Dairy-derived ingredients

Dairy-derived ingredients

Food and nutraceutical uses

Edited by
Milena Corredig

ELSEVIER

AMSTERDAM • BOSTON • HEIDELBERG • LONDON
NEW DELHI • NEW YORK • OXFORD • PARIS • SAN DIEGO
SAN FRANCISCO • SINGAPORE • SYDNEY • TOKYO
Woodhead Publishing is an imprint of Elsevier

WP
WOODHEAD PUBLISHING

Dairy-derived ingredients- Food and nutraceutical uses
Corredig

Woodhead Publishing Ltd.
An imprint of Elsevier
Abington Hall, Abington,
Cambridge CB21 6AH, England

ISBN: 978-1-84569-465-4

First Indian Print 2014

Indian ISBN: 978-93-5107-273-7

Published by Elsevier, a division of Reed Elsevier India Private Limited.

Registered Office: 305, Rohit House, 3 Tolstoy Marg, New Delhi-11 00 01
Corporate Office: 14th Floor, Tower 10B, DLF Cyber City, Phase-II,
Gurgaon - 122 002, Haryana

Digitally printed and bound at Avantika Printers Pvt. Ltd., New Delhi (India)

Contents

Contributor contact details

(* = main contact)

Chapter 1

Professor U. Kulozik
Food Process Engineering and Dairy Technology
Technische Universität München
Arcisstrasse 21
80333 München
Germany

E-mail: ulrich.kulozik@wzw.tum.de

Chapter 2

Dr Pierre Schuck
INRA
Agrocampus Ouest
UMR STLO
65, rue de Saint Brieuc
35042 Rennes cedex
France

E-mail: pierre.schuck@rennes.inra.fr

Chapter 3

L. Vercruysse, Professor J. Van Camp*, Professor K. Dewettinck
Department of Food Safety and Food Quality
Department of Crop Protection
Ghent University
Coupure Links 653
9000 Ghent
Belgium

E-mail: john.vancamp@ugent.be

Professor G. Smagghe and L. Vercruysse
Department of Crop Protection
Ghent University
Coupure Links 653
9000 Ghent
Belgium

Chapter 4

M. Sichien, N. Thienpont, E. Fredrick, T. Trung Le, Professor J. Van Camp and Professor K. Dewettinck*
Department of Food Safety and Food Quality (LA07)
Faculty of Bioscience Engineering
Ghent University
Coupure Links 653
9000 Ghent
Belgium

E-mail: Koen.Dewettinck@UGent.be

N. Thienpont
Department of Food Science and Technology
Faculty of Biosciences and Landscape Architecture
University College Ghent
Voskenslaan 270, 9000 Gent, Belgium

Chapter 5

Dr R. J. Durham
School of Natural Sciences
University of Western Sydney
Hawkesbury Campus
Locked Bag 1797
Sth Penrith, NSW 1797
Australia

E-mail: r.durham@uws.edu.au

Chapter 6

Dr P. X. Qi
United States Department of Agriculture
Agricultural Research Service
Eastern Regional Research Center
Dairy Processing and Products Research Unit
600 East Mermaid Lane
Wyndmoor, PA 19038
USA

E-mail: Phoebe.Qi@ars.usda.gov

Chapter 7

Dr I. Recio*, Dr F. J. Moreno and Professor R. López-Fandiño
Instituto de Fermentaciones Industriales (CSIC)
C/Juan de la Cierva, 3
28006 Madrid
Spain

E-mail: recio@ifi.csic.es

Chapter 8

T. Akhavan, S. Panahi, Professor G. H. Anderson and Dr B. L. Luhovyy*
Department of Nutritional Sciences
Faculty of Medicine
University of Toronto
150 College St., 329a FitzGerald Bldg
Toronto, ON
Canada
M5S 3E2

E-mail: bohdan.luhovyy@utoronto.ca

Chapter 9

Dr Riina Kekkonen*
Valio Ltd
R&D
PO Box 30
FIN-00039
Helsinki
Finland

E-mail: riina.kekkonen@valio.fi

Dr Katri Peuhkuri
Netnut
Espoo
Finland
Dr Riina Kekkonen and Dr Katri Peuhkuri
University of Helsinki
Institute of Biomedicine
Finland

Chapter 10

Professor H. J. Korhonen* and Dr P. Marnila
Biotechnology and Food Research
MTT Agrifood Research Finland
FIN-31600, Jokioinen
Finland

E-mail: hannu.j.korhonen@mtt.fi

Chapter 11

Dr P. Marnila* and Professor H. J. Korhonen
Biotechnology and Food Research
MTT Agrifood Research Finland
FIN-31600, Jokioinen
Finland

E-mail: pertti.marnila@mtt.fi

Chapter 12

Dr R. Hickey
Moorepark Food Research Centre
Teagasc
Moorepark
Fermoy, Co. Cork
Ireland

E-mail: rita.hickey@teagasc.ie

Chapter 13

Dr S. Mezouari* and Y. Pouliot
Department of Food Science and Nutrition
Laval University
Pavillon Paul-Comtois
Quebec City
Quebec
Canada
G1K 7P4

E-mail: mezouari4@yahoo.ca

Dr M. Britten
Food Research and Development Centre
Agriculture and Agri-Food Canada
3600 Casavant Boulevard West
St-Hyacinthe
QC
Canada
J2S 8E3

Chapter 14

Dr M. Corredig
Department of Food Science
University of Guelph
Guelph, Ontario
Canada
N1H 2W5

E-mail: mcorredi@uoguelph.ca

Chapter 15

Dr T. Huppertz
NIZO food research
P.O. Box 20
6710BA Ede
The Netherlands

E-mail: thom.huppertz@nizo.nl

Chapter 16

Professor A. L. Kelly*, Dr K. I. Kothari, Dr D. D. Voigt
Department of Food and Nutritional Sciences
University College Cork
Ireland

E-mail: a.kelly@ucc.ie

Dr T. Huppertz
NIZO Food Research
Ede
The Netherlands

Chapter 17

Professor M. A. Drake*, J. Wright, M. Whitson and M. Lloyd
Dept. Food, Bioprocessing, and Nutritional Sciences
Southeast Dairy Foods Research Center
North Carolina State University
Raleigh, NC 27695
USA

E-mail: mdrake@unity.ncsu.edu

Chapter 18

Dr M. Sibeijn
Process technology engineer
Dishman Netherlands B.V.
Veenendaal
The Netherlands

Dr J.A. Wouters*
NIZO Food Research
P.O. Box 20
6710 BA Ede
The Netherlands

E-mail: jeroen.wouters@nizo.nl

Chapter 19

Dr B. T. O'Kennedy
Moorepark Food Research Centre
Teagasc
Moorepark
Fermoy
Co. Cork
Ireland

E-mail: Brendan.okennedy@teagasc.ie

Chapter 20

T. P. Guinee
Moorepark Food Research Centre
Teagasc
Moorepark
Fermoy
Co. Cork
Ireland

E-mail: tim.guinee@teagasc.ie

Chapter 21

Dr P. J. Wilde
Institute of Food Research
Norwich
NR4 7UA
UK

E-mail: peter.wilde@bbsrc.ac.uk

Chapter 22

Dr C. M. Oliver and Dr M.-A. Augustin*
CSIRO Preventative Health Flagship
CSIRO Division of Food and Nutritional Sciences
671 Sneydes Road
Werribee
Victoria 3030
Australia

E-mail: maryann.augustin@csiro.au

Chapter 23

Dr P. M. Tomasula
United States Department of Agriculture
Agricultural Research Service
Eastern Regional Research Center
Dairy Processing and Products Research Unit
600 E. Mermaid Lane
Wyndmoor, PA 19038
USA

E-mail: Peggy.Tomasula@ars.usda.gov

Chapter 24

Dr A. H. J. Paterson*
School of Engineering and Advanced Technology
Massey University
Private Bag 11 222
Palmerston North
New Zealand

E-mail: a.paterson@massey.ac.nz

Dr S. J. Kellam
Hilmar Cheese Company
9001 N. Lander Avenue
PO BOX 910
Hilmar, CA 95324
USA

E-mail: skellam@hilmaringredients.com

Chapter 25

Professor Harjinder Singh
Riddet Institute
Massey University
Private Bag 11222
Palmerston North
New Zealand

E-mail: H.Singh@massey.ac.nz

Preface

This book is a great collection of chapters, written by acknowledged experts, providing up-to-date views on dairy ingredients manufacture, interactions and uses. Today milk can no longer be considered as a commodity, but rather as a source of value-added ingredients, both from the processing and the nutritional point of view. In the past few years there has been a fundamental change in the way we process and develop dairy products and ingredients.

Our knowledge of the structure and the structure–function changes of dairy components with processing has significantly improved over the past decade, and this has resulted in major advances in the production of ingredients. Membrane filtration, for example, has resulted in significant changes in the availability of whey, casein and milk protein concentrates and isolates, and the optimization of this unit operation has created a number of novel ingredients tailored to specific processing functionalities. Other processes, as for example, use of high pressure, have been studied in great detail, with possible commercial applications perhaps limited only by lack of initial capital investment. Because of the major progress made in chromatographic techniques and membrane separation, isolation of minor components in milk has become a reality, and studies on the functionality of these ingredients are no longer only an academic exercise but an industrial reality. For example, it is possible now to purify immunoglobulins from bovine colostrums and whey on a large scale, to produce concentrates as dietary supplements, and, with the application of high pressure processing to these products, it will be possible to minimize the denaturation of these molecules and therefore optimize their biological functionality.

The processing functionality of dairy ingredients continues to remain important, and therefore research efforts continue to be carried out to understand interactions and predict their behaviour during processing. In particular, there is an emerging trend to use dairy systems as platforms for the delivery of bioactive molecules in foods. Many milk components can function as encapsulating agents. Micro and nano-encapsulation can protect active components from the surrounding environment and tailor and control their release.

Although the study of the extraction and potential utilization of the various components in milk continues to be an important aspect of dairy research, the focus has shifted in the past decade from processing to nutritional functionality. A new generation of high-value, functional ingredients, not only for the food but also for the pharmaceutical market are being developed, as more and more evidence is brought forward of the beneficial effect of their consumption to human health.

The positive health-promoting aspects of dairy ingredients, whey, caseins, milk fat globule membrane material, oligosaccharides and glycoconjugates have been so far tested mostly *in vitro* or in simple animal models, and more work needs to be carried out to demonstrate their properties. However, the evidence continues to accumulate on antimicrobial, anti-inflammatory, hypocholesterolemic, hypertension controlling, anti-carcinogenic activities of milk-derived products. This book attempts to summarize our current knowledge in some of the most promising areas. In the future, milk ingredients may play a complementary role to the pharmacological drugs in the prevention and treatment of various chronic diseases. Epidemiological studies have linked the consumption of dairy products to alleviation of physiological characteristics associated with diabetes, cardiovascular diseases, metabolic syndrome and obesity; however, the results are often inconsistent, possibly because of the variety of composition of the dairy products consumed. Milk proteins are also considered one of the most important sources of bioactive peptides; however, their isolation and characterization still need to be developed, especially to evaluate the synergy occurring when consuming dairy products. In addition to simple enzymatic hydrolysis, an emerging field is biogenics, the study of the formation of bioactives as a result of tailored fermentations in dairy products. Another area that shows great promise is the study of the biological role of glycoproteins in milk and their potential for applications. Glycans from milk proteins may serve as recognition sites interacting with cell membranes, viruses and microorganisms. There is an increasing evidence that these molecules have beneficial effects within the gastrointestinal tracts, modulate the intestinal flora preventing attachment of pathogens and viruses, and interact directly with immune cells. These molecules could soon find therapeutic applications. Finally, the study of minor milk lipids derived from the milk fat globule membrane is also at its infancy.

As will become evident by reading the various chapters of this book, the future of the use of dairy ingredients is an interesting one and we are certainly at a turning point in the development of value-added components from milk. I am very grateful to the all the authors for their hard work on this project and their excellent contributions.

Milena Corredig

Part I

Modern approaches to the separation of dairy components and manufacture of dairy ingredients

Part I

Modern approaches to the separation of dairy components and manufacture of dairy ingredients

1

Novel approaches for the separation of dairy components and manufacture of dairy ingredients

U. Kulozik, Technische Universität München, Germany

Abstract: Novel approaches regarding the separation of dairy components are discussed in the light of continuous efforts to better understand boundary layer phenomena in membrane separation. For instance, deposit formation by proteins can be minimized only through crossflow filtration when the fluid dynamics along the membrane surface are under control and chemico-physical effects controlling adsorption are understood. Thus, even existing processes can be intensified by means of more appropriate processing and milieu conditions, together with selecting membranes with optimal membrane resistances relative to individual application. Hybrid processes and novel techniques such as membrane adsorption or electro assisted filtration are described as ways to expand the reach of current membrane technologies.

Key words: membrane filtration, boundary layer phenomena, membrane adsorption chromatography, dynamic membrane systems, electric field assisted membrane techniques.

1.1 Introduction

Fractionation of milk components has traditionally been a strong foundation for the production of a huge variety of dairy products. Centrifugation is applied to separate cream and skim milk. Coagulation is induced to separate protein from the aqueous phase in the manufacture of cheeses. Chromatographic techniques have been established on an industrial scale for the removal of lactose from milk to produce low lactose or lactose-free milk in Finland and other countries.

This chapter focuses mainly on more recent insights and new developments in research, and on the application of membrane technologies in the field of dairy production, with membranes applied alone or in combination with other processes.

Dairy components of relevance for separation and fractionation by means of membrane technologies are major and minor proteins in the first place and low molecular weight components such as ionic substances and lactose. All these components are of influence in membrane separation because of their molecular size or surface properties. Proteins, but also fat globule membrane components, are macromolecules prone to adsorption and formation of gel deposits on membrane surfaces. Ionic substances and lactose are small and therefore can pass through porous membranes. In contrast to that, they cannot pass through non-porous, homogeneous membranes, typical for reverse osmosis, and are therefore mainly responsible for the osmotic pressure of milk and whey. The osmotic pressure is a key intrinsic factor of the product influencing flux in reverse osmosis. The retention or passage of components nominally depends on the membrane pore size and structure. However, it is practically experienced that apart from size, other properties of molecules, such as charge, structure, and shape, play decisive roles in this respect. Beyond molecular weight, for instance, proteins vary in their surface charge and hydrodynamic volume or diameter as a function of pH and ionic composition of the aqueous phase environment. Depending on the physical properties of the membrane and other solutes, this often results in unspecific interactions of compounds with the membrane surface, causing adsorption or deposited layers, and these effects are difficult to predict.

An example is given in Fig. 1.1, where the impact of pH on the permeation of β-lactoglobulin through a microfiltration membrane is depicted. As can be seen, the permeation qualitatively follows the changes occurring to the quaternary structure of β-lactoglobulin, which, depending on pH, exists as a monomer, dimer or octamer. The practical implication of this effect in dairy technology is that β-lactoglobulin should have different permeation patterns in the filtration of acid or sweet whey if microfiltration is applied for the reduction of microorganisms in whey prior to the production of WPC products.

Another key factor is the ionic composition of the solution, which has an impact on the surface charge of colloidal particles such as proteins, including whey proteins and casein micelles. This is schematically shown in Fig. 1.2, where the effect of low or high ionic strengths on the electric double layer of the particles is described. The electrical double layer around charged particles (proteins carry a net charge depending on the composition of charged amino acids) is thin in high ionic environments, and high at lower ionic strengths. Thus, the hydrodynamic size of molecules is not constant – it rather depends on environmental factors such as pH and ionic composition, i.e. on electrical shielding effects and collapse of the hairy layer of protein residues protruding from the casein micelle into the outer phase.

In conclusion, despite its long history of application in dairy technology and strong impact on production methods, membrane separation is still a complex and not fully understood field because of various factors influen-

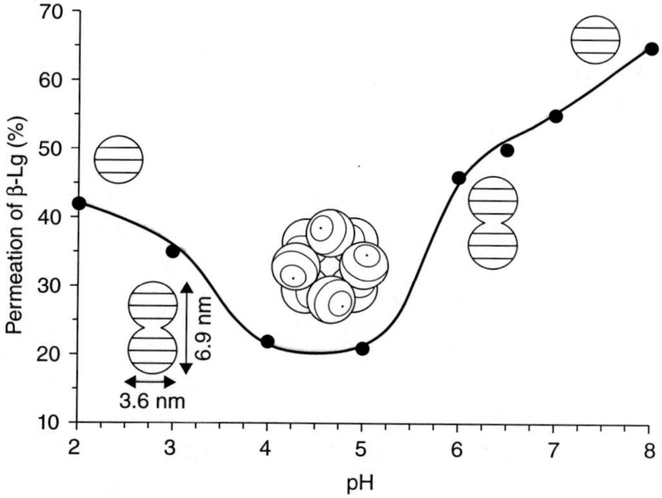

Fig. 1.1 Permeation (p) of β-lactoglobulin (molecular diameter approx. 3–4 nm) through a ceramic microfiltration membrane (nominal pore size 0.1 μm) as a function of pH. p is defined as the ratio of the β-lactoglobulin concentration found in the permeate relative to the concentration in the retentate ($p = c_{per}/c_{ret}$). Since the β-lactoglobulin is small in comparison to the nominal pore size, a permeation $p = 1$ could principally be expected.

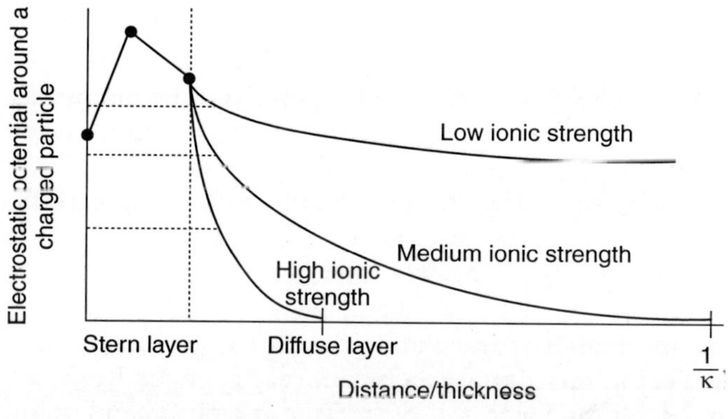

Fig. 1.2 Electrical double layer around a charged colloidal particle as a function of ionic strength. At higher levels of ionic concentration in the outer phase, the charge of particles is shielded at a closer distance from the surface such that the hydrodynamic radius is reduced.

cing separation results, factors that are less specific and not fully predictable when compared with other separation techniques. Nowadays, membrane separations are often combined with other technologies such as centrifugal or chromatographic separations, in hybrid processing modes.

1.2 State-of-the-art separation techniques

1.2.1 Interfacial effects in the separation of dairy components by means of membrane techniques

Membrane separation has become a widely-used unit operation in dairy products manufacture. The concentration of proteins from whey in the manufacture of so-called WPC products has been introduced during the early 1970s and has since been established also in speciality operations. Ultrafiltration is the most frequently used membrane process, mostly run in the crossflow processing mode. In contrast to dead-end filtration as it is applied in sterile filtration of biological media or in cake filtration in the beverage industry, the product to be concentrated is passed along the membrane surface to induce friction and, thus, a wall shear stress at the membrane surface. It is desirable to remove deposited particles from the surface of the membranes in order to avoid an overly strong effect of the deposited layer on permeation and retention. Therefore, the volume stream or fluid velocity is of great effect, as shown in an example in Fig. 1.3. It can be seen that the permeation increases linearly as a function of the transmembrane pressure difference when water is flowing across and permeating through the membrane. The permeation of protein solutions, on the other hand, is increasing less steeply and under-proportionally as the pressure goes up. The flux is determined by the membrane only at low transmembrane pressure differences (Δp_{TM}). Above approximately $\Delta p_{TM} = 3$ bar, depending on flow rate, the flux values flatten off. Beyond this point, often referred to as critical flux or critical Δp_{TM}, the permeation is controlled by a compressed, dense layer of deposited protein particles. These data demonstrate that, despite high wall shear stresses, the aim of preventing deposit formation is not fully reached and the deposit layer acts in fact as a secondary membrane which may even dominate the separation process.

The question arises in this context why the deposition of particles cannot be fully prevented. The answer lies in a dilemma situation of crossflow membrane filtration: Δp_{TM} and wall shear stress τ_W are dependent factors: the higher the wall shear stress (exerted by the volume flow along the membrane), the higher is the static pressure on the retentate side of the membrane. This, in turn, induces a high level of convective flow of solution towards the membrane as result of the transmembrane pressure Δp_{TM}. The retained particles accumulate at the membrane surface, thus generating a concentration increase in comparison to the bulk solution. This particularly applies to the region at the front end of the membrane, despite the fact that diffusional backflow and wall shear stress partially reduce the excess in concentration of retained particles. Further to that, the pressure loss along a flow path will create an inconsistent transmembrane pressure pattern along the membrane surface. These factors, in combination, impair the full control of processing conditions such that the membrane as a selective barrier is fully responsible for retention and permeation.

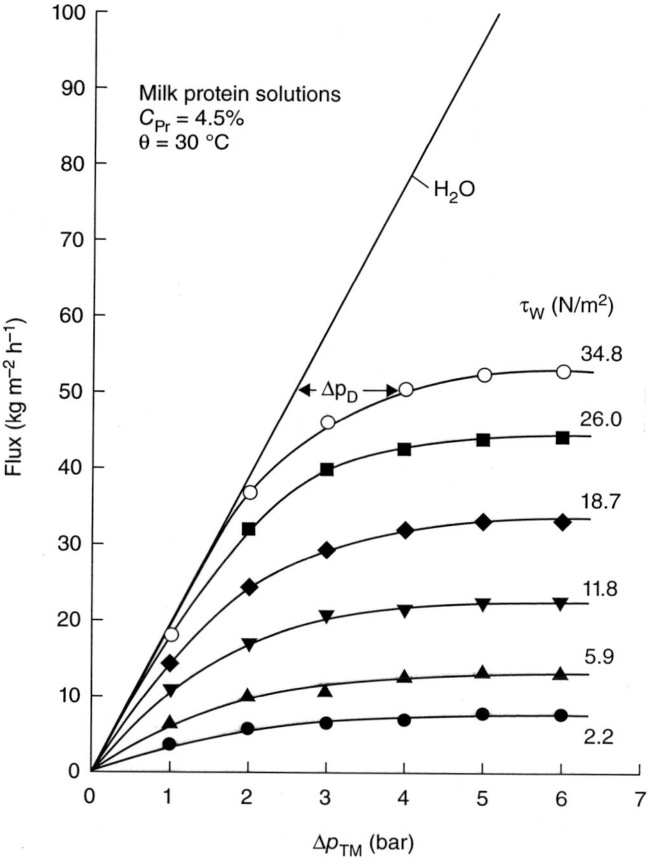

Fig. 1.3 Effect of transmembrane pressure difference Δp_{TM} and wall shear stress τ_W at a membrane surface on permeation rate in the ultrafiltration of milk protein solutions. The membrane material was poly sulfone (PS), nominal cut-off 25 kD.

An additional explanation for the difficulty of avoiding deposit formation is that dairy proteins, in particular casein micelles, are too large to be able to diffuse back from the membrane into the bulk solution and too small to be affected by lift forces imposed by the fluid stream. Calculations by Altmann and Ripperger[1] led to results presented in Fig. 1.4 explaining this situation in more detail. The diagram depicts the flux rate as a function of the size of the particles present in an aqueous model phase, water in this case. The flux curve can be divided into two regions: (i) at small particle sizes, the removal of deposited particles from the membrane surface is influenced by diffusion as a mechanism for particle removal. The smaller the dispersed particle, the higher the flux; (ii) the right wing of the graph, with an inverse slope as compared to the left side of the graph, can be

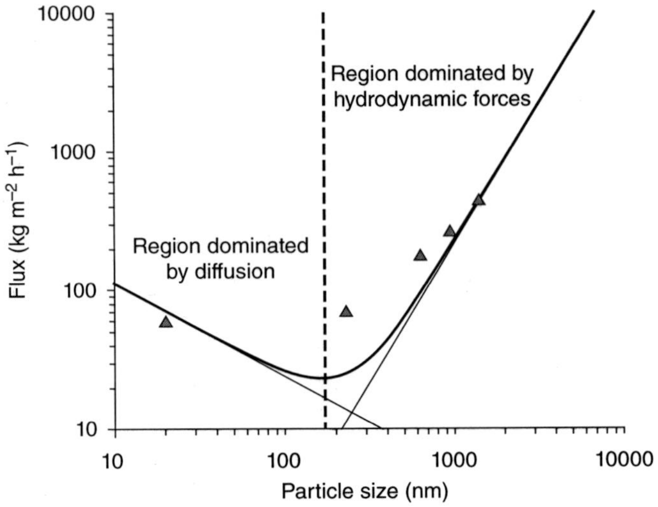

Fig. 1.4 Permeation rate as a function of particle size in model suspensions of Al_2O_3, TiO_2, SiO_2 and Kieselgur particles in sizes between 23–13 200 nm. The material, however, in this case is not of importance in relation to the size.[1]

explained by the fact that in this region, particles are affected by the shear force exerted by the fluid stream across the membrane surface. The larger the particles, the more effective they are influenced by hydrodynamic lift forces. It can be seen that the filtration of particles around 200 nm leads to a minimum of the curve. Neither diffusion nor lift forces are significantly affecting the particles deposited at the surface in this region. Considering these results, obtained for a model system consisting of inorganic particles in water, one can gain a basic understanding of why the deposition of casein micelles in milk, whose average size is in the area of 200 nm, can hardly be prevented in the cross flow filtration mode. Nevertheless, separation of casein micelles and native whey proteins has been investigated and successfully applied during recent years to obtain both protein fractions, without precipitation of either of them. This allows exploitation of their individual functional properties or shifting the ratio of both protein groups in order to influence the resulting textural profile of gels after acid or thermal coagulation.

A simplified processing scheme is shown as an insert in Fig. 1.5a. Skim milk is passed through a microfiltration unit equipped with a ceramic membrane (SCT membrane ZrO_2/Al_2O_3; nominal pore size 0.1 μm). The MF-permeate containing the whey proteins is then concentrated by an ultrafiltration unit concentrating the whey proteins. The UF-permeate containing the soluble low molecular weight substances of milk (lactose, minerals) at the original concentration, in turn, is recirculated to the feed tank

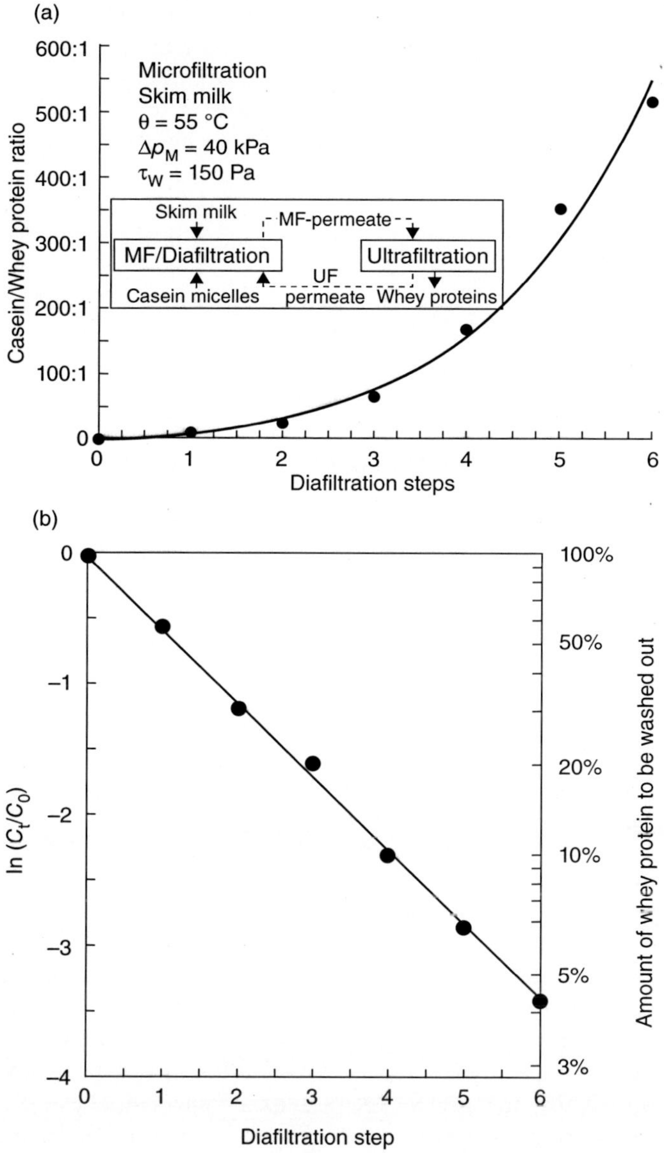

Fig. 1.5 Casein–whey protein ratio as a function of diafiltration steps in the fractionation of caseins and whey proteins by microfiltration from milk: (a) linear scale; (b) Logarithmic scale. The MF membrane used was a SCT membrane made of ZrO_2/Al_2O_3; the diafiltration was conducted using UF permeate from milk.

of the MF-unit to wash out the remaining whey proteins in a diafiltration mode. Using the UF permeate instead of water as a medium for diafiltration allows one to avoid a change in the native aqueous environment of the casein micelles. The number of volume turnovers to fully achieve a complete separation of the milk protein main fractions depends on the degree of permeation of the whey proteins through the MF-membrane. This is why the expected separation effect has to be defined in the form of the permeation value *p*, which is the ratio of this component in the retentate c_0 and in the permeate *c*:

$$p = c / c_0 \qquad [1.1]$$

Theoretically, i.e. assuming unlimited permeation, the permeate concentration of substances passing through is equally as high as in the retentate, i.e. $c = c_0$ and $p = 1$.

Since pressure-driven membrane processes principally separate according to molecule size relative to the pore size of the membrane, the separation efficiency is driven by the amount of filtrate, which convectively carries the permeating component through the membrane pores.

Figure 1.5 further demonstrates an experimental result from a fractionation process of casein micelles (medium size 150 nm) from whey proteins (medium size 3–4 nm) by means of microfiltration (SCT membrane ZrO_2/Al_2O_3, nominal cut-off: 0.1 μm). It is obvious that, despite optimal choice of the theoretically best membrane pore size, a diafiltration process is required to wash out the whey proteins completely. The number of diafiltration steps actually required depends on the real permeation value *p*, which is lower than the theoretical value of $p = 1$ due to unspecific retention effects, and is depicted as the x-axis, while the y-axis shows the casein–whey protein ratio, starting from approximately 8:2 (close to the origin), which is the original ratio of casein and whey proteins in milk.

As can be seen from Fig. 1.5, five to six diafiltration steps are required to reduce the content of whey proteins in the casein retentate by 95%, based on a permeation value $p = 0.5$, which is considerably lower than the theoretically assumed value of $p = 1$. This can be easily explained by the deposit formation of the retained casein micelles, as well as by an unspecific adsorption of whey proteins at the membrane surface. These observations confirm that we need to improve the understanding of the boundary layer effects at the membrane surface in order to be able to minimize their impact.

1.2.2 Separation of dairy components by means of chromatography

In contrast to its widespread use in the sugar industry, chromatography is a separation technique less common in the dairy industry. The use of chromatographic techniques for the fractionation of whey proteins, however, has offered new ways for the extraction of highly effective protein components of a natural milk source, which are products with an important role in promoting human health.[2] Together with the main fractions, β-lactoglobulin and α-lactalbumin, present in whey protein isolate often used in infant formula, immunoglobulins are also important food supplements. Glycomacropeptide plays a growing role as a protein source in phenylalanine-free food for phenyl-ketonuria patients. These are just a few examples indicating that the functional properties of individual protein fractions are appreciated more and more in the development of food systems with improved structures or with bio-functional properties. Chromatography has been continuously developed over the last twenty years to remove lactose from milk, to serve a latently existing market for products for lactose-intolerant consumers, across Europe and in developing dairy markets in Asia.[3] Meanwhile, the principle of chromatographic separation is being combined with the possibilities of other separation processes to improve throughput and to optimize the ratio of throughput and specificity, as will be shown below.

1.2.3 Separation of dairy components by centrifugal force

Beyond the standard separation of fat and skim milk, centrifugation as a unit operation is applied, in combination with membrane technology, to produce precipitated calcium salts from acid milk permeate.[4] According to an industrialized process, UF-permeates of whey or milk are concentrated by reverse osmosis to approximately 18 to 20% total solids. The pH is then adjusted to alkaline and the solution heated to promote the precipitation of calcium phosphate. While the reaction occurs instantaneously, a short holding time is applied to stabilize the precipitate, which subsequently is separated from the lactose solution either by centrifugation or cross-flow membrane filtration. The lactose stream is available for further downstream processing. The calcium phosphate slurry is further refined and concentrated by ultrafiltration with diafiltration to the desired end composition and lactose recovery rate.

1.3 Process intensification by hybrid processes

1.3.1 Thermal separation of whey proteins in combination with membranes

As mentioned above, chromatography can be applied on a large scale to separate proteins. Nevertheless, chromatography may be limited in terms of throughput, and also with regard to the cleaning process, which requires high amounts of chemicals to restore suitable conditions for the next production cycle. Therefore, alternative processes have been proposed to enable the separation of two originally equally-sized molecules by membrane separation, whereby the protein mixture is subjected to a thermal process under conditions, such as specific pH, ionic strength, protein and lactose concentration, which selectively affect individual proteins leading to their aggregation, while the target protein remains unaffected and native in solution.[5] The aggregated protein can then be easily removed from the solution by microfiltration. Figure 1.6 provides an example demonstrating this concept by an assessment of the reaction rate constant of the thermal aggregation process. The system heated is whey, i.e. comprised of β-lactoglobulin and α-lactalbumin, the two main whey proteins in the natural aqueous phase of milk. It can be seen by comparing the reaction rate constants, a measure for the rate of the aggregation, that the proteins react differently, depending on lactose and protein contents. β-Lactoglobulin, for instance, reacts relatively fast at low lactose and low protein contents, while α-lactalbumin reacts more slowly under the same conditions. Other regions on these graphs can be identified where the opposite is the case, or where both proteins are fast in aggregating. From these fundamental data, and from other results related to other thermal and milieu conditions, optimized thermal processes can be derived which allow one of the proteins to be converted into aggregates while the respective other protein remains native. Therefore, microfiltration can be applied to remove the aggregated protein as a so-called microparticulate (which has its own technological value), while the native protein can be obtained from the microfiltration permeate by ultrafiltration. Figure 1.7 depicts two examples related to this selective whey protein separation process.

1.3.2 Enzyme-induced crosslinking and separation of dairy proteins or peptides

Another example of the combination of other unit operations with membrane filtration is the selective upstream treatment of caseinomacropeptide (CMP) from sweet whey. The presence of CMP may sometimes be undesired; for instance, when a selective fractionation of whey proteins as described above is wanted. Tolkach and Kulozik[6] developed an enzyme assisted crosslinking process of CMP using transglutaminase (TG). A condition for the crosslinking of proteins or peptides is the presence of two amino acids, namely lysine and glutamine, between which TG creates isopeptide bonds. Thus, crosslinking the small CMP molecule is induced, generating peptide aggregates large enough to be removed by microfiltration from whey, as shown in Figures 1.8 and 1.9. The remaining whey proteins can be further fractionated either using chromatography or the processing options presented in Fig. 1.7.

1.3.3 Membrane adsorption chromatography for the separation of proteins and peptides from whey

As mentioned above, the permeation process in membrane technology is influenced by a variety of factors still not fully understood and, in addition,

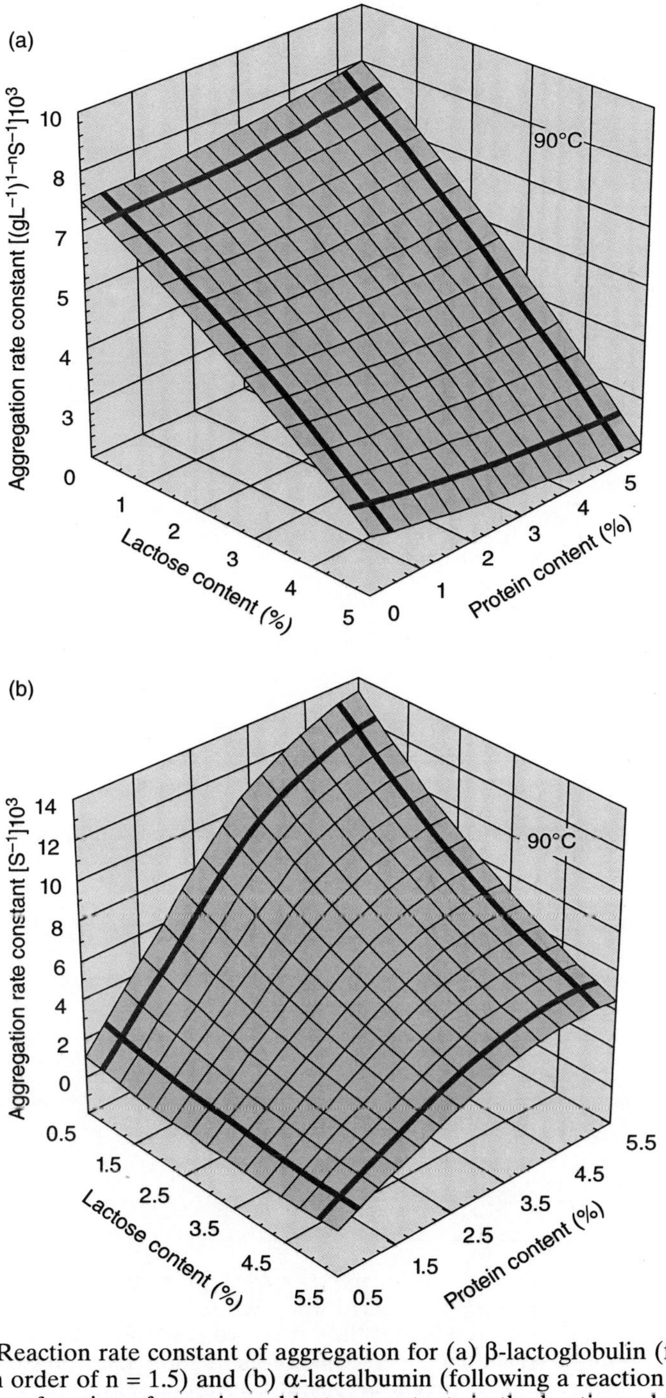

Fig. 1.6 Reaction rate constant of aggregation for (a) β-lactoglobulin (following a reaction order of n = 1.5) and (b) α-lactalbumin (following a reaction order of n = 1) as a function of protein and lactose contents in the heating solution at 90 °C.[5]

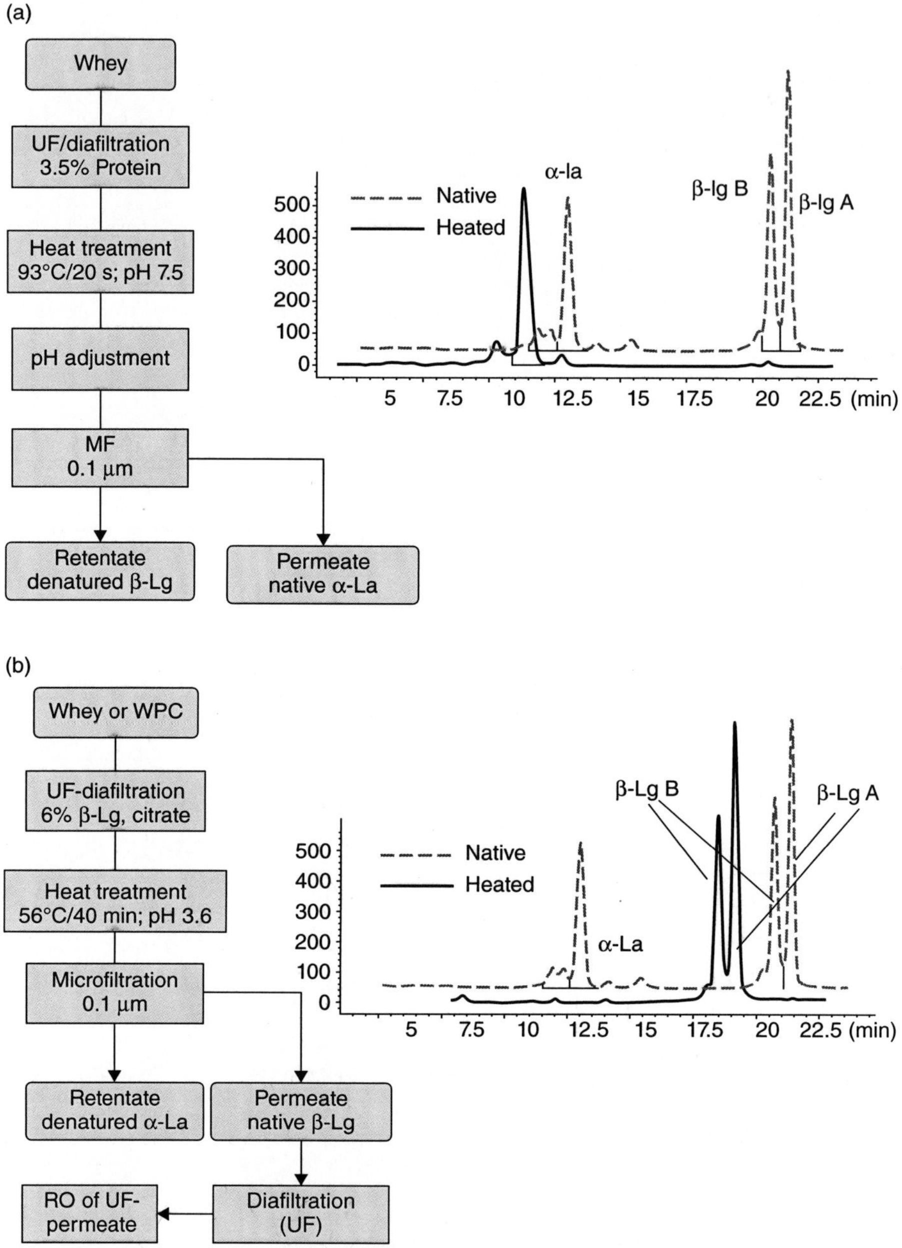

Fig. 1.7 Process diagrams for the separation of (a) β-lactoglobulin or (b) α-lactalbumin from pre-treated whey. Also shown are the respective chromatograms demonstrating that the native protein has hardly been affected by the upstream treatment. The dotted chromatogram lines show the whey protein composition prior to heat treatment, the solid lines the chromatograms after heat treatment.[5]

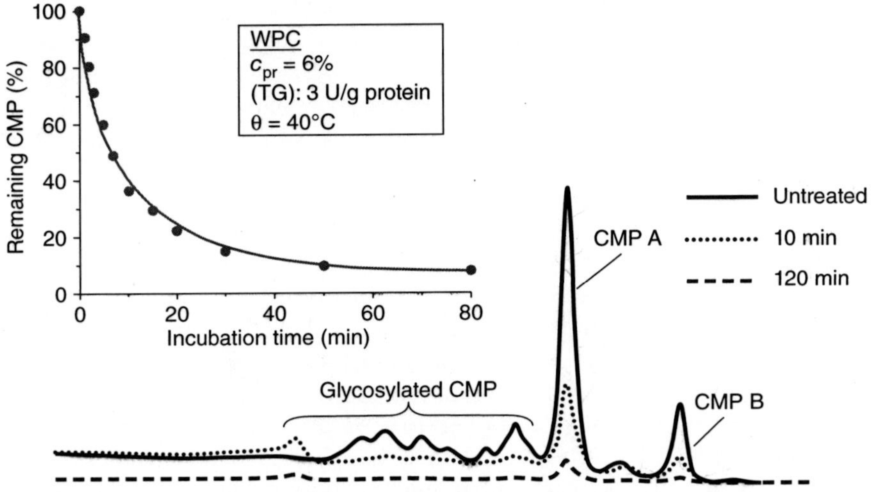

Fig. 1.8 Enzymatic crosslinking of caseinomacropeptide and removal from whey by microfiltration: Kinetics of CMP aggregation and whey protein HPLC chromatograms highlighting the loss of CMP over time. Both the glycosylated and the non-glycosylated parts of CMP are crosslinked and are therefore reduced in the chromatogram since they no longer remain soluble.

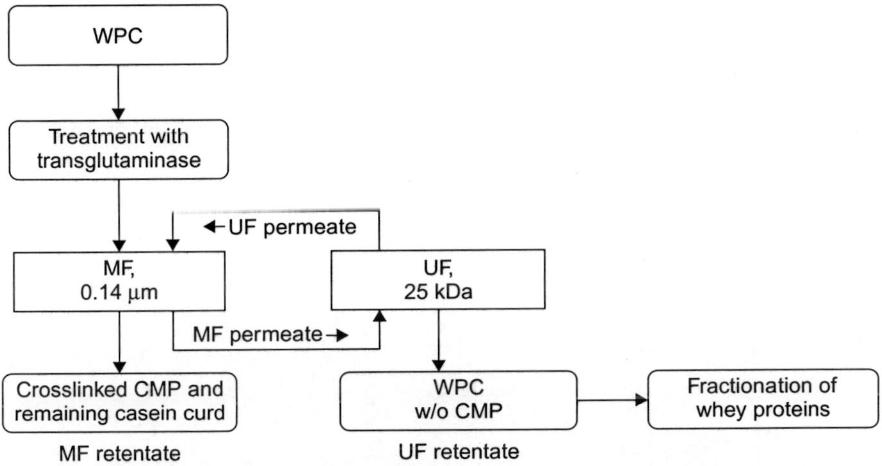

Fig. 1.9 Block diagram of a process using enzymatic crosslinking to aggregate CMP and microfiltration to remove the aggregates.

which are relatively unspecific. This is why separation tasks are often achieved in a more targeted way by chromatographic techniques. These techniques, however, are often limited by the low velocity of the diffusional processes which are responsible for the transport of substances into and out of the chromatographic beads. Therefore, novel hybrid systems have been developed combining the features of both processes, namely specificity of ligands in chromatography (quaternary ammonium ions in this case) and high throughput of membranes. Such systems are called membrane adsorption chromatography units (MAC), which consist of natural polymers, such as cellulose, as carrier material; these can be easily modified at their surfaces. Similar to adsorber beads, the surfaces can carry charged or otherwise adsorptive ligands (in this study, again using quaternary ammonium ions), which are convectively brought in contact with the substrates in the filtrate stream passing along them. Therefore, diffusional processes are much less limiting as compared to conventional chromatographic systems.

As could be shown in a study for the separation of fractions of caseinomacropetide (CMP), conventional ion exchange (IEX) chromatography and MAC can successfully be applied to separate the glycosylated from the unglycosylated parts, based on the fact that the glycosylated peptides are negatively charged. This might be of practical importance because of certain bioactivities and specific technological functionalities of CMP.[7] Without the MAC technique, this separation would be only possible by standard ion exchange chromatography (IEX). Figure 1.10 compares the separation results of MAC and conventional ion exchange chromatography. The diagrams depicts the elution curves detected by a UV detector as a function of elution time. As can be seen, the separation effect is very similar between

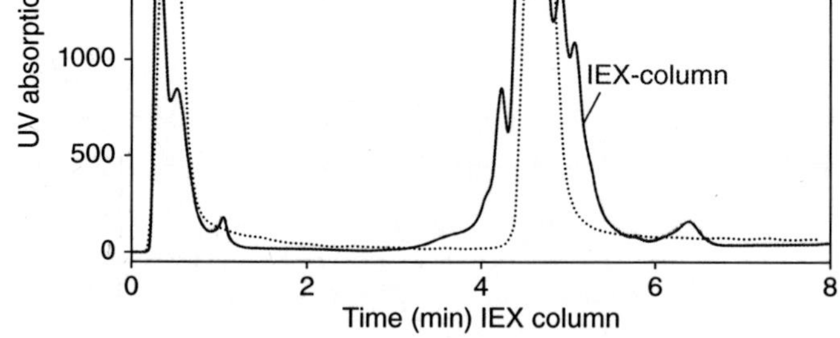

Fig. 1.10 Elution curves from a separation process of aglyco- and glyco-CMP from whey using ion exchange chromatography in comparison to membrane adsorption chromatography, both carrying quaternary ammonium groups as ligands.[8]

the systems. They both achieve a base line separation of peaks. The IEX chromatogram indicates that the glyco-CMP consists of various sub-fractions (see also Fig. 1.8) which could be further separated. However, the target here was to separate only the two bulk fractions, glyco- and aglyco-CMP. More importantly, it was shown that, in fact, MAC achieves the separation effect in a quarter of the time required by conventional IEX chromatography. It can be predicted that MAC technology will have further applications in the food industry or in pharmaceutical applications, since different ligands can be linked with the carrier surfaces and that a wide spectrum of separation tasks could successfully be solved by using MAC units.

1.4 Alternative membrane processing concepts

At the beginning of membrane technology, only polymeric materials were available and they are still dominating the market. Later on, ceramic or other inorganic materials have been introduced, which allow more extreme conditions of operation, cleaning and sterilization and have, partially, also more specific separation capabilities, and therefore have occupied a significant part of the membrane market. In addition, innovative technical solutions have been developed allowing a better control of processing conditions and, thus, of separation results, as will be shown below.

1.4.1 Membrane filtration with uniform transmembrane pressure difference

As was shown above, high wall-shear stresses in combination with sub-critical and uniform transmembranes pressure differences should yield better separation results. Different approaches have been taken to achieve improved situations with a better deposit control.

The first system was developed during the 1970s, achieving a uniform transmembrane pressure difference Δp_{TM} by means of pumping the permeate in a circular mode through the permeate side of the membrane modules.[9] The volume flow through the permeate side of the module can be adjusted such that the pressure loss is equivalent to the pressure loss along the retentate side of the membrane. By doing so, the pressure difference at each point along the membrane is the same, i.e. uniform. This results in a processing mode where, in contrast to conventional crossflow systems, volume flow rate and transmembrane pressure are no longer dependent factors delivering the same Δp_{TM}. This, in turn, allows the control of deposit formation through high levels of wall shear stress without simultaneously creating highly convective transportation of deposit-forming material towards the membrane surface. A processing scheme, as well as the related pressure situations on the retentate and permeate sides, is schematically depicted in Fig. 1.11 according to Sandblom.[9]

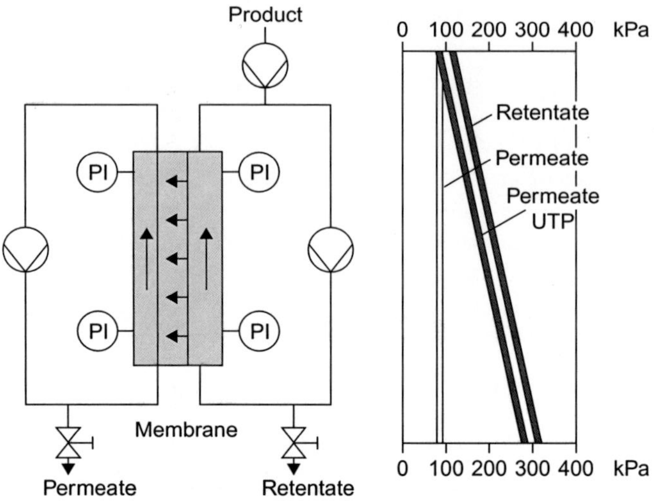

Fig. 1.11 Uniform Transmembrane Pressure (UTP) mode membrane filtration.[9] The pressure gradient on the retentate side of the membrane is equal to an induced pressure loss on the permeate side, thus creating the same transmembrane pressure along the membrane.

A result from a microfiltration experiment fractionating milk proteins is shown in Fig. 1.12, comparing the UTP mode with the classical crossflow mode. The permeation rate is strongly time-dependent in the case of crossflow MF, while for the UTP mode a much higher and constant flux can be obtained, indicating that the control of deposition of the retained casein micelles was more effective.

1.4.2 Gradient membranes

The positive effect described above stood behind the development of so-called gradient membranes made from ceramic materials, which have inbuilt gradients in permeation resistances. One of the concepts is based on a gradient in porosity within the membrane support layer (System Pall Exekia/France); another is based on the gradient in thickness of the selective front layer of the membrane (System TAMI/France). Either of the gradients can be established during membrane manufacture such that the permeation resistance along the flow path is inversely varied according to the linear decrease of static pressure on the retentate side. Provided that the respective gradient and the pressure loss on the retentate side are ideally compensating each other, the result is the same flux along the membrane. Figure 1.13 schematically depicts both gradient membrane concepts.

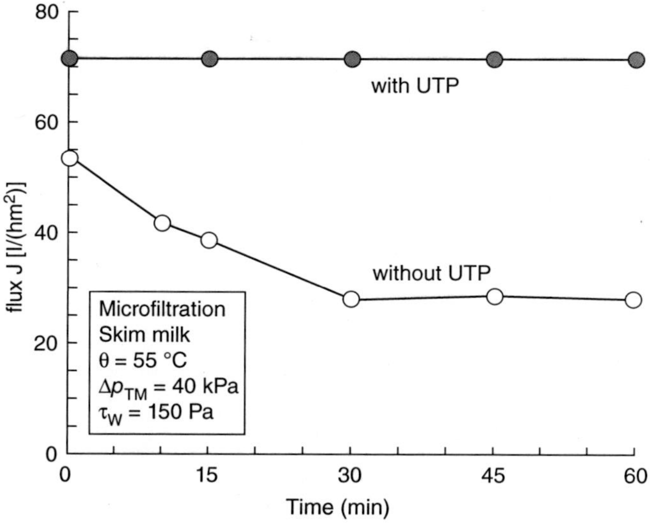

Fig. 1.12 Permeation rate in microfiltration-based fractionation of skim milk comparing crossflow and UTP modes using ceramic tubular membranes (ZrO_2/Al_2O_3).[10]

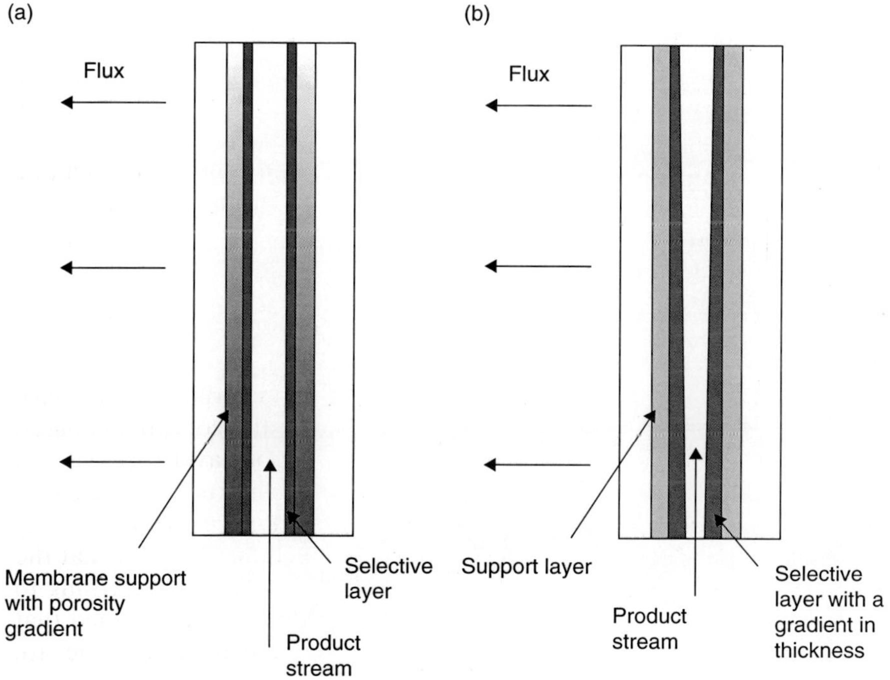

Fig. 1.13 Schematic depiction of membranes (a) with a gradient in membrane support porosity or (b) in the thickness of the selective layer at the membrane surface.

Gradient membranes should have the same effect (i.e. isoflux along the membrane) as the UTP processing mode, provided that the fluid properties (viscosity, concentration), which are varying according to the concentration through the stages of a larger membrane plant and which are responsible for the pressure loss, are compatible with the gradient of the respective membranes confronted with these variable properties of the retentate stream.

1.4.3 Dynamic membrane systems

Another approach in creating wall shear stress has been realized by the development of dynamic membrane systems with either rotating or oscillating membranes. The movement of the membrane, instead of high volume throughputs of the product, allows one to disentagle pressure difference and wall shear stress similar to the systems described above.

One of the first systems applying this concept has been the simple stirred vessel membrane cell, mainly for laboratory scale applications. A truly dynamic membrane system, the first of its kind, was the rotor–stator system by Sulzer (Winterthur/Switzerland) with membrane equipped cylindrical rotors and stators creating a circular gap between them, shear forces and Taylor vortices producing the required wall shear stress and removing deposited material. This concept has been later transferred to the development of flat sheet circular membrane systems, and deposit removal or control could be realized without large volume streams passing along the membrane surfaces. Figure 1.14 schematically shows one of the presently available systems with overlapping, rotating circular membranes. Figure 1.15 depicts results from a microfiltration experiment using an oscillating membrane system (PallSep; Pall/NJ) for the fractionation of milk proteins, i.e. whey proteins from caseins, with variations in the amplitude of the oscillation. As can be seen, the permeation is strongly dependent on the transmembrane pressure difference, despite the fact that a low volume stream is applied. This clearly shows that the system is operated in a

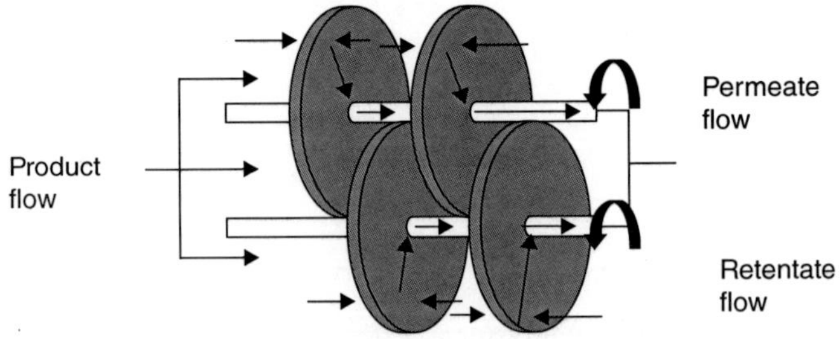

Fig. 1.14 Scheme of a dynamic membrane system with rotating membranes.

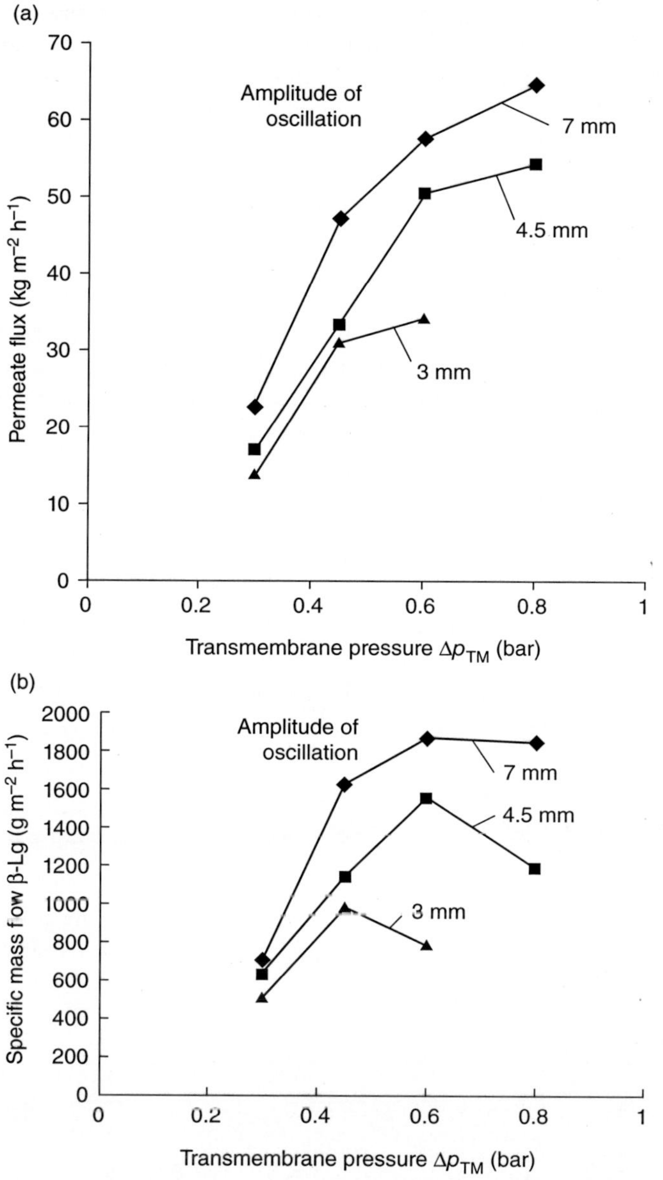

Fig. 1.15 (a) Permeation rate and (b) protein mass flow in MF using an oscillating membrane system for the fractionation of milk proteins.

membrane controlled region, where the deposit formation is negligible and therefore does not influence the permeation in a major way. On the other hand, the permeation is controlled by the deposited layer of retained protein particles at higher levels of the transmembrane pressure difference, in particular when the amplitude of oscillation is low. Figure 1.15b depicts the mass flow of smaller proteins, using β-lactoglobulin as an example, showing that an undesired retention can obviously be prevented as long as the critical value for Δp_{TM} is not exceeded.

1.4.4 Electric field assisted membrane techniques

Electrodialysis has long been applied in the dairy industry to separate ionic components, mainly from whey, using ion exchange membranes. Alternatively, charged particles can be removed from solutions using other kinds of membranes assisted by an electrical field, as has been demonstrated by Weigert *et al.*[11] The authors made use of the effect of an electrical field to enhance the effect of fluid dynamical lift forces in removing charged particles from membrane surfaces. The electrical field was polarized according to the net particle charge such that the direction of the forces was pointing away from the membrane surface. The results, depicted in Fig. 1.16, clearly show that submicron particles in a model system with anorganic (crystobalite) particles were effectively forced away from the membrane surface yielding a significantly higher flux rate as compared to the same filtration without an electrical field. Although these results were obtained for a non-food system, the potential of this principle for more complex solutions in dairy or other food applications exists and should be of interest to validate.

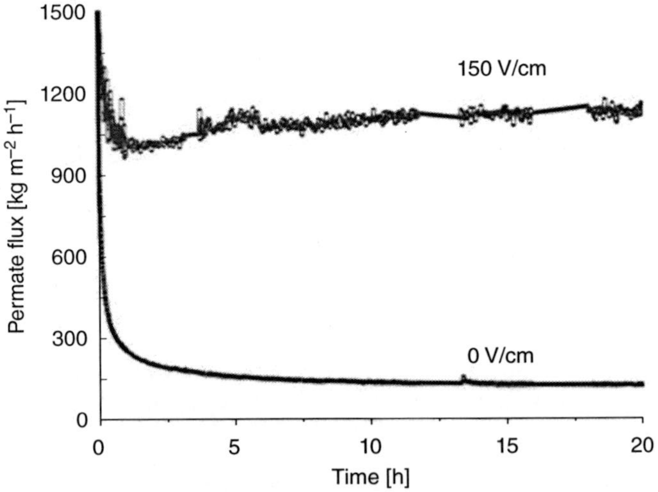

Fig. 1.16 Effect of an electrical field on the permate mass flow in crossflow microfiltration of anorganic crystobalite particles in comparison to conventional crossflow filtration.[11]

1.5 References

1 J. ALTMANN and S. RIPPERGER (1997): Particle deposition and layer formation at the crossflow microfiltration. *J. Membrane Sci.*, 124, 119–128

2 W. STROHMAIER (2003): Chromatographic fractionation of whey proteins. *Proceedings, IDF Conference on Advances in Fractionation and Separation: Processes for novel dairy applications*, Bruges, Belgium 7–12 Sept., pp. 81–93

3 P. JELEN and O. TOSSAVAINEN (2003): Low lactose and lactose free milk and dairy products – prospects, technologies and applications. *Australian J. Dairy Technol.* 58 (2), 161–165

4 C. F. BROTHERSEN, N. F. OLSON, and T. RICHARDSON (1982): Recovery of calcium phosphate from ultrafiltration permeates. *J. Dairy Sci.* 65, 17–23

5 A. TOLKACH and U. KULOZIK (2005): Optimization of thermal pre-treatment conditions for the separation of native α-lactalbumin from whey protein concentrates by means of selective denaturation of β-lactoglobulin. *J. Food Sci.*, 70 (9) 557–566

6 A. TOLKACH and U. KULOZIK (2005): Fractionation of whey proteins and caseinomacropeptide by means of enzymatic crosslinking and membrane separation techniques. *J. Food Eng.*, 67 (1–2) 13–20

7 C. THOMAE, I. KRAUSE, and U. KULOZIK (2006): Precipitation behaviour of caseinomacropeptides and their simultaneous determination with whey proteins by RP-HPLC. *Int. Dairy J.* 16 (4) 285–293

9 R. M. SANDBLOM, US Pat. 4 105 547, 1978

10 U. KULOZIK and M. KERSTEN (2001): Fraktionieren von Proteinen mittels Mikrofiltration. *Chemie Ingenieur Technik* 73, 1622–1625

11 T. WEIGERT, J. ALTMANN, and S. RIPPERGER (1999): Crossflow electrofiltration in pilot scale. *J. Membrane Sci.*, 159, 253–262

2

Understanding the factors affecting spray-dried dairy powder properties and behavior

P. Schuck, INRA, France

Abstract: In view of the increased development of filtration processes, the dairy industry requires greater understanding of the effects of spray drying on the quality of protein powders. Spray drying parameters have an important role in the quality of dairy powders during drying, storage and rehydration. The residence time of the droplet, and then of the powder, is so short that it is very difficult to study the mechanism of the structural change in the protein without fundamental research into relationships with the process/product interactions. Following an introduction to spray drying, this chapter covers four arcas, i.e. the principles of spray drying, the properties and qualities of spray-dried dairy powders, the effects of spray drying parameters on powder properties, and the control and improvement of powder properties.

Key words: spray drying, powders, dairy products, physical properties, rehydration.

2.1 Introduction

The purpose of the dehydration of milk and derivates is to stabilize these products for their storage and later use. Milk and whey powders are used mostly in animal feeds. With changes in agricultural policies (such as the implementation of the quota system and the dissolution of the price support system in the European Union), the dairy industry has been forced to look for better uses for dairy surplus and for the by-products of cheese (whey) produced from milk, and buttermilk produced from cream. Study of the reuse of protein fractions in terms of nutritional qualities and functionality has led us to believe that they could have several applications.

In the past 30 years, the dairy industry has developed new technological processes for extracting and purifying proteins (e.g. casein, caseinates, whey proteins, etc.) such as milk protein concentrate (MPC), milk protein isolate (MPI), whey protein concentrate (WPC), whey protein isolate (WPI), micellar casein concentrates (MCC) and isolates (MCI), whey concentrates, and selectively demineralized whey concentrates, mainly because of the emergence of filtration technology (e.g. microfiltration, ultrafiltration, nanofiltration and reverse osmosis). This recent emergence of new membrane separation techniques and improvements in chromatographic resins now provide the dairy technologist with several types of technique for the extraction and purification of almost all of the main milk proteins. (Maubois, 1991; Fauquant *et al.*, 1988; Pierre *et al.*, 1992, Schuck *et al.*, 1994a,b; Jeantet *et al.*, 1996).

The most frequently used technique for the dehydration of dairy products is spray drying. It became popular in the dairy industry in the 1970s, but at that time there were few scientific or technical studies on spray drying and, in particular, none on the effects of spray drying parameters or on the effects of the physico-chemical composition and microbiology of the concentrates on powder quality. Manufacturers acquired expertise in milk drying, and eventually in whey drying, processes through trial and error. Because of the variety and complexity of the mixes to be dried, a more rigorous method based on physico-chemical and thermodynamic properties has become necessary. Greater understanding of the biochemical properties of milk products before drying, water transfer during spray drying, the properties of powders and influencing factors is now essential for the production of milk powder. The lack of technical and economic information and of understanding of scientific methods prevents the manufacturer from optimizing his plant in terms of energy costs and powder quality.

A dairy powder is characterized not only by its composition (proteins, carbohydrates, fats, minerals and water) but also by its microbiological and physical properties (bulk and particle density, instant characteristics, flowability, floodability, hygroscopicity, degree of caking, whey protein nitrogen index, thermostability, insolubility index, dispersibility index, wettability index, sinkability index, rehydration time, free fat, occluded air, interstitial air, particle size, water activity, glass transition temperature, etc.) which form the basic elements of quality specifications, and there are well-defined test methods for their determination according to international standards (Pisecky, 1986, 1990, 1997; American Dairy Products Institute, 1990; Masters, 2002). These characteristics depend on drying parameters (type of tower spray drier, nozzles/wheels, pressure, agglomeration and thermodynamic conditions of the air, such as temperature, relative humidity and velocity), the characteristics of the concentrate before spraying (composition/physico-chemical characteristics, viscosity, thermo-sensitivity and availability of water) and storage conditions. Several scientific papers have been published on the effects of technological parameters on these properties (Baldwin *et al.*, 1980; Pisecky, 1980, 1981, 1986; De Vilder, 1986; Tuohy, 1989; Jeantet *et al.*, 2008a; Masters, 2002) (Fig. 2.1). Water content, water

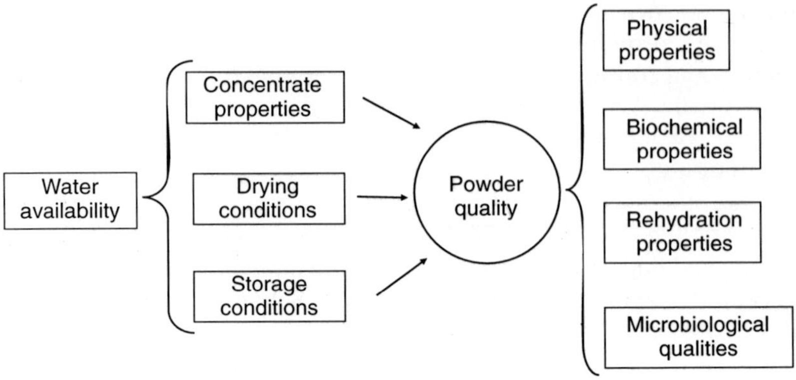

Fig. 2.1 Properties and qualities of powders.

dynamics and water availability are among the most important properties for all these powder properties and powder characteristics.

The aim of this chapter is to provide information on four areas, i.e. principles of spray drying, properties and qualities of spray-dried dairy powders, effects of spray drying parameters on powder properties, and control and improvement of powder properties.

2.2 Principles of spray drying

Drying is defined as the removal of a liquid (usually water) from a product by evaporation, leaving the solids in an essentially dry state. A number of different drying processes are in use in the dairy, food, chemical and pharmaceutical industries, such as spray-drying, fluid bed drying, roller drying, freeze-drying, microwave drying and superheated steam drying. Due to considerations related to drying economics and final product quality, the only processes of significance in milk protein powder manufacture are spray-drying and fluid bed drying (most often in combination). Only these two combined drying processes will be discussed in this section.

The basic principle of spray-drying is the exposure of a fine dispersion of droplets, created by means of atomization of preconcentrated milk products, to a hot air stream. According to Pisecky (1997), spray drying is an industrial process for the dehydration of a liquid by transforming the liquid into a spray of small droplets and exposing these droplets to a flow of hot air. The very large surface area of the spray droplets causes evaporation of the water to take place very quickly, converting the droplets into dry powder particles. The small droplet size created, and hence large total surface area, results in very rapid evaporation of water at a relatively low temperature, whereby heat damage to the product is minimized. (Refstrup,

2003) In fact, when a wet droplet is exposed to hot dry gas, variations in the temperature and the partial pressure of water vapor are spontaneously established between the droplet and the air:

- Heat transfer from the air to the droplet occurs under the influence of the temperature gradient;
- Water transfer occurs in the opposite direction, explained by variation in the partial pressure of water vapor between the air and the droplet surface.

Air is thus used both for fluid heating and as a carrier gas for the removal of water. The air enters the spray drier hot and dry, and leaves it wet and cool. Spray drying is a phenomenon of surface water evaporation maintained by the movement of capillary water from the interior to the surface of the droplet. As long as the average moisture is sufficient to feed the surface regularly, the evaporation rate is constant. If not, it decreases. The drying kinetics are related to three factors:

(i) The evaporation surface created by the diameter of the particles. Spraying increases the exchange surface (1 L of liquid sprayed in particles of 100 μm diameter develops a surface area of 60 m^2, whereas the surface area for one sphere of the same volume is approximately only 0.05 m^2).
(ii) The difference in the partial pressure of water vapor between the particle and the drying air. A decrease in the absolute humidity of the air and/or an increase in the air temperature tends to increase the difference in the partial pressure of water vapor between the particle and the drying air.
(iii) The rate of water migration from the center of the particle towards its surface. This parameter is essential for the quality of dairy powders. Indeed, it is important that there is always water on the surface of the product so that the powder surface remains at the wet bulb temperature for as long as possible. The rate of water migration depends on the water diffusion coefficient, which varies according to the biochemical composition, water content and droplet temperature. This is why calculation of this coefficient is complex and the mathematical models available are not easily exploitable by the dairy industry.

To define the components of a spray drying installation according to Masters (1991), Pisecky (1997) and Westergaard (2003), the main components of the spray drier shown in Fig. 2.2 are as follows:

- Drying chamber (Fig. 2.2, No. 7).
- Air filtration (Fig. 2.2, No. 17).
- Air heating system. The drying air can be heated in various ways: either indirectly by steam, oil, gas or hot oil, or directly by gas or electricity (Fig. 2.2, No. 5).

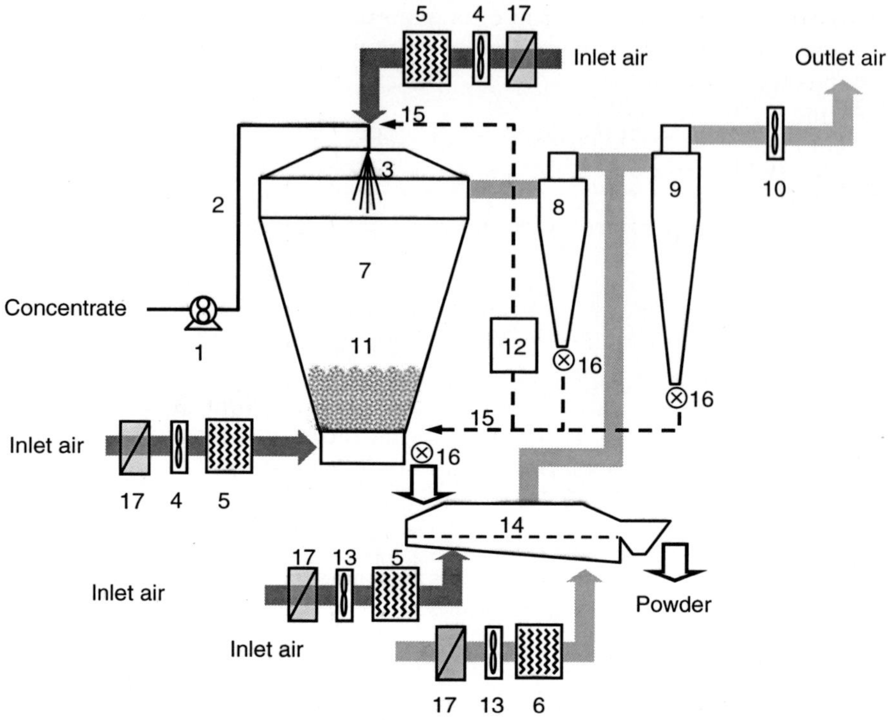

Fig. 2.2 Multiple effect spray-dryer; 1 Feed pump; 2 Feed flow; 3 Sprayer/Air disperser; 4 Inlet air fan; 5 Air heater; 6 Air cooler; 7 Drying chamber; 8 Primary cyclone; 9 Secondary cyclone; 10 Outlet air fan; 11 Integrated fluid bed; 12 Pressure conveyer system; 13 Vibro-fluidizer air fan; 14 Vibro-fluidizer; 15 Reincorporating fines; 16 Rotary valve; 17 Air filtration.

- Air distribution. The air flow chamber can be in co-current, counter-current or mixed mode. The system of air distribution is one of the most vital points in a spray dryer. There are various systems, depending on the plant design and the type of product to be produced. The most common system is where the air disperser is situated on top of the dryer ceiling and the atomizing device is placed in the middle of the air disperser, thus ensuring optimal mixing of the air and the atomized droplets.
- An atomizing device with a feed supply system such as a feed tank, feed pump, water tank, concentrate heater and atomizing device. The aim of atomizing the concentrate is to provide a very large surface area from which the evaporation of water can take place: the smaller the droplets, the greater the surface area; and the easier the evaporation, the greater the thermal efficiency obtained. The ideal situation from a drying point of view is a spray of droplets of the same size, where the drying time for all

particles would be the same to obtain equal moisture content. As mentioned previously, air distribution and atomization are key factors in the successful utilization of a spray-dryer. Atomization is directly responsible for many distinctive advantages offered by spray-drying: first, the very short drying time of the particles; second, the very short particle retention time in the hot atmosphere and low particle temperature (wet bulb temperature); and finally, the transformation of the liquid feed into a powder with long storage stability, ready for packing and transport.

In summary, the prime functions of atomization are:

- to produce a high surface-to-mass ratio, resulting in a high evaporation rate
- to produce particles of the desired shape, size and density.

There are three types of atomizing device: nozzle atomizer (pressure, pneumatic or sonic), rotary atomizer (wheel or disc) and combined atomizer (rotary and pneumatic) (Fig. 2.2, No. 3).

The basic function of pressure nozzles is to convert the pressure energy supplied by the high-pressure pump into kinetic energy in the form of a thin film, the stability of which is determined by the properties of the liquid, such as viscosity, surface tension, density and quantity per unit of time, and by the medium into which the liquid is sprayed.

The basic function of a pneumatic nozzle, also called a two-fluid nozzle, is to atomize the feed by high air velocity The energy available for atomization in two-fluid atomizers is independent of liquid flow and pressure. The necessary energy (kinetic) is supplied by compressed air. Two-fluid atomization is the only successful nozzle method for producing very small particles, especially from highly viscous liquids. It is not normally used in the drying of milk products.

In rotary atomizers, the liquid is accelerated continuously to the wheel edge by centrifugal forces produced by the rotation of the wheel. The liquid is distributed centrally and then extends over the wheel surface in a thin sheet and is discharged at high speed at the periphery of the wheel. The degree of atomization depends on peripheral speed, the properties of the liquid, and feed rate. According to Westergaard (2003), to select an optimal atomizer wheel, the liquid feed rate, peripheral speed and viscosity of the liquid should be taken into consideration.

- Powder recovery system. Separation of the dried product can be achieved by a primary discharge from the drying chamber followed by a secondary discharge from a particle collector (using a cyclone, bag filter or electrostatic precipitation), followed by total discharge from the particle collector and finishing with final exhaust air cleaning in a wet scrubber and dry filter.

> Authorities normally conclude that a powder loss of 250 mg m^{-3} (obtained classically by using only cyclones) is too high, and set a standard of 50 mg m^{-3} (obtained by using cyclones plus bag filters or cyclones plus wet scrubber which requires a final cleaning of the air). For these reasons, i.e. authorities' demand for reduced powder emission and powder producers' demand for lower energy consumption and reduced space requirements, a new powder recovery system has been developed, the cleanable-in-place (CIP-able) bag filter, which replaces the cyclones/bag filter (Westergaard, 2003).

According to Sougnez (1983), Masters (1991) and Pisecky (1997), the simplest types of installation are single-stage systems with a very short residence time (20–60 s). Thus, there is not enough residence time to obtain a real equilibrium between the relative humidity of the outlet air and water activity of the powder. The outlet temperature of the air must therefore be high, reducing the thermal efficiency of the single-stage spray drier. Generally, installations without any post-treatment system are suitable only for non-agglomerated powders not requiring cooling. If necessary, a pneumatic conveying system could be added to cool the powder while transporting the chamber fraction and the cyclone fraction to a single discharge point.

The two-stage drying system consists of limiting the spray drying process to a process with a longer residence time (several minutes) to provide a better thermodynamic balance. This involves a considerable reduction in the outlet air temperature (to increase the powder moisture content), and also an increase in the inlet air temperature, and in the concentrate and powder flow rate without risk of thermal denaturation due to the increase in moisture content of the powder during the first stage (thermal protection). A second, final drying stage is necessary to optimize the moisture content, by using an integrated fluid bed (static) or an external fluid bed (vibrating), the air temperatures of which are 15–25°C lower than with a single-stage system to improve and/or preserve the quality of the dairy powder (Fig. 2.2, Nos 11 and 14). Consequently, the surrounding air temperature at the critical drying stage and the particle temperature are also correspondingly lower, thus contributing to further improvement in economics. The integrated fluid bed can be either circular (e.g. Multi Stage Drier (MSD™) chamber) or annular (e.g. Compact Drier (CD) chamber). Two-stage drying has its limitations, but it can be applied to products such as skim milk, whole milk, pre-crystallized whey, caseinates, whey proteins and derivatives. The moisture content of the powder leaving the first stage is limited by the thermoplasticity of the wet powder, i.e. by its stickiness in relation to the water activity and the glass transition temperature (Roos, 2002). The moisture content must be close to 7–8, 9–10 and 2–3% for skim/whole milk, caseinate/whey protein and pre-crystallized whey powders, respectively. The two-stage drying techniques can be applied to the production of both non-agglomerated and agglomerated powders, but this technique is very suitable for the production of agglomerated powders, by separating the non-agglomerated particles from the agglomerates (i.e. collecting the cyclone fractions and re-introducing these fine fractions (called fines) into the wet zone around the atomizer of the chamber) (Fig. 2.2, No. 15).

Three-stage drying systems, with an internal fluid bed as a second stage in combination with an external vibrating fluid bed as a third-stage drier, first appeared at the beginning of the 1980s and were called Compact Drier Instantization (CDI) or MSD™. Today, they dominate the dairy powder industry (Fig. 2.2). Three-stage systems combine all the advantages of extended two-stage drying, using spray drying as the primary stage, fluid bed drying of a static fluid as the second drying stage and drying on an external vibrating fluid bed as the third drying stage. The final drying stage terminates with cooling to under the glass transition temperature. Evaporation performed at each stage can be optimized to achieve both gentle drying conditions and good thermal economy.

CD or CDI is suitable for producing both non-agglomerated and agglomerated powders of practically any kind of dried dairy product. It can also cope successfully with whey powders, fat-filled milk and whey products as well as caseinates, both non-agglomerated and agglomerated. It has a fat content limit of about 50% fat in total solids. Powder quality and appearance are comparable with those of products from two-stage drying systems but they have considerably better flowability and the process is more economical.

In comparison with CD, MSD™ can process an even wider range of products and can handle an even higher fat content. The main characteristic of MSD™ powder is due to very good agglomeration and mechanical stability, low particle size fractions (below 125 μm) and very good flowability. Optimization of the process allows considerable improvement in drying efficiency and the quality of the product obtained is generally better. The various advantages are:

- improved thermal efficiency: significant reduction in outlet air temperature, permitting an increase in inlet air temperature;
- reduced material obstruction: the capacity in one volume is two or three times greater than for a traditional unit;
- considerable reduction in powder emission to the atmosphere: reduction of the drying air flow and increase in powder moisture content together decrease the loss of fine particles in the outlet air;
- improved powder quality in relation to agglomeration level, solubility, dispersibility, wettability, particle size, density, etc.

There are other examples of drying equipment such as the 'tall form drier', the 'Filtermat™ drier', the 'Paraflash™ drier', the 'Tixotherm™ drier' and the Integrated Filter Dryer™ (IFD™). All these towers have characteristics related to the specific properties of the product being dried (e.g. high fat content, starch, maltodextrin, egg and hygroscopic products).

In terms of energy balance, Westergaard (2004) showed that energy consumption varies according to the drying processes. Energy consumption to produce 1 kg of powder from skim milk concentrate at 45% of total solids is 1595, 1350, 1280, 1038 and 960 Kcal for a one-stage spray dryer, a two-stage spray dryer, a two-stage spray dryer with high inlet air temperature, CDI and MSD™, respectively. This is explained by the increase in the number of drying stages simultaneously increasing the residence time, allowing increase in the inlet air temperature, concentrate flow rate and finally energy yield while preserving/improving powder quality (Bimbenet *et al.*, 2002; Jeantet *et al.*, 2008b).

2.3 Properties and qualities of dairy powders

According to Carić (2003), the properties of milk powders are categorized as physical, functional, biochemical, microbiological and sensory. There is a significant interrelationship between them, which affects the final quality. The physical and functional properties of milk powders are especially important when the powders are intended for recombining and in the manufacture of various food products. When intended for use as a food ingredient, milk powders should be light in color, free from off-flavors and easy to hydrate, disperse and dissolve in water. The basic properties that determine the quality of milk powder, and where defects are most likely, include powder structure, solubility, water content, scorched particles, flowability, floodability, oxidative changes, flavor, color and micro-organism contamination.

2.3.1 Physical properties

Powder structure

The physical structure of a milk powder can be defined as the way in which its chemical components are distributed and connected. Powder structure is very strongly affected by the drying technique. Powder produced by roller-drying has a compact structure of irregular shape with no occluded air. Roller-dried powder particles have a low bulk density (300–500 kg.m^{-3}), due to their irregular structure. The particles of spray-dried powder are spherical, with diameters in the range 10–250 μm. The particles contain occluded air and either large central vacuoles or smaller vacuoles which are distributed throughout the interior of the particles. The surface of spray-dried skim milk powder particles is usually wrinkled but it is smooth for high protein powders. The high inlet air temperature and large temperature differential between the hot air and the powder particles are the main causes of wrinkle formation, as is also the presence of lactose (Carić and Kaláb, 1987; Mistry *et al.*, 1992; Aguilar and Ziegler, 1993).

Particle size distribution

According to Carrić (2003), the particle size of a powder, which affects its appearance, reconstitution and flow characteristics, depends mainly on the atomization conditions and the viscosity of the concentrate; high atomizing pressure and low concentrate viscosity reduce particle size.

Powder density

Densities are classified into three groups: bulk (apparent) density, particle density and the density of the dry milk solids; all three are very much interrelated.

- Bulk density: Bulk density is regarded as the weight per unit volume and is expressed as $kg.m^{-3}$. It is a very important property, both from the point of view of cost and market requirements. Bulk density is currently determined by measuring the volume of 100 g of powder in a 250 mL graduated glass cylinder. The bulk density of milk powders is a very complex property, being the result of many other properties and being influenced by a number of factors such as feed concentration, feed temperature, feed foamability, milk preheating, age thickening, feed composition, type of atomizer, particle temperature history and particle size distribution (Pisecky, 1997). Bulk density depends also on particle density and occluded and interstitial air.
- Particle density: Particle density corresponds to the mass of particles (in grams) having a total volume of 1 cm^3. Particle density is influenced mainly by the amount of entrapped air. The processing factors that contribute significantly to particle density are viscosity and the incorporation of air into the concentrate prior to drying. The type of spray atomization affects air retention. Certain types of centrifugal spray-dried milk have more entrapped air than pressure spray products (Carrić, 2003).
  - Occluded air: The occluded air content is defined and quantified by the difference between the volume of a given mass of particles and the volume of the same mass of air-free milk solids. Many factors influence the occluded air content in powder particles, including incorporation of air into the feed, the system chosen for spray drying the concentrate, whipping action before and/or during atomization, properties of the feed and the ability of the feed to form a stable foam. The content and state of proteins might markedly affect stable foam formation, while fat has the opposite effect. High-fat concentrates are much less susceptible to foaming than skim milk. Undenaturated whey proteins in skim milk have a greater tendency to foam, which can be reduced by heat treatment, which causes protein

denaturation. Concentrates with low total solid contents foam more than more highly concentrated materials. An increase in temperature from 10 to 40°C increases the tendency to form foam for skim milk concentrate at 40 and 45% of total solids related to decrease in viscosity. Therefore, an increase in temperature from 10 and 40°C reduces the tendency to form foam for concentrate at 50% of total solids. This result could be explained by a thermal denaturation (Pisecky, 1997; Carrić, 2003).

  - Interstitial air: Interstitial air is defined as the difference between the volume of a given mass of particles and the volume of the same mass of tapped powder. This property depends primarily on the particle size distribution and the degree of agglomeration (Carrić, 2003).

- Density of the dry milk solids: This density expresses the density of solids without any air and is given by the composition of the power (Pisecky, 1997; Westergaard, 2004).

Flowability

Flowability is the ability of a powder to flow freely, like sand, without forming lumps, clusters or aggregates. Flowability can be measured as the time (in seconds) necessary for a given volume of powder to leave a rotary drum through given slits (Haugaard Sorensen *et al.*, 1978) or by the method developed by Carr (1965). Flowability depends also on particle size and shape, density and electrical charge. Large particles flow more easily than fines (particles with a diameter of <90 μm). Consequently, agglomeration is beneficial, as is uniformity of size. Moreover, according to Carrić (2003), a wide variation in particle size permits fines to occupy spaces between the large particles, which results in closer packing.

Rehydration

Most food additives are prepared in powder form and need to be dissolved before use. Water interactions in dehydrated products and dissolution are thus important factors in food development and formulation (Hardy *et al.*, 2002). Dissolution is an essential quality attribute of a dairy powder as a food ingredient (King, 1966). Many sensors and analytical methods such as the insolubility index (International Dairy Federation, 1988; American Dairy Products Institute, 1990), nuclear magnetic resonance (NMR) spectroscopy (Davenel *et al.*, 1997), turbidity, viscosity and particle size distribution (Gaiani *et al.*, 2006a) can now be used to study water transfer in dairy protein concentrates during rehydration. Using combinations of these methods, it is very easy to determine the different stages of the rehydration process (i.e. wettability, sinkability, dispersibility and solubility).

- Wettability is the ability (expressed as time in seconds) necessary for a given amount of powder to penetrate the still surface of water. In other words, wettability is the ability of a powder to absorb water on the surface and get wet (Haugaard Sorensen *et al.*, 1978). Generally, the wettability of powder particles depends on the surface activity of the particles, surface area, surface charge, particle size, density, porosity and presence of moisture-absorbing substances.

- Sinkability is the ability of powder particles to overcome the surface tension of water and sink into water after passing through the surface. Sinkability is expressed as milligrams of powder that sink per min per cm^2 of surface area. This property of powder is influenced by the forces that tend to submerge a particle on the surface and depends on the density of the particles, i.e. on the mass of the particles and the quantity of occluded air. Higher particle density and lower quantity of occluded air cause particles to sink (Carrić, 2003).
- Dispersibility reflects the ability of the wetted aggregates of powder particles to become uniformly dispersed when in contact with water. The effects of total heat treatment on casein during processing are of particular importance for good dispersibility. The dispersibility of milk powder can be improved by: (i) keeping the heat treatment on preheating to a minimum and (ii) minimizing the holding time and temperature of the concentrate.
- The insolubility index (ISI, in %), described by the IDF standard (International Dairy Federation, 1988) for skim milk, is the volume of sediment (for 50 mL) after rehydration (10 g of powder in 100 mL of distilled water, at 25°C), mixing (90 s, at 4000 $rev \cdot min^{-1}$) and centrifugation (300 s, at 160 g). With this method, the quantity of insoluble material can be determined but this IS is related only to the decrease in water transfer needed for rehydration. This index cannot differentiate between the truly insoluble (related to the thermal denaturation) and the falsly insoluble (related to the biochemical composition) (Schuck *et al.*, 1994b).

A modification of the rehydration parameters recommended by the IDF method (1988) resulted in marked variations in the ISI of micellar casein isolate (MCI) powder. The index decreased to 6.2 mL if the casein content of the sample of MCI powder was close to the casein content of milk (25 $g.L^{-1}$), the duration of stirring increased from 90 s to 900 s (ISI = 1.8 mL), the rehydration temperature increased from 24°C to 30°C (ISI = 7.2 mL), 40°C (ISI = 2.3 mL), 50°C (ISI = 0.9 mL%) or 60°C (ISI = 0.1 mL) and/or if the stirring velocity increased from 4000 to 10,000 $rev.min^{-1}$ (ISI = 0.8 mL). The replacement of the rehydration water by a saline solution (NaCl at 0.1 $mol.L^{-1}$), a microfiltrate or an ultrafiltrate did not change the ISI for MCI powder (Schuck *et al.*, 1994a,b). These results confirm that the ISI given by the IDF method (1988) results from a decrease in the water transfer towards the center of the particle and not from denaturation. Everything occurs as if the setting in contact with water created a high surface viscosity, slowing down the internal hydration of the MCI powder.

To determine the different steps of rehydration, various methods can be used, as follows:

- NMR spectroscopy is a technique for determining the rate of solution, the time required for complete reconstitution of powders and the transverse relaxation rate of reconstituted solutions (Davenel *et al.*, 1997). With this method, it is possible to differentiate between truly insoluble material and falsely insoluble material. (Schuck *et al.*, 1994b).
- For viscosity measurement, a rheometer can be used to obtain viscosity profiles. The blades of the agitator are placed at right angles to each other to provide good homogenization. Industrial dissolution processes usually include stirring at a constant speed and experiments have therefore been designed to provide a constant shear rate ($100\ s^{-1}$). The aqueous phase used is distilled water at a volume of 18 mL. The powder is dispersed in the rheometer cup 50 s after starting the rheometer. Dissolution is highly dependent on temperature and concentration (Gaiani *et al.*, 2005, 2006a).
- Experiments to provide turbidity profiles can be carried out in a 2 L vessel equipped with a four-blade 45° impeller rotating at 400 $rev.min^{-1}$. A double-walled jacket vessel maintains the temperature at 24°C. The turbidity sensor is placed 3 cm below the surface of the water and is positioned through the vessel wall to avoid disturbance during stirring. Turbidity changes accompanying powder rehydration are followed using a turbidity meter. The apparatus uses light in the near-infrared region (860 nm), the incident beam being reflected back at 180° by any particle in suspension in the fluid to the sensitive electronic sensor (Gaiani *et al.*, 2005).
- A laser light diffraction apparatus at a wavelength of 632.8 nm can be used to record particle size distributions. In the study of Gaiani *et al.* (2005, 2006a), the particle size distribution of particles was determined using a wet attachment, and the standard optical model presentation for particles dispersed in air was used. The results obtained corresponded to average diameters calculated according to the Mie theory. The criterion selected was d(50), meaning that the diameters of 50% of the particles were lower than this criterion (midpoint of cumulative volume distribution) (Gaiani *et al.*, 2005, 2006a).

Using a combination of the last three methods, it is possible to follow water transfer during rehydration and to obtain the wetting time, determined using the first peak of increased viscosity and turbidity, and the swelling time, determined using the second peak of viscosity in relation to the increase in particle size. The rehydration time can then be determined according to stabilization of the viscosity, turbidity and particle size values.

2.3.2 Physico-chemical properties

The aim of this short paragraph is to correlate certain thermodynamic information (moisture sorption isotherm, water activity (a_w), glass transition temperature (T_g) and state diagrams) to be able to understand the behavior of a powder at given temperatures and a_w conditions in terms of the quality of dairy ingredients and the rehydration behavior. From this thermodynamic information, it should be possible to anticipate the behavior of a powder under given temperature and a_w conditions (Schuck *et al.*, 2007).

Glass transition

Glass transition of dairy solids can be observed using differential scanning calorimetry (DSC). DSC measures any change in heat capacity that occurs over the glass transition temperature range (Roos, 2002). DSC is also the most common method for the determination of glass transition temperatures, taken from the onset or midpoint temperature of a change in heat capacity (Jouppila and Roos, 1994). Many sugars, including lactose, are transformed rapidly from the solid glassy state to a syrup-like, sticky liquid.

The glass transition of anhydrous lactose, as observed using DSC, has an onset temperature of 101°C, which is one of the highest temperatures measured for 'anhydrous' disaccharides (Jouppila and Roos, 1994). The glass transitions observed in milk solids are very close to those of pure lactose. However, if lactose is hydrolyzed, the T_g observed decreases dramatically, because of the much lower T_g of the galactose and glucose components. This also results in significant changes in the spray drying behavior and storage stability of lactose-hydrolyzed milk solids (Jouppila and Roos, 1994). Amorphous carbohydrates, including lactose and its hydrolysis products, are significantly plasticized by water, demonstrated by a rapidly decreasing T_g with increasing water content. The effect of water on the T_g of milk solids may be predicted using the Gordon–Taylor equation (Jouppila and Roos, 1994), or the Couchman–Karasz equation (Couchman and Karasz, 1978) according to Schuck *et al.* (2005a). Information regarding water plasticization can also be obtained from water sorption properties, which allow evaluation of the extent of water plasticization of dairy powders under various storage conditions.

Several studies have shown that the stickiness of dehydrated powders results from particle surface plasticization and concurrent decrease in viscosity, allowing the formation of liquid bridges between powder particles. It may be assumed that similar mechanisms control particle properties in the spray drying process. However, the process involves removal of the solvent and plasticizer, which has to occur at a rate competing between the position of the particle in the chamber and the formation of a dry surface to allow free flow of individual particles throughout the dehydration process.

Stickiness

Several techniques have been developed to characterize the stickiness behavior of food powders. All the tests are empirical in nature, and the techniques are still in development due to inaccuracies and the difficulties of application to real processing and handling situations. Due to the significant negative economic consequences in the industrial processing and handling of sticky products, investigations to find an accurate, simpler and cheaper technique to characterize the stickiness behavior of these types of product are still ongoing. For example, the glass transition temperature of skim milk solids, showing the stickiness and caking zone, is about 10°C above T_g or higher measured by DSC.

Some studies have tried to relate stickiness behavior and glass transition (Boonyai *et al.*, 2004, 2005). Various techniques and types of instrumentation have been developed to characterize the stickiness behavior of powder particles. The instrumental measurement concepts are generally based on the properties of food materials, such as resistance to shear motion, viscosity, optical properties and glass transition temperature. The first two properties provide a direct interpretation of stickiness behavior, whereas the measurements obtained from the latter can be indirectly correlated with stickiness. Stickiness characterization techniques may therefore be divided into direct and indirect techniques. The direct techniques can also be further classified as conventional, pneumatic and *in situ* techniques, according to the testing mechanism used. Conventional techniques involve mechanical movement, shearing or compression of the powder sample (propeller-driven methods, shear cell method, ampoule method or optical probe method), whereas pneumatic techniques involve the use of an air stream, with predetermined temperature and humidity, to blow or suspend the powder particles (fluidization test, cyclone stickiness test, blow test method). The *in situ* technique involves measurement of surface stickiness of single liquid droplets during hot air drying.

Water activity and sorption isotherm

The water activity (a_w) of dried milk products is largely correlated with moisture content and temperature. The composition and state of individual components, as influenced by various processing techniques, also play an important role. The composition of the solids more or less reflects the protein content. At low moisture content, characterized by $a_w < 0.2$, the casein is the main water absorber. Within the intermediate range of > 0.2 to < 0.6, sorption is dominated by the transformation of the physical state of lactose. Above this level, salts have a marked influence (Pisecky, 1997).

The water activity of milk powders consisting of non-fat milk solids and milk fat is controlled mainly by the moisture content expressed in non-fat solids since the fat has no influence. Thus, differences in a_w are due mostly to the state of proteins and the physical state of the lactose. The methods to determine a_w consist of putting the product in equilibrium with the surrounding atmosphere, then measuring the thermohygrometric characteristics of the air in equilibrium with the product. The a_w should be close to 0.2 at 25°C for optimal preservation (Efstathiou *et al.*, 2002). Water activity plays an important role during the dehydration process, and an understanding of sorption isotherms can provide valuable guidelines for the engineering design, control of the drying process (isothermal desorption curves) and storage stability (isothermal adsorption curves). In practical terms, all dairy powder isotherms found in the literature were obtained using final products as starting materials. The powder or the concentrate is exposed to air of defined relative humidity and brought to equilibrium, after which the moisture content is determined. Thus, the published isotherms are designated as adsorption or desorption.

Establishing equilibrium can very often take weeks. Many mathematical equations, both theoretical and empirical, have been reported in the literature to express the water sorption isotherms of dairy powders (Pisecky, 1997). With practical or theoretical sorption isotherms, the ideal moisture content can be determined for the optimal stabilization (at 0.2 a_w and at 25°C) of some dairy powders. For example, the corresponding moisture content must be close to 4%, 2–3% and 6%, for skim milk, whey and caseinate powders, respectively.

With T_g, water activity is one of the main factors governing many of the phenomena occurring during thermal dehydration, including:

- Easiness of water evaporation from a liquid droplet,
- Particle temperature history during the whole removal process,
- Moisture content equilibrium that can be achieved under given conditions at infinite residence time,
- Stickiness of the product (in relation to the T_g) and outlet conditions used for drying without occurrence of sticking (Pisecky, 1997).

2.4 Effects of spray drying parameters on powder properties

2.4.1 During spray drying

The properties of the final product are influenced by a number of factors, including the quality and composition of the raw milk and the processing conditions applied. As some of the factors are subject to both seasonal and daily variations, it is necessary to monitor those properties that might be affected by these variations and to make appropriate corrections to the operation parameters.

Two major properties, i.e. moisture content and bulk density, are good examples described below to show the strong relationships between process and products.

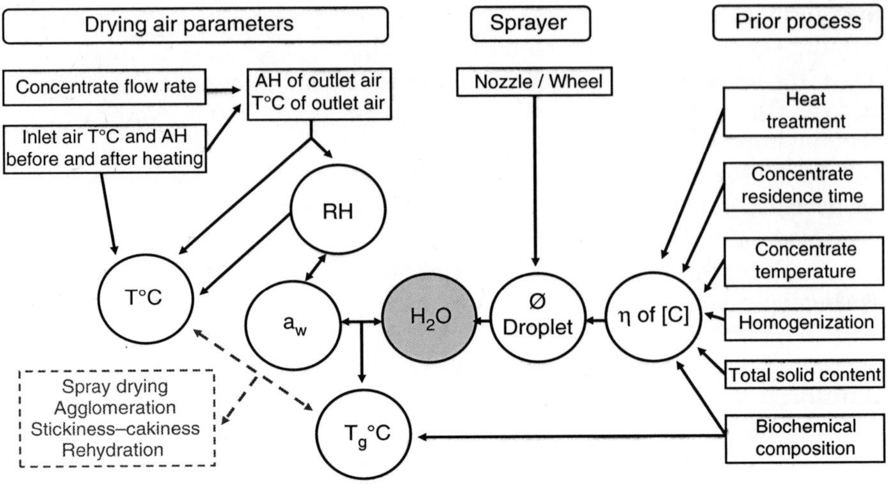

Fig. 2.3 Influence of various factors on powder moisture content. AH = Absolute humidity; RH = Relative humidity; H_2O = Moisture content; η = Viscosity; T°C = Temperature; T_g°C = Glass transition temperature; Ø = Size; [C] = Concentrate; a_w = water activity

Moisture content

Figure 2.3 shows the influence of various factors, such as the relative humidity (RH) of the outlet air, on powder moisture (Schuck *et al.*, 2008; Pisecky, 1997). As shown in this Figure, moisture content depends on the drying air parameters, on the spray dryer and on the previous process. Moisture content is influenced by the a_w, which depends on the RH of the outlet air. The relative humidity (RH) is influenced by the absolute humidity (AH) and temperature of the outlet air, which themselves depend on the concentrate flow rate, the inlet air temperature and AH before and after heating. The moisture content is also influenced by the size of the droplet, depending on the type of sprayer and viscosity. The viscosity of the concentrate depends on the heat treatment, temperature, residence time, homogenization, total solid content and biochemical composition. The temperature T of the droplet (and thus the powder) is influenced by the temperature and AH of the inlet air and by the RH, AH and temperature of the outlet air. The glass transition temperature (T_g) depends on biochemical composition, a_w and moisture content. Knowledge of the T and T_g makes it possible to monitor agglomeration, stickiness, cakiness and rehydration. Experiments reported by Schuck *et al.* (2008) show that the outlet air temperature is not always the optimum parameter to affect the moisture content of a dairy powder. The RH of the outlet air is the key parameter to optimize the moisture content and water activity of dairy powders.

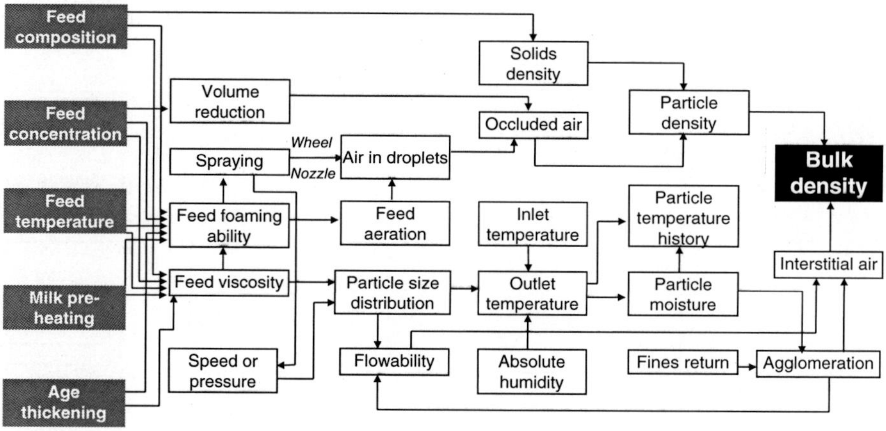

Fig. 2.4 Interrelationship of various drying parameters and physicochemical properties on bulk density (Pisecky, 1997; Carrić and Milanović, 2003).

Bulk density

Other important properties are bulk density and particle density. High bulk density is of economic importance in products for shipment in bulk, saving both packaging material and transport costs. Figure 2.4 shows the interrelationship of various drying parameters and physicochemical properties, and their relationship with bulk density (Pisecky, 1997; Carrić and Milanović, 2003). Apart from the product mass density, which is provided by its composition, two factors have a decisive influence on bulk density, namely the particle density (provided by product mass density and the occluded air content), and the interstitial air content (provided by particle size distribution and agglomeration).

Particle density depends on many factors. The composition of the solids plays an important part, first of all because it defines the product mass density. High protein content tends to reduce particle density as it increases the tendency of the feed to foam. This foaming can be suppressed somewhat by high heat pre-treatment, (denaturing the whey proteins), and also by high concentration combined with heating the feed.

To achieve high particle density, it is important to avoid any treatment that may incorporate air into the feed, such as excessive agitation. Rotary atomizers tend to incorporate air into the droplets, and pressure nozzles produce much higher particle density than rotary wheels. However, special vane-shaped rotary wheels are now available with less tendency to entrap air in product droplets.

The presence of air in the atomized droplets results in occluded air in the dried particles. Depending on drying conditions, or to be precise on the particle temperature during the drying process, any air bubbles initially present may expand and further reduce the particle density. Therefore, if

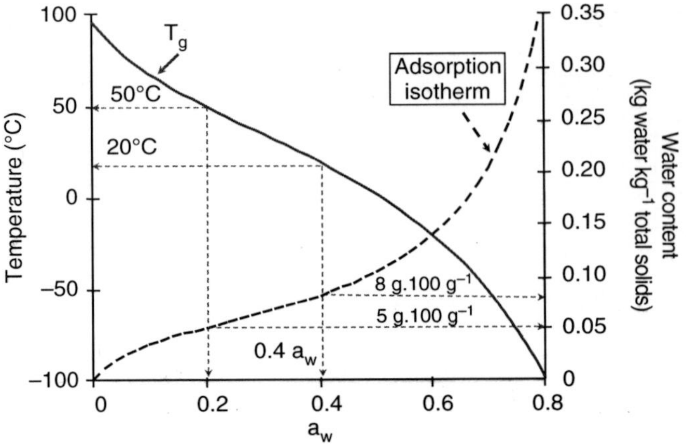

Fig. 2.5 Glass transition temperature (T_g) and water adsorption isotherm for skim milk solids and T_g curve of lactose (Roos, 2002).

all other conditions are the same, the two-stage or three-stage drying processes provide higher particle density than single-stage drying.

2.4.2 During storage

From the thermodynamic information (T_g, a_w, see Section 2.3.2), it should be possible to anticipate the behavior of a powder at a given temperature and under a_w conditions. Combination of the T_g and a_w curves in relation to the moisture content makes it possible to obtain a very interesting profile (Fig. 2.5) (Roos, 2002). From this curve, it is easy to determine the moisture content and the T_g at different a_w values in relation to a skim milk powder. For example, the moisture content and the T_g at 0.2 a_w, will be close to 5% and 50°C, respectively. If one wants an a_w close to 0.4, the moisture content and the T_g will be close to 8% and 20°C, respectively. However, for this powder some enzymatic and non-enzymatic reactions can begin at 0.4 a_w, which may increase deterioration and loss of quality. Moreover, it is not easy to stabilize this powder at a temperature lower than 20°C.

2.4.3 During rehydration

As explained in Section 2.3.1, Rehydration, the IDF method, NMR spectroscopy, turbidimetry, viscosimetry and granulometry provide valuable information regarding the mechanisms involved in dairy powder dissolution with constant stirring. However, to optimize the rehydration time, two parameters are very important, namely particle size, which depends on the spray drying conditions, and the powder structure, in particular the surface composition.

Effects of particle size

As expected, granulation has a positive effect on wetting. The wetting time is systematically better for granulated particles. This phenomenon is well known, with large particles forming large pores, high porosity and small contact angles between the powder surface and the penetrating water as fast wetting is enhanced (Freudig *et al.*, 1999). A surprising influence of granulation on the rehydration time was observed in the study of Gaiani *et al.* (2005). Depending on the nature of the protein, the granulation influence resulted in opposite effects. Whey protein isolate (WPI) rehydration time was shorter for granulated particles whereas it was shorter for non-granulated particles of micellar casein isolate (MCI). This was unexpected and could be explained by the controlling stage rate. The controlling stage for whey proteins is wetting. As granulation improves the wetting stage, the rehydration of whey powders is enhanced for granulated particles. In contrast, in this study, the controlling stage for casein proteins was dispersion. In fact, even with a shorter wetting time, a granulated powder is slower to rehydrate than a non-granulated powder (Gaiani *et al.*, 2005).

These results are not compatible with other studies on skim milk powder, in which it was generally accepted that a single particle size around 200 μm (Neff and Morris, 1968) or 400 μm (Freudig *et al.*, 1999) represented optimal dispersibility and sinkability. In fact, this optimal particle size depends on the composition of the dairy powder. If the industry wishes to optimize powder rehydration, it seems to be better to rehydrate granulated powders when the protein is whey and to rehydrate non-granulated powders when the protein is casein.

Effects of composition surface

Gaiani *et al.* (2006b) have shown that X-ray photoelectron spectroscopy (XPS) analysis yields essential information complementary to standard methods, providing greater understanding of rehydration by determination of the surface composition. A small change in powder surface composition or bulk composition can totally change the rehydration properties. It is worth noting that powders with less distinct compositions might present significant rehydration differences.

Table 2.1 shows the effects of storage on the surface composition of micellar casein powder and how lipids can migrate from the bulk to the surface. XPS has revealed that lipids are more likely to migrate toward the surface than lactose or proteins. We have demonstrated that the monolayer moisture capacity (X_m) determined by dynamic vapor sorption (DVS), which is decreased during storage, is probably due to the presence of surface lipids (formation of a hydrophobic surface) and that the presence of pores at the surface could be a route allowing lipid release. A fat layer could cover progressively the surface during storage, inducing a decrease of the monolayer capacity due to its hydrophobic nature.

Table 2.1 Monolayer moisture capacity (Xm), surface fat composition and wetting time modifications of micellar casein powders during storage at 20 and 50°C

Storage temperature (°C)	Storage duration (days)	Xm ($kg.kg^{-1}$)	Surface fat (%)	Wetting time (s)
20	0	0.0632	6	12
	15	0.0626	6	14
	30	0.0617	12	66
	60	0.0617	16	73
50	0	0.0632	6	12
	15	0.0574	6	15
	30	0.0537	13	68
	60	0.0524	17	265

The combination of these two methodologies (DVS and XPS) enhanced our understanding of lipid migration on a powder surface. This study also revealed the possibility of creating porosity inside the powder by controlled storage conditions (e.g. temperature, relative humidity, time). Further studies with powders containing more or less fat would therefore improve the understanding of fat release. In the future, XPS analysis could help to formulate dairy powders presenting better rehydration properties. Improving both the wetting time and the rehydration time could be one way to extend the applications of high protein content powders, the principal limitation of these powders being the low water transfer during rehydration (Gaiani *et al.*, 2007).

2.5 Control and improvement of powder properties

Due to the variety and complexity of the mixes to be dried, a more rigorous method based on physico-chemical and thermodynamic properties has now become necessary. Improved understanding of the biochemical properties of milk products before drying, water transfer during spray drying, the properties of powders and influencing factors has now become indispensable in the production of dairy powders. Lack of technical and economic information and of scientific methods prevents the manufacturer from optimizing his process in terms of energy costs and powder quality. Two approaches are necessary for dairy research into the spray drying of dairy products, one involving the products (availability of water related to the biochemical composition) and the second involving the process (knowledge and improvement of the drying parameters).

2.5.1 Availability of water

The aim of this section is to propose a new method (drying by desorption, using a thermohygrometer sensor) in order to determine major drying parameters according to food components in relation to their interactions with water (bound and free water) and linked to water transfer kinetics. Schuck and co-workers' studies (1998, 2009) have shown that drying by desorption is an excellent tool to determine and optimize the major spray-drying parameters in relation to biochemical composition according to water availability and desorption behavior (calculation of extra energy ΔE). The experimental device proposed by the authors differs from spray-drying equipment in terms of duration of drying, drying temperature, surface/volume ratio, etc. because the concentrate is dried in a cup and not in a droplet However, some computational tools have been developed to improve the method by taking these factors into account. Validation tests (>30 products) have indicated that this method could be applied to a wide range of food products and spray-dryer types. For reasons of calculation speed and reliability, this method has been computerized and it can already be used in the determination of parameters of spray drying for food products. The name of the new software is 'Spray Drying Parameter Simulation and Determination Software' ($SD^2P^{®}$), registered under the following identification: IDDN.FR.001.480002.002.R.P.2005.000.30100.

Analysis of the desorption curve (measured relative humidity versus time), combined with knowledge of the temperature, total solids, density and specific heat capacity of the concentrate, air flow rates, theoretical water content in relation to water activity and RH of the outlet air, current weather conditions, cost per kWh and percentage of drying in the integrated fluid allows determination of enthalpy, T, RH (including ΔE) for each inlet air, concentrate and powder flow rate, specific energy consumption, energy and mass balance, yield of the dryer and cost (in € or in $) to remove 1 kg of water or to produce 1 kg of powder. All these results are summarized in Fig. 2.6. This figure is a representation of the software delivery:

- air characteristics at the dryer/integrated fluid bed inlet and outlet (upper part),
- flow, energy and cost calculations (lower part) (Schuck *et al.*, 2009)

2.5.2 Process improvement

The aim of this section is to show the use of a thermohygrometric sensor, with some examples of such measurements (temperature, absolute (AH) and relative humidity (RH), dry air flow rate, water activity) through calculation of mass and absolute humidity to prevent sticking in the drying chamber and to optimize powder moisture and water activity in relation to the relative humidity of the outlet air.

It was demonstrated by Schuck *et al.* (2005b) that a thermohygrometer can be used to avoid sticking and to optimize water content and water activity in dairy powders. From these results, it could be seen that the

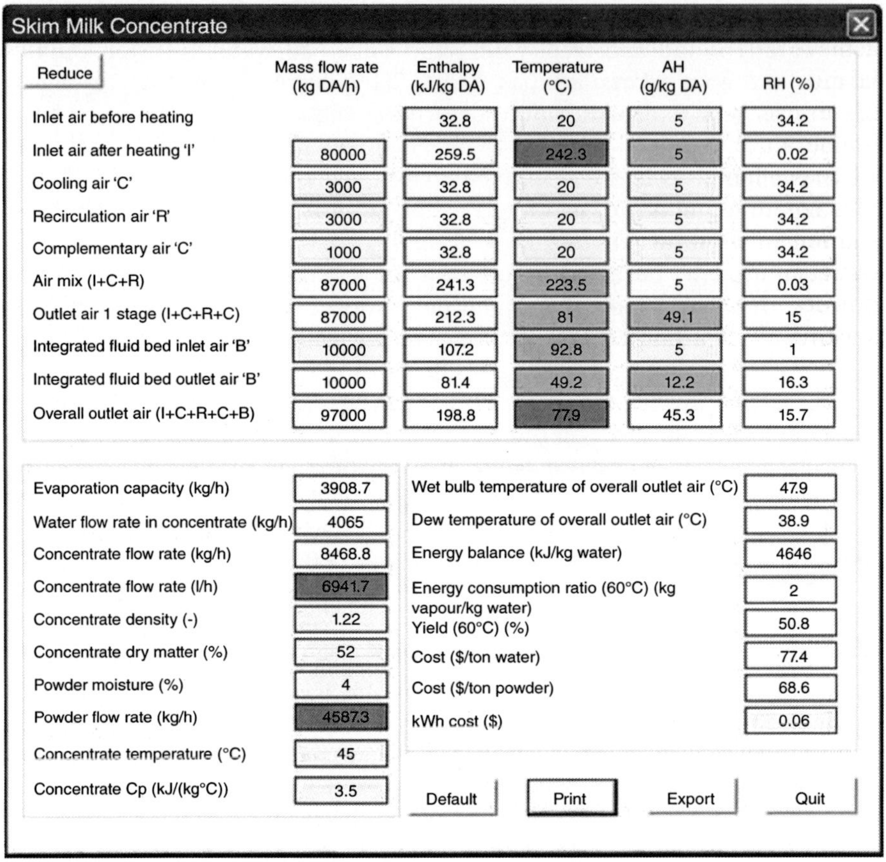

Skim Milk Concentrate

Reduce	Mass flow rate (kg DA/h)	Enthalpy (kJ/kg DA)	Temperature (°C)	AH (g/kg DA)	RH (%)
Inlet air before heating		32.8	20	5	34.2
Inlet air after heating 'I'	80000	259.5	242.3	5	0.02
Cooling air 'C'	3000	32.8	20	5	34.2
Recirculation air 'R'	3000	32.8	20	5	34.2
Complementary air 'C'	1000	32.8	20	5	34.2
Air mix (I+C+R)	87000	241.3	223.5	5	0.03
Outlet air 1 stage (I+C+R+C)	87000	212.3	81	49.1	15
Integrated fluid bed inlet air 'B'	10000	107.2	92.8	5	1
Integrated fluid bed outlet air 'B'	10000	81.4	49.2	12.2	16.3
Overall outlet air (I+C+R+C+B)	97000	198.8	77.9	45.3	15.7

Evaporation capacity (kg/h)	3908.7
Water flow rate in concentrate (kg/h)	4065
Concentrate flow rate (kg/h)	8468.8
Concentrate flow rate (l/h)	6941.7
Concentrate density (-)	1.22
Concentrate dry matter (%)	52
Powder moisture (%)	4
Powder flow rate (kg/h)	4587.3
Concentrate temperature (°C)	45
Concentrate Cp (kJ/(kg°C))	3.5

Wet bulb temperature of overall outlet air (°C)	47.9
Dew temperature of overall outlet air (°C)	38.9
Energy balance (kJ/kg water)	4646
Energy consumption ratio (60°C) (kg vapour/kg water)	2
Yield (60°C) (%)	50.8
Cost ($/ton water)	77.4
Cost ($/ton powder)	68.6
kWh cost ($)	0.06

Fig. 2.6 Parameters of spray drying calculated by the SD^{2}P® software.

calculated AH is systematically higher than the measured AH, because the calculated AH corresponded to the maximum theoretical value that could be reached. Calculation of AH by means of the mass balance is based on the hypothesis that the air circulating in the spray drier removes all the water from the concentrate. Thus, if the difference between calculated and measured absolute humidity of the outlet air is below 2 g of water per kg dry air (depending on the spray drier with regard to measurement accuracy), there is no problem of sticking in the spray dryer chambers, whatever the dairy concentrate used. On the other hand, sticking was observed in this study for differential AH above 2 g water per kg dry air, corresponding to lower water removal and consequently to favorable sticking conditions. The operator can follow the absolute humidity and anticipate a variation in drying parameters according to the differences between calculated and measured absolute humidity.

The operator can also follow the relative humidity in the outlet air. To achieve a dairy powder with the same water activity and moisture content, he must always maintain the same relative humidity in the outlet air according to each dairy product, whatever the spray drying conditions (inlet air temperature, relative and absolute humidity).

The changes in relative and absolute humidity (resulting from variations in absolute humidity of inlet air, total solid content of concentrate, crystallization rate, outlet air temperature, etc.) can be rapidly observed in the outlet air using a thermohygrometer before such changes significantly affect powder moisture, water activity and powder behavior with regard to sticking.

2.6 Conclusions and future trends

This chapter describes the effects of spray drying on the quality of dairy powders during drying and rehydration. It also demonstrates that the quality of these powders depends on the interactions between the biochemical composition and the process parameters.

Fig. 2.7 shows that the biochemical, microbiological and physical properties of a dairy powder and its recombined product depend on many parameters. For example:

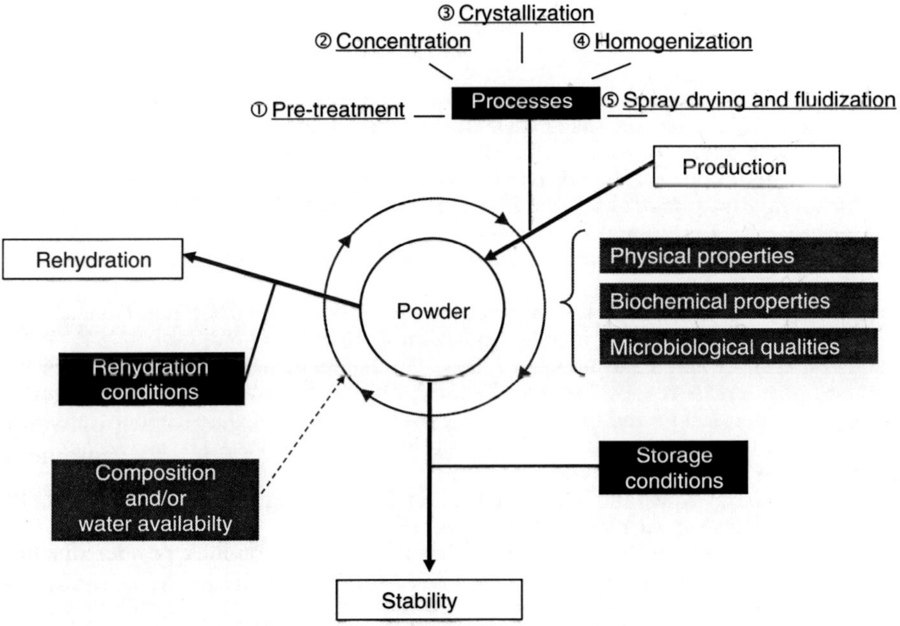

Fig. 2.7 Properties of dairy powders in relation to production, storage and rehydration.

- the pre-heat treatment process parameters, concentration by membrane filtration or by vacuum evaporation, crystallization, homogenization, spray drying and fluidization,
- the storage conditions (relative humidity, temperature, packaging, ...) to optimize stability over to time,
- the rehydration conditions (stirring conditions, temperature, concentration, ...) to improve water transfer to obtain the best quality recombined product from the corresponding powder.

Moreover, the biochemical composition (nature and content) and water availability interact in all the stages of production, stability and rehydration. The quality of a dairy powder can therefore be improved only if research is undertaken on the process–product interactions.

In conclusion, this chapter discusses the complexity of the spray drying process for dairy manufacturers who want to optimize their production. It is impossible to provide a simple mathematical model that takes into account all parameters simultaneously for the complex processes in the spray drying of dairy products. Furthermore, existing studies, essentially centered on skim milk, cannot easily be extrapolated to other dairy products. Study of more interaction processes, structures and functions of dairy products is necessary in order to increase our understanding of the mechanisms of water transfer, drying parameters, storage conditions and rehydration of dairy powders with thermodynamical and biochemical approaches.

2.7 References

AGUILAR CA and ZIEGLER GR (1993), 'Lactose crystallization in spray-dried milk powders exposed to isobutanol', *Food Struct*, 12, 43–50.

AMERICAN DAIRY PRODUCTS INSTITUTE (1990), *Standards for Grades of Dry Milk Including Methods of Analysis*, Chicago, American Dairy Products Institute.

BALDWIN AJ, BAUCKE AG and SANDERSON WB (1980), 'The effect of concentrate viscosity on the properties of spray dried skim milk powder', *N Z J Dairy Sci Technol*, 15, 289–297.

BIMBENET JJ, SCHUCK P, ROIGNANT M, BRULÉ G and MÉJEAN S (2002), 'Heat balance of a multistage spray-dryer: Principles and example of application', *Lait*, 82, 541–551.

BOONYAI P, BHANDARI B and HOWES T (2004), 'Stickiness measurement techniques for food powders: A review', *Powder Technol*, 145, 34–46.

BOONYAI P, BHANDARI B and HOWES T (2005), 'Measurement of glass–rubber transition temperature of skim milk powder by static mechanical test', *Drying Technol*, 23, 1499–1514.

CARIĆ M (2003), 'Types and Manufacture', in Roginsky H, *Encyclopedia of Dairy Sciences*, London, Academic Press, 1869–1874.

CARIĆ M and KALÁB M (1987), 'Effects of drying techniques on milk powders quality and microstructure: A review', *Food Microstruct*, 6, 171–180.

CARIĆ M and MILANOVIĆ (2003), 'Physical and functional properties of milk powders', in Roginsky H, *Encyclopedia of Dairy Sciences*, London, Academic Press, 1874–1880.

CARR RL JR (1965), 'Evaluating flow properties of solids', *Chem Eng*, 72, 163–168.

COUCHMAN PR and KARASZ FE (1978), 'A classical thermodynamic discussion of the effect of composition on glass transition temperatures', *Macromol* 11, 117–119.

DAVENEL A, SCHUCK P and MARCHAL P (1997), 'A NMR relaxometry method for determining the reconstitutability and the water-holding capacity of protein-rich milk powders', *Milchwissenschaft*, 52, 35–39.

DE VILDER J (1986), 'La fabrication de poudre de lait écrémé instantanée. I. Les caractéristiques physiques et chimiques', *Rev Agric*, 39, 865–877.

EFSTATHIOU T, FEUARDENT C, MÉJEAN S and SCHUCK P (2002), 'The use of carbonyl analysis to follow the main reactions involved in the process of deterioration of dehydrated dairy products: Prediction of most favourable degree of dehydration', *Lait*, 82, 423–439.

FAUQUANT J, MAUBOIS JL and PIERRE A (1988), 'Microfiltration du lait sur membrane minérale', *Tech Lait*, 1028, 21–23.

FREUDIG B, HOGEKAMP S and SCHUBERT H (1999), 'Dispersion of powders in liquid in a stirred vessel' *Chem Eng and Process*, 38, 525–532.

GAIANI C, BANON S, SCHER J, SCHUCK P and HARDY J (2005), 'Use of a turbidity sensor to characterize casein powders rehydration: Influence of some technological effects', *J Dairy Sci*, 88, 2700–2706.

GAIANI C, SCHER J, SCHUCK P, HARDY J, DESOBRY S and BANON S (2006a), 'The dissolution behaviour of native phosphocaseinate as a function of concentration and temperature using a rheological approach', *Int Dairy J*, 16, 1427–1434.

GAIANI C, EHRHARDT JJ, SCHER J, HARDY J, DESOBRY S and BANON S (2006b), 'Surface composition of dairy powders observed by X-ray photoelectronspectroscopy and effects on their rehydration properties', *Coll Surf B Biointer*, 49, 71–78.

GAIANI C, SCHUCK P, SCHER J, HARDY J, DESOBRY S and BANON S (2007), 'Dairy powder rehydration: Influence of proteins and some technological effects', *J Dairy Sci*, 90, 570–581.

HARDY J, SCHER J and BANON S (2002), 'Water activity and hydration of dairy powders', *Lait*, 82, 441–452.

HAUGAARD SORENSEN I, KRAG J, PISECKY J and WESTERGAARD V (1978), *Méthodes d'analyses desproduits laitiers déshydratés*, Copenhagen, Niro A/S.

INTERNATIONAL DAIRY FEDERATION (1988), *Dried Milk and Milk Products – Determination of Insolubility Index. IDF Standard 129A*, Brussels, International Dairy Federation.

JEANTET R, SCHUCK P, FAMELART MH and MAUBOIS JL (1996), 'Intérêt de la nanofiltration dans la production de poudres de lactosérum déminéralisées', *Lait*, 76, 283–301.

JEANTET R, CROGUENNEC T, MAHAUT M, SCHUCK P and BRULÉ G (2008a), *Les Produits Laitiers*, Paris, Technique et Documentation – Lavoisier,

JEANTET R, DUCEPT F, DOLIVET A, MÉJEAN S and SCHUCK P (2008b), 'Residence time distribution: A tool to improve spray-drying control', *Dairy Sci Technol*, 88, 31–43.

JOUPPILA K and ROOS YH (1994), 'Water sorption and time-dependent phenomena of milk powders', *J. Dairy Sci.* 77, 1798–1808.

KING N (1966), 'Dispersibility and reconstitutability of dried milk', *Dairy Sci Abs* 28, 105–118.

MASTERS K (1991), *Spray drying*, Essex, Longman Scientific & Technical.

MASTERS K (2002), *Spray drying in practice*, Charlottenlund SprayDryConsult International ApS.

MAUBOIS JL (1991), 'New applications of membrane technology in the dairy industry', *Aust J Dairy Technol* 46, 91–95.

MISTRY VV, HASSAN HN and ROBINSON DJ (1992), 'Effect of lactose and protein on the microsctructure of dried milk', *Food Struct* 11, 73–82.

NEFF E and MORRIS HA (1968), 'Agglomeration of milk powder and its influence on reconstitution properties', *J Dairy Sci*, 51, 330–338.

PIERRE A, FAUQUANT J, LE GRAËT Y, PIOT M and MAUBOIS JL (1992), 'Préparation de phosphocaséinate natif par microfiltration sur membrane' *Lait*, 72, 461–474.

PISECKY J (1980), 'Bulk density of milk powders.', *Aust J Dairy Technol*, 35, 106–111.

PISECKY J (1981), 'Technology of skimmed milk drying', *J Soc Dairy Technol*, 34, 57–62.

PISECKY J (1986), 'Standards, specifications and test methods for dry milk products', In MacCarthy D, *Concentration and Drying of Food*, London, Elsevier, 203–220.

PISECKY J (1990), '20 years of instant whole milk powder', *Scand Dairy Inf*, 4, 74–77.

PISECKY J (1997), *Handbook of Milk Powder Manufacture*, Copenhagen, Niro A/S.

REFSTRUP E (2003), 'Drying of Milk', in Roginsky H, *Encyclopedia of Dairy Sciences*, London, Academic Press, 860–871.

ROOS YH (2002), 'Importance of glass transition and water activity to spray drying and stability of dairy powders', *Lait*, 82, 478–484.

SCHUCK P, PIOT M, MÉJEAN S, FAUQUANT J, BRULÉ G and MAUBOIS JL (1994a), 'Déshydratation des laits enrichis en caséine micellaire par microfiltration; Comparaison des propriétés des poudres obtenues avec celles d'une poudre de lait ultra-propre', *Lait*, 74, 47–63.

SCHUCK P, PIOT M, MÉJEAN S, LE GRAËT Y, FAUQUANT J, BRULÉ G and MAUBOIS JL (1994b), 'Déshydratation par atomisation de phosphocaséinate natif obtenu par microfiltration sur membrane', *Lait*, 74, 375–388.

SCHUCK P, ROIGNANT M, BRULÉ G, DAVENEL A, FAMELART MH and MAUBOIS JL (1998), 'Simulation of water transfer in spray drying', *Drying Technol*, 16, 1371–1393.

SCHUCK P, BLANCHARD E, DOLIVET A, MÉJEAN S, ONILLON E and JEANTET R (2005a), 'Water activity and glass transition in dairy ingredients', *Lait*, 85, 295–304.

SCHUCK P, MÉJEAN S, DOLIVET A and JEANTET R (2005b), 'Thermohygrometric sensor: A tool for optimizing the spray drying process', *Innov. Food Sci Emerg Technol*, 6, 45–50.

SCHUCK P, MÉJEAN S, DOLIVET A, JEANTET R and BHANDARI B (2007), 'Keeping quality of dairy ingredients', *Lait*, 87, 481–488.

SCHUCK P, DOLIVET A, MÉJEAN S and JEANTET R (2008), 'Relative humidity of outlet air: Key parameter to optimize moisture content and water activity of dairy powder', *Dairy Sci Technol*, 88, 45–52.

SCHUCK P, DOLIVET A, MÉJEAN S, ZHU P, BLANCHARD E and JEANTET R (2009), 'Drying by desorption: A tool to determine spray drying parameters', *J Food Eng*, 94, 199–204.

SOUGNEZ M (1983), 'L'évolution du séchage par atomisation', *Chim Mag*, 1, 1–4.

TUOHY JJ (1989), 'Some physical properties of milk powders' *Irish J Food Sci Technol*, 13, 141–152.

WESTERGAARD V (2003), 'Dryer design', in Roginsky H, *Encyclopedia of Dairy Sciences*, London, Academic Press, 871–889.

WESTERGAARD V (2004), *Milk powder technology – Evaporation and spray-drying*, Copenhagen, A/S Niro A/S.

3

Production and enrichment of bioactive peptides derived from milk proteins

L. Vercruysse, J. Van Camp, K. Dewettinck and G. Smagghe, Ghent University, Belgium

Abstract: In this chapter, information is given on the production of bioactive peptides from milk proteins. After defining bioactive peptides, the necessity to release bioactive peptides from the parent protein is discussed. Then, the two main approaches for release, namely enzymatic hydrolysis and microbial fermentation are reviewed. Next, an overview is given of the main unit operations that can be applied for the enrichment and selective isolation of bioactive peptides from the protein hydrolysates, including membrane filtration, chromatography and precipitation. Attention is given mainly to ACE-inhibitory peptides, and to processes that show potential for application in the dairy industry.

Key words: milk, bioactive peptides, enzymatic hydrolysis, microbial fermentation, enrichment.

3.1 Introduction

3.1.1 Defining bioactive peptides

Food proteins are traditionally considered as sources of both energy and of building blocks for growth and maintenance of the body. However, recent data indicate that food proteins are also unique sources of bioactive sequences (Murray and FitzGerald, 2007). Biologically active peptides, also called bioactive peptides, exert physiological functions in the body resulting in beneficial effects towards health (Kitts and Weiler, 2003). Many bioactive peptides are latent in that they are inactive in the parent protein sequence and need to be released to exert their activity. There are two main approaches for releasing bioactive peptides. One approach is to use lactic acid bacteria (LAB) for fermentation. The other approach is to subject food proteins to *in vitro* hydrolysis by one or a combination of enzymes (Lopez-Fandino *et al.*, 2006). Bioactive peptides are short chains of amino acids, usually containing two to 20 amino acid residues per molecule. A wide range of activities has been described, including blood pressure lowering (angiotensin converting enzyme (ACE) inhibitory), opioid, antimicrobial, mineral binding, antioxidant, antithrombotic, immunomodulatory and anticancer effects. Many peptides are known to reveal multifunctional properties as they can exert more than one biological effect (Hartmann and Meisel, 2007).

3.1.2 Bioactive peptides in milk

Numerous food proteins of both animal and vegetable origin have been used to generate bioactive peptides. Plant protein sources of bioactive peptides include maize, soybean, wheat and rice (Yamamoto *et al.*, 2003; Murray and FitzGerald, 2007). Animal proteins, such as bovine, chicken and porcine muscle, contain bioactive peptides as well as fish proteins from bonito, sardine, tuna and salmon (Vercruysse *et al.*, 2005). However, the major source of bioactive peptides identified to date is milk. The most important is bovine milk, but also human, sheep (Papadimitriou *et al.*, 2007), goat, pig, buffalo and yak milk (Mao *et al.*, 2007) have been reported as sources of bioactive peptides (Minervini *et al.*, 2003). The main constituents of milk are casein and whey proteins. Bovine casein accounts for about 80% of the total milk proteins and is further divided into α-, β- and κ-casein. The major bioactive peptides derived from casein, exerting a role in the nervous, cardiovascular, digestive and immune system, have been reviewed by Silva and Malcata (2005). In addition, whey, which is a major product of the modern cheese and casein industry, is an important source of bioactive peptides exerting various activities (Smithers, 2008). The major whey proteins and peptides, such as β-lactoglobulin (β-Lg), α-lactalbumin and glycomacropeptide are known to positively influence human health (Chatterton *et al.*, 2006; Hernandez-Ledesma *et al.*, 2008). Milk, fermented milk, yoghurt and cheese, but also infant milk formulas (De Noni, 2008; Martin *et al.*, 2008) have been identified as sources of bioactive peptides. Milk proteins are attractive in the search for bioactive peptides as they are very well characterized and are available in large amounts at a high degree of purity and at a low price.

3.1.3 Necessity to release bioactive peptides from milk proteins

In past years, the protein fraction in milk has received great attention and several protein components exerting potential health-promoting activity have been identified and characterized. Intact milk proteins have been associated with a wide range of activities, including iron binding (e.g. lactoferrin) and antimicrobial effects (e.g. lactoperoxydase, lysozyme) (Haque and Chand, 2008). However, much of the biological activity of milk proteins is latent. The bioactive peptides have to be released from precursor milk proteins in one of the following ways: (a) enzymatic hydrolysis by digestive enzymes, (b) fermentation with proteolytic starter cultures, (c) proteolysis by enzymes derived from microorganisms or plants and (d) proteolysis by indigenous proteases present in milk (Korhonen and Pihlanto, 2007). Synergistic effects between intact bioactive milk proteins and bioactive milk-derived peptides are important as both compounds might be simultaneously present in the human gastrointestinal tract. Lopez-Exposito *et al.* (2008) indicated such a synergistic effect between lactoferrin and the antimicrobial peptide lactoferricin against *Escherichia coli* and *Staphylococcus epidermidis*.

3.2 Enzymatic hydrolysis

Enzymatic hydrolysis is the most common way to release bioactive peptides from milk proteins. One or a combination of enzymes can be used in a batch or continuous process. *In vitro* hydrolysis of milk proteins allows the selection of the protein substrate and enzyme specificity to optimize the yield of bioactive peptides.

Highly effective in the enzymatic release of bioactive peptides from milk proteins are pancreatic proteinases, especially trypsin. Other endoproteinases such as chymotrypsin, pepsin and thermolysin, and proteinase extracts derived from bacteria, moulds, plants and animals have also been used (Korhonen and Pihlanto, 2007). In comparison to digestive enzymes, microbial enzymes use different cleavage sites; thus, peptides liberated by these enzymes may differ from those liberated by digestive enzymes. Mizuno *et al.* (2004) found in a comparative study of nine commercial enzymes that an *Aspergillus oryzae* protease was most effective for the release of short and proline-rich anti-hypertensive peptides from casein. However, caseinates derived from milk of different species (bovine, sheep, goat, pig, buffalo, human) can show differences in susceptibility towards enzymatic hydrolysis, as demonstrated by Minervini *et al.* (2003) who used a partially purified serine proteinase from *Lactobacillus helveticus*. This result can be explained by differences in sequence identity, conformation, and relative proportions of individual caseins. Different species were shown to generate different types of peptides with ACE-inhibitory and antimicrobial activity (Minervini *et al.*, 2003). A combination of digestive enzymes will alter the digestive process and the peptides released, but will not necessarily result in a higher activity. Mullally *et al.* (1997) used gastric and pancreatic proteases to hydrolyze whey proteins in order to obtain ACE inhibitory peptides. Hydrolysis of α-lactalbumin and β-Lg with pepsin and trypsin separately resulted in similar ACE inhibitory activities. Similar activities were also obtained when using combinations of trypsin and pepsin; moreover, the order of addition of trypsin and pepsin had little effect on the ACE inhibitory values. However, the types of peptides present in the final hydrolysate may be affected by the order of enzyme addition (Mullally *et al.*, 1997). Gastro-intestinal digestion can be a way to release bioactive peptides from milk proteins. However, strong hydrolysis during gastro-intestinal digestion and absorption can limit the bioavailability of target peptides, which makes their applicability for systemic functions more difficult (Vermeirssen *et al.*, 2004a). The gastro-intestinal digestion process can be simulated in batch or semi-continuous reactor systems (Vermeirssen *et al.*, 2003).

The production of bioactive peptides using enzymatic hydrolysis can be studied using response surface methodology. Van der Ven *et al.* (2002) showed that the ACE inhibitory activity of whey protein hydrolysates could be controlled by systematical regulation of process conditions. Roufik *et al.* (2006) demonstrated that the *in vitro* digestibility of bioactive peptides derived from bovine β-Lg is not only influenced by pH, time of incubation, and enzyme/substrate ratio, but also by the length and the nature of the peptides. Also, other peptides present in the medium can influence the hydrolysis by gastro-intestinal enzymes, e.g. a higher degree of hydrolysis (DH) was obtained for pepsin and chymotrypsin when the opioid agonistic and ACE inhibitory peptide β-Lg f102-105 was combined with β-Lg f92-105, than when both peptides were hydrolyzed separately. It was suggested that in a mixture, peptides may form complexes that interact more efficiently with the hydrolyzing enzymes, compared with peptides alone (Roufik *et al.*, 2006). El-Zahar *et al.* (2005) also demonstrated that the enzymatic susceptibility towards pepsin of ovine β-Lg was significantly higher compared with that of bovine β-Lg. The effect was attributed to a slightly different tertiary structure and a higher surface hydrophobicity at pH 2 for ovine β-Lg. Especially, the more hydrophobic C-terminal end of ovine β-Lg may be more exposed at the surface of the molecule, making it susceptible to pepsin hydrolysis.

A more industrial approach to the production of bioactive peptides is enzymatic digestion in membrane reactors. Bouhallab *et al.* (1993) designed a continuous process in which enzymatic hydrolysis of casein was combined with purification of the bioactive peptides through the use of ultrafiltration. In addition, the immobilization of proteases can lead to a better hydrolysis efficiency. Different proteases, such as trypsin, chymotrypsin, papain, and cathepsins have been successfully immobilized on different supports. Immobilization has several advantages, such as capability of reuse, better control of hydrolysis rate and longer stability of the enzyme (Toldra *et al.*, 2005).

The amino acid sequence of milk proteins has been reported, and can be used to evaluate the presence of bioactive peptides by computer-aided models. *In silico* screening was used by Vermeirssen *et al.* (2004b) to rank individual milk proteins according to their potential to generate ACE-inhibitory peptides. A database containing the amino acid sequence of milk proteins was combined with a database containing the sequence and IC_{50}-

values of 498 ACE-inhibitory peptides reported in literature. Proteins were given weighed scores on the basis of high ACE-inhibitory activity (Score 1), as well as on short peptide size combined with high ACE-inhibitory activity (Score 2). Fifty different ACE-inhibitory peptides were identified in β-Lg, giving it the highest ranking among the milk proteins both by Scores 1 and 2. Other milk proteins with high scores were lactotransferrin and β-casein. A similar technique was developed by Dziuba *et al.* (2004). Again, a database containing protein and bioactive peptide sequences enables searching for protein fragments identical to bioactive peptides among protein amino acid sequences. Theoretical predictions and simulations are considered as an emerging tool in peptide science. Although these models give some insight into the number and type of active peptide sequences present in milk and other food proteins, because they do not take into account the release of active peptides from the source protein, their results have to be interpreted with caution (Minkiewicz *et al.*, 2008).

3.3 Microbial fermentation

Many industrially used dairy starter cultures are highly proteolytic. The peptides and amino acids degraded from milk proteins during fermentation contribute to the typical flavour, aroma and texture of dairy products. The proteolytic system of LAB can contribute to the liberation of bioactive peptides during fermentation. The cell-wall bound proteinases, which have a very broad substrate specificity, initiate the proteolytic attack and release of oligopeptides. In addition, transport systems specific for amino acids and peptides up to 18 amino acids are present in the LAB. Oligopeptides that cannot be transported into the cells can be further degraded after lysis of the bacterial cell, which allows the release of intracellular peptidases, such as di-, tri-, amino- or endopeptidases (Lopez-Fandino *et al.*, 2006). It is important to choose the right strains or combination of strains with optimal proteolytic activity and lysis at the right time, as the number of bioactive peptides relies on a balance between their formation and breakdown into inactive peptides and amino acids (Gobbetti *et al.*, 2000). In addition to a high number of LAB, some yeasts have also been used to ferment milk proteins in order to release bioactive peptides (see Table 3.1).

Fermented milk, and peptides isolated from fermented milk, have been demonstrated to have various biological activities. LeBlanc *et al.* (2002) detected immunostimulatory peptides in milk fermented with a *L. helveticus* strain. An antioxidative peptide, determined as Ala-Arg-His-Pro-His-Pro-His-Leu-Ser-Phe-Met, was isolated from κ-casein after fermentation with *Lactobacillus delbrueckii* subsp. *bulgaricus* IFO13953. Although the scavenging activity of the DPPH radical of the peptide was lower than butylated hydroxytoluene (BHT), the peptide showed about five times stronger antioxidative activity than BHT in the beta-carotene decolorization system

Table 3.1 Examples of micro-organisms used for the production of bioactive peptides

Micro-organism	Bioactivity	Representative references
Lactobacillus helveticus	ACE inhibition	Nakamura *et al.*, 1995a; LeBlanc *et al.*, 2002; Minervini *et al.*, 2003; Jauhiainen *et al.*, 2005
	Antioxidant	Singh and Chand, 2006
	Immunomodulatory	LeBlanc *et al.*, 2002
Lactobacillus casei	ACE inhibition	Ong and Shah, 2008
Lactobacillus plantarum	ACE inhibition	Gomez-Ruiz *et al.*, 2002
Lactobacillus acidophilus	ACE inhibition	Ryhanen *et al.*, 2001
Lactobacillus animalis	ACE inhibition	Hayes *et al.*, 2007
Lactococcus lactis subsp. *lactis*	Immunomodulatory and antihypertensive	Algaron *et al.*, 2004
Lactococcus lactis subsp. *cremoris*	ACE inhibition	Gobbetti *et al.*, 2000
Lactobacillus delbrueckii subsp. *bulgaricus*	Antioxidative	Ashar and Chand, 2004; Chobert *et al.*, 2005
	ACE inhibition	Gobbetti *et al.*, 2000
Streptococcus thermophilus	ACE inhibition	Chobert *et al.*, 2005; Tsai *et al.*, 2008
Saccharomyces cerevisiae	ACE inhibition	Nakamura *et al.*, 1995a
Klyveromyces marxianus var. *marxianus*	ACE inhibition	Belem *et al.*, 1999
Enterococcus faecalis	ACE inhibition	Miguel *et al.*, 2006; Quiros *et al.*, 2007
Leuconostoc mesenteroides subsp. *dextranicum*	ACE inhibition	Gomez-Ruiz *et al.*, 2002

(Kudoh *et al.*, 2001). Virtanen *et al.* (2007) investigated the production of antioxidant activity during fermentation of whey with 25 LAB strains. They indicated that the development of antioxidant activity was strain-specific with *Leuconostoc mesenteroides* ssp. *cremoris* strains, *Lactobacillus jensenii* (ATCC 25258) and *Lactobacillus acidophilus* (ATCC 4356) showing the highest activity. In addition, ACE inhibitory peptides released by fermentation of milk proteins have been reported in many studies and have been reviewed by Haque and Chand (2008). Nakamura *et al.* (1995a,b) identified two ACE inhibitory peptides, Val-Pro-Pro and Ile-Pro-Pro, derived from milk by fermentation with a combination of *L. helveticus* CP790 and *Saccharomyces cerevisiae*. These two peptides, of casein origin, exert *in vivo* antihypertensive activity in spontaneously hypertensive rats (SHR) and in hypertensive subjects (Takano, 1998). A fermented milk product containing these peptides was marketed by the Japanese Calpis company under the trade name Amiiru S. A European fermented milk drink with blood pressure lowering activity, Evolus from Valio Ltd, Finland, is produced by fermentation with another strain, namely *L. helveticus* LBK 16H but contains the same two active peptides (Tuomilehto *et al.*, 2004). Pihlanto-Leppala *et al.* (1998) studied the release of *in vitro* ACE inhibitory activity of whey and casein by fermentation with commercial yoghurt starters. They indicated that fermentation alone was not enough to generate ACE inhibitory activity, but a further digestion with pepsin and trypsin was necessary.

A number of studies has shown that during cheese making, a wide range of peptides are released and many of them have exhibited biological activities, such as ACE inhibitory activity. The origin of the milk and the manufacturing conditions, as well as the cheese-ripening time, determine the release of peptides and thus the activity (Korhonen and Pihlanto, 2003). In middle-aged Gouda cheese, a higher ACE inhibitory activity was detected compared to short-termed or long-termed ripened cheese. In ripened cheese, the ACE inhibitory activity probably results from a complex mixture of ACE inhibitory peptides. The activity increases during cheese maturation, but decreases when the proteolysis exceeds a certain level (Meisel *et al.*, 1997). Similarly, the free peptides from eight months-aged Gouda cheese exerted the strongest blood pressure lowering effect after oral administration in SHR, compared to 24 months-aged Gouda, Emmental, Blue, Camembert, Edam and Havarti cheese (Saito *et al.*, 2000). Ong and Shah (2008) studied the effect of probiotic adjuncts on ACE inhibitory activity of Cheddar cheeses. Cheeses made with the addition of *Lactobacillus casei* 279, *L. casei* LAFTI® L26 or *L. acidophilus* LAFTI® L10 had significantly higher ACE inhibitory activity than those without any probiotic adjunct. The probiotic adjuncts were added to improve proteolysis and enhance flavour during Cheddar cheese ripening. The two ACE inhibitory peptides Val-Pro-Pro and Ile-Pro-Pro have also been detected in various Swiss cheeses. The milk pretreatment, cultures, scalding conditions and ripening time were again identified as the key factors influencing the concentration of these two peptides in cheese. On average, the cheese varieties contained similar concentrations of the peptides as the commercial fermented milk products (Butikofer *et al.*, 2008).

Certain cell envelope-associated proteinases (CEP) of LAB can release bioactive health-beneficial peptides during milk fermentation. Hebert *et al.* (2008) characterized the peptide sequence in the hydrolysis pattern of α- and β-casein made by a CEP of *L. delbrueckii* subsp. *lactis* CRL 581. In this pattern, a series of potentially bioactive peptides could be detected. The antihypertensive tripeptide Ile-Pro-Pro can be released by *L. delbrueckii* subsp. *lactis* CRL 581; therefore, the micro-organism could be used as a functional cheese starter.

A combination of lactic acid fermentation and proteolysis of milk can be used to obtain a product which is more abundant in bioactive peptides than the traditional fermentation or proteolysis products of milk. Tsai *et al.* (2008) fermented milk using two LAB (*Streptococcus thermophilus* and *Lactobacillus bulgaricus*) and added flavourzyme at the beginning of the fermentation. This combination accelerated the production of peptides (32.8 $mg \cdot g^{-1}$ peptides for the combination in contrast to 5.8 $mg \cdot g^{-1}$ peptides for the lactic fermentation alone) and increased the ACE inhibitory activity (IC_{50} value of 0.266 $mg \cdot mL^{-1}$ for the combination compared to 0.515 $mg \cdot mL^{-1}$ for the lactic fermentation alone). Similarly, Chen *et al.* (2007) fermented milk with LAB and added a protease, prozyme 6, five hours after the beginning of fermentation. The prozyme-facilitated fermentation resulted in a higher ACE inhibitory activity compared to the non-facilitated fermentation (Chen *et al.*, 2007).

3.4 Enrichment of bioactive peptides

Various technologies have been applied for the enrichment and separation of bioactive peptides from the hydrolysates of milk proteins. However, commercial production has been limited by a lack of suitable large-scale technologies. Membrane filtration shows most potential to be used in pilot and industrial scale separations. Precipitation and chromatography are also available but are less frequently applied, mainly due to their lower selectivity and/or higher cost. The next sections will highlight a few recent evolutions in the use of membrane processing and chromatography for the separation and isolation of milk protein-derived bioactive peptides.

3.4.1 Membrane processing

Peptides can be separated from protein hydrolysates using suitable membrane technologies, such as nano- and ultrafiltration. In membrane reactors, hydrolysis of milk proteins is combined with enrichment of the bioactive peptides from the reaction mixture through the use of filtration. Ultrafiltration membrane reactors yield a uniform product with desired molecular mass. Stepwise filtration has been found useful to separate small peptides. Pihlanto-Leppala *et al.* (2000) fractionated ACE-inhibitory peptides from α-lactalbumin and β-Lg hydrolysates using membranes with molecular weight cut-off (MWCO) of 1 and 30 kDa. IC_{50}-values for ACE-inhibitory activities ranged from 109–837 $\mu g \cdot mL^{-1}$ in the <1 kDa fraction to 485–1134 $\mu g \cdot mL^{-1}$ in the 1–30 kDa fraction and 345–1733 $\mu g \cdot mL^{-1}$ in the original whey protein digests. The same technique was successfully used to enrich opioid peptides from α-lactalbumin and β-Lg hydrolysates (Pihlanto-Leppala *et al.*, 1996). From the studies of Pouliot *et al.* (1999, 2000), it became clear that peptide separation from a tryptic hydrolysate of whey protein using nano/ultra-filtration is influenced by the characteristics of the membrane and by the ionic strength and pH of the solution. Charge effects play an important role in the separation, as a negatively charged membrane with a MWCO of 1–5 kDa, gave relative concentration values in the permeate <100% for a group of acidic (negatively charged) peptides, while neutral and basic (positively charged) peptides showed values >100%. Addition of salt increased the concentration of negatively charged peptides in the permeate, contrary to neutral or positively charged peptides. The highest selectivity was obtained at pH 9 without salt addition (Pouliot *et al.*, 1999). Similar peptides can also be transferred differently through the membrane when present in different hydrolysates, as for β-Lg 76–82 and β-Lg 78–82 present in trypsin and chymotrypsin digests. This illustrates the importance of peptide–peptide interactions during membrane filtration (Pouliot *et al.*, 2000). Peptide–peptide interactions may also take place in the polarization layer of the membrane, and this can influence the transmission of bioactive peptides, as illustrated for the ACE-inhibitory peptide β-Lg f142-148 by Lapointe *et al.* (2003).

Electro-membrane filtration (EMF) involves the use of an electric field during pressure-driven filtration. Electrokinetic effects may help in orienting the transport of charged peptides towards the electrode of opposite charge, thereby improving the separation between neutral and charged peptides. Lapointe *et al.* (2006) applied EMF to selectively separate cationic peptides from a tryptic hydrolysate of β-Lg. A negatively charged nanofiltration membrane with a MWCO of 2.5 kDa was applied, whereby the electrode configuration in the EMF system was specifically set to favour transmission of positive peptides on the permeate side and increase the retention of negative peptides on the feed side. Applying an electric field of 5 V at pH 9 almost completely impeded the transmission of neutral peptides, while the impact on strong basic peptides was negligible. The ACE-inhibitory peptide β-Lg f142-148 with a pI = 9.80, due to the presence of a histidine and arginine residue, was mostly transmitted during EMF. A relative concentration in the permeate of 38% compared to 18% without applying an electric field, and 3.5% in the initial tryptic β-Lg hydrolysate, was obtained. Bargeman *et al.* (2002) hydrolyzed an α_{s2}-CN enriched solution by porcine pepsin A and successively fractionated the hydrolysate by EMF using a polysulphone membrane with a MWCO of 25 kDa. The relative content of α_{s2}-CN f(183–207), an antibacterial peptide with high positive net charge, increased from 7.5% in the feed solution to a maximum of 25% in the permeate. Conventional pressure-driven diafiltration using the same membrane did not result in the selective enrichment of this peptide, indicating that the electric field strength was essential for the isolation of positive charged peptides. Protection of the electrodes in the EMF-module by ion-exchange membranes has been suggested, in view of the electrolytic reactions that may occur at the cathode (permeate side) and at the anode, and that can result in acidification of the retentate and alkalisation of the permeate (Bargeman *et al.*, 2002; Lapointe *et al.*, 2006).

3.4.2 Chromatography

To date, chromatography has mainly been applied as an analytical tool for the characterization of milk protein derived bioactive peptides. Separation is thereby conventionally achieved by reverse-phase C18 HPLC with UV-detection (Vermeirssen *et al.*, 2005; Lapointe *et al.*, 2006), if necessary preceded by gel filtration or ion exchange chromatography for a first fractionation on the basis of molecular weight and/or charge (Lopez-Exposito *et al.*, 2006). Peptide fractions can be collected directly from the HPLC column and analyzed for peptide composition and bio-activity (Roufik *et al.*, 2006). Identified peptides can then be chemically synthesized and analyzed for bio-activity (Minervini *et al.*, 2003). A simultaneous separation and quantification of the ACE inhibitory peptides Val-Pro-Pro and Ile-Pro-Pro has been developed by Ferreira *et al.* (2007). Separation was obtained by RP-HPLC and quantification by UV detection at 214 nm in combination with calibration curves. Furthermore, for the detection and identification of ACE-inhibitory peptides, an on-line liquid chromatography–biochemical detection, coupled to electrospray mass spectrometry, is available. The method has been applied successfully for the rapid detection of ACE-inhibitory peptides in milk protein hydrolysates (van Elswijk *et al.*, 2003).

Chromatography can contribute to the purification of charged or hydrophobic peptides from milk protein hydrolysates. The antimicrobial peptide lactoferricin (Lfcin) has been purified from a pepsin hydrolysate of lactoferrin (Lf) by two-step hydrophobic chromatography, followed by reverse osmosis and freeze-drying. The resulting powder had a purity of more than 95% (Tomita *et al.*, 2002). Mineral binding peptides from milk have traditionally been isolated by selective precipitation methods using solvents and chelating agents (calcium and barium) at well defined pH and ionic strengths (Vegarud *et al.*, 2000). However, specific phosphopeptides released by the action of plasmin, e.g. β-CN-5P and β-CN-4P, can be purified from milk by a combination of TCA (5%) precipitation with gel filtration chromatography. The process can be made more efficient when purified β-casein (obtained by ion-exchange chromatography from sodium caseinate) is used as starting material for the plasmin hydrolysis (Andrews *et al.*, 2006).

Conventional chromatographic bed technology may cause problems in terms of fouling, long cycle times, and large pressure drops in the column. An alternative technology is absorptive membrane chromatography, which combines the use of chromatography with membrane filtration. Functional groups attached to the membranes in the column can be used to absorb functional components from dairy product streams. The strong acidic (cation exchange) membrane Sartobind S 15 cm^2 has been used for the purification of bovine lactoferricin (LfcinB) from pepsin-treated bovine lactoferrin (LfB). The latter was purified from sweet cheese whey using the same absorptive membrane technology (Plate *et al.*, 2006). At pH 7, LfcinB with a net positive charge binds to the membrane and can be isolated by a two-step salt gradient, i.e. 0.4 M sodium chloride for removal of other protein components followed by 2 M ammonium chloride for elution of LfcinB. For the same membrane, the binding capacity for LfcinB was higher than for lactoferrin. Possibly the size of the molecule was important, whereby the smaller LfcinB may have more access to the active sites of the membrane compared to LfB (Plate *et al.*, 2006).

Simulated moving bed systems have recently found access in the dairy industry for the isolation of functional ingredients. An example is continuous chromatographic separation (CSEP) technology that has been used for the large scale production of a glycomacropeptide-enriched whey protein isolate (DeSilva *et al.*, 2003). Furthermore, the properties of mesoporous silica materials make them suitable as adsorbants for the separation of food bioactive molecules. Their narrow pore size distributions and high surface areas provide selectivity in the separation process. Depending on the target molecule, the materials can be designed to achieve separations on the basis of size-exclusion, ion-exchange, hydrophobic or affinity interactions (Brady *et al.*, 2008). Zhao *et al.* (2002) employed C18 modified SBA-15 (Santa Barbara amorphous 15) as a substrate in HPLC to separate peptides and proteins, including lysozyme. Although mesoporous silica materials have promising properties, their use in industrial, large-scale separations is yet to be realized. Affinity chromatography is also a technology that shows potential in the future for isolation or selective enrichment of bioactive protein components from milk and dairy products. One example of a recent evolution is the successful immobilization of ACE on a glyoxyl–agarose gel (Megias *et al.*, 2006). Longer immobilization times increased the thermal stability of ACE at higher temperatures (60°C), although this was accompanied by a decrease in specific activity of the enzyme. The immobilized ACE is stable for several months at 4°C, and may be used for the determination of ACE-inhibitory activity as well as for the purification of ACE-inhibitory peptides (Megias *et al.*, 2006). Lund and Ardo (2004) have used immobilized metal affinity chromatography (IMAC) to isolate phosphopeptides from cheese. An iminodiacetic acid (IDA) Sepharose column was equilibrated with $FeCl_3$ and subsequently eluted with a gradient of ammonium dihydrogen phosphate. For a ten-month-old semihard Herrgard cheese, elution of the IMAC-Fe(III) column resulted in three distinct peaks corresponding to peptides carrying one, two and four phosphorylated serine residues.

3.4.3 Precipitation

Selective precipitation is a simple and efficient purification procedure, but can be used only in a few cases. Caseinophosphopeptides (CPP) are characterized by a particular functional group, a serine phosphate group -Ser(P)-Ser(P)-Ser(P)-Glu-Glu, which allows isolation by selective precipitation. Adamson and Reynolds (1995) succeeded in separating high purity CPP's from the acid clarified hydrolysate of sodium caseinate by a single-step precipitation with Ca^{2+} and ethanol.

3.5 Future trends

Extensive research carried out during the past 20 years has indicated the potential of milk proteins as a source of bioactive peptides. Because of their potential health benefits, there is a growing commercial interest in milk-based bioactive peptides in the context of health-promoting functional foods. However, the development of functional foods demands an integrated approach which faces challenges with regards to scientific and technological, as well as regulatory and marketing issues (Korhonen, 2002). In the future, bioactive peptides will be optimally exploited for human nutrition and health when knowledge in the different areas increases.

Further research on the production process of bioactive peptides should increase the availability of peptides on an industrial level. Initial research indicates that the immobilization of proteases can offer a great advantage as they can be used in continuous reaction systems in addition to increased protease stability and a more efficient use of proteases (Toldra *et al.*, 2005).

The production of large amounts of bioactive peptides may be possible by genetically transformed micro-organisms, for example by using food grade dairy fermentation bacteria as peptide production systems. Synthetic polynucleotides corresponding to bioactive peptides could be designed and molecularly cloned where after cloning vectors are needed for transport and uptake of the synthetic genes. Renye and Somkuti (2008) have reported such a cloning and expression of synthetic genes in lactic acid fermentation bacteria that encode mature bioactive peptides. They have successfully developed synthetic genes in *Streptococcus thermophilus* for the delivery and expression of an antimicrobial peptide (BL-11) and the C12 hypotensive peptide, which are normally derived from bovine milk proteins by enzymatic digestion.

Another possibility for generation of bioactive peptides may be achieved through the development of stable bovine colostrums concentrates or through gaining milk from hyperimmunized dairy cows (Moller *et al.*, 2008).

After peptide generation, processing technologies for the enrichment and purification of bioactive peptides are required. The downstream processes, such as membrane filtration and chromatography, available on laboratory scale, need to be transferred to pilot and industrial scales and need to be optimized to increase their efficiency of separation and purification.

3.6 Acknowledgement

This research is supported by a PhD grant for Lieselot Vercruysse from the Institute for the Promotion and Innovation by Science and Technology in Flanders (IWT). This project was also supported by the Special Research Fund of the Ghent University (No. 01102703). The financial support of grant G.0490.08 from the Fund for Scientific Research – Flanders (Fonds voor Wetenschappelijk Onderzoek (FWO) Vlaanderen) is gratefully acknowledged.

3.7 Sources of further information and advice

Korhonen H and Pihlanto A (2006), 'Bioactive peptides: Production and functionality', *Int Dairy J*, 16, 945–960.

3.8 References

ADAMSON N J and REYNOLDS E C (1995), 'Characterization of tryptic casein phosphopeptides prepared under industrially relevant conditions', *Biotechnol Bioeng*, 45, 196–204.

ALGARON F, MIRANDA G, LE BARS D and MONNET V (2004), 'Milk fermentation by *Lactococcus lactis* with modified proteolytic systems to accumulate potentially bio-active peptides', *Lait*, 84, 115–123.

ANDREWS A T, WILLIAMS R J H, BROWNSELL V L, ISGROVE F H, JENKINS K and KANEKANIAN A D (2006), 'beta-CN-5P and beta-CN-4P components of bovine milk proteose-peptone: Large scale preparation and influence on the growth of cariogenic microorganisms', *Food Chem*, 96, 234–241.

ASHAR M N and CHAND R (2004), 'Antihypertensive peptides purified from milks fermented with *Lactobacillus delbrueckii* ssp *bulgaricus*', *Milchwissenschaft-Milk Sci Int*, 59, 14–17.

BARGEMAN G, HOUWING J, RECIO I, KOOPS G H and VAN DER HORST C (2002), 'Electro-membrane filtration for the selective isolation of bioactive peptides from an alpha(s2)-casein hydrolysate', *Biotechnol Bioeng*, 80, 599–609.

BELEM M A F, GIBBS B F and LEE B H (1999), 'Proposing sequences for peptides derived from whey fermentation with potential bioactive sites', *J Dairy Sci*, 82, 486–493.

BOUHALLAB S, MOLLE D and LEONIL J (1993), 'Continuous hydrolysis of beta-casein in a membrane reactor – Preparation of a bioactive peptide', *Biotechnol Lett*, 15, 697–702.

BRADY R, WOONTON B, GEE M L and O'CONNOR A J (2008), 'Hierarchical mesoporous silica materials for separation of functional food ingredients – A review', *Innov Food Sci Emerg Technol*, 9, 243–248.

BUTIKOFER U, MEYER J, SIEBER R, WALTHER B and WECHSLER D (2008), 'Occurrence of the angiotensin-converting enzyme-inhibiting tripeptides Val-Pro-Pro and Ile-Pro-Pro in different cheese varieties of Swiss origin', *J Dairy Sci*, 91, 29–38.

CHATTERTON D E W, SMITHERS G, ROUPAS P and BRODKORB A (2006), 'Bioactivity of beta-lactoglobulin and alpha-lactalbumin – Technological implications for processing', *Int Dairy J*, 16, 1229–1240.

CHEN G W, TSAI J S and PAN B S (2007), 'Purification of angiotensin I-converting enzyme inhibitory peptides and antihypertensive effect of milk produced by protease-facilitated lactic fermentation', *Int Dairy J*, 17, 641–647.

CHOBERT J M, EL-ZAHAR K, SITOHY M, DALGALARRONDO M, METRO F, CHOISET Y and HAERTLE T (2005), 'Angiotensin I-converting-enzyme (ACE)-inhibitory activity of tryptic peptides of ovine beta-lactoglobulin and of milk yoghurts obtained by using different starters', *Lait*, 85, 141–152.

DE NONI I (2008), 'Release of beta-casomorphins 5 and 7 during simulated gastro-intestinal digestion of bovine beta-casein variants and milk-based infant formulas', *Food Chem*, 110, 897–903.

DESILVA K, STOCKMANN R and SMITHERS G W (2003), 'Isolation procedures for functional dairy components – Novel approaches to meeting the challenges', *Austr J Dairy Technol*, 58, 148–152.

DZIUBA J, NIKLEWICZ M, IWANIAK A, DAREWICZ M and MINKIEWICZ P (2004), 'Bioinformatic-aided prediction for release possibilities of bioactive peptides from plant proteins', *Acta Aliment*, 33, 227–235.

EL-ZAHAR K, SITOHY M, CHOISET Y, METRO F, HAERTLE T and CHOBERT J M (2005), 'Peptic hydrolysis of ovine beta-lactoglobulin and alpha-lactalbumin – Exceptional susceptibility of native ovine beta-lactoglobulin to pepsinolysis', *Int Dairy J*, 15, 17–27.

FERREIRA I M P L, ECA R, PINHO O, TAVARES P, PEREIRA A and ROQUE A C (2007), 'Development and validation of an HPLC/UV method for quantification of bioactive peptides in fermented milks', *J Liq Chromatogr Relat Technol*, 30, 2139–2147.

GOBBETTI M, FERRANTI P, SMACCHI E, GOFFREDI F and ADDEO F (2000), 'Production of angiotensin-I-converting-enzyme-inhibitory peptides in fermented milks started by *Lactobacillus delbrueckii* subsp *bulgaricus* SS1 and *Lactococcus lactis* subsp *cremoris* FT4', *Appl Environ Microbiol*, 66, 3898–3904.

GOMEZ-RUIZ J A, RAMOS M and RECIO I (2002), 'Angiotensin-converting enzyme-inhibitory peptides in Manchego cheeses manufactured with different starter cultures', *Int Dairy J*, 12, 697–706.

HAQUE E and CHAND R (2008), 'Antihypertensive and antimicrobial bioactive peptides from milk proteins', *Eur Food Res Technol*, 227, 7–15.

HARTMANN R and MEISEL H (2007), 'Food-derived peptides with biological activity: From research to food applications', *Curr Opin Biotechnol*, 18, 163–169.

HAYES M, STANTON C, SLATTERY H, O'SULLIVAN O, HILL C, FITZGERALD G F and ROSS R P (2007), 'Casein fermentate of *Lactobacillus animalis* DPC6134 contains a range of novel propeptide angiotensin-converting enzyme inhibitors', *Appl Environ Microbiol*, 73, 4658–4667.

HEBERT E M, MAMONE G, PICARIELLO G, RAYA R R, SAVOY G, FERRANTI P and ADDEO F (2008), 'Characterization of the pattern of alpha(s1)- and beta-casein breakdown and release of a bioactive peptide by a cell envelope proteinase from *Lactobacillus delbrueckii* subsp *lactis* CRL 581', *Appl Environ Microbiol*, 74, 3682–3689.

HERNANDEZ-LEDESMA B, RECIO I and AMIGO L (2008), 'beta-Lactoglobulin as source of bioactive peptides', *Amino Acids*, 35, 257–265.

JAUHIAINEN T, VAPAATALO H, POUSSA T, KYRONPALO S, RASMUSSEN M and KORPELA R (2005), '*Lactobacillus helveticus* fermented milk lowers blood pressure in hypertensive subjects in 24-h ambulatory blood pressure measurement', *Am J Hypertens*, 18, 1600–1605.

KITTS D D and WEILER K (2003), 'Bioactive proteins and peptides from food sources. Applications of bioprocesses used in isolation and recovery', *Curr Pharm Design*, 9, 1309–1323.

KORHONEN H (2002), 'Technology options for new nutritional concepts', *Int J Dairy Technol*, 55, 79–88.

KORHONEN H and PIHLANTO A (2007), 'Technological options for the production of health-promoting proteins and peptides derived from milk and colostrum', *Curr Pharma Design*, 13, 829–843.

KORHONEN H and PIHLANTO A (2003), 'Food-derived bioactive peptides – Opportunities for designing future foods', *Curr Pharma Design*, 9, 1297–1308.

KUDOH Y, MATSUDA S, IGOSHI K and OKI T (2001), 'Antioxidative peptide from milk fermented with *Lactobacillus delbrueckii* subsp *bulgaricus* IFO13953', *J Jpn Soc Food Sci Technol*, 48, 44–50.

LAPOINTE J F, GAUTHIER S F, POULIOT Y and BOUCHARD C (2006), 'Selective separation of cationic peptides from a tryptic hydrolysate of beta-lactoglobulin by electrofiltration', *Biotechnol Bioeng*, 94, 223–233.

LAPOINTE J F, GAUTHIER S F, POULIOT Y and BOUCHARD C (2003), 'Effect of hydrodynamic conditions on fractionation of beta-lactoglobulin tryptic peptides using nanofiltration membranes', *J Membrane Sci*, 212, 55–67.

LEBLANC J G, MATAR C, VALDEZ J C, LEBLANC J and PERDIGON G (2002), 'Immunomodulating effects of peptidic fractions issued from milk fermented with *Lactobacillus helveticus*', *J Dairy Sci*, 85, 2733–2742.

LOPEZ-EXPOSITO I, GOMEZ-RUIZ J A, AMIGO L and RECIO I (2006), 'Identification of antibacterial peptides from ovine alpha(s2)-casein', *Int Dairy J*, 16, 1072–1080.

LOPEZ-EXPOSITO I, PELLEGRINI A, AMIGO L and RECIO I (2008), 'Synergistic effect between different milk-derived peptides and proteins', *J Dairy Sci*, 91, 2184–2189.

LOPEZ-FANDINO R, OTTE J and VAN CAMP J (2006), 'Physiological, chemical and technological aspects of milk-protein-derived peptides with antihypertensive and ACE-inhibitory activity', *Int Dairy J*, 16, 1277–1293.

LUND M and ARDO Y (2004), 'Purification and identification of water-soluble phosphopeptides from cheese using Fe(III) affinity chromatography and mass spectrometry', *J Agric Food Chem*, 52, 6616–6622.

MAO X Y, NI J R, SUN W L, HAO P P and FAN L (2007), 'Value-added utilization of yak milk casein for the production of angiotensin-I-converting enzyme inhibitory peptides', *Food Chem*, 103, 1282–1287.

MARTIN M, WELLNER A, OSSOWSKI I and HENLE T (2008), 'Identification and quantification of inhibitors for angiotensin-converting enzyme in hypoallergenic infant milk formulas', *J Agric Food Chem*, 56, 6333–6338.

MEGIAS C, PEDROCHE J, YUST M D, ALAIZ M, GIRON-CALLE J, MILLAN F and VIOQUE J (2006), 'Immobilization of angiotensin-converting enzyme on glyoxyl-agarose', *J Agric Food Chem*, 54, 4641–4645.

MEISEL H, GOEPFERT A and GUNTHER S (1997), 'ACE-inhibitory activities in milk products', *Milchwissenschaft-Milk Sci Int*, 52, 307–311.

MIGUEL M, RECIO I, RAMOS M, DELGADO M A and ALEIXANDRE M A (2006), 'Antihypertensive effect of peptides obtained from *Enterococcus faecalis*-fermented milk in rats', *J Dairy Sci*, 89, 3352–3359.

MINERVINI F, ALGARON F, RIZZELLO C G, FOX P F, MONNET V and GOBBETTI A (2003), 'Angiotensin I-converting-enzyme-inhibitory and antibacterial peptides from *Lactobacillus helveticus* PR4 proteinase-hydrolyzed caseins of milk from six species', *Appl Environ Microbiol*, 69, 5297–5305.

MINKIEWICZ P, DZIUBA J, DAREWICZ M, IWANIAK A, DZIUBA M and NALECZ D (2008), 'Food peptidomics', *Food Technol Biotechnol*, 46, 1–10.

MIZUNO S, NISHIMURA S, MATSUURA K, GOTOU T and YAMAMOTO N (2004), 'Release of short and proline-rich antihypertensive peptides from casein hydrolysate with an *Aspergillus oryzae* protease', *J Dairy Sci*, 87, 3183–3188.

MOLLER N P, SCHOLZ-AHRENS K E, ROOS N and SCHREZENMEIR J (2008), 'Bioactive peptides and proteins from foods: Indication for health effects', *Eur J Nutr*, 47, 171 182.

MULLALLY M M, MEISEL H and FITZGERALD R J (1997), 'Angiotensin-I-converting enzyme inhibitory activities of gastric and pancreatic proteinase digests of whey proteins', *Int Dairy J*, 7, 299–303.

MURRAY B A and FITZGERALD R J (2007), 'Angiotensin converting enzyme inhibitory peptides derived from food proteins: Biochemistry, bioactivity and production', *Curr Pharma Design*, 13, 773–791.

NAKAMURA Y, YAMAMOTO N, SAKAI K, OKUBO A, YAMAZAKI S and TAKANO T (1995a), 'Purification and characterization of angiotensin I-converting enzyme-inhibitors from sour milk', *J Dairy Sci*, 78, 777–783.

NAKAMURA Y, YAMAMOTO N, SAKAI K and TAKANO T (1995b), 'Antihypertensive effect of sour milk and peptides isolated from it that arc inhibitors to angiotensin I-converting enzyme', *J Dairy Sci*, 78, 1253–1257.

ONG L and SHAH N P (2008), 'Release and identification of angiotensin-converting enzyme-inhibitory peptides as influenced by ripening temperatures and probiotic adjuncts in Cheddar cheeses', *LWT-Food Sci Technol*, 41, 1555–1566.

PAPADIMITRIOU C G, VAFOPOULOU-MASTROJIANNAKI A, SILVA S V, GOMES A M, MALCATA F X and ALICHANIDIS E (2007), 'Identification of peptides in traditional and probiotic sheep milk yoghurt with angiotensin I-converting enzyme (ACE)-inhibitory activity', *Food Chem*, 105, 647–656.

PIHLANTO-LEPPALA A, KOSKINEN P, PAAKKARI I, TUPASELA T and KORHONEN H (1996), 'Opioid whey protein peptides obtained by membrane filtration,' *IDF Bulletin*, 311, 36–38.

PIHLANTO-LEPPALA A, KOSKINEN P, PIILOLA K, TUPASELA T and KORHONEN H (2000), 'Angiotensin I-converting enzyme inhibitory properties of whey protein digests: Concentration and characterization of active peptides', *J Dairy Res*, 67, 53–64.

PIHLANTO-LEPPALA A, ROKKA T and KORHONEN H (1998), 'Angiotensin I converting enzyme inhibitory peptides derived from bovine milk proteins', *Int Dairy J*, 8, 325–331.

PLATE K, BEUTEL S, BUCCHOLZ H, DEMMER W, FISCHER-FRUHHOLZ S, REIF O, ULBER R and SCHEPER T (2006), 'Isolation of bovine lactoferrin, lactoperoxidase and enzymatically prepared lactoferricin from proteolytic digestion of bovine lactoferrin using adsorptive membrane chromatography', *J Chromatogr A*, 1117, 81–86.

POULIOT Y, GAUTHIER S F and L'HEUREUX J (2000), 'Effect of peptide distribution on the fractionation of whey protein hydrolysates by nanofiltration membranes', *Lait*, 80, 113–120.

POULIOT Y, WIJERS M C, GAUTHIER S F and NADEAU L (1999), 'Fractionation of whey protein hydrolysates using charged UF NF membranes', *J Membrane Sci*, 158, 105–114.

QUIROS A, RAMOS M, MUGUERZA B, DELGADO M A, MIGUEL M, ALEIXANDRE A and RECIO I (2007), 'Identification of novel antihypertensive peptides in milk fermented with *Enterococcus faecalis*', *Int Dairy J*, 17, 33–41.

RENYE J A and SOMKUTI G A (2008), 'Cloning of milk-derived bioactive peptides in *Streptococcus thermophilus*', *Biotechnol Lett*, 30, 723–730.

ROUFIK S, GAUTHIER S F and TURGEON S L (2006), 'In vitro digestibility of bioactive peptides derived from bovine beta-lactoglobulin', *Int Dairy J*, 16, 294–302.

RYHANEN E L, PIHLANTO-LEPPALA A and PAHKALA E (2001), 'A new type of ripened, low-fat cheese with bioactive properties', *Int Dairy J*, 11, 441–447.

SAITO T, NAKAMURA T, KITAZAWA H, KAWAI Y and ITOH T (2000), 'Isolation and structural analysis of antihypertensive peptides that exist naturally in Gouda cheese', *J Dairy Sci*, 83, 1434–1440.

SILVA S V and MALCATA F X (2005), 'Caseins as source of bioactive peptides', *Int Dairy J*, 15, 1–15.

SINGH B and CHAND R (2006), 'Antioxidative activity of bovine milk fermented with *Lactobacillus helveticus*', *Milchwissenschaft-Milk Sci Int*, 61, 63–65.

SMITHERS G W (2008), 'Whey and whey proteins – From "gutter-to-gold"', *Int Dairy J*, 18, 695–704.

TAKANO T (1998), 'Milk derived peptides and hypertension reduction', *Int Dairy J*, 8, 375–381.

TOLDRA F, MOLINA L, VINUESA S and ARISTOY M C (2005), 'Peptide generation from casein hydrolysis by immobilised porcine cathepsins', *Food Chem*, 92, 227–233.

TOMITA M, WAKABAYASHI H, YAMAUCHI K, TERAGUCHI S and HAYASAWA H (2002), 'Bovine lactoferrin and lactoferricin derived from milk: Production and applications', *Biochem Cell Biol*, 80, 109–112.

TSAI J S, CHEN T J, PAN B S, GONG S D and CHUNG M Y (2008), 'Antihypertensive effect of bioactive peptides produced by protease-facilitated lactic acid fermentation of milk', *Food Chem*, 106, 552–558.

TUOMILEHTO J, LINDSTROM J, HYYRYNEN J, KORPELA R, KARHUNEN M L, MIKKOLA L, JAUHIAINEN T, SEPPO L and NISSINEN A (2004), 'Effect of ingesting sour milk fermented using *Lactobacillus helveticus* bacteria producing tripeptides on blood pressure in subjects with mild hypertension', *J Human Hypertens*, 18, 795–802.

VAN ELSWIJK D A, DIEFENBACH O, VAN DER BERG S, IRTH H, TJADEN U R and VAN DER GREEF J (2003), 'Rapid detection and identification of angiotensin-converting enzyme inhibitors by on-line liquid chromatography–biochemical detection, coupled to electrospray mass spectrometry', *J Chromatogr A*, 1020, 45–58.

VAN DER VEN C, GRUPPEN H, DE BONT D B A and VORAGEN A G J (2002), 'Optimisation of the angiotensin converting enzyme inhibition by whey protein hydrolysates using response surface methodology', *Int Dairy J*, 12, 813–820.

VEGARUD G E, LANGSRUD T and SVENNING C (2000), 'Mineral-binding milk proteins and peptides: Occurrence, biochemical and technological characteristics', *Br J Nutr*, 84, S91–S98.

VERCRUYSSE L, VAN CAMP J and SMAGGHE G (2005), 'ACE inhibitory peptides derived from enzymatic hydrolysates of animal muscle protein: A review', *J Agric Food Chem*, 53, 8106–8115.

VERMEIRSSEN V, VAN CAMP J, DEVOS L and VERSTRAETE W (2003), 'Release of angiotensin I converting enzyme (ACE) inhibitory activity during in vitro gastrointestinal digestion: From batch experiment to semicontinuous model', *J Agric Food Chem*, 51, 5680–5687.

VERMEIRSSEN V, VAN CAMP J and VERSTRAETE W (2004a), 'Bioavailability of angiotensin I converting enzyme inhibitory peptides', *Br J Nutr*, 92, 357–366.

VERMEIRSSEN V, VAN CAMP J and VERSTRAETE W (2005), 'Fractionation of angiotensin I converting enzyme inhibitory activity from pea and whey protein in vitro gastrointestinal digests', *J Sci Food Agric*, 85, 399–405.

VERMEIRSSEN V, VAN DER BENT A, VAN CAMP J, VAN AMERONGEN A and VERSTRAETE W (2004b), 'A quantitative in silico analysis calculates the angiotensin I converting enzyme (ACE) inhibitory activity in pea and whey protein digests', *Biochimie*, 86, 231–239.

VIRTANEN T, PIHLANTO A, AKKANEN S and KORHONEN H (2007), 'Development of antioxidant activity in milk whey during fermentation with lactic acid bacteria', *J Appl Microbiol*, 102, 106–115.

YAMAMOTO N, EJIRI M and MIZUNO S (2003), 'Biogenic peptides and their potential use', *Curr Pharma Design*, 9, 1345–1355.

ZHAO J W, GAO F, FU Y L, JIN W, YANG P Y and ZHAO D Y (2002), 'Biomolecule separation using large pore mesoporous SBA-15 as a substrate in high performance liquid chromatography', *Chem Commun* 752–753.

4

Processing means for milk fat fractionation and production of functional compounds

M. Sichien, N. Thienpont, E. Fredrick, T. Trung Le, J. Van Camp and K. Dewettinck, Ghent University, Belgium

Abstract: This chapter discusses the various processing techniques applied to the major and minor components of milk fat and how these techniques can lead to the production of compounds with different technological and nutritional characteristics. After reviewing the global composition of bovine milk lipids, an overview is given of the methods for milk fat modification in order to enlarge the utilization range of milk fat in food products. The chapter then describes ways of isolating the polar lipids in dairy products with the aim of manufacturing added-value ingredients with special functionalities. The role of dairy lipids related to nutrition and health issues is covered in the last part of this chapter.

Key words: milk lipids, milk fat modification, polar lipids, technological and nutritional properties.

4.1 Introduction

Bovine milk can be considered as an oil-in-water emulsion in which the lipid fraction is mostly situated inside spherical droplets, the milk fat globules. Among the different natural fats and oils, bovine milk fat is considered as one of the most complex lipids due to the typical fatty acid pattern. This unique fatty acid profile affects to a great extent the physical and technological properties, making milk fat an intriguing substrate for physical and chemical modification. By these modification processes, it is possible to tailor milk fat for specific application areas without, or at least to a small extent, changing the unique organoleptic properties of milk fat. Furthermore, milk fat contains some components in smaller amounts, comprising the polar lipids, partial glycerides and sterols. These minor components can be found mainly in the milk fat globule membrane which is a thin layer surrounding the lipid droplets. The polar lipids in milk products are an interesting fraction from a technological and nutritional point of view. Because of their amphiphilic nature, milk polar lipids can be used as an emulsifying agent in various food products. Next to their attractive technological properties, the milk polar lipids are gaining more and more interest as a result of their health-promoting effects.

After giving an overview of the different major and minor components in milk fat and the factors affecting the global composition, the several modification processes applied to milk fat are covered in this chapter. This is by a discussion about the polar lipids in milk fat and their promising functionalities. In the last part of this chapter, the nutritional properties of dairy fats are reviewed.

4.2 Overview of milk lipids

The global composition of bovine milk lipids is depicted in Table 4.1. As for most other food fats, the main components of anhydrous milk fat are triglycerides (95–98%). The minor components of the milk lipids include mainly diglycerides (0.3–1.59%), monoglycerides (0.016–0.038%), phospholipids (PLs) (0.8–1.0%), sterols (0.22–0.41%) and free fatty acids (0.1–0.44%).

Table 4.1 Composition of bovine milk lipids (Kurtz, 1978 and Walstra *et al.*, 2006)

Class of lipid	% Total milk lipids (w/w)
Triglycerides	95–98
Diglycerides	0.3–1.59
Monoglycerides	0.016–0.038
Keto acid glycerides (total)	0.85–1.28
Hydroxy acid glycerides (total)	0.60–0.78
Lactonogenic glycerides	0.06
Neutral glyceryl ethers	0.016–0.020
Neutral plasmalogens	0.04
Free fatty acids	0.1–0.44
Phospholipids (total)	0.80–1.00
Sphingolipids (less sphingomyelin)	0.06
Sterols	0.22–0.41
Squalene	0.03
Carotenoids	5–10 E^{-4}
Vitamin A	6–9 E^{-4}
Vitamin D	0.85–2.1 E^{-6}
Vitamin E	0.0024
Vitamin K	1 E^{-4}

Table 4.2 Composition (weight% and mol%) of the major fatty acids in butterfat (Maniongui, 1991)

Fatty acid	Weight%	Mol%
4:0	3.70 ± 0.02	9.61 ± 0.05
6:0	2.45 ± 0.02	4.85 ± 0.04
8:0	1.36 ± 0.01	2.17 ± 0.01
10:0	2.88 ± 0.01	3.87 ± 0.01
12:0	3.21 ± 0.01	3.71 ± 0.01
14:0	11.19 ± 0.02	11.37 ± 0.02
14:1	1.05 ± 0.01	1.08 ± 0.01
15:0	1.27 +/− 0.05	1.35 +/− 0.05
16:0	27.37 ± 0.03	24.78 ± 0.03
16:1	1.65 ± 0.02	1.51 ± 0.02
18:0	13.91 ± 0.02	11.36 ± 0.02
18:1	27.99 ± 0.04	23.02 ± 0.03
18:2	2.05 ± 0.05	1.70 ± 0.04
18:3	0.99 ± 0.01	0.83 ± 0.01
20:0	0.19 ± 0.01	0.14 ± 0.01

4.2.1 Major components

Triglycerides consist of three fatty acids esterified on a glycerol backbone. Milk fat triglycerides can contain over 250 (Walstra *et al.*, 2006) to 400 (Jensen, 1992) different fatty acids. The major fatty acids are the even carbon-number linear fatty acids, which account for 95 mol% of the total fatty acids (Table 4.2) (Maniongui *et al.*, 1991). The broad range in chain length and degree of saturation of the fatty acids contributes to the wide melting range of milk fat. On the glycerol backbone, the fatty acids are not randomly distributed. The short-chain fatty acids are, for instance, primarily esterified on the sn-3 position. Although there is this somewhat stereospecific distribution of the fatty acids, an identification of all milk fat triglycerides is not feasible due to the highly complex fatty acid pattern. A milk fat mixture could contain up to 1331 different triglycerides (Walstra and Jennes, 1984). Therefore, the chemical composition of different milk fats is generally evaluated using the fatty acid profile.

Table 4.2 gives an overview of the major fatty acids present in milk fat in weight% and mol% terms (Maniongui *et al.*, 1991). About 70 mol% of the fatty acids is saturated. The most abundant fatty acid is palmitic acid (C16:0) followed by oleic acid (C18:1). Characteristic for milk fat is the occurrence of a significant amount of short-chain fatty acids (C4 to C10), *trans* fatty acids, like vaccenic acid (C18:1 *t*11) (Wolff *et al.*, 1995) and conjugated linoleic acid (CLA), generally C18:2 *c*9*t*11 (Collomb *et al.*, 2006). The short chain fatty acids are synthesized in the mammary gland and determine, to a large extent, the melting and crystallization properties of milk fat with respect to the physical properties. The presence and the amount of CLA and *trans* fatty acids is determined by both the rumen isomerization and biodegradation process of unsaturated fatty acids and the activity of Δ^9-desaturase in the mammary gland (Fritsche and Steinhart, 1998; Collomb *et al.*, 2006). Furthermore, the concentrations of branched fatty acids (Sonntag, 1979; Ha and Lindsay, 1991), hydroxy fatty acids (Jensen, 1992), keto fatty acids (Parks *et al.*, 1964) and odd length chain fatty acids are higher in dairy fats compared to vegetable fats.

4.2.2 Minor components

All non-triglyceride components present in the milk fat can be defined as minor components (Table 4.1). The concentrations of these minor components are dependent on the definition of milk fat (Vanhoutte, 2002). When milk fat refers to the apolar phase of milk, the concentrations of cholesterol and PLs will be higher than when milk fat is defined as the industrial product, anhydrous milk fat (AMF). The latter is the fat phase gained from butter after several centrifugal separations. PLs, cholesterol and cholesterol esters, which are mainly present in the milk fat globule membrane, (MFGM) can be found in lower concentrations in AMF due to incomplete exclusion of MFGM and as a result of their solubility in the triglyceride mixture.

In bovine milk, the PLs contribute to approximately 1% of the total milk lipid. The main PLs are phosphatidylcholine (PC), phosphatidylethanolamine (PE), phosphatidylserine (PS) and sphingomyelin (SM) (Jensen, 1973). Their fatty acid composition differs from the fatty acid profile of the triglycerides in milk fat. More long-chain unsaturated fatty acids are detected and no short-chain fatty acids are present (Jensen, 1973).

Free fatty acids (FFA), monoglycerides and diglycerides are present due to the unfinished synthesis of triglycerides or owing to lipolytic action. A high FFA content contributes to the hydrolytic rancidity in butter; especially the presence of C4 and C6 add strongly to this off-flavour. The C8 FFA does not contribute to rancid off-flavours since a rather low concentration is found in rancid butter, while C10 and C12 FFA give a distinct soapy and bitter taste in their usual concentration range in soapy rancid butters (Patton, 1973). The unsaponifiable fraction includes sterols, tocopherols, hydrocarbons, carotenoids, aliphatic alcohols, squalene and fat soluble vitamins (A, D, E and K). Most of the sterols are cholesterol (>95%) although traces of lanosterol, dihydrolanosterol and β-sitosterol are also present (Jensen, 1973).

4.2.3 Factors influencing the milk lipids composition

The chemical composition of milk fat is influenced primarily by feeding, the breed of the cow (genetics), its stage of lactation (physiological condition) and possible infections of the udder (mastitis). Actually, when all influencing factors are kept constant, differences in milk composition of different individuals can still be noticed (Walstra *et al.*, 2006).

Table 4.3 Chemical and physical properties of winter- and summer-fat (Bornaz *et al.*, 1993; Hettinga, 1996; Vanhoutte, 2002)

Samples	Fatty acids				Iodine number	Firmness (N)
	Volatile	Saturated	Mono-unsaturated	Poly-unsaturated		
Average of total	10.98	56.50	29.81	2.50	32.2	
Summer	9.49	58.82	33.53	3.14	36.8	1.5–3
Winter	12.45	59.15	26.15	1.86	27.7	7–9

The most important factor of variability is feeding. An excellent illustration of this is the variation between summer and winter milk fat (Vanhoutte, 2002). Table 4.3 illustrates that when cows change from stall feeding (silaged feed) in the winter to fresh grass during the summer months, a raise in iodine value, a change in fatty acid composition and a decrease of hardness can be detected. Moreover, adding trisaturated triglycerides to the diet alters the fatty acid composition of the milk fat. Due to the occurrence of biohydrogenation in the rumen, the relation between unsaturated fatty acids in feed and their concentration in the milk fat is less obvious (Grummer, 1991; Vanhoutte, 2002). The physical form of feed also has an effect. If unsaturated fatty acids should be present in a specific matrix that can escape biohydrogenation, it can result in a higher proportion of unsaturated fatty acids in milk fat (Walstra *et al.*, 2006). Moreover, several minor components, such as carotene and fat-soluble vitamins, are strongly affected by their content in the feed. The same counts for some trace elements.

The breed of the cow is mainly selected by man, according to the intended use and the local conditions, such as climate, feed, terrain and customs, resulting in some variability in milk composition. Concerning the lactation period, the proportion of short chains is, for example, low at the start and increases until at least the first 8 to 10 weeks of the lactation period (Palmquist *et al.*, 1993).

4.3 Milk fat modification

4.3.1 Introduction

Anhydrous milk fat (AMF) was produced as an excellent transport and storage form of the large stocks of butter during the 1980s in the European Union and the United States. By removing the water phase from butter, milk fat could be stored for a considerably longer period of time without significant loss of quality. Nowadays, milk fat has acquired a certain market share as a raw material for the confectionery and bakery industries, due to its unique organoleptic properties and its image as a natural ingredient (Kaylegian and Lindsay, 1995; Vanhoutte *et al.*, 2002a). However, AMF has disadvantages which limit its consumption and its use. Among these are negative health aspects due to its high concentration of saturated fatty acids and cholesterol and its high price compared to vegetable fats. Furthermore, the chemical composition of milk fat varies across different regions and seasons. These natural fluctuations, primarily due to feeding practices, result in milk fats with different physicochemical properties (Jensen *et al.*, 1991). In addition, the functionality of milk fat is often worse than that of vegetable fats such as margarines, restricting the potential use of milk fat in different food products such as puff pastry, cookies, and cold-spreadable products (Deffense, 1993). On the other hand, the much-appreciated flavour of AMF cannot be realized using these tailor-made vegetable products. Even addition of buttery aromas (e.g. diacetyl) seems unsatisfactory (Boudreau and Arul, 1993; Rajah, 1994).

The utilization range of milk fat in many food applications could be broadened by the application of various modification processes, such as hydrogenation, interesterification and fractionation. These modification techniques alter the chemical composition of milk fat, and therefore the physical and nutritional characteristics, making milk fat more suitable in a wider field of application or for tailored specific end uses (Kaylegian, 1999; Van Aken *et al.*, 1999).

In this part, we will discuss the technology of milk fat fractionation and interesterification, and the influence of process parameters in more detail. The effect of these modification techniques on the composition and the physicochemical properties of milk fat is also covered. Hydrogenation of milk fat will not be considered here as this modification technique is no longer applied in the dairy industry. This is because the increased levels of saturated and *trans* fatty acids in hydrogenated milk fat are not desirable from a nutritional point of view. Also, the quite high production costs and the negative attitude of consumers towards chemical modification of a fat with a natural image militate against its commercial application. Therefore, milk fat hydrogenation is replaced industrially by other modification processes (especially fractionation) for the production of milk fat with an increasing hardness, oxidative stability, and plastic range (De Greyt and Kellens, 2001; Timms, 2005).

4.3.2 Fractionation

Milk fat is composed of approximately 400 fatty acids and can thus contain, at the most, 400^3 triglycerides. Due to this very heterogeneous composition, milk fat exhibits a wide melting range, from about –40°C to 40°C, which provides the possibility of separating milk fat into a series of fractions having their own physical and chemical properties. It has been accepted that the melting profile of natural milk fat contains three major peaks, due

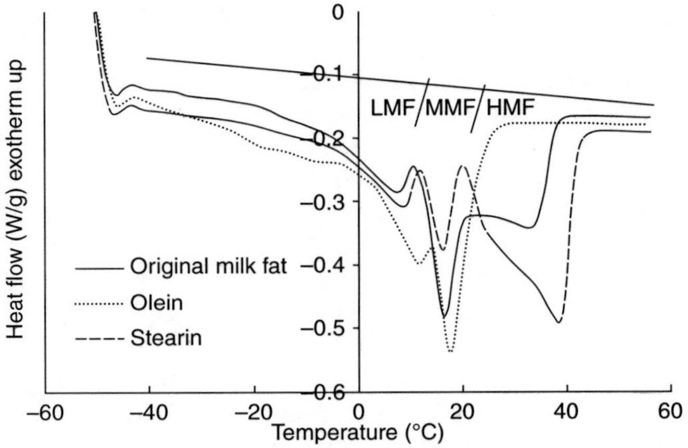

Fig. 4.1 Melting profile of milk fat, stearin and olein determined with a heating rate of 5°C/min (Vanhoutte *et al.*, 2002a).

to three major groups of triglycerides melting independently (Fig. 4.1): a low-melting fraction (LMF) which melts below 10°C, a middle-melting fraction (MMF) which melts from 10°C to 19°C, and a broad high-melting fraction (HMF) which melts above 20°C (Deffense, 1993; Kaylegian and Lindsay, 1995). Fractionation is the separation process of a fat into a higher melting solid fraction, the so-called 'stearin' and a lower melting liquid fraction, the so-called 'olein'. The olein fraction is enriched in short-chain triglycerides and long-chain unsaturated triglycerides, while the 'stearin' fraction contains more long-chain saturated triglycerides. Deffense (1993) and Vanhoutte *et al.* (2002a) demonstrated that, in the melting curve of the milk fat olein fraction, the peaks of LMF and MMF enlarge, while HMF is completely removed at a fractionation temperature of 18°C. Conversely, the stearin fraction is enriched in HMF and contains less MMF (Fig. 4.1). The increase of LMF in the high-melting fraction is explained by olein entrainment in the solid fraction, due to incomplete separation.

Fractionation of AMF is extensively reviewed by Kaylegian and Lindsay (1995) and, recently, more briefly by Augustin and Versteeg (2006). Different types of fractionation technologies have been employed and these technologies can be divided into two general groups:

(i) Fractionation methods based on crystallization and solubility behaviour of milk fat, including
   - Crystallization from melted milk fat, also known as dry fractionation (Deffense, 1993; Kellens, 1994; Lopez *et al.*, 2006; Tirtiaux, 1976) or as Lanza fractionation in the case that a detergent solution is used to facilitate the separation of the fractions (Jebson, 1970).

- Crystallization from milk fat dissolved in a solvent, also called wet fractionation (Boudreau and Arul, 1993; Hartel, 2001; Larsen and Samuelsson, 1979; Marangoni and Lencki, 1998).

(ii) Fractionation methods based on the solubility and volatility of the lipid constituents, including supercritical fluid fractionation (Kaufmann *et al.*, 1982; Arul *et al.*, 1987; Bhaskar *et al.*, 1993) and short-path distillation (Arul *et al.*, 1988; Campos *et al.*, 2003).

Dry fractionation
Crystallization from melted fat (or dry fractionation) is the most commonly employed and commercially exploited fractionation technique. This technique is very attractive for the dairy industry due to its reasonable cost of processing and its relatively simple equipment. Other advantages of dry fractionation are the preservation of the delicate flavour of milk fat, the absence of organic solvents or other processing aids, and the environmentally-friendly character confirming the 'natural' product image (Fatouh *et al.*, 2003). Today about 1000 tons per day of milk fat are fractionated with this fractionation process, predominantly in Europe.

Dry fractionation consists of the development of suitable sizes of crystals by controlled cooling and agitating of the molten fat, followed by the separation of the crystals from the remaining mother liquor by filtration or centrifugation. The chemical composition and physical characteristics of the fractions obtained are significantly affected by the equipment design, the related fractionation process, separation method and other process parameters such as the cooling and agitation rate, the residence time of the fat in the crystallizer and the fractionation temperature (Grall and Hartel, 1992). Currently, dry fractionation is commercially executed by two different processes, the Tirtiaux (Deffense, 1993; Tirtiaux, 1976) and the De Smet (Kellens, 1994) processes, both from Belgium. Both processes start with heating the milk fat to a temperature of 60–80°C to ensure that all the nuclei are destroyed. The Tirtiaux and De Smet process differ in the way the cooling step is carried out.

Tirtiaux has patented a slow-cooling process, usually carried out in vessels with gentle agitating. The fat is cooled following a special cooling pattern based on a constant temperature difference between the oil and the cooling liquid. Typically, the cooling rate is considerably decreased near the melting point, resulting in the formation of the desired β' crystals, which are easier to separate. The first cooling step with the formation of crystal nuclei takes place in special buffer tanks. From there, the oil is pumped into crystallizers in which the cooling is regulated by constant monitoring of the temperature of the milk fat in the crystallizer. The intention of this second step is the growth of very large, equally-sized crystals (optimal size of 200 to 300 μm) which are easy to separate from the oil slurry. The agitation rate can also be adjusted in order to obtain crystals of the desired size and shape. The major shortcoming of this process is the requirement of long crystallizing times (about 16 h to 20 h) to achieve the optimum crystallization.

In contrast, the cooling in the De Smet process is also carefully monitored but it is applied in a different way. The processes recommended by De Smet follow the principle of fast cooling by lowering stepwise the temperature of the cooling liquid in precise and precalculated steps. In that way, the milk fat is precooled for sufficient crystal nuclei formation and the final fractionation temperature is reached more quickly. Due to this cooling program, a high cooling surface per unit volume of fat is necessary in the industrial design of the concentric crystallizers of the De Smet process. Additionally, filters and centrifuges with a higher separation efficiency compared to those applied in the Tirtiaux process are necessary as the fat crystals in the De Smet process are much more heterogeneous in size, resulting in a considerable higher filter cost. However, the De Smet process achieves its optimum crystallization in much less time (6 h to 8 h) than the Tirtiaux process.

The next stage in this modification process is the separation step, resulting in the actual production of a solid product and a liquid fraction. The efficiency of this step has an enormous impact on the properties and yields of the resulting fractions. In the dairy industry, separation of the milk fat slurry is mostly accomplished by filtration. There are two different filtration techniques used; vacuum filtration on belt or drum filters and membrane press filtration. Vacuum filtration separates the liquid first by gravity and then the entrained olein is further removed under vacuum pressure. Although the separation can be carried out continuously, this technique does not permit a very thorough separation and a considerable amount of the olein fraction stays entrained in the filter cake. Classically, a continuous belt Florentine vacuum (50 to 200 mbar) filter, shown in Fig. 4.2, can be used in the Tirtiaux process since very large, equally-sized crystals are formed, which are easy to separate. In this type of filter, filtration takes place on the upper, horizontal surface of a rotating, stainless-steel perforated belt, which has a vacuum chamber beneath it (Illingworth, 2002). Better separation efficiency is achieved by using membrane press filters, introduced during the eighties in the fractionation industry. The application of pressure to the cake in the press chamber brings about a lower level of liquid entrainment in the filter cake. The higher pressure in the chambers of the membrane filter is generated by inflating the watertight membranes with air, reducing the chamber volumes. After deflating the membranes, the filter cakes can be batch-wise discharged into a tank by gravity. These principles of operation are illustrated in Fig. 4.3. Today, the membrane filter press is the most suitable filtration technique in milk fat dry fractionation and its usage is growing increasingly for all the crystallization processes. Especially in the case of the De Smet crystallization process, membrane filters are much preferred. Standard membrane filter presses normally operate at a maximum pressure of 4–8 bars, which is more than

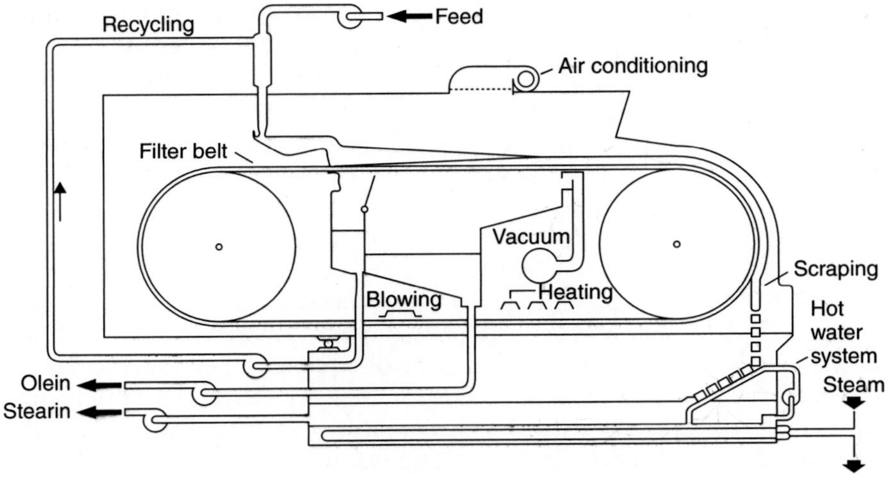

Fig. 4.2 A schematic diagram of a continuous belt Florentine vacuum filter (Illingworth, 2002).

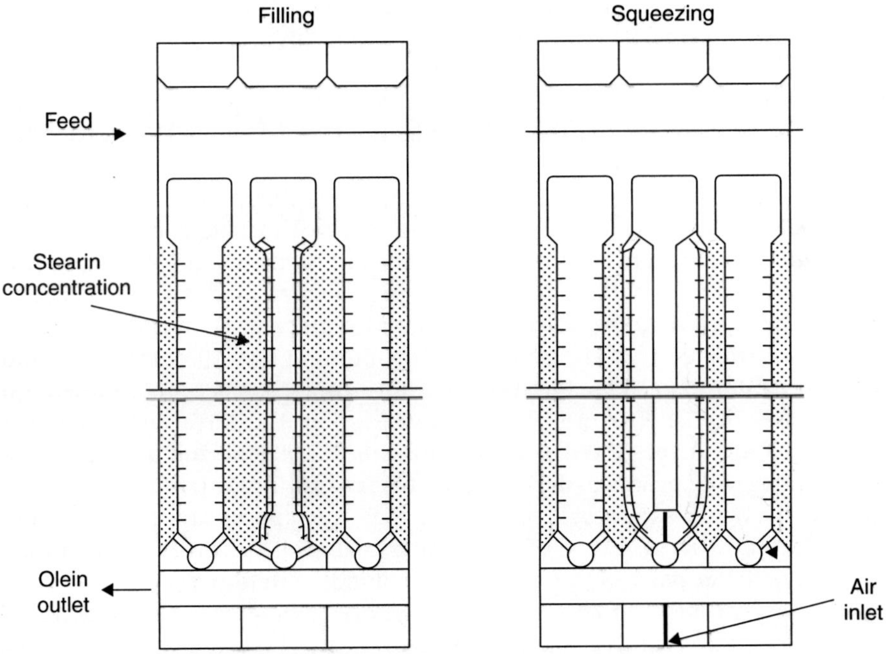

Fig. 4.3 Intersection of membrane plates during filling and squeezing (Calliauw, 2008).

sufficient in most cases. During recent years, new high-pressure membrane filter presses have been developed to operate at pressures of up to 30 bar. Those high-pressure membrane filter presses make use of hydraulic fluid as a pressure medium instead of air used for the standard membrane filter presses (Gibon and Tirtiaux, 2002). Higher pressures are necessary to meet the required specifications, especially in the production of high-purity olein fractions.

Milk fat fractions can also be separated by use of centrifugal forces such as with nozzle and screen centrifuges. As yet, only a few commercial installations are currently known to exist. For instance, Westfalia developed and launched in 1997 a nozzle centrifuge as an alternative to membrane filtration for AMF fractionation. In this piece of equipment, the fat crystals are separated in a centrifugal field due to the density difference of olein and stearin. During the operation, crystals are drawn outwards where they are discharged through backwardly reacting nozzles by the centrifugal force. This type of separation can be done continuously and in an enclosed system, the latter being very important in preventing the possible deterioration of milk fat (fractions) when exposed to oxygen or light. However, as stated, only a few installations have been used on an industrial scale, possibly due to a separation efficiency which is not as good as those of the membrane press, and due to complicated operation handling operations (Deffense, 2000; Dijkstra, 2007; Illingworth, 2002).

The fractionation process, and more predominantly the crystallization and filtration properties, are greatly affected by a number of process parameters such as agitation rate, cooling rate and fractionation temperature. The effect of these parameters on milk fat fractionation and the properties of the resulting fractions have been investigated several times (Deffense, 1987; Patience *et al.*, 1999; Vanhoutte *et al.*, 2002a, 2003). For instance, Vanhoutte *et al.* (2003) concluded that higher fractionation temperatures result in a lower stearin yield and a higher melting point of the olein fraction. Fractionation temperature also influences the cake density and thus the filtration characteristics to a certain extent, probably related to differences in crystal size and/or morphology. At lower temperatures, more compacted clumps of smaller crystals are formed, resulting in less oil entrapment and a denser filter cake. The agitation rate also has a considerable impact on the crystal size and/or morphology and thus on the filtration behaviour. Herrera and Hartel (2000) noticed that samples crystallized at the highest shear rates will produce the narrowest particle size distribution and the smallest crystals, as a result of an enhanced nucleation rate (Herrera and Hartel, 2000) and breakage of the crystals by the shear forces of the impeller (Patience *et al.*, 1999). On the other hand, agitation ensures effective heat transfer on the cooled surfaces of the crystallization vessel and avoids local overheating and subsequent melting of the crystals.

Some minor components of the milk fat can also alter the fractionation properties. Vanhoutte *et al.* (2002b) observed that high levels of free fatty acids (from inadequate heat treatment or microbial contamination) or high levels of partial glycerides (from a deodorization process) have a negative influence on the filtration quality of the resulting crystal suspension. In the small-scale fractionation experiments, lower filtration rates and an increased oil inclusion resulting in a softer stearin fraction were detected.

The presence of phospholipids changes negatively both the crystallization kinetics and the filtration properties. Increased induction times and worse filtration qualities leading also to a softer stearin, were observed at higher concentrations of added phospholipids. These results emphasise the importance of good quality raw materials for fractionation of AMF and an adequate removal of the membranous material (containing a lot of phospholipids) during the production of AMF.

By fractionation, two fractions are produced from the original milk fat, with modified chemical and physical properties. However, for applications in a broad range for food products, the separation of AMF into sharply defined fractions usually happens in a multi-stage process. In such a multistage process, one or both of the fractions are remelted and refractionated but with different process parameters and cooling patterns. Most commonly, the olein of the first step is fractionated again at a fractionation temperature that is lower than the temperature of the first step. This process can then be repeated. In the dairy industry, most of the fractionation processes of AMF happen in three steps, although seven steps have been reported in literature. Figure 4.4 presents some fractionation routes applied to milk fat, together with the melting characteristics of the fractions obtained (Deffense, 1993; Illingworth, 2002; Kaylegian, 1999; Kaylegian and Lindsay, 1995).

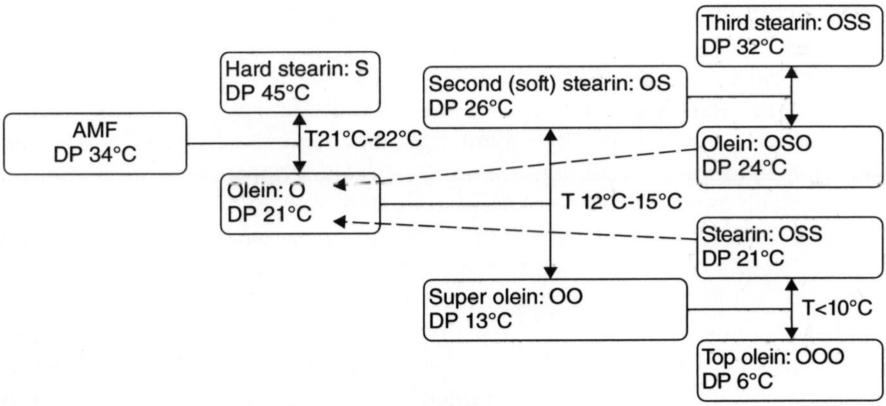

Fig. 4.4 Example of a multistep fractionation of milk fat, with optional recycling of some fractions. Abbreviations: DP, dropping point; S, stearin; O, olein (adapted from Illingworth, 2002).

The typical three-step fractionation process of milk fat creates thus five fractions, all with different melting characteristics and, as a result, with other application areas. A first field of application of the different fractions is the bakery industry, due to the pleasant flavour, and the richness and tenderness of milk fat. For instance, the hard stearin fractions may be used as the fat phase in laminated pastries such as puff pastry, Danish pastries and croissants, which are flaky products containing successive layers of dough and fat. The fat phase in these products must remain solid during production, but must not be too brittle otherwise the fat will break and will not create a separate layer. In cookies and biscuits, lower melting fractions (olein fractions) are required as the fat has to be liquid to prevent gluten interactions resulting in the typical, short texture. Also, in confectionery products such as in chocolate, stearins can be used. These high melting fractions are more compatible with cocoa butter, making it possible to use more milk fat. Additionally, these fractions act as a fat bloom inhibitor in chocolate. Stearin can also be used as a component of cold-spreadable butter and as hardstock for margarine production in place of hydrogenated vegetable fats. The various olein fractions can be used to improve the spreadability of butter and dairy spreads at refrigerator temperature. A first way to make spreadable butter is by blending the cream and 15% to 25% super olein, followed by the cream ripening and churning process. The second possibility to create the desired melting profile of cold-spreadable butter is recombining 70–80% of the top olein fraction with higher melting fractions, followed by emulsification of the fat and water phase in a scraped surface heat exchanger and texturisation (Burgess, 2001; Deffense, 1993; Hartel and Kaylegian, 2001; Kaylegian, 1999).

Lanza fractionation

The so-called Lanza or detergent fractionation is based on a technique patented by Fratelli Lanza in 1905. In the 1970s, Tetra Pak (formerly Alfa-Laval) marketed the technique as the Lipofrac process (Fjaervoll, 1969). The crystallization step is very similar to that used in dry fractionation but it differs in the way in which the olein is separated from the stearin. Separation is not accomplished by means of a mechanical treatment, but by an aqueous solution of detergent (sodium lauryl sulphate) containing an electrolyte such as magnesium sulphate, followed by centrifugation. The detergent covers the surface of the fat crystals that have been precipitated from the melt. As a result, the wetted crystals become hydrophilic and will move from the oil phase to the aqueous phase. When the detergent solution is added, it is essential to stir strongly in order to disintegrate the crystal agglomerates which could trap the olein and thereby avoid the wetting of some crystals. The olein phase is generally separated from the aqueous phase (with the stearin crystals) via centrifuges. Due to a higher difference in specific density, centrifuges with a lower separation power can be used compared to those required in the dry fractionation process. The water/

crystal phase is then heated to melt the crystals, followed by a second centrifugation step to complete the separation.

With this type of fractionation, much smaller crystals are required, allowing much shorter crystallization times. Other benefits are an improved olein yield and better separation efficiency.

In the 1970s and early 1980s, Lanza fractionation of AMF was executed on a commercial scale in New Zealand (Illingworth, 2002; Norris *et al.*, 1971). However, this process is, at the moment, ousted by dry fractionation as a result of world-wide resistance from the consumers against any treatment of milk fat with chemical additives, even if only as a processing aid. A number of countries have even prohibited by law the use of detergent-separated fractions in fat-rich dairy products. The higher energy consumption and effluent problems as a result of the use of the wetting agent also make Lanza fractionation unattractive for AMF fractionation.

Solvent fractionation

Another possibility for separating milk fat into fractions with different melting profiles is based on dissolution in an appropriate solvent followed by crystallization at a defined fractionation temperature. The so-called solvent or wet fractionation was one of the first techniques used for milk fat fractionation and it was already being used in the late 1950s (Bhalerao *et al.*, 1959; Hartel and Kaylegian, 2001). The presence of the solvent influences largely the crystallization kinetics, resulting from the lower viscosity of the solvent–milk fat mixture compared with the milk fat itself. In fact, the crystallization time is very rapid and often in the order of 30 min (compared with the 24 h in the Tirtiaux process), and the crystallization can even be executed in a scraped-surface tubular heat exchanger. Either apolar (e.g. pentane, hexane) or polar (e.g. acetone, ethanol, isopropanol) solvents may be used and the solvent type also affects the crystallization behaviour of the milk fat. Larsen and Samuelsson (1979) observed that crystallization in polar solvents yields rod-shaped crystals which form compact aggregates, while the crystals formed in apolar solvents are not well-defined and have a gel-like structure. When using polar solvents, the separation efficiency is thus improved compared to dry fractionation where the crystals have a spherulite structure, allowing a higher inclusion of the olein fraction in the crystals. The dilution of the interstitial olein by the solvent and washing the filter cake with fresh solvent also influence the separation efficiency in a positive way. The separation is usually executed via vacuum filtration or less commonly via centrifugation. The milk fat fractions obtained are then freed from the solvent by distillation. However, despite the advantages of faster crystallization, improved olein yields and superior separation sharpness, solvent separation is never used on an industrial scale predominantly due to the loss of aroma and flavour compounds of the fractions obtained. Additionally, the high investment and operation costs, the need for safety precautions and the potential toxicological concerns about solvent residues ensure that solvent fractionation of AMF is not found in commercial applications (Augustin and Versteeg, 2006; Dijkstra, 2007; Kaylegian and Lindsay, 1995; Wright *et al.*, 2000).

Supercritical fluid fractionation

Supercritical fluid fractionation is based on the solubility of milk fat in a supercritical fluid (instead of a solvent as in the wet fractionation process) and on volatility properties of the milk fat components. A supercritical fluid is any substance at a temperature and pressure above its thermodynamic critical point (i.e. the temperature above which liquid cannot be formed simply by increase in pressure) and such fluids combine liquid-like and gas-like behaviour in an interesting manner, making them attractive extraction agents (Arul *et al.*, 1987; Bhaskar *et al.*, 1993). The densities of supercritical fluids are similar to those of liquids, resulting in good solvent power, while the high diffusion rates due to their low viscosity and diffusivity values ensure that they can generally penetrate AMF faster than liquid solvents. In spite of the poor solubility of triglycerides in supercritical carbon dioxide, the most attractive fractionating solvent is CO_2 as a consequence of its inexpensive, non-toxic and non-flammable character. Furthermore, supercritical CO_2 does not pose any toxicological concerns about solvent residues and has an environmentally friendly image (Boudreau and Arul, 1993). The density, and thus the solvent power, of supercritical CO_2 can be manipulated by varying the process conditions of fractionation, which makes it possible to selectively separate multiple fractions in one step. The solvent power is enhanced when the temperature and the pressure increase. At low pressure, only the low-molecular weight triglycerides are extracted. As the pressure increases at a constant temperature, the long-chain triglycerides become more soluble and can be extracted. Because the fractionation of AMF is primarily based on differences in the molecular weight of the triglycerides, the fractions obtained are more markedly different in fatty acid and triglyceride composition compared to melt crystallization. On the other hand, the fractions obtained by supercritical CO_2 fractionation show less distinct differences in melting characteristics, thereby restricting the use of supercritical extraction in the production of fractions for specific food applications. Together with the high initial capital costs, this explains why supercritical fluid fractionation of AMF is applied only on a laboratory scale, but not yet on an industrial scale (Augustin and Versteeg, 2006; Bhaskar *et al.*, 1998).

Short-path distillation

Short-path distillation consists of the evaporation of molecules into a vacuum. Since short-path distillation is founded on variations in the volatility of the milk fat triglycerides, this fractionation process has a very high degree of molecular weight and carbon number separation, even more than supercritical fluid extraction. The distillate (the olein fraction) is enriched in short-chain and medium-chain fatty acids and depleted in long-chain fatty acids, while the opposite trends are observed in the retentate (the stearin fraction). In comparison with other processes, a higher concentration of unsaturated long-chain fatty acids is noticed in the solid fraction than in the liquid fraction (Arul *et al.*, 1988; Boudreau and Arul, 1993; Campos *et al.*, 2003). The chemical and physical properties are also affected by the applied distillation temperature (Boudreau and Arul, 1991). Campos *et al.* (2003) characterized the fraction properties at distillation temperatures from 125°C to 250°C. They found an augmented SFC-profile of the retentate and an increasing yield of the distillate from 0.3% to 42.7% when the distillation temperature was increased.

Short-path distillation thus offers the possibility of fractionating milk fat into fractions with unique chemical and physical characteristics. However, it has not been used in the industry as a result of thermal damage to the milk fat (decomposing and polymerizing of the unsaturated fatty acids), loss in flavour and high thermal demands (Boudreau and Arul, 1991).

4.3.3 Interesterification

Interesterification consists of a rearrangement of the acyl groups within and among triglycerides, which can consequently lead to changes in the different physical and functional characteristics of the interesterified fat. The distribution of the fatty acids on the glycerol molecule is thus modified, although the fatty acid composition of the fat is retained.

Since the reaction rate of interesterification is very slow, the chemical reaction should be carried out in the presence of a catalyst. Two types of catalyst are currently in use, namely inorganic catalysts (chemical interesterification) and biologically derived enzymes (enzymatic interesterification), both with their own benefits and drawbacks.

Chemical interesterification

Chemical interesterification involves the use of chemicals such as alkali metals (sodium, potassium and their alloys) and their corresponding hydroxides and alkoxides to accelerate the redistribution of the fatty acids. A first paper considering this modification process applied to milk fat appeared in the late 1950s (Weihe and Greenbank, 1958; Weihe, 1961). The authors concluded that interesterification leads to an increase in the melting point of milk fat. Greater chances of increases were established by directed interesterification than by random interesterification. In directed interesterification, the temperature of the reaction is controlled so that certain higher melting triglycerides crystallize and are no longer able to participate in further reactions. In order to re-establish the equilibrium, the reaction is driven to form these higher melting triglycerides. Furthermore, larger increases in melting point were identified in the presence of a solvent.

Huyghebaert *et al.* (1994), Rodrigues and Gioielli (2003) and Rousseau *et al.* (1996) showed that interesterification has also a considerable influence on the solid fat content profile of the modified milk fat. They observed a reduced solid fat content at lower temperatures and an increased solid fat content at higher temperatures. This smoother SFC-profile and the changed melting point are the result of a changed triglyceride composition of the milk fat. Interesterification increases the level of lower molecular weight triglycerides, consisting 22 to 30 carbon atoms, explained by randomization of the initially asymmetrically distributed short-chain fatty acids. The amount of higher molecular weight triglycerides (C_{44}–C_{54}) is also increased, explaining the higher melting point of randomized milk fat. The most predominant increase is observed for the C_{50} triglyceride. Furthermore, Huyghebaert *et al.* (1994) used the C_{38}/C_{50} ratio as a parameter of the interesterification reaction. During the reaction, the amount of mid-fraction and especially the C_{38} content is reduced and the C_{50} content increases.

One of the major shortcomings of chemical interesterification is the loss of the desired, buttery flavour due to several side reactions. The produced short-chain and medium-chain fatty acid methyl esters result in, respectively, a fruity and a bitter flavour. Postrefining is essential to remove these unwanted off-flavours (Huyghebaert *et al.*, 1994). Furthermore, the chemical catalyst is a hazardous product, requiring very careful handling. On the other hand, the technology and equipment for chemical interesterification is already well-established, as it has been applied for a long period of time.

Enzymatic interesterification

Enzymatic interesterification makes use of lipases (glycerol ester hydrolase, EC 3.1.1.3) to speed up the exchange of fatty acids on the glycerol molecule. Lipase is an enzyme that catalyzes the complete or partial hydrolysis of tri-, di- and monoglycerides, but under appropriate conditions the biocatalyst can promote ester formation as well. An interesting phenomenon of enzymes in general is their possible selectivity to accelerate only a well-specified reaction. Based on their specificity, three main classes of lipases used for interesterification can be distinguished: non-specific, 1,3-specific and fatty acid-specific. During an interesterification reaction with a non-specific lipase, products are formed with a random fatty acid distribution similar to products resulting from a chemically catalyzed interesterification. The 1,3-specific lipases act only at outer positions on the glycerol backbone and do not change the fatty acid profile at the *sn*-2 position. The fatty acid-specific lipases cause only the exchange of a specific fatty acid, regardless of its position (Macrae, 1983).

The lipases can be used in a free powdered form or immobilized on diverse substrates. Immobilization of lipases can positively affect their recoverability, reusability and stability, making the interesterification process economically more feasible. Interesterification can also be completed using a solvent to reduce the reaction viscosity.

Several studies have been performed to characterize the effect of enzymatic catalyzed interesterification on the physico-chemical properties of milk fat (Kalo *et al.*, 1986, 1990; Chmura *et al.*, 2008; Marangoni and Rousseau, 1998). They all observed a modified triglyceride profile which was very similar to that resulting from chemical interesterification. Interesterification causes a decrease in the content of middle-chain triglycerides (C_{32}–C_{42}) and an increase in the amount of higher molecular weight triglycerides ($>C_{42}$). Despite the use of 1,3-specific lipases, no positional specificity was observed in the different studies. This can probably be explained by the non-enzymatic acyl migration which is a migration of the fatty acids to the neighbour position on the glycerol backbone resulting in a redistribution of the fatty acids at the *sn*-2 position as well. This changed triglyceride profile has a considerable effect on the melting characteristics of the interesterified fat. Enzymatic interesterification also results in a decrease of the solid fat content at lower temperatures, and above 25°C it generally leads to an increase. This smoother, more linearized SFC-profile was also observed by Marangoni and Rousseau (1998), although they found that the SFC drops upon enzymatic interesterification at all temperatures. Lower SFC values at every temperature may result from an increase in water content in the reaction medium yielding higher amounts of mono- and diglycerides (Kalo *et al.*, 1986).

Compared with chemical interesterification, the enzymatic catalyzed process has a lower impact on the typical flavour components of milk fat. Another major benefit is its qualification as a 'green' chemistry solution due to lower amount of product losses and waste products. The milder reaction condition also leads to a more environmentally friendly process. On the other hand, the biocatalyst costs more and requires a longer reaction time and better reaction control, which may result in increased processing costs (De Greyt and Kellens, 2001).

Applications of interesterified milk fat

The modifications of the physical properties creates the possibility of using interesterified milk fat in a wider range of food products. For instance, Timms and Parekh (1980) reported that interesterified milk fat seems to be better suited to chocolate than unmodified milk fat. Moreover, Huyghebaert *et al.* (1994) stated that the texture of recombined butter made from interesterified milk fat has a better spreadability at 5°C than the original butter. However, the product is not fully spreadable and can be further improved by removing the high melting triglycerides by fractionation.

Neither chemical nor enzymatic interesterification of milk fat has yet been carried out on a commercial scale. The major drawback for its industrial exploitation is the high operating cost, which is mainly the result of considerable oil losses. The process is commercially viable only if this high operating cost is offset against the added value of the interesterified product. In the case of milk fat, this seems very doubtful, because interesterification has a negative impact on the desirable buttery flavour (De Greyt and Kellens, 2001).

4.4 Milk fat minor components

4.4.1 Downstream processing of polar lipids

The polar lipids in dairy products, which comprise PLs and sphingolipids, are predominantly situated in the MFGM membrane (60–70%) and the lipoprotein particles. The MFGM membrane surrounds the fat globules, stabilizes the fat droplets in the milk serum, and protects them from enzymatic degradation by lipases (Danthine *et al.*, 2000). During the processing of milk (e.g. churning), destabilization of the fat globules and disruption of the MFGM occurs. Then this biological membrane becomes no longer associated with the fat globules and the MFGM fragments preferentially migrate to the aqueous phases such as buttermilk and butter serum (McPherson and Kitchen, 1983). As the polar lipids of the MFGM are closely associated with a variety of specific membrane proteins, of which xanthin oxidase/dehydrogenase, butyrophilin and adipophilin are the most abundant (Mather, 2000), the MFGM proteins will probably co-migrate with the polar lipids during dairy processing. Corredig *et al.* (2003), Roesch *et al.* (2004) and Rombaut *et al.* (2007) have confirmed that dairy products with a high polar lipid content are also enriched in MFGM proteins. Generally, the most promising dairy products for further purification of MFGM components are products with a high polar lipid content on either a dry matter or a total lipid base (e.g. buttermilk, butterserum and whey), because water can easily be removed by filtration, evaporation or drying. Nonetheless, solvents for extraction and fractionation are often required to separate neutral and polar lipids. Various techniques have been used for purification of MFGM fragments, such as coagulation, filtration, centrifugation and conventional or supercritical extraction techniques or combinations thereof (Rombaut and Dewettinck, 2006).

Laboratory applications for MFGM isolation

Generally, laboratory MFGM isolation methods consist of four major steps (Singh, 2006), namely the separation of the fat globules by centrifugation, followed by several washing steps to remove contaminants (Mangino and Brunner, 1975; Kanno and Kim, 1990; Ye *et al.*, 2002; Fong *et al.*, 2007), the disruption of the fat globules to release the MFGM and finally, the recovery of the MFGM material. An overview of the various washing solutions that have been used for MFGM isolation is given in the review of Dewettinck *et al.* (2008). Churning, agitation at reduced temperatures, freeze–thaw cycles, extractions with polar aprotic solvents, bile salts, or nonionic detergents are often used to release the MFGM from the triglyceride fat core from the washed fat globules into the aqueous phase (Dowben *et al.*, 1967; Harrison *et al.*, 1975; Snow *et al.*, 1980; Patton, 1982; Dapper *et al.*, 1987). The final step involves the collection of the released MFGM material by ultracentrifugation, freeze-drying or microfiltration. The membrane material can also be obtained by precipitation at low pH or by 'salting out' with ammonium sulfate and recovering the precipitated material by centrifugation (Anderson and Brooker, 1974; Kanno and Kim, 1990; Rombaut *et al.*, 2006a; Morin *et al.*, 2007a). The above mentioned procedures are laboratory applications for the isolation of MFGM material from untreated milk, which are summarized in Fig. 4.5.

MFGM isolation from industrial sources

Rich sources of MFGM material such as buttermilk, butter serum and whey are still considered as low value by-products originating from dairy processing. Generally, these products have a low stability to oxidation due to their higher polyunsaturated fatty acid content, and a high variability between batches. Furthermore, the processing steps affect the interactions of the serum proteins with the MFGM. The lack of whey proteins recovered in the permeate illustrates the strong binding of whey proteins with MFGM components. The resulting aggregates are too large to pass through the membrane pores during microfiltration and are recovered in the retentate (Ward *et al.*, 2006). Corredig and Dalgleish (1997) and Morin *et al.* (2007b) showed that these interactions most likely to occur during heating the MFGM isolates, even at temperatures as low as 65°C. However, to date, only limited data are available on the influence of processing on the MFGM-interactions, so more research should be undertaken to better understand them and to control the manufacturing conditions to obtain products of consistent quality.

In spite of this, many authors have used buttermilk as a raw material for the isolation of MFGM fragments. Tangential micro- and ultrafiltration techniques seem to be appropriate for MFGM purification from dairy by-products, because of their ability to separate particles from dissolved solutes. Filtration conditions such as pH, temperature, type of membrane material, pore size and the type of buttermilk will influence the permeation of the different components of the buttermilk through the membrane (Morin *et al.*, 2004, 2006; Rombaut *et al.*, 2007). The isolation of polar lipids by membrane filtration is based on the selective removal of casein, whey proteins, lactose and minerals from the concentrate. Sometimes the filtration is combined with the stepwise addition of water, for further washing out of undesirable components such as lactose, whey-proteins and minerals (¼ diafiltration). However, microfiltration alone cannot achieve optimal separation of MFGM lipids and proteins from caseins and whey proteins.

Buttermilk and whey are rich sources of MFGM material, but they also may contain significant amounts of skim milk components, such as whey proteins and caseins. The casein micelles can cause difficulties during the purification of polar lipids by cross flow membrane filtration. They have sizes similar to the MFGM fragments and filtration solely cannot remove these micelles. Sachdeva and Buchheim (1997) achieved a tenfold increase in polar lipid concentration (70–77% recovery) by removing the caseins by acid and rennet coagulation prior to filtration. Another strategy is reduction

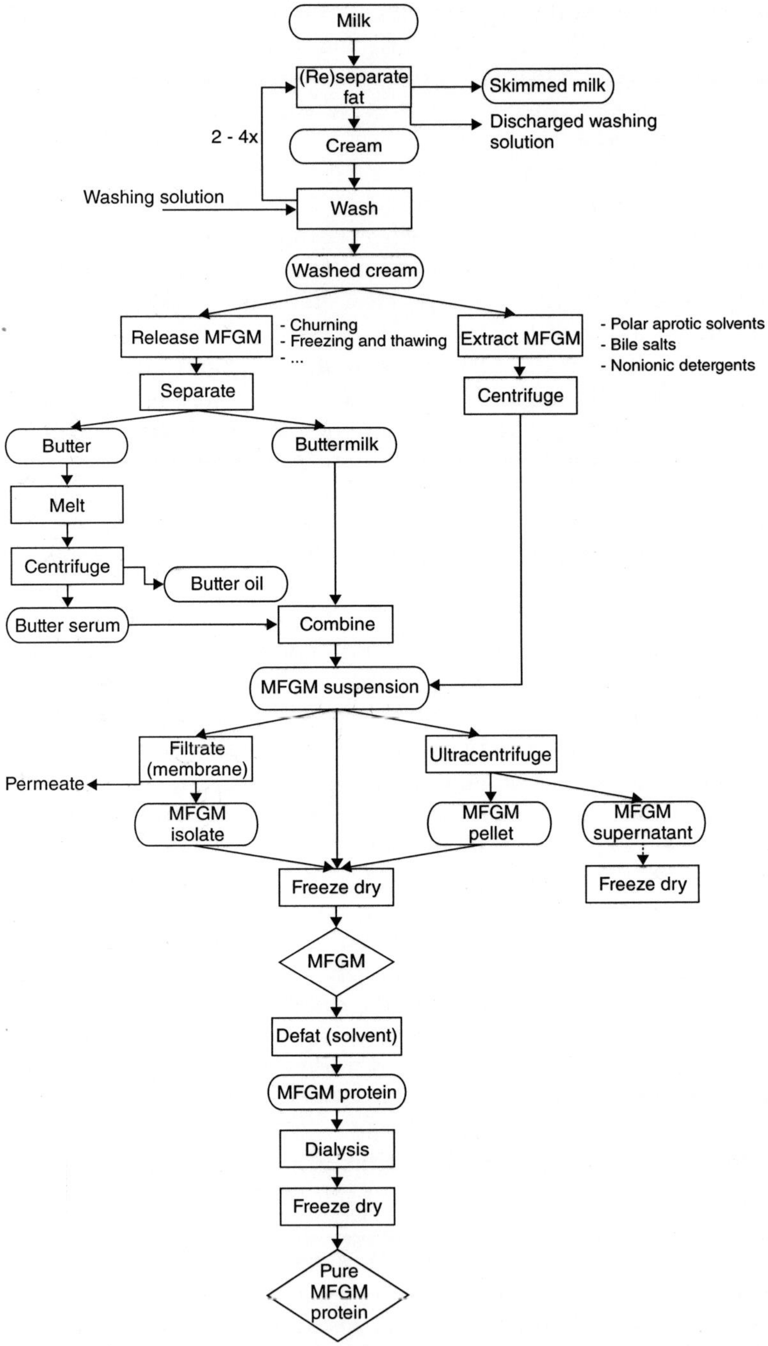

Fig. 4.5 Summary of isolation methods of MFGM (Dewettinck *et al.*, 2008).

of the casein micelle diameter, which allows it to be transferred through the membrane and recovered in the permeate. Corredig *et al.* (2003) and Roesch *et al.* (2004) used sodium citrate to dissociate the casein micelles into submicelles, followed by microfiltration to collect MFGM material in the retentate. In this way, the fat and casein content were increased in the retentate and permeate respectively. An enriched fraction from buttermilk containing about 80% of MFGM material with a significantly reduced amount of caseins was obtained (Corredig *et al.*, 2003). The application of several diafiltration steps with deionized water resulted in a reduction of the casein contamination in the retentate, but also in a loss of MFGM material (Rombaut *et al.*, 2006a). Roesch and Corredig (2002) described an enzymatic treatment with trypsin and chymotrypsin to destroy the casein micelles. This procedure reduced the retention of caseins during microfiltration, leading to a phospholipid-enriched retentate.

Rombaut *et al.* (2006a) added citrate to butter serum. These authors reported that 44% of the polar lipids were lost during filtration due to blocking and fouling of the filter membrane with MFGM particles. Addition of sodium citrate agent caused dispersion of not only casein micelles but also of MFGM fragments. Morin *et al.* (2006) compared the fractionation of regular buttermilk and whey buttermilk, the aqueous fraction obtained by churning of whey cream which is separated from cheese whey, by ceramic microfiltration. The latter contains similar amounts of PLs to regular buttermilk and the absence of caseins helps to minimize separation problems. However, the variability of the lipid content of whey cream results in considerable variations in the lipid content of whey buttermilk. Careful standardization of the whey cream prior to churning should be considered.

Rombaut *et al.* (2007) used acid buttermilk cheese whey, the aqueous fraction obtained by acidification of sweet-cream buttermilk, as a starting material for concentration of MFGM components by microfiltration. Acid buttermilk cheese whey is relatively rich in MFGM fragments and, as in whey buttermilk, the casein micelles are absent. The authors reported that the membrane material and pore size have a great influence on filtration characteristics and MFGM retention. The best results were obtained with negatively charged membranes with a pore size less than 0.15 μm to avoid excessive MFGM losses. MFGM retention was also affected by the pH and at optimized conditions, 98% of the polar lipids from the acid buttermilk whey could be recovered in the retentate. A thermocalcic aggregation (* see page 90) whey pre-treatment for whey clarification resulted in low permeate fluxes and high retention of ash and whey proteins, which makes it less suitable as a MFGM purification technique.

Another approach is the use of cream washed with skim milk ultrafiltrate to remove the caseins prior to churning, which results in a protein reduction of 74% compared with normal cream buttermilk (Morin *et al.*, 2007a). This procedure induced losses of MFGM material and a yield reduction of the process, but it significantly improved the permeation flux, the concentration of MFGM material and the removal of skim milk proteins.

*Thermocalcic aggregation is a sequestration technique of residual lipids, lipoproteins, and MFGM fragments; it was originally developed for the clarification of sweet cheese whey. After the coagulation of the polar lipid components by calcium addition and moderate heating, these aggregates can be separated by settling, filtration, or centrifugation. Thermocalcic aggregation is often applied in the manufacture of whey protein concentrates as a pre-treatment prior to UF, because the residual fat, consisting mainly of small fat globules, liposomes and MFGM particles, would lead to rapid fouling and flux decline of the membranes. Around 80% of the β-lactoglobulin and 85% of the α-lactalbumin remain soluble in the clarified whey, and the resulting sediment contains merely all polar lipids (Fauquant *et al.*, 1985; Pierre *et al.*, 1992). Theodet and Gandemer (1994) concentrated the polar lipids of Emmenthal whey by thermocalcic aggregation followed by microfiltration, and reported an increase in polar lipid concentration from 0.33 g to 1.10 g/100 g dry matter. Rombaut *et al.* (2006b) observed a similar rate of increase of 1.84 g to 7.95 g/100 g dry matter using buttermilk whey instead of Emmenthal whey. The combination of membrane techniques, thermocalcic aggregation, enzymatic hydrolysis and demineralization, results in much higher degrees of purification, as observed by Baumy *et al.* (1990). These authors obtained a concentrate containing 16.79 g/100 g dry matter starting from whey, and combined with solvent extraction and precipitation even higher concentrations were reported. The hexane–isopropanol extraction resulted in a polar lipid concentration of 36.8 g/100 g dry matter, whereas subsequent acetone precipitation yielded 76.7 g/100 g dry matter.

In general, the coagulation and filtration techniques do not seriously alter the polar lipid/total lipid ratio. These processes do not exhibit selective effects on lipids, so solvent extraction and fractionation techniques are required to increase this ratio. Supercritical CO_2 (SCO_2) is excellently suited as an extraction solvent due to its low critical parameters (31.1°C, 73.8 bar), relatively low cost, non-toxicity, chemical inertness and non-flammability. Furthermore, the resulting products are free from organic solvent residues. It is therefore an environmentally benign alternative to conventional lipid fractionation techniques, based on toxic solvents. Supercritical fluid extraction can be used to extract lipid and lipid-soluble materials from complex matrices. Astaire *et al.* (2003) developed a two-step procedure to increase concentrations of the polar MFGM lipids by microfiltration and supercritical fluid extraction (SFE) from buttermilk. They removed the triglycerides present in microfiltration-processed buttermilk powder using SCO_2 extraction, thereby increasing the polar lipid content from 31.02 g to 83.15 g/100 g total lipids. The resulting MFGM isolate contained a significantly reduced concentration of non-polar lipids and an increased concentration of polar lipids derived from the MFGM. Most likely, the inclusion of a second SCO_2 extraction step, with addition of a modifier such as ethanol, would exclusively remove the polar lipids, thereby resulting in a 100% pure fraction. This was carried out on ground soybean by Montanari *et al.* (1999). A first SCO_2 extraction step of soybean flakes was performed to extract the oil, leaving the PLs in the defatted soybean flakes, after which the PLs were recovered by a second SCO_2/ethanol extraction step resulting in a PL rich fraction. The PLs will not effectively dissolve in CO_2, and thus a suitable co-solvent such as ethanol is added to enhance their solubility. These results are remarkable, as the traditional extraction with hexane leaves about 50% of the total PLs in the spent seed matrix, while the other 50% are extracted into the hexane.

4.4.2 Technological functionalities

PLs make up approximately 1% of milk lipids, and they play an important role in the milk structural system. PLs are amphiphilic, and in milk they are present mostly in the MFGM, where they stabilize the fat globule against coalescence. Due to their dipolar nature, PLs in general, and milk PLs in particular, are considered to have good emulsifying properties.

The forms of milk PLs used in the food industry have evolved from by-products of the dairy processing industry which have a high content of MFGM to the use of isolated and fractionated ingredients which are concentrated in the PL content. Compared to skim milk powder, buttermilk powder has better emulsifying properties since it contains a higher concentration of PL. Because of that, the former has been being replaced by the latter in the bakery industry. Buttermilk is also used to replace partially skim milk in dairy products such as yoghurt and cheese due to its lower cost and nutritional properties. Adding a certain amount of liquid buttermilk, which is concentrated by several means, reduces the free oil under baking and increases the yield of pizza cheese by increase moisture content without adversely affect functional properties of the cheese (Govindasamy-Lucey *et al.*, 2007; Mistry *et al.*, 1996). Similar findings were observed with production of Mozzarella cheese (Poduval and Mistry, 1999). Whey buttermilk, resulting from churning of whey cream in the manufacture of whey butter, gave higher emulsifying properties and lower foaming ability compared with sweet or cultured buttermilk, probably due to a higher ratio of PL to protein (Sodini *et al.*, 2006).

MFGM isolated from untreated milk, which contains 25–30% PL depending on the isolation methods, has a good emulsifying capacity. One percentage of MFGM material isolated from untreated milk is enough to stabilize an emulsion made from 25% milk fat and the emulsion stability increases with increasing concentration of the MFGM (Kanno, 1989). The fat globule size ranges from 0.9–17 μm in diameter (Kanno *et al.*, 1991). The effects of emulsifying time, homogenization, temperature, and pH on emulsifying properties, size and the surface of globules were also investigated in this study. The emulsion had the highest viscosity and was the least stable at pH around 5, which is near the isoelectric point of MFGM, 4.9 (Kanno, 1989; Kanno *et al.*, 1991). However, the emulsifying properties of MFGM material isolated from industrial buttermilk are not always superior compared to that of the buttermilk (Corredig and Dalgleish, 1997; Wong and Kitts, 2003b). This varying observation on emulsifying properties has been attributed to the difference in protein composition of the isolated MFGM material. Attachment of β-lactoglobulin, and to a lesser extent α-lactalbumin and κ-casein, to the MFGM during heating of milk is well-known (Ye *et al.*, 2004). As a result, MFGM material isolated from industrial sources is different in composition compared to that of the material isolated from untreated milk. Along with proteins, PLs contribute a major part of the milk fat globule membrane. However, there is limited information available on the effect of processing treatments on this moiety. Whether or not the changes in PL composition or chemical structure, or alteration in interaction of PLs together or with other components of the membrane or of the milk serum phase, would change the technological properties of the MFGM isolate is still to be investigated. Corredig and Dalgleish (1997) concluded that heating the MFGM isolate strongly affected the technological properties of these membrane fractions, caused by association of the whey proteins with the MFGM. Using capillary electrophoresis, bovine serum albumin and other proteins were shown to interact with PS and PC when they were incubated together (Bo and Pawliszyn, 2006; Hu *et al.*, 2001).

MFGM material isolated from raw milk is currently used mainly for research purposes while potential applications today relate to the enriched PL fractions isolated from the industrial sources. As summarized in the previous section, many researchers have been successful in isolating MFGM fragments, which contain high concentrations of PLs and specific membrane proteins, from by-products of dairy industrial processing. This opens opportunities to make use of such cheap sources to manufacture added-value ingredients with special functionalities. However, the treatment during milk processing affects the composition of the materials obtained as well as their technological functionalities. Further studies are still needed on this aspect and certain adaptation of dairy processing procedures would probably be needed in order to protect technological and nutritional properties of the MFGM which is concentrated in the by-products after processing of milk or cream (Ward *et al.*, 2006).

Soybean and egg yolk lecithin have a long tradition of being employed in food products, as well as others such as cosmetic products, animal feeds, medicinal and agrochemical products, paints and lubricants. However, the use of PL from dairy sources is limited to pharmaceutical and cosmetic applications due to technological difficulties in extracting and hence high prices of the prepared isolates (Miura *et al.*, 2006). Soy lecithin does not contain SM and contains very limited concentrations of PS, while PL fractionated from dairy sources contains about 24–32% SM and 3.2–12% PS

(Burling and Graverholt, 2008; Miura *et al.*, 2006; Thompson and Singh, 2006). The cost issue which limits applications of bovine milk lecithin would become less important as its beneficial properties in both technological as well as nutritional aspects are fully accounted.

Several dairy functional ingredients enriched in PLs have been launched on the market. Lacprodan-20 and -75 from Arla Foods Ingredients amba (Denmark) contain 20 and 75% PLs, respectively, and Phospholac 600, produced by Fonterra Cooperative Group Ltd. (New Zealand) contains more than 70% of PLs. One interesting result from the use of such enriched PL isolate has been reported from the research group of Snow Brand Milk Products (Japan), where bovine milk lecithin (85% PL) was found to stabilize while soy lecithin (95% PL) solidified the cream reconstituted from butter oil (Miura *et al.*, 2006). Among the PL species, PC but not PE or SM, regardless of their origin, is the component to exert the emulsifying effect (Miura *et al.*, 2004, 2006).

Several forms of liposomes, unilamellar, multilamellar, and multivesicular, can be created from Phospholac 600 using microfluidization, and these PL dispersions are quite stable towards oxidation (Thompson and Singh, 2006). The capacity to produce liposomal formulations gives rise to opportunities to use MFGM PLs to improve moisturizing effects, and to encapsulate and protect sensitive cosmetic ingredients or bioactive ingredients in functional foods. The capacity of SM to improve the epidermal function is one of great attraction for employing milk PLs in cosmetic and pharmaceutical applications (Haruta *et al.*, 2008).

Compared to triglycerides, PLs of milk contain a high concentration of unsaturated fatty acids, so they are very susceptible to oxidation. The rate of milk fat oxidation is well-known to be accelerated in the presence of oxidizing factors such as oxygen, light, heat and catalysts such as metal ions. Oxidation is, in almost all cases, an important issue to deal with when phospholipid-rich ingredients are used. However, PLs can be pro-oxidative or anti-oxidative to milk fat, depending on processing conditions and intrinsic factors of the matrices where PLs are incorporated (Chen and Nawar, 1991a,b; King, 1963; Lee and Choe, 2008). The antioxidation ability of PLs depends both on their head groups and their fatty acid composition (Nwosu *et al.*, 1997). An amount of 0.1–0.2% buttermilk powder inhibits lipid oxidation in a simple peroxidizing model system and the reducing activity of the buttermilk is mainly attributed to sulfhydryl groups (of proteins). These groups may scavenge peroxide and hydroxyl radicals, and sequester both Fe^{2+} and Fe^{3+} (Wong and Kitts, 2003a).

The nutritional properties of PL will be highlighted in the following section. For applications of milk PL, or dairy ingredients enriched in PL, in functional foods or any other products, one could utilize their technological functionalities, beneficial nutritional properties or their protective effects (e.g. in the form of liposome) or any combination of these three advantages.

4.5 Nutritional properties of dairy fats

A large amount of research data with focus on the nutritional properties of dairy lipids is now available. Nevertheless, additional data are still needed to gain better insights in the role of dairy lipids in nutrition and health, and to make appropriate recommendations for consumers in terms of dairy fat intake, and to producers for modification of dairy lipid composition and production of modified food products.

Currently, the nutrition debate on dairy lipids has a high focus on the relationship between dairy fat intake and cardiovascular disease (Steijns, 2008; Tholstrup, 2006). The ratio total cholesterol:HDL cholesterol in serum is now considered as a more sensitive and specific risk factor for coronary heart disease (CHD) compared with total cholesterol (Mensink *et al.*, 2003). CHD risk is lowered when saturated fatty acids (SFAs) and *trans*-monounsaturated fatty acids are replaced by *cis*-unsaturated fatty acids. In the case of individual SFAs, myristic acid (C14 : 0) and palmitic acid (C16 : 0) show limited change in total cholesterol:HDL cholesterol when replacing 1 E% from carbohydrate, and are considered to have a negative effect on CHD risk (Mensink *et al.*, 2003; Steijns, 2008). Although myristic and palmitic acids as well as *trans*-monounsaturated fatty acids are important constituents of dairy fats, their effects on CHD risk when present in ingested dairy products are still far from understood. Milk fat consumption shows less pronounced effects on serum lipids than expected from its fat content, and epidemiological data frequently lack an association between milk consumption and CHD. Emerging issues that receive or need further attention in CHD research are: (i) the role of short and medium chain fatty acids and sphingolipids in milk and milk products; (ii) the influence of specific non-fat components in milk and milk products, e.g. calcium, proteins, peptides and vitamin K; (iii) the type of dairy product used to carry dairy fat, e.g. cheese consumption may have more favourable effects on serum lipoprotein levels compared to butter at equal saturated fat intake; (iv) the influence of stereospecific distribution of fatty acids in triglycerides; and (v) the influence of dairy constituents on additional biomarkers for CHD risk, e.g. small and dense LDL particles (Pfeuffer and Schrezenmeir, 2006; Haug *et al.*, 2007; Tholstrup, 2006; Steijns, 2008). A more complete scientific update on the relationship between dairy fats and cardiovascular disease has recently been published (Lock *et al.*, 2008).

Next to effects on cardiovascular disease, effects on other disorders are also being evaluated, such as insulin resistance and type 2 diabetes, obesity, and cancer (for reviews, see Mensink, 2006; Collomb *et al.*, 2006; Pfeuffer and Schrezenmeir, 2006). The milk fat globule membrane in milk containing a high level of polar lipids has also received significant attention in view of its potential bioactive effects (for a review, see Dewettinck *et al.*, 2008). For instance, it is believed that PC supports liver recovery from chemical or viral damages, and has a protective effect on the gastrointestinal mucosa against toxic attacks. Furthermore, some PLs are digested in the gastrointestinal tract to metabolic breakdown products that might possess antibacterial and antiviral activity.

4.6 Acknowledgement

The financial support of grant G.0490.08 from the Fund for Scientific Research – Flanders (Fonds voor Wetenschappelijk Onderzoek (FWO) Vlaanderen) is gratefully acknowledged.

4.7 References

ANDERSON M and BROOKER B E (1974), 'Loss of material during the isolation of milk fat globule membrane', *J Dairy Sci*, 58, 1442–1448.

ARUL J, BOUDREAU A, MAKHLOUF J, TARDIF R and BELLAVIA T J (1988), 'Fractionation of anhydrous milk fat by short-path distillation', *J Am Oil Chem Soc*, 65, 1642–1646. DOI:10.1007/BF02912569.

ARUL J, BOUDREAU A, MAKLOUF J, TADRIF R and SAHASRABUDHE M R (1987), 'Fractionation of milk fat by supercritical carbon dioxide', *J Food Sci*, 52, 1231–1236. DOI:10.1111/j.1365-2621.1987.tb14050.x.

ASTAIRE J C, WARD R, GERMAN J B and JIMÉNEZ-FLORES R (2003), 'Concentration of polar MFGM lipids from buttermilk by microfiltration and supercritical fluid extraction', *J Dairy Sci*, 86, 2297–2307.

AUGUSTIN M A and VERSTEEG C (2006), 'Milk fat: Physical, chemical and enzymatic modification', in Fox P F and McSweeney P L H, *Advanced Dairy Chemistry Volume 2, Lipids 3rd Edition*, New York, Springer, 293–332. DOI:10.1007/0-387-28813-9_8.

BAUMY J J, GESTIN L, FAUQUANT J, BOYAVAL E and MAUBOIS J L (1990), 'Technologies de purification des phospholipides du lactosérum', *Process*, 1047, 29–33.

BHALERAO V R, JOHNSON O C and KUMMEROW F A (1959), 'Effect of thermal oxidative polymerization on the growth-promoting value of some fractions of butterfat', *J Dairy Sci*, 42, 1057–1062.

BHASKAR A R, RIZVI S S H and SHERBON J W (1993), 'Anhydrous milk fat fractionation with continuous countercurrent supercritical carbon dioxide', *J Food Sci*, 58, 748–752. DOI:10.1111/j.1365-2621.1993.tb09350.x.

BHASKAR A R, RIZVI S S H, BERTOLI C, FAY L B and HUG B (1998), 'A comparison of physical and chemical properties of milk fat fractions obtained by two processing technologies', *J Am Oil Chem Soc*, 75, 1249–1264. DOI:10.1007/s11746-998-0170-0.

BO T and PAWLISZYN J (2006), 'Characterization of phospholipid–protein interactions by capillary isoelectric focusing with whole-column imaging detection', *Anal Biochem*, 350, 91–98. DOI:10.1016/j.ab.2005.11.039.

BOUDREAU A and ARUL J (1991), 'Physical and chemical modification of milk fat. 2.1 Fractionation', in *Monograph on Utilizations of Milk Fat*, Bulletin, No. 260, International Dairy Federation, Brussels, 7–15.

BOUDREAU A and ARUL J (1993), 'Cholesterol reduction and fat fractionation technologies for milk fat: An overview', *J Dairy Sci*, 76, 1772–1781.

BURGESS K J (2001), 'Milk fats as ingredients', *Int J Dairy Technol*, 54, 56–60. DOI:10.1046/j.1471-0307.2001.00015.x.

BURLING H and GRAVERHOLT G (2008), Milk – A new source for bioactive phospholipids for use in food formulations, *Lipid Technol,* 20, 229–231. DOI:10.1002/lite.200800058.

CALLIAUW G (2008), *Molecular interactions affecting the phase composition during dry fractionation of palm olein*, PhD thesis, Ghent University, Belgium, 275p.

CAMPOS R J, LITWINENKO J W and MARANGONI A G (2003), 'Fractionation of milk fat by short-path distillation', *J Dairy Sci*, 86, 735–745.

CHEN Z and NAWAR W W (1991a), 'Prooxidative and antioxidative effects of phospholipids on milk fat', *J Am Oil Chem Soc*, 68, 938–940. DOI:10.1007/BF02657539.

CHEN Z Y and NAWAR W W (1991b), 'Role of Milk Fat Globule Membrane in Autoxidation of Milk Fat', *J Food Sci*, 56, 398–401. DOI:10.1111/j.1365-2621.1991.tb05289.x.

CHMURA M, STANIEWSKI B, PANFIL-KUNCEWICZ H, SZPENDOWSKI J and ZAWADZKA J (2008), 'The effects of enzymatic interesterification on selected physicochemical characteristics of milk fat', *Milchwissenschaft*, 63, 37–40.

COLLOMB M, SCHMID A, SIEBER R, WECHSLER D and RYHANEN E L (2006), 'Conjugated linoleic acids in milk fat: Variation and physiological effects', *Int Dairy J*, 16, 1347–1361. DOI:10.1016/j.idairyj.2006.06.021.

CORREDIG M and DALGLEISH D G (1997), 'Isolates from industrial buttermilk: Emulsifying properties of materials derived from the milk fat globule membrane', *J Agric Food Chem*, 45, 4595–4600. DOI:10.1021/jf970531f.

CORREDIG M, ROESCH R R and DALGLEISH D G (2003), 'Production of a novel ingredient from buttermilk', *J Dairy Sci*, 86, 2744–2750.

DANTHINE S, BLECKER C, PAQUOT M, INNOCENTE N and DEROANNE C (2000), 'Progress in milk fat globule membrane research: A review', *Lait*, 80, 209–222. DOI:10.1051/lait:2000120.

DAPPER C H, VALIVULLAH H M and KEENAN T W (1987), 'Use of polar aprotic solvents to release membranes from milk lipid globules', *J Dairy Sci*, 70, 760–765.

DE GREYT W and KELLENS M (2001), 'Improvement of the nutritional and physicochemical properties of milk fat', in Gunstone F D, *Structure and Modified Lipids*, New York, Marcel Dekker, 285–312.

DEFFENSE E (1987), 'Multi-step butteroil fractionation and spreadable butter', *Fett Wiss Technol*, 89, 502–507.

DEFFENSE E (1993), 'Milk fat fractionation today: A review', *J Am Oil Chem Soc*, 70, 1193–1201. DOI:10.1007/BF02564225.

DEFFENSE E (2000), 'Dry fractionation technology in 2000', *Eur J Lipid Sci Technol*, 102, 234–236. DOI:10.1002/(SICI)1438-9312(200003)102:3<234::AID-EJLT234>3.0.CO;2-N.

DEWETTINCK K, ROMBAUT R, THIENPONT N, LE T T, MESSENS K and VAN CAMP J (2008), 'Nutritional and technological aspects of milk fat globule membrane material', *Int Dairy J*, 18, 436–457. DOI:10.1016/j.idairyj.2007.10.014.

DIJKSTRA A J (2007), 'Modification processes and food uses', in Gunstone F D, Harwood J L and Dijkstra A J, *The Lipid Handbook – 3rd edition*, London, CRC Press, 263–354. DOI:10.1007/BF02640487.

DOWBEN R M, BRUNNER J R and PHILPOTT D E (1967), 'Studies on milk fat globule membranes', *Biochim Biophys Acta*, 135, 1–10. DOI:10.1016/0005-2736(67)90002-8.

FATOUH A E, SINGH R K, KOEHLER P E, MAHRAN G A, EL-GHANDOUR M A and METWALLY A E (2003), 'Chemical and thermal characteristics of buffalo butter oil fractions obtained by multi-step dry fractionation', *Lebensm Wiss Technol*, 36, 483–496. DOI:10.1016/S0023-6438(03)00044-6.

FAUQUANT J, VIECO E, BRULE G and MAUBOIS J L (1985), 'Clarification des lactosérums doux par agrégation thermocalcique de la matière grasse résiduelle', *Lait*, 65, 1–20. DOI:10.1051/lait:1985647-6481.

FJAERVOLL A (1969), '(Butter oil and butterfat fractionation)', *Svenska Mejeritidende*, 61, 491–496.

FONG B Y, NORRIS C S and MACGIBBON A K H (2007), 'Protein and lipid composition of bovine milk-fat-globule membrane', *Int Dairy J*, 17, 275–288. DOI:10.1016/j.idairyj.2006.05.004.

FRITSCHE J and STEINHART H (1998), 'Analysis, occurrence and physiological properties of trans fatty acids (TFA) with particular emphasis on conjugated linoleic acid isomers (CLA) – a review', *Fett/Lipid*, 100, 190–210.

GIBON V and TIRTIAUX A (2002), 'Latest trends in dry fractionation', *Lipid Technol*, 14, 33–36.

GOVINDASAMY-LUCEY S, LIN T, JAEGGI J J, MARTINELLI C J, JOHNSON M E and LUCEY J A (2007), 'Effect of type of concentrated sweet cream buttermilk on the manufacture, yield, and functionality of pizza cheese', *J Dairy Sci*, 90, 2675–2688. DOI:10.3168/jds.2006-681.

GRALL D S and HARTEL R W (1992), 'Kinetics of butterfat crystallization', *J Am Oil Chem Soc*, 69, 741–747. DOI:10.1007/BF02635909.

GRUMMER R R (1991), 'Effect of feed on the composition of milk fat', *J Dairy Sci*, 72, 1996–2001.

HA J K and LINDSAY R C (1991), 'Volatile branched fatty acids and phenolic compounds in aged Italian cheese flavors', *J Food Sci* 56, 1241–1247. DOI:10.1111/j.1365-2621.1991.tb04743.x.

HARRISON R, HIGGINBOTHAM J D and NEWMAN R (1975), 'Sialoglycopeptides from bovine milk fat globule membrane', *Biochim Biophys Acta*, 389, 449–463. DOI:10.1016/0005-2736(75)90156-X.

HARTEL R W (2001), *Crystallization in Foods*, Gaithersburg, MD, Aspen Publications.

HARTEL R W and KAYLEGIAN K E (2001), 'Advances in milk fat fractionation – technology and applications', in Garti N and Sato K, *Crystallization processes in fats and lipid systems*, New York, Marcel Dekker, 381–427.

HARUTA Y, KATO K and YOSHIOKA T (2008), 'Dietary phospholipid concentrate from bovine milk improves epidermal function in hairless mice', *Biosci Biotechnol Biochem*, 72, 2151–2157. DOI:10.1271/bbb.80212.

HAUG A, HOSTMARK A T and HARSTAD O M (2007), 'Bovine milk in human nutrition – a review', *Lipids Health Dis*, 6, 25–31. DOI:10.1186/1476-511X-6-25.

HERRERA M L and HARTEL R W (2000), 'Effect of processing conditions on physical properties of a milk fat model system: Microstructure', *J Am Oil Chem Soc*, 77, 1197–1204. DOI:10.1007/s11746-000-0186-2.

HU S, ZHANG L and DOVICHI N J (2001), 'Characterization of the interaction between phospholipid and protein by capillary electrophoresis with laser-induced fluorescence detection', *J Chromatogr A*, 924, 369–375. DOI:10.1016/S0021-9673(01)00988-8.

HUYGHEBAERT A, VERHAEGHE D and DE MOOR H (1994), 'Fat products using chemical and enzymatic interesterification', in Moran D P J and Rajah K K, *Fats in Food Products*, Glasgow, Blackie Academic and Professional, 319–342.

ILLINGWORTH D (2002), 'Fractionation of fats', in Marangoni A G and Narine S S, *Physical Properties of Lipids*, New York, Marcel Dekker, 411–447.

JEBSON R S (1970), 'Fractionation of milkfat into high and low melting point components', *Proceedings of XVIII International Dairy Congress Vol. 1E*, Sydney, 240.

JENSEN R G (1973), 'Composition of bovine milk lipids', *J Am Oil Chem Soc*, 50, 186–192. DOI:10.1007/BF02640488.

JENSEN R G (1992), 'Fatty acids in milk and dairy products', in Chow C K, *Fatty Acids in Foods and Their Health Implications*, New York, Marcel Dekker, 109–124.

JENSEN R G, FERRIS A M and LAMMI-KEEFE C J (1991), 'The composition of milk fat', *J Dairy Sci*, 74, 3228–3243.

KALO P, HUOTARI H and ANTILA M (1990), '*Pseudomanas fluorescens* lipase-catalyzed interesterification of butter fat in absence of a solvent', *Milchwissenschaft*, 45, 281–285.

KALO P, PARVIAINEN P, VAARA K, ALI-YRRKÖ S and ANTILA M (1986), 'Changes in the triglyceride composition of butter fat induced by lipase and sodium methoxide catalyzed interesterification reactions', *Milchwissenschaft*, 41, 82–85.

KANNO C (1989). 'Emulsifying properties of bovine milk fat globule membrane in milk fat emulsion: Conditions for the reconstitution of milk fat globules', *J Food Sci*, 54, 1534–1539. DOI:10.1111/j.1365-2621.1989.tb05153.x.

KANNO C and KIM D-H (1990), 'A simple procedure for the preparation of bovine milk fat globule membrane and a comparison of its composition, enzymatic activities, and electrophoretic properties with those prepared by other methods', *Agr Biol Chem*, 54, 2845–2854.

KANNO C, SHIMOMURA Y and TAKANO E (1991), 'Physicochemical properties of milk fat emulsions stabilized with bovine milk fat globule membrane', *J Food Sci*, 56, 1219–1223. DOI:10.1111/j.1365-2621.1991.tb04738.x.

KAUFMANN V G, BIERNOTH G, FREDE E, MERK W, PRECHT D and TIMMEN H (1982), 'Fraktionierung von Butterfett durch extraktion mit uberkritischem CO_2', *Milchwissenschaft*, 37, 92–96.

KAYLEGIAN K E (1999), 'The production of specialty milk fat ingredients', *J Dairy Sci*, 82, 1433–1439.

KAYLEGIAN K E and LINDSAY R C (1995), *Handbook of Milkfat Fractionation Technology and Application*, Champaign, AOCS Press.

KELLENS M (1994), 'Developments in fractionation technology', *Lecture presented at the Symposium on Fractional Crystallisation of Fats, organised by the Society of Chemical Engineers*, London (9th March).

KING R L (1963), 'Oxidation of milk fat globule membrane material. II. Relation of ascorbic acid and membrane concentrations', *J Dairy Sci*, 46, 267–274.

KURTZ F E (1978), 'The lipids of milk: Composition and properties', in Webb B H, Johnson A H and Alford J A, *Fundamentals of Dairy Chemistry*, (2nd ed.) Westport, The Avi Publishing Company, 125–219.

LARSEN N E and SAMUELSSON E G (1979), 'Some technological aspects on fractionation of anhydrous butter-fat', *Milchwissenschaft*, 34, 663–665.

LEE Y and CHOE E (2008), 'Singlet oxygen quenching effects of phosphatidylcholine in emulsion containing sunflower oil', *J Food Sci*, 73, C506–C511. DOI:10.1111/j.1750-3841.2008.00852.x.

LOCK A, DESTAILLATS F, KRAFT J and GERMAN B (2008), 'Introduction to the proceedings of the symposium "Scientific update on Dairy Fats and Cardiovascular Diseases"', *J Am Coll Nutr*, 27, 720S–722S.

LOPEZ C, BOURGAUX C, LESIEUR P, RIAUBLANC A and OLLIVON M (2006), 'Milk fat and primary fractions obtained by dry fractionation. 1. Chemical composition and crystallisation properties', *Chem Phys Lipids*, 144, 17–33. DOI:10.1016/j.chemphyslip.2006.06.002.

MACRAE A R (1983), 'Lipase-catalysed interesterification of oils and fats', *J Am Oil Chem Soc*, 60, 243A–246A. DOI:10.1007/BF02543502.

MANGINO M E and BRUNNER J R (1975), 'Molecular weight profile of fat globule membrane proteins', *J Dairy Sci*, 58, 313–318.

MANIONGUI C, GRESTI J, BUGAUT M, GAUTHIER S and BÉZARD J (1991), 'Determination of bovine butterfat triacylglycerols by reversed-phase liquid chromatography and gas chromatography', *J Chromatogr*, 543, 81–103. DOI:10.1016/S0021-9673(01)95757-7.

MARANGONI A G and LENCKI R W (1998), 'Ternary phase behavior of milk fat fractions', *J Agric Food Chem*, 46, 3879–3884. DOI:10.1021/jf9801668.

MARANGONI A G and ROUSSEAU D (1998), 'Chemical and enzymatic modification of butterfat and butterfat–canola oil blends', *Food Res Int*, 31, 595–599. DOI:10.1016/S0963-9969(99)00033-2.

MATHER I H (2000), 'A review and proposed nomenclature for major proteins of the milk-fat globule membrane', *J Dairy Sci*, 83, 203–247.

MCPHERSON A V and KITCHEN B J (1983), 'Review of the progress of dairy science – The bovine milk fat globule membrane – Its formation, composition, structure and behavior in milk and dairy products', *J Dairy Res*, 50, 107–133.

MENSINK R P, ZOCK P L, KESTER A D and KATAN M B (2003), 'Effects of dietary fatty acids and carbohydrates on the ratio of serum Total to HDL cholesterol and on serum lipids and apolipoproteins: A meta-analysis of 60 controlled trials', *Am J Clin Nutr*, 77, 1146–1155.

MENSINK R P (2006), 'Dairy products and the risk to develop type 2 diabetes or cardiovascular disease', *Int Dairy J*, 16, 1001–1004. DOI:10.1016/j.idairyj.2005.10.013.

MISTRY V V, METZGER L E and MAUBOIS J L (1996), 'Use of ultrafiltered sweet buttermilk in the manufacture of reduced fat Cheddar cheese', *J Dairy Sci*, 79, 1137–1145.

MIURA S, MUTOH T, SHIINOKI Y and YOSHIOKA T (2006), 'Emulsifying properties of phospholipids in the reconstitution of cream using butter oil', *Eur J Lipid Sci Technol*, 108, 898–903. DOI:10.1002/ejlt.200600109.

MIURA S, TANAKA M, SUZUKI A and SATO K (2004), 'Application of phospholipids extracted from bovine milk to the reconstitution of cream using butter oil', *J Am Oil Chem Soc*, 81, 97–100. DOI:10.1007/s11746-004-0863-1.

MONTANARI L, FANTOZZI P, SNYDER J M and KING J W (1999), 'Selective extraction of phospholipids from soybeans with supercritical carbon dioxide and ethanol', *J Supercrit Fluids*, 14, 87–93. DOI:10.1016/S0896-8446(98)00110-7.

MORIN P, BRITTEN M, JIMÉNEZ-FLORES R and POULIOT Y (2007a), 'Microfiltration of buttermilk and washed cream buttermilk for concentration of milk fat globule membrane components', *J Dairy Sci*, 90, 2132–2140. DOI:10.3168/jds.2006-832.

MORIN P, JIMÉNEZ-FLORES R and POULIOT Y (2004), 'Effect of temperature and pore size on the fractionation of fresh and reconstituted buttermilk by microfiltration', *J Dairy Sci*, 87, 267–273.

MORIN P, JIMÉNEZ-FLORES R and POULIOT Y (2007b), 'Effect of processing on the composition and microstructure of buttermilk and its milk fat globule membranes', *Int Dairy J*, 17, 1179–1187. DOI:10.1016/j.idairyj.2007.03.010.

MORIN P, POULIOT Y and JIMÉNEZ-FLORES R (2006), 'A comparative study of the fractionation of regular buttermilk and whey buttermilk by microfiltration', *J Food Eng*, 77, 521–528. DOI:10.1016/j.jfoodeng.2005.06.065.

NORRIS R, GRAY I K, MCDOWELL A K R and DOLBY R M (1971), 'The chemical composition and physical properties of fractions of milk fat obtained by a commercial fractionation process', *J Dairy Res*, 38, 79–82.

NWOSU C, BOYD L and SHELDON B (1997), 'Effect of fatty acid composition of phospholipids on their antioxidant properties and activity index', *J Am Oil Chem Soc*, 74, 293–297. DOI:10.1007/s11746-997-0139-4.

PALMQUIST D L, BEAULIEU A D and BARBANO D M (1993), 'Feed and animal factors influencing milk fat composition', *J Dairy Sci*, 76, 1753–1771.

PARKS O W, SCHWARTZ D P, KATZ L and KEENEY M (1964), 'Isolation and characterization of methyl ketone precursor in butter fat', *J Lipid Res*, 5, 232–235.

PATIENCE D B, HARTEL R W and ILLINGWORTH D (1999), 'Crystallization and pressure filtration of anhydrous milk fat: Mixing effects', *J Am Oil Chem Soc*, 76, 585–594. DOI:10.1007/s11746-999-0008-4.

PATTON S (1973), 'Origin of the milk fat globule', *J Am Oil Chem Soc*, 50, 178–185.

PATTON S (1982), 'Release of remnant plasma membrane from milk fat globules by Triton X-100', *Biochim Biophys Acta*, 688, 727–734. DOI:10.1016/0005-2736(82)90285-1.

PFEUFFER M and SCHREZENMEIR J (2006), 'Milk and the metabolic syndrome', *Obes Rev*, 8, 109–118. DOI:10.1111/j.1467-789X.2006.00265.x.

PIERRE A, LEGRAET Y, FAUQUANT J, PIOT M, DURIER C and KOBILINSKY A (1992), 'Optimization of physicochemical factors for whey clarification', *Lait*, 72, 405–420. DOI:10.1051/lait:1992429.

PODUVAL V S and MISTRY V V (1999), 'Manufacture of reduced fat Mozzarella cheese using ultrafiltered sweet buttermilk and homogenized cream', *J Dairy Sci*, 82, 1–9.

RAJAH K K (1994), 'Milk fat developments', *J Soc Dairy Technol*, 47, 81–83.

RODRIGUES J N and GIOIELLI L A (2003), 'Chemical interesterification of milkfat and milkfat–corn oil blends', *Food Res Int*, 36, 149–159. DOI:10.1016/S0963-9969(02)00130-8.

ROESCH R R and CORREDIG M (2002), 'Production of buttermilk hydrolyzates and their characterization', *Milchwissenschaft*, 57, 376–380.

ROESCH R R, RINCON A and CORREDIG M (2004), 'Emulsifying properties of fractions prepared from commercial buttermilk by microfiltration', *J Dairy Sci*, 87, 4080–4087.

ROMBAUT R and DEWETTINCK K (2006a), 'Properties, analysis and purification of milk polar lipids', *Int Dairy J*, 16, 1362–1373. DOI:10.1016/j.idairyj.2006.06.011.

ROMBAUT R, DEJONCKHEERE V and DEWETTINCK K (2006a), 'Microfiltration of butter serum upon casein micelle destabilization', *J Dairy Sci*, 89, 1915–1925.

ROMBAUT R, DEJONCKHEERE V and DEWETTINCK K (2007), 'Filtration of milk fat globule membrane fragments from acid buttermilk cheese whey', *J Dairy Sci*, 90, 1662–1673. DOI:10.3168/jds.2006-587.

ROMBAUT R, VAN CAMP J and DEWETTINCK K (2006b), 'Phospho- and sphingolipid distribution during processing of milk, butter and whey', *Int J Food Sci Technol*, 41, 435–443. DOI:10.1111/j.1365-2621.2005.01091.x.

ROUSSEAU D, FORESTIERE K, HILL A R and MARANGONI A G (1996), 'Restructuring butterfat through blending and chemical interesterification. 1. Melting behavior and triacylglycerol modifications', *J Am Oil Chem Soc*, 73, 963–972. DOI:10.1007/BF02523403.

SACHDEVA S and BUCHHEIM W (1997), 'Recovery of phospholipids from buttermilk using membrane processing', *Kieler Milchw Forsch*, 49, 47–68.

SINGH H (2006), 'The milk fat globule membrane – A biophysical system for food applications', *Curr Opin Colloid Interface Sci*, 11, 154–163. DOI:10.1016/j.cocis.2005.11.002.

SNOW L D, DOSS R C and CARRAWAY K L (1980), 'Cooperativity of the concanavalin A inhibition of bovine milk fat globule membrane 5′-nucleotidase. Response to extraction of nucleotidase and of putative cytoplasmic surface coat components', *Biochim Biophys Acta*, 611, 333–341.

SODINI I, MORIN P, OLABI A and JIMÉNEZ-FLORES R (2006), 'Compositional and functional properties of buttermilk: A comparison between sweet, sour, and whey Buttermilk', *J Dairy Sci*, 89, 525–536.

SONNTAG N O V (1979), 'Composition and characteristics of individual fats and oils', in Swern D, *Bailey's Industrial Oil and Fat Products, Vol. 1, 4th ed.*, New York, John Wiley and Sons, 289–477.

STEIJNS J M (2008), 'Dairy products and health: Focus on their constituents or on the matrix?', *Int Dairy J*, 18, 425–435. DOI:10.1016/j.idairyj.2007.11.008.

THEODET C and GANDEMER G (1994), 'Fate of lipids during whey defatting process', *Lait*, 74, 281–295. DOI:10.1051/lait:1994423.

THOLSTRUP T (2006), 'Diary products and cardiovascular disease', *Curr Opin Lipidol*, 17, 1–10.

THOMPSON A K and SINGH H (2006), 'Preparation of liposomes from milk fat globule membrane phospholipids using a microfluidizer', *J Dairy Sci*, 89, 410–419.

TIMMS R E (2005), 'Fractional crystallisation – the fat modification process for the 21st century', *Eur J Lipid Sci Technol*, 107, 48–57. DOI:10.1002/ejlt.200401075.

TIMMS R E and PAREKH J V (1980), 'The possibilities for using hydrogenated, fractionated or interesterified milk fat in chocolate', *Lebensm Wiss Technol*, 13, 177–181.

TIRTIAUX F (1976), 'Le fractionnement industriel des corps gras par crystallisation dirigée – procédé Tirtiaux', *Oléagineux*, 31, 279–285.

VAN AKEN G A, TEN GROTENHUIS E, VAN LANGEVELDE A J and SCHENK H (1999), 'Composition and crystallization of milk fat fractions', *J Am Oil Chem Soc*, 76, 1323–1331. DOI:10.1007/s11746-999-0146-8.

VANHOUTTE B (2002), *Milk fat crystallization: Fractionation and texturisation*, PhD thesis, Ghent University, Belgium, 368p.

VANHOUTTE B, DEWETTINCK K, VANLERBERGHE B and HUYGHEBAERT A (2002a), 'Monitoring milk fat fractionation: Effect of agitation, temperature and residence time on physical properties', *J Am Oil Chem Soc*, 79, 1169–1176. DOI:10.1007/s11746-002-0622-3.

VANHOUTTE B, DEWETTINCK K, VANLERBERGHE B and HUYGHEBAERT A (2003), 'Monitoring milk fat fractionation: Filtration properties and crystallization kinetics', *J Am Oil Chem Soc*, 80, 213–218. DOI:10.1007/s11746-003-0679-z.

VANHOUTTE B, FOUBERT I, DUPLACIE F, HUYGHEBAERT A and DEWETTINCK K (2002b), 'Effect of phospholipids on isothermal crystallization and fractionation of milk fat', *Eur J Lipid Sci Technol*, 104, 738–744. DOI:10.1002/1438-9312(200211)104:11<738::AID-EJLT738>3.0.CO;2-F.

WALSTRA P and JENNES R (1984), *Dairy Chemistry and Physics, 5. Lipids*, New York, John Wiley and Sons, 58–97.

WALSTRA P, WOUTERS J T M and GEURTS T J (2006), *Dairy Technology: Principles of Milk Properties and Processes, 2nd edition. Chapter 2: Milk Components*, New York, CRC press, 27–105.

WARD R E, GERMAN J B and CORREDIG M (2006), 'Composition, applications, fractionation, technological and nutritional significance of milk fat globule membrane material', in Fox P F and McSweeney P L H, *Advanced Dairy Chemistry, Volume 2, Lipids, 3rd Edition*, New York, Springer, 213–244. DOI:10.1007/0-387-28813-9_6.

WEIHE H D and GREENBANK G R (1958), 'Properties of interesterified butter oil', *J Dairy Sci*, 41, 703.

WEIHE H D (1961), 'Interesterified butter oil', *J Dairy Sci*, 44, 944–947.

WOLFF R L, BAYARD C C and FABIEN R J (1995), 'Evaluation of sequential methods for determination of butterfat fatty acid composition with emphasis on trans-18:1 acids. Application of seasonal variations in French butters', *J Am Oil Chem Soc*, 72, 1471–1483. DOI:10.1007/BF02577840.

WONG P Y Y and KITTS D D (2003a), 'Chemistry of buttermilk solid antioxidant activity', *J Dairy Sci*, 86, 1541–1547.

WONG P Y Y and KITTS D D (2003b), 'A comparison of the buttermilk solids functional properties to nonfat dried milk, soy protein isolate, dried egg white, and egg yolk powders', *J Dairy Sci*, 86, 746–754.

WRIGHT A J, MCGAULEY S E, NARINE S S, WILLIS W M, LENCKI R W and MARANGONI A G (2000), 'Solvent effects on the crystallization behavior of milk fat fractions', *J Agric Food Chem*, 48, 1033–1040. DOI:10.1021/jf9908244.

YE A, SINGH H, TAYLOR M W and ANEMA S (2002), 'Characterization of protein components of natural and heat-treated milk fat globule membranes', *Int Dairy J*, 12, 393–402. DOI:10.1016/S0958-6946(02)00034-1.
YE A, SINGH H, TAYLOR M W and ANEMA S (2004), 'Interactions of whey proteins with milk fat globule membrane proteins during heat treatment of whole milk', *Lait*, 84, 269–283. DOI:10.1051/lait:2004004.

5

Modern approaches to lactose production

R. J. Durham, University of Western Sydney, Australia

Abstract: This chapter reviews recent improvements in the purification and crystallisation of lactose from whey. The properties of lactose and traditional methods of manufacture are discussed, within this framework methods for whey and permeate pretreatments are examined including; calcium precipitation, nanofiltration, ion exchange, electrodialysis, adsorptive decolourisation and ion exclusion chromatography. The fundamentals of lactose crystallisation are explored together with the impact of impurities on lactose crystal growth rate, purity, shape and size distribution. Innovations in lactose crystallisation and specialised pharmaceutical products are discussed including spray dried and anhydrous lactose, and dry powder inhalers. Post handling, separation, washing, milling and drying and impacts on lactose quality are described. Finally standards for hygiene and quality control of lactose manufacture, applications and functionality of lactose are covered in the last part of this chapter.

Key words: lactose, production, manufacture, whey, calcium, nanofiltration, ion exchange, chromatography, crystallisation, pharmaceutical, edible.

5.1 Introduction

Lactose is the major carbohydrate found in the milk of mammals, whereby cows' milk contains about 4.4–5.2% lactose and human milk 6.5–7.5% (Ganzle *et al.*, 2008). Lactose is typically recovered from whey, a by-product from cheese or casein manufacture. Recovery has mostly relied on the low solubility of lactose, using crystallisation as the basis for separation and purification. Early methods involved crystallisation from concentrated whole whey, which may have been adjusted to pH 6.8 and boiled to precipitate whey proteins and calcium. It was concentrated to 60% solids in single-effect vacuum pans, then the lactose was allowed to crystallise. The resulting crude lactose crystals were then washed and dried in rotary driers

(Weisberg, 1954). Purity and yields were low. Advances in membrane technology in the 1970s led to vast improvements in whey processing, with reductions in membrane costs throughout the 1980–1990s, enabling the efficient recovery of whey proteins as a valuable by-product from whey, creating protein-free permeates for lactose manufacture. Improvements in lactose manufacture have addressed purification as a key determinant of process efficiency, with research and development focused on calcium precipitation, nanofiltration, ion exchange, decolourisation and chromatography. Research on lactose crystallisation has been aimed at better understanding of crystallisation at the molecular level, for better control over lactose crystal purity, shape and size distribution. There has also been much research directed at understanding the anomeric character of lactose and transformations during drying, milling and storage, and consequences on lumping and post-handling. Finally, the development of sophisticated pharmaceutical lactose products, such as dry powder inhalers, has been the sharp edge of lactose research, directed at developments in particle engineering for fine control over lactose morphology and performance of lactose as a drug delivery device.

5.1.1 Properties of lactose

Lactose is a disaccharide with a molecular formula of $C_{12}H_{22}O_{11}$, composed of D-galactose and D-glucose linked by a β-1,4 glycosidic bond. The aldehyde group of the glucose moiety remains free and is able to be oxidised; therefore lactose is a reducing sugar (Rosenthal, 1991). Lactose has an anomeric centre at the C1 carbon of the glucose moiety, with the hydroxyl group able to be positioned in an equatorial (β-anomer) or axial (α-anomer) direction relative to the pyranose ring, see Fig. 5.1.

Lactose in solution exists in dynamic equilibrium between α and β anomeric forms. Mutarotation rates are affected by pH, temperature, ionic concentration and solvent. The solubility of the α and β anomers differ

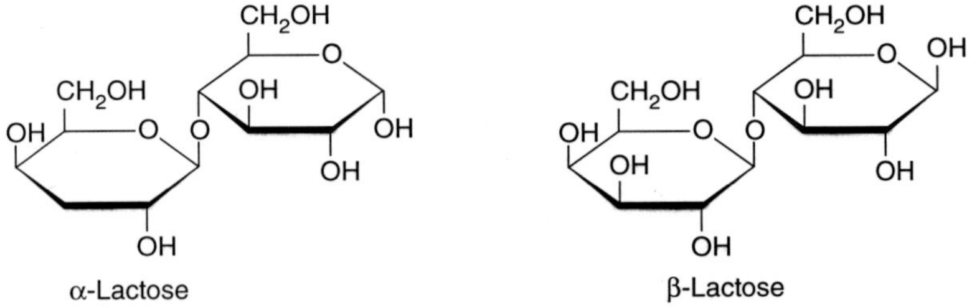

Fig. 5.1 The chemical structure of the α- and β-lactose.

significantly, for example the solubility of α-lactose monohydrate is 7 g/100 g water at 15°C, β anhydrous is 50 g/100 g water, while an equilibrium mixture of α and β is 17 g/100 g water at 15°C (Hartel and Shastry, 1991). The equilibrium lactose solubility increases with temperature and is determined by the mutarotation rate below 93.5°C. Above this temperature, β-lactose solubility determines the overall solubility. Equations describing the equilibrium solubility (g lactose/100g water) of anhydrous lactose were developed by Lifran 2007 (Eqs 5.1 and 5.2), using data by Hudson (1904), Herrington (1934), Foremost Foods (1970) and Visser (1982).

$$\text{Lactose solubility } C_S = e^{2.389+0.028T} \text{ at } T < 93.5°C \quad [5.1]$$

$$\text{Lactose solubility } C_S = e^{3.569+0.028T} \text{ at } T > 93.5°C \quad [5.2]$$

(Lifran, 2007)

Lactose normally crystallises as α-lactose monohydrate, containing one molecule of water for each molecule of lactose. The crystals are very hard and dissolve slowly, and if their size is over 10–14 μm, they can be detected sensorially, creating a defect known as sandiness (Ganzle *et al.*, 2008). α-Lactose monohydrate crystals normally form a tomahawk shape during crystallisation (see Fig. 5.2). The tomahawk shape arises from inhibition of growth of the {011} face, by the attachment and inclusion of

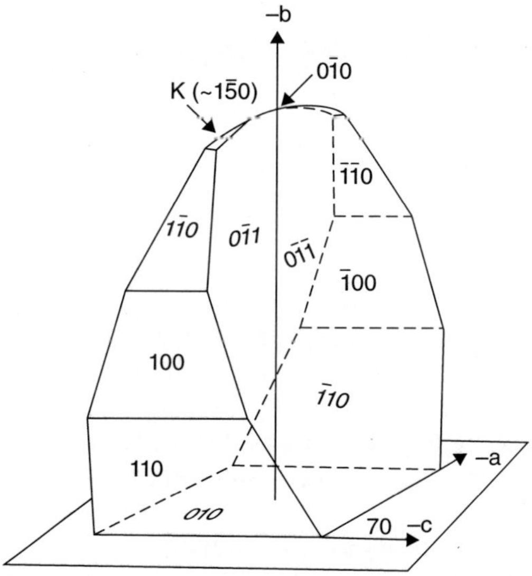

Fig. 5.2 Tomahawk shape of α-lactose monohydrate crystals (van Kreveld and Michaels, 1965).

β-lactose at this face inhibiting further growth (Dincer *et al.*, 1999; Raghavan *et al.*, 2000). β-lactose is the predominant crystalline form for crystallisations conducted above 93°C. Amorphous lactose forms when lactose is quickly dried and consists of a glassy state mixture of α and β lactose. Amorphous lactose is very hygroscopic, with a strong tendency to recrystallise into α-lactose monohydrate, forming hard lumps during storage.

The properties of crystalline lactose are affected by its polymorphic structure, which to a large extent is defined by the anomeric ratio of β to α. Different polymorphs of lactose have different densities, hygroscopicities, melting points, solubilities and speeds of dissolution. Consequently, the crystal form of lactose in dairy powders determines the physical properties of the powder; including flowability, hygroscopicity, solubility and caking.

5.1.2 Current lactose manufacturing processes

The manufacture of edible lactose typically commences with the ultrafiltration of whey to produce whey permeate as the feedstock to lactose processing. The whey permeate may be partially demineralised by calcium precipitation and/or nanofiltration. The permeate is then concentrated in a falling film multiple-effect evaporator to 60–70% TS, taking about 6 hours to fill large, specially designed crystallisation tanks. Crystallisation takes place with slow controlled cooling to maximise the size of the lactose crystals, with gentle mixing, taking up to 20 hours. The large lactose crystals are separated in decanter washers, creating a mother liquor by-product containing 20–30% of the lactose and 90% of the ash. The lactose crystals are washed in a refiner washer with cold water (1:1 water to lactose ratio). Wash water is often recycled to maximise the yield and minimise waste. The crystals are dried in a fluidised bed drier and milled to the desired size. The yields are in the range 65–75%. The edible lactose process is shown on Fig. 5.3, showing comments on the typical issues associated with each production step.

Pharmaceutical grade lactose is manufactured by further refining of edible lactose. This involves redissolving the lactose crystals to 60–65% solids and treating the solution with activated carbon, which adsorbs a number of impurities including riboflavin and peptides (Kellam, 1996). The carbon is removed by flocculation and filtration, and is typically discarded. The concentrated lactose is then crystallised, with recycling to maximise the yield of pharmaceutical grade lactose, see Fig. 5.4.

Traditional lactose processes continue to challenge manufacturers despite improvements over the years. Impurities in the lactose-rich whey permeate constrain the efficiency of the process; evaporator fouling by calcium phosphate can reduce operation times to 4–6 hours between cleaning cycles, creating high volumes of cleaning wastes. Purification of lactose

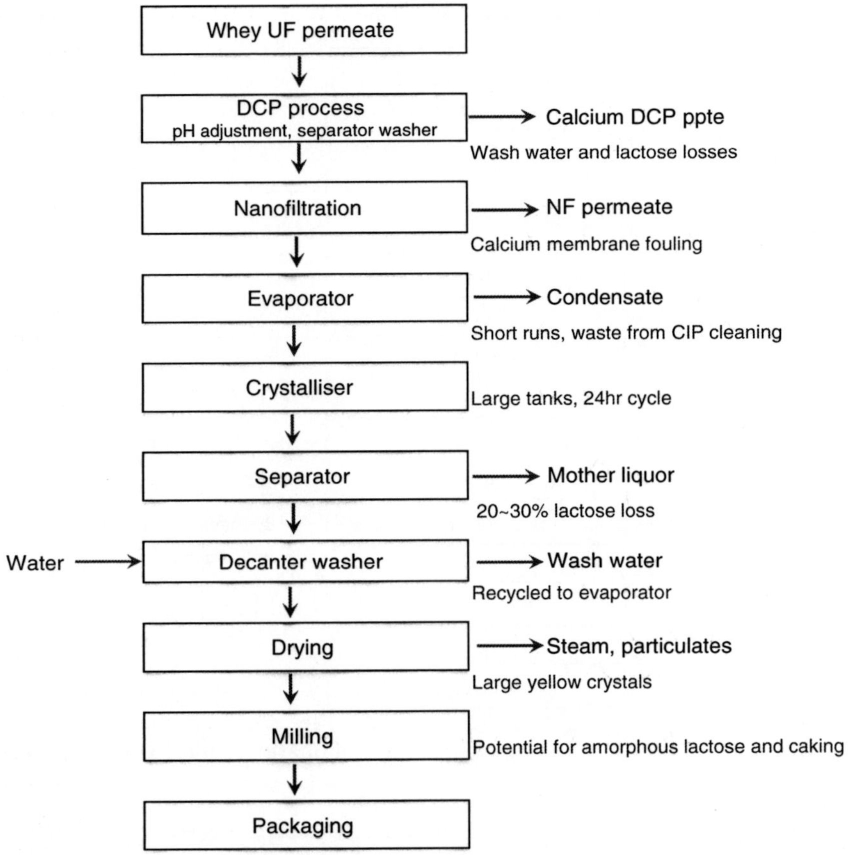

Fig. 5.3 Process chart for the manufacture of edible grade lactose.

is dependant on slow crystal growth and washing procedures. Impurities affect crystal growth and yield, which are aggravated by recycling mid-streams, required to improve yields and efficiencies. Further losses occur after crystallisation in the decanter separator, with high wash-water demands and yield loss. The large crystals need to be milled and this can lead to caking and lumping during storage of the milled product (Listiohadi *et al.*, 2005b, c). Entrained impurities and yellow colour reduce the grade and value of the lactose crystals. The yield is 65–75% with best practise. Greater losses can result from poor process control and variable feed quality. Pharmaceutical lactose production requires a second crystallisation, with higher purities achieved by activated carbon and filter aids, which impose further costs and wastage. The mother liquor presents a significant waste disposal problem for lactose manufacturers with BOD_5

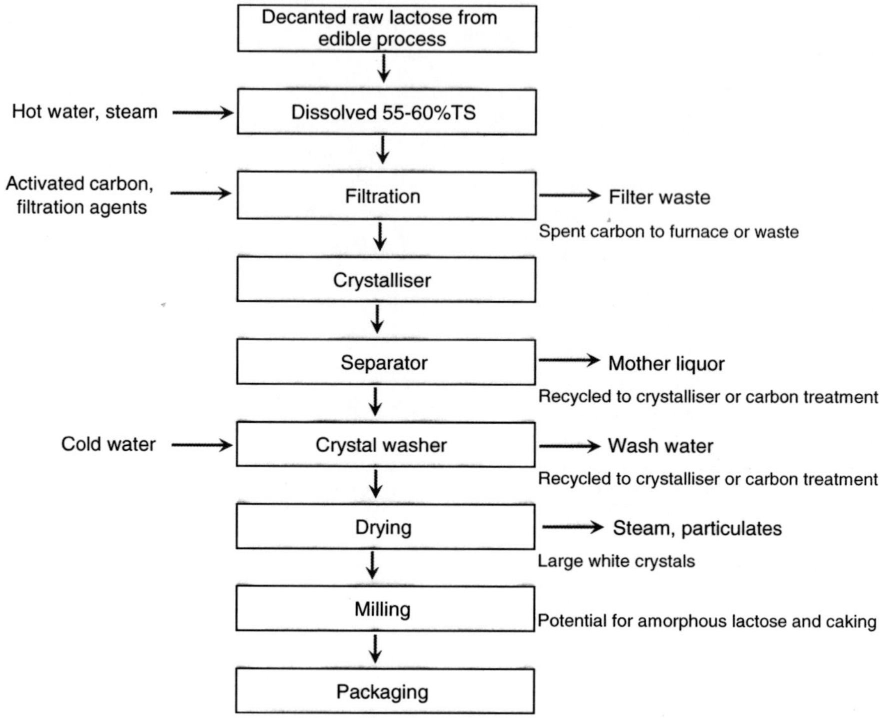

Fig. 5.4 Process chart for the manufacture of pharmaceutical grade lactose.

at 160 000 mg/L. If the mother liquor is not quickly used as stockfeed by farmers close to the factory, its disposal can present serious environmental problems.

5.2 Whey and permeate pretreatments

5.2.1 Whey – feedstock for lactose manufacture

The feedstock for lactose manufacture is ultrafiltered whey permeate, predominantly from sweet cheese whey. Whey contains 95% of the original water, most of the lactose, 20% of the protein, and traces of fat. There are two main types of whey, sweet whey and acid whey, see Table 5.1. Sweet whey is produced from rennetted cheese, such as mozzarella or cheddar. Sweet whey has higher pH and lower calcium phosphate levels compared with acid whey. Acid whey is produced from lactic acid fermentation for fresh cheese or from hydrochloric acid casein production.

Table 5.1 Composition of milk and milk permeate; rennetted sweet cheese whey and permeate; lactic whey; hydrochloric acid whey and permeate (Durham, 2000)

Component		Milk	Milk permeate	Cheese whey	Cheese whey permeate	Lactic whey	HCl acid whey	HCl whey permeate
Total solids	g/100 g	12.9	5.6	6.7	5.5	6.0	5.1	5.8
Lactose	g/100 g	4.8	4.7	4.8	4.7	3.9	4.4	4.3
Protein	g/100 g	3.4	0.33	0.6	0.05	0.72	0.73	0.06
Total N	g/100 g	0.53	0.052	0.094	0.008	0.113	0.114	0.009
NPN	g/100 g	–	0.032	0.039	0.003	0.042	0.034	0.03
Fat	g/100 g	3.5	<0.01	0.25	<0.01	0.06	0.05	<0.01
Ash	g/100 g	0.75	0.46	0.54	0.51	0.72	0.6	0.56
Citric acid	g/100 g	0.18		0.16	1.49			
Sodium	ppm	410	370	433	430	400	370	
Potassium	ppm	1300	1400	1460	1445	1530	1200	
Calcium	ppm	1100	280	420	375	1140	1200	
Magnesium	ppm	100	80	82	79	89	90	
Phosphorus	ppm	890	330	440	411	900	680	
Riboflavin	ppm	1.4	1.1	1.2	1.0	1.2	1.4	
pH		6.8	6.6	6.1	6.0	4.0	4.7	

During the formation of *acid whey*, the pH is lowered (by lactic fermentation or HCl) below the isoelectric point of casein to form the curd. The colloidal calcium phosphate is solubilised and the casein micelle structure is disrupted; consequently the casein proteins aggregate releases calcium phosphate into the whey (Brown, 1988). During the formation of *sweet whey*, the added rennet cleaves κ-casein on the surface of the casein micelle, destabilising the casein complex, releasing casein-macropeptide into the whey, while the hydrophobic α, β paracasein micelles flocculate together, retaining the calcium phosphate within the curd structure. The composition of lactic fermented wheys is further complicated by the presence of microbial metabolites such as lactic acid and higher levels of non-protein nitrogen.

Whey is ultrafiltered for the manufacture of whey protein concentrate or whey protein isolate. Ultrafiltration processes need to be monitored and controlled, as leaky membranes can lead to elevated protein levels in the permeate which can cause downstream problems during lactose manufacture. Diafiltration can be employed during ultrafiltration to produce higher protein products (e.g. WPC85). This can result in variable concentration of permeate to the lactose plant, particularly at the start-up and end of runs. Measures to control calcium fouling during ultrafiltration include pre-heating and holding the whey at elevated temperatures to allow the calcium to precipitate (Hiddink *et al.*, 1981). However, this impacts upon protein functionality; therefore ultrafiltration is now normally conducted at <15°C.

5.2.2 Calcium and phosphorous precipitation in whey and permeate

Cheese whey permeate contains about 375 ppm calcium and 411 ppm phosphorus. At these concentrations whey and permeates are supersaturated with calcium and phosphate, with respect to the solubility product of the various calcium precipitates – dicalcium phosphate dihydrate 2.6×10^{-7} mol/L, dicalcium phosphate 1.2×10^{-47} mol/L, hydroxyapatite 10^{-58} mol/L (Walstra and Jenness, 1984). Consequently, whey permeate is thermodynamically unstable and precipitation of some form of calcium phosphate will take place sooner or later (Schmidt and Both, 1987).

Calcium phosphate precipitation is not rapid and involves a number of stages. Initially the solution becomes opalescent with the formation of hydrated amorphous calcium phosphate. This is followed by one of the precursor phases – dicalcium phosphate dihydrate (DCPD), dicalcium phosphate (DCP), tricalcium phosphate (TCP) or octacalcium phosphate (OCP), with the structure and composition depending on the pH, temperature and concentration. These precursors eventually recrystallise into hydroxyapatite (HAP) (Feenstra and de Bruyn, 1979). The fact that this process is slow and continues through stages is central to the problem of calcium phosphate removal.

Calcium phosphate precipitation is favoured by concentration, elevated temperatures and alkaline pH. Heating whey ultrafiltrate causes the formation of finely dispersed precipitate and reduced soluble calcium (Tessier and Rose, 1958). Heat induces changes in the salt balance between the colloidal phase and serum phase, favouring the formation of dicalcium phosphate dihydrate. Alkaline pH favours calcium phosphate precipitation, with HPO_4^{2-} the dominant species above pH 8, related to the pKa of phosphoric acid at 2.2, 7.2 and 12.3 (Brule *et al.*, 1978).

During the manufacture of lactose, calcium phosphate precipitation is inevitable during heating and concentration, forming a hard, shell-like material on the heating surfaces of the evaporator. Lactose evaporators can have operational cycles as short as 4–6 hours, with frequent cleaning required to remove calcium precipitates fouling the heating surfaces.

5.2.3 Precipitation treatments to remove calcium and phosphorus

Rapid removal of calcium and phosphate prior to heating and concentration is the preferred option for lactose processing. Researchers have employed heat treatments and pH adjustments (Hayes *et al.*, 1974; Hickey *et al.*, 1980; Hobman, 1984) and additives such as calcium (Hayes, 1982; Karleskind *et al.*, 1995), polyphosphates (Evans, 1980; Hiddink *et al.*, 1981; Vembu and Rathinam, 1997) and EDTA (Evans, 1980; Ramachandra Rao *et al.*, 1994) to force the equilibrium one way or another, see Table 5.2.

Early work was aimed at precipitating calcium in whey prior to ultrafiltration to form calcium aggregates which remained in suspension rather than fouling the membrane (Hayes *et al.*, 1974; Matthews *et al.*, 1978). These

Table 5.2 Whey and UF permeate pretreatments to reduce calcium and phosphate fouling

Feed material	pH	Other additives	Heating temp (°C)	Heating time	Separation	Comments	Reference
HCl casein whey	4.4	0	80	15 s	nil	higher UF flux	Hayes *et al.*, 1974
HCl casein whey	5.9	0	80	15 s	nil	double UF flux	Hayes *et al.*, 1974
Cheese whey	6.2	0	80	15 s	nil	no effect on flux	Hayes *et al.*, 1974
Acid casein whey	5.7	0	80	15 s	nil	higher UF flux	Matthews *et al.*, 1978
Acid casein whey	5.7	0	50	1 hr	nil	higher UF flux	Matthews *et al.*, 1978
HCl casein whey	6.2	0	80	15 s	nil	higher UF flux	Hickey *et al.*, 1980
Cheese whey	6.2	0	81	15 s	nil	lower flux	Hickey *et al.*, 1980
UF perm. 40–45%TS	6	0	82–93	30–90 min	centrifuged	Ca and P removed	Pederson, 1980
Acid w.perm. 33–40%TS	7	EDTA, polyphosphate		50		lower ash in lactose	Evans *et al.*, 1982
UF permeate	7.4	0.022 M Ca	nil	nil	centrifuged	36% Ca, 65% P removed	Hayes, 1982
UF permeate	9	0.065 M Ca	20	nil	centrifuged	70% Ca, 80% P removed	Brothersen *et al.*, 1982
Lactic permeate	5.5	0.024 M Na_2CO_3	80	20 min	centrifuged	75% Ca, 90% P removed	Hobman, 1984
Lactic permeate	6.7	0	70	8 min	centrifuged	80% Ca, 65% P removed	Hobman, 1984
UF permeate	6.9–7.2	0	75–80	20–30 min	ultrafiltered		GEA, 2008

treatments were found to have a more marked impact on lactic whey ultrafiltration due to the higher calcium levels that would otherwise cause extensive fouling of the membranes.

Subsequent work on whey permeates for lactose production employed combinations of pH adjustments, heat treatments and centrifugal separation. Some researchers (Brothersen *et al.*, 1982; Hayes, 1982; Hobman, 1984) also added calcium to push the equilibrium towards greater removal of phosphorous and calcium. Calcium addition also precipitated citrate, lipids, proteins and lactose. It is worth noting that calcium precipitation is the basis of the Steffen process used for the separation of lactose from whey (Olano *et al.*, 1977). There are also patents describing the use of EDTA and polyphosphates to sequester calcium in concentrated UF permeate (Evans *et al.*, 1982), whereby the addition of 2×10^{-4} g/100 g EDTA to concentrated whey permeate produced lower ash lactose.

Calcium precipitation is retarded by the presence of permeate peptides (Halbert *et al.*, 2000; Schmidt and Both, 1987). Studies have shown that the addition of milk or whey proteins to permeate before heat treatment reduces the precipitation of calcium salts (Brule *et al.*, 1978). Schmidt and Both (1987) found that ultrafiltration permeate is much less stable than that of the corresponding milk or whey, as the protein in milk stabilises the calcium and phosphate in solution. This is an issue when leaky ultrafilter membranes increase protein contamination of permeates, impacting upon the performance of calcium phosphate precipitation processes.

The most successful calcium precipitation process has been the GEA Westphalia de-calcification process (GEA, 2008a). In this process, whey permeate is adjusted to pH 6.7 by the addition of NaOH and heated to 75–80°C for 20–30 mins. The resulting calcium precipitate is stabilised by a short holding time, then separated by cross-flow ultrafiltration with diafiltration. The slurry is then spray dried into a fine powder. There are some variations whereby the calcium precipitate is separated in a decanter washer and then dried; however, calcium driers require special care as the calcium precipitate is highly abrasive. The milk calcium can be sold at a premium for use in dietary supplements as a natural milk product (Harju, 2001).

5.2.4 Nanofiltration treatments to remove monovalent ions

Nanofiltration has been widely applied to demineralise and concentrate whey permeate for increased efficiency in lactose manufacture. Yield increases have been reported at 8–10% (Guu and Zall, 1992) and 6% (Mikkonen *et al.*, 2001), along with reduced lactose solids lost in the mother liquor. Mikkonen *et al.* (2001) also found that nanofiltration reduced viscosity, enabling higher solids concentration and higher yields. However, some researchers found the effect on final lactose purity to be small, as polyvalent ions and riboflavin continue to be retained in the crystallising liquor (Mikkonen *et al.*, 2001).

The rejection characteristics of nanofiltration membranes depends on the true charge of the membrane and the concentration of the salts in solution (Baticle *et al.*, 1997). Neutral molecules larger than 200–300 g/mole are rejected by the membrane, while inorganic ions are rejected on the basis of their charge, valency, degree of hydration, concentration and their interactions with other inorganic ions present in solution. The commonly available thin film polymeric nanofiltration membranes have ionisable carboxyl and amine functional groups and tend to carry a negative surface charge at neutral pH (Childress and Elimelech, 1996); therefore they actively reject multivalent anions (such as colourants, phosphates and citrates) and favour the permeation of monovalent ions (chloride, sodium and potassium) (Suarez *et al.*, 2006; Cuartas-Uribe *et al.*, 2007). If the feed has a very high content of multivalent anions, the permeation of monovalent anions is favoured to maintain electroneutrality, thus negative rejection of chloride ions can occur due to their small size and charge (Suarez *et al.*, 2006).

Multivalent cations such as calcium and magnesium tend to be rejected due to their larger hydrated size compared to monovalent cations. van der Horst *et al.* (1995) observed retentions of 75% Ca^{2+} and 45% Na^+ from $CaCl_2$/NaCl mixtures. Calcium can also be involved in phosphate complexes and citrate chelation, which further reduces the availability of free Ca^{2+}. Studies on the effect of pH on membrane rejection have shown that membrane rejection characteristics can be manipulated due to the amphoteric nature of polymeric nanofiltration membranes. For example, calcium rejection can be enhanced when the feed is acidified below the isoelectric point of the membrane, i.e. <pH 3, thereby achieving calcium rejection up to 95% (Durham *et al.*, 2003). This effect can be employed for recycling spent ion exchange brines used for softening, whereby nanofiltration can be used to separate the calcium from spent brine regenerant (see Section 5.2.6).

Studies on the operating conditions of nanofiltration of whey permeate have established that 18–20 Bar trans-membrane pressure and 12–15°C maximises flux, minimises lactose transmission and maximises mineral transfer (Durham, 2000). Lower operating temperatures, although associated with lower fluxes, are preferable to higher temperatures due to microbial risks in the range 25–55°C, plus increased propensity for calcium precipitation (Rice *et al.*, 2006). Furthermore, the lactose content of the permeate increases with temperature, with the rate of lactose permeation higher at higher temperatures (Atra *et al.*, 2005). Increased demineralisation rates are possible by diafiltration (Cuartas-Uribe *et al.*, 2007).

Membrane fouling is a common problem for nanofiltration, particularly with calcium and phosphate being concentrated at the surface of the membrane (Alkhatim *et al.*, 1998; Tay *et al.*, 2002). Additionally, peptides and proteins can absorb onto the membrane and contribute to fouling. Fouling leads to increased membrane cleaning costs, waste disposal and is associated with shortened membrane life. Removal of calcium and phosphate prior to nanofiltration can alleviate many of these problems, either by calcium phosphate precipitation as described in Section 5.2.3 or by softening ion exchange described in Section 5.2.6.

5.2.5 Electrodialysis for removal of monovalent ions

Electrodialysis demineralisation relies on transport of ionic species across permeable ion exchange selective membranes under the influence of a DC electric current. The electric field acts as the driving force and charged membranes perform the separation (Bazinet, 2004). The charged membranes can carry cation or anion ionisable groups, much like ion exchange resins. The membranes are available as monopolar (cation or anion) or bipolar membranes (Bazinet, 2004). Monopolar membranes are permeable to only one type of ion. Cation and anion exchange membranes can be stacked alternately and employed for demineralisation. Bipolar membranes are composed of three parts: an anion exchange layer, a cation exchange layer, and a hydrophilic transition layer at their junction. These membranes can carry out dissociation of water in the presence of an electric field (Bazinet, 2004).

Electrodialysis demineralisation has been employed for many years, principally for the demineralisation of whey for the manufacture of infant formula, but it is less reported for lactose manufacture. As with nanofiltration, monovalents are preferentially removed, enhancing the Ca:Na ratio, favourable for manufacture of demineralised whey for infant formula. Partial (50%) demineralisation is economically achievable, while higher levels of demineralisation requires a disproportionate increase in recirculating time and electrical power (Batchelder, 1987). The process can suffer from fouling, with calcium deposits forming on the cation membrane and protein fouling of the anion membrane (Ayala-Bribiesca *et al.*, 2006). Recent patents describe the use of softening ion exchange of whey to remove divalent cations and anions prior to electrodialysis, thus achieving improved process efficiency by reducing calcium fouling (Noel, 2002).

Newer applications of bipolar membranes have been the focus of much recent research, including for the production of organic acids (Bazinet, 2004), acid casein manufacture (Balster, 2006), peptide fractionation (Poulin *et al.*, 2008), delipidisation of whey protein (Lin Teng Shee *et al.*, 2007), deflavouring whey protein (Crowley, 2006) and waste water processing (Sleigh *et al.*, 2006); and may offer opportunities in lactose processing not yet realised.

5.2.6 Ion exchange demineralisation or softening of permeates

Ion exchange demineralisation is well established for the production of demineralised whey for use in infant formula (Herve, 1974; Houldsworth, 1980; Berghoffer and Scheibl, 1986; Jonsson and Arph, 1987). Ion exchange demineralisation of whey permeate prior to crystallisation has been described by Gonzales *et al.* (2008) for the manufacture of modified whey powder with superior organoleptic qualities, whereby ion exchange was used to remove 95% chloride, 90% of sodium and potassium and 60–70% of calcium. Ion exchange softening has also found increased acceptance for sugar juice processing (cane and beet) for sugar manufacture (Rousset, 1997; Kochergin and Tzschatzsch, 2005; Gula *et al.*, 2007) and offers comparable benefits for lactose manufacture.

Ion exchange demineralisation employs mixed bed or sequential cation and anion resin filled columns. The process can be operated for complete demineralisation, whereby the resin is regenerated with concentrated alkali and acids, or for softening decalcification, whereby the resin is regenerated with salt solutions. The control of the separation is based on the relative affinities of ions to the ion exchange resin as follows:

Strong cation resin: $Ce^{3+} > Al^{3+} \gg Ba^{2+} > Sr^{2+} > Ca^{2+} > Mg^{2+} \gg Ag^{+} > Rb^{+} > K^{+} > NH^{4+} > Na^{+} > H^{+} > Li^{+}$

Strong base anion resin: citrate $> SCN^{-} > I^{-} > S_2O_3^{2-} > CrO_4^{2-} > SO_4^{2-} > HPO_4^{2-} > NO^{3-} > Cl^{-} > HCO_3^{-} > H_2PO_4^{-} > HCO_2^{-} > BrO_3^{-} > ClO_3^{-} > F^{-} > OH^{-}$

(Haddad and Jackson, 1990)

There are a number of difficulties associated with ion exchange processing; short running times between regeneration, high consumption of regenerant chemicals, salt waste from regenerant chemicals, high water requirements to rinse resins and remove NaOH in particular, plus pH fluctuations when regenerated with strong acids and bases (Jonsson and Arph, 1987).

Innovations to ion exchange processing of whey permeate has been reported by Durham (2000) and Durham *et al.* (1998). This process is based on the ion exchange softening or decalcification of whey permeate. The key to this process is the recovery of monovalent salts from the nanofilter permeate for use as regeneration brine. Moreover, water can also be recovered from the permeate streams for resin rinsing. This work was aimed at addressing the cost and wastage associated with regenerant chemicals, resulting in a cleaner, more cost-efficient softening process. The softened whey permeate has many advantages for lactose processing: nanofiltration with improved flux and no fouling; improved efficiencies for subsequent ion exclusion purification; resulting in longer operation times between cleaning cycles; and minimal fouling during evaporation.

Studies on the operating conditions of ion exchange softening have established that calcium removal is greater with treating single strength whey permeate, otherwise high sodium and potassium concentrations in concentrated whey permeate create a competing regeneration effect during ion exchange (Durham *et al.*, 2007). Further studies confirmed that the temperature effect on ion exchange was minimal; therefore operating at low temperature was better from a microbiological perspective. It was also observed that protein or peptides can adsorb onto ion exchange resin and negatively impact on ion exchange efficiency. Therefore it is important to monitor ultrafilter membrane integrity to prevent peptides entering the permeate, to maintain process stability for the ion exchange softening of permeates.

This softening process has been patented by Durham *et al.* (1998), covering a four-step process of ion exchange, nanofiltration, recovery of whey brine for the regeneration of the ion exchange column and subsequent purification of lactose by ion exclusion. The process has been licensed by Groupé Novasep. The process was developed as a pretreatment for ion exclusion chromatography for lactose purification; however, it has also been employed as a softening pretreatment to ion exchange demineralisation, with improvement in process and cost efficiencies (Novasep, 2005; Friesland Foods Domo, 2007).

5.2.7 Chromatography purification of lactose

The softening pretreatment patented by Durham *et al.* (1998) creates a softened nanofiltered whey permeate that can be further purified by ion exclusion chromatography to produce a purified lactose stream suitable for direct crystallisation into pharmaceutical grade lactose or conversion into lactose derivatives (Durham *et al.*, 2007).

Ion exclusion chromatography is a method whereby sugars (non-ionic components) can be separated from minerals (ionic components) on long, resin-filled columns. In this process, discrete quantities of a sugar/salt mixture such as softened whey permeate, are loaded onto long columns filled with a strong cation resin and eluted with water. The resin is saturated with counter ions in the same composition as the feed, so there is no net ion exchange between the resin and the feed. The ionic material is excluded by the resin and quickly moves through the column (see Fig. 5.5). The non-ionic substances partition between the eluent phase and the occluded resin liquid and pass slowly through the column, moving into and out of the resin beads as they traverse the length of the column. The ionic materials emerge first, followed by the non-ionic substances.

Ion exclusion separations are improved when the resin is in the monovalent cation form. Monovalent resins have a stronger Donnan potential so that ionic components are more strongly excluded, ensuring that they quickly pass down the column and separate from the non-ionic components. There is no internal ion exchange during elution. Monovalent separations have a single, narrow, mineral peak. The separation of the sugar is also enhanced as the minerals move faster down the column, allowing the resin beads to swell after the minerals pass, more able to absorb the sugar and delay its progress. To achieve monovalent separations, the process relies on pre-treatments such as ion exchange to remove calcium from the feed prior to ion exclusion (Durham *et al.*, 2007).

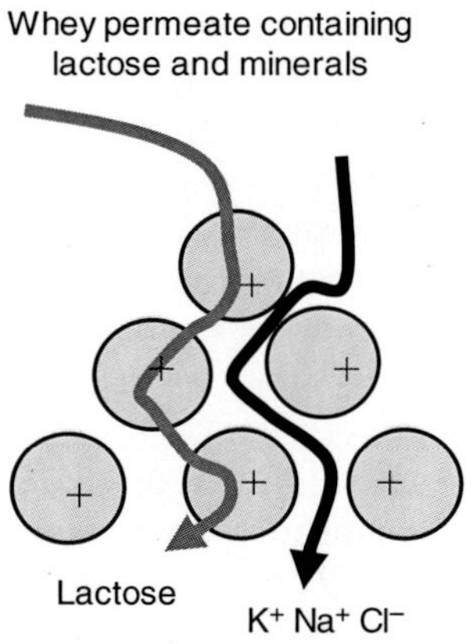

Fig. 5.5 Schematic of ion exclusion resin showing Donnan exclusion of minerals from the charged resin particles and delayed path of lactose permeation through the resin beads (Durham *et al.*, 2004).

There are a number of advantages associated with ion exclusion chromatography. The resin does not require regeneration; therefore no downtime for regeneration, no need for expensive regenerant chemicals, and no accompanying impact of spent regenerant salts on the environment. The elution solvent is water, which can be recovered by recycling the evaporator condensate and reverse osmosis permeate. Anionic and cationic species can be separated in a single pass with up to 99% demineralisation achieved. This process has found widespread application for the recovery of sucrose from beet and cane molasses (Ehrenhuber, 2005; Suhr and Schulze, 2003; Lancrenon, 1997; Chertudi, 1991) and offers similar benefits to lactose processing.

Research on chromatographic purification for lactose manufacture can be traced to Harju and Heikkilae (1989), who describe a chromatographic process which recovers lactose from the mother liquor, similar to chromatographic separation of beet sugar molasses being applied commercially in the sugar industry (Harju and Heikkilae, 1989; Harju, 2007). In this process, only the mother liquor is treated by ion exclusion to increase overall lactose yield.

More recently, the Ion Exclusion Lactose (IEL) process by Durham *et al.* (2007) describes a technique whereby the entire lactose stream is recovered by chromatography as purified lactose. In this process, the ion exchanged, softened, nanofiltered whey permeate is separated by ion exclusion chromatography. The mineral peak contains the ionic fraction (potassium, sodium, chloride, non-protein nitrogen, etc.). The lactose peak is essentially pure lactose. The purity, yield and throughput is determined by column length, flowrate and injection size. The resultant purified lactose stream is suitable for direct crystallisation into pharmaceutical grade lactose or conversion into derivatives. This process captures the entire lactose stream, achieving an environmental footprint smaller than traditional technology, while avoiding the production of mother liquor. Costings of chromatographic purification of lactose reported by Theobald (2007) show a 2-year return on investment for a plant processing 1 million L UF permeate per day, with a 20% increase yield of lactose, optimal lactose purity, lower water and chemical consumption and lower effluent rejection.

5.2.8 Decolourisation by activated carbon or resin adsorption

Whey and permeate solutions also contain coloured compounds, such as riboflavin and Maillard browning products, which are responsible for yellow and brownish colours in crude and edible grade lactose. Riboflavin is sparingly soluble in water, up to 100–250 μg/ml at room temperature (Association of Vitamin Chemists, 1966). In concentrated permeates, riboflavin tends to co-crystallise with lactose at low temperatures, and is difficult to wash away. Consequently, the final crystallisation temperature can be affected by the amount of riboflavin present in the concentrate.

Maillard browning products are formed by reactions between free amino groups of peptides (particularly lysine) and reducing sugars such as lactose. This reaction is favoured by elevated temperatures and long holding times, as occur during evaporation and filling the crystallisation tanks, which can take many hours. Leakage of primary amino groups through the ultrafilter membranes can add to this problem. These coloured compounds can be removed by activated carbon or adsorbent resins.

Activated carbon is made from a variety of carbonaceous materials including wood, coal, peat, nut shells, lignite, bone and petroleum residues. The material is dehydrated and carbonised in a kiln in the absence of oxygen, which drives off the volatile matter and produces a rudimentary pore structure. The 'charcoal' is then activated by thermal or chemical processes (OMRI, 2002). Activated carbon is employed in the manufacture of pharmaceutical grade lactose. The process involves redissolving (edible grade) lactose crystals and treating the solution with granular or powdered activated carbon to absorb a range of solutes including riboflavin, proteose peptones, Maillard browning products and a variety of proteins. Montilla *et al.* (2005) found 65–92% colour reduction using 5 mg/L activated carbon. Kellam (1998) reports that extra protein can be absorbed onto the activated carbon by temporarily adjusting the liquor pH. The carbon can be removed from the liquor by flocculation and filtration, and is typically discarded (Kellam, 1998). Activated carbon can be recycled by thermal regeneration, but this is mainly done off-site by specialised companies. Activated carbon has a number of disadvantages, including costly operating expenses, complex thermal regeneration requirements (or waste disposal costs), the risk of microbial growth if conducted at ambient temperatures, and health risks to workers if the carbon is not handled safely (Gula and Paillat, 2005).

Adsorptive ion exchange resins can also be applied to remove coloured compounds. Ion exchange decolourisation involves a number of phenomena such as ion exchange (mostly anionic interactions regenerated by chloride ions); hydrogen bond interactions between colourant and resin; steric interactions whereby slow flowrates and macroporous resins maximise retention of coloured compounds; or by hydrophobic absorption of non-polar solutes requiring non-polar regeneration (Gula and Paillat, 2005). Studies by Pragono *et al.* (2008) on batch columns using macroreticular adsorbent resin, found more than 90% of riboflavin could be removed from a 15% TS lactose solution containing 8–9 mg/L riboflavin without nearing saturation. These studies also demonstrated that sorption followed a pseudo-first-order kinetic model in batch adsorption column system, and that the mechanism for riboflavin adsorption relies on hydrophobic interaction (Pragono *et al.*, 2008).

Adsorption resins have been widely applied in sugar processing for decolourisation of beet and cane juice prior to crystallisation (Broadhurst and Rein, 2003; Coco *et al.*, 2008; Garcia Agudo *et al.*, 2002; Gula and Paillat, 2005; Rein *et al.*, 2007; Serpen *et al.*, 2007; Singhal, 2005). However, resin-based decolourisation is less widely reported for whey and permeate processing. Resin based decolourisation has advantages over activated carbon, as the resin can be regenerated and, when combined with membrane recycling of the spent regeneration brine, the waste can be reduced by 90% (Gula and Paillat, 2005), or 75–80% (Coco *et al.*, 2008).

5.3 Lactose crystallisation processes

5.3.1 Fundamentals of lactose crystallisation

Industrial crystallisation of lactose is usually performed by slow batch cooling of evaporated permeate, producing large lactose crystals that can be separated, washed and dried. The mother liquor from which lactose is crystallised, comprises a complex mixture of minerals, peptides and organic acids. Consequently, crystallisation functions as a purification and separation process, as much as a crystal recovery step. Crystallisation rates rely on a complex set of interactions determined by lactose solubility, supersaturation, α/β lactose mutarotation, nucleation, temperature, and agitation, and is affected by the type and concentration of impurities in solution (Randolph and Larson, 1988).

Lactose crystallisation commences with the formation of nuclei in a super-saturated solution. At high supersaturation, spontaneous nucleation (primary) can occur. Below this region is the intermediate zone, where nucleation will occur only if seed crystals are present, known as secondary nucleation. Next is the metastable zone, where nucleation does not occur, but seed crystals will grow (Lifran, 2007), see Fig. 5.6. Nucleation rates depend on supersaturation, agitation, impurities, seed crystal number and

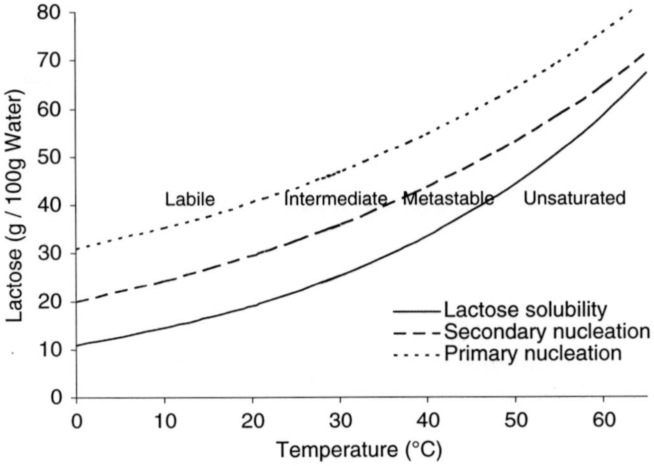

Fig. 5.6 Solubility regions for lactose, showing primary and secondary nucleation thresholds (Lifran, 2007).

size (Kauter, 2003). There are difficulties in accurate measurement of the kinetic parameters for nucleation. Studies by Shi *et al.* (1990), and Liang *et al.* (1991) calculated parameters of nucleation kinetics for continuous lactose crystallisation using Mixed Suspension, Mixed Product Removal (MSMPR) population balance technique. However, Butler (1998) found that these studies ignored the large numbers of small crystals below 10 µm. Butler (1998) subsequently developed an equation representing a closer estimate of nucleation rate, see Equation 5.3.

$$C_{SNT} = e^{2.992+0.0196T} \text{ at } T < 60°C \quad [5.3]$$

(Butler, 1998)

The principle driving force for crystallisation is supersaturation. This is achieved by slowly cooling the concentrated lactose solution (65–70% TS) from 70–75°C to 20–10°C. Cooling crystallisation is suited to lactose due to its low solubility (at 22 g/100 g at 25°C, compared with sucrose at 200 g/100 g at 25°C). During lactose crystallisation, the nuclei begin to grow, lactose molecules diffuse towards the crystal, adsorb on the surface and incorporate into the lattice (Hartel and Shastry, 1991). As the crystals grow, α-lactose is depleted and β-lactose mutarotates into α-lactose which continues to be crystallised into α-monohydrate. Concurrently, the solids concentration of the syrup reduces, and cooling is continued maintaining supersaturation and crystal growth. Lactose crystallisation is thus a complex interrelated phenomenon, determined by the kinetics of mutarotation, nucleation and crystal growth, described in a kinetic model by Mimouni (2009).

Existing practices for cooling crystallisation of lactose take 20–24 hours, including 6 hours to fill the crystallising tank with evaporated whey permeate and 14–18 hours crystallisation. Lactose crystallisation is conducted in specially designed lactose crystallisers, with large cooling surfaces for controlled cooling. Stirring agitator blades ensure that the crystals remain suspended in solution, allowing the crystals to grow but not be damaged by the gentle agitation (Terlet, 2008). Agitation has been found to markedly increase the overall rate of crystallisation by increasing the surface area available for crystallisation from the supersaturated solution (Haase and Nickerson, 1966).

Optimum lactose crystals are 200–300 μm and tomahawk shaped (Ostergaard, 1988). Crystal size distribution (CSD) is one of the main variables governing the performance of downstream processes such as separation, washing and drying. It has been reported that crystals of uniform shape and size, growing under the same environment, do not grow at the same rate and this is known as growth rate dispersion (Hartel, 2001). The resulting CSD affects the end product performance, such as bulk density and flow properties (Vu *et al.*, 2004b). Large crystals with a narrow size distribution, with the least mother liquor inclusion, which can be easily washed, produce the highest quality lactose (Shi *et al.*, 2006).

5.3.2 Effect of impurities on lactose crystallisation

The components naturally present in whey and permeate have been extensively studied to determine their effect on crystallisation of lactose (Bhargava and Jelen, 1996; Guu and Zall, 1991; Jelen and Coulter, 1973; Michaels and van Kreveld, 1966; Nickerson and Moore, 1974; Smart, 1988; Smart and Smith, 1991; Visser, 1980, 1984). Some impurities have been found to promote growth, others not to affect it, and yet others to inhibit crystal growth. Impurities influence crystal growth in three different ways: impurities can change the equilibrium solubility, therefore supersaturation; impurities can adsorb onto the crystal faces inhibiting the growth by limiting docking of more growth units; impurities can adsorb on the crystal surface, inhibiting growth, and become included within the crystal lattice, modifying the crystal habit (Mullin, 2001).

Bhargava and Jelen (1996) examined the effect of mineral salts on lactose growth, following on from earlier studies by Jelen and Coulter (1973). They confirmed that some salts accelerated growth (LiCl, $CaCl_2$ and $MgSO_4$) while other salts slowed growth (K_2HPO_4 and KCl). The effect of the different minerals is complex and depends on the species involved. Some minerals increase lactose solubility; for example, calcium increases lactose solubility due to the formation of lactose-$CaCl_2$ complexes (Herrington, 1934). Other mineral impurities may accelerate the surface deposition step of crystallisation, by lowering lactose solubility and accelerating the mutarotation (Jelen and Coulter, 1973). Guu and Zall (1991) found that potassium, in particular, decreased lactose growth and yields. Bhargava and Jelen (1996) also found that lactose crystal growth was slowed in the presence of whey permeate, reflecting the presence of potassium in whey permeate.

Lactose phosphate has been found to strongly retard lactose crystal growth (Visser, 1988). This has been attributed to the ability of lactose phosphate to dock at the growth sites of all lactose faces, preferentially onto the fastest growing faces {010} and {110}. Further studies on the mechanisms of lactose phosphate by Lifran (2007) found that lactose phosphate is incorporated into the crystal since its structure is comparable to lactose and it is anionic at the pH of crystallisation. Consequently, lactose phosphate monovalent anions form a single layer on the crystal surface, creating a negative charge and forming a barrier against further adsorption. The presence of only 60 ppm lactose phosphate was found to be sufficient to cut crystal growth rate by 50% compared to growth rates in ultra pure lactose (Lifran *et al.*, 2006a). Lactose phosphate also decreased the yield, the final median size, and modified the shape of some crystals. Lactose phosphate was found to be concentrated in pharmaceutical grade lactose at levels of 270 to 400 ppm, being concentrated into the lactose during repeated crystallisations. Theories on the origin of lactose phosphate have shown links to cheesemaking processes and the metabolism of lactic acid bacteria, with 10 fold increases in lactose phosphate found in cheese and whey 24 hours after manufacture, compared with the content in fresh milk (Lifran *et al.*, 2006b).

Riboflavin has been found to retard the growth of lactose crystals due to its ability to dock onto the crystal growing surface and block growth sites, retarding the growth of that particular face (Michaels and van Kreveld, 1966; Smart and Smith, 1991). However, riboflavin is mainly adsorbed at the surface of the crystals and can be washed off, as opposed to lactose phosphate which is integrated within the crystal structure (Lifran, 2007).

β-Lactose has also been found to impact on α-lactose crystal growth. The asymmetric tomahawk shape of α-lactose monohydrate crystals arise from the growth inhibition caused by incorporation of the β isomer in the {011} sector of the crystal (Dincer *et al.*, 1999; Raghavan *et al.*, 2000). Studies have shown that growth under conditions that restrict β-lactose formation result in needle-like crystals (Raghavan *et al.*, 2000).

The effect of impurities on lactose quality is magnified when crystallisation is accelerated, as would occur under highly supersaturated conditions or rapid cooling, with the impurities present in the mother liquor entrained in the lactose crystals. Smart and Smith (1991) found calcium phosphate and calcium sulphate salts co-precipitated with the lactose when their solubility had been exceeded during crystallisation. Impurity analysis of lactose crystals by Butler (1998) also revealed entrapped calcium phosphate and adsorbed riboflavin. Consequently, rapid crystallisation is not desirable when it results in the incorporation of large amounts of impurities within the crystal.

5.3.3 Innovations in lactose crystallisation

Lactose is generally crystallised in cooling batch crystallisers in a highly empirical process which can be unpredictable, with difficulties in optimising for quality, recovery, efficiency, capital cost and energy consumption (Shi *et al.*, 2006). The control of lactose crystallisation has been the subject of much recent research, with a range of crystallisation parameters being the focus for improvement, including; cooling profile (Shi *et al.*, 2006), seeding (Guu and Zall, 1991; Jayaprakasha *et al.*, 1994; Vu *et al.*, 2004a), sonocrystallisation (Bund and Pandit, 2007a; Vu *et al.*, 2003), continuous crystallisation (Henningfield and Dinesen, 2003); recycling mother liquor (Cloidt and Lehmann, 2007), crystallisation from solvents and ternary systems (Bouchard *et al.*, 2008; Brito and Giulietti, 2007; Bund and Pandit, 2007b; Lifran *et al.*, 2005) and permeate crystalliser driers.

A process employing a controlled cooling profile to achieve lactose crystals with uniform size and narrow size distribution has been patented by Shi *et al.* (2006). This patent recommends that the concentrated lactose permeate be initially rapidly cooled from 80°C to 50°C and then held at 50°C to encourage nucleation, thereafter gradually cooled to 30°C to assist crystal growth and maximise yield. The resulting crystals with narrow size distribution can be more efficiently washed and separated. This process relies on primary nucleation as the source of seeds and control of the cooling profile to manage the size distribution of the resulting crystals.

Seeding is rarely employed for control of industrial crystallisation of edible lactose, yet there has been consistent research demonstrating the value of seeding for improved yield and control over crystal size distribution. Ideally, seeded crystallisation is conducted within the metastable solubility zone; that is where seed crystals grow but nucleation does not occur. This allows for better control of the crystallisation process, with the crystal size distribution controlled by seeding and controlled cooling (Lifran, 2007). Guu and Zall (1991) and Jayaprakasha *et al.* (1994) both showed that seeding increased the relative gross yield. Vu *et al.* (2004a) found that seeding could speed up crystallisation growth rate with more consistent product quality, achieving high yields of crystals with narrow size distributions compared to self-nucleated batch crystallisation.

A dynamic optimization model developed by Vu *et al.* (2006) has been used to investigate and compare seeded and non-seeded batch cooling, semi-batch and continuous cooling crystallisation of lactose, to identify conditions for producing uniform size distribution at the highest yield. Semi-batch seeded crystallisation was found to achieve higher yields in shorter times, compared with batch cooling. Semi-batch mode involves cooling in accordance with a defined temperature profile, while syrup is continually fed into the tank until it is filled. Continuous feeding of fresh syrup maintains the high driving force for crystallisation and constrains temperature drops (Vu *et al.*, 2006). These results also suggest that evaporative batch operation is faster than batch cooling; however, its implementation is more complicated and nucleation is difficult to control.

A continuous evaporative crystallisation apparatus for lactose has been patented by Henningfield and Dinesen (2003), whereby the concentrated permeate or lactose is simultaneously concentrated and crystallised whilst being conveyed along a horizontal evaporative crystalliser with a hollow screw shaft for agitation and cooling. The resulting lactose crystals are recovered from the crystalline slurry (at 75% TS).

Another recent patent in lactose crystallisation describes a process whereby a part of the desugarised permeate (mother liquor) is fed back (recycled) to the lactose concentrate prior to the evaporation step and reduces yield loss from the mother liquor. The process includes demineralisation of the permeate by nanofiltration, electrodialysis or ion exchange to maintain purity to the crystalliser (Cloidt and Lehmann, 2007).

Lactose crystals with uniform size distribution have been produced using drop-based microfluidic crystallisers, whereby supersaturated solutions of lactose are expelled dropwise from narrow tubing at 100 μm, 150 μm or 300 μm I.D., creating crystals in the size range close to that of the tubing diameter. Dombrowkski *et al.* (2007) report CSD at 16% compared to 40% for bulk crystallisation, while drops containing single crystals had 7% CSD.

Sonocrystallisation has been investigated to initiate nucleation and accelerate the growth of lactose crystals (Bund and Pandit, 2007a; Vu *et al.*, 2003, 2004b). Sonocrystallisation thus provided a solution to the difficulties associated with long induction times associated with natural nucleation of lactose (Raghavan *et al.*, 2001). Studies have shown that ultrasound enhances nucleation, with higher ultrasound levels resulting in higher crystal yields and smaller size particles, due to the high number of nuclei that have been generated, and also that agglomerates are broken-up (Vu *et al.*, 2004b). Sonocrystallisation has also been combined with solvent addition and cooling to produce fine seed crystals for use in controlling bulk crystallisation of lactose (Vu *et al.*, 2003, 2004b).

The use of solvents for accelerated crystallisation to control crystal shape and size distribution, has also been the subject of research in a number of studies over the years (Olano *et al.*, 1977; Singh *et al.*, 1991; Lifran *et al.*, 2002, 2004b; Bund and Pandit, 2007a). Bund and Pandit (2007a) found that alcoholic crystallisation increased the yield of lactose from concentrated deproteinated paneer whey (containing 19.5% w/v lactose) up to 90% with 65–85% ethanol at 7°C. The method described by Lifran *et al.* (2002) involves staged alcohol (preferably ethanol) addition for demineralisation and then controlled crystallisation, producing free-flowing pharmaceutical grade, rosette shaped crystals with unique properties. Work by Vu *et al.* (2004b) found that solvent addition and sonocrystallisation could be used to produce very fine crystals for pharmaceutical applications or seed lactose, thus avoiding the problems associated with amorphous lactose created during size reduction milling (hammer, ball or jet mill).

Finally, recent literature describes a number of permeate crystalliser driers used for the production of free-flowing dried permeate. These processes employ special evaporators capable of making permeate pastes at very high concentrations (78–86% TS) yet avoiding heat damage. Evaporation is followed by rapid crystallisation to create very fine lactose crystals which are dried in specially constructed driers to produce a free flowing powder. The *Niro Thixotherm* process is named after the thixotropic nature of highly concentrated crystallised lactose. The permeate is dried in a *Rosinaire*™ paddle dryer with a screw conveyor to cure and crystallise the permeate followed by drying and cooling in a combined back-mix/plug-flow fluid bed drier (Pisecky, 2005). The *Anhydro Paraflash* process utilises a *Paraflash* forced circulation flash evaporator to concentrate the permeate to 78% solids, accompanied by continuous crystallisation of lactose and drying in a spin flash dryer (Andersen, 2005). The *Relco Keller* drying system concentrates the permeate to 78% solids, followed by triple C (cooling concentrator crystalliser), then an air lift drier, an integrated two-stage fluid bed (Keller, 2007). Carlisle Process Systems also developed new approaches for drying acid whey and permeates which are prone to stickiness (Boersen and Veldmeijer, 2003), described by Neville (2006), in a modern, purpose-built whey and permeate drying plant in the Netherlands.

5.4 Post handling – separation, washing and drying crystals

5.4.1 Decanting and washing crystals

The crystalline lactose slurry exiting the crystalliser requires separation and washing to remove the mother liquor and prepare for drying. There are a number of different types of separators available; solid bowl decanter (horizontal shaft centrifugal decanter), screen bowl decanter (worm/screen centrifuge) (Brinkman, 1976; Gesterkamp, 1994; Mans, 1994; Bylund, 2003; TEMA, 2007), and hydrocyclones (Zeevalkink *et al.*, 2005). Decanter separators are deployed in series, with the crystal slurry first being centrifuged in the solid bowl decanter to separate the mother liquor and raw lactose. Efficient operation of the decanter centrifuge is facilitated by large crystals (200–300 μm) with narrow size distribution, ensuring a free-flowing crystal slurry, from which the mother liquor can be drained and separated. Too many fines will clog the separator and large amounts of mother liquor will be retained, with the resulting crystals containing higher levels of impurities and higher levels of moisture impacting on drier running costs.

The raw lactose is then washed and centrifuged in a series of screen bowl centrifuges. The quantity of wash water required to wash the crystals depends on the CSD, the number of wash steps and the counter current recycling of wash water (Mans, 1994). Wash water corresponding to 10% of the feed quantity has been reported by Ostergaard (1988) and van den

Bos (1987), while Bylund (2003) report 1:1 wash water ratio with the lactose processed. The wash water can be recycled back to an earlier step in the process to recover lactose and minimise losses from washing. The purity of the wash water impacts on the purity of the lactose; thus, higher grades of lactose require demineralised water or RO permeate, while potable water would suffice for edible grades of lactose. After decantation, the lactose is 88–90% TS and requires drying.

5.4.2 Drying lactose crystals

Moist lactose crystals exiting decanter separators are dried in a fluidised bed drier to produce a free-flowing powder with a final moisture content of 4.5–5.5% (in α-lactose monohydrate). This applies to both edible grade lactose and pharmaceutical grade lactose that has been further processed by secondary purification and crystallisation. The fluidised bed drier is limited by a product temperature of 93°C, otherwise β-lactose will be formed from the residual amorphous lactose. Above 100°C, further moisture will be driven out, forming anhydrous unstable α-lactose (Brinkman, 1976). The fines from the drier exhaust should be collected in a bag filter to maintain air quality in the surrounding area (GEA, 2008b). Drying time impacts on product quality: if the lactose is dried too fast then amorphous lactose on the surface of the crystals dries as amorphous glass and increases the risk of caking during storage. The amorphous lactose should be given sufficient time to crystallise as α-lactose monohydrate during drying.

The air used to dry the crystals should be dehumidified and monitored for RH% (GEA, 2008b), to prevent caking problems during storage. The lactose should also be cooled prior to packing, otherwise when placed in a cool store, moisture will migrate from the hot region within the centre of the bag to the cool region at the surface; and if there is enough moisture present in the powder, capillary condensation between particles will cause liquid bridging and dissolution, which eventually leads to caking and lumping as the moisture is redistributed when the temperature has equilibrated (Paterson and Bronlund, 2009).

5.4.3 Grinding and milling

Lactose crystals are normally milled and sieved to a range of size specifications. There is a range of lactose sizes available on the market: for example, 60 Mesh (40–65% 250 μm), 100 Mesh (60–80% 100 μm), 200 Mesh (75–85% 75 μm), 350 mesh (>65% 45 μm), achieved by milling and sieving. Milling has the potential to cause changes in the crystalline lactose. Otsuka *et al.* (1991) found that the crystal water of α-monohydrate was converted into absorbed water by mechanical stress, and that isomerisation of lactose occured in the solid state during grinding. Crystallinity decreases with grinding time, resulting in the formation of amorphous lactose with the release of the water of crystallinity (Otsuka *et al.*, 1991). Amorphous lactose is very susceptible to elevated humidity levels, with lumps of black mould found to occur within the sacks of lactose when there have been insufficient control over humidity during drying and milling. Furthermore, there is also an increased propensity to lump and cake as the amorphous lactose recrystallises into α-lactose monohydrate during storage (Listiohadi *et al.*, 2005a).

5.5 Specialised pharmaceutical products

5.5.1 Spray-dried lactose

Spray-dried lactose consists of spherical particles of small α-lactose monohydrate crystals, glued together in an amorphous lactose matrix (Vanderbist *et al.*, 1998). It is commonly used as an excipient for direct compression tableting, filler and flowing agent in pharmaceutical preparations (Patel and Chen, 2008). Its spherical shape is advantageous for use in high-speed tableting machines used by pharmaceutical manufacturers.

Spray-dried lactose is manufactured from redissolved pharmaceutical grade lactose, which has been pre-crystallised by flash cooling and rapid mixing in specialised tanks to encourage a shower of very fine crystals (Jensen, 1987). The lactose slurry is atomised in the spray drier and the semi-dried particles are subsequently processed in vibratory fluidised bed driers to encourage secondary crystallisation into α-lactose monohydrate. An agglomerated powder can be produced by recycling the fines into the drying chamber (Patel and Chen, 2008). Agglomerated powders have improved flowability, reduced dustiness and instant properties.

Pre-crystallisation has a number of advantages in that it is possible to process concentrates up to 60% solids with improved process efficiency if the lactose has been subject to crystallisation. It is not possible to spray dry non-crystallised concentrate over 42–45% due to risk of blockages (Jensen, 1987). Pre-crystallisation into α-lactose monohydrate also creates a product that is less hygroscopic and more stable, with improved flowability and drug dispersion properties (Schuck and Dolivet, 2002).

Spray-dried agglomerated lactose typically contains 9–12% β-lactose, present in an amorphous matrix which is usually 15–20 wt.% of the total product (Bolhuis *et al.*, 2004). Controlling the degree of crystallisation in spray-dried lactose is the key factor for controlling many of its properties, including density and porosity (Trivedi and Axe, 2001), solubility (Blagden *et al.*, 2007; Pokharkar *et al.*, 2006), flowability (Fitzpatrick *et al.*, 2007) and sorption characteristics (Chiou and Langrish, 2008; Lehto *et al.*, 2006). The degree of pre-crystallisation, the inlet air temperature and the particle temperatures throughout the dryer have been established to be important in determining the final crystallinity of the product, storage stability and propensity to cake (Chiou and Langrish, 2008).

5.5.2 Anhydrous lactose

Anhydrous lactose, containing high levels of crystalline β-lactose, is used as a pharmaceutical excipient for wet granulation and direct compression, and for Direct Powder Inhalation (DPI) formulations. β-Lactose is more soluble than α-lactose monohydrate and has low water content, which is advantageous if the pharmaceutical active is sensitive to moisture (Vanderbist *et al.*, 1998). Roller-dried anhydrous lactose also yields more reproducible results as a carrier for DPI, due to its highly crystalline, anhydrous nature (Chow *et al.*, 2007).

Anhydrous lactose is manufactured by redissolving pharmaceutical grade lactose in purified water then roller-drying between two heated counter-rotating drums. The metal rollers are heated from inside with condensing steam (120–160°C) (Shilton, 2002). The product to be dried is spread as a thin film onto the surface of the hot drum and after one revolution is scraped off by knives (Vanderbist *et al.*, 1998). Since the lactose is dried at <93°C, it comprises more than 70% crystalline β-lactose (Lerk *et al.*, 1983). The flowability of the roller-dried lactose can be improved by sieving to remove fines (Kussendrager and Walsma, 2006). There are also patents indicating that improved flowability and compressibility can be achieved by an agglomeration process whereby the anhydrous lactose primary particles are treated by a wet granulation step and dried, to obtain anhydrous lactose agglomerates (Kussendrager and Walsma, 2006).

There are also patents describing the manufacture of anhydrous stable lactose by extrusion, whereby dry lactose is introduced into an extruder heated above 100°C, with 161–164°C exemplified in the patent (van Leverink, 1981). The resulting extruded lactose yields an anhydrous product high in β-lactose, which can be used for tableting and other pharmaceutical applications requiring anhydrous lactose.

5.5.3 Dry powder inhalers

The highest value pharmaceutical grade lactose with the most exacting physical specifications is Dry Powder Inhaler (DPI) lactose, a carrier for drugs used for the treatment of asthma and chronic obstructive pulmonary disease, administered locally to the lungs and airways by inhalation. The micronised drug (<5 μm) is evenly mixed with very fine micronised lactose particles (1–5 μm) and a coarser grade lactose (typically 30 to 100 μm). The fine lactose adheres to the 'drug active' to aid dispersion, while the coarse lactose plays a role in the aerosol, carrying the drug into the airways. During inhalation, the drug has to be 'stripped' from the lactose so that it can be inhaled through the airways into the lungs (Williams, 2007).

The behaviour of particles is the foundation on which dry powder inhaler performance is built. Powder engineering advances include micronisation and powder blending, controlled solvent crystallisation, spray drying, spray freeze drying, particle formation from liquid dispersion systems, supercritical fluid processing and particle coating (Chow *et al.*, 2007). Research has also been directed towards understanding how the specific properties of formulation components are directly linked to aerosolisation performance of DPI systems (Young *et al.*, 2007). Hickey *et al.* (2007) has also reviewed static and dynamic methods of characterising dry powder inhaler formulations with the intention of developing standard methods.

5.6 Lactose standards and specifications

5.6.1 Standards for lactose

The standard for edible grade lactose is defined by the Food Chemicals Codex (2004) as a white to creamy white, crystalline powder, which functions as a nutritive sweetener, processing aid, humectant and texturiser (see Table 5.3). The standards for pharmaceutical grade lactose are governed by the pharmacopoeia; European Pharmacopeoia (PhEur), Japanese Pharmacopoeia (JP) and US Pharmacopoeia–National Formulary (USP–NF). In recent years, these have been harmonised, with ten of the eleven chapters covering excipient monographs signed off as interchangeable in 2006 (US Pharmacopeia, 2005). The Pharmacopeial Discussion Group (PDG) continues to meet and discuss further harmonisation with the ultimate goal of globally workable standards.

Table 5.3 Lactose specifications for edible grade lactose according to Food Chemicals Codex[a] (2006) and pharmaceutical grade lactose according to US Pharmacopoeia[b]

Analysis	Edible[a]	Pharmaceutical[b]
α-Lactose monohydrate		
Lactose content (anhydrous basis)	98.0–100.5%	99.80%
Ash (sulphated, anhydrous basis)	<0.3%	<0.1%
Water content ('as is' basis)		
Loss on drying at 120°C for 16 hours	4.5–5.5%	
Loss on drying at 80°C for 2 hours		<0.5%
Karl Fischer		4.5–5.5%
pH	4.5–7.5	
Total plate count cfu/g		<100
Anhydrous lactose		
Lactose content (anhydrous basis)	98.0–100.5%	>99.8%
Ash (sulphated, anhydrous basis)	<0.3%	<0.1%
Water content ('as is' basis)		
Loss on drying at 120°C for 16 hours	<1.0%	
Loss on drying at 80°C for 2 hours		<0.1%
Karl Fischer		<1.0%

5.6.2 Hygiene and quality control

Lactose manufacturers are required to apply accredited quality management practises (HACCP; ISO 9000; ISO 22000) for the control and management of lactose manufacturing processes. The manufacture of lactose should be conducted in accordance with the Codex General Principles of Food Hygiene, and comply with the Principles for Microbiological Criteria for Food (Codex, 2001). Manufacturers of pharmaceutical excipients will also need to comply with upcoming legislation covering Good Manufacturing Practises, being implemented in the light of recent Public Health concerns (McGlue, 2008).

Microbiological specifications for pharmaceutical grade lactose require <100 cfu/g total plate count. The Food Chemicals Codex (2004) does not specify standards for edible grade lactose. Manufacturers quote a range of maximum microbiological specifications: <500 cfu/g for Hilmar refined edible (Hilmar, 2008); <1000 cfu/g for Lactose India edible (Lactose India, 2008); <2500 cfu/g for Hilmar natural lactose (Hilmar, 2008) and <10,000 cfu/g for Davisco edible grade lactose (Davisco, 2008). High microbial counts have implications in food applications, increasing the risk of mishandling. These numbers can be compared to the standards for a major application for lactose, powdered infant formula, at <1000 cfu/g (FSANZ, 2008).

There are a number of microbial hazards naturally associated with the lactose process. The raw material is a by-product from cheese or casein manufacture, high in nutrients and perishable, and should be treated with care. The evaporator is operated at conditions whereby thermophilic bacteria can multiply. The extended crystallisation times are within the temperature range for microbial growth of 60°C to 20°C. Crystalline α-lactose monohydrate contains 5% crystalline water which may become available for mould growth should amorphous lactose form during milling, or if the lactose is exposed to high humidities during drying, milling and storage.

Operational prerequisite programs or good manufacturing practises, include hygienic plant design, sanitiser foot baths, protective clothing, regular cleaning and sanitation programs, filtration of air into factory, and proper management and documentation of the process. Critical control points include: adequate pasteurisation of the whey prior to ultrafiltration; holding the evaporated concentrate at 75°C for at least 15 min prior to initiating cooling to pasteurise and kill microorganisms (this carries a risk of Maillard browning when filling extremely large tanks, as the entire tank is generally held at elevated temperatures prior to cooling); and the control of relative humidity of drying and milling conditions to control the risk of mould growth during subsequent storage.

5.7 Mother liquor

Mother liquor is the liquor remaining after lactose crystallisation and contains significant quantities of nitrogen (peptides, non-protein nitrogen),

Table 5.4 Dry weight composition of mother liquor and grandmother liquor (Friend *et al.*, 2004)

Component	Mother liquor g/kg mean ± SD n = 88		Grandmother liquor g/kg mean ± SD n = 11	
Total solids	276	±19	352	±29
Fat	3	±2	4	±4
Protein (Nx6.38)	23	±5	37	±6
Ash	183	±26	229	±36
Salt	81	±33	86	±17
Sodium	19	±12	27	±4
Potassium	25	±13	63	±10
Calcium	23	±13	20	±2
Chloride	50	±14	62	±19
Magnesium	4	±1	4	±1
Phosphorous	16	±8	24	±2
Sulphur	3	±3	4	±2
Lactose	670	±36	558	±31
Lactic acid	12	±13	24	±11
Citric acid	50	±9	63	±8

minerals, vitamins and lactose. Mother liquor is also known as delactosed permeate (DLP) or delac. Although the composition is highly variable from day-to-day and plant-to-plant, mother liquor will typically contain 28–30% solids, of which two thirds is lactose. This represents significant loss of lactose yield. The composition of mother liquor is shown in Table 5.4.

The disposal of mother liquor is a major problem; it has limited use as a stockfeed due to its perishable nature, whereby animals need to live close to the factory. Animal feeding also needs infrastructure; transport and feed lots. In New Zealand, a product based on mother liquor is used as a potassium-rich, liquid feed supplement with 42% solids, known as *Proliq* (Dexcel, 2007), while the same name is used in Australia for a foliar nutrient (FarmOz, 2000). Mother liquor can be also used as a fertiliser by metering it into irrigation; however, it has a very high BOD_5, and has been found to kill fish if leakage occurs directly into streams.

The best way to deal with mother liquor is to limit its production. There are several patents and processes describing the recovery of the mother liquor to increase the yield of lactose and minimise the mother liquor by-product. Friend *et al.* (2004), describe recrystallisation of mother liquor, increasing lactose yield and creating grandmother liquor, with the ash content as a proportion of lactose at 27% in mother liquor and 41% in grandmother liquor. Cloidt and Lehmann (2007) describe a process whereby a part of the mother liquor is returned to the lactose concentrate prior to the evaporation step, for re-purification and crystallisation to increase the yield of lactose. Harju and Heikkilae (1989) describe a process whereby the mother liquor is treated by ion exclusion, similar to cane and beet molasses desugarisation used for increasing sugar yields (Harju, 2007; Rousset, 1997; Chertudi, 1991; Saska and Lancrenon, 1994).

5.8 Applications and functionality of lactose

Lactose has number of properties that make it a unique energy supply for the growing infant, with low glycaemic index ensuring steady energy release, prebiotic functions for a healthy gut plus low cariogenicity, it is an ideal energy source for the suckling infant (Rutishauser, 2002). Lactose also has a range of functions that make it a unique sugar in a wide range of food products, low sweetness, reducing sugar properties and stable protective functions. The major applications of lactose include standardisation of milk for skim milk powder, infant formula, milk chocolate and other confectionery, baked goods and as a carrier base for dry food ingredient mixes. Below is a list summarising the functionality and applications of lactose:

Physiological – Glycaemic Index 46, ensuring gradual release of energy controlled by lactase activity, providing a gradual and prolonged energy supply and constant blood glucose levels.

Prebiotic function – encourages the growth of beneficial bacteria, reducing pathogenic bacteria proliferation, producing antimicrobials and enzymes that release bioactive peptides ensuring a healthy intestine.

Sweetness – 30–35% of the sweetening power of sucrose, it has synergistic effects with other food ingredients, including sweetness enhancement and sweetness suppression. Beneficial in masking the bitter aftertaste of saccharine or enhancing the milky flavour in caramels. Lactose can be used in a range of foods, including savoury mixes, confectionery, fruit fillings and bakery products, to accentuate natural flavours without being cloying.

Colour/flavour carrying properties – a strong affinity for absorbing substances such as flavour volatiles and pigments, for bulking and free flow.

Free flow agent – α-lactose monohydrate is non-hygroscopic and can be used in powdered foods as a free flow agent or anti-caking agent. The relatively low solubility of lactose allows lactose to be used in instantising, or increasing the dispersibility of dry food mixes.

Protein stabilising properties – protects the solubility of casein during spray drying, maintaining its functional properties, e.g. coffee whiteners.

Crystallisation control – alters the crystallisation behaviour of other sugars and can be used to control crystallisation in food formulations. In formulations containing both lactose and sucrose, all crystals remain smaller and the tendency of sucrose crystals to combine together is reduced. This reduces sandiness and yields a softer, smoother crystalline mass, with improved mouth-feel and chewiness.

Reducing properties – reducing sugar able to react with proteins, peptides and amino acids by the Maillard reaction to form compounds that are highly flavoured and golden brown. e.g. creating the brown crumb colour of baked products.
Emulsifying properties – aids in the emulsification of shortenings, improving the volume, grain and tenderness of bakery products and reduces the amounts of sugar and fat needed.
Selective fermentation – is not fermented by bakers yeast so it can be added to bread for brown crusts and production of caramelised flavours.
Chocolate and chocolate products – substitution for sucrose in milk chocolate gives chocolate a shorter, less sweet flavour; adsorbs volatiles enhancing the flavour; participates in the Maillard reaction during conching, contributing to typical chocolate flavours and formation of brown pigments; lowers tempered viscosity assisting the release of entrained air bubbles during moulding.
Processed means – can be incorporated into brines for processed meas such as hams, to increase the concentration of salt and phosphate in the aqueous phase, without sweetening.

Pharmaceutical grade lactose is further refined and converted into a range of products: milled lactose crystals, spray-dried lactose, anhydrous β-lactose and micronised lactose. It is mainly used in pharmaceuticals as a tableting excipient, but also as a dry powder inhaler. Pharmaceutical lactose and edible lactose are also used as raw material for lactose derivatives such as lactulose, lactobionic acid, lactitol and galacto-oligosaccharides, or hydrolysed to liquid lactose syrup.

5.9 Current and future trends

Global production of all types of lactose (edible and pharmaceutical) was 870 000 tonnes in 2006, up from 724 000 tonnes in 2002 (Affertsholt, 2007). The major lactose producers are located in Europe (46%), United States (37%) and New Zealand (13%) (Affertsholt, 2007). In recent years, lactose manufacturers have undergone amalgamations and takeovers resulting in a few major companies dominating the market, such as DMV Fonterra, Friesland Foods Domo, Leprino Foods, Hilmar Ingredients, Meggle, Lactalis and Arla Foods. Larger and more sophisticated companies are able to invest in infrastructure and support the manufacture of high value pharmaceutical lactose products and derivatives, thus meeting the requirements of the pharmaceutical market with security of quality and supply.

While the volumes of lactose production has continued to grow, there have been significant price fluctuations in 2006–2007, see Fig. 5.7. These price fluctuations, should be viewed in the context of the preceding 10 years of stable prices. The rapid price increases in 2006 have been attributed to drought and biofuels (competing for land); the growth of China and India;

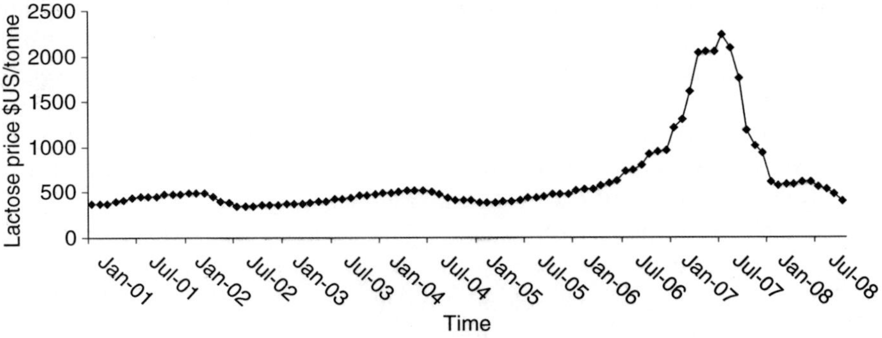

Fig. 5.7 Lactose prices $US/tonne from 2001–2008 (data from Gould, 2008).

and the growing demand for lactose for infant nutrition, pharmaceuticals and milk standardisation. However, the practises of futures market traders have also been implicated in the price shocks across all the commodity sectors (Baum, 2008). In the end, the price rises were not sustainable, with the resulting correction bringing the lactose price back to pre-2006 levels, as have prices for all dairy commodities in recent times.

These price challenges have reinforced the commodity nature of edible lactose and dried permeate. Manufacturers will need to respond to this by continuing to develop higher value lactose products, such as specialist high-grade pharmaceutical lactose products and lactose derivatives, which are less vulnerable to the vagaries of commodity prices. The market for these products has shown significant growth rates of up to 20% p.a. in recent years (Affertsholt, 2008). High-value fractions and derivatives can be marketed into the health and 'wellness' industry and for use in dietary supplements, sports nutrition, clinical nutrition and functional foods (Affertsholt, 2008), to become the main driving force for market development of lactose and lactose products into the future.

5.10 Sources of further information and advice

International dairy organisations

IDF-FIL International Dairy Federation. Website: http://www.fil-idf.org

IDF Symposium on Lactose and Lactose Derivatives, 14–16 May 2007, Moscow, Russia. Website: http://www.lactose.ru/

International Dairy Foods Association. Website: http://www.idfa.org/

National dairy organisations

EuroMilk – European Dairy Association / European Whey Products Association. Website: http://www.euromilk.org/

American Dairy Products Institute. Website: http://www.adpi.org/

U.S. Dairy Export Council (USDEC). A non-profit, independent membership organisation that represents the export trade interests of U.S. milk producers, proprietary processors, dairy cooperatives, and export traders. Website: http://www.usdec.org/

Dairy Australia. Research and development, information, national marketing and international trade development. Website: http://www.dairyaustralia.com.au

Pharmaceutical organisations

International Pharmaceutical Excipients Council (IPEC). Global association of manufacturers and users of pharmaceutical excipients. The association consists of three closely related but independent associations [IPEC Americas, IPEC Europe and IPEC Japan (JPEC)]. Website: www.ipec.org

United States Pharmacopeia (USP). An official public standards-setting authority for all prescription and over-the-counter medicines. USP sets standards for the quality, purity, strength, and consistency of these products – critical to public health. USP's standards are recognised and used in more than 130 countries around the globe. Acquired the rights for maintaining *The Food Chemicals Codex* in 2006. Website: www.usp.org

Major lactose manufacturers

Campina DMV International. Joint venture with Fonterra in 2007, Joint venture with Friesland Foods in Dec 2008. Address: NCB-laan 80, 5462 GE Veghel, The Netherlands. Websites: http://www.dmv-international.com http://www.dmv-fonterra-excipients.com/ http://www.campina.com/english/Products/Ingredients/Lactose.aspx

Friesland Foods-Domo. Joint venture with Warrnambool Cheese and Butter in 2008. Address: Hanzeplein 25, 8017 JD Zwolle, The Netherlands. Websites: http://www.domo.nl/ http://www.domo.nl/pharma/ http://www.lactose.com/

Hilmar Ingredients. Address: 9001 N. Lander Avenue Hilmar, California 95324 USA. Website: http://www.hilmarcheese.com/Lactose.cms

Molkerei Meggle. Address: Megglestraße 6–12, 83512 Wasserburg am Inn, Germany. Websites: www.meggle-pharma.com/ http://www.meggle-foodingredients.de/Commodities-Angebot.42.0.html

Sheffield Pharma Ingredients. Division of Kerry Bio-Science, a Kerry Group company. Strategic marketing agreement with Foremost Farms USA. Address: 158 State Highway 320, Norwich, NY 13815 USA. Website: http://www.sheffield-products.com/index.php?/content/view/23/56/

5.11 References

AFFERTSHOLT T (2007) Market developments and industry challenges for lactose and lactose derivatives, *IDF Lactose Symposium*, 14–16 May, Moscow.

AFFERTSHOLT T (2008) *The World Market for Whey and Lactose Ingredients 2007–2010*, 3A Business Consulting, 30 October 2008. NPI Centre, [Available at http://www.npicenter.com/anm/templates/newsATemp.aspx?articleid=22523 andzoneid=8 Accessed 29-12-2008].

ALKHATIM H S, ALCAINA M I, SORINO E, LORA J, ARNAL J (1998) Treatment of whey effluents by nanofiltration membranes, *Desalination*, 119, 177–184.

ANDERSEN G (2005) The Anhydro Paraflash process for production of whey and permeate powder, *Innovations in Food Technology*, 27 (20), 22–23.

ASSOCIATION OF VITAMIN CHEMISTS (1966) *Methods of Vitamin Assay*, N.Y., Wiley Interscience.

ATRA R, VATAI G, BEKASSY-MOLNAR E, BALINT A (2005) Investigation of ultra and nanofiltration for utilisation of whey protein and lactose, *Journal of Food Engineering*, 67, 325–332.

AYALA-BRIBIESCA E, ARAYA-FARIAS M, POURCELLY G, BAZINET L (2006) Effect of concentrate solution pH and mineral composition of a whey protein diluate solution on membrane fouling formation during conventional electrodialysis, *Journal of Membrane Science*, 280 (1, 2), 790–801.

BALSTER J (2006) *Membrane module and process development for monopolar and biopolar membrane electrodialysis*, PhD thesis, University of Twente, The Netherlands.

BATCHELDER B T (1987) Electrodialysis applications in whey processing, *Bulletin of the IDF*, 212 (4), 84–90.

BATICLE P, KIEFER C, LAKHCHAF N, LARBOT A, LECLERC O, PERSIN M, SARRAZIN J (1997) Salt filtration on gamma alumina nanofiltration membranes fired at two different temperatures, *Journal of Membrane Science*, 135, 1–8.

BAUM C (2008) What's going on with commodity prices? *Sydney Morning Herald. Business Day*. May 15. [Available at http://business.smh.com.au/business/whats-going-on-with-commodity-prices-20080515-2egu.html Accessed 11-11-2008].

BAZINET L (2004) Electrodialytic phenomena and their applications in the dairy industry: A review *Critical Reviews in Food Science & Nutrition*, 44 (7, 8), 525–544.

BERGHOFFER E, SCHEIBL A (1986) Technical fractionation of whey by ion exchange chromatography. I Literature survey. II Experiments. *Milchwissenscraft*, 41 (11), 700–703 and 41 (12), 774–779.

BHARGAVA A, JELEN P (1996) Lactose solubility and crystal growth as affected by mineral impurities, *Journal of Food Science*, 61, 180–184.

BLAGDEN N, DE MATAS M, GAVAN P T, YORK P (2007) Crystal engineering of active pharmaceutical ingredients to improve solubility and dissolution rates. *Advanced Drug Delivery Reviews*, 59 (7), 617–630.

BOERSEN A C, VELDMEIJER L (2003) *Method and Device for Producing Whey Powder*. PCT Patent WO 2003-NL00948; EP1581063 (A1); US2006088642 (A1); EP1581063 (A0); AU2003295282 (A1) NL1022291C (C2).

BOLHUIS G K, KUSSENDRAGER K, LANGRIDGE J (2004) New developments in spray dried lactose, *Pharmaceutical Technology*, June, 26–31.

BOUCHARD A, JOVANOVIC N, HOFLAND G W, JUSKOOT W, MENDES E, CROMMELIN D J A, WITKAMP G-J (2008) Supercritical fluid drying of carbohydrates: Selection of suitable excipients and process conditions, *European Journal of Pharmaceutics and Biopharmaceutics*, 68, 781–794.

BRINKMAN G E (1976) New ideas for the utilisation of lactose – principles for lactose manufacture, *J. Soc. Dairy Technology*, 29 (2), 101–107.

BRITO A B N, GIULIETTI M (2007) Study of lactose crystallisation in water–acetone solutions, *Crystallisation Research Technology*, 6, 583–588.

BROADHURST H A, REIN P W (2003) Modeling adsorption of cane-sugar solution colorant in packed-bed ion exchangers, *AIChE Journal*, 49, 2519–2532.

BROTHERSEN C F, OLSON N F, RICHARDSON T (1982) Recovery of calcium phosphate from ultrafiltration permeates, *Journal of Dairy Science*, 65 (1), 17–23.

BROWN R J (1988) Milk coagulation and protein denaturation. Ch. 11, In: Wong N P, Jenness R, Keeney M, Marth E H (eds). *Fundamentals of Dairy Chemistry.* 3rd Edition, Van Nostrand Reinhold. New York.

BRULE G, REAL DEL SOL E, FAUQUANT J, FIAUD C (1978) Mineral salts stability in aqueous phase of milk: Influence of heat treatments, *Journal of Dairy Science*, 61, 1225–1232.

BUND R K, PANDIT A B (2007a) Sonocrystallisation: Effect on lactose recovery and crystal habit, *Ultrasonics Sonochemistry*, 14, 143–152.

BUND R K, PANDIT A B (2007b) Rapid lactose recovery from buffalo whey by use of anti-solvent ethanol, *Journal of Food Engineering*, 82, 333–341.

BUTLER B (1998) *Modelling industrial lactose crystallisation.* PhD Thesis. University of Queensland, St. Lucia, Qld, Australia.

BYLUND G (2003) *Dairy Processing Handbook*, 2nd Edition, Tetra Pak Processing Systems, S-221 86, Lund, Sweden.

CHERTUDI K P (1991) The Tasco chromatographic separator at Twin Falls factory, *International Sugar Journal*, 93 (1106), 28–32.

CHILDRESS A E, ELIMELECH M (1996) Effect of solution chemistry on the surface charge of polymeric reverse osmosis and nanofiltration membranes, *J. Membrane Science*, 119, 253–268.

CHIOU D, LANGRISH T A G (2008) The effect of temperature on the crystallinity of lactose powders produced by spray drying, *Journal of Food Engineering*, 86 (2), 288–293.

CHOW A H L, TONG H H Y, CHATTOPADHYAY P, SHEKUNOV B Y (2007) Particle engineering for pulmonary drug delivery, *Pharmaceutical Research*, 24 (3), 411–437.

CLOIDT R, LEHMANN H (2007) *Process for the preparation of lactose from whey*, Patent application EP1869984.

COCO M, GARCÍA M T, MATO S, CARTÓN Á, GONZÁLEZ G (2008) Evolution of colorants in sugarbeet juices during decolorization using styrenic resins, *Journal of Food Engineering*, 89, 429–434.

CODEX (2001) *Codex standard for sugars*, Codex Standard 212-1999 [Amd 1-2001].

CROWLEY C (2006) *Preparation of pumpable, edible composition using electrodialysis.* United States Patent Application Publication, US 2006/0024413 A1. Kraft Foods Holdings Inc.

CUARTAS-URIBE B, VINCENT-VELA M C, ALVAREZ-BLANCO S, ALCAINA-MIRANDA M I, SORIANO-COSTA E (2007) Nanofiltration of sweet whey and prediction of lactose retention as a function of permeate flux using the Kedem–Spieglar Donnan Steric Partioning models, *Separation and Purification Technology*, 56, 38–46.

DAVISCO (2008) Refined Edible lactose, [Available at http://www.daviscofoods.com/Lactose/ Accessed 28-12-2008].

DEXCEL (2007) Feeding Mother Liquer (Proliq) to dairy cows, *Dexcel Farmfact 5–21* [Available at http://www.dexcel.co.nz/data/usr/5-21%20Feeding%20Mother%20Liquer%20(Proliq)%20to%20dairy%20cows.pdf Accessed 28-12-2008].

DINCER T D, PARKINSON G M, ROHL A L, OGDEN M I (1999) Crystallisation of α-lactose monohydrate from DMCO solution: Influence of β-lactose, *Journal of Crystal Growth*, 205, 368–374.

DOMBROWSKI R D, LITSTER J D, WAGNER N J, HE Y (2007) Crystallisation of alpha-lactose monohydrate in a drop based microfluidic crystallizer. *Chemical Engineering Science*, 62, 4802–4810.

DURHAM R J (2000) *Development of a process for the purification of lactose from whey.* PhD Thesis, UWS Hawkesbury, Richmond, NSW Australia.

DURHAM R J, HOURIGAN J A, SLEIGH R W (2007) New approach for high purity lactose – utilising the by-products from whey. *IDF Lactose Symposium*, 14–16 May, Moscow.

DURHAM R J, HOURIGAN J A, SLEIGH R W, JOHNSON R J (1998) *Process for the Purification of Nutrients from Food Process Streams*. PCT Patent WO99/04903; Australia 726559; New Zealand 502015; United States Patent 6475390.

DURHAM R J, SLEIGH R W, HOURIGAN J A (2004) Pharmaceutical lactose: A new whey with no waste, *Australian Journal Dairy Technology*, 59 (2), 138.

DURHAM R J, SLEIGH R W, HOURIGAN J A (2003) Nanofiltration for recovery of spent ion exchange brines, *'IMSTEC' 03, 5th International Membrane, Science and Technology Conference*, University of NSW, Kensington, NSW.

EHRENHUBER B (2005) Molasses desugarizing with betaine production, *Zuckerindustrie*, 130 (6), 463–469.

EVANS E W (1980) Whey research, *J. Society Dairy Technology*, 33 (3), 95.

EVANS J W, YOUNG G C, STAGER C W (1982) *Production of stable lactose product*. United States Patent 4342604.

FARMOZM (2000) *Proliq Foliar Nutrient* [Available at http://www.farmoz.com.au/label/farmoz/PROLIQ_16103556.pdf Accessed 29-12-2008].

FEENSTRA T P, DE BRUYN P L (1979) Formation of calcium phosphates in moderately supersaturated solutions, *J. Physical Chemistry*, 83 (4), 475–479.

FITZPATRICK J J, BARRY K, CERQUEIRA P S M, IQBAL T, O'NEILL J, ROOS Y H (2007) Effect of composition and storage conditions on the flowability of dairy powders, *International Dairy Journal*, 17 (4), 383–392.

FOOD CHEMICALS CODEX (2004) *Committee on Food Chemicals Codex, Food and Nutrition Board, Institute of Medicine (U.S)*, 5th edition, Published by National Academies Press.

FOREMOST FOODS (1970) *Lactose: Chemical applications. Lactose: Pharmaceutical applications*, Trade literature, *Foremost Foods*, 111 Pine St. San Francisco, CA94111, USA.

FRIEND M A, KAISER A G, PILTZ J W, SILLENCE M N, JOLLIFFE S K (2004) Use of delactosed whey permeate as a supplement for cattle on cereal straw based diet, *Australian Journal of Experimental Agriculture*, 44, 833–840.

FRIESLAND FOODS DOMO (2007) *Friesland Foods Domo to develop technology together with Novasep Process*, Nov 19 2007 [Available at http://www.frieslandfoods.com/en/domo/news/Pages/Default.aspx?guid=e3c3396c-3010-47d6-91ca-061466931065andiid=7 Accessed 29-12-2008].

FSANZ (2008) *Food standards code, Standard 1.6.1, Microbiological Limits For Food*, Food Standards Australia New Zealand, Commonwealth of Australia, Canberra, Published by Anstat Pty Ltd.

GANZLE M G, HAASE G, JELEN P (2008) Lactose: Crystallization, hydrolysis and value-added derivatives, *International Dairy Journal*, 18, 685–694.

GARCIA AGUDO J A, GARCIA CUBERO M T, BENITO G G, MIRANDA M P (2002) Removal of coloured compounds from sugar solutions by adsorption onto anionic resins: Equilibrium and kinetic study, *Separation and Purification Technology*, 29, 199–205.

GEA (2008a) *De-calcification of UF permeate, GEA Process Engineering Inc* [Available at http://www.geafiltration.com/filtration_library/de-calcification_UF_permeate.pdf Accessed 4-11-2008].

GEA (2008b) *Niro lactose production process, GEA Process Engineering Inc* [Available at http://www.niroinc.com/drying_dairy_food/Niro_lactose_production_process.asp Accessed 4-11-2008].

GESTERKAMP M (1994) Lactose: Sugar from milk, *Food Technology in New Zealand*, Feb, 16–20.

GONZALES O D-J G, ROCKLAGE B, BERNAUER H-P (2008) *Modified whey powder and process for producing the same*, Canadian Patent Application CA 2620529 A1 2008/08/07.

GOULD B W (2008) Weekly lactose price, University of Wisconsin dairy marketing and risk management program, *Agricultural and Applied Economics* UW Madison. [Available at http://future.aae.wisc.edu/data/monthly_values/by_area/23 Accessed 11-11-2008].

GULA F, BURRIS B, LANCRENON X (2007) The use of ion exchange resins for beet sugar juice processing, *International Sugar Journal*, 109 (1300), 235–241.

GULA F, PAILLAT D (2005) Decolorization of refinery liquors: A technical and economic comparison between the different systems using activated carbon or resins, *International Sugar Journal*, 107 (1276), 235–240.

GUU M Y K, ZALL R R (1992) Nanofiltration concentration effect on the efficacy of lactose crystallisation, *Journal of Food Science*, 57 (3), 735–739.

GUU M Y K, ZALL R R (1991) Lactose crystallisation: Effects of minerals and seeding, *Process Biochemistry*, 26, 167–172.

HAASE G, NICKERSON T A (1966) Kinetic reactions of alpha and beta lactose. II Crystallisation, *Journal of Dairy Science*, 49, 757–761.

HADDAD P R, JACKSON P E (1990) *Ion Chromatography, Principles and Applications*, Journal of Chromatography Library, Vol 46. Elsevier. Amsterdam.

HALBERT C, O'KENNEDY B T, HALLIHAN A, KELLY P M (2000) Stabilisation of calcium phosphate using denatured whey proteins, *Milchwissenchaft*, 55, 386–389.

HARJU M (2001) Milk sugars and minerals as ingredients, *International Journal of Dairy Technology*, 54 (2), 61–63.

HARJU M (2007) Chromatographic separation of lactose and its applications in the dairy industry, *IDF Lactose Symposium*, 14–16 May, Moscow.

HARJU M E, HEIKKILAE H O (1989) A process of recovering lactose from whey, European Patent Application EP 0 311 977 A2.

HARTEL R W (2001) *Crystallisation in Foods*, Aspen Publication, Aspen Publishers, Gaithersburg, MD.

HARTEL R W, SHASTRY A V (1991) Sugar crystallisation in food products, *Critical Reviews in Food Science and Nutrition*, 1 (1), 49–112.

HAYES J F (1982) Calcium and phosphorous removal from milk and whey permeates, *Australian Dairy Technology Review Conference* (Working Papers) Glenormiston Victoria, Australia.

HAYES J F, DUNKERLEY J A, MULLER L L (1974) Studies on whey processing by ultrafiltration. 2. Improving permeation rates by preventing fouling, *Australian Journal Dairy Technology*, 29, 132–140.

HENNINGFIELD T D, DINESEN R A (2003) *Process and plant for evaporative concentration and crystallisation of a viscous lactose containing aqueous liquid*, Patent WO/2003/086091.

HERRINGTON B L (1934) Some physico-chemical properties of lactose. VI. The solubility of lactose in salt solutions, the isolation of a compound of lactose and calcium chloride, *Journal Dairy Science*, 17, 805–814.

HERVE D (1974) Whey treatment by ion exchange resins, *Process Biochemistry*, Feb–Mar, 16–17.

HICKEY A J, MANSOUR H M, TELKO M J (2007) Physical characterization of component particles included in dry powder inhalers. I. Strategy review and static characteristics, *Journal of Pharmaceutical Sciences*, 96 (5), 1282–1301.

HICKEY M W, HILL R D, SMITH B R (1980) Investigations into the ultrafiltration and reverse osmosis of wheys. I. The effects of certain pretreatments. *NZ Journal Dairy Science and Technology*, 15, 109–121.

HIDDINK J, DE BOER R, NOOY P F C (1981) Effects of various pretreatments on the ultrafiltration of sweet cheese whey at about 55°C, *Milchwissenscraft*, 36 (11), 657–663.

HILMAR (2008) *Lactose specifications*, [Available at http://www.hilmarcheese.com/Lactose.cms Accessed 28-12-2008].

HOBMAN P G (1984) Review of the processes and products for utilisation of lactose in deproteinated milk serum, *Journal Dairy Science*, 67 (11), 2654–2679.

HOULDSWORTH D W (1980) Demineralisation of whey by means of ion exchange and electrodialysis, *Journal Society of Dairy Technology*, 33 (2), 45–51.

HUDSON C S (1904) The hydration of milk sugar in solution, *Journal of the American Chemical Society*, 30 (2), 1767–1783.

JAYAPRAKASHA H M, NISSEN B H, PATEL R S, SINGH S, RENNER E (1994) Effect of physico chemical properties of whey concentrates on lactose crystallisation, *Indian Journal of Dairy and Biosciences*, 5, 30–33.

JELEN P, COULTER S T (1973) Effects of certain salts and other whey substances on the growth of lactose crystals, *Journal of Food Science*, 38, 1186–1189.

JENSEN G K (1987) New drying technologies, *Bulletin of the IDF*, 212 (6), 27–37.

JONSSON H, ARPH S O (1987) Ion exchange for the demineralisation of cheese whey, *Bulletin of the IDF*, 212 (4), 91–98.

KARLESKIND D, LAYE I, MEI F-L, MORR C V (1995) Chemical pretreatment and microfiltration for making delipidised whey protein concentrate, *Journal Food Science*, 60 (2), 221–226.

KAUTER M D (2003) *The effects of impurities on lactose crystallization*, PhD Thesis, University of Queensland, St. Lucia, Qld, Australia.

KELLAM S (1998) *The manufacture of lactose*, The Lactose Company of New Zealand Ltd [Available at http://www.nzic.org.nz/ChemProcesses/dairy/3F.pdf Accessed 29-12-2008].

KELLER A K (2007) Practical considerations for profitable production of lactose and permeate powders, *IDF Lactose Symposium*, 14–16 May, Moscow.

KOCHERGIN V, TZSCHATZSCH O (2005) Ion exchange thin juice softening using fractal technology, *International Sugar Journal*, 107 (1278), 342–352.

KUSSENDRAGER K D, WALSMA B (2006) *Anhydrous lactose agglomerates and the preparation thereof*, PCT Patent Application WO/2006/068484.

LACTOSE INDIA (2008) Lactose Edible Grade [Available at http://lactoseindia.com/spec3.html Accessed 28-12-2008].

LANCRENON X (1997) Innovative separation technologies: The development of new applications in the Nth American sugar industry. Paper presented at the *29th Biennial Meeting American Society of Sugar Beet Technologies*, March, 1997.

LEHTO V-P, TENHO M, VAHA-HEIKKILA K, HARJUNEN P, PAALLYSAHO M, VALISAARI J, NIEMELA P, JARVINEN K (2006) The comparison of seven different methods to quantify the amorphous content of spray-dried lactose. *Powder Technology*, 167 (2), 85–93.

LERK C F (1983) Physico-pharmaceutical properties of lactose. In: *Proceedings of the 22nd International Colloquium on Industrial Pharmacy*, Ghent, pp. 59–89.

LIANG B, SHI Y, HARTEL R W (1991) Growth-rate dispersion effects on lactose crystal size distributions from a continuous cooling crystallizer, *Journal of Food Science*, 56, 848–854.

LIFRAN E (2007) *Influence of lactose phosphate on the crystallisation of lactose*, PhD Thesis, Centre for Plant and Food Science, University of Western Sydney, Richmond, Australia.

LIFRAN E V, VU T T L, DURHAM R J, HOURIGAN J A, SLEIGH R W (2005) Crystallisation Kinetics of Lactose in the presence of Lactose Phosphate. Paper No. 71. *Chemeca 2005. 33rd Australasian Chemical Engineering Conference*. 25–28 September 2005, Brisbane, Australia.

LIFRAN E V, VU T T L, DURHAM R J, HOURIGAN J A, SLEIGH R W (2006a) Crystallisation kinetics of lactose in the presence of lactose phosphate, *Powder Technology*, 179 (1–2), 43–54.

LIFRAN E V, VU T T L, DURHAM R J, HOURIGAN J A, SLEIGH R W (2006b) Lactose phosphate and its origin in cheese-making: A case study, *Aust J Dairy Tech.*, 61 (2), 163–166.

LIFRAN E, SLEIGH R W, JOHNSON R L, STEELE R J, HOURIGAN J A, DALZIEL S M (2002) *Method for purification of lactose*. PCT Patent WO 2002/50089 A1. Australia 2002218862, CA 2442181, EP 2001271379, JP 2002551982, NZ 526975, US 10451347.

LIN TENG SHEE F, ANGERS P, BAZINET L (2007) Delipidation of a whey protein concentrate by electroacidification with bipolar membranes, *Journal of Agricultural and Food Chemistry*, 55 (10), 3985–3989.

LISTIOHADI Y D, HOURIGAN J A, SLEIGH R W, STEELE R J (2005c) Effect of milling on the caking behaviour of lactose, *Australian Journal Dairy Technology*, 60 (3), 214–224.

LISTIOHADI Y D, HOURIGAN J A, SLEIGH R W, STEELE R J (2005b) Properties of lactose and its caking behaviour, *Australian Journal Dairy Technology*, 60 (1), 33–52.

LISTIOHADI Y D, HOURIGAN J A, SLEIGH R W, STEELE R J (2005a) Role of amorphous lactose in the caking of alpha-lactose monohydrate powders, *Australian Journal Dairy Technology*, 60 (1), 19–32.

MANS J (1994) Dietrich's Milk Products Inc, Sweet success, *Dairy Foods*, April, 42–46.

MATTHEWS M E, DOUGHTY R K, SHORT J L (1978) Pretreatment of acid casein wheys to improve processing rates in ultrafiltration. *NZ J. Dairy Sci. Technol.* 13, 216–220.

MCGLUE K (2008) European Excipient Regulation: Implications for manufacturers and suppliers, *Innovations in Pharmaceutical Technology*, 74–76. [Available at http://www.ipec-europe.org/docs/pdf/news/IPT.pdf Accessed 29-12-2008].

MICHAELS A S, VAN KREVELD A (1966) Influences of additives on growth rates in lactose crystals, *Netherlands Milk and Dairy Journal*, 20, 163–181.

MIKKONEN H, HELAKORPI P, MYLLYKOSKI L, KEISKI R L (2001) Effect of nanofiltration on lactose crystallisation, *Milchwissenschaft*, 56 (6), 307–310.

MIMOUNI A, SCHUCK P, BOUHALLAB S (2009) Isothermal batch crystallization of alpha-lactose: A kinetic model combining mutarotation, nucleation and growth steps, *International Dairy Journal*, 19, 129–136.

MONTILLA A, CASTILLO M D, DEL SANZ M L, OLANO A (2005) Egg shell as catalyst of lactose isomerisation to lactulose, *Food Chemistry*, 90 (4), 883–890.

MULLIN J W (2001) *Crystallisation*, 4th Edition, Butterworth-Heinemann, Oxford.

NEVILLE J (2006) Developments in whey protein and lactose permeate production processes and their relationship to specific product attributes, *International Journal of Dairy Technology*, 59 (2), 67–69.

NICKERSON T A, MOORE E E (1974) Factors influencing lactose crystallisation, *Journal of Dairy Science*, 57, 1315–1319.

NOEL R (2002) *Method of processing whey for demineralization purposes*. United States Patent 6383540.

NOVASEP (2005) *Novasep Process, Applexion and Australian Join Forces in Whey Fractionation*, Pharmaceutical Technology.com [Available at http://www.pharmaceutical-technology.com/contractors/purification/novasep/press3.html Accessed 28-12-2008].

OLANO A, BERNHARD R A, NICKERSON T A (1977) Alteration of in the ratio of α and β lactose co-crystallised from organic solvents, *J. Food. Sci.*, 42 (4), 1066–1072.

OMRI (2002) Activated carbon processing, National Organic Standards Board Technical Advisory Panel Review, *Organic Materials Review Institute for the USDA National Organic Program* [Available at http://www.omri.org/AC_processing.pdf Accessed 29-12-2008].

OSTERGAARD B (1988) Lactose from UF permeate, *North European Food and Dairy Journal*, 5, 145–156.

OTSUKA M, OHTNAI H, KANENIWA N, HIGUCHI S (1991) Isomerisation of lactose in solid state by mechanical stress during grinding, *J Pharm Pharmacol*, 43, 148–153.

PATEL K C, CHEN X D (2008) Drying of aqueous lactose solutions in a single stream dryer, *Food and Bioproducts Processing*, 86, 185–197.

PATERSON A H J, BRONLUND J E (2009) The practical implications of temperature induced moisture migration in bulk lactose, *Journal of Food Engineering*, 91, 85–90.

PEDERSON H T (1980) *Treatment of whey*, US Patent 4202909.

PISECKY J (2005) Spray drying in the cheese industry, *International Dairy Journal*, 15, 531–536.

POKHARKAR V B, MANDPE L P, PADAMWAR M N, AMBIKE A A, MAHADIK K R, PARADKAR A (2006) Development, characterization and stabilization of amorphous form of a low T_g drug. *Powder Technology*, 167 (1), 20–25.

POULIN J F, AMIOT J, BAZINET L (2008) Impact of feed solution flow rate on peptide fractionation by electrodialysis with ultrafiltration membrane, *Journal of Agricultural and Food Chemistry*, 56 (6), 2007–2011.

PRAGONO R, VU L T T, HOURIGAN J A, DURHAM R J, SLEIGH R W (2008) Sorption kinetics of riboflavin from diluted lactose solution in batch systems, *Proceedings of Chemeca 2008*, 28 Sept, Newcastle, Australia.

RAGHAVAN S L, RISTIC R I, SHEEN D B, SHERWOOD J N (2001) The bulk crystallisation of α-lactose monohydrate from aqueous solution, *Journal of Pharmaceutical Soc.*, 90 (7), 823–832.

RAGHAVAN S L, RISTIC R I, SHEEN D B, SHERWOOD J N, TROWBRIDGE L, YORK P (2000) Morphology of crystals of alpha-lactose hydrate grown from aqueous solution, *Journal of Physical Chemistry B*, 104, 12256–12262.

RAMACHANDRA RAO H G, LEWIS M J, GRANDISON A S (1994) Effect of soluble calcium of milk on fouling of ultrafiltration membranes, *J. Sci. Food Agric.*, 65, 249–256.

RANDOLPH A D, LARSON M A (1988) *Theory of Particulate Processes, Analysis and Techniques of Continuous Crystallisation*, 2nd Edition, Academic Press, San Diego.

REIN P W, BENTO L S M, CORTES R (2007) The direct production of white sugar in a cane sugar mill, *International Sugar Journal*, 109 (1301), 286–299.

RICE G, KENTISH S, O'CONNOR A, STEVENS G, LAWRENCE N, BARBER A (2006) Fouling behaviour during the nanofiltration of dairy ultrafiltration permeate, *Desalination*, 199, 239–241.

ROSENTHAL I (1991) *Milk and Dairy Products*, VCH Pub. Inc., New York.

ROUSSET F (1997) New developments in chromatographic desugarisation of beet molasses. Applexion. *British Sugar EuroTechLink 97.*, York. England.

RUTISHAUSER I H E (2002) Infant Nutrition, In: *Food and Nutrition*, Wahlqvist, M.L. (Ed), Allen and Unwin, Australia.

SASKA M, LANCRENON X (1994) Applications of continuous chromatographic separation in the sugar industry. Part III. Desugarisation of cane molasses. *International Sugar Journal*, 96 (1150), 403–410.

SCHMIDT D G, BOTH P (1987) Studies on the precipitation of calcium phosphate. I. Experiments in the pH range 5.3 to 6.8 at 25°C and 50°C in the absence of additives, *Neth. Milk Dairy J.* 41, 105–120.

SCHUCK P, DOLIVET A (2002) Lactose crystallisation: Determination of α-lactose monohydrate in spray dried dairy products, *Lait*, 82 (4), 413–421.

SERPEN A, ATAC B, GOKMEN V (2007) Adsorption of Maillard reaction products from aqueous solutions and sugar syrups using adsorbent resin, *Journal of Food Engineering*, 82, 342–350.

SHI Y, LIANG B, HARTEL R W (1990) Crystallization kinetics of alpha-lactose monohydrate in a continuous cooling crystallizer, *Journal of Food Science*, 55, 817–820.

SHI Y, LIANG B, HARTEL R W (2006) *Crystal refining technologies by controlled crystallization*, United States Patent Application 200610128953 A1.

SHILTON N (2002) *Food Process Engineering – Drying, University College Dublin*, [Available at http://www.ucd.ie/food/html/homepage/newdrying.doc Accessed 28-12-2008].

SINGH R K, NIELSEN S, CHAMBERS J V (1991) Lactose crystallization from whey permeate in an ethanol–water mixture, *Journal of Food Science*, 56 (4), 935–937.

SINGHAL S (2005) Brine recovery at a sugar refinery, *International Sugar Journal*, 107 (1280), 462–466.

SLEIGH R W, ARAL H, SIMONS L, SANTANGELO R, STEELE R J, PALMOWSKI L (2006) Sustainable management of salty waste streams from cheese manufacture, *Aust J Dairy Tech.*, 61 (2), 193–195.

SMART J B (1988) Effect of whey components on the rate of crystallisation and solubility of alpha-lactose monohydrate, *New Zealand Journal of Dairy Science and Technology*, 23 (4), 275–289.

SMART J B, SMITH J M (1991) Effect of selected compounds on the rate of alpha-lactose monohydrate crystallisation, crystal yield and quality, *International Dairy Journal*, 1, 41–53.

SUAREZ E, LOBO A, ALVAREZ S, RIERA F, ALVAREZ R (2006) Partial demineralisation of whey and milk ultrafiltration permeate by nanofiltration at pilot plant scale, *Desalination*, 198, 274–281.

SUHR M R, SCHULZE B C (2003) Technology in use in the modern beet sugar factory, *International Sugar Journal*, 105 (1250), 52.

TAY M R, LIU J, SUN D D (2002) Effect of solution physico-chemistry on the charge property of nanofiltration membranes, *Water Research*, 36, 585–598.

TEMA (2007) *Siebtechnik centrifuges for chemicals and minerals*. TEMA BV, The Netherlands. [Available at http://www.tema.info/pdf/centrcal.pdf Accessed 29-12-2008].

TERLET (2008) *Tanks and Vessels – Dairy. Crystallisation tank for the production of lactose*, Terlet International The Netherlands, [Available at http://mpe.frontis.nl/fileadmin/images_mpe/terlet/downloads/pdf_tank/UK/T_V_Dairy.pdf Accessed 29-12-2008].

TESSIER H, ROSE D (1958) Calcium ion concentration in milk, *Journal of Dairy Science*, 41, 351–359.

THEOBALD J (2007) Purification processes for lactose, *IDF Lactose Symposium*, 14–16 May, Moscow.

TRIVEDI P, AXE L (2001) Ni and Zn sorption to amorphous versus crystalline iron oxides: Macroscopic studies. *Journal of Colloid and Interface Science*, 244 (2), 221–229.

U S PHARMACOEPEIA (1995) *Pharmaceutical lactose standards*, USP 23, NF 18, USP Convention, Inc. Washington.

U S PHARMACOPEIA (2005) *Pharmacopeial Discussion Group (PDG) convenes to discuss future harmonisation and approve new monographs and general chapters, USP Pharmacopeia Media Advisory*, 19 Dec [Available at http://vocuspr.vocus.com/vocuspr30/Newsroom/ViewAttachment.aspx?SiteName=uspharmandEntity=PRAssetandAttachmentType=FandEntityID=99425andAttachmentID=6465bac6-4d2e-4818-8d9f-1fe5cffbd2de Accessed 5-10-2008].

VAN DEN BOS M J (1987) Backgrounds of the technologies used for the production of lactose, *Bulletin of the IDF*, 212 (4), 99–102.

VANDERBIST F, MAES P, BAUDIER P (1998) *Dry powder inhaler excipient, process for its preparation and pharmaceutical compositions containing it*, PCT Patent Application WO/1998/050015.

VAN DER HORST H C, TIMMER J M K, ROBERTSON T, LEENDERS J (1995) Use of nanofiltration for concentration and demineralisation of cheese whey: A mass transport model, *Journal of Membrane Science*, 104 (3), 205–218.

VAN KREVELD A, MICHAELS A S (1965) Measurements of crystal growth of α-lactose, *Journal of Dairy Science*, 48, 259–264.

VAN LEVERINK J (1981) *Extrusion process for the preparation of anhydrous stable lactose*, US Patent 4280997.

VEMBU R, RATHINAM V (1997) *Separation of minerals from whey permeate*, US Patent 5639501.

VISSER R A (1980) A natural crystal growth retarder in lactose, *Netherlands Milk and Dairy Journal*, 34, 255–275.

VISSER R A (1982) Supersaturation of α-lactose in aqueous solutions in mutarotation equilibrium, *Netherlands Milk and Dairy Journal*, 36, 89–101.

VISSER R A (1984) Experiments for tracing growth retarders in lactose, *Netherlands Milk and Dairy Journal*, 38, 107–133.

VISSER R A (1988) Crystal growth retarding of alpha-lactose hydrate by sugar phosphates, a continued study, *Netherlands Milk and Dairy Journal*, 42, 449–468.

VU L T T, DURHAM R J, HOURIGAN J A, SLEIGH R W (2003) Fine seed preparation for crystallisation processes. *Chemeca 2003, 31st Australasian Chemical Engineering Conference.* Adelaide, South Australia.

VU L T T, DURHAM R J, HOURIGAN J A, SLEIGH R W (2004a) Dynamic modelling and optimisation of α-lactose monohydrate seeded batch cooling crystallisation. *Proceedings of the 14th European Symposium on Computer Aided Process Engineering (ESCAPE).* Paper NO 82, Vol. 15. May 16–19, 2004; Lisbon, Portugal.

VU L T T, DURHAM R J, HOURIGAN J A, SLEIGH R W (2004b) The uses of ultrasonic power and solvents in lactose cooling crystallisation (Paper 86), *Chemeca 2004, 32nd Australasian Chemical Engineering Conference.* Sept 26–29, Sydney, Australia.

VU L T T, DURHAM R J, HOURIGAN J A, SLEIGH R W (2006) Dynamic modelling optimisation and control of lactose crystallisations: Comparison of process alternatives, *Separation and Purification Technology*, 48, 159–166.

WALSTRA P, JENNESS R (1984) *Dairy Chemistry and Physics.* John Wiley and Sons. New York.

WEISBERG S M (1954) Recent progress in the manufacture and use of lactose: A review, *Journal of Dairy Science*, 37 (9) 1106–1115.

WILLIAMS A (2007) Modern challenges for lactose and derivatives, *IDF Lactose Symposium*, 14–16 May, Moscow.

YOUNG P M, TRAINI D, COATES M (2007) Recent advances in understanding the influence of composite-formulation properties on the performance of dry powder inhalers, *Physica B-Condensed Matter*, 394 (2), 315–319.

ZEEVALKINK H, KEIZER M, VAN OOSTEN H (2005) *Method and device for separating solid from milk or a milk product*, PCT Patent Application WO/2005/110070.

Part II

Biological functionality of dairy components and nutraceutical applications

6

Studies of the biological function and structure of casein micelles, and future implications

P. X. QI, Agricultural Research Service, USDA, USA

Abstract: At the heart of the milk system are the colloidal casein–calcium–transport complexes termed the casein micelles. Despite extensive research in more than half a century, accomplished by the application of a wide range of physical techniques such as light, neutron, and X-ray scattering, and Electron Microscopy (EM), the molecular details of casein micelles and the main contributing forces that stabilize them in milk remain controversial and still sustain interest and effort in the dairy science research community. Among various proposed molecular models, two main conflicting theories about the internal structure of casein micelles have emerged. One places the emphasis on the four main protein constituents, α_{s1}-, α_{s2}-, β- and κ-casein in the micelles, while the other proposes that inorganic calcium phosphate nano-clusters are the dominant players in holding the micelles together. In this chapter, casein micelles are examined in the light of recent advances in understanding protein–protein interactions (associations) and protein structure–function relationships. The biological significance of casein micelles, in relation to their unique construct, allows for an efficient transit through the mammary secretory apparatus and this is critically assessed, in addition to the existing overwhelming amount of evidence supporting the argument that proteinaceous complexes act as the formative agents in the synthesis of casein micelles in mammary tissue; in other words, that protein–protein interactions are paramount in the formation and stabilization of casein micelles.

Key words: casein micelles, structure, protein–protein interaction, biological function, TEM.

6.1 Introduction

The appearance of milk is that of a creamy white fluid. The lubricity and taste of milk are based upon three unique biological structures: the colloidal calcium–protein complexes, often termed as casein micelles, the milk fat globules with their limiting membrane, and the milk sugar, lactose. The complexity of these structures is necessitated by the fact that milk, in most species, is in essence predominantly water. It is the accommodation of these ingredients to an aqueous environment that forms the basis for the structure of milk at the molecular level and calls for the unique secretory process: milk synthesis (Patton, 2004).

It is generally believed (Farrell, 1999; Farrell *et al.*, 2003a) that the casein micelles are responsible for transporting and delivering the otherwise insoluble inorganic calcium and phosphate, indispensable nutrients for the neonates. Although these colloids have been the subject of extensive research for more than half a century (Noble and Waugh, 1965; Waugh *et al.*, 1970; Payens, 1979; Schmidt, 1982; Griffin and Roberts, 1985; Farrell and Thompson, 1988; Holt, 1992; Hansen *et al.*, 1996; Walstra, 1999; Horne, 2002; Tuinier and de Kruif, 2002; Qi, 2007; McMahon and Oommen, 2008; Fox and Brodkorb, 2008), the structural details of the casein micelles on the molecular level remain elusive and still sustain interest and effort (Farrell *et al.*, 2006a; Qi, 2007; Fox and Brodkorb, 2008; McMahon and Oommen, 2008). Past biochemical and physical studies of these colloids have focused on the size and properties of the colloids, their protein and mineral composition, and the central building block of the micelles. Conflicting views on the structure of the casein micelles have arisen from various interpretations of the core data published so far. In the light of recent advancements in the field of structural biology and systems biology, this manuscript will focus on the implications of protein–protein interactions in an attempt to discern the biologically competent route for the formation, and possibly stabilization, of the casein micelles.

The central theme of protein primary sequence to structure to function that dominated the field of molecular biology and structural biology for the past few decades is currently being challenged and debated by many (Meier *et al.*, 2008; Szilagyi *et al.*, 2008; Wright and Dyson, 1999). It has recently been recognized that the family of natively unfolded (Weinreb *et al.*, 1996) or intrinsically unstructured (Wright and Dyson, 1999; Dyson and Wright, 2002, 2005; Tompa, 2002) or intrinsically disordered proteins (Dunker *et al.*, 2001; Uversky, 2002; Bracken *et al.*, 2004; Radivojac *et al.*, 2007) is rapidly expanding. They are often found in eukaryotes (Dunker *et al.*, 2008) and are believed to be involved in multiple-partner binding sites, post-translational modifications, and alternative splicing that may lead to protein aggregation and amyloidogenesis (Uversky, 2008). These proteins, or sometimes long stretches of sequence in one protein, have been categorized as 'non-folding' and contain different amino acid sequences from those that fold into globular 3D structures. Caseins have been implicated as a prime example of this class of proteins (Farrell *et al.*, 2006b,c).

6.2 Brief review of proposed models for casein micelles

6.2.1 Protein composition of skim milk

The dominant protein constituents in skim milk are present in an aggregate form: the casein micelle (Fig. 6.1). This unique supramolecular aggregate

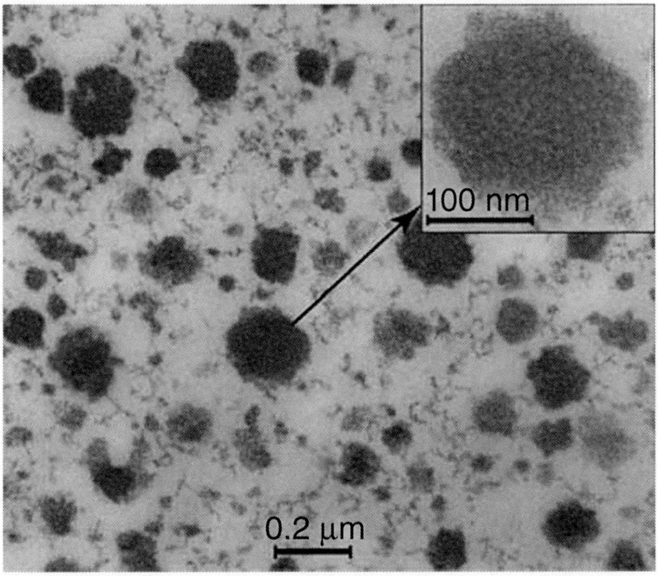

Fig. 6.1 Transmission Electron Micrograph of bovine milk casein micelles fixed with glutaraldehye and stained with uranyl acetate and lead citrate (pH 7.0). Inset at the upper right shows an enlarged single micelle.

imparts the opalescence characteristic of skim milk. The chief function of the micelle is to fluidize the casein molecules and solubilize the calcium and phosphate (Farrell and Thompson, 1988). In general, when milks that contain >2% protein are analyzed, the accompanying inorganic phosphate and calcium levels found would, by themselves, yield insoluble precipitates (apatite or brushite, depending upon the pH). Conversely, in the absence of these salts, the casein components, as a result of their open structures, would have a high viscosity. The formation of the casein colloidal complexes, micelles, averts these two problems.

Four major casein components in cows' milk, α_{s1}-, α_{s2}-, β- and κ-casein, have been well characterized in the past (Swaisgood, 2003). Caseins have been found to be homologous to these proteins in their gene and primary protein sequences in all species examined to date (Farrell *et al.*, 2004). However, the proportions of the various caseins vary widely. For example, β-casein is the primary casein in human milk, and in goats' milk it comprises 40% to 50% of the casein. In goats' milk there is also a high degree of variance in casein proportions in different animals, which appears to be genetically controlled (Table 6.1). Despite the variations in casein components, all species competently form colloidal casein micelles for the transport of calcium and phosphate. At the ultra-structural level, particularly in terms of the average size, the casein micelles of most species such as goat

Table 6.1 Casein distribution (%) in various milks

Milk	α_{s1}-	α_{s2}-	β-	κ-
Goat	5 to 17	6 to 20	50	15
Cow	38	10	40	12
Human	Trace	Trace	70	27

(Pierre *et al.*, 1998) and mouse (Burgoyne and Duncan, 1998) appear to be quite similar.

The α_{s1}-, α_{s2}-, and β-caseins are precipitated by calcium (calcium-sensitive) binding to their phosphoserine residues at the concentrations of protein and calcium found in most milks (Fox and Brodkorb, 2008). However, κ-casein is not only soluble in calcium (calcium-insensitive), but also interacts with and stabilizes the calcium insoluble caseins to initiate formation of the stable colloidal state. It has been generally recognized by studies (Hill and Wake, 1969; Dalgleish, 1980; de Kruif and Zhulina, 1996; Carroll and Farrell, 1983). Using numerous chemical, enzymatic and immunological techniques, that while the majority of the κ-casein resides on the surface of the casein micelles, other caseins might occur there as well (Dalgleish, 1998; Horne, 2006). In all models for casein structure, κ-casein is thought to predominate on the micellar surface (Hill and Wake, 1969). In milk clotting in the neonate stomach, the enzyme chymosin (rennin) specifically cleaves one bond in κ-casein to initiate aggregation of the micelles. It has been clearly demonstrated recently (Shekar *et al.*, 2006) by κ-casein gene null mutation experiments in mice, that κ-casein is essential for the assembly of the casein micelles and for lactation to occur *in vivo*.

6.2.2 The submicelle theory of casein structure

For many years, the most accepted theory (Farrell, 1988; Schmidt, 1982) of the structure of the casein micelle was that it was composed of spherical aggregates of the caseins (submicelles) held together by calcium–phosphate linkages. The submicelle hypothesis has been historically supported by biochemical and biophysical studies (Kumosinski *et al.*, 1987; Stothart, 1989; Kakalis *et al.*, 1990; Jang and Swaisgood, 1990) on the individual casein components and reconstitution of micelles from their component caseins (Griffin *et al.*, 1988; Slattery, 1979), as well as electron microscopy of the micelles themselves (Carroll *et al.*, 1968) and partially disrupted micelles. Early studies on the calcium-depleted caseins demonstrated that in the absence of calcium they formed rather large aggregates, traditionally termed as 'submicelles' (Walstra, 1990; Stothart, 1989; Pepper and Farrell, 1982; Jang and Swaisgood, 1990), and that these aggregates formed colloidal complexes in the presence of added calcium. Gradual removal of calcium by dialysis or EDTA treatment showed the emergence of rather uniform submicellar structures, as visualized by physical techniques and freeze fracture electron microscopy (Lin *et al.*, 1972; Knoop *et al.*, 1979). These aggregated particles had physical properties similar to the aggregates found in whole casein preparations in the absence of calcium and were considered as submicellar in nature. Based on these data, Schmidt proposed the 'submicelle' model. In summary, this model begins with 'specific and productive' casein–casein interactions, traps calcium phosphate and forms the colloidal complexes.

This theory has been later challenged by concepts (Holt, 1992; Horne, 1998) arising from the study of the casein–calcium–phosphate interactions, the micelles themselves, and physical chemical studies of the structure–function relationships and calculations from polymer condensation theory.

6.2.3 Casein micelle models with an internal gel matrix

Two more recently proposed models for the casein micelle have emerged that refute the notion of discrete submicellar structures within the micelle. The first experiment to indicate departure from the submicelle theory was that of Rose and Colvin (1966) who, from X-ray diffraction, postulated that the electron-dense particles were granules of colloidal calcium phosphate instead of proteins. During a series of studies on casein–calcium–phosphate interactions, Holt and coworkers (Holt *et al.*, 1998; Holt, 1998) discovered that the phosphopeptide fraction of β-casein could bind to and stabilize calcium–phosphate aggregates, resulting in the formation of nanoclusters of a discrete size and composition: without the peptides, the calcium phosphate structures would grow randomly and precipitate. This discovery led De Kruif and Holt to propose that such nanoclusters are the centerpiece of the casein micelle structure (de Kruif *et al.*, 2002). The formation of nanoclusters with a radius of 2.3 nm would drive micelle formation by randomly binding phosphoproteins, causing an inverted micelle, and then more proteins could coat this new hydrophobic surface and, in turn, bind more calcium phosphate until a size limited colloid is formed. There are about 800 of these amorphous calcium phosphate nanoclusters in an average sized casein micelle (~100 nm in diameter). This nanocluster model is supported by the rheomorphic theory of casein structure (Holt and Sawyer, 1993). In this view, the unstructured proteins form about the amorphous inorganic species and their function of binding to the calcium phosphate gives rise to their structure; hence no specific protein secondary structures or protein–protein interactions are invoked, except that a surface position for κ-casein is required. However, it is not yet clear what is the cell assembly mechanism that causes this to happen.

The casein micelle model proposed by Horne (1998, 2002, 2006) considers the surface chemistry of the individual caseins and their assembly behavior, and concludes that protein–protein interactions are indeed important, but in essence the model retains the rheomorphic concept. In this view, the amphiphilic nature of the caseins causes them to act more as block copolymers of alternating charge and hydrophobicity, that is, a charged phosphopeptide loop on the N-terminal and a hydrophobic train toward the C-terminal for β-casein, the reverse being true for κ-caseins, in which an N-terminal hydrophobic train is followed by a charged loop. In the case of α_{s1}-casein, it ends with a final C-terminal hydrophobic train. In this model, the hydrophobic interactions among various caseins are mostly considered, and the growth of the calcium phosphate nanoclusters begins the process of micelle formation, but it is limited by binding to the phosphopeptide loop regions of the caseins. Once bound to the amorphous inorganic matrix, further protein–protein interactions are related to the hydrophobic blocks, and polymerization proceeds by repeating the entire process. Micelle formation leads to an internal gel-like structure with embedded nanoclusters of calcium and phosphate, and the reaction of κ-casein, which contains only one phosphoserine residue, limits micellar growth by acting as a dead-end capping unit, in analogy with the growth of synthetic polymers.

Interestingly, these two distinctly different views of the internal structure of the casein micelles–submicelles versus gel matrices arise essentially from the same biochemical and physical chemical databases. For a further exposition on their similarities and differences, see the excellent review of Pieter Walstra (1999). A recent review by McManhon and Oommen (2008) suggested an interlocked lattice model (to reconcile the combined action between casein-calcium phosphate aggregates and casein polymer chains as the main building block) that maintains the integrity and stability of the casein micelles.

6.3 Synthesis and secretion of caseins

6.3.1 Cell physiology

The evolution of the mammary gland, presumably from external sweat glands, has yielded a great variety of exterior appearances in many species (Patton, 2004), but at the tissue level there is a common organizational theme as shown in Fig. 6.2a. Mammary secretory cells are epithelial in nature and are arranged in alveoli which are connected to ductal tissue. The secretory epithelial cells (SEC) are surrounded by a layer of myoepithelial cells, which are able to contract and expel milk into the ducts in response to the hormone oxytocin. The alveoli are highly vascularized to ensure a constant flow of the metabolic precursors needed for milk synthesis and secretion. Finally the vascularized alveoli are embedded in an extracellular matrix. This matrix not only supports the cells, but also, through cell–cell interactions, is responsible for the full expression of the genes that control milk synthesis (Patton, 2004).

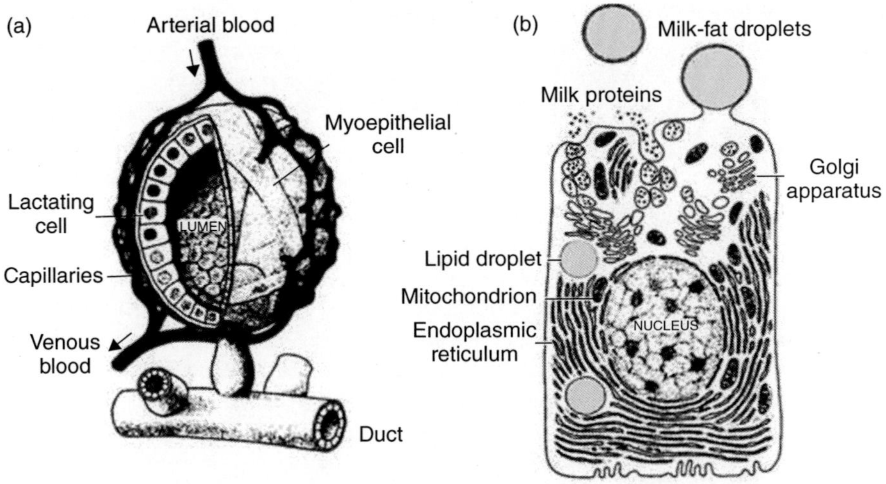

Fig. 6.2 Cell physiology of lactating mammary gland. (a) A single alveolus consisting of lactating epithelial cells (SEC) surrounding the lumen. (b) A typical lactating cell indicating active secretion of protein and lipid by distinct mechanisms; reprinted with permission of *Scientific American* and S. Patton.

6.3.2 Protein synthesis and secretion: Overview

Adaptation of milk components to their ultimate aqueous environment begins during secretion. Lipid and protein synthesis are partitioned from the start. Amino acids and their metabolic precursors are actively transported into the SEC and assembled into proteins on the ribosomes of the highly developed rough endoplasmic reticulum (Patton, 2004). All milk proteins of mammary origin have conserved leader sequences which cause insertion of the nascent proteins into the lumen of the endoplasmic reticulum (ER), shown in Fig. 6.2b. The proteins are then transported through the Golgi apparatus (Patton, 2004; Farrell, 1988), as shown in Fig. 6.2b; presumably the globular proteins of milk are folded during this period. In the Golgi apparatus, the caseins, which are the major milk proteins in most species, appear to be spherical complexes of about 10 nm in diameter. The caseins are phosphorylated by a calcium-activated membrane-bound kinase to begin the process of calcium transport (Bingham and Farrell, 1974). A membrane associated ATPase delivers calcium to the vesicles (Bingham *et al.*, 1993). The gradual intercalation of calcium, casein, and phosphate into the submicellar structures (or complexes formed through specific casein–casein interactions) leads to the formation of casein micelles and insures the effective transport of these vital minerals. This process can be visualized in Fig. 6.3 (top), where small submicellar or particles of casein–casein complexes are seen in the secretory vesicles nearest the trans Golgi. Through the binding of calcium and the accretion of phosphate, the

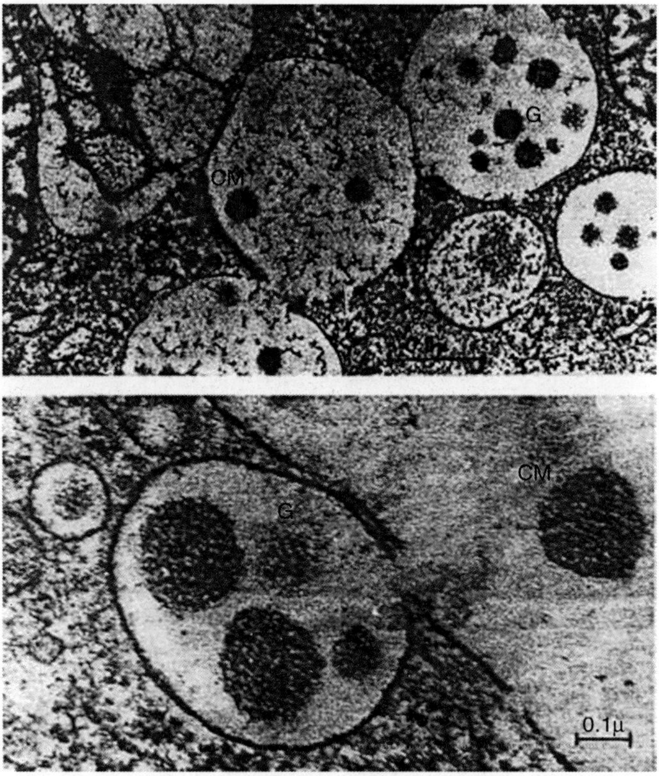

Fig. 6.3 Formation of casein micelles (CM) within Golgi vesicles (G) and depicting the aggregation of small submicellar particles into large micelles (top). A Golgi vesicle (G) about to discharge its contents into the alveolar lumen (bottom); a casein micelle (CM) is already present in the lumen; reprinted from Farrell *et al.* (2006a).

colloidal casein micelles are formed and finally secreted by reverse pinocytosis (Fig. 6.3, bottom). Overall, this view strongly supports the involvement of proteinaceous complexes in the synthesis and secretion of casein micelles though it remains unclear if any structural rearrangements or changes occur during and after secretion.

6.3.3 Protein synthesis and secretion: Details in the ER lumen

The process of casein secretion within the lumen of the ER has not been studied in specific detail (Farrell *et al.*, 2006a); however, reference to recent information on cell biology and protein folding in other systems may shed more light on the issues of casein micelle formation. The issue of quality control of protein folding has become a widely researched area in cell biology (Meusser *et al.*, 2005). The process of endoplasmic reticulum associated degradation (ERAD) has been found to occur in many secretory systems. As noted above, conserved leader sequences of casein cause insertion of the nascent proteins into the lumen of the ER. Before the newly secreted proteins can traffic beyond the ER, they must fold into their final dispositions, and multicomponent systems must assemble. Failure to properly associate or fold leads to the unfolded protein response, tagging and removal by the ERAD system. Overall, the environment of the ER lumen would be conducive to the signature functional property of the caseins (self-association) so initial casein–casein interactions would naturally occur here. While the ER lumen serves as a calcium storage center, the free calcium ion concentration fluctuates between 1 and 3 μM, a concentration well below the binding constants for the unphosphorylated caseins. Presumably, proper association of the caseins helps them to escape degradation and move on to the Golgi for processing. It has been our contention that conserved sequences of the individual caseins give rise to selected secondary structural elements and that these elements lead to self-association of the caseins without classical protein folding (Farrell *et al.*, 2002b, 2003a): that is to say, caseins contain little or no tertiary structure and proceed directly from secondary structure formation to quaternary structures exhibiting both rigid and flexible elements. This is in line with the basic tenet of structural biology that protein structure gives rise to biological function – the Anfinsen hypothesis (Anfinsen, 1973). Recent development in the field of protein folding has shed light on understanding protein self-assembly and aggregation as delineated by Jaenicke and Lilie (2000). Self-associations are those on-line and productive reactions which lead to competent biological protein assemblies such as amyloid (Jahn *et al.*, 2006), whereas non-productive aggregations lead to often mis-assembled protein complexes. The latter would be targeted for the ERAD process. The question now arises as to which of the many known *in vitro* reactions of the caseins studied to date are important in this context. It may be of interest to examine those 'productive' self-associations that may lead to the formation of casein micelles and those 'non-productive' aggregations formed through various protein–protein interactions.

The self-association of bovine β-casein has been studied by many (for example de Kruif and Grinberg, 2002; O'Connell *et al.*, 2003). The most commonly accepted mechanism is that the β-casein molecules form rather spherical polymers of limited size, following a critical micelle pathway. Interestingly, the dimensions of this polymer are rather fixed, but its molecular weight is highly dependent on ionic strength and temperature. In species such as human, where β-casein is the predominate protein, this process could be viewed as a prominent on-line self-association in the ER lumen. It must be noted that the unphosphorylated form of human β-casein has nearly the same propensity for self association as its phosphorylated forms (Bu *et al.*, 2003). The weight average molecular weight of bovine

Table 6.2 Weight average molecular weights of selected caseins and mixtures by analytical ultracentrifugation at 37°C

Casein or mixture	Weight average molecular weight	Weight average polymeric size	Rotor speed (rpm)
α_{S1}-Casein[b]	56000	Dimer	12000
β-Casein[c]	1250000	52 mer	3000
RCM κ-Casein[d]	3040000	160 mer	3000
1.5 α_{S1}-:1 RCM κ-[d]	316000	15 mer	3000
4 α_{S1}-:1 RCM κ-[c]	92400	Tetramer	6000
4β-:1 RCM κ-[c]	1010000	43 mer	3000
1β-:1α_{S1}-[c]	213000	Nonamer	3000
RCM whole casein[c]	110000	Hexamer	6000

[a] All data were obtained at 37°C, pH 6.75 in 25 mM PIPES (disodium piperazine-N,N′-bis(2-ethane sulfonic acid)) with 80 mM KCl to mimic milk salt conditions in the mammary gland in the absence of calcium: The rotor speeds were appropriate to the weight average molecular weight as previously described (Malin *et al.*, 2005; Farrell *et al.*, 2003b). The protein SH groups were reduced and carboxymethylated (RCM).
[b] (Malin *et al.*, 2005).
[c] (Farrell *et al.*, 2006a).
[d] (Farrell *et al.*, 2003b).

β-casein at 37°C in the absence of calcium is 1250000, as shown in Table 6.2. This is almost three times the estimated value for the submicellar structures (or casein–casein complexes) and could potentially lead to ER stress and ERAD tagging.

Studies on the polymerization of α_{s1}-casein had previously shown (Alaimo *et al.*, 1999) that the molecule exhibits a progressive consecutive association to dimers, tetramers, hexamers, etc., and that this process is highly dependent on pH and ionic strength. Most early studies (Schmidt, 1982) on the polymerization of α_{s1}-casein were conducted at or below 25°C. Because of its hydrophobic nature, it was expected that polymerization would be accentuated at 37°C. This clearly is not the case, as shown in Table 6.2: all three major genetic variants depolymerize (Malin *et al.*, 2005) and behave essentially as dimers. The dimer formation is centered on its C-terminal half, with a strong selective interaction between residues 136 and 160 of each monomer, and there appears to be no involvement of N-terminal hydrophobic train of α_{s1}-casein at normal ionic strength and 37°C. In addition, the latter region is positively charged and could participate in phosphate binding. This appears to contradict one of the modes of crosslinking postulated by Horne (1998) as part of the casein gel network. Finally, it has been shown (Bingham *et al.*, 1972) that both native and dephosphorylated α_{s1}-caseins undergo similar aggregation and precipitation reactions. Thus, in the ER lumen, the α_{s1}-casein molecule would not be greatly polymerized, and the higher order polymers observed *in vitro* are the products of an aggregation process, which would likely be 'non-productive'.

Of all the caseins, the self-association behavior of α_{s2}-casein has been studied the least (Euston and Horne, 2005). Horne (1998) has speculated that it will form linear micelles similar to α_{s1}-casein.

κ-Casein is the calcium-soluble stabilizing protein of the casein micelles: it is also the only casein whose disulfide bonds play a significant role in casein structure. As isolated from milk, the protein displays a unique laddered disulfide bonded pattern in SDS gel electrophoresis, with sizes ranging from dimer to octamer and above (Farrell *et al.*, 2003b). In analogy with the use of cleavable disulfide reporter groups, it can be stated that the nearest neighbor to a κ-casein molecule is another κ-casein molecule, as shown in the Schmidt model (Schmidt, 1982). In essence, the monomeric κ-casein molecules depicted in the Horne model (Horne, 1998) are not found.

The source of the laddered disulfide pattern of κ-casein is unknown. In most secretory systems, the enzyme protein disulfide isomerase (PDI) occurs on the inner membrane of the ER, and acts as both a chaperone and a catalyst for the rearrangement of disulfide bonds. However, this enzyme may not be responsible for the linear polymer pattern. For this pattern to form, the κ-casein monomers would have to be in a queue at or near the PDI. Theκ-casein's ability to polymerize may inhibit the action of PDI and thus allow it to retain the SH character in the ER. Therefore, the studies on the polymerization of reduced κ-casein that demonstrate an association involving a critical micelle model similar to that of β-casein (Schmidt, 1982) may be relevant. However, as was the case for α_{s1}-casein, these studies were primarily conducted below 25°C. At 37°C, reduced carboxymethylated κ-casein (RCM-κ) forms large, stranded amyloid bodies (Farrell *et al.*, 2003b); the latter, as seen in Table 6.2, are clearly aggregations and not a part of competent casein secretion. So the κ-casein must either be SH capped or self associate with other caseins for the successful transit through the ER lumen. At a ratio of 1.5 α_{s1}-:1 RCM κ-casein, amyloid formation is inhibited and moderate molecular weight complexes of 316000 are formed at 37°C (Table 6.2). Increasing the ratio of α_{s1}-casein to 4:1 substantially reduces the complexes to 92400. Bovine β-casein, while limiting amyloid formation, does not have the same effect of reducing the weight average molecular weight as α_{s1}-casein (Table 6.2). Since the κ- and β-caseins seem to share a similar self-association mechanism, it would appear that in mixed associations the reduced κ-casein can be inserted interchangeably into the self-association reaction of β-casein, but the resulting overall size is somewhat smaller than that of β-casein alone (Table 6.2). Such large complexes might not allow the associated proteins to escape ERAD and move on to the Golgi apparatus. In fact, Chanat *et al.* (1999) have studied the transport of caseins from the ER to the Golgi apparatus in mammary epithelial cells. Their data suggest that for animals with high casein content, α_{s1}-casein must interact with the other caseins for efficient transport to the Golgi. In cells that completely lack α_{s1}-casein, the accumulation of β-casein (or κ–β mixtures?) is observed in the ER. In the long term, this causes ER stress, activates the ERAD system and impedes secretion. Recent work by Shekar *et al.* (2006) has demonstrated that κ-casein is essential for the formation of casein micelles, as well as for lactation to occur.

To test the efficacy of α_{s1}-casein at reducing the size of β-casein aggregates *in vitro*, 1:1 mixtures of the two proteins were studied by analytical ultracentrifugation at 37°C. The weight average molecular weights of the complexes were speed dependent, increasing with decreasing speed, indicating strong hydrophobic interactions; see Table 6.2. The weight average molecular weight of the 1:1 complexes was 213000, which represents a six-fold reduction of the β-casein aggregate or a four-fold increase over that of the α_{s1}-casein alone. This result is of importance because there are few studies of the mixed associations of these two caseins at 37°C and conditions close to that of the ER. Here, the α_{s1}-casein acts to diminish the size of either the β- or κ-casein aggregates; in this sense it may be considered a molecular detergent for the other caseins. Thus, these *in vitro* data confirm the *in vivo* observations that α_{s1}-casein can reduce aggregated species and allow the associated particles to escape the ER. For human milk, the small amount of α-caseins present may help reduce the aggregates of β-casein: also, the net casein content in human milk is only 17% of that in bovine milk (Table 6.1), so smaller aggregates would be favored (Dev *et al.*, 1994, Sood *et al.*, 1997). Finally, RCM-derived whole bovine casein, with the standard ratios of the four caseins, has a weight average molecular weight of 110000 at 37°C under the same conditions of pH and in the absence of calcium. From all of these biological and physical chemical studies it would appear that for the competent synthesis and secretion of casein: preformed casein complexes of the size of the putative casein submicelles (or casein–casein complexes) must form through protein–protein interactions – triggered by conserved protein sequences – and emerge from the ER for efficient transit via secretory vesicles to the Golgi apparatus.

6.3.4 Protein synthesis and secretion: Details in the Golgi vesicles

From the discussion above, it seems most likely that the individual casein molecules undergo significant self-association in the ER and are then transported in vesicles to the cis face of the Golgi apparatus. In this region, three significant events occur in the process of casein secretion. The first event is most likely an increase in calcium concentration, accomplished by an ATPase-driven pump (Bingham *et al.*, 1993). The second most likely next step is the phosphorylation of the associated caseins by a membrane-associated casein kinase which uses calcium-ATP as substrate and is specific for Ser residues, preceded at the n + 2 position by Glu or a serine phosphate (SerP) residue (Farrell *et al.*, 2004). The casein kinase (Bingham and Farrell,

1974) responsible for this reaction has not been purified, but in Golgi preparations the enzyme requires a surprisingly high calcium ion concentration (K_M~20 mM) so that, at the time of phosphorylation, calcium ions may be almost immediately bound to the caseins (K_D~5 mM). The third and often overlooked step is that one of the byproducts of the kinase reaction, ADP, retains bound calcium, so that when this is converted by membrane associated diphosphatases to phosphate and AMP (Farrell *et al.*, 1992), both calcium and phosphate are released near the interior membrane surface where the casein proteins are still being phosphorylated. Studies by West and Clegg (1983) showed that phosphorylation of casein is still proceeding in large Golgi vesicles as is, most likely, calcium transport. Thus, both calcium and phosphate may automatically be bound to the casein–casein complexes, as seen in Fig. 6.3, prior to micelle formation. In general, when milks that contain >2% protein are analyzed, the accompanying inorganic phosphate and calcium levels found, yield insoluble precipitates (apatite or brushite) in the absence of casein. But the question arises, are the concentrations of these compounds ever high enough or concentrated enough to form nanoclusters in the Golgi vesicles? Additionally, would the energy gained by coating these precipitating clusters be sufficient to depolymerize the preformed casein complexes in the manner suggested by the Horne model? Veis (2005) has suggested that, in general, mineralization in mammalian systems such as collagen and dentin matrices is directed and controlled (assembled) by the structural proteins present: inorganic direction appears to be limited to simpler systems such as the crystalline shells of corals.

Based on the average composition of the colloidal caseinate (Farrell, 1988), the average concentrations of calcium and phosphate within the colloidal complexes are 18.7 and 15.2 mM, respectively. In turn the average concentration of casein within the casein micelles is 1 mM, but the average casein molecule has 6.5 phosphate groups, thus the concentration is 6.5 mM SerP. Moreover, all four caseins have selected areas of positive surface (Swaisgood, 2003) which may bind phosphate after calcium binds to the protein as suggested for α_{s1}-casein (Malin *et al.*, 2005). In most of the models available on the structure of the casein micelles, the binding details of phosphate have been largely neglected. The prominent positive patches of the caseins are: 1–10 of α_{s1}- (+6); 165–199 of α_{s2}- (+11); 97–113 of β- (+6) and 97–116 of κ-casein (+6); these, too, average out to 6.5 mM. Assuming each serine phosphate binds one calcium, which in turn binds one inorganic phosphate, this would result in a double layer. Further, assuming each positive charge in the above mentioned areas then binds one phosphate and one calcium in a second double layer, then the possible concentrations of unbound calcium and phosphate within the micelle is further reduced. It can be seen in a molecular dynamics study (Farrell *et al.*, 2002a) that both the smaller calcium ions and the larger chloride ions bind to the peptide and then associated ions tend to form a charged double layer about the peptide chain. In this case there is limited attraction between the aqueous calcium and chloride ions, but complexes between calcium and phosphate would be more numerous. *In vivo*, the formation of such a double layer of calcium phosphate would reduce the unbound (uninfluenced) concentrations of these two ions to 5.7 and 2.2 mM respectively. This is far from the concentrations used by Holt and coworkers (Holt *et al.*, 1998, Holt, 1998) to form nanoclusters in vitro (37 mM calcium, 30 mM phosphate and 3 mM phosphopeptide). In this same context the supramolecular aggregate would have 65 negative charges (Malin *et al.*, 2005) due to clusters of SerP groups and 65 positive charges due to clusters of basic amino acids: in total it could carry 260 ions as a simple double layer. The standard nanocluster has about 355 ions at its core (Holt *et al.*, 1998; Holt, 1998). The coalescence of two casein–casein complexes, with their bound (influenced) ions, would then terminate any possible calcium phosphate growth and begin micelle formation. It may be that nanoclusters then have an interesting and informative chemistry but actually represent a process which is an inorganic aggregation similar to amyloid formation by κ-casein, rather than an on-line productive biological process. Thus, from the point of view of the synthesis and secretion of casein micelles, the Schmidt model may be representative of the biological process, although the stoichiometry of the inorganic 'cement' is probably incorrect, based upon the latest physical chemical data (de Kruif and Holt, 2003; Dalgleish *et al.*, 2004), which indicates that a type of apatite is the most likely candidate for the molecular structure within the micelles.

6.4 Studies on the structure of casein micelles

Clearly, the proteinaceous complexes, formed through specific casein–casein interactions with an average size of ~10 nm, not necessarily spherical (Figs 6.1 and 6.3), play a major role in the on-line formation of casein micelles in mammary tissue. Additionally, it would appear that α_{s1}-casein acts as a type of detergent to limit the size of these submicellar particles in order to defeat the unfolded response and escape the ERAD system. The formation of these controlled aggregates (productive association) allows for and facilitates transfer from the ER to the Golgi apparatus. Once present in the Golgi, presumably, micelle formation subsequently occurs. Past research on understanding the detailed structure of the micelles has centered on using electron microscopy, neutron scattering and X-ray scattering techniques. Atomic Force Microscopy (AFM) (Uricanu *et al.*, 2004; Gebhardt *et al.*, 2006) has only recently been applied to the study of the casein micelles.

6.4.1 Electron microscopy

Electron microscopy (EM) represents a powerful tool for elucidation of biological ultrastructures, as seen in Figs 6.1 and 6.4. The problem with this

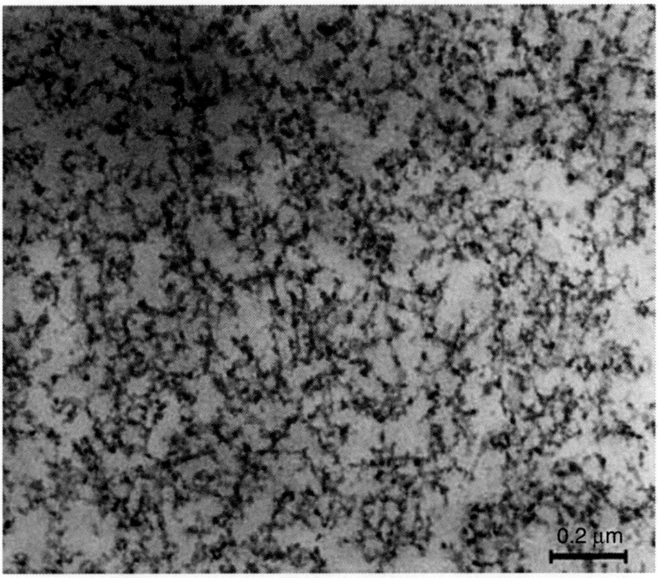

Fig. 6.4 Transmission Electron Micrograph of sodium caseinate in imidazole buffer (pH 7.0), fixed with glutaraldehyde and stained with uranyl acetate and lead citrate.

technique lies in the fixatives and metal staining used to accentuate the particular features they react with and visualize, usually at the expense of other features. In contrast, when uranyl oxalate is used as a positive stain for proteins (McMahon and McManus, 1998; McMahon and Oommen, 2008), a more uniform distribution of material is seen because the stain is binding to the caseins, particularly the SerP, and accentuating the protein distribution. However, more details can be visualized from our studies of casein micelles and sodium caseinate (Figs 6.1 and 6.4). Similar strand-like protein structures with a 'knot' about 10 nm (diameter) can be found in skim milk (Fig. 6.1) and sodium caseinate, a casein product depleted of calcium. In addition, Fig. 6.3 shows the presence of casein–casein aggregates in rat Golgi vesicles. Clearly, EM images are influenced by the stains used in the experiments. One is tempted to employ the scientific dialectic here and say that neither the thesis (submicelles) nor the antithesis (no submicelles) is correct but that synthesis is needed. Walstra (1999) has proposed that the submicelles re-emerge in EM representations of products such as cheese. Work on a variety of cheeses (Tunick *et al.*, 1997) demonstrates the dynamic nature of the submicellar structures, loosely defined as casein–casein complexes of the cheese protein matrix.

6.4.2 Micelle dissociation studies

Studies on micelle dissociation were among the first to indicate the existence of submicelles or casein–casein complexes, and Schmidt (1982) drew heavily on these and on the reconstitution studies (Schmidt and Payens, 1976). Clearly, the Walstra hypothesis on micelle equilibria is at play in these experiments (Walstra, 1999). Those components in rapid equilibrium will quickly exchange and yield one result, while those slow to equilibrate will accentuate another feature. It has also been suggested that calcium binding to caseinates must precede phosphate binding (Visser *et al.*, 1979). Temperature plays another role in that the aggregation of α_{s1}-casein is, as noted above, accentuated at lower temperatures.

Finally, Holt (1998) studied the effect of κ-casein on micelle dissociation: he expected that added κ-casein would cause dissociation of the micelles, and it did not. The experiment conducted was similar to what Talbot and Waugh (1970) termed micelle transformation. In all previous studies of κ-casein content versus size, the more κ-casein present, the smaller the micelles (Sood *et al.*, 2003). Addition of purified κ-casein to micelles causes a shift to smaller sizes, not complete dissociation. It should also be noted that the purified κ-casein used in these experiments represents an SH-capped laddered polymer and not a reactive reduced monomeric species as discussed above (Table 6.2).

6.4.3 Scattering studies

Both small-angle X-ray scattering (SAXS) (Kumosinski *et al.*, 1988; Holt *et al.*, 2003; Marchin *et al.*, 2007; Pignon *et al.*, 2004) and small angle neutron scattering (SANS) (Stothart and Cebula, 1982; Hansen *et al.*, 1996) have been applied to the casein micelles. Both these techniques provide information on the electron density of the sample relative to the solvent. Because of technical limitations regarding the wavelength of the radiation relative to the total particle size, SAXS methodologies, in essence, provide a viewing window on the micelles. The data then must be interpreted in terms of the density of the average particles observed within the micelles. To circumvent this problem it has been common to study first the sodium caseinate (protein aggregates), which is totally contained within the experimental q-scale, and so determine its scattering density. Comparison of the density difference is then made with that observed for the window on the micelle. When this is done carefully, good inferences into the nature of the particles within the overall micelle structure can be made. Using these concepts and enhanced experimental techniques, the two SANS studies came to very similar conclusions that the micelles have within them particles with electron densities (scattering centers) similar to those found for the sodium caseinates. For the SAXS data, the electron density difference for the sodium caseinate is extremely low relative to globular proteins (9.9 e/nm^3 for casein versus 67 e/nm^3 for α-lactalbumin) and the particles within the micelles have this same low electron density: similar calculations can be done for the SANS data. Where these calculations differ is in the mathematical models used to fit the data for the proteins. Basically, the scattering centers within the micelles display a good deal of heterogeneity leading Hansen *et al.* (1996) to conclude a polydisperse distribution of submicelles, while Stothart and Cebula (1982) postulated submicelles of a more closely packed nature. Kumosinski *et al.*(1988) fitted their data for sodium caseinate to a somewhat lopsided sphere within a sphere: basically, a spherical hydrophobic core and a loose hydrophilic shell reminiscent of Schmidt's submicelle model (Schmidt, 1982). However, the data could be fitted well only for reformed synthetic micelles when there was significant overlap among the casein molecules contained in adjacent submicelles. These studies arrived at the same conclusion: proteinaceous complexes formed through specific casein–casein interactions exist within the casein micelles.

Holt *et al.* (1998) studied calcium phosphate nanoclusters with both SANS and SAXS, and this led them to speculate that the calcium phosphate 'clusters rather than putative submicelles could be solely responsible for the heterogeneous structure revealed by electron microscopy, neutron scattering and X-ray scattering.' Recent studies on casein micelles by Marchin *et al.* (Pignon *et al.*, 2004; Marchin *et al.*, 2007) using SAXS/USAXS, combined with the more modern microscopic technique of cryo-TEM, also questioned the existence of submicelles and suggested that the micelles are dominated by calcium phosphate nano-clusters of 2.5 nm in diameter, which is in general agreement with Holt's model.. However, the details of the internal structure and subsequent reorganization upon the removal of serum calcium phosphate or of β-casein remain uncharacterized.

Arguably, we raise the prospect that casein–casein interactions might be as prevalent as the possible existence of inorganic calcium phosphate nanoclusters within the micelles. Despite the fact that the debate over the molecular details involving the formation and stabilization of casein micelles is likely to continue, few would doubt the complex and dynamic nature in the delicate balance of hydrophobic (casein–casein) interactions and electrostatic (casein–casein and casein–calcium–phosphate) interactions. The fast advancement in the field of protein folding research, theoretical and experimental, combined with emerging bioinformatics methodologies will, without doubt, significantly advance our understanding of the growing class of proteins such as casein which inherently lack a well-defined three-dimensional (3D) structure.

6.5 Future trends

Past research on casein micelles has provided us with a wealth of information for consideration of possible applications aimed at improving human health. Various attributes of casein micelles, including their size and relative stability, make them an ideal candidate for the currently ever-growing arena of nanotechnology. It has recently been demonstrated (Semo *et al.*, 2007; Sozer and Kokini, 2009) that casein micelles may be manipulated to act as an effective delivery vehicle for minerals, vitamins, bioactive and nutraceutical compounds through nanoencapsulation and nanoparticulation processes. The lack of unifying safety regulations and generally negative public perception of manufactured nanomaterials for food applications will undoubtedly highlight the natural and superior advantages of casein micelles.

6.6 Acknowledgements

The author would like to thank Dr Harold M. Farrell, Jr. for fruitful discussions and acknowledge Dr Peter H. Cooke for his technical assistance in obtaining the high resolution TEM images presented in this work.

6.7 References

ALAIMO, M. H., WICKHAM, E. D. and FARRELL, H. M., JR. (1999) Effect of self-association of α_{s1}-casein and its cleavage fractions α_{s1}-casein(136–196) and α_{s1}-casein(1–197), on aromatic circular dichroic spectra: Comparison with predicted models. *Biochim. Biophys. Acta*, 1431, 395–409.

ANFINSEN, C. B. (1973) Principles that govern the folding of protein chains. *Science*, 181, 223–30.

BINGHAM, E. W. and FARRELL, H. M., JR. (1974) Casein kinase from the Golgi apparatus of lactating mammary gland. *J. Biol. Chem.*, 249, 3647–51.

BINGHAM, E. W., FARRELL, H. M., JR. and CARROLL, R. J. (1972) Properties of dephosphorylated α_{s1}-casein. Precipitation by calcium ions and micelle formation. *Biochemistry*, 11, 2450–4.

BINGHAM, E. W., MCGRANAGHAN, M. B., WICKHAM, E. D., LEUNG, C. T. and FARRELL, H. M., JR. (1993) Properties of [Ca^{2+} + Mg^{2+}]-adenosine triphosphatases in the Golgi apparatus and microsomes of the lactating mammary glands of cows *J. Dairy Sci.*, 76, 393–400.

BRACKEN, C., IAKOUCHEVA, L. M., ROMERO, P. R. and DUNKER, A. K. (2004) Combining prediction, computation and experiment for the characterization of protein disorder. *Curr. Opin. Struct. Biol.*, 14, 570–6.

BU, H., HU, Y., SOOD, S. M. and SLATTERY, C. W. (2003) Comparison of native and recombinant non-phosphorylated human β-casein: Further evidence for a unique β-casein folding pattern. *Arch. Biochem. Biophys.*, 415, 213–20.

BURGOYNE, R. D. and DUNCAN, J. S. (1998) Secretion of milk proteins. *J. Mammary Gland Biol. Neoplasia*, 3, 275–86.

CARROLL, R. J. and FARRELL, H. M., JR. (1983) Immunological approach to location of κ-casein in the casein micelle by electron microscopy. *J. Dairy Sci.*, 66, 679–86.

CARROLL, R. J., THOMPSON, M. P. and NUTTING, G. C. (1968) Glutaraldehyde fixation of casein micelles for microscopy. *J. Dairy Sci.*, 51, 1903–8.

CHANAT, E., MARTIN, P. and OLLIVIER-BOUSQUET, M. (1999) $\alpha_{(S1)}$-Casein is required for the efficient transport of β- and κ-casein from the endoplasmic reticulum to the Golgi apparatus of mammary epithelial cells. *J. Cell Sci.*, 112, 3399–412.

DALGLEISH, D. G. (1980) A mechanism for the chymosin-induced flocculation of casein micelles. *Biophys. Chem.*, 11, 147–55.

DALGLEISH, D. G. (1998) Casein micelles as colloids: Surface structures and stabilities. *J. Dairy Sci.*, 81, 3013–8.

DALGLEISH, D. G., SPAGNUOLO, P. A. and GOFF, H. D. (2004) A possible structure of the casein micelle based on high-resolution field-emission scanning electron microscopy. *Int. Dairy J.*, 14, 1025–31.

DE KRUIF, C. G. and GRINBERG, V. Y. (2002) Micellisation of β-casein. *Colloids Surface A*, 210, 183–90.

DE KRUIF, C. G. and HOLT, C. (2003) Casein micelle structure, functions and interactions. In Fox, P. F. and McSweeney, P. L. H. (Eds.) *Advanced Dairy Chemistry: Proteins*. New York, NY, Kluwer Academic / Plenum Publishers.

DE KRUIF, C. G., TUINIER, R., HOLT, C., TIMMINS, P. A. and ROLLEMA, H. S. (2002) Physicochemical study of κ- and β-casein dispersions and the effect of cross-linking by transglutaminase. *Langmuir*, 18, 4885–91.

DE KRUIF, C. G. and ZHULINA, E. B. (1996) κ-Casein as a polyelectrolyte brush on the surface of casein micelles. *Colloid Surface A*, 117, 151–9.

DEV, B. C., SOOD, S. M., DEWIND, S. and SLATTERY, C. W. (1994) κ-Casein and β-casein in human milk micelles: Structural studies. *Arch. Biochem. Biophys.*, 314, 329–36.

DUNKER, A. K., LAWSON, J. D., BROWN, C. J., WILLIAMS, R. M., ROMERO, P., OH, J. S., OLDFIELD, C. J., CAMPEN, A. M., RATLIFF, C. M., HIPPS, K. W., AUSIO, J., NISSEN, M. S., REEVES, R., KANG, C., KISSINGER, C. R., BAILEY, R. W., GRISWOLD, M. D., CHIU, W., GARNER, E. C. and OBRADOVIC, Z. (2001) Intrinsically disordered protein. *J. Mol. Graph. Model.*, 19, 26–59.

DUNKER, A. K., SILMAN, I., UVERSKY, V. N. and SUSSMAN, J. L. (2008) Function and structure of inherently disordered proteins. *Curr. Opin. Struct. Biol.*, 18, 756–64.

DYSON, H. J. and WRIGHT, P. E. (2002) Coupling of folding and binding for unstructured proteins. *Curr. Opin. Struct. Biol.*, 12, 54–60.

DYSON, H. J. and WRIGHT, P. E. (2005) Intrinsically unstructured proteins and their functions. *Nat. Rev. Mol. Cell. Biol.*, 6, 197–208.

EUSTON, S. R. and HORNE, D. S. (2005) Simulating the self-association of caseins. *Food Hydrocolloids*, 19, 379–86.

FARRELL, H. M., JR. (1988) Physical equilibria: Proteins. In Wong, N. B. (Ed.) *Fundamentals of Dairy Chemistry, 3rd Edition*. Reinhold, NY, Van Nostrand.

FARRELL, H. M., JR. (1999) Milk, composition and synthesis. In Knobil, E. and Neill, J. D. (Eds.) *Encyclopedia of Reproduction*. San Diego, CA, Academic Press, Inc.

FARRELL, H. M., JR., BROWN, E. M., HOAGLAND, P. D. and MALIN, E. L. (2003a) Higher order structures of the caseins: A paradox? In Fox, P. F. and McSweeney, P. L. H. (Eds.) *Advanced Dairy Chemistry – 1: Proteins, 3rd edition*. New York, NY, Kluwer Academic / Plenum Publishers.

FARRELL, H. M., JR., COOKE, P. H., WICKHAM, E. D., PIOTROWSKI, E. G. and HOAGLAND, P. D. (2003b) Environmental influences on bovine κ-casein: Reduction and conversion to fibrillar (amyloid) structures. *J. Protein Chem.*, 22, 259–73.

FARRELL, H. M., JR., JIMENEZ-FLORES, R., BLECK, G. T., BROWN, E. M., BUTLER, J. E., CREAMER, L. K., HICKS, C. L., HOLLAR, C. M., NG-KWAI-HANG, K. F. and SWAISGOOD, H. E. (2004) Nomenclature of the proteins of cows' milk–sixth revision. *J. Dairy Sci.*, 87, 1641–74.

FARRELL, H. M., JR., KUMOSINSKI, T. F., MALIN, E. L. and BROWN, E. M. (2002a) The caseins of milk as calcium binding proteins. In Vogel, H. J. (Ed.) *Methods Mol. Biol.* Totwa, NJ, Humana Press, Inc.

FARRELL, H. M., JR., LEUNG, C. T. and WICKHAM, E. D. (1992) Distribution of ADPase activity in the lactating rat mammary gland and its possible role in an ATP cycle in the Golgi apparatus. *Arch. Biochem. Biophys.*, 292, 368–75.

FARRELL, H. M., JR., MALIN, E. L., BROWN, E. M. and QI, P. X. (2006a) Casein micelle structure: What can be learned from milk synthesis and structural biology? *Curr. Opin. Colloid Interface Sci.*, 11, 135–47.

FARRELL, H. M., JR., QI, P. X., BROWN, E. M., COOKE, P. H., TUNICK, M. H., WICKHAM, E. D. and UNRUH, J. J. (2002b) Molten globule structures in milk proteins: Implications for potential new structure–function relationships. *J. Dairy Sci.*, 85, 459–71.

FARRELL, H. M., JR., QI, P. X. and UVERSKY, V. N. (2006b) New views of protein structure: Applications to the caseins. In Fishman, M. L., Qi, P. X. and Wicker, L. (Eds.) *Advances in Biopolymers: Molecules, Clusters, Networks and Interactions*. Washington, D. C., American Chemical Society.

FARRELL, H. M., JR., QI, P. X. and UVERSKY, V. N. (2006c) New views of protein structure: Implication for potential new protein structure–function relationships. In Fishman, M. L., Qi, P. X. and Wicker, L. (Eds.) *Advances in Biopolymers: Molecules, Clusters, Networks and Interactions*. Washington, D. C., American Chemical Society.

FARRELL, H. M., JR. and THOMPSON, M. P. (1988) Caseins as calcium binding proteins. In Thompson, M. P. (Ed.) *Calcium binding proteins*. Boca Raton, FL, CRC Press.

FOX, P. F. and BRODKORB, A. (2008) The casein micelle: Historical aspects, current concepts and significance. *Int. Dairy J.*, 18, 677–84.

GEBHARDT, R., DOSTER, W., FRIEDRICH, J. and KULOZIK, U. (2006) Size distribution of pressure-decomposed casein micelles studied by dynamic light scattering and AFM. *Eur. Biophys. J.*, 35, 503–9.

GRIFFIN, M. C., LYSTER, R. L. and PRICE, J. C. (1988) The disaggregation of calcium-depleted casein micelles. *Eur. J. Biochem.*, 174, 339–43.

GRIFFIN, M. C. and ROBERTS, G. C. (1985) A ^{1}H-n.m.r. study of casein micelles. *Biochem. J.*, 228, 273–6.

HANSEN, S., BAUER, R., LOMHOLT, S. B., BRUUNQVIST, K., PEDERSEN, J. S. and MORTENSEN, K. (1996) Structure of casein micelles studied by small-angle neutron scattering. *Eur. Biophys. J.*, 24, 143–7.

HILL, R. J. and WAKE, R. G. (1969) Amphiphile nature of κ-casein as the basis for its micelle stabilizing property. *Nature*, 221, 635–9.

HOLT, C. (1992) Structure and stability of bovine casein micelles. *Adv. Protein Chem.*, 43, 63–151.

HOLT, C. (1998) Casein micelle substructure and calcium phosphate interactions studied by Sephacryl colume chromatography. *J. Dairy Sci.*, 81, 2994–3003.

HOLT, C., DE KRUIF, C. G., TUINIER, R. and TIMMINS, P. A. (2003) Substructure of bovine casein micelles by small-angle X-ray and neutron scattering. *Colloids Surface A*, 213, 275–84.

HOLT, C. and SAWYER, L. (1993) Caseins as rheomorphic proteins: Interpretation of primary and secondary structures of the α_{S1}-, β- and κ-caseins. *J. Chem. Soc. Faraday Trans.*, 89, 2683–92.

HOLT, C., TIMMINS, P. A., ERRINGTON, N. and LEAVER, J. (1998) A core–shell model of calcium phosphate nanoclusters stabilized by β-casein phosphopeptides, derived from sedimentation equilibrium and small-angle X-ray and neutron-scattering measurements. *Eur. J. Biochem.*, 252, 73–8.

HORNE, D. S. (1998) Casein interactions: Casting light on the black boxes, the structure of dairy products. *Int. Dairy J.*, 8, 171–7.

HORNE, D. S. (2002) Casein structure, self-assembly and gelation. *Curr. Opin. Colloid Interface Sci.*, 7, 456–61.

HORNE, D. S. (2006) Casein micelle structure: Models and muddles. *Curr. Opin. Colloid Interface Sci.*, 11, 148–53.

JAENICKE, R. and LILIE, H. (2000) Folding and association of oligomeric and multimeric proteins. *Adv. Protein Chem.*, 53, 329–401.

JAHN, T. R., PARKER, M. J., HOMANS, S. W. and RADFORD, S. E. (2006) Amyloid formation under physiological conditions proceeds via a native-like folding intermediate. *Nat. Struct. Mol. Biol.*, 13, 195–201.

JANG, H. D. and SWAISGOOD, H. E. (1990) Characteristics of the interaction of calcium with casein submicelles as determined by analytical affinity chromatography. *Arch. Biochem. Biophys.*, 283, 318–25.

KAKALIS, L. T., KUMOSINSKI, T. F. and FARRELL, H. M., JR. (1990) A multinuclear, high-resolution NMR study of bovine casein micelles and submicelles. *Biophys. Chem.*, 38, 87–98.

KNOOP, A. M., KNOOP, E. and WIECHEN, A. (1979) Sub-structure of synthetic casein micelles. *J. Dairy Res.*, 46, 347–50.

KUMOSINSKI, T. F., PESSEN, H., FARRELL, H. M., JR. and BRUMBERGER, H. (1988) Determination of the quaternary structural states of bovine casein by small-angle X-ray scattering: Submicellar and micellar forms. *Arch. Biochem. Biophys.*, 266, 548–61.

KUMOSINSKI, T. F., PESSEN, H., PRESTRELSKI, S. J. and FARRELL, H. M., JR. (1987) Water interactions with varying molecular states of bovine casein: 2H NMR relaxation studies. *Arch. Biochem. Biophys.*, 257, 259–68.

LIN, S. H., LEONG, S. L., DEWAN, R. K., BLOOMFIELD, V. A. and MORR, C. V. (1972) Effect of calcium ion on the structure of native bovine casein micelles. *Biochemistry*, 11, 1818–21.

MALIN, E. L., BROWN, E. M., WICKHAM, E. D. and FARRELL, H. M., JR. (2005) Contributions of terminal peptides to the associative behavior of alpha s1-casein. *J. Dairy Sci.*, 88, 2318–28.

MARCHIN, S., PUTAUX, J. L., PIGNON, F. and LEONIL, J. (2007) Effects of the environmental factors on the casein micelle structure studied by cryo transmission electron microscopy and small-angle X-ray scattering/ultra small-angle X-ray scattering. *J. Chem. Phys.*, 126, 045101.

MCMAHON, D. J. and MCMANUS, W. R. (1998) Rethinking casein micelle structure using electron microscopy. *J. Dairy Sci.*, 81, 2985–93.

MCMAHON, D. J. and OOMMEN, B. S. (2008) Supramolecular structure of the casein micelle. *J. Dairy Sci.*, 91, 1709–21.

MEIER, S., BLACKLEDGE, M. and GRZESIEK, S. (2008) Conformational distributions of unfolded polypeptides from novel NMR techniques. *J. Chem. Phys.*, 128, 052204.

MEUSSER, B., HIRSCH, C., JAROSCH, E. and SOMMER, T. (2005) ERAD: The long road to destruction. *Nat. Cell Biol.*, 7, 766–72.

NOBLE, R. W., JR. and WAUGH, D. F. (1965) Casein micelles. Formation and structure. I. *J. Am. Chem. Soc.*, 87, 2236–45.

O'CONNELL, J. E., GRINBERG, V. Y. and DE KRUIF, C. G. (2003) Association behavior of β-casein. *J. Colloid Interface Sci.*, 258, 33–9.

PATTON, S. (2004) *Milk: Its remarkable contribution to human health and well-being*, New Brunswick, NJ, Transaction Publishers.

PAYENS, T. A. (1979) Casein micelles: The colloid-chemical approach. *J. Dairy Res.*, 46, 291–306.

PEPPER, L. and FARRELL, H. M., JR. (1982) Interactions leading to formation of casein submicelles. *J. Dairy Sci.*, 65, 2259–66.

PIERRE, A., MICHEL, F., LE GRAET, Y. and ZAHOUTE, L. (1998) Casein micelle size in relation with casein composition and α_{s1}- α_{s2}- β- and κ-casein contents in goat milk. *Lait*, 78, 591–605.

PIGNON, F., BELINA, G., NARAYANAN, T., PAUBEL, X., MAGNIN, A. and GESAN-GUIZIOU, G. (2004) Structure and rheological behavior of casein micelle suspensions during ultrafiltration process. *J. Chem. Phys.*, 121, 8138–46.

QI, P. X. (2007) Studies of casein micelle structure: The past and the present. *Le Lait*, 87, 363–83.

RADIVOJAC, P., IAKOUCHEVA, L. M., OLDFIELD, C. J., OBRADOVIC, Z., UVERSKY, V. N. and DUNKER, A. K. (2007) Intrinsic disorder and functional proteomics. *Biophys. J.*, 92, 1439–56.

ROSE, D. and COLVIN, J. R. (1966) Internal structure of casein micelles from bovine milk. *J. Dairy Sci.*, 49, 351–5.

SCHMIDT, D. G. (1982) Association of caseins and casein micelle structure. In Fox, P. F. (Ed.) *Developments in Dairy Chemistry: 1*. London, UK, Applied Science Publishers.

SCHMIDT, D. G. and PAYENS, T. A. (1976) Micellar aspects of casein. In Matijevic, E. (Ed.) *Surface and Colloid Science*. New York, NY, John Wiley.

SEMO, E., KESSELMAN, E., DANINO, D. and LIVNEY, Y. D. (2007) Casein micelle as a natural nano-capsular vehicle for nutraceuticals. *Food Hydrocolloids*, 21, 936–42.

SHEKAR, P. C., GOEL, S., RANI, S. D., SARATHI, D. P., ALEX, J. L., SINGH, S. and KUMAR, S. (2006) κ-Casein-deficient mice fail to lactate. *Proc. Natl. Acad. Sci. USA*, 103, 8000–5.

SLATTERY, C. W. (1979) A phosphate-induced sub-micelle-micelle equilibrium in reconstituted casein micelle systems. *J. Dairy Res.*, 46, 253–8.

SOOD, S. M., ERICKSON, G. and SLATTERY, C. W. (2003) κ-Casein interactions in the suspension of the two major calcium-sensitive human β-caseins. *J. Dairy Sci.*, 86, 2269–75.

SOOD, S. M., HERBERT, P. J. and SLATTERY, C. W. (1997) Structural studies on casein micelles of human milk: Dissociation of β-casein of different phosphorylation levels induced by cooling and ethylenediaminetetraacetate. *J. Dairy Sci.*, 80, 628–33.

SOZER, N. and KOKINI, J. L. (2009) Nanotechnology and its applications in the food sector. *Trends Biotechnol.*, 27, 82–9.

STOTHART, P. H. (1989) Subunit structure of casein micelles from small-angle neutron-scattering. *J. Mol. Biol.*, 208, 635–8.

STOTHART, P. H. and CEBULA, D. J. (1982) Small-angle neutron scattering study of bovine casein micelles and sub-micelles. *J. Mol. Biol.*, 160, 391–5.

SWAISGOOD, H. E. (2003) Chemistry of the caseins. In Fox, P. F. and McSweeney, P. L. H. (Eds.) *Advanced Dairy Chemistry: Proteins*. New York, NY, Kluwer Academic / Plenum Publishers.

SZILAGYI, A., GYORFFY, D. and ZAVODSZKY, P. (2008) The twilight zone between protein order and disorder. *Biophys. J.*, 95, 1612–26.

TALBOT, B. and WAUGH, D. F. (1970) Micelle-forming characteristics of monomeric and covalent polymeric kappa caseins. *Biochemistry*, 9, 2807–13.

TOMPA, P. (2002) Intrinsically unstructured proteins. *Trends Biochem. Sci.*, 27, 527–33.

TUINIER, R. and DE KRUIF, C. G. (2002) Stability of casein micelles in milk. *J. Chem. Phys.*, 117, 1290–5.

TUNICK, M. H., COOKE, P. H., MALIN, E. L., SMITH, P. W. and HOLSINGER, V. H. (1997) Reorganization of casein submicelles in mozzarella cheese during storage. *Int. Dairy J.*, 7, 149–55.

URICANU, V. I., DUITS, M. H. and MELLEMA, J. (2004) Hierarchical networks of casein proteins: An elasticity study based on atomic force microscopy. *Langmuir*, 20, 5079–90.

UVERSKY, V. N. (2002) Natively unfolded proteins: A point where biology waits for physics. *Protein Sci.*, 11, 739–56.

UVERSKY, V. N. (2008) Amyloidogenesis of natively unfolded proteins. *Curr. Alzheimer Res.*, 5, 260–87.

VEIS, A. (2005) Enhanced: A Window on Biomineralization. *Science*, 307, 1419–20.

VISSER, J., SCHAIER, R. W. and VAN GORKOM, M. (1979) The role of calcium, phosphate and citrate ions in the stabilization of casein micelles. *J. Dairy Res.*, 46, 333–5.

WALSTRA, P. (1990) On the stability of casein micelles. *J. Dairy Sci.*, 73, 1965–79.

WALSTRA, P. (1999) Casein sub-micelles: Do they exist? *Int. Dairy J.*, 9, 189–92.

WAUGH, D. F., CREAMER, L. K., SLATTERY, C. W. and DRESDNER, G. W. (1970) Core polymers of casein micelles. *Biochemistry*, 9, 786–95.

WEINREB, P. H., ZHEN, W., POON, A. W., CONWAY, K. A. and LANSBURY, P. T., JR. (1996) NACP, a protein implicated in Alzheimer's disease and learning, is natively unfolded. *Biochemistry*, 35, 13709–15.

WEST, D. W. and CLEGG, R. A. (1983) Casein kinase activity in rat mammary gland Golgi vesicles. Phosphorylation of endogenous caseins. *Eur. J. Biochem.*, 137, 215–20.

WRIGHT, P. E. and DYSON, H. J. (1999) Intrinsically unstructured proteins: Re-assessing the protein structure–function paradigm. *J. Mol. Biol.*, 293, 321–31.

7

Glycosylated dairy components: Their roles in nature and ways to make use of their biofunctionality in dairy products

I. Recio, F. J. Moreno and R. López-Fandiño, Instituto de Fermentaciones Industriales (CSIC), Spain

Abstract: This chapter deals with the most important proteins that are naturally glycosylated in milk, including i) κ-casein and caseinmacropeptide, ii) glycosylated whey proteins and iii) glycoproteins associated with the milk-fat globule membrane. Data based on structure–function and biological role of these glycoproteins are discussed in detail, with special attention to the health promoting activities that may be mediated by the carbohydrate portion. For instance, glycans from milk proteins may serve as recognition sites interacting with viruses, microorganisms and cell membrane lectins. In this sense, this chapter underlines the importance of carrying out further research on milk biologically active glycoproteins, as it could allow expanding their potential commercial applications in the food and pharmaceutical industries.

Keywords: milk proteins, glycosylation, caseinmacropeptide, whey, milk-fat globule membrane.

7.1 Introduction: Glycosylation of milk proteins

Glycan moieties that are covalently bound to proteins play a significant role in defining the physicochemical and biological properties of such proteins. Carbohydrates can be attached to proteins in two major ways, resulting in the so-called *N*-glycans and *O*-glycans. *N*-glycans consist of a carbohydrate moiety *N*-linked to the amide nitrogen of Asp side chains within an Asn-X-Ser/Thr motif, where X corresponds to any amino acid except for Pro, although there is also the option of having Cys instead of Ser/Thr (Vance *et al.*, 1997). Due to their common biosynthetic pathway, all *N*-glycans share a well-defined trimannosyl pentasaccharide core with two molecules of *N*-acetylglucosamine ($Man_3GlcNAc_2$) and can be divided into three classes: i) high mannose-type glycans, containing only two GlcNAc, a variable number of Man and sometimes also glucose (Glc) residues; ii) complex-type glycans, exhibiting variable numbers of GlcNAc, galactose (Gal), fucose (Fuc) and sialic acid, and sometimes also GalNAc residues in addition to the pentasaccharide core; and iii) hybrid-type glycans, combining the characteristic features of both high-mannose and complex-type species (Dalpathado and Desaire, 2008; Geyer and Geyer, 2006). Depending on the number of branches attached, complex-type glycans can be also subdivided to bi-, tri- and tetra-antennary structures (Nilsson, 1994). *O*-glycans are formed by glycosylation of the hydroxy oxygen of Ser and Thr side chains, generally in sequence regions of high hydroxyamino acid density.

The mammary gland naturally secretes *N*- or *O*-glycosylated proteins and glycosylation is undoubtedly one of the most important post-translational event for functional proteins. It is essential for the stability of many proteins and is often necessary for correct protein folding (3D), conformation, solubility, intracellular transport or tissue targeting. Glycosylation is also required for the biological activity of several proteins, e.g. antibodies. Likewise, since many pathogens use carbohydrate-binding proteins to attach to cells and initiate infection processes (Zopf and Roth, 1996), a number of milk glycoproteins, particularly those containing sialylated oligosaccharides, have been shown both to inhibit adhesion of pathogens and separate bacteria already attached on the host tissue, as will be explained below.

The purpose of this chapter is to review the newest data based on structure–function relationships and the biological role of the most important milk glycoproteins, including i) κ-casein and caseinmacropeptide, ii) glycosylated whey proteins and iii) glycoproteins associated with the milk-fat globule membrane.

7.2 κ-casein and caseinmacropeptide

κ-casein is the only glycosylated component among the four milk caseins. In this protein, glycosylation, as well as all the other post-translational modifications, is found exclusively in the C-terminal portion of the molecule. The 64 C-terminal amino acids of κ-casein can be released by chymosin or pepsin cleavage between Phe^{105}-Met^{106}, during the manufacture of cheese or during digestion in the stomach, producing a heterogeneous group of polypeptides known as caseinmacropeptide (CMP) (Delfour *et al.*, 1965). CMP is abundant and comprises, approximately, 15–20% (w/w) of the protein in bovine milk whey. CMP is a highly soluble, acidic peptide, with an isoelectric point of 4–5. It is heat stable and, as it is the case of κ-casein, it can exist in several forms due to extensive post-translational modifications such as glycosylation, phosphorylation and genetic variance. Table 7.1 summarises the sites of genetic variant substitution, glycosylation and phosphorylation, which contribute to this heterogeneity in bovine

Table 7.1 Amino acid residues of bovine CMP, genetic variants and postranslational modification sites involved in its heterogeneity. The nomenclature is based on the amino acid sequence of κ-casein whole mature protein

Amino acid sequence (dominant variant A)	Genetic polymorphism	Glycosilation sites	Phosphorilation sites
κ-CN f(106–169) Met-Ala-Ile-Pro-Pro-Lys-Lys-Asn-Gln-Asp-Lys-Thr-Glu-Ile-Pro-Thr-Ile-Asn-Thr-Ile-Ala-Ser-Gly-Glu-Pro-Thr-Ser-Thr-Pro-Thr-Thr-Glu-Ala-Val-Glu-Ser-Thr-Val-Ala-Thr-Leu-Glu-Asp-Ser-Pro-Glu-Val-Ile-Glu-Ser-Pro-Pro-Glu-Ile-Asn-Thr-Val-Gln-Val-Thr-Ser-Thr-Ala-Val	A/F) T^{135} T^{136} D^{148} I^{153} S^{155} B/C = D) T^{135} I^{136} A^{148} I^{153} S^{155} B2) T^{135} I^{136} A^{148} T^{153} S^{155} E) T^{135} T^{136} D^{148} I^{153} G^{155} G/H) I^{135} T^{136} D^{148} I^{153} S^{155}	T^{121} T^{131} T^{133} T^{135}/T^{136} S^{141} T^{142} T^{165}	S^{127} S^{149}
Mercier *et al.* (1973)	Alexander *et al.* (1988) Coolbear *et al.* (1996) Prinzenberg *et al.* (1999)	Claverol *et al.* (2003) Holland *et al.* (2004) Minkiewicz *et al.* (1996) Mollé & Léonil (1995) Pisano *et al.* (1994) Vreeman *et al.* (1986)	Claverol *et al.* (2003) Holland *et al.* (2004) Mercier (1981) Minkiewicz *et al.* (1996) Mollé & Léonil (1995) Talbo *et al.* (2001)

CMP. The structures of ovine and caprine CMP have been reviewed by Manso and López-Fandiño (2004).

As shown in Table 7.1, up to 11 genetic variants of κ-casein are known, but variants A and B dominate in bovine milk. These differ by two amino acid substitutions: Thr^{136} in variant A and Ile^{136} in variant B, and Asp^{148} in variant A and Ala^{148} in variant B. In addition, there are three phosphorylation sites. Ser^{149} is fully phosphorylated and Ser^{127} is the second most important phosphorylation site.

Glycosylated forms represent only about 50% of the total κ-casein or bovine CMP, which, in this case, is referred to as glycomacropeptide (Vreeman *et al.*, 1986). Five different mucin-type carbohydrate chains, composed of a sialic acid, *N*-acetylneuraminic acid (NeuAc), galactose (Gal) and *N*-acetylgalactosamine (GalNAc), have been identified in κ-casein (Saito and Itoh, 1992): (i) monosaccharide GalNAc-O-R, (ii) disaccharide Galβ1-3GalNAc-O-R, (iii) trisaccharide NeuAcα2-3 Galβ1-3GalNAc-O-R,

(iv) trisaccharide Galβ1-3(NeuAc α2-6)GalNAc-O-R, (v) tetrasaccharide NeuAcα2-3Galβ1-3(NeuAc α2-6)GalNAc-O-R.

These are *O*-glycosidically linked to the peptide through threonine residues. Thr^{121}, Thr^{131}, Thr^{133}, Thr^{136} (variant A), and Thr^{142} seem to be accepted as the most important glycosylation sites, but also Thr^{165}, Thr^{135} and Ser^{141} have been proposed as potential glycosylation sites (Kanamori *et al.*, 1980; Pisano *et al.*, 1994; Takeuchi *et al.*, 1985; Zevaco and Ribadeau-Dumas, 1984). A more recent identification of the modified sites, which involved the previous separation of the glycoforms by two-dimensional electrophoresis, revealed a distinct hierarchy in the *O*-glycosylation of κ-casein that implies an ordered addition of glycans to Thr^{131}, Thr^{142} and Thr^{133} (Holland *et al.*, 2005). Microheterogeneity occurs due to variations in the extent and type of oligosaccharides linked. The post-translational modifications result in 14 glycosylated forms of κ-casein variant A and CMP, in addition to the identified aglyco-forms (Mollé and Léonil, 1995).

CMP is a multifunctional peptide with many possible biological applications (for reviews, see Abd El-Salam *et al.*, 1996; Brody, 2000; Dziuba and Minkiewicz, 1996; Manso and López-Fandiño, 2004; Thomä-Worringer *et al.*, 2006). Structure-activity studies have pointed out the importance of certain elements of the peptide for the different biological functions. In particular, those activities that rely on interactions with cell components are linked to the content and structure of the sugar residues, while others depend on the primary structure and can be exerted by smaller peptides encrypted in the molecule. The value of CMP as a multifunctional bioactive compound is supported by the finding in human plasma of bovine and human CMP at physiologically active concentrations (Chabance *et al.*, 1995, 1998). In fact, *in vivo* studies showed that the peptide is released intact from the stomach and undergoes only partial hydrolysis by pancreatic enzymes (Fosset *et al.*, 2002; Ledoux *et al.*, 1999), although the digestion profile may vary depending on the degree of glycosylation (Boutrou *et al.*, 2008). Furthermore, ovine CMP had a very long life when incubated in human or guinea-pig plasma (Qian *et al.*, 1995). In addition, several active fragments of κ-casein were found in the blood stream of humans and rats, which provides evidence for the resistance of the active sequences, either generated *in vitro* or *in vivo*, to further digestion (Chabance *et al.*, 1998; Fosset *et al.*, 2002).

The functional properties of CMP make it an interesting ingredient for use in the development of novel foods with health-promoting benefits. On top of its solubility and heat stability, bovine CMP displays a good emulsifying capacity, with a maximum activity observed at alkaline pH and a minimum in the pH range 4.5 to 5.5 (Martín-Diana *et al.*, 2005). However, the stability of heat-treated CMP stabilised emulsions deteriorates after 24 h of storage (Chobert *et al.*, 1989; Minkiewicz *et al.*, 1996; Moreno *et al.*, 2002). CMP modified by conjugation with disaccharides or fatty acids may exhibit improved functionality and even enhanced biological activities

(Moreno *et al.*, 2002; Wong *et al.*, 2006). Further information about the technological functional properties of CMP and its incorporation into food matrices can be found in Thomä-Worringer *et al.* (2006).

Various methods have been described for the isolation of CMP, mainly by means of chromatography or ultrafiltration, although the latter is preferred due to ease of scaling up and low cost. Most of the ultrafiltration methods make use of the pH-dependence of the molecular weight of CMP to influence its permeation behaviour and to separate it from the other whey proteins (Kawasaki *et al.*, 1993a). Such pH-induced changes in the molecular weight of CMP are attributed to self-associations through non-covalent interactions that lead to the formation of oligomers at neutral pH, which partially dissociate at acidic pH (Kawasaki *et al.*, 1993a; Xu *et al.*, 2000), although disagreement on this matter exists (Mikkelsen *et al.*, 2005a; Minkiewicz *et al.*, 1996). Other methods use the high thermal stability of CMP in comparison with whey proteins, which denature and aggregate at 90°C, to facilitate its separation by ultrafiltration (Martin-Diana *et al.*, 2002; Metwally *et al.*, 2001). As an alternative to the separation of CMP from the whey proteins, it is possible to treat micellar casein with chymosin to obtain a solution rich in CMP, before concentration by microfiltration or diafiltration (Thomä and Kulozik, 2004).

Regarding its recombinant production, human non-glycosylated CMP was produced in *Escherichia coli* with high yield (Liu *et al.*, 2008), while yeasts were used to produce human glycosylated CMP, although it was significantly less glycosylated than bovine CMP (Kim *et al.*, 2005). Transgenic rabbits were also treated to express high levels of low-Phe κ-casein in their milk. In the recombinant protein, four out of the five Phe were mutated, but it kept its micelle formation ability and was digestible by chymosin (Baranyi *et al.*, 2007).

In view of the different processes available for the production of CMP and of the importance of its structure (in particular glycosylation) in its various biological activities, it is necessary to know whether the methodology used to obtain the CMP influences its structure and activity (Li and Mine, 2004a; Lieske *et al.*, 2004; Thomä-Worringer *et al.*, 2006). Similarly, storage and technological processes, such as heating, may affect the glycosylation degree or the chemical stability of CMP (Lieske *et al.*, 2004).

7.2.1 Bioactivity of glycosylated caseinmacropeptide

The sialic acid content of CMP is very important in terms of bioactivity. Large amounts of this carbohydrate are found in the brain and in the central nervous system in the form of gangliosides and glycoproteins, which contribute to the functioning of cell membranes and membrane receptors, and to normal brain development. An *in vivo* experiment with laboratory animals has shown that the exogenous administration of sialic acid increased the production of ganglioside sialic acid in the brain, improving learning ability, an effect that could also be achieved with dietary CMP (Wang *et al.*, 2001, 2004). In fact, sialic acid from a CMP source fed to piglets, in amounts similar to those present in mature sow milk, for 35 days, was associated with faster learning, higher concentration of protein-bound sialic acid in the frontal cortex and higher mRNA levels of two learning-related genes (Wang *et al.*, 2007).

κ-casein and CMP interact with toxins, viruses and bacteria, exerting health-promoting activities that are strongly mediated by the carbohydrate fraction, as many pathogens and enterotoxins adhere to cells by recognising carbohydrate receptors (Dziuba and Minkiewicz, 1996). Human κ-casein prevented the adhesion of *Helicobacter* to sections of stomach tissue (Strömquist *et al.*, 1995). CMP inhibits the binding of cholera toxins to their oligosaccharide receptors on cell walls (Kawasaki *et al.*, 1992; Oh *et al.*, 2000) and protects cells from infection by influenza virus (Kawasaki *et al.*, 1993b). The carbohydrate structure and, in particular NeuAc, is considered crucial for the binding ability of CMP to *Salmonella enteriditis* and enterohemorrhagic *Escherichia coli* 0157: H7 (Nakajima *et al.*, 2005). However, the antiadhesive activity of CMP shows a strong strain-dependent behaviour, which stresses the influence on the results of the existence of CMP forms varying in the degree of glycosylation, the lack of knowledge of the cellular receptors for different strains of pathogenic and non pathogenic bacteria, and the use of different assay formats and cell lines, (Brück *et al.*, 2002, 2003a, 2006a; Rhoades *et al.*, 2005). *In vivo*, CMP was shown to reduce *Escherichia coli* induced diarrhoea in infant rhesus monkeys and it was observed that CMP supplementation even increased zinc absorption (Brück *et al.*, 2003b; Kelleher *et al.*, 2003).

Sialic acid containing substances, including human and bovine CMP, promote the growth of the *Bifidobacterium* genus: *B. breve*, *B. bifidum*, and *B. infantis* (Azuma *et al.*, 1984; Idota *et al.*, 1994; Metwally *et al.*, 2001). Supplementation of milk with 2% CMP increased the counts of *B. lactis* when compared with unsupplemented milk (Janer *et al.*, 2004). The ability of CMP to nourish healthy gut microflora points at its potential as a prebiotic in functional foods, or as a supplement in infant formulae to simulate the beneficial bacteriological effects of breast milk, effects that are supported by its anti-adhesive capacity to pathogens (Brück *et al.*, 2002, 2003a, 2006a). However, when the prebiotic effect of CMP-supplemented infant formulae was tested in healthy term infants, it could not be demonstrated that CMP altered the gastrointestinal flora making it closer to that of breast-fed infants. A possible explanation of the apparent discrepancy between this result and those previously observed *in vitro*, in batch and continuous culture models, is that the prebiotic effect is seen only with low starting populations of the beneficial microbiota (Brück *et al.*, 2006b). Furthermore, to exert desirable effects on the gut microflora, the antiadhesive effect would need to be selective for pathogens and potentially harmful organisms over the desirable probiotic flora (Rhoades *et al.*, 2005).

Dietary intake of CMP as a source of NeuAc was found to increase the sialic acid content of the saliva in piglets that influences its viscosity and protective properties (Wang *et al.*, 2004). Because of its anti-adhesive properties, CMP inhibits the adhesion of cariogenic bacteria such as *Streptococcus mutans*, *S. sanguis* and *S. sobrinus* to the oral cavity (Neeser *et al.*, 1988, 1994; Vacca Smith *et al.*, 2000) and modulates the composition of the dental plaque microbiota (Guggenheim *et al.*, 1999; Schupbach *et al.*, 1996). Various structurally different glycosylated CMP compounds prevent the adhesion to different degrees (Neeser *et al.*, 1988). As part of common personal hygiene products, CMP can prevent dental caries by helping to control acid formation in the dental plaque, in turn reducing hydroxyapatite dissolution from tooth enamel and promoting remineralisation (Aimutis, 2004).

Glycosidic CMP forms containing sialic acid could also regulate food intake and gastrointestinal functions through the stimulation of the release of cholecystokinin (CCK) from intestinal receptors in the duodenum of animals and humans (Beucher *et al.*, 1994; Yvon *et al.*, 1994). CCK controls food intake and digestion by regulating the release of pancreatic enzymes and slowing down intestinal contractions (Pedersen *et al.*, 2000). However, in contrast with reports stating that CMP could suppress appetite, according to Ney *et al.* (2008) CMP did not reduce food intake in growing mice. Whey protein isolate (WPI)-fed Wistar rats experienced a significant reduction in body weight gain when compared to casein-fed animals, but inclusion of CMP did not offer a significant benefit in weight control as compared to WPI alone, although it decreased abdominal fat accumulation (Royle *et al.*, 2008). In humans, a study conducted with adults over a short-time period suggested that CMP (at doses of 0.4 and 2.0 g) had no effect on food intake or on subjective indicators of satiety (Gustafson *et al.*, 2001). Furthermore, no remarkable effects of CMP (0.8 g) were found on CCK release in ten men and ten women, although it was suggested that CMP could play a role in compensatory energy intake regulation (Burton-Freeman, 2008). In another recent paper, CMP-enriched WPI (with a dose of 27 g/day of CMP for 6 months) had no additional effect on the weight loss produced by a high-protein meal (Keogh and Clifton, 2008). In the two latter studies, it was concluded that the most likely explanation for the lack of activity was that insufficient CMP was provided, as compared with the doses used in the rat studies (Keogh and Clifton, 2008). In particular, in food preparations with low protein, higher doses of CMP may be required to achieve the CCK release and satiety effects reported in animals (Burton-Freeman, 2008).

κ-casein exerts immunosuppressive effects, with the inhibitory activity depending on the CMP portion (Otani *et al.*, 1992; Otani and Monnai, 1993; Otani and Hata, 1995). CMP inhibits mitogens from inducing the proliferation of lymphocytes and it can even induce apoptosis of certain lymphocytes (Matin and Otani, 2000; Otani *et al.*, 1995). The immunomodulatory effects of CMP, such as the inhibition of mitogen-induced proliferative responses (Otani *et al.*, 1995) and the enhancement of the proliferation and phagocitic activity of human macrophages (Li and Mine, 2004b), were reported to depend both on the polypeptide portion and on the presence of NeuAc. In contrast, Mikkelsen *et al.* (2005b) stated that sialic acid was not important for the immunomodulatory activity of the four sialylated milk proteins: κ-casein, CMP, lactoferrin or PP3. This discrepancy could be attributed to the dependency of these effects on the immunological responsiveness of the mouse strain used for the cell proliferation assay, the cell type and the mitogen employed to stimulate proliferation (Gauthier *et al.*, 2006; Mikkelsen *et al.*, 2005b; Otani *et al.*, 2005).

Monnai *et al.* (1998) showed that dietary CMP had a significant suppressing activity towards the production of IgG antibodies specific to dietary ovalbumin, helping to down-regulate the immune response to antigens in newborn mammals. Immunisation or feeding of mice with κ-casein gave rise to CMP-specific IgG1, IgG2 and IgM antibodies, whereas CMP *per se* lacked immunogenicity independently of the route of presentation (Mikkelsen *et al.*, 2006). While these results suggest that CMP is not likely to cause allergy, Pizzano *et al.* (2005) reported that at least one patient who had outgrown allergy to cows' milk, presented serum IgE that exclusively recognised bovine κ-casein glycoforms.

κ-casein and CMP may influence downstream events in the immune inflammatory response through the modulation of cytokine production, suppression of interleukin receptors expression, or induction of receptor antagonists (Monnai and Otani, 1997; Otani and Monnai, 1995; Otani *et al.*, 1996). *In vitro* assays showed that κ-casein inhibited all proinflammatory cytokines in lipopolysaccharide-stimulated murine dendritic cells, whereas less marked effects of CMP and lactoferrin were seen, and that these activities were independent of sialic acid (Mikkelsen *et al.*, 2005b). In fact, CMP displayed anti-inflammatory activity in rats with hapten-induced colitis and so its use was proposed in immunosuppressive foods and for the management of inflammatory bowel disease (Daddaoua *et al.*, 2005). The mechanism of action may involve downregulation of interleukin 17 (Th17) and regulatory T cells (Requena *et al.*, 2008).

7.2.2 Nutritional and biological activities that do not depend on glycosylation

Regardless of the sugar content, CMP has a unique amino acid profile, as it contains no aromatic amino acids (Phe, Trp, Tyr), no Cys, Arg or His, but it is rich in branched-chain amino acids (Ile and Thr) and low in Met. These characteristics make it a useful component that can be formulated into a variety of acceptable low Phe foods and beverages (Lim *et al.*, 2007). CMP is suitable for nutrition in cases of phenylketonuria (PKU), a genetic disorder caused by a deficiency of hepatic phenylalanine hydroxylase, with severe neuropsychological consequences. However, since CMP contains limiting amounts of the indispensable amino acids: Arg, His, Leu, Met, Trp and Try, it has to be supplemented to provide an adequate source of protein for individuals with PKU. In support of the utilisation of CMP in the PKU diet, Ney *et al.* (2008) observed similar growth, with significantly lower concentrations of Phe, in plasma and brain of a mouse model of PKU, compared to an amino acid diet. However, because of its high content of Thr, CMP can cause hyperthreoninemia (Fanaro and Vigi, 2002; Rigo *et al.*, 2001), and it has been postulated that the application in infant formula of whey protein concentrates low or free from CMP would give plasma amino acid profiles that more closely resemble that of breast milk-fed infants (lower Thr, higher Trp and Cys) (Mallee and Steijns, 2007; Sandström *et al.*, 2008).

Certain biological activities can be exerted by non-glycosylated κ-casein or CMP, or by shorter fragments within their structure, which can be released by specific enzymatic hydrolysis *in vitro* or *in vivo*. For instance, several short sequences, generated by enzyme digestion, possess antibacterial properties against different Gram-positive and Gram-negative bacteria (López-Expósito *et al.*, 2006; Malkoski *et al.*, 2001).

The region 106–116 of κ-casein is analogous to the fragment 400–411 of fibrinogen γ-chain and thus, CMP can inhibit fibrinogen binding to its platelet receptors, which leads to platelet aggregation and the formation of thrombi (Jollès *et al.*, 1978, 1986; Rutherfurd and Gill, 2000). A number of peptides deriving from the sequence 106–116 of κ-casein that can be produced by tryptic hydrolysis were shown to inhibit platelet aggregation *in vitro* (Léonil and Mollé, 1990; Manso *et al.*, 2002; Maubois *et al.*, 1991; Quian *et al.*, 1995). Interestingly, the antithrombotic activity of bovine, ovine and human CMP, as well as bovine κ-casein fragments 106–116 and 112–116, was achieved *in vivo* at lower doses than those suspected from the *in vitro* platelet aggregation data (1 mg kg^{-1}). CMP was antithrombotic in the guinea-pig (Bal dit Sollier *et al.*, 1996).

Peptides with *in vitro* angiotensin I converting enzyme (ACE)-inhibitory activity were liberated from bovine, ovine and caprine CMP by proteolysis with trypsin (Manso and López-Fandiño, 2003). In addition, CMP and its tryptic peptides exhibited, *in vitro*, relaxant actions on endothelium intact aortic rings, and *in vivo* studies revealed a blood pressure lowering effect of CMP and its tryptic hydrolysate on spontaneously hypertensive rats (Miguel *et al.*, 2007). Provided that their efficacy is proved in humans, CMP and its tryptic peptides, by virtue of their combined cardiovascular effects, could be applied in the development of functional foods to regulate blood circulation.

7.3 Glycosylated whey proteins

Whey contains components that not only provide nutrition, but can also prevent and attenuate disease, or augment conventional therapies, when delivered in amounts that exceed normal dietary intakes. Several reviews have covered the health properties of whey proteins and their clinical implications (Krissansen, 2007; Marshall, 2004). These biological activities are due mainly to the amino acid composition, to the presence of specific clusters of amino acids, or to certain conformations; and, in some cases, these activities are also related to the glycan groups attached to the protein molecule. This section will deal with whey proteins that are naturally glycosylated in milk and their biological activities in the organism. Special attention will be paid to the importance that glycan moieties may play on protein conformation and physiological function.

7.3.1 Lactoferrin

Lactoferrin is an iron-binding glycoprotein (about 80 kDa) of the transferrin family, found mainly in the milk of various mammals and in other exocrine secretions such as tears, saliva, synovial fluids and blood (Lonnerdal, 2003). The concentration of lactoferrin in milk varies widely among species: lactoferrin in human milk is the second most abundant whey protein, next to α-lactalbumin, whereas bovine milk contains much lower amounts of lactoferrin. The concentration of lactoferrin changes from 2–5 mg/ml in colostrum to 0.1–0.3 mg/ml in mature bovine milk. The structure of lactoferrin has been studied in detail (reviewed recently by Legrand *et al.*, 2008). Lactoferrin is composed of two homogeneous globular lobes (the N- and C-lobes). These two lobes share 40% identity, which is believed to be the result of duplication of an early ancestral gene (Baker *et al.*, 2002). Each molecule of lactoferrin is capable of binding two molecules of Fe^{3+} and this binding affects its structure. Each lobe comprises two sub-lobes or domains, which form a cleft where the ferric ion is tightly bound in synergistic cooperation with a carbonate anion. In the iron-free form (apo-lactoferrin), the N-lobe has an 'opened' conformation and the C-lobe has a 'closed' conformation. In the ion-saturated form (holo-lactoferrin), both lobes are in their 'closed' conformations. (Kanyshkova *et al.*, 2001).

Multiple molecular mass forms of bovine lactoferrin have been isolated from colostrum and mature milk. The high and low molecular mass forms have masses of 84 and 81 kDa and are designated as lactoferrin-a and -b, respectively. It has been suggested that the difference is due to the difference in the glycan moieties. The content of the high molecular mass form in colostrum and mature milk is less than half of that of lactoferrin-b. Bovine lactoferrin posses five potential *N*-linked glycosylation sites (Asn^{233}, Asn^{281}, Asn^{368}, Asn^{476}, and Asn^{545}), of which four are always utilised, but Asn^{281} is glycosylated only in lactoferrin-a (Wei *et al.*, 2000). The carbohydrate moieties of bovine lactoferrin consist of high-mannose type glycans, and the relative proportions of these glycans vary with the period of lactation. In bovine lactoferrin, Asn^{233} and Asn^{545} are linked to high-mannose type glycans and Asn^{368} and Asn^{476} with heterogeneous complex-type structures (Wei *et al.*, 2001). In addition, the glycan containing GalNAc is only located at Asn^{476} (Wei and Nishimura, 2003), and *N*-acetylglucosamine (GlcNAc) is present at Asn^{368} (Moore *et al.*, 1997).

LF exerts various biological activities, which can be grouped under the heading 'host defence'. These activities would include antibacterial, antiviral and antitumour activities; regulation of cell growth and differentiation;

anti-inflammatory activities; and modulation of the immune response (reviewed by Legrand *et al.*, 2005, 2008; Valenti and Antonini, 2005; Wakabayashi *et al.*, 2006; Ward *et al.*, 2005; Yamauchi *et al.*, 2006). However, other less-known activities have also been reported for this protein, such as proteolytic activity (Hendrixson *et al.*, 2003), inhibition of bacterial biofilms (Singh *et al.*, 2002) and promotion of bone growth (Cornish *et al.*, 2004).

The significance of glycosylation for lactoferrin is not completely understood. Most of the glycosylation sites are highly exposed on the protein surface, but there is one for which this is not the case, and that is likely to affect protein dynamics and/or function. Asn^{545}, in cow, buffalo, sheep and goat lactoferrins, lies in a surface cleft between the two domains of the C-lobe, where, by providing additional interactions with both domains, it may help to stabilise the 'closed state' (Jabeen *et al.*, 2005; Moore *et al.*, 1997). Crystal structures showed that the carbohydrate chain attached to Asn^{545} appears to contribute to interdomain interactions and may modulate iron release from the C-lobe (Moore *et al.*, 1997). Although it has been suggested that glycan moieties in lactoferrin could be involved in antimicrobial functions (Tomita *et al.*, 1994) and in a decrease of the immunogenicity of the protein (Spik *et al.*, 1988), it has been demonstrated only that glycosylation protects lactoferrin against tryptic hydrolysis. The major tryptic cleavage site reported for bovine lactoferrin is Lys^{282} (Shimazaki *et al.*, 1993), which is located in the contiguous position to Asn^{281}. This explains why the high molecular mass form of bovine lactoferrin, lactoferrin-a, where Asn^{281} is glycosylated, is less susceptible to tryptic hydrolysis than bovine lactoferrin-b (van Veen *et al.*, 2004). Due to this resistance, it would be expected that bovine lactoferrin-a might also be superior in offering protection in the mammary gland and the intestinal tract of the newborn. However, even though bovine lactoferrin-a is about 10 times more resistant to trypsin than bovine lactoferrin-b, it is still much more sensitive to trypsin than human lactoferrin (which is 100-fold more resistant). Because human lactoferrin is also less susceptible to pepsin, it is clear that human lactoferrin has more resistance to survive in the environment of the gastrointestinal tract, and therefore it should be considered when looking for oral applications of lactoferrin. For human lactoferrin, it has also been demonstrated that glycosylation has little effect on protein affinity but confers resistance to hydrolysis. A comparative study between native and deglycosylated human lactoferrin showed that their binding properties for iron, lysozyme and bacterial lipopolysaccharide are identical but the loss of carbohydrate increases the sensitivity to proteolysis (Vanberkel *et al.*, 1995). Finally, in order to gain more insight into the latest understanding of the human health benefits of lactoferrin, the reader may be referred to Chapter 11 of this book.

7.3.2 α-Lactalbumin

α-Lactalbumin is, quantitatively, the second most important protein in bovine whey, representing approximately 20% (w/w) of the total whey protein, and it is fully synthesised in the mammary gland. In human milk, α-lactalbumin is the most abundant protein. It functions as a galactosyltransferase co-enzyme in lactose synthesis. The crystal structure has been solved, revealing α-helical and β-sheet domains and four disulfide bonds. The molecular mass of the bovine protein is 14178 Da and it is a metalloprotein with a high-affinity binding site for Ca^{2+}, although other divalent ions can also interact. In fresh milk, a small part of the native protein occurs in the glycosylated form (Barman, 1970). The glycan moiety is composed mainly of GlcNAc, GalNAc, and Man, and contains low quantities of fucose (Fuc), galactose (Gal) and NeuAc (Tilley *et al.*, 1991). In bovine α-lactalbumin, Asn^{45} is reported to be the single point of attachment of carbohydrate groups, although one more potential glycosylation site is present at Asn^{74}. The glycan moiety does not seem to be involved in the role of α-lactalbumin in lactose biosynthesis, since the glycosylated and nonglycosylated forms are equally active (Barman, 1970).

In addition to its activity as co-factor in the synthesis of lactose, α-lactalbumin has been found to exert other biological activities. α-Lactalbumin is especially rich in Cys and Trp. It is known that Trp and its metabolites regulate several neurobehavioural effects such as appetite, sleeping–waking rhythm and pain perception (Heine *et al.*, 1995). Other studies have found that α-lactalbumin has a prebiotic effect, as demonstrated by using *in vitro* (Brück *et al.*, 2003a; Petschow and Talbott, 1991) and human studies (Bettler and Kullen, 2007; Brück *et al.*, 2006b). *In vitro*, a significant reduction in the count of *Salmonella* and enteropatogenic *Escherichia coli* was observed in media supplemented with α-lactalbumin (Brück *et al.*, 2003a). It has also been observed that α-lactalbumin and its hydrolysates decrease association and internalisation of pathogenic bacteria to intestinal cells (Caco-2) (Brück *et al.*, 2006a; Peterson *et al.*, 1998). However, the mechanism by which α lactalbumin exerts this activity remains unclear. These authors suggested that this whey protein might disturb cellular metabolism by increasing cell permeability. It is likely that its prebiotic effect is related to the high Cys content of the protein, or to a possible antimicrobial effect of the protein and its hydrolysates rather than to the glycan chains of the molecule. As described in Section 7.2.1 for CMP, bacterial binding is mediated by adhesins and capsular material on the bacterial cell surface or by fimbriae or pili, which are specific for various ceramide and ganglioside glycoconjugates of the epithelial cell membranes (Neeser *et al.*, 1995). It is therefore considered that exogenous substances containing the same carbohydrate residues, such as NeuAc, might competitively inhibit bacterial adhesion to intestinal cells and inhibit colonization. However, as only a small proportion of α-lactalbumin is glycosylated, it does not seem likely that the prebiotic effect is primarily caused by the glycosylated fraction of the protein.

An additional activity for α-lactalbumin, not related to the glycan moieties of the molecule but to a conformational change, is the antitumoral activity. Partly unfolded apo-α-lactalbumin forms an active complex with oleic acid known as HAMLET (human α-lactalbumin made lethal to tumor cells). The *in vitro* and *in vivo* activity and the mechanism of action of this molecular complex have been recently reviewed (Hallgren *et al.*, 2008).

7.3.3 Immunoglobulins

Immunoglobulins (Ig) constitute a complex group of glycoproteins present in the serum, tissue fluids and milk of all mammals. Colostrum is particularly rich in immunoglobulins, but their concentration declines during normal lactation. In the cow, the major class of immunoglobulins is IgG1, with concentrations of about 48 g/l in colostrum and 0.6 g/l in mature milk; while in human milk the predominant class of immunoglobulins is IgA, with levels of about 17 g/l in colostrum and about 1 g/l in milk. They play the role of antibodies in response to antigenic or immunogenic stimuli, such as bacteria and viruses, and thus provide protection against microbial infections. The technologies to produce specific immunoglobulins and the clinical studies undertaken to determine their efficacies against human and animal infections have recently been reviewed (Mehra *et al.*, 2006), and the biological basis for their health benefits will be considered in Chapter 10 of this book.

In terms of quaternary structure, immunoglobulins are monomers or polymers of a four-chain molecule, consisting of two light polypeptide chains (with molecular masses in the range of 25 kDa) and two heavy chains (50–70 kDa) (Fig. 7.1). Antibodies form a Y-shaped structure consisting of two identical Fab 'arms', and an Fc stem. The Fab regions contain the antigen-binding sites, while the elimination of the antigen is mediated by the Fc through the activation of host immune mechanisms. All Igs contain carbohydrates at conserved positions in the constant regions of the heavy chains. IgG has a *N*-linked biantennary structure at Asn^{297} on each heavy chain, and crystallographic studies (Deisenhofer, 1981) have shown that the two CH2 domains of the Fc stem do not interact by protein–protein contacts, but instead through an interstitial region that is formed by oligosaccharides (Fig. 7.1). In addition, up to 30% of human and rabbit IgGs possess also Fab-associated carbohydrates. IgA1 and IgD molecules also contain several *O*-linked sugars (Wright and Morrison, 1997).

Structural roles for glycosylation in Igs are well documented (Krotkiewski, 1999; Rudd *et al.*, 2001; Wright and Morrison, 1997). Sugar chains influence physicochemical properties of Igs in a manner similar to other types of glycoproteins, that is: (i) increasing solubility in polar solvents, (ii) increasing thermal resistance, (iii) protecting against protease digestion, and (iv) being responsible, to some exent, for the conformation of Ig molecules. The role of glycosylation in Ig secretion has been extensively studied with inhibitors of *N*-glycosylation (e.g. tunicamycin) or by mutation of Asn^{297}. Secretion of aglycosylated species was totally abolished in the case

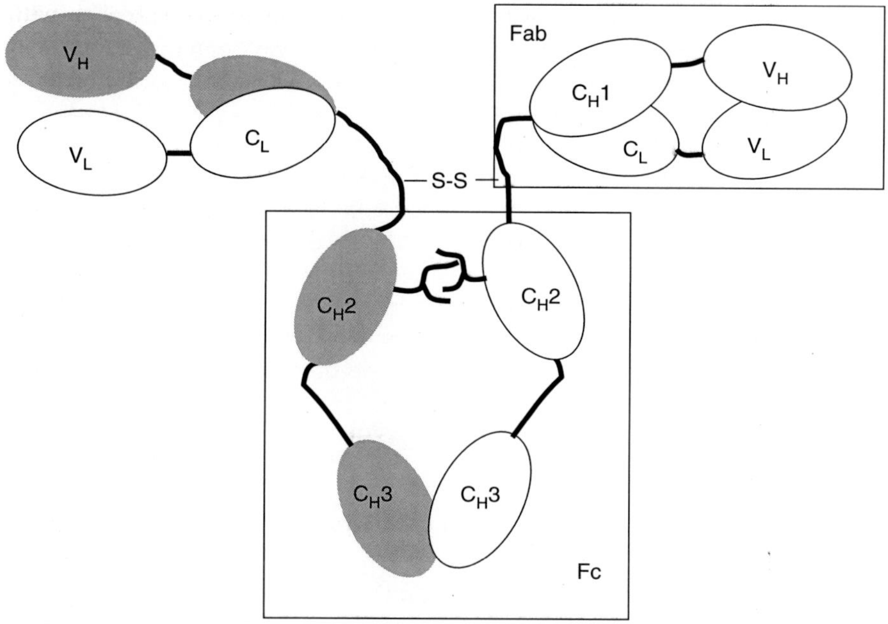

Fig. 7.1 Schematic representation of an IgG1 molecule, composed of two identical light chains (each composed of two domains, labelled VL and CL), and two identical heavy chains (each with four domains, labelled VH, CH1, CH2, and CH3). A flexible hinge separates Fabs from Fc. Amino acid residues in the VH and VL domains form the antigen combining sites. Effector functions, such as complement activation and binding to cytotoxic cells, are mediated by the Fc region. The carbohydrates between the CH2 domains are represented by forked structures. Reproduced from Wright and Morrison (1997), with permission from Elsevier.

of IgE, while secretion of IgG and IgD was largely unaffected and IgM and IgA secretion was reduced to a moderate extent (Taylor and Wall, 1988). There is some discussion about the role of the carbohydrate moiety in the clearance of antibodies from circulation, and it has been suggested that the carbohydrate residues contribute only partly to the survival of IgG *in vivo* (Wawrzynczak *et al.*, 1992).

N-glycans attached at the conserved position Asn^{297} of the CH2 domain of the Fc stem contribute to several biological activities of Igs (see reviews by Krotkiewski, 1999; Rudd, *et al.*, 2001; Wright and Morrison, 1997): (i) they influence the activation via the classical pathway, which starts with the binding of C1q (the first component of the complement cascade) to the CH2 domain of an IgG molecule; (ii) they contribute to the induction of antibody-dependent cellular cytotoxicity; (iii) they influence the binding of IgGs to Fc receptors; and (iv) they contribute to the rapid elimination of antigen–antibody complexes from circulation. In contrast, Fab glycosylation either influences antigen binding in a positive way (enhancing binding) or diminishes binding of antigen, depending on the structure and specificity of the antibody and the structure of the antigen.

It should be noted that almost all key molecules involved in the innate and adaptive immune response (including Igs, cytokines, cytokine receptors, complements, etc.) are glycoproteins. Changes in glycans or glycopeptide may have a significant role in a variety of human immune-related diseases, such as rheumatoid arthritis, autoimmune disease, Wiskott–Aldrich syndrome, infection diseases and cancer (Zhang, 2006). For example, in rheumatoid arthritis, agalactosylated glycoforms of aggregated IgG may induce association with the mannose-binding lectin and contribute to the pathology (Rudd *et al.*, 2001).

7.3.4 Antimicrobial milk enzymes: Lactoperoxidase and lysozyme

Lactoperoxidase is a member of the family of mammalian peroxidases and it is one of the most abundant enzymes in mature milk: it represents approximately 1% (w/w) of the total protein pool in whey (Reiter and Perraudin, 1991). Lactoperoxidase utilizes hydrogen peroxide to oxidize specific substrates, resulting in the production of potent antimicrobial reaction products. This mechanism also requires the presence of a second substrate, such as thiocyanate or a halogen, in order to form the highly reactive end-products, thiocyanate anion (SCN-), I-, or Br- (Kussendrager and van Hooijdonk, 2000).

Bovine lactoperoxidase is a heme-containing glycoprotein that consists of a single polypeptide chain containing 612 amino acid residues, and has a molecular weight of 78 kDa (Kussendrager and van Hooijdonk, 2000). The lactoperoxidase molecule has a carbohydrate content ranging from 6.4 to 11.5%, and possesses five potential *N*-glycosylation sites. All five of the Asn residues (at positions 6, 112, 222, 258, and 349) appear to be glycosylated, except for Asn^{112}, which is found to be 85% glycosylated. The nature of the glycans has been studied by mass spectrometry. Results indicated that the oligosaccharides attached to Asn^{222} are solely Man, those at Asn^{6} and Asn^{222} are complex oligosaccharides, and those at Asn^{258} and Asn^{349} contain a mixture of high mannose, complex and hybrid structures (Wolf *et al.*, 2000). Although previous studies had indicated the presence of sialic acid (Watanabe *et al.*, 1998), its content could not be confirmed by mass spectrometry.

The biological function of the lactoperoxidase system is predominantly the defence against microbial infections, and many *in vitro* studies have shown its bacteriostatic and bactericidal effect against a broad spectrum of microorganisms (reviewed by de Wit and van Hooydonk, 1996). Other reported biological functions of lactoperoxidase include antiviral activity (Yamaguchi *et al.*, 1993), and protection against H_2O_2-mediated peroxidation (Reiter and Perraudin, 1991). The lactoperoxidase system is commercially used as a natural preservative (de Wit and van Hooydonk, 1996).

While it is known that lactoperoxidase is a glycosylated protein, most attention has been focused on its amino acid sequence, the heme linkage to the protein (Thanabal and Lamar, 1989) and the binding of SCN- and I- to the lactoperoxidase molecule (Modi *et al.*, 1989). However, as it occurs for other glycosylated enzymes, it is to be expected that the glycan moieties could play an important role in the biological function of this enzyme. It is already known that carbohydrate structures play a role in dimer formation in other mammalian peroxidases (Moguilevsky *et al.*, 1991). Although previous data indicated that there are no significant differences in the enzymatic activity of the various lactoperoxidase fractions (Paul and Ohlsson, 1985), the establishment of glycosylation-function relationships for lactoperoxidase needs further study.

Lysozyme is known to be a potent antibacterial enzyme that cleaves the β-1,4 glycosidic linkages between *N*-acetylmuramic acid and GlcNAc, resulting in the lysis of the bacterial cell wall. Besides this well-known muramidase-based mechanism, more recently there are also evidences of a non-enzymatic antibacterial mechanism of native and denatured lysozymes. The antibacterial properties and the different modes of action of lysozyme and modified molecules have been reviewed by various authors (Floris *et al.*, 2003; Ibrahim *et al.*, 2002; Masschalck and Michiels, 2003). Lysozyme is one of the most abundant proteins in human milk (0.4 g/L) but its concentration in bovine milk is about 3000 times lower (0.13 mg/L). In addition, human lysozyme, as compared to bovine lysozyme, appears to be more stable (Chandan *et al.*, 1965). Because of its abundance in hen egg white (1–3 g/L), many structural studies have been performed with lysozyme from this source. It is known that the active site of hen egg white lysozyme consists of six subsites A, B, C, D, E, and F, and contains six consecutive sugar residues (Jollès, 1996). There is little information about the importance of the carbohydrate residues in human or bovine milk lysozyme on their activity. From studies with recombinant human lysozyme, it has been concluded that glycosylated and nonglycosylated forms of mutant lysozyme have an enzymatic activity similar to normal human milk lysozyme (Horst *et al.*, 1991).

7.4 Glycoproteins associated with the milk-fat globule membrane (MFGM)

Milk contains a wide number of minor proteins, many of which are heavily glycosylated, distributed mainly in the whey fraction and the fat globule membrane. Although the importance and significance of most of them remain poorly understood, minor milk glycoproteins are becoming of increasing interest in nutritional and pharmacological studies as they may possess interesting physiological properties (Fox and Kelly, 2003).

The MFGM is a rich source for membrane, cytoplasmic or secreted proteins or glycoproteins, which account for only 1–2% of the total protein fraction in milk (Riccio, 2004). Following proteomic studies, up to 120 proteins have been recently identified in the bovine MFGM (Fong *et al.*, 2007; Reinhardt and Lippolis, 2006; Smolenski *et al.*, 2007; Vanderghem *et al.*, 2008). As their functions are not yet very well known, they have been grouped into eight broad functional categories, based primarily on functions listed in the Swiss-Prot database (http://expasy.org/sprot/). These general functions include membrane/protein trafficking, cell signalling or fat transport/metabolism, among others (Reinhardt and Lippolis, 2006).

The contribution of MGFM components to health and disease raises a controversial debate (Dewettinck *et al.*, 2008; Michalski and Januel, 2006). On the one hand, Spitsberg (2005) proposed the use of MFGM as a potential nutraceutical due to the accumulated evidences that some MFGM components acted as health beneficial factors, such as the cholesterolemia-lowering factor, inhibitors of cancer cell growth or gastrointestinal pathogens, vitamin binders, bactericidal agents, etc. On the other hand, bovine MFGM has been also been considered to be a component of milk with adverse effects in relation to certain human pathologies such as autism or coronary heart disease (CHD) (Riccio, 2004). Regarding the latter, contradictory epidemiological studies have been published on bovine milk consumption and the risk of CHD (Moss and Freed, 2003; Warensjo *et al.*, 2004), indicating that these results should be taken with caution (Spitsberg, 2005).

7.4.1 Butyrophilin (BTN)

Butyrophilin (BTN) belongs to the Ig superfamily and it is the most abundant protein in bovine MFGM as it constitutes ~20–43% of the total MFGM protein fraction (Mather, 2000). The sequenced mature primary structure contains 500 amino acids with two site-specific *N*-glycosylation sites, Asn^{55} and Asn^{215} (Table 7.2). Its glycan moieties have been fully characterised and comprise bi-, tri-, and tetra-antennary high-mannose sugar chains (Sato *et al.*, 1995). Furthermore, half of the Asn^{55}-linked oligosaccharides contain the unusual GalNAcβ1-4GlcNAc structure, previously detected in bovine CD36 (see Section 7.4.4.) (Nakata *et al.*, 1993). This structure represents 37% of the total oligosaccharides content in bovine BTN (Sato *et al.*, 1993).

BTN seems to be essential for the regulation of milk fat globule secretion (Mather and Keenan, 1998), with data supporting that BTN plays a direct role in the assembly, transport or secretion of lipid droplets (Ogg *et al.*, 2004). To date, it remains unknown if the glycan moieties of BTN exert some specific biological activity or effect on the human health. In this sense, a role of BTN on the development or suppression of experimental

Table 7.2 Summary of structural and functional features of the major glycoproteins of the bovine milk-fat globule membrane

Protein name	Family	Sequence accession number (ExPASy)	Mr of mature protein (kDa)	pI	Content in milk (% of the total MFGM protein fraction)	Glycan moieties	Glycosylation sites identified	Function	Health aspects	References
Butyrophilin	Immunoglobulin superfamily	P18892	66–67	5.0–5.4	20–43	Bi-, tri- and tetra-antennary high-mannose *N*-linked glycans also containing Gal, GlcNAc, GalNAc and sialic acid.	Asn^{29}, Asn^{189}.	– Milk fat globule secretion.	– Induces or modulates Experimental Allergic Encephalomyelitis – Suppresses Multiple Sclerosis – Influences pathogenesis of autistic behaviour	Heid *et al.* (1983); Sato *et al.* (1995); Mather (2000); Riccio (2004); Mañá *et al.* (2004); Fong *et al.* (2007); Dewettinck *et al.* (2008)
Mucin 1	Mucin family	Q8WML4	120–220	<5	1.2	Major *O*-glycans and *N*-glycans consist of GalNAc, Gal, GlcNAc, Man, Fuc and sialic acid	*O*-sites not determined. Putative *N*-sites: Asn^{139}, Asn^{179}, Asn^{219}, Asn^{362}, Asn^{438}	– Protection from physical damage. – Functions in the development of ducts and cavities during embryogenesis.	– Protective effect against rotavirus infection and invasive gut pathogens (important role of glycan moieties).	Patton & Patton (1990); Patton (1994); Patton *et al.* (1995); Mather (2000); Pallesen *et al.* (2001); Kvistgaard *et al.* (2004); Liu *et al.* (2005); Dewettinck *et al.* (2008)

Table 7.2 *Cont'd*

Protein name	Family	Sequence accession number (ExPASy)	Mr of mature protein (kDa)	pI	Content in milk (% of the total MFGM protein fraction)	Glycan moieties	Glycosylation sites identified	Function	Health aspects	References
Mucin 15		Q8MI01	115–140	<4.5	1.5		*O*-sites not determined. *N*-sites Asn^{7}, Asn^{48}, Asn^{56}, Asn^{71}, Asn^{99}, Asn^{115}, Asn^{124}, Asn^{131}, Asn^{139}, Asn^{152}, Asn^{191}	– Marker of the secretory and ductal epithelium	– Unknown	Pallesen *et al.* (2002); Riccio (2004); Pallesen *et al.* (2007)
PAS 6/7	Cadherin superfamily	Q95114	47 (PAS 7) and 50–52 (PAS 6)	5.6–7.6	17–22	*O*-linked glycans composed of Gal, GalNAc, GlcNAc and Fuc (and sialic acid in PAS 7). High-mannose biantennary *N*-glycans also containing Gal, GlcNAc, Fuc and sialic acid.	Ser^{9}, Asn^{41}, Asn^{209} (PAS 6) Thr^{16}, Asn^{41} (PAS 7)	– Ca-dependent adhesive properties. – Phospholipid-binding.	– Role in epithelization, cell polarization, cell movement and rearrangement, neurite outgrowth. – Synaptic activity in the central nervous system. – Phospholipid-blocking anticoagulant.	Mather *et al.* (1980); Kim *et al.* (1992); Hvarregaard *et al.* (1996); Kim *et al.* (1998); Seok *et al.* (2001); Shi & Gilbert (2003); Shi *et al.* (2004); Riccio (2004)

Content in

CD 36	CD36 family	P26201	75–88	<7	2–5	High mannose-type, hybrid-type, and bi-, tri-, and tetra-antennary complex-type *N*-glycans also containing Gal, GlcNAc, GalNAc, Fuc and sialic acid.	Asn^{78}, Asn^{101}, Asn^{171}, Asn^{204}, Asn^{234}, Asn^{246}, Asn^{320}, Asn^{416}	– Fatty acid transporter – Macrophages marker. – Involvement in phagocytosis by apoptotic neutrophils.	– Unknown	Greenwalt *et al.* (1990); Berglund *et al.* (1996); Riccio (2004); Dewettinck *et al.* (2008)
Proteose-peptone component 3 (PP3)	PP3/GlyCAM-1 family	P80195	18–30	6.3	Not determined. (~300 mg/L in milk)	Biantennary complex-type *N*-glycans with, Man, Fuc, GlcNAc, GalNAc, Gal and NeuAc. *O*-glycans containing GalNAc, Gal, GlcNAc and sialic acid.	Asn^{77}, Thr^{16}, Thr^{86}, Ser^{60}	– Control of the solubility of the calcium phosphate not bound to casein micelles. – Inhibition of lipolytic activities.	– Stimulation of the growth of bifidobacteria. – Growth-inhibitory activity against some Gram+ and Gram– bacteria. – Mitogenic activities on MARK 3 hybridoma cells related to the sialic acid content.	Cartier *et al.* (1990); Mati *et al.* (1993); Sorensen & Petersen (1993); Etienne *et al.* (1994); Girardet *et al.* (1995); Coddeville *et al.* (1998); Kjeldsen *et al.* (2003); Campagana *et al.* (2004); Fong *et al.* (2007); Dewettinck *et al.* (2008)

autoimmune encephalomyelitis (EAE), a multiple sclerosis-like disease, has been described. Bovine BTN can modulate the encephalitogenic T-cell response to myelin oligodendrocyte glycoprotein (MOG), which is a minor component of the myelin membrane in the central nervous system and also a candidate auto-antigen in human multiple sclerosis, through molecular mimicry (Berer *et al.*, 2007; Guggenmos *et al.*, 2004; Stefferl *et al.*, 2000). Alternatively, Mañá *et al.* (2004) have reported a tolerance induction mechanism, also caused by the molecular mimicry between bovine BTN and MOG, which suppressed the clinical manifestations of EAE in mice treated with BTN, before or after immunisation with MOG.

Despite some epidemiological studies (Lauer, 1997; Malosse and Perron, 1993; Malosse *et al.*, 1992) having given some support to the hypothesis of a link between milk consumption and multiple sclerosis (Butcher, 1986), recent reports have pointed out that the recommendation of the removal of milk from the diet of patients suffering multiple sclerosis is not warranted on current evidence (Rutter, 2006). Thus, further investigation must be addressed to elucidate the role of BTN and other milk products, such as vitamins D or B12, in protecting or increasing the risk of multiple sclerosis.

Finally, an association between autism and BTN has been also established. Vojdani *et al.* (2002) found, in children with autism, some cross-reactivity between several neuron-specific antigens, such as MOG and the unglycosylated BTN peptide 89–109, suggesting a role for common antibodies against brain cross-reactive food antigens and infectious agents in the pathogenesis of autistic behaviour.

7.4.2 Mucin-like glycoproteins

Mucins belong to a heterogeneous family of heavily *O*-glycosylated, high-molecular weight glycoproteins present on the apical surface of many epithelial cells (Patton, 2001). Three different mucins, MUC1, MUCX and MUC15 have been described in the milks of several species including bovine (Liu *et al.*, 2005; Pallesen *et al.*, 2002; Patton *et al.*, 1995). Whilst there is relatively extensive information on MUC1 and MUC15, much less is known about MUCX. Conversely to the classical mucins, these glycoproteins are integral membrane components as they are transferred to the MFGM upon its secretion from mammary lactating cells. However, they can also behave as soluble proteins under certain conditions, and various soluble forms of MUC1 lacking the membrane anchor have been described (Mather, 2000).

The mature secreted bovine MUC1 comprises 558 amino acids, giving a calculated average mass of 55.7 kDa (Pallesen *et al.*, 2001). The difference between this mass value and those estimated by SDS-PAGE for the most abundant polymorphic forms of MUC1 (120–220 kDa, Table 7.2), as well as their PAS-reagent stainability, which is specific for glycoproteins (Mather,

2000), indicate an extensive glycosylation. Bovine MUC1 contains around 50% (w:w) of carbohydrate, 30% of which is sialic acid and the rest corresponds to Fuc, Gal, Man, GlcNAc and GalNAc (Pallesen *et al.*, 2001; Snow *et al.*, 1977). The primary structure of the mature bovine MUC15 consists of 307 amino acids (Pallesen *et al.*, 2002) and, as is the case for MUC1, the large divergence between the calculated average molecular mass, 33.3 kDa, and the estimated 115–140 kDa from the electrophoretic mobility on SDS-PAGE gels can be explained by the heavy glycosylation. Thus, it is estimated that the carbohydrate fraction constitutes 65% of the total molecular weight of MUC15, with glycans composed of Fuc, Gal, Man, GalNAc, GlcNAc and sialic acid (Pallesen *et al.*, 2007).

The determination of α-linked (1-3, 1-6 or 1-2) mannose residues, together with the identification of several *N*-glycosylation potential sites both in MUC1 and MUC15 (Table 7.2), indicate that bovine MFGM mucins also contain *N*-glycans in addition to the abundant *O*-linked glycans (Pallesen *et al.*, 2001, 2002). The presence of sialic acids α2-6, and to a lesser degree α2-3, linked to core Galβ1-3GalNAc or Galβ 1-4GlcNAc *O*-linked structures, as well as unsubstituted Galβ 1-3GalNAc, was determined in the three mucins. However, substantial differences have also been described in relation to their glycan moieties; as examples, bovine MUC1 possesses a much lower content of GlcNAc than bovine MUC15 (Pallesen *et al.*, 2007), and MUCX seems to contain more complex carbohydrate structures than MUC1 (Liu *et al.*, 2005).

The physiological significance of mucin-like milk glycoproteins is still not totally elucidated. In general, mucins are proposed to be involved in many important biological processes, including barrier functions on epithelial surfaces, immune responses, or protection of exposed cells from physical damage and invasive pathogenic microorganisms (Moniaux *et al.*, 2001). Mucin like milk glycans are believed to be involved in some of these biological functions. For instance, the high sialic acid content of the mucins confers a strong negative charge to epithelial cell surfaces, which may prevent wall-to-wall adherence and serve to keep ducts and alveoli open (Patton, 1999). Thereby, MUC1 might be a factor in the virtual non-existence of mammary tumours in the cow as it could prevent cancer cell adhesion and aggregation (Patton, 1999). Also, it has been shown that mucins from the human MFGM could protect the breast-fed infants from infectious diseases as they could act as a decoy for bacteria preventing the attachment of pathogens to the gastrointestinal mucosa (Hamosh *et al.*, 1999; Peterson *et al.*, 1998). This is supported by the fact that mucins were shown to be very resistant to degradation in the stomach and large fragments of these glycoproteins were detected in faeces of breast milk-fed infants (Patton, 1994; Peterson *et al.*, 2001). Milk mucins from human MFGM have been also shown to prevent adherence of S-fimbriated *Escherichia coli* to buccal epithelial cells (Schroten *et al.*, 1992) and to have antiviral properties (Habte *et al.*, 2008, 2007; Yolken *et al.*, 1992). In fact, in one of these studies, the removal of sialic acid from the milk mucin resulted in the loss of antirotaviral activity (Yolken *et al.*, 1992).

As occurred for human κ-casein (Section 7.2.1), high molecular weight mucin-like components from bovine MFGM showed a high inhibitory activity towards *Helicobacter pylori* adhesion, which was significantly decreased after the removal of sialic acids (Hirmo *et al.*, 1998). Wang *et al.* (2001), in a further study, also demonstrated the inhibition of *Helicobacter pylori* infection in a BALB/cA mouse model by both bovine non-defatted and defatted MFGM fractions, indicating that the major role in the anti-*Helicobacter pylori* activity was played by the protein component of MFGM. MUC1 from bovine milk was also shown to inhibit efficiently the neuraminidase-sensitive rotavirus RRV strain, whereas the neuraminidase-resistant rotavirus Wa strain was unaffected (Kvistgaard *et al.*, 2004).

7.4.3 Periodic acid/Schiff (PAS 6/7)

Glycoprotein isoforms PAS 6 and PAS 7, also known under the common name 'lactadherin', are major peripheral components of the MFGM of several species, although they are also found in a wide range of tissues and body fluids (Butler *et al.*, 1980; Mather, 2000). Bovine PAS 6/7 share a common polypeptide chain of 45.6 kDa in their mature form (Aoki *et al.*, 1995; Mather *et al.*, 1993), and the differences observed in their SDS-PAGE molecular masses and isoelectric points (Table 7.2) are due to the glycosylation pattern.

The carbohydrate content for PAS 6 and PAS 7 was estimated to be 7% and 5.5%, respectively (Kim *et al.*, 1992). Hvarregaard *et al.* (1996) identified the *N*- and *O*-glycosylation sites, as well as the monosaccharide composition, which was previously reported by Kim *et al.* (1992), for both isoforms (Table 7.2). PAS 6/7 contain similar *O*-glycans, comprising equimolar ratios of Fuc, GlcNAc and Gal, whereas an additional *O*-sialylated form including NeuAc was also detected in PAS 7 (Hvarregaard *et al.*, 1996). Four biantennary complex-type *N*-glycans based on the core structure: Manα1-6(Manα1-3)Manβ1-4GlcNAcβ1-4GlcNAc, bound to additional residues of Fucα1-6, GlcNAcβ1-2, Galβ1-4 or β1-3 and NeuAcα were characterised in PAS 6. In a similar study, Seok *et al.* (2001) determined in PAS 7 nine different biantennary *N*-linked sugar chains comprised of the same core structure linked to Fucα1-6, GlcNAcβ1-2, Galβ1-4, and NeuAcα2-6. An additional *N*-glycan with a unique structure containing a terminal Manα1-6 and two N-acetyllactosamine (Galβ1-4GlcNAc) units linked to a Manα1-3 arm was also described in PAS 7 (Seok *et al.*, 2001).

The structure, topology, and tissue distribution of PAS 6/7 suggests that they may function as adhesive proteins (Mather, 2000). In fact, they belong to the cadherin superfamily, a major class of membrane proteins with prominent roles in cell adhesion, and the regulation of tissue organisation and morphogenesis (Pettitt, 2005). The bovine PAS 6/7 N-terminal region consists of two epidermal growth factor (EGF)-like domains (amino acids

1–44 and 45–89) (Andersen *et al.*, 2000; Hvarregaard *et al.*, 1996). The second EGF domain of bovine PAS 6/7 contains an Arg-Gly-Asp cell-adhesion (unglycosylated) sequence in an extended loop which binds to integrin receptors on cell surfaces (Andersen *et al.*, 1997, 2000; Hvarregaard *et al.*, 1996). Thus, it has been considered that these proteins could have an important physiological role in apoptosis and endocytosis events involving the PAS 6/7 integrin receptors (Andersen *et al.*, 2000). Bovine PAS 6/7 also bind to phospholipids, preferentially phosphatidyl-L-Ser, through their C-termini that resemble blood clotting factors V and VIII (Shi *et al.*, 2004). Consequently, bovine PAS 6/7 may function as potent anticoagulants by competing with blood coagulation proteins for phospholipid binding sites *in vitro* (Shi and Gilbert, 2003). Furthermore, the ability of bovine PAS 6/7 to inhibit thrombosis and hemostasis in mice by binding to platelets with partial or complete phosphatidylserine exposure has recently been reported (Shi *et al.*, 2008).

Human milk PAS 6/7 prevent symptomatic rotavirus-induced infection *in vitro* and *in vivo* (Kvistgaard *et al.*, 2004; Newburg *et al.*, 1998; Peterson *et al.*, 1998; Yolken *et al.*, 1992). Yolken *et al.* (1992) demonstrated that the sialic acid residues of human PAS 6/7 play an important role in their antiviral action. Rotavirus infection occurs in the small intestine and is restricted to the upper and middle part of the villi in mature enterocytes (Kvistgaard *et al.*, 2004). Human PAS 6/7 survive to the stomach digestion and are present in the gastric aspirates of breast-fed infants (Peterson *et al.*, 1996). Results obtained in a cohort of clinical studies with 200 infants determined an increased risk of suffering a symptomatic rotavirus infection when the human milk PAS 6/7 concentration was low (Newburg *et al.*, 1998). Kvistgaard *et al.* (2004) showed that, unlike human PAS 6/7, bovine PAS 6/7 did not inhibit rotavirus infection *in vitro*. A possible explanation for this dissimilar behaviour could lay in the differences between the glycan moieties of human and bovine PAS 6/7. Unfortunately, the glycosylation pattern and carbohydrate composition of human PAS 6/7 has still to be elucidated (Kvistgaard *et al.*, 2004). In contrast to these results, Kanamaru *et al.* (1999) observed that a high-molecular weight glycoprotein fraction from bovine milk containing MUC1, PAS 6/7 and an unidentified 80-kDa protein was effective in inhibiting the replication of human rotavirus *in vitro*. Despite the fact that this fraction was not further purified, these authors suggested that bovine PAS 6/7 might be responsible for the inhibitory activity.

7.4.4 Cluster of Differentiation (CD36)

CD36 is a ubiquitous glycoprotein widely expressed on epithelial cells from the mammary gland or on endothelial cells of adipocytes, monocytes, macrophages, platelets and numerous organs such as heart, liver and pancreas (Greenwalt and Mather, 1985; Mather, 2000). Bovine CD36 has a 52.8 kDa polypeptide backbone that is heavily glycosylated to give rise to a final molecular mass of 75 to 88 kDa, depending on its source (Greenwalt *et al.*, 1990). The CD36 amino acid sequence is well conserved across species and a high homology, from 83 to 87%, is found for human, mouse, rat and hamster CD36 as compared to the bovine counterpart (Mather, 2000). Berglund *et al.* (1996) determined that bovine MFGM CD36 contains around 24% (w/w) carbohydrate (not including the amount of sialic acid) and eight *N*-glycosylation sites occupied with glycans containing Man, Gal, GlcNAc, GalNAc, Fuc and sialic acid (Table 7.2). Previously, Nakata *et al.* (1993) characterised 16 different Asn-linked sugar chains in bovine MFGM CD36 consisting of high mannose-type, hybrid-type, and bi-, tri-, and tetra-antennary complex-type oligosaccharides. About 28% of the total oligosaccharides contained the unusual GalNAcβ1-4GlcNAc group, which was also detected in bovine BTN (Sato *et al.*, 1995).

CD36 is considered a multifunctional protein as it may act as a long-chain fatty acid transporter (Ibrahimi *et al.*, 1996), or as a signal-transmission molecule in human platelets (Huang *et al.*, 1991) and monocytes (Schuepp *et al.*, 1991). Also, it plays a significant role in oxidised low-density lipoprotein uptake *in vitro* and *in vivo* (Endemann *et al.*, 1993; Luangrath *et al.*, 2008) and it can be involved in platelet aggregation and activation by binding thrombospondin (Asch *et al.*, 1987) or collagen fibrils (Tandon *et al.*, 1989), as well as in the macrophage recognition and phagocytosis of apoptotic neutrophils (Savill *et al.*, 1992). In addition, it may also bind to specific proteins on the surface of erythrocytes parasitised with malaria (Oquendo *et al.*, 1989). Nevertheless, the specific biological function of bovine MFGM CD36 is yet unknown and it is not clear why this glycoprotein is expressed on the surface of mammary epithelial cells and the MFGM (Mather, 2000; Wilcox *et al.*, 2002). The *N*-linked oligosaccharides appeared to be important for the stability of the conformation of human and bovine CD36 as both their resistance to proteolytic cleavage and their solubility in non-ionic detergent were dependent on the presence of *N*-linked oligosaccharides (Greenwalt *et al.*, 1991). However, it still remains to determine if the glycan moieties of CD36 have any physiologically-relevant function.

7.4.5 Component-3 of the proteose peptone fraction (PP3)

Bovine PP3 is a native phosphorylated glycoprotein with an apparent molecular mass of 28 kDa and whose concentration in milk is around 300 mg/L. Its high thermostability enables it to remain soluble after heat coagulation of whey proteins and to appear within the proteose peptone fraction (Girardet and Linden, 1996). Considering that PP3 does not derive from the enzymic hydrolysis of caseins, the term proteose peptone is considered a misnomer and alternative terms for PP3 such as 'lactophorin' (Kanno, 1989), 'GlyCAM-1' (Glycosylation-dependent Cell Adhesion Molecule 1) (Groenen et al., 1995) or 'lactoglycophorin' (Girardet and Linden, 1996) have been proposed. Early immunological studies showed PP3 to react with antibodies raised against the MFGM, suggesting that this glycoprotein could be associated with the membrane or share one or several common carbohydrate epitopes with MFGM glycoproteins (Nejjar *et al.*, 1986; Paquet *et al.*, 1988). Later on, Sorensen *et al.* (1997) demonstrated that PP3 is also present in the MFGM fraction, as they ruled out any cross-reaction between PP3 and the glycan moieties of other MFGM proteins. Despite this immunological confirmation, it is still debatable whether PP3 is inherently part of the MFGM or is present as the result of protein–protein interactions (Fong *et al.*, 2007).

Bovine PP3 possesses a single peptidic chain of 135 amino acid residues with five phosphorylation sites at Ser^{29}, Ser^{31}, Ser^{38}, Ser^{40} and Ser^{46}, one *N*-glycosylation site at Asn^{77} and three *O*-glycosylation sites at Thr^{16}, Thr^{86} (Sorensen and Petersen, 1993) and Ser^{60} (Kjeldsen *et al.*, 2003) (Table 7.2). PP3 exhibits a great microheterogeneity of the glycan structures, and the structures of eight different *N*-linked biantennary oligosaccharide chains at Asn^{77}, containing a common octasaccharide core comprised by GlcNAc, Fuc and Man units linked to GalNAc, Gal or NeuAc as terminal units, were determined by NMR (Girardet *et al.*, 1995). Three *O*-glycan structures, (i) GalNAc, (ii) Galβ1-3GalNAc, (iii) Galβ1-4GlcNAcβ1-6(Galβ1-3)GalNAc, linked to the Thr^{86}, were also characterized (Coddeville *et al.*, 1998), and Kjeldsen *et al.* (2003) identified a tetrasaccharide, NeuNAc-Hex-(NeuNAc)-HexNAc, linked both to Thr^{16} and Ser^{60}.

Although the exact biological role of PP3 remains unknown, several biological functions, attributed to the apoglycoprotein region, have been proposed. Thus, the phosphorylated N-terminus of the PP3 may have the ability to bind iron (Bernos *et al.*, 1997) and calcium ions, thus controlling the solubility of the calcium phosphate that is not directly bound to casein micelles (Sorensen and Petersen, 1993). Hydrophobic peptides (from 1 to 5 kDa) of bovine PP3 stimulated the growth of bifidobacteria (Etienne *et al.*, 1994), whereas the amphipathic α-helix region, spanning residues from 98 to the C-terminal 135, was responsible for the inhibition of lipolytic activities (Cartier *et al.*, 1990). Within this amphipathic region, the fragment 113–135, called lactophoricin, displayed growth-inhibitory activity against some Gram-positive (*Streptococcus thermophilus*) and Gram-negative (*Salmonella St Paul* and *Pseudomonas aeruginosa*) bacteria (Campagna *et al.*, 2004). Regarding the biological significance of the glycan moieties, mitogenic activities of caprine, ovine and bovine PP3 on MARK 3 hybridoma cells were related to the sialic acid content (Mati *et al.*, 1993). On the other hand, the cDNA sequence of bovine PP3 has a high homology with a gene coding for GlyCAM-1 in mice and rats (Johnsen *et al.*, 1995). GlyCAM-1 is a sulphated glycoprotein which is indirectly involved in the immune response by acting as an endothelial cell surface ligand for the L-selectin leukocyte. This fact has suggested that PP3 belongs to a family of GlyCAM type molecules that might play a (unknown) role in the immune defence of mammals (Girardet and Linden, 1996).

7.5 Conclusions and future trends

It is recognised that glycans affect the activities of the protein to which they are attached by modulating their functions. Oligosaccharide groups present in glycoproteins impact upon the folding and conformational stability of the polypeptide scaffolds. In addition, carbohydrate-based biomolecular interactions also play key roles in pathological processes, including tumour metastasis, viral and bacterial infection and inflammation. Consequently, the biological and biomedical significance of glycoproteins makes studies of their functions one of the most important fields of research in the post-genomic era. In this field, advances in mass spectrometry instrumentation are being of great utilility in the characterisation of recombinant proteins, including post-translational modifications such as disulfide bonds, glycosylation and phosphorylation. For gaining better understanding of the molecular basis of the function of glycoproteins in biological processes, several chemical processes are being developed. These chemical approaches include chemical methods for probing carbohydrates, detection of proteins using synthetic carbohydrate probes, the construction of homogeneous glycoproteins and carbohydrate microarrays for the rapid analysis of carbohydrate-recognition events (Park *et al.*, 2008).

Despite the important body of knowledge gathered so far about milk glycoproteins, the development of milk glycoprotein-based functional foods is limited by the absence of enough clinical data on their physiological effect and further studies should be carried out to generate the data for a dose/response ratio valid for humans.

Regarding the bioactivity of the glycosylated compounds, deeper structure/activity relationships need to be established to link the role of the carbohydrate moiety in the biological functions. On the basis of the findings published to date, it is thought that further research on milk biologically active minor glycoproteins would allow expanding their potential commercial applications in the food and pharmaceutical industries.

7.6 Acknowledgements

This work was financially supported by projects Consolider-Ingenio 2010. CSD2007-0063 from the Spanish Ministry of Education and Science (MEC), S-505/AGR/0153-01 from the Autonomous Community of Madrid and PIF-SIALOBIOTIC 200870F0101 from CSIC.

7.7 References

ABD EL-SALAM MH, EL-SHIBINI S and BUCHHEIM W (1996) Characteristics and potential uses of the casein macropeptide. *Int Dairy J* 6, 327–341.

AIMUTIS WR (2004) Bioactive properties of milk proteins with particular focus on anticariogenesis. *J Nutr* 134, 989S-995S.

ALEXANDER LJ, STEWART AF, MACKINLAY AG, KAPELISKAYA TV, TKACH TM and GORODETSKY SI (1988) Isolation and characterization of the bovine κ-casein gene. *Eur J Biochem* 178, 395–401.

ANDERSEN MH, BERGLUND L, RASMUSSEN JT and PETERSEN TE (1997) Bovine PAS-6/7 binds alpha(V)beta(5) integrin and anionic phospholipids through two domains. *Biochemistry* 36, 5441–5446.

ANDERSEN MH, GRAVERSEN H, FEDOSOV SN, PETERSEN TE and RASMUSSEN JT (2000) Functional analyses of two cellular binding domains of bovine lactadherin. *Biochemistry* 39, 6200–6206.

AOKI N, KISHI M, TANIGUCHI Y, ADACHI T, NAKAMURA R and MATSUDA T (1995) Molecular cloning of glycoprotein antigens MGP57/53 recognized by monoclonal antibodies raised against bovine milk fat globule membrane. *Biochim Biophys Acta – General Subjects* 1245, 385–391.

ASCH AS, BARNWELL J, SILVERSTEIN RL and NACHMAN RL (1987) Isolation of the thrombospondin membrane-receptor. *J Clin Invest* 79, 1054–1061.

AZUMA N, YAMAUCHI K and MITSUOKA T (1984) Bifidus growth-promoting activity of a glycomacropeptide derived from human κ-casein. *Agr Biol Chem* 48, 2159–2162.

BAKER EN, BAKER HM and KIDD RD (2002) Lactoferrin and transferrin: Functional variations on a common structural framework. *Biochem Cell Biol* 80, 27–34.

BAL DIT SOLLIER C, DROUET L, PIGNAUD G, CHEVALIER C, CAEN JP, FIAT AM, IZQUIERDO C and JOLLÈS P (1996) Effect of κ-casein derived peptides on platelet aggregation and on thrombus formation in the guinea pig. *Thromb Res* 81, 427–437.

BARANYI M, HIRIPI L, SZABÓ L, CATUNDA AP, HARSÁNYI I, KOMÁROMY P and BÖSZE Z (2007) Isolation and some effects of functional, low-phenylalanine κ-casein expressed in the milk of transgenic rabbits. *J Biotechnol* 128, 383–392.

BARMAN TE (1970) Purification and properties of bovine milk glyco-α-lactalbumin. *Biochim Biophys Acta* 214, 242–244.

BERER K, STARCH M, MATHER I and LININGTON C (2007) Regulation through mimicry: The autologous milk protein butyrophilin determines susceptibility to myelin oligodendrocyte glycoprotein-induced experimental autoimmune encephalomyelitis. *Clin Immunol* 123, S11-S11.

BERGLUND L, PETERSEN TE and RASMUSSEN JT (1996) Structural characterization of bovine CD36 from the milk fat globule membrane. *Biochim Biophys Acta – Gene Structure and Expression* 1309, 63–68.

BERNOS E, GIRARDET JM, HUMBERT G and LINDEN G (1997) Role of the O-phosphoserine clusters in the interaction of the bovine milk alpha(s1)-, beta-, kappa-caseins and the PP3 component with immobilized iron(III) ions. *Biochim Biophys Acta – Protein Structure and Molecular Enzymology* 1337, 149–159.

BETTLER J and KULLEN MJ (2007). Infant formula enriched with α-lactalbumin has a prebiotic effect in healthy term infants. *J Pediatr Gastroenterol Nutr* 44, 197–197.

BEUCHER S, LEVENEZ F, YVON M and CORRING T (1994) Effects of gastric digestive products from casein on CCK release by intestinal-cells in rat. *J Nutr Biochem* 5, 578–584.

BOUTROU R, JARDIN J, BLAIS A, TOMÉ D and LÉONIL J (2008) Glycosylations of κ-casein-derived caseinomacropeptide reduce its accessibility to endo- but not exointestinal brush border membrane peptidases. *J Agric Food Chem* 56, 8166–8173.

BRODY EP (2000) Biological activities of bovine glycomacropeptide. *Br J Nutr* 84, S39–S46.

BRÜCK WM, GRAVERHOLT G and GIBSON GR (2002) Use of batch culture and a two-stage continuous culture system to study the effect of supplemental α-lactalbumin and glycomacropeptide on mixed populations of human gut bacteria. *FEMS Microbiol Ecol* 41, 231–237.

BRÜCK WM, GRAVERHOLT G and GIBSON GR (2003a) A two-stage continuous culture system to study the effect of supplemental α-lactalbumin and glycomacropeptide on mixed populations of human gut bacteria challenged with enteropathogenic *Escherichia coli* and *Salmonella* serotype *Typhimurium*. *J Appl Microbiol* 95, 44–53.

BRÜCK WM, KELLEHER SL, GIBSON GR, GRAVERHOLT G and LÖNNERDAL B (2006a) The effects of α-lactalbumin and glycomacropeptide on the association of Caco-2 cells by enteropathogenic *Escherichia coli, Salmonella typhimurium* and *Shigella flexneri*. *FEMS Microbiol Lett* 259, 158–162.

BRÜCK WM, KELLEHER SL, GIBSON GR, NIELSEN KE, CHATTERTON DEW and LÖNNERDAL B (2003b) rRNA probes used to quantify the effects of glycomacropeptide and α-lactalbumin supplementation on the predominant groups of intestinal bacteria of infant rhesus monkeys challenged with enteropathogenic *Escherichia coli*. *J Pediatr Gastroenterol Nutr* 37, 273–280.

BRÜCK WM, REDGRAVE M, TUOHY KM, LONNERDAL B, GRAVERHOLT G, HERNELL O and GIBSON GR (2006b) Effects of bovine α-lactalbumin and casein glycomacropeptide-enriched infant formulae on faecal microbiota in healthy term infants. *J Pediatr Gastroenterol Nutr* 43, 673–679.

BURTON-FREEMAN BM (2008) Glycomacropeptide (CMP) is not critical to whey-induced satiety, but may have a unique role in energy intake regulation through cholecystokinin (CCK). *Physiol Behav* 93, 379–387.

BUTCHER PJ (1986) Milk consumption and multiple-sclerosis – an etiologic hypothesis. *Med Hypotheses* 19, 169–178.

BUTLER JE, PRINGNITZ DJ, MARTENS CL and CROUCH N (1980) Bovine-associated mucoprotein .1. distribution among adult and fetal bovine-tissues and body-fluids. *Differentiation* 17, 31–40.

CAMPAGNA S, MATHOT AG, FLEURY Y, GIRARDET JM and GAILLARD JL (2004) Antibacterial activity of lactophoricin, a synthetic 23-residues peptide derived from the sequence of bovine milk component-3 of proteose peptone. *J Dairy Sci* 87, 1621–1626.

CARTIER P, CHILLIARD Y and PAQUET D (1990) Inhibiting and activating effects of skim milks and proteose-peptone fractions on spontaneous lipolysis and purified lipoprotein–lipase activity in bovine-milk. *J Dairy Sci* 73, 1173–1177.

CHABANCE B, JOLLÈS P, IZQUIERDO C, MAZOYER E, FRANCOUAL C, DROUET L and FIAT AM (1995) Characterization of an antithrombotic peptide from κ-casein in newborn plasma after milk ingestion. *Br J Nutr* 3, 583–590.

CHABANCE B, MARTEAU P, RAMBAUD JC, MIGLIORE-SAMOUR D, BOYNARD M, PERROTIN P, GUILLET R, JOLLÈS P and FIAT AM (1998) Casein peptide release and passage to the blood in humans during digestion of milk or yogurt. *Biochimie* 80, 155–165.

CHANDAN RC, PARRY RM and SHAHANI KM (1965) Purification and some properties of bovine milk lysozyme. *Biochim Biophys Acta* 110, 389–398.

CHOBERT JM, TOUATI A, BERTRANDHARB C, DALGALARRONDO M and NICOLAS MG (1989) Solubility and emulsifying properties of κ-casein and its caseinomacropeptide. *J Food Biochem* 13, 457–473.

CLAVEROL S, BURLET-SCHILTZ O, GAIRIN JE and MONSARRAT B (2003) Characterization of protein variants and post-translational modifications: ESI-MSn analyses of intact proteins eluted from polyacrylamide gels. *Mol Cell Proteomics* 2, 483–493.

CODDEVILLE B, GIRARDET JM, PLANCKE Y, CAMPAGNA S, LINDEN G and SPIK G (1998) Structure of the O-glycopeptides isolated from bovine milk component PP3. *Glycoconj J* 15, 371–378.

COOLBEAR KP, ELGAR DF and AYERS JS (1996) Profiling of genetic variants of bovine κ-casein macropeptide by electrophoretic and chromatographic techniques. *Int Dairy J* 6, 1055–1068.

CORNISH J, CALLON KE, NAOT D, PALMANO KP, BANOVIC T, BAVA U, WATSON M, LIN JM, TONG PC, CHEN Q, CHAN VA, REID HE, FAZZALARI N, BAKER HM, BAKER EN, HAGGARTY NW, GREY AB and REID IR (2004) Lactoferrin is a potent regulator of bone cell activity and increases bone formation *in vivo*. *Endocrinology* 145, 4366–4374.

DADDAOUA A, PUERTA V, ZARZUELO A, SUÁREZ MD, SÁNCHEZ DE MEDINA F and MARTÍNEZ-AUGUSTIN O (2005) Bovine glycomacropeptide is anti-inflammatory in rats with hapten-induced colitis. *J Nutr* 135, 1164–1170.

DALPATHADO DS and DESAIRE H (2008) Glycopeptide analysis by mass spectrometry. *Analyst* 133, 731–738.

DEISENHOFER J (1981) Crystallographic refinement and atomic models of a human fc fragment and its complex with fragment-b of protein-a from Staphylococcus-aureus at 2.9-A and 2.8-A resolution. *Biochemistry* 20, 2361–2370.

DELFOUR A, JOLLES J, ALAIS C and JOLLES P (1965) Caseino-glycopeptides: Characterization of a methionin residue and of the N-terminal sequence. *Biochem Biophys Res Commun* 19, 452–455.

DEWETTINCK K, ROMBAUT R, THIENPONT N, LE TT, MESSENS K and VAN CAMP J (2008) Nutritional and technological aspects of milk fat globule membrane material. *Int Dairy J* 18, 436–457.

DE WIT JN and VAN HOOYDONK ACM (1996) Structure, functions and applications of lactoperoxidase in natural antimicrobial systems. *Neth Milk Dairy J* 50, 227–244.

DZIUBA J and MINKIEWICZ P (1996) Influence of glycosylation on micelle-stabilizing ability and biological properties of C-terminal fragments of cows' κ-casein. *Int Dairy J* 6, 1017–1044.

ENDEMANN G, STANTON LW, MADDEN KS, BRYANT CM, WHITE RT and PROTTER AA (1993) CD36 is a receptor for oxidized low-density-lipoprotein. *J Biol Chem* 268, 11811–11816.

ETIENNE L, GIRARDET JM and LINDEN G (1994) Growth promotion of bifidobacterium-animalis by bovine-milk proteose-peptone. *Lait* 74, 313–323.

FANARO S and VIGI V (2002) Protein quality and quantity in infant formulas. A critical look. *Minerva Pediatr* 54, 203–209.

FLORIS R, RECIO I, BERKHOUT B and VISSER S (2003) Antibacterial and antiviral effects of milk proteins and derivatives thereof. *Curr Pharm Design* 9, 1257–1275.

FONG BY, NORRIS CS and MACGIBBON AKH (2007) Protein and lipid composition of bovine milk-fat-globule membrane. *Int Dairy J* 17, 275–288.

FOSSET S, FROMENTIN G, GIETZEN DW, DUBARRY M, HUNEAU JF, ANTOINE JM, LANG V, MATHIEU-CASSERON F and TOME D (2002) Peptide fragments released from Phe-caseinomacropeptide *in vivo* in the rat. *Peptides* 23, 1773–1781.

FOX PF and KELLY AL (2003) Developments in the chemistry and technology of milk proteins 2. Minor milk proteins. *Food Aust* 55, 231–234.

GAUTHIER SF, POULIOT Y and SAINT-SAUVEUR D (2006) Immunomodulatory peptides obtained by the enzymatic hydrolysis of whey proteins. *Int Dairy J* 16, 1315–1323.

GEYER H and GEYER R (2006) Strategies for analysis of glycoprotein glycosylation. *Biochim Biophys Acta – Proteins and Proteomics* 1764, 1853–1869.

GIRARDET JM and LINDEN G (1996) PP3 component of bovine milk: A phosphorylated whey glycoprotein. *J Dairy Res* 63, 333–350.

GIRARDET JM, CODDEVILLE B, PLANCKE Y, STRECKER G, CAMPAGNA S, SPIK G and LINDEN G (1995) Structure of glycopeptides isolated from bovine milk component PP3. *Eur J Biochem* 234, 939–946.

GREENWALT DE and MATHER IH (1985) Characterization of an apically derived epithelial membrane glycoprotein from bovine-milk, which is expressed in capillary endothelia in diverse tissues. *J Cell Biol* 100, 397–408.

GREENWALT DE, TANDON NN and JAMIESON GA (1991) Role of carbohydrate on the conformation of human and bovine GPIV (CD36). *Thromb Haemost* 65, 1153–1153.

GREENWALT DE, WATT KWK, SO OY and JIWANI N (1990) PAS-IV, an integral membrane-protein of mammary epithelial-cells, is related to platelet and endothelial cell-CD36 (GP-IV). *Biochemistry* 29, 7054–7059.

GROENEN MAM, DIJKHOF RJM and VANDERPOEL JJ (1995) Characterization of a GLY-CAM1-like gene (glycosylation-dependent cell-adhesion molecule-1) which is highly and specifically expressed in the lactating bovine mammary-gland. *Gene* 158, 189–195.

GUGGENHEIM B, SCHMID R, AESCHLIMANN JM, BERROCAL R and NEESER JR (1999) Powdered milk micellar casein prevents oral colonization by *Streptococcus sobrinus* and dental caries in rats: A basis for the caries-protective effect of dairy products. *Caries Res* 33, 446–454.

GUGGENMOS J, SCHUBART AS, OGG S, ANDERSSON M, OLSSON T, MATHER IH and LININGTON C (2004) Antibody cross-reactivity between myelin oligodendrocyte glycoprotein and the milk protein butyrophilin in multiple sclerosis. *J Immunol* 172, 661–668.

GUSTAFSON DR, MCMAHON DJ, MORREY J and NAN R (2001) Appetite is not influenced by a unique milk peptide: Caseinomacropeptide (CMP). *Appetite* 36, 157–163.

HABTE HH, DE BEER C, LOTZ ZE, TYLER MG, KAHN D and MALL AS (2008) Inhibition of human immunodeficiency virus type 1 activity by purified human breast milk mucin (MUC1) in an inhibition assay. *Neonatology* 93, 162–170.

HABTE HH, KOTWAL GJ, LOTZ ZE, TYLER MG, ABRAHAMS M, RODRIQUES J, KAHN D and MALL AS (2007) Antiviral activity of purified human breast milk mucin. *Neonatology* 92, 96–104.

HALLGREN O, AITS S, BREST P, GUSTAFSSON L, MOSSBERG A-K, WULLT B and SVANBORG C (2008) Apoptosis and tumor cell death in response to HAMLET (human α-lactalbumin made lethal to tumor cells). *Adv Exp Med Biol* 606, 217–240.

HAMOSH M, PETERSON JA, HENDERSON TR, SCALLAN CD, KIWAN R, CERIANI RL, ARMAND M, MEHTA NR and HAMOSH P (1999) Protective function of human milk: The milk fat globule. *Semin Perinatol* 23, 242–249.

HEID HW, WINTER S, BRUDER G, KEENAN TW and JARASCH ED (1983) Butyrophilin, an apical plasma membrane-associated glycoprotein characteristic of lactating mammary-glands of diverse species. *Biochim Biophys Acta* 728, 228–238.

HEINE W, RADKE M and WUTZKE KD (1995) The significance of tryptophan in human-nutrition. *Amino Acids* 9, 191–205.

HENDRIXSON DR, QIU J, SHEWRY SC, FINK DL, PETTY S, BAKER EN, PLAUT AG and ST GEME JW, 3RD (2003) Human milk lactoferrin is a serine protease that cleaves Haemophilus surface proteins at arginine-rich sites. *Mol Microbiol* 47, 607–617.

HIRMO S, KELM S, IWERSEN M, HOTTA K, GOSO Y, ISHIHARA K, SUGURI T, MORITA M, WADSTROM T and SCHAUER R (1998) Inhibition of *Helicobacter pylori* sialic acid-specific haemagglutination by human gastrointestinal mucins and milk glycoproteins. *FEMS Immunol Med Microbiol* 20, 275–281.

HOLLAND JW, DEETH H and ALEWOOD PF (2004) Proteomic analysis of κ-casein microheterogeneity. *Proteomics* 4, 743–752.

HOLLAND JW, DEETH H and ALEWOOD PF (2005) Analysis of *O*-glycosylation site occupancy in bovine κ-casein glycoforms separated by two dimensional gel electrophoresis. *Proteomics* 5, 990–1002.

HORST M, HARTH N and HASILIK A (1991) Biosynthesis of glycosylated human lysozyme mutants. *J Biol Chem* 266, 13914–13919.

HUANG MM, BOLEN JB, BARNWELL JW, SHATTIL SJ and BRUGGE JS (1991) Membrane glycoprotein-iv (CD36) is physically associated with the fyn, lyn, and yes protein-tyrosine kinases in human platelets. *Proc Natl Acad Sci U S A* 88, 7844–7848.

HVARREGAARD J, ANDERSEN MH, BERGLUND L, RASMUSSEN JT and PETERSEN TE (1996) Characterization of glycoprotein PAS-6/7 from membranes of bovine milk fat globules. *Eur J Biochem* 240, 628–636.

IBRAHIM HR, AOKI T and PELLEGRINI A (2002) Strategies for new antimicrobial proteins and peptides: Lysozyme and aprotinin as model molecules. *Curr Pharm Design* 8, 671–693.

IBRAHIMI A, SFEIR Z, MAGHARAIE H, AMRI EZ, GRIMALDI P and ABUMRAD NA (1996) Expression of the CD36 homolog (FAT) in fibroblast cells: Effects on fatty acid transport. *Proc Natl Acad Sci U S A* 93, 2646–2651.

IDOTA T, KAWAKAMI H and NAKAJIMA I (1994) *Bifidobacterium* growth promoting activity effect of N-acetylneuraminic acid containing substances. *Biosci Biotechnol Biochem* 58, 1720–1722.

JABEEN T, SHARMA S, SINGH N, BHUSHAN A and SINGH TP (2005) Structure of the zinc-saturated C-terminal lobe of bovine lactoferrin at 2.0 A resolution. *Acta Crystallogr D Biol Crystallogr* 61, 1107–1115.

JANER C, PELÁEZ C and REQUENA T (2004) Caseinomacropeptide and whey protein concentrate enhance *Bifidobacterium lactis* growth in milk. *Food Chem* 86, 263–267.

JOHNSEN LB, SORENSEN ES, PETERSEN TE and BERGLUND L (1995) Characterization of a bovine mammary-gland PP3 cDNA reveals homology with mouse and rat adhesion molecule GLYCAM-1. *Biochim Biophys Acta – Gene Structure and Expression* 1260, 116–118.

JOLLÈS P (1996) *Lysozymes: Model Enzymes in Biochemistry and Biology*, Basel, Birkhäuser Verlag.

JOLLÈS P, LÉVY-TOLEDANO S, FIAT AM, SORIA C, GILLENSEN D, THOMAIDIS A, DUNN FW and CAEN J (1986) Analogy between fibrinogen and casein. Effect of an undecapeptide isolated from κ-casein on platelet function. *Eur J Biochem* 158, 379–382.

JOLLÈS P, LOUCHEUX-LEFEBVRE MH and HENSCHEN A (1978) Structural relatedness of κ-casein and fibrinogen γ-chain. *J Mol Evol* 11, 271–277.

KANAMARU Y, ETOH M, SONG XG, MIKOGAMI T, HAYASAWA H, EBINA T and MINAMOTO N (1999) A high-Mr glycoprotein fraction from cows' milk potent in inhibiting replication of human rotavirus *in vitro*. *Biosci Biotechnol Biochem* 63, 246–249.

KANAMORI M, KAWAGUCHI N, IDUKI F and DOI H (1980) Attachment sites of carbohydrate moieties to peptide chain of bovine κ-casein from normal milk. *Agric Biol Chem* 44, 1855–1861.

KANNO C (1989) Purification and separation of multiple forms of lactophorin from bovine-milk whey and their immunological and electrophoretic properties. *J Dairy Sci* 72, 883–891.

KANYSHKOVA TG, BUNEVA VN and NEVINSKY GA (2001) Lactoferrin and its biological functions. *Biochem Moscow* 66, 1–7.

KAWASAKI Y, ISODA H, SHINMOTO H, TANIMOTO M, DOSAKO S, IDOTA T and NAKAJIMA I (1993b) Inhibition by κ-casein glycomacropeptide and lactoferrin of influenza virus hemmaglutination. *Biosci Biotechnol Biochem* 57, 1214–1215.

KAWASAKI Y, ISODA H, TANIMOTO M, DOSAKO S, IDOTA T and AHIKO K (1992) Inhibition by lactoferrin and κ-casein glycomacropeptide of binding of Cholera toxin to its receptor. *Biosci Biotechnol Biochem* 56, 195–198.

KAWASAKI Y, KAWAKAMI H, TANIMOT M, DOSAKO S, TOMIZAWA A, KOTAKE M and NAKAJIMA I (1993a) pH-dependent molecular weight changes of kappa-casein glycomacropeptide and its preparation by ultrafiltration. *Milchwissenschaft* 48, 191–196.

KELLEHER SL, CHATTERTON D, NIELSEN K and LONNERDAL B (2003) Glycomacropeptide and α-lactalbumin supplementation of infant formula affects growth and nutritional status in infant rhesus monkeys. *Am J Clin Nutr* 77, 1261–1268.

KEOGH JB and CLIFTON P (2008) The effect of meal replacements high in glycomacropeptide on weight loss and markers of cardiovascular disease risk. *Am J Clin Nutr* 87, 1602–1605.

KIM DH, KANNO C and MIZOKAMI Y (1992) Purification and characterization of major glycoproteins, PAS-6 and PAS-7, from bovine-milk fat globule-membrane. *Biochim Biophys Acta* 1122, 203–211.

KIM Y-J, PARK S, OH Y-K, KANG W, KIM HS and LEE EY (2005) Purification and characterisation of human caseinmacropeptide produced by a recombinant *Saccharomyces cerevisiae. Protein Expr Purif* 41, 441–446.

KIM DH, AZUMA N, TANAKA H and KANNO C (1998) Structures of the *N*-linked sugar chains in the PAS-6 glycoprotein from the bovine milk fat globule membrane. *Glycoconj J* 15, 361–369.

KJELDSEN F, HASELMANN KF, BUDNIK BA, SORENSEN ES and ZUBAREV RA (2003) Complete characterization of posttranslational modification sites in the bovine milk protein PP3 by tandem mass spectrometry with electron capture dissociation as the last stage. *Anal Chem* 75, 2355–2361.

KRISSANSEN GW (2007) Emerging health properties of whey proteins and their clinical implications. *J Am Coll Nutr* 26, 713S–723S.

KROTKIEWSKI H (1999) Carbohydrate moiety of immunoglobulins in health and pathology. *Acta Biochim Polonica* 46, 341–350.

KUSSENDRAGER KD and VAN HOOIJDONK ACM (2000) Lactoperoxidase: Physico-chemical properties, occurrence, mechanism of action and applications. *Brit J Nutr* 84, S19–S25.

KVISTGAARD AS, PALLESEN LT, ARIAS CF, LOPEZ S, PETERSEN TE, HEEGAARD CW and RASMUSSEN JT (2004) Inhibitory effects of human and bovine milk constituents on rotavirus infections. *J Dairy Sci* 87, 4088–4096.

LAUER K (1997) Diet and multiple sclerosis. *Neurology* 49, S55-S61.

LEDOUX N, MAHÉ S, DUBARRY M, BOURRAS M, BENAMOUZIG R and TOMÉ D (1999) Intraluminal immunoreactive caseinomacropeptide after milk protein digestion in humans. *Nahrung* 43, 196–200.

LEGRAND D, ELASS E, CARPENTIER M and MAZURIER J (2005) Lactoferrin: A modulator of immune and inflammatory responses. *Cell Mol Life Sci* 62, 2549–2559.

LEGRAND D, PIERCE A, ELASS E, CARPENTIER M, MARILLER C and MAZURIER J (2008) Lactoferrin structure and functions. *Adv Exp Med Biol* 606, 163–194.

LÉONIL J and MOLLÉ D (1990) Liberation of tryptic fragments from caseinmacropeptide of bovine κ-casein involved in platelet function. Kinetic study. *Biochem J* 271, 247–252.

LI EW and MINE Y (2004a) Comparison of chromatographic profile of glycomacropeptide from cheese whey isolated using different methods. *J Dairy Sci* 87, 174–177.

LI EW and MINE Y (2004b). Immunoenhancing effects of bovine glycomacropeptide and its derivatives on the proliferative response and phagocytic activities of human macrophage like cells, U937. *J Agric Food Sci* 52, 2704–2708.

LIESKE B, KONRAD G and KLEINSCHMIDT TH (2004) Isolation of caseinmacropeptide from rennet whey by a multi-stage ultrafiltration process. II. Influence of pH and heating on the carbohydrate moiety of glycomacropeptide. *Milchwissenschaft* 59, 291–294.

LIM K, VAN CALCAR SC, NELSON KL, GLEASON ST and NEY DM (2007) Acceptable low-phenyalanine foods and beverages can be made with glycomacropeptide from cheese whey for individuals with PKU. *Mol Genet Metab* 92, 176–178.

LIU C, ERICKSON AK and HENNING DR (2005) Distribution and carbohydrate structures of high molecular weight glycoproteins, MUC1 and MUCX, in bovine milk. *J Dairy Sci* 88, 4288–4294.

LIU F, LIAO L and CHEN J (2008) Preparation of unglycosylated human caseinomacropeptide by engineering DAB *Escherichia coli*. *J Agric Food Chem* 56, 889–893.

LONNERDAL B (2003) Nutritional and physiologic significance of human milk proteins. *Am J Clin Nutr* 77, 1537S–1543S.

LÓPEZ-EXPÓSITO I, MINERVINI F, AMIGO L and RECIO I (2006) Identification of anitbacterial peptides from bovine κ-casein. *J Food Prot* 69, 2992–2997.

LUANGRATH V, BRODEUR MR, RHAINDS D and BRISSETTE L (2008) Mouse CD36 has opposite effects on LDL and oxidized LDL metabolism *in vivo*. *Arterioscler Thromb Vasc Biol* 28, 1290–1295.

MALKOSKI M, DASHPER SG, O'BRIEN-SIMPSON NM, TALBO GH, MACRIS M, CROSS KJ and REYNOLDS EC (2001) Kappacin, a novel antibacterial peptide from bovine milk. *Antimicrob Agents Chemother* 45, 2309–2315.

MALLEE L and STEIJNS J (2007) Whey protein concentrates from acidic whey: Benefits for use in infant formulas. *Agro Food Ind Hi-Tech* 18, XXIV–XXVI.

MALOSSE D and PERRON H (1993) Correlation-analysis between bovine populations, other farm-animals, house pets, and multiple-sclerosis prevalence. *Neuroepidemiology* 12, 15–27.

MALOSSE D, PERRON H, SASCO A and SEIGNEURIN JM (1992) Correlation between milk and dairy product consumption and multiple-sclerosis prevalence – a worldwide study. *Neuroepidemiology* 11, 304–312.

MAÑÁ P, GOODYEAR M, BERNARD C, TOMIOKA R, FREIRE-GARABAL M and LINARES D (2004) Tolerance induction by molecular mimicry: Prevention and suppression of experimental autoimmune encephalomyelitis with the milk protein butyrophilin. *Int Immunol* 16, 489–499.

MANSO MA and LÓPEZ-FANDIÑO R (2003) Angiotensin I converting enzyme-inhibitory activity of bovine, ovine and caprine κ-casein macropeptides and their tryptic hydrolyzates. *J Food Prot* 66, 1686–1692.

MANSO MA and LÓPEZ-FANDIÑO R (2004) κ-casein macropeptides from cheese whey: Physicochemical, biological, nutritional, and technological features for possible uses. *Food Rev Int* 20, 329–355.

MANSO MA, ESCUDERO C, ALIJO M and LÓPEZ-FANDIÑO R (2002) Platelet aggregation inhibitory activity of bovine, ovine and caprine κ-casein macropeptides and their tryptic hydrolysates. *J Food Prot* 65, 1992–1996.

MARSHALL K (2004) Therapeutic applications of whey proteins. *Alt Medicine Rev* 9, 136–156.

MARTÍN-DIANA AB, FRAGA MJ and FONTECHA J (2002) Isolation and characterisation of caseinmacropeptide from bovine, ovine, and caprine cheese whey. *Eur Food Res Technol* 214, 282–286.

MARTÍN-DIANA AB, FRÍAS J and FONTECHA J (2005) Emulsifying properties of whey protein concentrate and caseinomacropeptide of cow, ewe and goat. *Milchwissenschaft* 60, 363–367.

MASSCHALCK B and MICHIELS CW (2003) Antimicrobial properties of lysozyme in relation to foodborne vegetative bacteria. *Crit Rev Microbiol* 29, 191–214.

MATHER IH (2000) A review and proposed nomenclature for major proteins of the milk-fat globule membrane. *J Dairy Sci* 83, 203–247.

MATHER IH and KEENAN TW (1998) Origin and secretion of milk lipids. *J Mammary Gland Biol Neoplasia* 3, 259–273.

MATHER IH, TAMPLIN CB and IRVING MG (1980) Separation of the proteins of bovine milk-fat-globule membrane by electrofocusing with retention of enzymatic and immunological activity. *Eur J Biochem* 110, 327–336.

MATHER IH, BANGHART LR and LANE WS (1993) The major fat-globule membrane-proteins, bovine components 15/16 and guinea-pig GP-55, are homologous to

MFG-E8, a murine glycoprotein containing Epidermal growth factor-like and factor-V/VIII-like sequences. *Biochem Mol Biol Int* 29, 545–554.

MATI A, MOULTIMATI F, GIRARDET JM, FOKOU E, BELLEVILLENABET F, NABET P and LINDEN G (1993) Mitogenic activity of hydrophobic fractions of proteose peptone from cows, ewes and goats milk measured with mark-3 hybridoma culture. *J Dairy Res* 60, 443–448.

MATIN MA and OTANI H (2000) Release of cytotoxic glycopeptides from human acid casein fraction by the action of stomach proteinases. *Michwissenschaft* 55, 6–10.

MAUBOIS JL, LÉONIL J, TROUVÉ R and BOUHALLAB S (1991) Les peptides du lait à activité physiologique III. Peptides du lait à effet cardiovasculaire: activités antithrombotique et antihypertensive. *Lait* 71, 249–255.

MEHRA R, MARNILA P and KORHONEN H (2006) Milk immunoglobulins for health promotion. *Int Dairy J* 16, 1262–1271.

MERCIER JC (1981) Phosphorylation of caseins. Present evidence for an amino acid triplet code post-translationally recognized by specific kinases. *Biochimie* 63, 1–17.

MERCIER JC, BRIGNON G and RIBADEAU-DUMAS B (1973) Primary structure of bovine κ B casein. Complete sequence. *Eur J Biochem* 35, 222–235.

METWALLY MM, EL-SHIBINY S, DIEB SME, EL-ETRIBY HMM and ASSEM FA (2001) Large scale preparation and growth promoting effects on *Bifidobacterium* of glycomacropeptide from sweet whey. *Egypt J Dairy Sci* 29, 29–36.

MICHALSKI MC and JANUEL C (2006) Does homogenization affect the human health properties of cow's milk? *Trends Food Sci Technol* 17, 423–437.

MIGUEL M, MANSO MA, LÓPEZ-FANDIÑO R, ALONSO MJ and SALAICES M (2007) Vascular effects and antihypertensive properties of κ-casein macropeptide. *Int Dairy J* 17, 1473–1477.

MIKKELSEN TL, BACKMAN S, SØRENSEN ES, BARKHOLT V and FRØKIÆR H (2005b) Sialic acid-containing milk proteins show differential immunomodulatory activities depending on sialic acid. *J Agric Food Chem* 53, 7673–7680.

MIKKELSEN TL, FRØKIÆR H, TOPP C, BONOMI F, IAMETTI F, PICARIELLO G, FERRANTI P and BARKHOLT V (2005a) Caseinomacropeptide self-association is dependent on whether the peptide is free or restricted in κ-casein. *J Dairy Sci* 88, 4228–4238.

MIKKELSEN TL, RASMUSSEN E, OLSEN A, BARKHOLT V and FRØKIÆR H (2006) Immunogenicity of κ-casein and glycomacropeptide. *J Dairy Sci* 893, 824–830.

MINKIEWICZ P, SLANGEN CJ, LAGERWERF FM, HAVERKAMP J, ROLLEMA HS and VISSER S (1996) Reversed-phase high-performance liquid chromatographic separation of bovine κ-casein macropeptide and characterization of isolated fractions. *J Chromatogr A* 743, 123–135.

MODI S, BEHERE DV and MITRA S (1989) Binding of thiocyanate to lactoperoxidase – H-1 and N-15 nuclear magnetic-resonance studies. *Biochemistry* 28, 4689–4694.

MOGUILEVSKY N, GARCIAQUINTANA L, JACQUET A, TOURNAY C, FABRY L, PIERARD L and BOLLEN A (1991) Structural and biological properties of human recombinant myeloperoxidase produced by Chinese-hamster ovary cell-lines. *Eur J Biochem* 197, 605–614.

MOLLÉ D and LÉONIL J (1995) Heterogeneity of the bovine κ-caseinomacropeptide, resolved by liquid chromatography on-line with electrospray ionization mass spectrometry. *J Chromatogr A* 708, 223–230.

MONIAUX N, ESCANDE F, PORCHET N, AUBERT JP and BATRA SK (2001) Structural organization and classification of the human mucin genes. *Front Biosci* 6, D1192–D1206.

MONNAI M and OTANI H (1997) Effect of bovine κ-caseinoglycopeptide on secretion of interleukin-1 family cytokines by P388D1 cells, a line derived from mouse monocyte/macrophage. *Milchwissenschaft* 52, 192–196.

MONNAI M, HORIMOTO Y and OTANI H (1998) Immunomodificatory effect of dietary bovine κ-caseinoglycopeptide on serum antibody levels and proliferative responses of lymphocytes in mice. *Milchwissenschaft* 53, 129–132.

MOORE SA, ANDERSON BF, GROOM CR, HARIDAS M and BAKER EN (1997) Three-dimensional structure of diferric bovine lactoferrin at 2.8 angstrom resolution. *J Mol Biol* 274, 222–236.

MORENO FJ, LÓPEZ-FANDIÑO R and OLANO A (2002) Characterization and functional properties of lactosyl caseinomacropeptide conjugates. *J Agric Food Chem* 50, 5179–5184.

MOSS M and FREED D (2003) The cow and the coronary: Epidemiology, biochemistry and immunology. *Int J Cardiol* 87, 203–216.

NAKAJIMA K, TAMURA N, KOBAYASHI-HATTORI K, YOSHIDA T, HARA-KUDO Y, IKEDO M, SUGITA-KONISHI Y and HATTORI M (2005) Prevention of intestinal infection by glycomacropeptide. *Biosci Biotechnol Biochem* 69, 2294–2301.

NAKATA N, FURUKAWA K, GREENWALT DE, SATO T and KOBATA A (1993) Structural study of the sugar chains of CD36 purified from bovine mammary epithelial-cells – occurrence of novel hybrid-type sugar chains containing the Neu5Ac-α-2-]6GalNAc-β-1-]4GlcNAc and the Man-α-1-]2Man-α-1-]3Man-α-1-]6Man groups. *Biochemistry* 32, 4369–4383.

NEESER JR, CHAMBAZ A, DEL VEDOVO S, PRIGENT MJ and GUGGENHEIM B (1988) Specific and nonspecific inhibition of adhesion of oral actinomyces and streptococcii to erythrocytes and polystyrene by caseinoglycopeptide derivatives. *Infect Immun* 56, 3201–3208.

NEESER JR, GOLLIARD M, WOLTZ A, ROUVET M, GILLMANN ML and GUGGENHEIM B (1994) *In vitro* modulation of oral bacterial adhesion to saliva-coated hydroxyapatite beads by milk casein derivatives. *Oral Microbiol Immunol* 9, 193–201.

NEESER JR, GRAFSTROM RC, WOLTZ A, BRASSART D, FRYDER V and GUGGENHEIM B (1995) A 23 kDa membrane glycoprotein bearing NeuNAc alpha 2–3Gal beta 1–3GalNAc O-linked carbohydrate chains acts as a receptor for Streptococcus sanguis OMZ 9 on human buccal epithelial cells. *Glycobiology* 5, 97–104.

NEJJAR Y, PAQUET D, GODBILLON G and LEDEAUT JY (1986) Immunological relationship between the hydrophobic fraction of proteose-peptone and the milk-fat globule-membrane of bovine-milk. *Int J Biochem* 18, 893–900.

NEWBURG DS, PETERSON JA, RUIZ PALACIOS GM, MATSON DO, MORROW AL, SHULTS J, GUERRERO MD, CHATURVEDI P, NEWBURG SO, SCALLAN CD, TAYLOR MR, CERIANI RL and PICKERING LK (1998) Role of human-milk lactadherin in protection against symptomatic rotavirus infection. *Lancet* 351, 1160–1164.

NEY DM, HULL AK, VAN CALCAR SC, LIU X and ETZEL MR (2008) Dietary glycomacropeptide supports growth and reduces the concentrations of phenylalanine in plasma and brain in a murine model of phenylketonuria. *J Nutr* 138, 316–322.

NILSSON B (1994) Analysis of protein glycosylation by mass spectrometry. *Mol Biotechnol* 2, 243–280.

OGG SL, WELDON AK, DOBBIE L, SMITH AJH and MATHER IH (2004) Expression of butyrophilin (Btn1a1) in lactating mammary gland is essential for the regulated secretion of milk-lipid droplets. *Proc Natl Acad Sci U S A* 101, 10084–10089.

OH S, WOROBO RW, KIM B, RHEEM S and KIM S (2000) Detection of cholera toxin-binding activity of κ-casein macropeptide and optimization of its production by the response surface methodology. *Biosci Biotechnol Biochem* 64, 516–522.

OQUENDO P, HUNDT E, LAWLER J and SEED B (1989) CD36 directly mediates cytoadherence of plasmodium-falciparum parasitized erythrocytes. *Cell* 58, 95–101.

OTANI H and HATA I (1995) Inhibition of proliferative responses of mouse spleen lymphocytes and rabbit Peyer's patch cells by bovine milk caseins and their digests. *J Dairy Res* 62, 339–348.

OTANI H and MONNAI M (1993) Inhibition of proliferative responses of mouse spleen lymphocytes by bovine κ-casein digests. *Food Agric Immunol* 5, 219–229.

OTANI H and MONNAI M (1995) Induction of an interleukin-1 receptor antagonist-like component produced from mouse spleen cells by bovine κ-caseinoglycopeptide. *Biosci Biotechnol Biochem* 59, 1166–1168.

OTANI H, HORIMOTO Y and MONNAI M (1996) Suppression of interleukin-2 receptor expression on mouse CD4+T cells by bovine κ-caseinoglycopeptide. *Biosci Biotechnol Biochem* 60, 1017–1019.

OTANI H, HORIMOTO Y, MATIN MA, OHNUKI H and KAWAHARA T (2005) Effect of bovine κ-caseinoglycopeptide on mitogen-induced proliferative responses of spleen cells from different strains of mice. *Milchwissenschaft* 60, 245–248.

OTANI H, MONNAI M and HOSONO A (1992) Bovine κ-casein as inhibitor of the proliferation of mouse splenocytes induced by lipopolysaccharide stimulation. *Milchwissenschaft* 47, 512–515.

OTANI H, MONNAI M, KAWASAKI Y, KAWAKAMI H and TANIMOTO M (1995) Inhibition of mitogen-induced proliferative responses of lymphocytes by bovine κ-caseinoglycopeptides having different carbohydrate chains. *J Dairy Res* 62, 349–357.

PALLESEN LT, ANDERSEN MH, NIELSEN RL, BERGLUND L, PETERSEN TE, RASMUSSEN LK and RASMUSSEN JT (2001) Purification of MUC1 from bovine milk-fat globules and characterization of a corresponding full-length cDNA clone. *J Dairy Sci* 84, 2591–2598.

PALLESEN LT, BERGLUND L, RASMUSSEN LK, PETERSEN TE and RASMUSSEN JT (2002) Isolation and characterization of MUC15, a novel cell membrane-associated mucin. *Eur J Biochem* 269, 2755–2763.

PALLESEN LT, PEDERSEN LRL, PETERSEN TE and RASMUSSEN JT (2007) Characterization of carbohydrate structures of bovine MUC15 and distribution of the mucin in bovine milk. *J Dairy Sci* 90, 3143–3152.

PAQUET D, NEJJAR Y and LINDEN G (1988) Study of a hydrophobic protein-fraction isolated from milk proteose-peptone. *J Dairy Sci* 71, 1464–1471.

PARK S, LEE M-L and SHIN I (2008) Chemical tools for functional studies of glycans. *Chem Soc Rev* 37, 1579–1591.

PATTON S (1994) Detection of large fragments of the human-milk mucin MUC-1 in feces of breast-fed infants. *J Pediatr Gastroenterol Nutr* 18, 225–230.

PATTON S (1999) Some practical implications of the milk mucins. *J Dairy Sci* 82, 1115–1117.

PATTON S (2001) MUC1 and MUC-X, epithelial mucins of breast and milk. In Newburg DS, *Bioactive Components of Human Milk*, New York, Kluwer Academic / Plenum Publishers, 35–45.

PATTON S, GENDLER SJ and SPICER AP (1995) The epithelial mucin, MUC1, of milk, mammary gland and other tissues. *Biochim Biophys Acta – Reviews on Biomembranes* 1241, 407–423.

PATTON S and PATTON RS (1990) Genetic polymorphism of PAS-I, the mucin-like glycoprotein of bovine milk fat globule membrane. *J Dairy Sci* 73, 3567–3574.

PAUL KG and OHLSSON PI (1985) The chemical structure of lactoperoxidase. In *The Lactoperoxidase System, Chemistry and Biological Significance*, Pruitt KM and Tenovuo JO, editors. New York, Marcel Dekker, pp. 15–29.

PEDERSEN NLR, NAGAIN-DOMAINE C, MAHE S, CHARIOT J, ROZE C and TOME D (2000) Caseinomacropeptide specifically stimulates exocrine pancreatic secretion in the anesthetized rat. *Peptides* 21, 1527–1535.

PETERSON JA, HENDERSON TR, SCALLAN C, KIWAN R, MEHTA NR, TAYLOR MR, CERIANI RL and HAMOSH M (1996) Human milk fat globule (MFG) glycoproteins are present in gastric aspirates of human milk-fed preterm infants. *Pediatr Res* 39, 739–739.

PETERSON JA, PATTON S and HAMOSH M (1998) Glycoproteins of the human milk fat globule in the protection of the breast-fed infant against infections. *Biol Neonate* 74, 143–162.

PETERSON JA, SCALLAN CD, CERIANI RL and HAMOSH M (2001) 'Structural and functional aspects of three major glycoproteins of the human milk fat globule membrane', in Newburg DS, *Bioactive Components of Human Milk*, New York, Kluwer Academic / Plenum Publishers, 179–187.

PETSCHOW BW and TALBOTT RD (1991) Response of bifidobacterium species to growth promoters in human and cow milk. *Pediatr Res* 29, 208–213.

PETTITT J. THE CADHERIN SUPERFAMILY (December 29, 2005), WormBook, ed. *The C. elegans Research Community*, WormBook, doi/10.1895/wormbook.1.50.1.

PISANO A, PACKER NH, REDMOND JW, WILLIAMS KL and GOOLY AA (1994) Characterization of O-linked glycosylation motifs in the glycopeptide domain of bovine κ-casein. *Glycobiology* 4, 837–844.

PIZZANO R, NICOLAI MA, MANZO C, GIANNATTASIO M and ADDEO F (2005) Human IgE binding to the glycosidic moiety of bovine κ-casein. *J Agric Food Chem* 53, 7971–7975.

PRINZENBERG EM, KRAUSE I and ERHARDT G (1999) SSCP analysis at the bovine CSN3 locus discriminates six alleles corresponding to known protein variants (A, B, C, E, F, G) and three new DNA polymorphisms (H, I, A(1)). *Anim Biotechnol* 10, 49–62.

QIAN ZY, JOLLÈS P, MIGLIORE-SAMOUR D, SCHOENTGEN F and FIAT AM (1995) Sheep κ-casein peptides inhibit platelet aggregation. *Biochim Biophys Acta* 1244, 411–417.

REINHARDT TA and LIPPOLIS JD (2006) Bovine milk fat globule membrane proteome. *J Dairy Res* 73, 406–416.

REITER B and PERRAUDIN JP (1991) Lactoperoxidase: Biological functions. In *Peroxydases in Chemistry and Biology*, J. Everse, K. E. Everse and M. B. Grisham (eds), Boca Raton, CRC Press. pp. 143–180.

REQUENA P, DADDAOUA A, MARTÍNEZ-PLATA E, GONZÁLEZ M, ZARZUELO A, SUÁREZ MD, SÁNCHEZ DE MEDINA F and MARTÍNEZ-AUGUSTIN O (2008). Bovine glycomacropeptide ameliorates experimental rat ileitis by mechanisms involving downregulation of interleukin 17. *Br J Pharmacol* 154, 825–832.

RHOADES JR, GIBSON GR, FORMENTIN K, BEER M, GREENBERG N and RASTALL RA (2005) Caseinglycomacropeptide inhibits adhesion of pathogenic *Escherichia coli* strains to human cells in culture. *J Dairy Sci* 88, 3455–3459.

RICCIO P (2004) The proteins of the milk fat globule membrane in the balance. *Trends Food Sci Technol* 15, 458–461.

RIGO J, BOEHM G, GEORGI G, JELINEK J, NYAMBUGABO K, SAWATZKI G and STUDZINSKI F (2001) An infant formula free of glycomacropeptide prevents hyperthreoninemia in formula-fed preterm infants. *J Pediatr Gastroenterol Nutr* 32, 127–130.

ROYLE PJ, MCINTOSH GH and CLIFON PM (2008) Whey protein isolate and glycomacropeptide decrease weight gain and alter body composition in male Wistar rats. *Br J Nutr* 100, 88–93.

RUDD PM, ELLIOTT T, CRESSWELL P, WILSON IA and DWEK RA (2001) Glycosylation and the immune system. *Science* 291, 2370–2376.

RUTHERFURD KJ and GILL HS (2000) Peptides affecting coagulation. *Br J Nutr* 84, S99–S102.

RUTTER ERF (2006) Multiple sclerosis and milk: To drink or not to drink? *Int J Dairy Technol* 59, 223–228.

SAITO T and ITOH T (1992) Variations and distributions of O-glycosidically linked sugar chains in bovine κ-casein A. *J Dairy Sci* 75, 1768–1774.

SANDSTRÖM O, LÖNNERDAL B, GRAVERHOLT G and HERNELL O (2008) Effects of α-lactalbumin-enriched formula with different levels of glycomacropeptide on infant nutrition. *Am J Clin Nutr* 87, 921–928.

SATO T, FURUKAWA K, GREENWALT DE and KOBATA A (1993) Most bovine-milk fat globule-membrane glycoproteins contain asparagine-linked sugar chains with GalNAc-β-1-]4GlcNAc groups. *J Biochem* 114, 890–900.

SATO T, TAKIO K, KOBATA A, GREENWALT DE and FURUKAWA K (1995) Site-specific glycosylation of bovine butyrophilin. *J Biochem* 117, 147–157.

SAVILL J, HOGG N, REN Y and HASLETT C (1992) Thrombospondin cooperates with CD36 and the vitronectin receptor in macrophage recognition of neutrophils undergoing apoptosis. *J Clin Invest* 90, 1513–1522.

SCHROTEN H, LETHEN A, HANISCH FG, PLOGMANN R, HACKER J, NOBISBOSCH R and WAHN V (1992) Inhibition of adhesion of S-fimbriated *Escherichia-coli* to epithelial-cells by meconium and feces of breast-fed and formula-fed newborns – mucins are the major inhibitory component. *J Pediatr Gastroenterol Nutr* 15, 150–158.

SCHUEPP BJ, PFISTER H, CLEMETSON KJ, SILVERSTEIN RL and JUNGI TW (1991) CD36-mediated signal transduction in human monocytes by anti-CD36 antibodies but not by anti-thrombospondin antibodies recognizing cell membrane-bound thrombospondin. *Biochem Biophys Res Commun* 175, 263–270.

SCHUPBACH P, NEESER JR, GOLLIARD M, ROUVET M and GUGGENHEIM B (1996) Incorporation of caseinoglycomacropeptide and caseinophosphopeptide into the salivary pellicle inhibits adherence of mutans streptococci. *J Dent Res* 75, 1779–1788.

SEOK JS, SHIMODA M, AZUMA N and KANNO C (2001) Structures of the N-linked sugar chains in PAS-7 glycoprotein sharing the same protein core with PAS-6 glycoprotein from the bovine milk fat globule membrane. *Biosci Biotechnol Biochem* 65, 901–912.

SHI J, PIPE SW, RASMUSSEN JT, HEEGAARD CW and GILBERT GE (2008) Lactadherin blocks thrombosis and hemostasis *in vivo*: Correlation with platelet phosphatidylserine exposure. *J Thromb Haemost* 6, 1167–1174.

SHI JL and GILBERT GE (2003) Lactadherin inhibits enzyme complexes of blood coagulation by competing for phospholipid-binding sites. *Blood* 101, 2628–2636.

SHI JL, HEEGAARD CW, RASMUSSEN JT and GILBERT GE (2004) Lactadherin binds selectively to membranes containing phosphatidyl-L-serine and increased curvature. *Biochim Biophys Acta – Biomembranes* 1667, 82–90.

SHIMAZAKI K, TANAKA T, KON H, OOTA K, KAWAGUCHI A, MAKI Y, SATO T, UEDA Y, TOMIMURA T and SHIMAMURA S (1993) Separation and characterization of the C-terminal half molecule of bovine lactoferrin. *J Dairy Sci* 76, 946–955.

SINGH PK, PARSEK MR, GREENBERG EP and WELSH MJ (2002) A component of innate immunity prevents bacterial biofilm development. *Nature* 417, 552–555.

SMOLENSKI G, HAINES S, KWAN FYS, BOND J, FARR V, DAVIS SR, STELWAGEN K and WHEELER TT (2007) Characterisation of host defence proteins in milk using a proteomic approach. *J Proteome Res* 6, 207–215.

SNOW LD, COLTON DG and CARRAWAY KL (1977) Purification and properties of major sialoglycoprotein of milk-fat globule membrane. *Arch Biochem Biophys* 179, 690–697.

SORENSEN ES and PETERSEN TE (1993) Phosphorylation, glycosylation and amino-acid-sequence of component PP3 from the proteose peptone fraction of bovine-milk. *J Dairy Res* 60, 535–542.

SORENSEN ES, RASMUSSEN LK, MOLLER L and PETERSEN TE (1997) The localization and multimeric nature of component PP3 in bovine milk: Purification and characterization of PP3 from caprine and ovine milks. *J Dairy Sci* 80, 3176–3181.

SPIK G, CODDEVILLE B and MONTREUIL J (1988) Comparative study of the primary structures of sero-, lacto- and ovotransferrin glycans from different species. *Biochimie* 70, 1459–1469.

SPITSBERG VL (2005) Bovine milk fat globule membrane as a potential nutraceutical. *J Dairy Sci* 88, 2289–2294.

STEFFERL A, SCHUBART A, STORCH M, AMINI A, MATHER I, LASSMANN H and LININGTON C (2000) Butyrophilin, a milk protein, modulates the encephalitogenic T cell response to myelin oligodendrocyte glycoprotein in experimental autoimmune encephalomyelitis. *J Immunol* 165, 2859–2865.

STRÖMQUIST M, FALK P, BERGSTRÖM S, HANSSON L, LÖNNERDAL B, NORMARK S and HERNELL O (1995) Human milk κ-casein and inhibition of *Helicobacter pylori* adhesion to human gastric mucosa. *J Pediatr Gastroenterol Nutr* 21, 288–296.

TAKEUCHI M, TSUDA E, YOSHIKAWA M, SASAKI R and CHIBA H (1985) Fractionation and characterisation of 9 subcomponents of bovine κ-casein A. *Agric Biol Chem* 49, 2269–2276.

TALBO GH, SUCKAU D, MALKOSKI M and REYNOLDS EC (2001) MALDI-PSD-MS analysis of the phosphorylation sites of caseinomacropeptide. *Peptides* 22, 1093–1098.

TANDON NN, KRALISZ U and JAMIESON GA (1989) Identification of glycoprotein-IV (CD36) as a primary receptor for platelet-collagen adhesion. *J Biol Chem* 264, 7576–7583.

TAYLOR AK and WALL R (1988) Selective removal of alpha heavy-chain glycosylation sites causes immunoglobulin a degradation and reduced secretion. *Mol Cell Biol* 8, 4197–4203.

THANABAL V and LAMAR GN (1989) A nuclear overhauser effect investigation of the molecular and electronic-structure of the heme crevice in lactoperoxidase. *Biochemistry* 28, 7038–7044.

THOMÄ C and KULOZIK U (2004) Methods of obtaining isolated caseinomacropeptide from milk and whey and functional properties. *Bull IDF* 389, 74–77.

THOMÄ-WORRINGER C, SØRENSEN J and LÓPEZ-FANDIÑO R (2006) Health effects and technological features of caseinomacropeptide. *Int Dairy J* 16, 1324–1333.

TILLEY CA, SINGER A, HARRIS-BRANDTS M and MOSCARELLO MA (1991) The major oligosaccharide of bovine α-lactalbumin carries terminal β-linked GalNAc. *Glycoconjugate J* (Abstr. 11th Int. Symp. On Glycoconjugates, 30 June-5 July, Toronto, Canada) 8, 249–250.

TOMITA S, MATSUE M, MATSUYAMA J and KIYOSAWA I (1994) Agglutination of bacterial-cells of Clostridium-innocuum, Bifidobacterium-longum, and Micrococcus-luteus by lactoferrin and ovotransferrin. *Biosci Biotechnol Biochem* 58, 722–726.

VACCA SMITH AM and BOWEN WH (2000) The effects of milk and κ-casein on salivary pellicle formed on hydroxyapatite discs *in situ*. *Caries Res* 34, 88–93.

VALENTI P and ANTONINI G (2005) Lactoferrin: An important host defence against microbial and viral attack. *Cell Mol Life Sci* 62, 2576–2587.

VAN BERKEL PHC, GEERTS MEJ, VAN VEEN HA, KOOIMAN PM, PIEPER FR, DEBOER HA and NUIJENS JH (1995) Glycosylated and unglycosylated human lactoferrins both bind iron and show identical affinities towards human lysozyme and bacterial lipopolysaccharide, but differ in their susceptibilities towards tryptic proteolysis. *Biochem J* 312, 107–114.

VANCE BA, WU W, RIBAUDO R, SEGAL DM and KEARSE KP (1997) CD69 heterogeneity results from usage of both typical (N-X-T) and atypical (N-X-C) N-linked glycosylation motifs. *J Allergy Clin Immunol* 99, 1869–1869.

VANDERGHEM C, BLECKER C, DANTHINE S, DEROANNE C, HAUBRUGE E, GUILLONNEAU F, DE PAUW E and FRANCIS F (2008) Proteome analysis of the bovine milk fat globule: Enhancement of membrane purification. *Int Dairy J* 18, 885–893.

VAN VEEN HA, GEERTS MEJ, VAN BERKEL PHC and NUIJENS JH (2004) The role of N-linked glycosylation in the protection of human and bovine lactoferrin against tryptic proteolysis. *Eur J Biochem* 271, 678–684.

VOJDANI A, CAMPBELL AW, ANYANWU E, KASHANIAN A, BOCK K and VOJDANI E (2002) Antibodies to neuron-specific antigens in children with autism: Possible cross-reaction with encephalitogenic proteins from milk, Chlamydia pneumoniae and Streptococcus group A. *J Neuroimmunol* 129, 168–177.

VREEMAN HJ, VISSER S, SLANGEN CJ and VAN RIEL JAM (1986) Characterization of bovine κ-casein fractions and the kinetics of chymosin-induced macropeptide release from carbohydrate-free and carbohydrate containing fractions determined by high performance gel permeation chromatography. *Biochem J* 240, 87–97.

WAKABAYASHI H, YAMAUCHI K and TAKASE M (2006) Lactoferrin research, technology and applications. *Int Dairy J* 16, 1241–1251.

WANG B, STAPLES A, SUN Y, KARIM M and BRAND-MILLER J (2004) Effect of dietary sialic acid supplementation on saliva content in piglets. *Asia Pac J Clin Nutr* 13, S75.

WANG B, YU B, KARIM M, HU HH, SUN Y, MCGREEVY P, PETOCZ P, HELD S and BRAND-MILLER J (2007) Dietary sialic acid supplementation improves learning and memory in piglets. *Am J Clin Nutr* 85, 561–569.

WANG X, HIRMO S, WILLEN R and WADSTROM T (2001) Inhibition of *Helicobacter pylori* infection by bovine milk glycoconjugates in a BALB/cA mouse model. *J Med Microbiol* 50, 430–435.

WARD PP, PAZ E and CONNEELY OM (2005) Multifunctional roles of lactoferrin: A critical overview. *Cell Mol Life Sci* 62, 2540–2548.

WARENSJO E, JANSSON JH, BERGLUND L, BOMAN K, AHREN B, WEINEHALL L, LINDAHL B, HALLMANS G and VESSBY B (2004) Estimated intake of milk fat is negatively associated with cardiovascular risk factors and does not increase the risk of a first acute myocardial infarction. A prospective case-control study. *Br J Nutr* 91, 635–642.

WATANABE S, VARSALONA F, YOO YC, GUILLAUME JP, BOLLEN A, SHIMAZAKI K and MOGUILEVSKY N (1998) Recombinant bovine lactoperoxidase as a tool to study the heme environment in mammalian peroxidases. *FEBS Lett* 441, 476–479.

WAWRZYNCZAK EJ, CUMBER AJ, PARNELL GD, JONES PT and WINTER G (1992) Blood clearance in the rat of a recombinant mouse monoclonal-antibody lacking the N-linked oligosaccharide side-chains of the CH2 domains. *Mol Immunol* 29, 213–220.

WEI Z and NISHIMURA T (2003) Comparative study of glycans in lactoferrin-a and lactoferrin-b. In Shahidi F, Ho C-T, Watanabe S and Osawa T, *Food Factors in Health Promotion and Disease Prevention*, New York, Oxford University Press, 410–418.

WEI Z, NISHIMURA T and YOSHIDA S (2000) Presence of a glycan at a potential N-glycosylation site, Asn-281, of bovine lactoferrin. *J Dairy Sci* 83, 683–689.

WEI Z, NISHIMURA T and YOSHIDA S (2001) Characterization of glycans in a lactoferrin isoform, lactoferrin-a. *J Dairy Sci* 84, 2584–2590.

WILCOX CP, JANOLINO VG and SWAISGOOD HE (2002) Isolation and partial characterization of CD36 from skim milk. *J Dairy Sci* 85, 1903–1908.

WOLF SM, FERRARI RP, TRAVERSA S and BIEMANN K (2000) Determination of the carbohydrate composition and the disulfide bond linkages of bovine lactoperoxidase by mass spectrometry. *J Mass Spectrom* 35, 210–217.

WONG PYY, NAKAMURA S and KITTS DD (2006) Functional and biological activities of casein glycomacropeptide as influenced by lipohilization with medium and long chain fatty acid. *Food Chem* 97, 310–317.

WRIGHT A and MORRISON SL (1997) Effect of glycosylation on antibody function: Implications for genetic engineering. *Trends Biotechnol* 15, 26–32.

XU Y, SLEIGH R, HOURIGAN J and JOHNSON R (2000) Separation of bovine immunoglobulin G and glycomacropeptide from dairy whey. *Process Biochem* 36, 393–399.

YAMAGUCHI Y, SEMMEL M, STANISLAWSKI L, STROSBERG AD and STANISLAWSKI M (1993) Virucidal effects of glucose-oxidase and peroxidase or their protein conjugates on human-immunodeficiency-virus type-1. *Antimicrob Agents Chemother* 37, 26–31.

YAMAUCHI K, WAKABAYASHI H, SHIN K and TAKASE M (2006) Bovine lactoferrin: Benefits and mechanism of action against infections. *Biochem Cell Biol* 84, 291–296.

YOLKEN RH, PETERSON JA, VONDERFECHT SL, FOUTS ET, MIDTHUN K and NEWBURG DS (1992) Human-milk mucin inhibits rotavirus replication and prevents experimental gastroenteritis. *J Clin Invest* 90, 1984–1991.

YVON M, BEUCHER S, GUILLOTEAU P, LE HUEROU-LURON I and CORRING T (1994) Effects of caseinomacropeptide (CMP) on digestion regulation. *Reprod Nutr Dev* 34, 527–537.

ZEVACO C and RIBADEAU-DUMAS B (1984) A study on the cabohydrate binding sites of κ-casein using high performance liquid chromatography. *Milchwissenschaft* 39, 206–210.

ZHANG XL (2006) Roles of glycans and glycopeptides in immune system and immune-related diseases. *Curr Med Chem* 13, 1141–1147.

ZOPF D and ROTH S (1996) Oligosaccharide anti-infective agents. *Lancet* 347, 1017–1021.

8

Application of dairy-derived ingredients in food intake and metabolic regulation

T. Akhavan, S. Panahi, G. H. Anderson, B. L. Luhovyy, University of Toronto, Canada

Abstract: This chapter discusses the role of milk and dairy products, and their ingredients in obesity and the regulation of food intake and components of metabolic syndrome. In addition to protein (whey and casein), fat (saturated, mono- and poly-unsaturated fatty acids) and carbohydrate (lactose), milk contains biologically active substances such as immunoglobulins, enzymes, antimicrobial peptides, oligosaccharides, hormones, cytokines and growth factors. Each of these may affect food intake and metabolic regulation through a large number of physiologic mechanisms. Thus, their actions may explain the positive health associations between more frequent dairy consumption, a healthier body weight, and decreased risk of developing the metabolic syndrome.

Key words: milk, dairy, food intake, metabolic syndrome, functional foods.

8.1 Introduction

Metabolic syndrome is defined by the presence of a combination of three or more of the following: (a) abdominal circumference >102 cm (40 inches) for men and 88 cm (35 inches) for women; (b) hypertension; (c) hyperglycemia; (d) dyslipidemia (elevated triacylglycerols (TG) and low levels of high density lipoproteins (HDL) in the blood) (Torpy *et al.*, 2006). An American heart association, The National Heart, Lung and Blood Institute (Grundy *et al.*, 2005), has set the criteria of blood pressure as ≥130/85 mmHg; fasting plasma glucose ≥5.6 mmol/L (100 mg/dL); triglycerides >1.7 mmol/L (150 mg/dL), HDL cholesterol <1.03 mmol/L (40 mg/dL) in men or <1.3 mmol/L (50 mg/dL) in women. Most of the metabolic disorders occur in the presence of insulin resistance (Bonora *et al.*, 1998). In addition, metabolic syndrome is further characterized by prothrombotic and proinflammatory states and is often associated with fatty liver, cholesterol gallstones, obstructive sleep apnea, gout, depression, musculoskeletal disease and polycystic ovarian syndrome (Grundy, 2008).

Impaired fasting glucose, impaired glucose tolerance and type 2 diabetes are the most frequently identified risk factors for the metabolic syndrome. In addition, high cholesterol levels, particularly high low-density lipoprotein (LDL) and low HDL are risk factors for the development of (CVD). These risk factors combine to markedly increase the risk of cardiovascular disease CVD, the leading cause of morbidity and mortality in developed countries (Isomaa *et al.*, 2001).

A dietary and lifestyle approach to the prevention and management of the metabolic syndrome is recognized as the first preferred strategy to improve metabolic functions. Thus, it is important to identify foods, food components and dietary practices that contribute to a healthy body weight and reduce risk factors for the metabolic syndrome and other obesity-related co-morbidities.

For this purpose, dairy products and ingredients are of interest for several reasons. First, milk and dairy products are readily available, good sources of high quality proteins and relatively inexpensive. Second, dairy products and components, particularly dairy proteins, are known to suppress short-term food intake, increase subjective satiety and stimulate the mechanisms known to signal satiation and satiety (Aziz and Anderson, 2007), and do so beyond what is attributable to their energy content alone. Third, positive associations have been found between dairy consumption and the maintenance of healthy body weights (Barr *et al.*, 2000; Phillips *et al.*, 2003). Fourth, calcium (Zemel *et al.*, 2004; Heaney *et al.*, 2002), medium-chain triglycerides (St-Onge and Jones, 2003) and conjugated linoleic acid (CLA) (Wang and Jones, 2004) in dairy products have been established as factors modulating lipid metabolism and energy expenditure. Finally, the reduced consumption of milk and other dairy products in the last few decades has been associated with a simultaneous rise in metabolic disease rates such as obesity, dysglycemia and dyslipidemia. Therefore, consumption of milk and dairy products has the potential to be an effective part of countermeasures against obesity, the metabolic syndrome and related diseases.

The objective of this chapter is to examine the role of milk and milk derived components as modulators of obesity, food intake and the metabolic syndrome.

8.2 Dairy-derived ingredients

In addition to protein (whey and casein), fat (saturated, mono- and polyunsaturated fatty acids) and carbohydrate (lactose), milk contains biologically active substances such as immunoglobulins, enzymes, antimicrobial peptides, oligosaccharides, hormones, cytokines and growth factors. The health benefits of dairy products may be due, in part, to each of these components.

8.2.1 Milk proteins and peptides

The primary effect of milk on both food intake and metabolic regulation originates from milk proteins, which is the main focus of this chapter. Casein and whey proteins are high quality proteins accounting for 80% and 20% of cow milk proteins, respectively. A large majority of casein is consumed in dairy products but whey protein, the by-product of cheese manufacturing, is under-utilized.

Bioactive peptides are both naturally present in milk and produced through the action of bacteria in fermented dairy products such as yogurt, sour milk and cheese. Milk proteins are considered the most important food source of bioactive peptides and an increasing number of these peptides has been identified in milk protein hydrolysates and fermented dairy products. At least 26 bioactive peptides are encrypted in the primary structure of milk proteins and many of them have been isolated from dairy products, including sour milk, yogurts, and cheeses (Haque *et al.*, 2009). Many bioactive peptides are now considered to be multifunctional peptides, such as α-lactorphin, which is a tetrapeptide originally released from milk protein α-LA by enzymatic hydrolysis, has opioid activity, but also inhibits the angiotensin-converting enzyme (ACE) to have an antihypertensive effect (Meisel, 2004).

Bovine casein (CN) consists of α_{s1}-CN (12–15 g/L of skim milk), α_{s2}-CN (3–4 g/L), β-CN (9–11 g/L), and κ-CN (2–4 g/L) (Farrell *et al.*, 2004). Caseins, when digested by proteolytic enzymes, produce bioactive peptides that affect immune, digestive, cardiovascular and nervous systems (Korhonen and Pihlanto, 2006). Although caseinomacropeptide (CMP) or the glycosylated form of CMP called glycomacropeptide (GMP) is present in κ-casein, it becomes a whey constituent because it is hydrolyzed by chymosin during cheese-making into para-κ-casein, which remains in the curd, and into CMP, which is removed with the whey liquid.

Bovine whey protein includes beta-lactoglobulin (β-LG) (2–4 g/L of skim milk), alpha-lactalbumin (α-LA) (0.6–1.7 g/L), lactoferrin (0.02–0.1 g/L), serum albumin (0.4 g/L), immunoglobulins and secretory components (0.47–0.85 g/L) (Farrell *et al.*, 2004). Many of. whey protein's health-promoting properties have been attributed to peptides released during digestion. Peptides derived from whey have a number of physiological functions including the modulation of blood pressure, inflammatory processes, hyperglycemia and systems regulating food intake. However, these actions are not limited only to whey proteins and peptides, but are also due to synergism between whey proteins and other whey constituents such as calcium. The effect of milk with calcium in attenuating weight and fat gain and in reducing blood pressure is much stronger than that of calcium supplementation alone (Zemel *et al.*, 2004).

Whey proteins (α-LA, β-LG) are precursors of ACE-inhibitory peptides called lactokinins (Hartmann and Meisel, 2007), which have antihypertensive and potentially anti-obesogenic activities. The peptides, α-lactorphin and β-lactorphin (Meisel, 2004) affect adipocyte lipogenesis due to their ACE-inhibitory activities and may also reduce food intake via peripheral opioid receptors, similar to casein and soy protein hydrolysates (Pupovac and Anderson, 2002).

The differences in the physical properties of casein and whey contribute to their functionality in food systems and to their physiological effects when ingested, including their effect on food intake, which will be discussed later in this chapter. Whey protein is rapidly digested (a fast protein), whereas casein is more slowly digested (a slow protein). The classification of whey and casein as 'fast' and 'slow' proteins is based on their contribution to protein synthesis and their effect on plasma amino acid concentrations (Boirie *et al.*, 1997). In humans, the intake of whey (0.45 g/kg body weight) results in a fast, but short and transient, increase in plasma amino acids that peak in 40 minutes to 2 hours after its ingestion and returns to baseline values after 3 to 4 hours. Casein, in contrast, consistent with its slow gastric emptying, results in plasma amino acid concentrations that rise more slowly and are lower, but sustain a prolonged plateau lasting for at least 7 hours after its consumption (Boirie *et al.*, 1997).

High protein diets, high in whey proteins, may reduce fat deposition and improve insulin sensitivity (Ebringer *et al.*, 2008; Luhovyy *et al.*, 2007; Pfeuffer and Schrezenmeir, 2007). Whey leads to higher pre-meal insulin concentrations than casein (Dangin *et al.*, 2001) and may contain the predominant insulin secretagogue (a substance which causes another substance to be secreted) in milk because the insulin response after preloads of 25 g carbohydrate with 18.2 g of whey protein was 50% higher than after milk or cheese (Nilsson *et al.*, 2004). Addition of whey to a meal containing rapidly digested and absorbed carbohydrates, stimulated greater plasma insulin concentrations (+57% after lunch) and reduced postprandial blood glucose (–21% over 180 min response) in subjects with type 2 diabetes (Frid *et al.*, 2005).

Amino acids may be the primary factor accounting for the insulinotropic effect of milk protein. Healthy subjects who ingested a mixture of leucine, isoleucine, valine, lysine and threonine had glycemic and insulinemic responses similar to those after whey ingestion (Nilsson *et al.*, 2007), suggesting that the branched-chain amino acids (BCAA) of whey are the major determinants of insulinemia as well as reduced glycemia resulting from the whey drink. However, the whey drink stimulated incretin (gastric inhibitory peptide (GIP) and glucagon-like peptide -1 (GLP-1)) response, while the BCAA mixture did not, suggesting that the action of whey is not related simply to the amino acid content, but presumably is due to the synergistic action of amino acids and peptides. The authors concluded that whey-induced hyperinsulinemia occurs by two or even more separate pathways, including one connected to the significant increment in certain amino acids, but the other connected through the incretins, which are believed to interact with bioactive peptides derived from proteins (Nilsson *et al.*, 2007).

8.2.2 Milk minerals

In the food patterns, milk and dairy products contribute >10% of the requirements of many nutrients, especially minerals including calcium, phosphorus, magnesium, zinc and potassium (Weaver, 2009). Bioavailability of minerals and trace elements including calcium, magnesium, zinc, selenium and iron is improved by dairy proteins and peptides (Vegarud *et al.*, 2000). The role of many of these minerals in the etiology and treatment of metabolic disorders has been a subject of interest (Scholz-Ahrens and Schrezenmeir, 2006). Compared with other foods, milk and dairy products are a good source of dietary calcium which not only plays an important role in the regulation of body weight (Loos *et al.*, 2004), but may also improve lipid profiles (Karanja *et al.*, 1987; Denke *et al.*, 1993). In addition, an inverse relationship between calcium intake and blood pressure has been consistently reported from epidemiological studies (Kromhout *et al.*, 1985; Jorde and Bonaa,, 2000). A recent review showed that inadequate calcium/dairy intake may increase the risk of positive energy balance and the risk of developing metabolic syndrome (Major *et al.*, 2008).

8.2.3 Milk fats

The overall health benefits of milk fat may be more beneficial than reflected by current opinion (German and Dillard, 2004, 2006). Association between dairy fat consumption and increased risk for coronary heart disease (CHD) has led to the reduction of milk fat in many dairy products and the appearance of fat-free dairy products. However, studies linking dairy consumption with elevated serum cholesterol levels and CHD mortality (Turpeinen, 1979) have limitations, as reviewed recently (Nestel, 2008). Although the saturated fat content of milk has been found to increase blood lipids, some recent studies have reported that a milk supplemented diet decreased blood lipids (St-Onge *et al.*, 2000, Samuelson *et al.*, 2001), suggesting that milk may contain substances that counterbalance the effect of the saturated fatty acids (FA) in milk (Pfeuffer and Schrezenmeir, 2000). Furthermore, the FA in milk have several valuable metabolic functions.

Fat assists in the absorption of essential nutrients and fat-soluble vitamins A, D, E, and K, and plays an important antioxidant role. In addition, the amount and quality of fat in the diet is of importance in the development of metabolic syndrome. Most raw cows' milk contains about 3.5% fat, but the level varies widely, depending on many factors such as breed, stage of lactation, season, nutritional status, type of feed, health and age of the animal, intervals between milking and the point during milking when the sample is taken (Fox and McSweeney, 1998). Processing has led to many variations in the fat composition of milk available for consumption, ranging from whole milk to skim milk, and more recently to milk with increased omega-3 FA (Visioli *et al.*, 2000). On average, milk contains 33 g/L total fat, where TG account for about 95% of this lipid fraction. Other milk lipids are diacylglycerols (about 2% of the lipid fraction), cholesterol (less than 0.5%), phospholipids (about 1%), and free fatty acids (FFA) accounting for less than 0.5% of total milk lipids. Nomenclature of FA classifies them as short-chained (≤C4:0), medium-chained (C6:0–C10:0) and long-chained (≥C12:0) (Molkentin, 2000). Increased levels of FA in milk may result in off-flavours in milk and dairy products (Ebringer *et al.*, 2008). Cow milk fat contains up to 400 distinct fatty acids, the major proportion comprising twelve of these, of which 65% are saturated, 29.8% mono-unsaturated, and 3.2% poly-unsaturated (Fox and McSweeney, 1998). Although much of the details of the composition and biological properties of the major FA in bovine milk have been described, the FA are the least constant and most variable component of milk (German and Dillard, 2006). Milk lipids are aggregated into milk fat globules (MFG) in natural milk; however, MFG are broken down during homogenization or heat treatment. Although it has little effect on fat composition, the nutritional effects associated with MFG are expected to be lost (Argov *et al.*, 2008, Michalski, 2007).

Fat is an important regulator of intake and metabolic responses via the ileal brake which leads to delayed gastric emptying, and frequency of jejunal contractions, increased small intestinal transit time and reduced gastric and pancreatic secretion (Maljaars *et al.*, 2008). Both FA chain length and saturation predetermine the magnitude of the ileac brake. Milk fat, which has 32.2% (% weight) of C18:1 plus C18:2 (2.4%) of total lipids in whole milk (Fox and McSweeney, 1998) may contribute to the satiety value of milk. A recent double-blinded, randomized, crossover study with healthy subjects showed that when shea oil, high in C18:0, canola oil, high in C18:2 and safflower oil, high in C18:2 FA were infused into the ileum, canola oil and safflower oil increased fullness, reduced hunger and increased cholecystikinin (CCK) secretion, although no effect on food intake was found (Maljaars *et al.*, 2009). This finding suggests that the dairy products, where these FA naturally occur, may contribute to the ileal brake phenomenon while fat-free dairy products do not.

Milk poly-unsaturated FA are precursor eicosanoids and affect inflammatory and thrombogenic processes and blood pressure regulation (Vessby, 2003). Among the saturated FA, lauric acid (C12:0) is present in bovine and human milk at approximately 3.1% of milk fat (Fox and McSweeney, 1998). While a meta-analysis based on reported intakes of individual saturated FA found that lauric acid was the most potent total- and HDL cholesterol-raising saturated FA, this effect is proportionally higher for HDL than for LDL. Moreover, the replacement of the different saturated FA with carbohydrates led to an increase in serum triacylglycerol concentrations to the same extent (Mensink *et al.*, 2003). Myristic acid (C14:0) makes up 9.5% of the fat content of bovine milk and 5.1% of human milk (Fox and McSweeney, 1998). Dairy fat is a major source of myristic acid and high intake (10% of total energy) of this ingredient in human dietary interventions was associated with high LDL cholesterol and apolipoprotein-B levels, and a low HDL to LDL ratio (Zock *et al.*, 1994). However, very few studies have been conducted with lower (realistic) intakes or by taking into consideration the position of the FA on the triacylglycerol. In the case of myristic acid, in milk TG, it exists in the *sn-2* position, which is the only bioavailable form (Dabadie *et al.*, 2006). In a dietary intervention for five weeks where the main source of myristic acid (1.2% and 1.8% total energy) was milk, an improvement in lipid profiles was shown (Dabadie *et al.*, 2005).

Palmitic acid (C16:0) accounts for 26.3% and 20.2% of milk fat in bovine and human milk, respectively (Fox and McSweeney, 1998). A comparison between the effects of dietary laurate–myristate and the effects of palmitic acid in humans with normal lipid concentrations showed that palmitic acid lowers serum cholesterol levels (Sundram *et al.*, 1994). Also, replacement of dietary laurate–myristate with palmitate–oleate has a beneficial effect on an important index of thrombogenesis, the ratio of thromboxane to prostacyclin in plasma (Ng *et al.*, 1992). Furthermore, stearic acid is present in bovine and human milk at 14.6% and 5.9% of milk fat, respectively (Fox and McSweeney, 1998). Dietary stearic acid decreases plasma and liver cholesterol concentrations by reducing intestinal cholesterol absorption through altering the microflora populations that synthesize secondary bile acids (Cowles *et al.*, 2002).

In addition to these saturated FA, milk fat provides CLA and other components such as sphingomyelin and other sphingolipids, etheric lipids, and butyric acids that have health benefits. Sphingolipids are highly bioactive compounds that have profound effects on cell regulation. Per capita sphingolipid consumption in the United States is estimated to be approximately 115–140 g/year (or 0.3–0.4 g/day) and is primarily received from dairy products (Vesper *et al.*, 1999). Animal studies have shown that feeding sphingolipids inhibits colon carcinogenesis, reduces serum LDL and elevates HDL (Vesper *et al.*, 1999).

Milk and meat products from ruminants are a good source of CLA in the human diet (Ebringer *et al.*, 2008). There are two predominant isomers of CLA, and they include *cis*-9, *trans*-11 CLA and *trans*-10, *cis*-12 CLA. Conjugated linoleic acid in animal fat was reported to reduce adiposity, improve plasma lipoprotein concentrations and reduce inflammatory responses in laboratory animals. However, in order to cause such meaningful changes in humans, the daily intake of CLA should be very high (equivalent to 3.5 kg of cheese per day) to take in 10 g of CLA (MacRae *et al.*, 2005). On the other hand, benefits have not been found in a number of trials of dietary interventions (Nestel, 2008). For example, dairy products enriched with as much as 3 g of CLA daily did not affect serum lipid profiles in moderately overweight subjects (Naumann *et al.*, 2006). However, the potential mechanism of action of CLA on human adiposity are limited and understanding it may provide more certainty to the results of controlled studies (Brown and McIntosh, 2003, Brown *et al.*, 2004). Both human and animal studies have shown that there are some concerns about CLA-isomers and their safety when they are used as dietary supplements. For instance, the CLA *trans*-10, *cis*-12 isomer may produce liver hypertrophy and insulin resistance via a redistribution of fat deposition that resembles lipodystrophy (Larsen *et al.*, 2003).

Milk fat also contains medium-chained FA. Caproic acid (C6:0) is present in bovine and human milk at approximately 1% and 0.1% of milk fat, respectively. Caprylic acid (C8:0) and capric acid (C10:0) are present at approximately 0.3% of human and 1.2% of cow milk fat, respectively (German and Dillard, 2004). These three FA have similar biological activities and have not been associated with the risk of coronary artery disease when consumed as part of a diet (German and Dillard, 2004). A study with healthy men and women suggests that the medium-chained triacylglycerides in the diet may reduce body weight and fat more than long-chained triglycerides (Tsuji *et al.*, 2001).

Among short-chained FA, butyric acid (C4:0), the shortest saturated FA, is present in ruminant milk fat and human milk at 2 to 5% and 0.4% by weight, respectively (German and Dillard, 2004). No other common food fat contains this FA. Butyric acid produces butter's distinctive flavour and occurs as TG in 5 to 6% of butter fat; it is found in rancid butter and parmesan cheese. It is a potent antineoplastic agent and its action is improved through synergy with other milk fat components (Parodi, 1999).

Of the mono-unsaturated FA, oleic acid (C18:1) is relatively high in milk (approximately 8 g/L). Oleic acid, like other unsaturated FA, lowers plasma cholesterol, LDL and triacylglycerol concentrations. The concentration of poly-unsaturated FA, including linoleic and α-linolenic acids is 2 g/L in bovine milk (German and Dillard, 2004). These FA are converted to arachidonic acid and eicosapentaenoic acid (EPA), which have protective effects against dyslipidemia and CVD (Angerer and von Schacky, 2000).

The fat from sheep and goat milk, in comparison with cows' milk fat, is richer in medium-chained TG (FA C6:0–10:0) (Sampelayo *et al.*, 2007), which has led to the suggestion that it may confer more health benefits. Medium-chained TG are absorbed through the portal system without re-synthesis of TG in intestinal cells, and are subjected predominantly to β-oxidation in the liver, and are not stored as fat (Tsuji *et al.*, 2001).

In summary, individual FA yield extremely diverse effects on risk factors for type 2 diabetes and CVD. There is reasonably good understanding of the effects of individual FA on blood lipids and lipoproteins; however, much remains to be learned about these individual FA with respect to other risk factors such as platelet function, blood pressure, endothelial function, development of atherosclerosis and insulin resistance, and other risk factors related to metabolic syndrome. There is a growing appreciation of the health benefits of mono-unsaturated and poly-unsaturated FA. A greater understanding of the roles played by milk fat in the pathogenesis of insulin resistance and metabolic syndrome is required.

8.2.4 Milk carbohydrates

The primary source of milk carbohydrate is lactose, an oligosaccharide composed of glucose and galactose. It is hydrolyzed by the lactase enzyme and absorbed in the small intestine. Insufficient activity of this enzyme results in lactose intolerance. Lactose is dissolved in whey, the serum phase of fluid milk. While the normal pasteurization used for fluid milk has no significant effect on lactose, the higher temperatures used for ultra high temperature pasteurization, to extend the shelf life of milk, or spray drying can cause isomerization of lactose and prevents digestive discomfort in lactose intolerant individuals. Milk lactose had not been found to have any negative effects on metabolic regulation and coronary artery disease.

Compared to sucrose and glucose, lactose has a lower glycemic response, and therefore contributes to the low glycemic effect of milk. In addition, when consumed by obese subjects, lactose (56 g) results in lower glycemic and insulin response and lower food intake at 180 min (Bowen *et al.*, 2006b). Lactose also confers health benefits because it is an essential substrate for fermentation for many probiotic bacteria, including those involved in the processing of dairy products. While many consumers tend to avoid dairy products with lactose due to intolerance, in fact, it was shown that lactose maldigesters can consume at least one cup of milk without experiencing symptoms, and that tolerance can be further improved by consuming the milk with a meal or choosing yogurt or hard cheeses (Byers and Savaiano, 2005).

Tagatose, a hexose epimer of fructose, occurs naturally in dairy products in small amounts, but is increased by high-temperature treatment, i.e. in sterilized milk (Troyano *et al.*, 1991). Tagatose is also commercially produced by isomerization of galactose and has proven to be of interest in management of metabolic disease. Tagatose reduces postprandial hyperglycaemia and hyperinsulinaemia, decreases food intake and in a 14-month trial reduced body weight and increased HDL-cholesterol (Lu *et al.*, 2008). This dairy-derived ingredient is now undergoing phase 3 of clinical trial under FDA-affirmed protocol.

8.3 Relationship between milk and chronic diseases

Excess body fat, primarily abdominal fat, is a major predictor of the risk of developing conditions of the metabolic syndrome. Therefore, the role of dairy and milk consumption in body weight control is reviewed first, followed by an examination of their effect on conditions of the metabolic syndrome.

8.3.1 Milk and obesity

There is currently no clear consensus that dairy consumption plays a significant role in the prevention and treatment of obesity based on epidemiological evidence and randomized controlled trials. Milk and dairy consumption has been associated with both increased and decreased energy intake and body weight, perhaps explained by the wide range of fat contents and energy densities of dairy products, and the change in fat content of consumed products with time. In one epidemiological study conducted in Bogalusa, LA, USA, more frequent consumption of dairy products was associated with higher energy and saturated fat intake, suggesting that higher energy intake was largely due to the intake of high fat dairy products (Ranganathan *et al.*, 2005). In contrast, other epidemiological studies with adults in the USA (Pereira *et al.*, 2002; Loos *et al.*, 2004) report an inverse association between dairy intake and adiposity. Similarly, a large cross-sectional study conducted in Portugal found that high milk consumption was associated with a lower body mass index (BMI) in the entire population, with the exception of no effect in post-menopausal women (Marques-Vidal *et al.*, 2006).

Prospective studies of the relationship between dairy and adiposity in adults (Elwood *et al.*, 2005; Drapeau *et al.*, 2004; Rajpathak *et al.*, 2006) have also given inconsistent findings. A 20-year prospective analysis of 45–59-year-old men in South Wales showed that individuals whose milk consumption was above the median had a significantly lower BMI compared to those consuming less milk (Elwood *et al.*, 2005). Similarly, in the Quebec Family Study, changes in dietary patterns of a sample of adults were followed for up to 6 years and found that among 51 food categories, increased consumption of fruit and milk (skimmed and partly skimmed) were the only food patterns inversely associated with body weight change (Drapeau *et al.*, 2004). However, in a later examination of the relationship between dairy and body weight over 12 years in the USA, men with the largest increase in total high fat dairy intake gained slightly more weight than men who decreased intake the most (Rajpathak *et al.*, 2006). Results from intervention trials also fail to provide conclusions on the impact of a diet rich in dairy products on adiposity (Zemel *et al.*, 2004; Gunther *et al.*, 2005; Harvey-Berino *et al.*, 2005; Lin *et al.*, 2000). A secondary review of nine randomized trials provided little support for an effect of dairy supplementation in reducing body weight or fat mass; however, none of the reviewed studies were specifically designed or powered to address this issue (Barr, 2003). In a placebo controlled study with 32 obese adults maintained for 24 weeks on balanced deficit diets (500 kcal/d deficit) and randomized to a standard diet (400 to 500 mg of dietary calcium/d supplemented with placebo), a high-calcium diet (standard diet supplemented with 800 mg of calcium/d), or high-dairy diet (1200 to 1300 mg of dietary calcium/d supplemented with placebo) a significant decrease in body weight and fat loss from the trunk region was found in those subjects following high-dairy diet (Zemel *et al.*, 2004).

8.3.2 Milk and cardiovascular disease

Based on some ecological studies, milk has become identified as a risk factor for CVD due to its saturated FA (Seely, 1981; Grant, 1998). However, the results from prospective studies are in contrast to those from ecological studies. A comprehensive review of ten cohort studies provided no evidence that milk adversely affects CVD and, in contrast, found that drinking milk may contribute a small, but worthwhile reduction in heart disease and stroke risk (Elwood *et al.*, 2004), an observation confirmed in a later report (Elwood *et al.*, 2005), and for ischaemic heart disease (Elwood *et al.*, 2004), As summarized recently, the contribution of milk consumption to total dietary fat, and the intake of saturated FA by a majority of subjects was relatively small (Elwood *et al.*, 2004). Based on a study of foodstuffs purchased by 224 women in the United Kingdom, milk fat contributed around 5% of the fat in their total food purchases (Lip *et al.*, 1995).

8.3.3 Milk and the metabolic syndrome

Dietary patterns characterized by increased dairy consumption have a strong inverse association with insulin resistance, risk of Type 2 diabetes and CVD among overweight adults. The Dietary Approaches to Stop Hypertension (DASH) diet, comprising low-fat dairy among other components (fruits and vegetables), showed ability to reduce blood pressure among 429 adults (Appel *et al.*, 1997). Results obtained from 88 517 women aged 34 to 59 following the DASH diet from 1980 through to 2004 showed that adherence to this diet provides lower incidences of CVD and stroke (Fung *et al.*, 2008).

In the Health Professionals Follow-up Study, male participants with no history of diabetes, CVD or cancer at baseline were followed for twelve years. Each serving per day increase in total dairy intake was associated with a 9% lower risk of Type 2 diabetes and for low-fat dairy intake there was a 12% reduced risk (Choi *et al.*, 2005). Another ten-year prospective follow-up study of dairy intake and the risk of Type 2 diabetes in 37 183 middle-aged or older women without a history of diabetes, CVD, and/or cancer at baseline found that higher low-fat dairy intake may lower the risk of Type 2 diabetes (Liu *et al.*, 2006). The ten-year cumulative incidence of insulin resistance syndrome and its association with dairy consumption, measured by diet history interview, showed that dairy consumption was inversely associated with the incidence of insulin resistance among individuals who were overweight at baseline but not among lean individuals

(Pereira *et al.*, 2002). Daily consumption of dairy product was associated with a 21% lower likelihood of insulin resistance syndrome, both in blacks and whites or men and women (Pereira *et al.*, 2002).

Dairy consumption was inversely associated with the risk of metabolic syndrome in a cross-sectional study conducted in Iran (Azadbakht *et al.*, 2005). In France, dairy intake was inversely associated with the frequency of the metabolic syndrome in men, but not women (Mennen *et al.*, 2000). Men who consumed more than one serving of dairy products per day had a 40% lower prevalence of the metabolic syndrome. This association may be indicated by the heptadecanoic acid and pentadecanoic acid (15:0) content of milk. These biomarkers in plasma were significantly and inversely correlated with TG, insulin, pro-insulin and leptin; suggesting that milk fat reduces risk factors of acute myocardial infarction (Warensjo *et al.*, 2004). In another study, intake of milk fat and cream, and the plasma concentration of these same biomarkers of consumption, was inversely associated with glucose levels at fasting and an oral glucose tolerance test (Smedman *et al.*, 1999). In addition, in 70-year-old men, inverse associations between intake of milk products and BMI, waist circumference, LDL:HDL ratio, HDL, TG, and fasting plasma glucose were found, whereas relationships to HDL and apolipoprotein A-I tended to be positive (Smedman *et al.*, 1999).

Not all studies have reported positive associations between dairy consumption and the metabolic syndrome, but negative findings are remarkably few. High dairy consumption by a Dutch elderly population was not associated with more favourable responses in components of the metabolic syndrome, except for a modest association with lower blood pressure (Snijder *et al.*, 2007). The British Women's Heart and Health Study found that women 60 to 79 years old (n = 111) who never drank milk had lower insulin resistance scores, triacylglycerol levels and BMI, higher HDL and suffered less frequently from diabetes (Lawlor *et al.*, 2005).

8.3.4 Milk and dysglycemia

The glycemic index (GI), which is a measure of the effects of carbohydrates on blood glucose levels, has important implications for controlling blood glucose in both healthy individuals and those with Type 2 diabetes (Wolever and Miller, 1995); however, it overlooks the contribution of insulin to metabolic disorders (Hollenbeck and Coulston, 1991) because the glucose response is not a good indicator of insulin response when the food consumed contains protein (Gannon *et al.*, 1988; Holt *et al.*, 1997).

Milk consumption with carbohydrate foods greatly modifies the glycemic response to their ingestion. Lower glycemic responses have been found after high milk breakfasts compared with high fat breakfasts or high fiber breakfasts in both healthy participants and individuals with Type 2 diabetes (Schrezenmeir *et al.*, 1989) and when milk and whey is added to carbohydrate meals which are equi-carbohydrate to a white bread reference (Nilsson *et al.*, 2004; Ostman *et al.*, 2001). However, the insulin response to milk was not different from that of the reference food (Nilsson *et al.*, 2004; Ostman *et al.*, 2001), showing a large dissociation between the GI and insulin index (II) for healthy normal subjects. The insulintropic effect of milk is seen with both whole and skim milk, indicating that the dissociation between the GI and II is not related to the fat content of milk, but is due to the protein content (Ostman *et al.*, 2001; Pfeuffer and Schrezenmeir, 2007). Even the addition of milk to a low GI mixed meal elicits an insulintropic effect and reduced glycemic response (Liljeberg Elmstahl and Bjoorck, 2001).

The increase in insulin after consumption of milk proteins has been attributed to the BCAA because of a correlation between postprandial insulin and early increments in plasma amino acids, primarily leucine, valine, lysine, and isoleucine (Nilsson *et al.*, 2004). Therefore, a role of bioactive peptides or BCAA in the enhanced insulin response through releasing higher GIP and GLP-1 concentrations has been suggested as an underlying possible mechanism (Nilsson *et al.*, 2004; Frid *et al.*, 2005).

8.4 Dairy components and their effect on satiety and food intake regulation

Many dairy components stimulate regulatory systems that are involved in the brain's regulation of food intake and energy balance. These mechanisms are reviewed briefly as a background to describe the contributions of dairy components to intake regulation.

8.4.1 The role of the central nervous system in satiety and food intake regulation

Both long- and short-term food intake mechanisms are regulated by the hypothalamus in response to neural, metabolic and hormonal signals entering the brain (Schwartz *et al.*, 2000). Short- and long-term pathways regulating food intake overlap in the central nervous system (CNS), where all the signals are integrated and translated into food intake control (Schwartz *et al.*, 2000). Two major long-term food intake regulatory hormones are leptin and insulin (Badman and Flier, 2005). Both leptin (produced from adipose tissues) and insulin (produced from pancreatic β-cells) concentrations are influenced by the adipose tissue mass to promote satiety. Leptin inhibits AMP-activated protein kinase (AMPK) activity in the hypothalamus and its inhibition is necessary for the satiety – induced effect of leptin (Minokoshi *et al.*, 2008).

Insulin is a hormone involved in both short- and long-term food intake regulations. Although insulin resistance is associated with an increased body weight and obesity, insulin contributes to long-term energy balance in normal weight individuals, and hyperinsulinemia appears to have a transient protective effect on overeating (Abou Samra *et al.*, 2007). In the short-term, insulin response is a stronger correlate of satiety and food intake suppression than any of the gastrointestinal satiety hormones (Abou Samra *et al.*, 2007). In addition to modifying the glycemic response, plasma concentrations of insulin strongly associate with short-term satiety and decreased food intake (Akhavan and Anderson, 2007).

8.4.2 The role of gut hormones in satiety and food intake regulation

Satiety refers to the afferent signalling that leads to cessation of eating. Satiety can be induced through physiological processes, including satiety-signaling pathways from the gastrointestinal tract to the CNS. More than 20 different regulatory peptide hormones are released in the gastrointestinal system. Many of them are recognized to be involved in the regulation of food intake and are sensitive to the gut nutrient content and composition (Murphy and Bloom, 2006). Primary mediators of satiety signaling form in the gastrointestinal tract, including CCK, GLP-1, peptide tyrosine tyrosine (PYY), and GIP (D'Alessio, 2008).

CCK, a satiety hormone involved in controlling food intake, is released postprandially, particularly after fat and protein ingestion (Liddle *et al.*, 1985). CCK is produced by the endocrine cells (I-cells) of the duodenum and upper jejunum of the small intestine. CCK induces satiety in a dose-responsive manner in animals both through central mechanisms after being transported across the blood–brain barriers and peripheral mechanisms through abdominal vagal nerves (Reidelberger *et al.*, 2004). The effects of CCK on satiety and meal size are synergized by the action of leptin (Barrachina *et al.*, 1997).

GLP-1 is produced from L-cells located primarily in the ileum and colon. GLP-1 receptors are expressed not only in pancreatic islet cells, but in the stomach peripheral nerves and specific regions in the brain (D'Alessio, 2008). GLP-1 is involved in stimulating insulin secretion and inhibiting hyperglycemia (D'Alessio and Vahl, 2004). In addition, GLP-1 inhibits gastric emptying and suppresses food intake. Like CCK, GLP-1 is also dependent on leptin to induce satiety (Williams *et al.*, 2006).

PYY, also secreted from the L-cells of the gastrointestinal tract, is released into the circulation in response to ingesting a meal, and produces satiety. Like CCK and GLP-1, the satiety-induced effect of PYY is likely dependant on leptin signalling (Moran *et al.*, 2005).

Lastly, GIP is released from K cells in the duodenum after food ingestion and it may play an important role in obesity development since it is known that GIP-receptor-knockout mice are resistant to obesity when fed a high-fat diet (Murphy and Bloom, 2006).

8.4.3 Dairy components and their effect on satiety and food intake regulation

In Section 8.2.3 we hypothesized the possible mechanisms of dairy fat ingestion on metabolic responses, including satiety and food intake via the ileal brake. However, there is considerable evidence that the effect of milk and dairy products on satiety and food intake is mainly mediated through the effect of dairy proteins on the release of satiety hormones. Proteins are more satiating than either carbohydrates or fats (Anderson *et al.*, 2004). Nevertheless, various quantities and types of proteins have different impacts on metabolic regulation, weight loss and weight maintenance. Dietary proteins regulate satiety and metabolic function by the combined actions of the intact protein, encrypted peptides and amino acids on gastrointestinal and central pathways. Dairy proteins also suppress short-term food intake and can contribute to reduced energy intake and eventually body weight through their effects on many hormones that signal satiety.

Milk proteins increase CCK concentrations in plasma (Bowen *et al.*, 2006b; Hall *et al.*, 2003). Whey proteins are the precursors of many bioactive peptides that have potential to exert their actions through gut satiety mechanisms. This has been confirmed when whey protein increased CCK more than casein (Hall *et al.*, 2003), although in another study, whey protein resulted in similar CCK responses compared with soy and gluten proteins (Bowen *et al.*, 2006b), perhaps due to differences in the GMP content of the whey product derived from making cheese. GMP stimulates CCK release (Hall *et al.*, 2003; Liddle *et al.*, 1985), and blocking CCK-A receptors prevents suppression of food intake in rats given casein, in which the CMP is encrypted (Pupovac and Anderson, 2002). Although carbohydrate and fat are more potent stimulators of GLP-1 than protein (Brubaker and Anini, 2003), milk proteins stimulate GLP-1 release independently of carbohydrate and fat (Aziz and Anderson, 2007), possibly a response that is primarily due to the action of whey. A high protein breakfast (58% of total energy) consisting mainly of dairy products enriched with whey protein isolate resulted in higher GLP-1 concentrations over 3 hours than a high carbohydrate breakfast (19% of total energy from protein) consisting mainly of plain yogurt (predominantly casein) (Blom *et al.*, 2006). Similarly, when protein preloads of 50 g were given with an additional 200 kcal from fat and carbohydrate, whey protein ingestion resulted in higher plasma concentrations of GLP-1 than casein for up to 3 hours in humans (Hall *et al.*, 2003). However, since whey is a 'fast protein' and casein is a 'slow protein', and plasma amino acid concentrations begin to fall 3 hours after whey ingestion, but remain elevated for at least 7 hours after casein ingestion, casein may have a stronger effect than whey at a later time (Boirie *et al.*, 1997). In support of this, plasma GLP-1 concentrations fell substantially 2 hours after the administration of isoenergetic whey, whey hydrolysate and casein hydrolysate solutions, but continued to increase only after the casein solution (Calbet and Holst, 2004). Plasma GLP-1 concentration was independent of the degree of protein fractionation (Calbet and Holst, 2004).

Whey stimulates GIP response (+80%) in healthy subjects while BCAA mixtures did not have such an effect (Nilsson *et al.*, 2007), suggesting that release of bioactive peptides present in whey and released during digestion are the primary stimulators of GIP secretion (Nilsson *et al.*, 2007). In contrast, PYY in plasma increased after intra-gastric administration of whey protein or whey peptide hydrolysate to healthy subjects, but it was independent of the degree of protein fractionation (Calbet and Holst, 2004). The effect of whey compared with other proteins has not been reported.

Milk proteins also reduce plasma ghrelin, the orexogenic hormone. Complete milk protein, casein, whey or GMP, given by gavage as preloads to rats, reduced plasma ghrelin concentrations similarly at 30 min when compared to the water control (Peng, 2005). In humans, whey protein isolate and calcium caseinate suppressed ghrelin concentrations similar to lactose, but more than glucose over 3 hours, an effect that was correlated with lower subsequent energy intake (Bowen *et al.*, 2006b). Whey is not unique among proteins in its effect on ghrelin because whey, soy and gluten proteins decreased ghrelin similarly over 3 hours (Bowen *et al.*, 2006a).

It remains unclear, however, if the effects of dairy ingredients, especially proteins, on food intake, subjective satiety and intake regulatory mechanisms in humans are obtained from usual serving sizes of dairy products. The effects described have been observed in short-term experiments and when those ingredients were consumed in much higher amounts.

8.5 Dairy-based functional foods

Functional dairy products are defined as products produced from either milk enriched with a functional component, or ingredients originating from milk (Mattila-Sandholm and Saarela, 2003). Based on their main health effects, functional dairy products can be divided into products providing: (i) gastrointestinal health and general well-being including pro-, pre- and syn- biotics, and products with reduced lactose; (ii) cardiovascular health including antihypertensive and cholesterol-lowering effects; (iii) other benefits including antiosteoporotic, sleep-modulating, immuno-enhancing effects. There are some examples of recent advances in this area. In 2008, the Irish dairy and ingredient group, Carbery Food Ingredients, developed a hydrolyzed whey protein ingredient, Optipep, which is able to increase the insulin response compared to native whey protein and could be beneficial post-exercise, especially for athletes (Starling, 2008). Several functional dairy products have been patented in the form of powdered milk, liquid milk or milk slice for eradicating functional constipation or preventing and inhibiting hyperlipemia (Wang *et al.*, 2008). These products contain 0.01 to 15% functional oligosaccharide from low-polymerized seminose, xylo-oligosaccharide, fructooligosaccharides, galactooligosaccharide, stachyose, raffinose, isomaltooligosaccharide, maltooligosaccharide, soybean oligosaccharide, chitooligosaccharides, oligofructose, and/or oligosaccharide. Another approach to the use of complex carbohydrates for novel functional dairy products is the example of low-fat mozzarella cheese where barley beta-glucan concentrate is used as a fat replacer (Vithanage *et al.*, 2008). Such strategies reach two goals. The added polysaccharide, beta-glucan, is firstly known to improve glycemic control and reduce serum cholesterol levels, and secondly, to reduce the fat content of the milk product at the cost of an ingredient with similar technologic properties.

Some functional products are aimed to target specific biomarkers associated with certain health conditions. For example, the ingredient derived from the mushroom *Panellus serotinus*, is claimed to enhance the level of adiponectin in ZF rats, and authors propose to use this substance in the wide variety of foods including dairy products to prevent obesity and related diseases (Yanagida and Nagao, 2008).

The abundance of milk whey as a by-product of cheese-making had a strong impact on the production of whey-based beverages starting in the 1970s. Since whey has a high content of lactose and may not be consumed by the population with lactose intolerance, the specific cultures of bacteria were chosen to produce functional products with high nutritional value and acceptable sensory characteristics, including non-alcoholic beverages and alcoholic beverages such as whey beer and whey wine with up to 1.5% of alcohol (Jelicic *et al.*, 2008).

The world market for functional foods where dairy products play a significant role is estimated between 33 to 61 billion US$. The United States is the largest market segment, followed by Europe and Japan, and these three account for up to 90% of the total sales for functional foods (Siro *et al.*, 2008). However, more research, especially clinical trials, are required to substantiate the effectiveness of functional dairy products for the prevention or mitigation of obesity and the metabolic syndrome among different populations. This may be illustrated by the recent rejection of health claims related to dairy products and healthy body weight by European Food Safety Authority (EFSA). Three portions of dairy foods per day are claimed to promote a healthy body weight in children and adolescents, a portion being specified as 200 ml milk, 28 g cheese or 125 ml yogurt. EFSA found that of the five cross-sectional plus the seven prospective observations presented, only four studies (two cross-sectional and two prospective studies) support an association between the consumption of dairy foods and BMI or body fat estimates in children and adolescents. The EFSA Panel noted that cross-sectional studies provide only an association and do not provide sufficient evidence of a cause and effect relationship. The two prospective studies included only a small sample size (53 and 92 subjects) and covered a limited age range (two to 13 years). Finally, EFSA concluded that a cause and effect relationship is not established between the daily consumption of dairy foods (milk, cheese and yogurt) and a healthy body weight in children and adolescents. The main flaw of this claim, as it was concluded by EFSA, is the lack of intervention studies to specifically examine any causal relationship, and there are insufficient data to establish a specific level or frequency of consumption of dairy foods associated with any specific effect on body weight (EFSA, 2008). Therefore, more intervention studies are needed to establish the effect of dairy products, as well as dairy-derived ingredients, on the risk factors associated with metabolic syndrome.

8.6 Conclusion

Dairy-derived ingredients have potential to improve physiological characteristics associated with obesity, Type 2 diabetes, and CVD. However, their mechanism of action in bringing about effects on food intake and the components of the metabolic syndrome action requires elucidation.

8.7 Notation

α-LA alpha-lactalbumin, AMPK AMP-activated protein kinase, β-LG beta-lactoglobulin, BMI body mass index, CN casein, BCAA branched-chain amino acids, CVD cardiovascular disease, CMP caseinomacropeptide, CNS central nervous system, CCK cholecystikinin, CLA conjugated linoleic acid, CHD coronary heart disease, EPA eicosapentaenoic acid, EFSA European Food Safety Authority, FA fatty acids, FFA free fatty acids, GIP gastric inhibitory peptide, GI glycemic index, GLP-1 glucagon-like peptide -1, GMP glycomacropeptide, HDL high density lipoproteins, MFG milk fat globules, PYY peptide tyrosine-tyrosine, TG triacylglycerols

8.8 References

ABOU SAMRA, R., WOLEVER, T. M. S. and ANDERSON, G. (2007) Enhanced food intake regulatory responses after a glucose drink in hyperinsulinemic men. *International Journal of Obesity*, 31, 1222–1231. DOI:10.1038/sj.ijo.0803565

AKHAVAN, T. and ANDERSON, G. H. (2007) Effects of glucose-to-fructose ratios in solutions on subjective satiety, food intake, and satiety hormones in young men. *American Journal of Clinical Nutrition*, 86, 1354–1363.

ANDERSON, G. H., TECIMER, S. N., SHAH, D. and ZAFAR, T. A. (2004) Protein source, quantity, and time of consumption determine the effect of proteins on short-term food intake in young men. *Journal of Nutrition*, 134, 3011–3015.

ANGERER, P. and VON SCHACKY, C. (2000) n-3 Polyunsaturated fatty acids and the cardiovascular system. *Current Opinion in Lipidology*, 11, 57–63.

APPEL, L. J., MOORE, T. J., OBARZANEK, E., VOLLMER, W. M., SVETKEY, L. P., SACKS, F. M., BRAY, G. A., VOGT, T. M., CUTLER, J. A., WINDHAUSER, M. M., LIN, P. H. and KARANJA, N. (1997) A clinical trial of the effects of dietary patterns on blood pressure. *New England Journal of Medicine*, 336, 1117–1124.

ARGOV, N., LEMAY, D. G. and GERMAN, J. B. (2008) Milk fat globule structure and function: Nanoscience comes to milk production. *Trends in Food Science & Technology*, 19, 617–623. DOI:10.1016/j.tifs.2008.07.006

AZADBAKHT, L., MIRMIRAN, P., ESMAILLZADEH, A. and AZIZI, F. (2005) Dairy consumption is inversely associated with the prevalence of the metabolic syndrome in Tehranian adults. *American Journal of Clinical Nutrition*, 82, 523–530.

AZIZ, A. and ANDERSON, G. H. (2007) The Effect of Dairy Components on Food Intake and Satiety: Mechanisms of Actions and Implications for the Development of Functional Foods. In Saarela, M. (Ed.) *Functional Dairy Products*. Cambridge, UK, Woodhead Publishing Limited.

BADMAN, M. K. and FLIER, J. S. (2005) The gut and energy balance: Visceral allies in the obesity wars. *Science*, 307, 1909–14. DOI:10.1126/science.1109951

BARR, S. I., MCCARRON, D. A., HEANEY, R. P., DAWSON-HUGHES, B., BERGA, S. L., STERN, J. S. and OPARIL, S. (2000) Effects of increased consumption of fluid milk on energy and nutrient intake, body weight, and cardiovascular risk factors in healthy older adults. *Journal of the American Dietetic Association*, 100, 810–817. DOI:10.1016/S0002-8223(00)00236-4

BARR, S. I. (2003) Increased dairy product or calcium intake: Is body weight or composition affected in humans? *Journal of Nutrition*, 133, 245S-248S.

BARRACHINA, M. D., MARTINEZ, V., WANG, L. X., WEI, J. Y. and TACHE, Y. (1997) Synergistic interaction between leptin and cholecystokinin to reduce short-term food intake in lean mice. *Proceedings of the National Academy of Sciences of the United States of America*, 94, 10455–10460.

BLOM, W. A. M., LLUCH, A., STAFLEU, A., VINOY, S., HOLST, J. J., SCHAAFSMA, G. and HENDRIKS, H. F. J. (2006) Effect of a high-protein breakfast on the postprandial ghrelin response. *American Journal of Clinical Nutrition*, 83, 211–220.

BOIRIE, Y., DANGIN, M., GACHON, P., VASSON, M. P., MAUBOIS, J. L. and BEAUFRERE, B. (1997) Slow and fast dietary proteins differently modulate postprandial protein accretion. *Proceedings of the National Academy of Sciences of the United States of America*, 94, 14930–14935.

BONORA, E., KIECHL, S., WILLEIT, J., OBERHOLLENZER, F., EGGER, G., TARGHER, G., ALBERICHE, M., BONADONNA, R. C. and MUGGEO, M. (1998) Prevalence of insulin resistance in metabolic disorders: The Bruneck Study. *Diabetes*, 47, 1643–9.

BOWEN, J., NOAKES, M. and CLIFTON, P. M. (2006a) Appetite regulatory hormone responses to various dietary proteins differ by body mass index status despite similar reductions in ad libitum energy intake. *Journal of Clinical Endocrinology and Metabolism*, 91, 2913–2919. DOI:10.1210/jc.2006-0609

BOWEN, J., NOAKES, M., TRENERRY, C. and CLIFTON, P. M. (2006b) Energy intake, ghrelin, and cholecystokinin after different carbohydrate and protein preloads in overweight men. *Journal of Clinical Endocrinology and Metabolism*, 91, 1477–1483. DOI:10.1210/jc.2005-1856

BROWN, J. M. and MCINTOSH, M. K. (2003) Conjugated linoleic acid in humans: Regulation of adiposity and insulin sensitivity. *Journal of Nutrition*, 133, 3041–3046.

BROWN, J. M., BOYSEN, M. S., CHUNG, S., FABIYI, O., MORRISON, R. F., MANDRUP, S. and MCINTOSH, M. K. (2004) Conjugated linoleic acid induces human adipocyte delipidation – Autocrine/paracrine regulation of MEK/ERK signaling by adipocytokines. *Journal of Biological Chemistry*, 279, 26735–26747. DOI:10.1074/jbc.M401766200

BRUBAKER, P. L. and ANINI, Y. (2003) Direct and indirect mechanisms regulating secretion of glucagon-like peptide-1 and glucagon-like peptide-2. *Canadian Journal of Physiology and Pharmacology*, 81, 1005–1012. DOI:10.1139/Y03-107

BYERS, K. G. and SAVAIANO, D. A. (2005) The myth of increased lactose intolerance in African-Americans. *Journal of the American College of Nutrition*, 24, 569S-573S.

CALBET, J. A. L. and HOLST, J. J. (2004) Gastric emptying, gastric secretion and enterogastrone response after administration of milk proteins or their peptide hydrolysates in humans. *European Journal of Nutrition*, 43, 127–139. DOI:10.1007/s00394-004-0448-4

CHOI, H. K., WILLETT, W. C., STAMPFER, M. J., RIMM, E. and HU, F. B. (2005) Dairy consumption and risk of Type 2 diabetes mellitus in men – A prospective study. *Archives of Internal Medicine*, 165, 997–1003.

COWLES, R. L., LEE, J. Y., GALLAHER, D. D., STUEFER-POWELL, C. L. and CARR, T. P. (2002) Dietary stearic acid alters gallbladder bile acid composition in hamsters fed cereal-based diets. *Journal of Nutrition*, 132, 3119–3122.

DABADIE, H., MOTTA, C., PEUCHANT, E., LERUYET, P. and MENDY, F. (2006) Variations in daily intakes of myristic and alpha-linolenic acids in sn-2 position modify lipid profile and red blood cell membrane fluidity. *British Journal of Nutrition*, 96, 283–289. DOI:10.1079/BJN20061813

DABADIE, H., PEUCHANT, E., BERNARD, M., LERUYET, P. and MENDY, F. (2005) Moderate intake of myristic acid in sn-2 position has beneficial lipidic effects and enhances DHA of cholesteryl esters in an interventional study. *Journal of Nutritional Biochemistry*, 16, 375–382. DOI:10.1016/j.jnutbio.2005.01.010

D'ALESSIO, D. (2008) Intestinal hormones and regulation of satiety: The case for CCK, GLP-1, PYY, and Apo A-IV. *Journal of Parenteral and Enteral Nutrition*, 32, 567–568. DOI:10.1177/0148607108322401

D'ALESSIO, D. A. and VAHL, T. P. (2004) Glucagon-like peptide 1: Evolution of an incretin into a treatment for diabetes. *American Journal of Physiology–Endocrinology and Metabolism*, 286, E882–E890. DOI:10.1152/ajpendo.00014.2004

DANGIN, M., BOIRIE, Y., GARCIA-RODENAS, C., GACHON, P., FAUQUANT, J., CALLIER, P., BALLEVRE, O. and BEAUFRERE, B. (2001) The digestion rate of protein is an independent regulating factor of postprandial protein retention. *American Journal of Physiology–Endocrinology and Metabolism*, 280, E340–E348.

DENKE, M. A., FOX, M. M. and SCHULTE, M. C. (1993) Short-term dietary calcium fortification increases fecal saturated fat-content and reduces serum-lipids in men. *Journal of Nutrition*, 123, 1047–1053.

DRAPEAU, V., DESPRES, J. P., BOUCHARD, C., ALLARD, L., FOURNIER, G., LEBLANC, C. and TREMBLAY, A. (2004) Modifications in food-group consumption are related to long-term body-weight changes. *American Journal of Clinical Nutrition*, 80, 29–37.

EBRINGER, L., FERENCIK, M. and KRAJCOVIC, J. (2008) Beneficial Health Effects of Milk and Fermented Dairy Products. *Folia Microbiologica*, 53, 378–394.

EFSA (2008) Scientific Opinion of the Panel on Dietetic Products Nutrition and Allergies on a Request from the European Commission on Dairy Foods and Healthy Body Weight. *The EFSA Journal,* 786, 1–10.

ELWOOD, P. C., PICKERING, J. E., HUGHES, J., FEHILY, A. M. and NESS, A. R. (2004) Milk drinking, ischaemic heart disease and ischaemic stroke. II. Evidence from cohort studies. *European Journal of Clinical Nutrition*, 58, 718–724. DOI:10.1038/sj.ejcn.1601869

ELWOOD, P. C., STRAIN, J. J., ROBSON, P. J., FEHILY, A. M., HUGHES, J., PICKERING, J. and NESS, A. (2005) Milk consumption, stroke, and heart attack risk: Evidence from the Caerphilly cohort of older men. *Journal of Epidemiology and Community Health*, 59, 502–505. DOI:10.1136/jech.2004.027904

FARRELL, H. M., JIMENEZ-FLORES, R., BLECK, G. T., BROWN, E. M., BUTLER, J. E., CREAMER, L. K., HICKS, C. L., HOLLAR, C. M., NG-KWAI-HANG, K. F. and SWAISGOOD, H. E. (2004) Nomenclature of the proteins of cows' milk – Sixth revision. *Journal of Dairy Science*, 87, 1641–1674.

FOX, P. F. and MCSWEENEY, P. L. H. (1998) *Dairy Chemistry and Biochemistry*, London; New York, Blackie Academic & Professional.

FRID, A. H., NILSSON, M., HOLST, J. J. and BJORCK, I. M. (2005) Effect of whey on blood glucose and insulin responses to composite breakfast and lunch meals in Type 2 diabetic subjects. *American Journal of Clinical Nutrition*, 82, 69–75.

FUNG, T. T., CHIUVE, S. E., MCCULLOUGH, M. L., REXRODE, K. M., LOGROSCINO, G. and HU, F. B. (2008) Adherence to a DASH-Style diet and risk of coronary heart disease and stroke in women. *Archives of Internal Medicine*, 168, 713–720.

GANNON, M. C., NUTTALL, F. Q., NEIL, B. J. and WESTPHAL, S. A. (1988) The insulin and glucose responses to meals of glucose plus various proteins in Type II diabetic subjects. *Metabolism*, 37, 1081–8.

GERMAN, J. B. and DILLARD, C. J. (2004) Saturated fats: What dietary intake? *American Journal of Clinical Nutrition*, 80, 550–559.

GERMAN, J. B. and DILLARD, C. J. (2006) Composition, structure and absorption of milk lipids: A source of energy, fat-soluble nutrients and bioactive molecules. *Critical Reviews in Food Science and Nutrition*, 46, 57–92. DOI:10.1080/10408690590957098

GRANT, W. B. (1998) Milk and other dietary influences on coronary heart disease. *Alternative Medicine Review*, 3, 281–94.

GRUNDY, S. M. (2008) Metabolic syndrome pandemic. *Arteriosclerosis Thrombosis and Vascular Biology*, 28, 629–636. DOI:10.1161/atvbaha.107.151092

GRUNDY, S. M., CLEEMAN, J. I., DANIELS, S. R., DONATO, K. A., ECKEL, R. H., FRANKLIN, B. A., GORDON, D. J., KRAUSS, R. M., SAVAGE, P. J., SMITH, S. C., SPERTUS, J. A. and COSTA, F. (2005) Diagnosis and management of the metabolic syndrome – An American Heart Association/National Heart, Lung, and Blood Institute scientific statement – Executive summary. *Circulation*, 112, E285-E290. DOI:10.1161/CIRCULATIONAHA.105.169405

GUNTHER, C. W., LEGOWSKI, P. A., LYLE, R. M., MCCABE, G. P., EAGAN, M. S., PEACOCK, M. and TEEGARDEN, D. (2005) Dairy products do not lead to alterations in body weight or fat mass in young women in a 1-y intervention. *American Journal of Clinical Nutrition*, 81, 751–756.

HALL, W. L., MILLWARD, D. J., LONG, S. J. and MORGAN, L. M. (2003) Casein and whey exert different effects on plasma amino acid profiles, gastrointestinal hormone secretion and appetite. *British Journal of Nutrition*, 89, 239–248. DOI:10.1079/BJN2002760

HAQUE, E., CHand R and KAPILA S (2009) Biofunctional Properties of Bioactive Peptides of Milk Origin. *Food Reviews International*, 25, 15. DOI:10.1080/87559120802458198

HARTMANN, R. and MEISEL, H. (2007) Food-derived peptides with biological activity: From research to food applications. *Current Opinion in Biotechnology*, 18, 163–169. DOI:10.1016/j.copbio.2007.01.013

HARVEY-BERINO, J., GOLD, B. C., LAUBER, R. and STARINSKI, A. (2005) The impact of calcium and dairy product consumption on weight loss. *Obesity Research*, 13, 1720–1726.

HEANEY, R. P., DAVIES, K. M. and BARGER-LUX, M. J. (2002) Calcium and weight: Clinical studies. *Journal of the American College of Nutrition*, 21, 152S-155S.

HOLLENBECK, C. B. and COULSTON, A. M. (1991) The clinical utility of the glycemic index and its application to mixed meals. *Canadian Journal of Physiology and Pharmacology*, 69, 100–107.

HOLT, S. H. A., MILLER, J. C. B. and PETOCZ, P. (1997) An insulin index of foods: The insulin demand generated by 1000-kJ portions of common foods. *American Journal of Clinical Nutrition*, 66, 1264–1276.

ISOMAA, B., ALMGREN, P., TUOMI, T., FORSEN, B., LAHTI, K., NISSEN, M., TASKINEN, M. R. and GROOP, L. (2001) Cardiovascular morbidity and mortality associated with the metabolic syndrome. *Diabetes Care*, 24, 683–9.

JELICIC, I., BOTANIC, R. and TRATNIK, L. (2008) Whey based beverages – New generation of dairy products. *Mljekarstvo*, 58, 17.

JORDE, R. and BONAA, K. H. (2000) Calcium from dairy products, vitamin D intake, and blood pressure: The Tromso study. *Am J Clin Nutr*, 71, 1530–1535.

KARANJA, N., MORRIS, C. D., ILLINGWORTH, D. R. and MCCARRON, D. A. (1987) Plasma lipids and hypertension: Response to calcium supplementation. *American Journal of Clinical Nutrition*, 45, 60–65.

KORHONEN, H. and PIHLANTO A (2006) Bioactive peptides: Production and functionality *International Dairy Journal*, 16, 16. DOI:10.1016/j.idairyj.2005.10.015

KROMHOUT, D., BOSSCHIETER E. B. and COULANDERR C. D. (1985) Potassium, calcium, alcohol intake and blood pressure: The Zutpgen Study. *American Journal of Clinical Nutrition*, 41, 1299–1304.

LARSEN, T. M., TOUBRO, S. and ASTRUP, A. (2003) Efficacy and safety of dietary supplements containing CLA for the treatment of obesity: Evidence from animal and human studies. *Journal of Lipid Research*, 44, 2234–2241. DOI:10.1194/jlr.R300011-JLR200

LAWLOR, D. A., EBRAHIM, S., TIMPSON, N. and SMITH, G. D. (2005) Avoiding milk is associated with a reduced risk of insulin resistance and the metabolic syndrome: Findings from the British Women's Heart and Health Study. *Diabetic Medicine*, 22, 808–811.

LIDDLE, R. A., GOLDFINE, I. D., ROSEN, M. S., TAPLITZ, R. A. and WILLIAMS, J. A. (1985) Cholecystokinin bioactivity in human plasma. Molecular forms, responses to feeding, and relationship to gallbladder contraction. *Journal of Clinical Investigation*, 75, 1144–1152.

LILJEBERG ELMSTAHL, H. and BJOORCK, I. (2001) Milk as a supplement to mixed meals may elevate postprandial insulinaemia. *European Journal of Clinical Nutrition*, 55, 994–999.

LIN, Y. C., LYLE, R. M., MCCABE, L. D., MCCABE, G. P., WEAVER, C. M. and TEEGARDEN, D. (2000) Dairy calcium is related to changes in body composition during a two-year exercise intervention in young women. *Journal of the American College of Nutrition*, 19, 754–760.

LIP, G. Y., MALIK, I., LUSCOMBE, C., MCCARRY, M. and BEEVERS, G. (1995) Dietary fat purchasing habits in whites, blacks and Asian peoples in England–implications for heart disease prevention. *International Journal of Cardiology*, 48, 287–293.

LIU, S., CHOI, H. K., FORD, E., SONG, Y., KLEVAK, A., BURING, J. E. and MANSON, J. E. (2006) A prospective study of dairy intake and the risk of Type 2 diabetes in women. *Diabetes Care*, 29, 1579–84. DOI:10.2337/dc06-0256

LOOS, R. J. F., RANKINEN, T., LEON, A. S., SKINNER, J. S., WILMORE, J. H., RAO, D. C. and BOUCHARD, C. (2004) Calcium intake is associated with adiposity in black and white men and white women of the HERITAGE Family Study. *Journal of Nutrition*, 134, 1772–1778.

LU, Y., LEVIN, G. V. and DONNER, T. W. (2008) Tagatose, a new antidiabetic and obesity control drug. *Diabetes Obes Metab*, 10, 109–34. DOI:10.1111/j.1463–1326.2007.00799.x

LUHOVYY, B. L., AKHAVAN, T. and ANDERSON, H. (2007) Whey proteins in the regulation of food intake and satiety. *Journal of the American College of Nutrition*, 26, 704S–712S.

MACRAE, J., O'REILLY, L. and MORGAN, P. (2005) Desirable characteristics of animal products from a human health perspective. *Livestock Production Science*, 94, 95–103. DOI:10.1016/j.livprodsci.2004.11.030

MAJOR, G. C., CHAPUT, J. P., LEDOUX, M., ST-PIERRE, S., ANDERSON, G. H., ZEMEL, M. B. and TREMBLAY, A. (2008) Recent developments in calcium-related obesity research. *Obesity Reviews*, 9, 428–445. DOI:10.1111/j.1467-789X.2007.00465.x

MALJAARS, J., ROMEYN, E. A., HADDEMAN, E., PETERS, H. P. and MASCLEE, A. A. (2009) Effect of fat saturation on satiety, hormone release, and food intake. *American Journal of Clinical Nutrition*, 89, 1019–1024. DOI:10.3945/ajcn.2008.27335

MALJAARS, P. W. J., PETERS, H. P. F., MELA, D. J. and MASCLEE, A. A. M. (2008) Ileal brake: A sensible food target for appetite control. A review. *Physiology & Behavior*, 95, 271–281. DOI:10.1016/j.physbeh.2008.07.018

MARQUES-VIDAL, P., GONCALVES, A. and DIAS, C. M. (2006) Milk intake is inversely related to obesity in men and in young women: Data from the Portuguese Health Interview Survey 1998–1999. *International Journal of Obesity*, 30, 88–93. DOI:10.1038/sj.ijo.0803045

MATTILA-SANDHOLM, T. and SAARELA, M. (2003) Functional Dairy Products, Boca Raton, USA, CRC Press; Cambridge, UK, Woodhead Publishing.

MEISEL, H. (2004) Multifunctional peptides encrypted in milk proteins. *Biofactors*, 21, 55–61.

MENNEN, L. I., LAFAY, L., FESKENS, E. J. M., NOVAK, M., LEPINAY, P. and BALKAU, B. (2000) Possible protective effect of bread and dairy products on the risk of the metabolic syndrome. *Nutrition Research*, 20, 335–347.

MENSINK, R. P., ZOCK, P. L., KESTER, A. D. M. and KATAN, M. B. (2003) Effects of dietary fatty acids and carbohydrates on the ratio of serum total to HDL cholesterol and on serum lipids and apolipoproteins: A meta-analysis of 60 controlled trials. *American Journal of Clinical Nutrition*, 77, 1146–1155.

MICHALSKI, M. C. (2007) On the supposed influence of milk homogenization on the risk of CVD, diabetes and allergy. *British Journal of Nutrition*, 97, 598–610. DOI:10.1017/S0007114507657900

MINOKOSHI, Y., SHIUCHI, T., LEE, S., SUZUKI, A. and OKAMOTO, S. (2008) Role of hypothalamic AMP-kinase in food intake regulation. *Nutrition*, 24, 786–90. DOI: 10.1016/j.nut.2008.06.002

MOLKENTIN, J. (2000) Occurrence and biochemical characteristics of natural bioactive substances in bovine milk lipids. *British Journal of Nutrition*, 84, S47–S53.

MORAN, T. H., SMEDH, U., KINZIG, K. P., SCOTT, K. A., KNIPP, S. and LADENHEIM, E. E. (2005) Peptide YY(3–36) inhibits gastric emptying and produces acute reductions in food intake in rhesus monkeys. *American Journal of Physiology–Regulatory Integrative and Comparative Physiology*, 288, R384–R388. DOI:10.1152/ajpregu.00535.2004

MURPHY, K. G. and BLOOM, S. R. (2006) Gut hormones and the regulation of energy homeostasis. *Nature*, 444, 854–9. DOI:10.1038/nature05484

NAUMANN, E., CARPENTIER, Y. A., SAEBO, A., LASSEL, T. S., CHARDIGNY, J. M., SEBEDIO, J. L. and MENSINK, R. P. (2006) Cis-9, trans-11 and trans-10, cis-12 conjugated linoleic acid (CLA) do not affect the plasma lipoprotein profile in moderately overweight subjects with LDL phenotype B. *Atherosclerosis*, 188, 167–174. DOI:10.1016/j.atherosclerosis.2005.10.019

NESTEL, P. J. (2008) Effects of Dairy Fats within Different Foods on Plasma Lipids. *Journal of the American College of Nutrition*, 27, 735S–740S.

NG, T. K., HAYES, K. C., DEWITT, G. F., JEGATHESAN, M., SATGUNASINGAM, N., ONG, A. S. and TAN, D. (1992) Dietary palmitic and oleic acids exert similar effects on serum cholesterol and lipoprotein profiles in normocholesterolemic men and women. *Journal of the American College of Nutrition*, 11, 383–390.

NILSSON, M., HOLST, J. J. and BJORCK, I. M. E. (2007) Metabolic effects of amino acid mixtures and whey protein in healthy subjects: Studies using glucose-equivalent drinks. *American Journal of Clinical Nutrition*, 85, 996–1004.

NILSSON, M., STENBERG, M., FRID, A. H., HOLST, J. J. and BJORCK, I. M. (2004) Glycemia and insulinemia in healthy subjects after lactose-equivalent meals of milk and other food proteins: The role of plasma amino acids and incretins. *American Journal of Clinical Nutrition*, 80, 1246–1253.

OSTMAN, E. M., ELMSTAHL, H. and BJORCK, I. M. E. (2001) Inconsistency between glycemic and insulinemic responses to regular and fermented milk products. *American Journal of Clinical Nutrition*, 74, 96–100.

PARODI, P. W. (1999) Conjugated linoleic acid and other anticarcinogenic agents of bovine milk fat. *Journal of Dairy Science*, 82, 1339–1349.

PENG, X. (2005) Milk proteins, glycomacropeptide, and regulation of short-term food intake in rats (M.Sc. dissertation). Canada: Unibersity of Toronto Available from: Dissertations & Theses @ University of Toronto. Accessed May 12, 2009, Publication Number: AAT MR07448.

PEREIRA, M. A., JACOBS, D. R., VAN HORN, L., SLATTERY, M. L., KARTASHOV, A. I. and LUDWIG, D. S. (2002) Dairy consumption, obesity, and the insulin resistance syndrome in young adults – The CARDIA study. *Journal of the American Medical Association*, 287, 2081–2089.

PFEUFFER, M. and SCHREZENMEIR, J. (2000) Bioactive substances in milk with properties decreasing risk of cardiovascular diseases. *British Journal of Nutrition*, 84, S155–S159.

PFEUFFER, M. and SCHREZENMEIR, J. (2007) Milk and the metabolic syndrome. *Obesity Reviews*, 8, 109–118. DOI:10.1111/j.1467-789X.2006.00265.x

PHILLIPS, S. M., BANDINI, L. G., CYR, H., COLCLOUGH-DOUGLAS, S., NAUMOVA, E. and MUST, A. (2003) Dairy food consumption and body weight and fatness studied longitudinally over the adolescent period. *International Journal of Obesity*, 27, 1106–1113. DOI:10.1038/sj.ijo.0802370

PUPOVAC, J. and ANDERSON, G. H. (2002) Dietary peptides induce satiety via cholecystokinin-A and peripheral opioid receptors in rats. *Journal of Nutrition*, 132, 2775–2780.

RAJPATHAK, S. N., RIMM, E. B., ROSNER, B., WILLETT, W. C. and HU, F. B. (2006) Calcium and dairy intakes in relation to long-term weight gain in US men. *American Journal of Clinical Nutrition*, 83, 559–566.

RANGANATHAN, R., NICKLAS, T. A., YANG, S. J. and BERENSON, G. S. (2005) The nutritional impact of dairy product consumption on dietary intakes of adults (1995–1996): The Bogalusa heart study. *Journal of the American Dietetic Association*, 105, 1391–1400. DOI: 10.1016/j.jada.2005.6.024

REIDELBERGER, R. D., HERNANDEZ, J., FRITZSCH, B. and HULCE, M. (2004) Abdominal vagal mediation of the satiety effects of CCK in rats. *American Journal of Physiology–Regulatory Integrative and Comparative Physiology*, 286, R1005–R1012. DOI:10.1152/ajpregu.00646.2003

SAMPELAYO, M. R. S., CHILLIARD, Y., SCHMIDELY, P. and BOZA, J. (2007) Influence of type of diet on the fat constituents of goat and sheep milk. *Small Ruminant Research*, 68, 42–63. DOI:10.1016/j.smallrumres.2006.09.017

SAMUELSON, G., BRATTEBY, L. E., MOHSEN, R. and VESSBY, B. (2001) Dietary fat intake in healthy adolescents: Inverse relationships between the estimated intake of saturated fatty acids and serum cholesterol. *British Journal of Nutrition*, 85, 333–341. DOI:10.1079/BJN2000279

SCHOLZ-AHRENS, K. E. and SCHREZENMEIR, J. (2006) Milk minerals and the metabolic syndrome. *International Dairy Journal*, 16, 1399–1407. DOI:10.1016/j.idairyj.2006.06.009

SCHREZENMEIR, J., TATO, F., TATO, S., KUSTNER, E., KRAUSE, U., HOMMEL, G., ASP, N. G., KASPER, H. and BEYER, J. (1989) Comparison of glycemic response and insulin requirements after mixed meals of equal carbohydrate content in healthy, Type-1, and Type-2 diabetic men. *Klinische Wochenschrift*, 67, 985–994.

SCHWARTZ, M. W., WOODS, S. C., PORTE, D., JR., SEELEY, R. J. and BASKIN, D. G. (2000) Central nervous system control of food intake. *Nature*, 404, 661–71. DOI: 10.1038/35007534

SEELY, S. (1981) Diet and coronary-disease – a survey of mortality-rates and food-consumption statistics of 24 countries. *Medical Hypotheses*, 7, 907–918.

SIRO, I., KAPOLNA, E., KAPOLANA, B. and LUGASI, A. (2008) Functional food. Product development, marketing and consumer acceptance – a review. *Appetite*, 51, 11. DOI: 10.1016/j.appet.2008.05.060

SMEDMAN, A. E. M., GUSTAFSSON, I. B., BERGLUND, L. G. T. and VESSBY, B. O. H. (1999) Pentadecanoic acid in serum as a marker for intake of milk fat: Relations between intake of milk fat and metabolic risk factors. *American Journal of Clinical Nutrition*, 69, 22–29.

SNIJDER, M. B., VAN DER HEIJDEN, A., VAN DAM, R. M., STEHOUWER, C. D. A., HIDDINK, G. J., NIJLELS, G., HEINE, R. J., BOUTER, L. M. and DEKKER, J. M. (2007) Is higher dairy consumption associated with lower body weight and fewer metabolic disturbances? The Hoorn Study. *American Journal of Clinical Nutrition*, 85, 989–995.

STARLING, S. (2008) Carbery Boosts Whey Response [http://www.dairyreporter.com/Products/Carbery-boosts-whey-response], Decision News Media SAS.

ST-ONGE, M. P. and JONES, P. J. H. (2003) Greater rise in fat oxidation with medium-chain triglyceride consumption relative to long-chain triglyceride is associated with lower initial body weight and greater loss of subcutaneous adipose tissue. *International Journal of Obesity*, 27, 1565–1571. DOI:10.1038/sj.ijo.0802467

ST-ONGE, M. P., FARNWORTH, E. R. and JONES, P. J. H. (2000) Consumption of fermented and nonfermented dairy products: Effects on cholesterol concentrations and metabolism. *American Journal of Clinical Nutrition*, 71, 674–681.

SUNDRAM, K., HAYES, K. C. and SIRU, O. H. (1994) Dietary palmitic acid results in lower serum cholesterol than does a lauric–myristic acid combination in normolipemic humans. *American Journal of Clinical Nutrition*, 59, 841–846.

TORPY, J. M., LYNM, C. and GLASS, R. M. (2006) JAMA patient page. The metabolic syndrome. *The Journal of the American Medical Association*, 295, 850. DOI:10.1001/jama.295.7.850

TROYANO, E., OLANO, A., FERNÁNDEZ-DÍAZ, M., SANZ, J. and MARTÍNEZ-CASTRO, I. (1991) Gas chromatographic analysis of free monosaccharides in milk. *Chromatographia*, 32, 379–382. DOI:10.1007/BF02321437

TSUJI, H., KASAI, M., TAKEUCHI, H., NAKAMURA, M., OKAZAKI, M. and KONDO, K. (2001) Dietary medium-chain triacylglycerols suppress accumulation of body fat in a double-blind, controlled trial in healthy men and women. *Journal of Nutrition*, 131, 2853–2859.

TURPEINEN, O. (1979) Effect of cholesterol-lowering diet on mortality from coronary heart disease and other causes. *Circulation*, 59, 1–7.

VEGARUD, G. E., LANGSRUD, T. and SVENNING, C. (2000) Mineral-binding milk proteins and peptides; Occurrence, biochemical and technological characteristics. *British Journal of Nutrition*, 84, S91-S98.

VESPER, H., SCHMELZ, E. M., NIKOLOVA-KARAKASHIAN, M. N., DILLEHAY, D. L., LYNCH, D. V. and MERRILL, A. H. (1999) Sphingolipids in food and the emerging importance of sphingolipids to nutrition. *Journal of Nutrition*, 129, 1239–1250.

VESSBY, B. (2003) Dietary fat, fatty acid composition in plasma and the metabolic syndrome. *Current Opinion in Lipidology*, 14, 15–19.

VISIOLI, F., RISE, P., PLASMATI, E., PAZZUCCONI, F., SIRTORI, C. R. and GALLI, C. (2000) Very low intakes of N-3 fatty acids incorporated into bovine milk reduce plasma triacylglycerol and increase HDL-cholesterol concentrations in healthy subjects. *Pharmacological Research*, 41, 571–576.

VITHANAGE, C. J., MISHRA, V. K., VASILJEVIC, T. and SHAH, N. P. (2008) Use of beta-glucan in development of low-fat Mozzarella cheese. *Milchwissenschaft-Milk Science International*, 63, 420–423.

WANG, Y. W. and JONES, P. J. H. (2004) Conjugated linoleic acid and obesity control: efficacy and mechanisms. *International Journal of Obesity*, 28, 941–955. DOI:10.1038/sj.ijo.0802641

WANG, Y., WANG, X. and WU, D. (2008) Functional health dairy product, e.g. for preventing and inhibiting hyperlipemia, comprises oligosaccharide from low-polymerized seminose, xylo-oligosaccharide, fructooligosaccharides, and/or galactooligosaccharide. *GUANGZHOU BOKAI BIOLOGY TECHNOLOGY CO LTD (GUAN-Non-standard).*

WARENSJO, E., JANSSON, J. H., BERGLUND, L., BOMAN, K., AHREN, B., WEINEHALL, L., LINDAHL, B., HALLMANS, G. and VESSBY, B. (2004) Estimated intake of milk fat is negatively associated with cardiovascular risk factors and does not increase the risk of a first acute myocardial infarction. A prospective case-control study. *British Journal of Nutrition*, 91, 635–642. DOI:10.1079/BJN20041080

WEAVER, C. M. (2009) Should dairy be recommended as part of a healthy vegetarian diet? Point. *American Journal of Clinical Nutrition*, 89 (suppl), 1S–4S. DOI:10.3945/ajcn.2009.26736O

WILLIAMS, D. L., BASKIN, D. G. and SCHWARTZ, M. W. (2006) Leptin regulation of the anorexic response to glucagon-like peptide-1 receptor stimulation. *Diabetes*, 55, 3387–93. DOI:10.2337/db06-055

WOLEVER, T. M. S. and MILLER, J. B. (1995) Sugars and blood-glucose control. *American Journal of Clinical Nutrition*, 62, S212–S227.

YANAGIDA, A. and NAGAO, K. (2008) Agents useful as food/beverage products for increasing adiponectin levels, and treating life-style related diseases, e.g. obesity, diabetes and arteriosclerosis, comprise substance derived from mushroom of Panellus genus. *KOKURITSU DAIGAKU HOJIN SAGA DAIGAKU (KOKU-Non-standard).*

ZEMEL, M. B., THOMPSON, W., MILSTEAD, A., MORRIS, K. and CAMPBELL, P. (2004) Calcium and dairy acceleration of weight and fat loss during energy restriction in obese adults. *Obesity Research*, 12, 582–590. DOI:10.1038/oby.2004.67

ZOCK, P. L., DEVRIES, J. H. M. and KATAN, M. B. (1994) Impact of myristic acid versus palmitic acid on serum lipid and lipoprotein levels in healthy women and men. *Arteriosclerosis and Thrombosis*, 14, 567–575.

9

Bioactive milk protein and peptide functionality

R. Kekkonen, Valio Ltd, Finland and University of Helsinki, Institute of Biomedicine, Finland; and K. Peuhkuri, Netnut, Finland and University of Helsinki, Institute of Biomedicine, Finland

Abstract: Bioactive peptides (BPs), encrypted in the amino acid sequences of protein, are potential health-enhancing nutraceuticals. Numerous bioactivities, including antithrombotic, antilipemic, antioxidative, immunomodulatory, antimicrobial, opiate, osteoprotective, anticariogenic, and growth-promoting properties, have been described for BPs, mostly *in vitro* or in animal models. The clinical evidence for antihypertensive effects is convincing and the evidence for the regulation of satiety, food intake and obesity-related metabolic disorders is increasing. In order to exert health effects, peptides have to be absorbed from the intestine and must reach the target cells in sufficient concentrations or act via receptors and cell signalling in the gut without absorption.

Key words: cardioprotective peptides, immunomodulatory peptides, opioid peptides, gastroprotective peptides, body composition.

9.1 Introduction

Interest in the health aspects of bioactive compounds found in milk and dairy products is increasing. The number, potency, and importance of bioactive compounds in milk, especially in milk proteins, are probably greater than previously thought. Whereas some of these compounds are normal milk components, others emerge during digestive or fermentation processes.

Besides intact proteins, such as alpha-lactalbumin (α-lactalbumin), beta-lactoglobulin (β-lactoglobulin), growth factors, immunoglobulins and lactoferrin, increasing interest is focused on peptides which may be released from proteins by enzymatic activity. Bioactive peptides (BPs) may be encrypted in the amino acid sequence of a larger protein and usually consist of 3–20 amino acids. Encrypted peptides can be released in three ways: *in vivo* during digestion by digestive enzymes, such as trypsin; *in vivo* during digestion by microbial enzymes; or *in vitro* during food processing or ripening by isolated or microbial enzymes.

Concentrates of these peptides are potential health-enhancing nutraceuticals for food and pharmaceutical applications. Numerous bioactivities, including antihypertensive, antithrombotic, antilipemic, antioxidative, immunomodulatory, antimicrobial, opiate, osteoprotective, anticariogenic and growth-promoting properties, have been described for bioactive bovine peptides. Furthermore, there has been growing evidence of the role of dairy proteins even in the regulation of satiety, food intake and obesity-related metabolic disorders. This chapter provides an overview of the beneficial physiological effects of bioactive protein components in milk and also discusses their potential health effects.

9.2 Major bioactive milk proteins and peptides

The average protein content of milk is 3.2%, of which approximately 80% is casein (2.7 g/100 g milk) and the rest is whey. The concentration of caseins and whey proteins in milk is presented in Table 9.1. In addition to these main milk proteins, a small number of minor proteins and peptides with physiological activities (Table 9.1) are natural constituents of milk. These compounds are secreted from blood into milk or are synthesized in the mammary gland.

Milk proteins contain a wide variety of biologically active peptides. These peptides, which are encrypted within the sequence of the parent proteins, can be released by enzymatic proteolysis. BPs have been obtained from casein as well as whey proteins from *in vivo* digests after feeding the precursor protein, or from *in vitro* digestion with proteolytic enzymes during food processing. Typically, they are inactive within the sequence of the precursor proteins before proteolysis. Once produced, BPs may act in the body as regulatory compounds with hormone-like activity.

Bioactive peptides may be liberated from food proteins throughout the whole gastrointestinal tract, and they can display bioactivity in the small and large intestine (Fig. 9.1). However, since most food proteins are degraded during their transit through the small intestine, and because microbial activity occurs predominantly in the large intestine, the release by microbial enzymes during digestion applies to proteins when reaching this site of the intestine. In addition, intestinal digestive enzymes and microbial enzymes in the gut, or in food, use different cleavage sites. As a result, peptides liberated by microbial enzymes may differ from those released by digestive enzymes.

It is well known that di- and tripeptides are easily absorbed in the intestine (Hara *et al.*, 1984). However, less is known about the absorption of BPs of higher molecular weight in the intestine. In order to exert

Table 9.1 Concentration and biological functions of bovine milk proteins

Protein	Concentration (g/l)	Bioactive peptide	Function
Total caseins	26.0		
α	13.0	α-Casein exorphin	Opioid agonist
		Casokinins, lactorphins	ACE-inhibitory
		Immunopeptides	Immunomolulatory
		Casocidin, Isracidin	Antimicrobial
		Phosphopeptides	Mineral carriers
β	9.3	β-Casomorphins	Opioid agonist
		Casokinins, lactorphins	ACE-inhibitory
		Immunopeptides	Immunomolulatory
		Phosphopeptides	Mineral carriers
κ	3.3	Casoxins, lactorphins	Opioid antagonist
		Immunopeptides	Immunomolulatory
		Casoplatelins	Antithrombotic
Total whey	6.3		
β-Lactoglobulin	3.2	β-Lactorphin	Opioid agonist
		Lactokinins	ACE-inhibitory
α-Lactalbumin	1.2	α-Lactorphin	Opioid agonist
		Lactokinins	ACE-inhibitory
Immunoglobulins (A, M, G)	0.7		
Serum albumin	0.4		
Lactoferrin and transferrin	0.1	Lactoferricin	Antimicrobial
		Casoplatelins	Opioid antagonist
Lactoperoxidase	0.03		
Lysozyme	0.0004		
Proteose-peptone	<1		
Glycomacropeptide	0.5–1.5		
Miscellaneous minor proteins	0.8		
Thyrotropin-releasing hormone (THR)			
Luteinizing hormone-releasing hormone (LHRH)			
Somatostatin (SIH)			
Gastrin-releasing peptide (GRP)			
Calcitonin			
Adrenocorticotropic peptide (ACTH)			
Insulin			
Growth factors, e.g.			
EGF			
Relaxin			
Prolactin			
Thyroid stimulating hormone (TSH)			
Enzymes, e.g. plasmin			

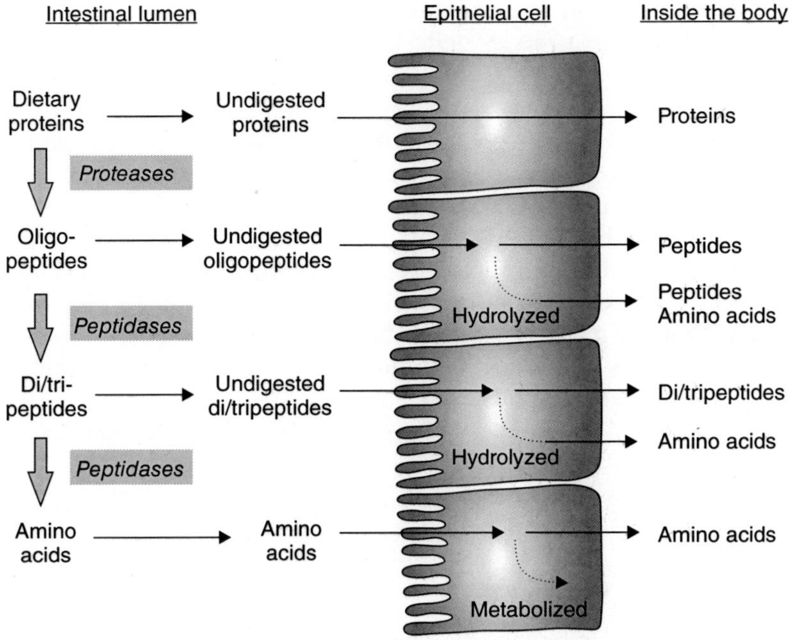

Fig. 9.1 The structure of dietary proteins and peptides change during the process of digestion and absorption in the intestine and epithelial cells.

antihypertensive effects, for example, peptides have to be absorbed from the intestine and must reach the target cells in blood vessels in sufficient concentrations. Thus, bioactive food components may be degraded during digestion, may not be absorbed, or may not reach adequate concentrations in blood and other target tissues.

Some proteins, such as lactoferrin and immunoglobulins, have been shown to, at least partly, escape degradation in the small intestine (Drescher *et al.*, 1999; Roos *et al.*, 1995). They may be absorbed as whole proteins, or act directly in the intestinal tract or via receptors and cell signalling in the gut without absorption.

9.2.1 Safety

Because bioactive proteins and peptides are natural components of milk and dairy products, their degradation in the gastrointestinal tract is assumed not to differ from any other dietary proteins and peptides. Among BPs, the ACE-inhibitory peptides and antihypertensive peptides have been extensively studied worldwide. Furthermore, safety and toxicity studies have been conducted mostly with antihypertensive tripeptides derived from milk proteins. In these studies on experimental animal models or in clinical trials, no adverse effects were reported even with the highest dose tested (Dent *et al.*, 2007; Pins and Keenan, 2006; Jauhiainen *et al.*, 2005; Mizuno *et al.*, 2005; Nakamura *et al.*, 2005). An interesting observation is that even high daily doses had no clinical signs nor biologically meaningful effects on systolic or diastolic blood pressure, the pulse rate, or clinical pathology (serum chemistry or haematology) in adult normotensive volunteers (for review, see Bernard *et al.*, 2005).

Dairy proteins have been a part of the human diet for a long time. Although there is no theoretical safety concern about the natural ingredients of dairy products, accumulative clinical data from longer interventions and clinical use of products containing BPs will be valuable in expanding knowledge of their safety, as the duration of the published interventions is only a few months at the most.

9.3 Health-promoting properties of bioactive milk proteins and peptides

9.3.1 Effects on the cardiovascular system

Antihypertensive effects

Angiotensin I-converting enzyme (ACE) is a key enzyme in the regulation of blood pressure. Within the renin-angiotensin system, ACE catalyzes the conversion from angiotensin I to angiotensin II, which is a very potent vasoconstrictor. Consequently, inhibition of synthesis of angiotensin II lowers blood pressure. Moreover, since ACE also degrades bradykinin, which has vasodilatory properties, the widely-used ACE inhibitor antihypertensive pharmaceuticals reduce hypertension by at least two mechanisms.

Dairy products may act as natural functional foods with ACE inhibitory activity and blood pressure lowering capacity. Meisel (1997) has listed ACE inhibitory activities in various dairy products. Low activity was found in products with a low degree of proteolysis, such as yoghurt, quark, and fresh cheeses. Ripened cheese contained more activity, but this was dependent on the degree of maturation.

The antihypertensive effect of orally administered doses of alpha-s1- and beta-casein (αs1- and β-casein) hydrolysates in spontaneous hypertensive rats has been reported by Yamamoto *et al.* (1994), Nakamura *et al.* (1995) and Maeno *et al.* (1996). Tripeptides Val-Pro-Pro (VPP) and Ile-Pro-Pro (IPP) were identified as antihypertensive components which were produced by fermenting milk proteins with *Lactobacillus helveticus* and *Saccharomyces cerevisiae*. A sour milk product fermented with these microbes and containing these tripeptides (Calpis™) reduced arterial blood pressure in rats and humans (Hata *et al.*, 1996; Masuda *et al.*, 1996). Accordingly, Seppo *et al.* (2003) and Jauhiainen *et al.* (2005) found moderate or significant reduction in the blood pressure of hypertensive subjects after 21- and

10-week-interventions with daily doses of sour milk products (Evolus®), compared to a control group. Evolus® contains bioactive VPP and IPP peptides that can pass the intestinal tract, and can, after absorption, inhibit the production of angiotensin-II in the blood, which suggests a mild pharmacological antihypertensive effect.

Other peptides may be equally effective. In fact, a placebo-controlled study conducted by Ashar and Chand (2004) demonstrated the effect of 'Dahi', a fermented milk product containing ACE-inhibitory peptide Ser-Lys-Val-Tyr-Pro, on hypertensive subjects. The product was produced by fermenting milk with *L.delbrueckii* ssp.*bulgaricus*, *Str.thermophilus* and *Lc.lactis* susp.*lactis* biovar *diacetylactis*. A significant decline in the systolic blood pressure was recorded after two and four weeks from the beginning of the trial. In addition, β-lactoglobulin and α-lactalbumin have been shown to release several ACE-inhibitory peptides after digestion with proteolytic enzymes (Hernández-Ledesma *et al.*, 2007; Pihlanto-Leppälä *et al.*, 2000; Mullally *et al.*, 1997).

Milk peptides may exert antihypertensive effects also through other mechanisms. Maes *et al.* (2004) investigated the endothelin-converting enzyme-inhibitory (ECE) properties of whey-derived peptides. Tryptic digest of β-lactoglobulin was found to inhibit the endothelin-1 (ET-1) release from endothelial cells *in vitro*. Endothelin is an even stronger vasoconstrictor than angiotensin II. The membrane-bound ECE converts the precursor endothelin into active endothelin. Thus, inhibition of ECE leads to lower blood pressure. Whereas Perpetuo *et al.* (2003) demonstrated a stimulation of bradykinin activity, Sipola *et al.* (2002) found enhancement of endothelium-derived nitric oxide production and Nurminen *et al.* (2000) reported enhancement of the vasodilatory action of milk peptides of binding to opiate receptors.

To exert an antihypertensive effect after oral ingestion, active peptides must be absorbed in an intact form from the intestine and, further, be resistant to degradation by plasma peptidases in order to reach the target sites. Proline-containing peptides, such as VPP and IPP, are generally resistant to degradation by digestive enzymes. In fact, VPP can be transported intact through the cell layer via paracellular and transcellular routes, although a significant amount of peptide is hydrolyzed to amino acids by intracellular peptidases (Satake *et al.*, 2002). Masuda *et al.* (1996) detected VPP and IPP in the abdominal aorta of SHR rats after oral administration of a sour milk product containing these tripeptides. On the other hand, *in vitro* results by Walsh *et al.* (2004) indicated that a tryptic digest of β-lactoglobulin, which is a potent inhibitor of ACE according to the study of Mullally *et al.* (1997), is probably not sufficiently stable in the gastrointestinal tract and in the serum to act as a hypotensive agent in humans.

Hence, only a few of the large number of *in vitro* potential antihypertensive milk peptides have proven clinically effective in animal and human trials. The effect is achieved after continuous ingestion of products containing effective peptides, since the blood pressure is gradually reverted to the pre-trial level after the intervention period. Further studies are still needed to better understand the mechanisms behind the blood pressure-reducing effects of milk peptides and also to demonstrate the long-term physiological effects induced by the consumption of such peptides.

Antithrombotic effects

Antithrombotic drugs are used for reducing platelet aggregation and for enhancing fibrinolysis in order to prevent the development of thrombosis. A large number of similarities have been reported between the clotting of blood and the clotting of milk. Structural homologies between bovine κ-casein and the human fibrinogen γ-chain were found by Jollés *et al.* (1978). The κ-casein fragment called casopiastrin has displayed antithrombotic activity by inhibiting fibrinogen binding to platelets *in vitro* (Manso *et al.*, 2002; Chabance *et al.*, 1997; Fiat *et al.*, 1993). Moreover, Bal dit Sollier *et al.* (1996) proved some antithrombotic activity *in vivo* after parenteral administration in guinea pigs. Another peptide called casoplatelin inhibits both the aggregation of activated platelets and the binding of the human fibrinogen γ-chain to its receptor on the platelet surface (Jollés *et al.*, 1986).

The human lactoferrin-derived peptide, KRDS, holds structural similarities to the fibrinogen α-chain and has inhibited platelet aggregation in experimental models both *in vitro* and *in vivo* (Drouet *et al.*, 1990; Mazoyer *et al.*, 1990; Raha *et al.*, 1988). However, although no pharmacological data on the clinical significance of milk peptides with antithrombotic activity is available, such peptides have been detected in the plasma of newborn babies after breastfeeding or ingestion of cow-milk-based infant formula, which indicates absorption of antithrombotic peptides into the blood (Chabance *et al.*, 1995).

Hypocholesterolemic activity

Whey proteins have been reported to exhibit a greater hypocholesterolemic effect, in comparison with casein or soybean proteins, in rats and healthy men (Kawase *et al.*, 2000; Minehira *et al.*, 2000; Zhang and Beynen, 1993). Fermented acidophilus milk had a hypocholesterolemic effect in human subjects during a three-week intervention (Ashar and Prajapati, 2000). By contrast, Dent *et al.* (2001) did not report any effect of whey in postmenopausal women.

On the contrary, dietary casein actually tended to increase serum and hepatic lipids in animal studies when comparing the effect of different dietary protein supplements (Aoyama and Wada, 1999; Van der Meer *et al.*, 1988). In a recent screening by Nass *et al.* (2008), however, two of the αs1-casein hydrolysates activated the MAP-kinase signalling of the cholesterol metabolizing enzyme 7alpha-hydroxylase (7α-hydroxylase), even if no effect on 7α-hydroxylase expression was seen, which thus suggests some hypocholesterolemic effect after all.

Nagaoka *et al.* (2001) identified four peptide sequences from a β-lactoglobulin hydrolysate, lactostatin, which inhibited cholesterol absorption by Caco-2 cells *in vitro*. In animal studies, Ile-Ile-Ala-Glu-Lys-corresponding fragments f(71–75) demonstrated an even more powerful effect on the serum cholesterol level than exhibited by β-sitosterol, a known agent for reducing cholesterol absorption. This effect was, at least in part, due to an enhancement of faecal steroid excretion, but the accurate mechanisms of effect remain to be clarified. Recent data have indicated that lactostatin is capable of inducing gene transcription of human cholesterol 7α-hydroxylase in a human liver cell line (Morikawa *et al.*, 2007). This activity involves calcium and MAPK-dependent signalling pathways.

β-Lactotensin, a neurotensin NT2 agonist derived from β-lactoglobulin, has been found to exhibit hypocholesterolemic activity after administration of 100 mg/kg p.o. to mice for two days (Yamauchi *et al.*, 2003). This hypocholesterolemic effect was reported to be due to neurotensin NT2 receptor agonist action.

Bovine lactoferrin (1%) mixtured with a standard commercial mouse diet reduces plasma or hepatic triacylglycerol and cholesterol (Takeuchi *et al.*, 2004). In addition, during the intervention with a normal diet, plasma HDL-cholesterol levels increased significantly. However, no significant suppressive effects on the lipid metabolism of the high-fat diet fed mice were noticed. The authors supposed this action to be due to the suppressive effects on lipid metabolism and the effects on micellar solubility of cholesterol.

The effects of dietary proteins are thus due to the interaction between protein and hepatic lipogenesis, in addition to which inhibition of cholesterol absorption has been demonstrated. However, most of the claimed hypocholesterolemic effects of BPs have been observed *in vitro* or in animal models, and human clinical data remains limited. The accurate mechanisms remain to be clarified, but it is evident that the action of BPs is miscellaneous.

Antioxidative effects

Reactive oxygen species (ROS) are produced as part of normal cell metabolism and are involved in a variety of normal regulatory systems. A high level of ROS can disturb the redox balance and promote oxidative stress in cells.

Whey has potent antioxidant activity, likely by contributing cystein-rich proteins that aid in the synthesis of glutathione (GSH). GSH is comprised of glycine, glutamate and cystein. Dietary intake of cystein, or its precursor methionine, has a powerful effect on the synthesis of GSH (for review, see Wu *et al.*, 2004). Nonetheless, in addition to the amount of dietary cystein, also the entire protein composition influences oxidative stress (Blouet *et al.*, 2007; Mariotti *et al.*, 2004).

Several *in vitro* studies have shown the antioxidative capacity of whey proteins and peptides (for review, see Bounous, 2000). The antioxidative activity of unhydrolyzed β-lactoglobulin, however, is not remarkable *in vitro* (Liu *et al.*, 2007; Elia *et al.*, 2006), although several supplements of whey fractions have exhibited antioxidative activity *in vivo* in rats (Zommara *et al.*, 2002, 1998).

Ingestion of a whey-based cysteine donor in a product named *Immunocal* (HMS90) increased the circulating lymphocyte GSH concentration in healthy young adults by over 30% (Lands *et al.*, 1999). The increase in GSH was linear with the amount of whey protein ingested (Zavorsky *et al.*, 2007). If oxidative stress has been imposed during physical exercise, inflammation or use of alcohol, varying doses of whey protein supplementation prevented oxidative stress both in experimental and clinical trials (Elia *et al.*, 2006; Oner *et al.*, 2006; Middleton *et al.*, 2004; Rosaneli *et al.*, 2002). Furthermore, antioxidative effects of the *Immunocal* product have been demonstrated in several clinical groups, such as asthma, cystic fibrosis, cancer, and HIV-infected patients (Tozer *et al.*, 2008; Lothian *et al.*, 2006; Moreno *et al.*, 2006; Grey *et al.*, 2003; Micke *et al.*, 2002, 2001). However, in a small trial by Brown *et al.* (2004), no antioxidative effects of whey were found in participants of a fitness course.

Antioxidative properties of milk peptides have not been systematically studied. Pazos *et al.* (2006) compared the activity *in vitro* and found lactoferrin to be a more powerful antioxidant than bovine serum albumin or β-lactoglobulin, which were quite equal in their antioxidative capacity. Hernández-Ledesma and co-workers (2005) investigated the antioxidant activity of hydrolysates of β-lactoglobulin and α-lactalbumin and demonstrated that the presence of tryptophan (Trp), tyrosine (Tyr), and methionine (Met), as well as their position within the peptide sequence, are the most determinant factors on the antioxidative activity of peptides (Hernández-Ledesma *et al.*, 2007). The high antioxidant activity of these amino acids may be explained by the capacity of the indolic and phenolic groups to serve as hydrogen donors. L-Tyr has been found to be an effective antioxidant in different *in vitro* assays (Gülçin, 2007).

Lactoferrin is a member of the iron-binding transferrin protein family which inhibits the formation of reactive oxygen species. It is an efficient iron chelator and decreases the intensity of peroxidation processes *in vitro* and *in vivo* (Tsubota *et al.*, 2008; Chodaczek *et al.*, 2007; Shoji *et al.*, 2007; Kruzel *et al.*, 2006; Sandomirsky *et al.*, 2003; Raghuveer *et al.*, 2002). However, Konishi *et al.* (2006) found no markers of altered iron metabolism during eight weeks of lactoferrin supplementation in patients with chronic hepatitis, suggesting that the inhibitory effect of lactoferrin supplementation on lipid peroxidation may be based on several mechanisms.

Casein and casein-derived peptides are not considered to be as effective antioxidants as whey proteins. *In vitro* casein and casein hydrolysates inhibit lipid peroxidation and scavenge free radical intermediates, at least partly dose-dependently (Sakanaka *et al.*, 2005; Diaz and Decker, 2004; Rival *et al.*, 2001). Ingestion of casein significantly protected rats from iron and exercise-induced oxidative damage (Zunquin *et al.*, 2006), but the effects were not equal to those of soy protein or casein supplemented with methionine or cystein in rats and mice (Park *et al.*, 2005; Alhamdan and Grimble, 2003; Madani *et al.*, 2000; Taniguchi *et al.*, 2000). By contrast, a casein-supplemented diet had no antioxidative effects in mice in the trial by Elia *et al.* (2006). Rankin and co-workers (2006), instead, found no difference between casein and whey supplementation in trained athletes. Of casein protein hydrolysates, the Glu-Leu sequence is important for the radical scavenging activity (Suetsuna *et al.*, 2000).

To summarize, the results of some animal and clinical studies suggests that fermented milk products containing BPs do indeed have an antioxidative effect. Further studies are needed to elucidate the potential clinical benefits and significance of the antioxidative peptides.

9.3.2 Effects on the immune system

The immune system is a complex and highly developed system consisting of innate immunity and adaptive or acquired immunity. The innate immunity provides immediate defence against infections and pathogens in a non-specific manner. The adaptive immune system, on the other hand, is composed of highly specialized, systemic cells and processes that eliminate or prevent infections by inducing a complex interplay between antigens, lymphocytes, and antibody production.

Diet is known to play an important role in the body's defence mechanisms. Research concerning the role of functional peptides on the immune system is quite recent but seems promising. The two main activities being studied are the immunomodulatory (stimulation of immune system) and antimicrobial (inhibition of micro-organisms) effects.

Most of the studies on the effects of milk proteins and peptides on the proliferation and maturation of lymphocytes, macrophage activity, cytokine and antibody production and natural killer cells have been conducted *in vitro* (for review, see Möller *et al.*, 2008; Baldi *et al.*, 2005; Cross and Gill, 2000). Fewer studies have been carried out on the immune function in animals and humans. *In vivo*, lymphocyte activity in mice was increased by dietary supplements of milk proteins (Wong and Watson, 1995), α-lactalbumin (Bounous *et al.*, 1983), lactoferrin (Debbabi *et al.*, 1998), and κ-casein-based glycomacropeptide (Otani *et al.*, 1996). An increased amount of immunoglobulin A (IgA) secreting cells and an increased IgA concentration have been observed by administration of fermented milk protein (LeBlanc *et al.*, 2002; Matar *et al.*, 2001; Perdigon *et al.*, 1994).

Phagocytosing capacity was enhanced in peritoneal macrophages in mice by oral application of casein hydrolysate for five days (Kazlauskaite *et al.*, 2005). Despite that, the immunostimulatory effects were not intense enough to prevent mice from experimentally induced inflammation. Moreover, daily intake of bovine colostrum for seven days in healthy adults did not enhance the expression of phagocytosis receptors on neutrophils and monocytes (He *et al.*, 2001).

According to Bounous *et al.* (1985, 1983), the quality of dietary protein has an effect on the activity on B-lymphocytes and antibody production, in particular. In these studies, the immune responses of mice on a α-lactalbumin diet were up to almost five times higher than those of mice fed on a corresponding casein diet (Bounous *et al.*, 1983). Enhanced cell-mediated immune responses followed by systemic immunization with ovalbumin were also reported by Wong and Watson (1995) in an intervention study, in which mice were fed with whey proteins or other dietary proteins for five to eight weeks. Subsequent to that, Low *et al.* (2003) observed boosted intestinal tract antibody responses to orally-administered antigens in mice, as well. In a mouse study by Otani *et al.* (2003), a caseinphosphopeptide preparation, consisting mainly of bovine αs-casein and β-casein, increased faecal and intestinal IgA responses towards orally ingested lipopolysaccharides (LPS), in addition to which the spleen cells produced larger amounts of IgA, interleukin 5 (IL-5), and IL-6 than the cells of mice on the control diet.

The ability of lactoferrin to modulate the function of several immune cells has been shown *in vitro* and *in vivo*. Intraperitoneally injected lactoferrin augmented peritoneal macrophage activities and enhanced the production of superoxide anion and nitric oxide by peritoneal macrophages (Wakabayashi *et al.*, 2003); even an oral dose boosted cytokine production (Wakabayashi *et al.*, 2006). In experimental studies, orally ingested lactoferrin has augmented natural killer cell and lymphocyte activity in mice (Kuhara *et al.*, 2000; Wang *et al.*, 2000), and trends to this effect have also been observed in healthy volunteers and in patients with chronic hepatitis (Ishii *et al.*, 2003; Yamauchi *et al.*, 1998). Once exposed to stimuli, T-lymphocytes can produce cytokines and immunoglobulins which modulate the developing immune response, and can, in addition, up-regulate the expression and number of receptor molecules on the cell surface. Indeed, both oral intake and gavaged lactoferrin in animals has increased the production of cytokines and immunoglobulins in the intestine (Togawa *et al.*, 2002; Kuhara *et al.*, 2000; Wang *et al.*, 2000; Debbabi *et al.*, 1998). Sfeir *et al.* (2004) compared different modes of oral administration of lactoferrin regarding the regulatory effect on intestinal and systemic immune response. While no differences were noticed between gastric intubation, single buccal doses and continuous oral doses in the diet, lactoferrin in the drinking water had only minor effects.

Antimicrobial properties of milk have been widely acknowledged for years. These properties are mainly due to immunoglobulins, lactoferrin, lactoperoxidase, and lysozyme. Among the most potent antimicrobial peptides are lactoferrin and its fragments (for review, see Brock, 2002). More recently, other whey proteins have been considered as potential precursors of bacterial fragments. Antibacterial fragments have been derived also from αs1-, αs2- and κ-casein (for review, see Lahov and Regelson, 1996). These peptides have been found to be active against a broad range of pathogenic organisms, e.g. Escherichia, Helicobacter, Listeria, Salmonella and Staphylococcus, as well as viruses, yeasts and some fungi.

To summarize the complex framework of the immune system and BPs, most of the studies on the effects of BPs have been conducted *in vitro* with somewhat conflicting results. By contrast, fewer studies exist on the immune function in animals and only some have been carried out on humans. In order to establish the physiological significance of multifunctional immunomodulatory peptides, further fundamental research, as well as clinical trials, are needed in this field. The immunomodulatory effects of lactoferrin are promising.

9.3.3 Effects on the nervous system

Some milk peptides, the so-called opioid peptides, have an affinity for an opiate receptor and opiate-like effects. Several types of opioid-agonists as well as antagonist peptides have been characterized, but the major opioid peptides are fragments of β-caseins (Table 9.2). In addition to dietary intake, opioid peptides are formed in the gut as a result of *in vivo* hydrolysis of milk proteins.

Once absorbed into the blood, some of these peptides can cross the blood–brain barrier, travel to the brain and various other organs, and elicit pharmacological properties similar to opium or morphine. Many opioid peptide transport systems have been characterized at the functional level, and these are expressed differentially at different sites of the body (for review, see Ganapathy and Miyauchi, 2005). Individual receptors are responsible for specific physiological effects, i.e. the μ-receptor for emotional behaviour and suppression of intestinal motility and the κ-receptor for sedation and food intake.

κ-casein-derived casoxin and lactoferrin have antagonistic effects, whereas the effects of β-casein derived β-casomorphins and morphiceptin and α-casein-derived exorphins are agonistic. Casomorphins, as opioid

Table 9.2 Some opioid peptides derived from bovine milk proteins

Protein precursor	Bioactive peptide	Bioactivity
α-, β-Casein	Casomorphins, e.g. β-casomorphin 5, morphiceptin, α-Casein exorphin	Opioid agonists
κ-Casein	Casoxins e.g. casoxin 4, casoxin A	Opioid antagonists
α-Lactalbumin	α-Lactorphin	Opioid agonists
β-Lactoglobulin	β-Lactorphin	Opioid agonists
Lactoferrin	Lactoferroxins	Opioid antagonists

ligands, modulate social behaviour (Pansepp *et al.*, 1984), induce analgesic behaviour (Matthies *et al.*, 1984), prolong gastrointestinal transit time by inhibiting intestinal peristalsis and motility (Schulte-Frohlinde *et al.*, 2000; Becker *et al.*, 1990), exert antidiarrhoeal action (Daniel *et al.*, 1990), and modulate the amino acid transport (Ermisch *et al.*, 1989). Furthermore, opioid-like milk peptides play a regulatory role on the appetite by modifying endocrine activity of the pancreas and by increasing insulin output, at least in experimental trials (Schusdziarra *et al.*, 1983). Effects of systemic intraperitoneal administration of β-casomorphins on learning and memory have been shown in mice by Sakaguchi *et al.* (2006, 2003). Low doses improved the impairment of short and long-term memory resulting from cholinergic dysfunction by agonisting μ-receptor. Intraperitoneal injection of lactoferrin suppressed psychological stress in rats, possibly by activating opioid receptors via nitric oxide synthase activation (Kamemori *et al.*, 2004).

In addition to opioid receptor agonist and antagonist properties, milk proteins may have other mechanisms to affect the nervous system. For example, the intake of amino acid tryptophan affects serotonin synthesis and the intake of tyrosine influences the synthesis of dopamine and noradrenaline (for review, see Riedel *et al.*, 2002). Serotonergic stimulation by a single oral dose of α-lactalbumin in women complaining of premenstrual symptoms mildly impaired premenstrual memory decline (Schmitt *et al.*, 2005). An increase in dietary tryptophan and stimulation of serotonin function by an oral dose of α-lactalbumin increased the ability to cope with stress in high stress-prone subjects (Markus *et al.*, 2000).

9.3.4 Effects on the gastrointestinal tract

Gastric mucosal integrity is maintained by a balance between endogenous aggressive factors (hydrochloric acid, pepsin, bile and pancreatic enzymes) and a number of defence mechanisms including the gastric mucosal barrier, mucus secretion, gastric microcirculation, cellular regeneration and endogenous protective agents. These protective mechanisms are modulated by a wide range of substances (for review, see Martin and Wallace, 2006). Of the mediators, nitric oxide, eicosanoids, protease-activated receptors and cytokines are endogenous factors which are considered potential targets for therapeutic effects of dietary proteins.

There are several studies on the effects of whey proteins on gastric mucosa, showing different protective mechanisms. A single dose of whey protein isolate protected rats against ethanol and stress-induced gastric mucosal injury with the same potency as an antiulcer agent (Matsumoto *et al.*, 2001). Casein, instead, had no protective effect. Rosaneli *et al.* (2002) suggested this effect to be, at least partly, dependent on the stimulation of glutathione synthesis, as previously demonstrated *in vitro* by Hiraishi *et al.* (1994). Daily gavage doses of whey protein ameliorated gastric and hepatic oxidative damage in ethanol-induced gastric ulcers by reducing inflammation by inhibiting neutrophil infiltration and by reducing the generation of cytotoxic free radicals (Jahovic *et al.*, 2005). Activated neutrophils may enhance the activity of cytotoxic enzymes, e.g. myeloperoxidase, by the release of oxidants (Kettle and Winterbourn, 1997). Thus, by reducing the accumulation of activated neutrophils, the formation of reactive oxidants decreases. On the other hand, Rosaneli *et al.* (2004) reported whey proteins to stimulate mucus and bicarbonate production by affecting prostaglandin synthesis. In addition, Clarke *et al.* (2002) found the protective effect to be due to growth factors (such as insulin-like growth factor-1 and transforming growth factor-beta) of whey protein concentrate.

The known gastroprotective components of whey include α-lactalbumin and lactoferrin (Ushida *et al.*, 2003; Rosaneli *et al.*, 2002; Matsumoto *et al.*, 2001). α-Lactalbumin stimulates mucin production and secretion in gastric mucus-producing cells, and, contrary to the results of Rosaneli *et al.* (2004), this effect is inherent to endogenous prostaglandin synthesis (Ushida *et al.*, 2007). Lactoferrin has protected gastric mucosa against nonsteroidal anti-inflammatory drugs, helicobacter infection, and some viruses in animals, and there is even some evidence to this effect from preclinical studies (Bortoli *et al.*, 2007; Dial *et al.*, 2005, 2000; Sato *et al.*, 1996). Dial *et al.* (2005) assumed the mechanism to be related to the ability of lactoferrin to prevent neutrophil migration into the gut epithelium and also to its other properties, including the local antibiotic effect and possible stimulation of intestinal components of the immune system. However, no evidence has been found implying that gastroprotective mechanisms of whey components could be based on altered prostaglandin synthesis or the action of nitric oxide.

Evidence on the protective effect of caseins on gastric mucosa is scanty (Claustre *et al.*, 2002; Matsumoto *et al.*, 2001; Koo, 1994). However, β-casomorphins have induced a strong release of mucin in the perfused jejunum of the rat through the activation of the enteric nervous system and opioid receptors (Trompette *et al.*, 2003; Claustre *et al.*, 2002). Zoghbi *et al.* (2006) found this to be due to direct stimulation of intestinal goblet cells and the activation of μ-receptors, indicating a possible protective effect of caseins on the mucosa as well.

To conclude, gastroprotective effects of BPs have been studied in experimental models. Whey proteins, especially α-lactalbumin and lactoferrin, exert gastroprotective effects by several mechanisms, such as stimulation of glutathione synthesis and induction of mucin production. The clinical relevance of these effects remains to be studied.

9.3.5 Effects on body composition

In this chapter, the effects of BPs on body composition, i.e. on growth, as well as bone, fat and muscle tissues, are briefly reviewed.

Growth and height

According to the 'milk hypothesis' put forward by Bogin (1998), greater consumption of milk during infancy and childhood will result in taller adult stature. Observational studies indicate a distinctly more positive association between the consumption of cows' milk and linear growth, but in intervention studies from well-nourished populations, the association is not as clear (Adebamowo *et al.*, 2008; Rich-Edwards *et al.*, 2007; Rogers *et al.*, 2006; Berkey *et al.*, 2005; Wiley 2005; Hoppe *et al.*, 2004a, b; Okada, 2004; Merrilees *et al.*, 2000; Bonjour *et al.*, 1997; Cadogan *et al.*, 1997; Baker *et al.*, 1980). More studies are needed to identify to which of the milk components this result is attributable. It is not likely to be the protein quantity, since the effect is evident also in well-nourished populations whose protein intake exceeds the basic physiological need considerably. Since milk protein, for example, is more efficient than protein in meat (Grillenberger *et al.*, 2003, Hoppe *et al.*, 2004c), the quality of protein is a more feasible factor. Potential candidates include BPs, amino acids, the insulin-like growth factor-1 (IGF-1) in cows' milk, or milk minerals, including calcium. Casein is more potent to stimulate the release of insulin, which is a broad anabolic stimulator (Nilsson *et al.*, 2004), but, on the other hand, Hoppe (2006) has some evidence in favour of whey.

Osteoprotective effects

Since casein phosphopeptides (CPP) can bind and solubilize minerals, they have been considered physically beneficial for bone health and for the prevention of osteoporosis. Several cell culture experiments and animal studies have detected the effect of dietary CPP on intestinal calcium absorption, calcium balance, and bone formation, with conflicting results (Mora-Gutierrez *et al.*, 2007; Ferraretto *et al.*, 2003, 2001; Erba *et al.*, 2002; Tsuchita *et al.*, 2001, 1996, 1993; Bennett *et al.*, 2000; Saito *et al.*, 1998; Hansen *et al.*, 1996; Reynolds *et al.*, 1995; Scholz-Ahrens *et al.*, 1990). In general, studies using tracers have revealed positive effects of CPP on calcium bioavailability, whereas most of the balance studies have failed to find any effect of CPP addition to diet. The number of human trials remains scanty. Hansen *et al.* (1997) found the addition of CPP to have no influence on calcium and zinc absorption in healthy young adults who were given a phytate-containing test meal. Neither Narva *et al.* (2003) nor Teucher *et al.* (2006) reported any bone boosting effect by CPP in milk, fermented milk, or drinkable calcium lactate drink in healthy adults or postmenopausal women.

The results on CPP's anticariogenic effect is more consistent. CPP can exert its anticariogenic effect by promoting recalcification of tooth enamel, whereas glycomacropeptide (GMP) derived from κ-casein seems to act by inhibiting the adhesion and growth of plaque-forming bacteria on oral mucosa (Reynolds *et al.*, 2003; Malkoski *et al.*, 2001; Brody, 2000).

Obesity and fat tissue

Diet is the most important contributing factor to obesity. Observational and follow-up studies have revealed some beneficial effects of ingestion of milk and dairy products on obesity and body fat stores, even though conflicting results have been published (Marques-Vidal *et al.*, 2006; Murakami *et al.*, 2006; Rajpathak *et al.*, 2006; Rosell *et al.*, 2006; Azadbakht *et al.*, 2005; Choi *et al.*, 2005; Liu *et al.*, 2005; Mirmiran *et al.*, 2005; Drapeau *et al.*, 2004; Newby *et al.*, 2003; Pereira *et al.*, 2002; Lin *et al.*, 2000). Most of the published interventions are designed to reveal the effect of dairy calcium, not of the intake of milk or milk proteins. Zemel and co-workers (2004a, b, 2005a, b), however, have published several studies indicating that supplemental intake of milk decreases body fat tissue and increases fat degradation. This has been confirmed by Melanson *et al.* (2005) and Gunther *et al.* (2005a, b), but not by Bowen *et al.* (2005), Thompson *et al.* (2005), Boon *et al.* (2005) or Harvey-Berino *et al.* (2005). In animal models, whey-based diets have proved to be more efficient than casein-based diets (Pilvi *et al.*, 2007, Royle *et al.*, 2007).

Dietary proteins may have a satiating effect and may also affect appetite in the postprandial as well as the post-absorptive state. The results are, however, somewhat conflicting and the role of whey and casein proteins, in particular, requires further research (Burton-Freeman, 2008; Diepvens *et al.*, 2008; Keogh and Clifton, 2008; Bowen *et al.*, 2007, 2006a, b; Luhovyy *et al.*, 2007; Lejeune *et al.*, 2005; Sanggaard *et al.*, 2004; Westerterp-Plantega *et al.*, 2004; Aziz and Anderson, 2003; Hall *et al.*, 2003; Hira *et al.*, 2003; Gustafson *et al.*, 2001). In addition, the effects have been observed in short-term experiments and when protein and peptide are consumed in higher amounts than are contained in the usual servings of dairy products. The physiological mechanism behind these effects is presumed to be due to postprandial changes in hunger/satiety hormones, such as cholecystokinin, glucagon-like peptide-1, peptide YY, and ghrelin. Even the renin-angiotensin system may partly contribute to the effects on fat tissue, since angiotensin II promotes adipocyte growth and preadipocyte recruitment (Engeli *et al.*, 2003).

Effects on muscle protein synthesis

Muscle protein synthesis (MPS) and muscle protein breakdown are simultaneous, ongoing processes. Protein consumption, and the accompanying hyperaminoacidemia, stimulates an increase in MPS and a small suppression of muscle protein breakdown. In addition to protein quantity, there is evidence supporting the role of protein quality on MPS and protein balance (MPS minus muscle protein breakdown). Milk proteins are effective in supporting exercise-induced lean body mass gain (Roy, 2008; Phillips *et al.*, 2005; Rankin *et al.*, 2004). The field of nutrition in MPS, lean body mass, muscle force, endurance, or on exercise performance, is greatly studied among athletes. There are, however, conflicting results of the clinical efficacy of whey proteins, as recently reviewed by Tang and Phillips (2009), Hayes and Cribb (2008), and Ha and Zemel (2003). The number of trials with casein supplements remains smaller.

However, metabolically active skeletal muscle can offset the morbidities associated with the sarcopenia of ageing, such as Type 2 diabetes, decline in aerobic fitness and the reduction in metabolic rate, which can lead to fat mass accumulation. Ageing has been characterized by a decreased sensitivity of muscle protein synthesis to branched-chain amino acids, especially leucine. Whey proteins are a rich source of essential amino acids that rapidly elevate plasma amino acids, whereby leucine supplementation may correct this defect in old rats and elderly humans (Koopman *et al.*, 2008; Rieu *et al.*, 2007; Katsanos *et al.*, 2006; Paddon-Jones *et al.*, 2006). Long-term research with defined health outcomes is currently insufficient for specifying an optimal quality and quantity of dietary protein in the elderly. Further work is certainly required, especially considering the possibility that simple dietary and exercise strategies could improve the maintenance of skeletal muscle mass, resulting in a likely decrease in the overall burden of a number of diseases and improved quality of life as we age.

After reviewing the reports on the effects of BPs on body composition, it is clear that growing interest is focused on this field, because bone and muscle strength, lipid-lowering effects as well as an increased metabolic rate are all beneficial for ageing populations. Evidence for the benefit of BPs is accumulating but evaluation of their clinical relevance should be based on long-term clinical studies in target populations.

9.4 Future trends

Addition of bioactive peptides to milk products or other foods, or ingesting them as a natural part of fermented dairy products, seems to have the potential to offer specific health benefits to consumers. Most of the claimed physiological properties of bioactive milk components have been carried out *in vitro* or in animal models. While there is a demanding need for further basic research to clarify why these peptides have physiological effects and for clinical research to confirm these effects in humans, new products containing bioactive peptides are now commercially available. Food and pharmaceutical companies are actively considering ways to exploit bioactive peptides in both human nutrition and in health promotion.

9.5 Sources of further information and advice

An increasing number of foods sold in the EU bear nutrition and health claims. In December 2006, EU decision-making bodies adopted the

Regulation on the Use of Nutrition and Health Claims for foods. This Regulation lays down harmonised EU-wide rules for the use of health or nutritional claims on foodstuffs based on nutrient profiles. It aims to enable consumers to make informed and meaningful choices about the foods they buy, and to ensure fair competition and promote innovation in the food industry.

http://www.efsa.europa.eu/EFSA/efsa_locale-1178620753812_1178621456747.htm

Process for the Assessment of Scientific Support for Claims on Foods (PASSCLAIM) builds on the principles defined within the previous EU project 'Functional Food Science in Europe' (FUFOSE). The main thrust of the FUFOSE Consensus Document was a scheme to base claims for functional foods on solid scientific evidence. FUFOSE suggested that any claim for 'enhanced function' and 'reduced risk of disease' should be scientifically justified. The PASSCLAIM document delivers criteria to assess the scientific support for claims on foods. It is a consensus document that has been borne of wide and intensive consultation among diverse stakeholders including academic experts, representatives of public interest groups, regulators, and the food industry.

http://europe.ilsi.org/activities/ecprojects/PASSCLAIM/passpubs.htm

9.6 References

ADEBAMOWO CA, SPIEGELMAN D, BERKEY CS, DANBY W, ROCKETT H, COLDITZ GA, WILLET W, and HOLMES MD, (2008). Milk consumption and acne in teenaged boys. *J Am Acad Dermatol*, 58, 787–93.

ALHAMDAN AA, and GRIMBLE RF, (2003). The effect of graded levels of dietary casein, with or without methionine supplementation, on glutathione concentration in unstressed and endotoxin-treated rats. *Int J Vitam Nutr Res*, 73, 468–77.

AOYAMA Y, and WADA M, (1999). Effects of dietary protein type on the response of lipid metabolism to orotic acid in rats. *Biosci Biotechnol Biochem*, 63, 2005–8.

ASHAR M, and CHAND R, (2004). Fermented milk containing ACE-inhibitory peptides reduces blood pressure in middle-aged hypertensive subjects. *Milchwissenschaft*, 59, 363–6.

ASHAR MN, and PRAJAPATI JB, (2000). Verification of hypocholesterolemic effect of fermented milk on human subjects with different cholesterol levels. *Folia Microbiol*, 45, 263–8.

AZADBAKHT L, MIRMIRAN P, ESMAILLZADEH A, and AZIZI F, (2005). Dairy consumption is inversely associated with the prevalence of the metabolic syndrome in Tehranian adults. *Am J Clin Nutr*, 82, 523–30.

AZIZ A, and ANDERSON GH, (2003). Exendin-4, a GLP-1 receptor agonist, interacts with proteins and their products of digestion to suppress food intake in rats. *J Nutr*, 133, 2326–30.

BAKER IA, ELWOOD PC, HUGHES J, JONES M, MOORE F, and SWEETNAM PM, (1980). A randomised controlled trial of the effect of the provision of free school milk on the growth of children. *J Epidemiol Community Health*, 34, 31–4.

BALDI A, IOANNIS P, CHIARA P, ELEONORA F, ROUBINI C, and VITTORIO D, (2005). Biological effects of milk proteins and their peptides with emphasis on those related to the gastrointestinal ecosystem. *J Dairy Res*, 72, 66–72.

BAL DIT SOLLIER C, DROUET L, PIGNAUD G, CHEVALLIER C, CAEN J, FIAT AM, IZQUIERDO C, and JOLLES P, (1996). Effect of kappa-casein split peptides on platelet aggregation and on thrombus formation in the guinea-pig. *Thromb Res*, 81, 427–37.

BECKER A, HEMPEL G, GRECKSCH G, and MATTHIES H, (1990). Effects of beta-casomorphin derivatives on gastrointestinal transit in mice. *Biomed Biochim Acta*, 49, 1203–7.

BENNETT T, DESMOND A, HARRINGTON M, MCDONAGH D, FITZGERALD R, FLYNN A, and CASHMAN KD, (2000). The effect of high intakes of casein and casein phosphopeptide on calcium absorption in the rat. *Br J Nutr*, 83, 673–80.

BERKEY CS, ROCKETT HR, WILLETT WC, and COLDITZ GA, (2005). Milk, dairy fat, dietary calcium, and weight gain: A longitudinal study of adolescents. *Arch Pediatr Adolesc Med*, 159, 543–50.

BERNARD BK, NAKAMURA Y, BANDO I, and MENNEAR JH, (2005). Studies of the toxicological potential of tripeptides (L-valyl-L-prolyl-L-proline and L-isoleucyl-L-prolyI-L-proline): II. Introduction. *Int J Toxicol*, 24 (Suppl 4), 5–11.

BLOUET C, MARIOTTI F, AZZOUT-MARNICHE D, MATHÉ V, MIKOGAMI T, TOMÉ D, and HUNEAU JF, (2007). Dietary cysteine alleviates sucrose-induced oxidative stress and insulin resistance. *Free Radic Biol Med*, 42, 1089–97.

BOGIN B, (1998). Milk and human development: An eassay on the 'milk hypothesis'. *Antropologia Portuguesa*, 15, 23–36.

BONJOUR JP, CARRIE AL, FERRARI S, CLAVIEN H, and SLOSMAN D, (1997). Calcium-enriched foods and bone mass growth in prepubertal girls: A randomized, double-blind, placebo-controlled trial. *J Clin Invest*, 99, 1287–94.

BOON N, HUL GB, VIGUERIE N, SICARD A, LANGIN D, and SARIS WH, (2005). Effects of three diets with various calcium contents on 24-h energy expenditure, fat oxidation, and adipose tissue message RNA expression of lipid metabolism-related proteins. *Am J Clin Nutr*, 82, 1244–52.

BORTOLI DE N, LEONARDI G, CIANCIA E, MERLO A, BELLINI M, COSTA F, MUMOLO MG, RICCHIUTI A, CRISTIANI F, SANTI S, ROSSI M, and MARCHI S, (2007). Helicobacter pylori eradication: A randomized prospective study of triple therapy versus triple therapy plus lactoferrin and probiotics. *Am J Gastroenterol*, 102, 951–6.

BOUNOUS G, (2000). Whey protein concentrate (WPC) and glutathione modulation in cancer treatment. *Anticancer Res*, 20, 4785–92.

BOUNOUS G, LÉTOURNEAU L, and KONGSHAVN PA, (1983). Influence of dietary protein type on the immune system of mice. *J Nutr*, 113, 1415–21.

BOUNOUS G, SHENOUDA N, KONGSHAVN PA, and OSMOND DG, (1985). Mechanism of altered B-cell response induced by changes in dietary protein type in mice. *J Nutr*, 115, 1409–17.

BOWEN J, NOAKES M, and CLIFTON PM, (2005). Effect of calcium and dairy foods in high protein, energy-restricted diets on weight loss and metabolic parameters in overweight adults. *Int J Obes*, 29, 957–65.

BOWEN J, NOAKES M, and CLIFTON PM, (2006a). Appetite regulatory hormone responses to various dietary proteins differ by body mass index status despite similar reductions in ad libitum energy intake. *J Clin Endocrinol Metab*, 91, 2913–9.

BOWEN J, NOAKES M, and CLIFTON PM, (2007). Appetite hormones and energy intake in obese men after consumption of fructose, glucose and whey protein beverages. *Int J Obes*, 31, 1696–703.

BOWEN J, NOAKES M, TRENERRY C, and CLIFTON PM, (2006b). Energy intake, ghrelin, and cholecystokinin after different carbohydrate and protein preloads in overweight men. *J Clin Endocrinol Metab*, 91, 1477–83.

BROCK JH, (2002). The physiology of lactoferrin. *Biochem Cell Biol*, 80, 1–6.

BRODY EP, (2000). Biological activities of bovine glycomacropeptide. *Br J Nutr*, 84 (Suppl 1), 39–46.

BROWN EC, DISILVESTRO RA, BABAKNIA A, and DEVOR ST, (2004). Soy versus whey protein bars: Effects on exercise training impact on lean body mass and antioxidant status. *Nutr J*, 3, 22.

BURTON-FREEMAN BM, (2008). Glycomacropeptide (GMP) is not critical to whey-induced satiety, but may have a unique role in energy intake regulation through cholecystokinin (CCK). *Physiol Behav*, 93, 379–87.

CADOGAN J, EASTELL R, JONES R, and BARKER ME, (1997). Milk intake and bone mineral acquisition in adolescent girls: Randomised, controlled intervention trial. *BMJ*, 315, 1255–60.

CHABANCE B, JOLLÈS P, IZQUIERDO C, MAZOYER E, FRANCOUAL C, DROUET L, and FIAT AM, (1995). Characterization of an antithrombotic peptide from kappa-casein in newborn plasma after milk ingestion. *Br J Nutr*, 73, 583–90.

CHABANCE B, QIAN ZY, MIGLIORE-SAMOUR D, JOLLÈS P, and FIAT AM, (1997). Binding of the bovine caseinoglycopeptide to the platelet membrane glycoprotein GPIb alpha. *Biochem Mol Biol Int*, 42, 77–84.

CHODACZEK G, SAAVEDRA-MOLINA A, BACSI A, KRUZEL ML, SUR S, and BOLDOGH I, (2007). Iron-mediated dismutation of superoxide anion augments antigen-induced allergic inflammation: Effect of lactoferrin. *Postepy Hig Med Dosw*, 61, 268–76.

CHOI HK, WILLETT WC, STAMPFER MJ, RIMM E, and HU FB, (2005). Dairy consumption and risk of Type 2 diabetes mellitus in men: A prospective study. *Arch Intern Med*, 165, 997–1003.

CLARKE J, BUTLER R, HOWARTH G, READ L, and REGESTER G, (2002). Exposure of oral mucosa to bioactive milk factors reduces severity of chemotherapy-induced mucositis in the hamster. *Oral Oncol*, 38, 478–85.

CLAUSTRE J, TOUMI F, TROMPETTE A, JOURDAN G, GUIGNARD H, CHAYVIALLE JA, and PLAISANCIE P, (2002). Effects of peptides derived from dietary proteins on mucus secretion in rat jejunum. *Am J Physiol Gastrointest Liver Physiol*, 283, G521–528.

CROSS ML, and GILL HS, (2000). Immunomodulatory properties of milk. *Br J Nutr*, 84, S81–9.

DANIEL H, VOHWINKEL M, REHNER G, (1990). Effect of casein and beta-casomorphins on gastrointestinal motility in rats. *J Nutr*, 120, 252–7.

DEBBABI H, DUBARRY M, RAUTUREAU M, and TOMÉ D, (1998). Bovine lactoferrin induces both mucosal and systemic immune response in mice. *J Dairy Res*, 65, 283–93.

DENT MP, O'HAGAN S, BRAUN WH, SCHAETTI P, MARBURGER A, and VOGEL O, (2007). A 90-day subchronic toxicity study and reproductive toxicity studies on ACE-inhibiting lactotripeptide. *Food Chem Toxicol*, 45, 1468–77.

DENT SB, PETERSON CT, BRACE LD, SWAIN JH, REDDY MB, HANSON KB, ROBINSON JG, and ALEKEL DL, (2001). Soy protein intake by perimenopausal women does not affect circulating lipids and lipoproteins or coagulation and fibrinolytic factors. *J Nutr*, 131, 2280–7.

DIAL EJ, DOHRMAN AJ, ROMERO JJ, and LICHTENBERGER LM, (2005). Recombinant human lactoferrin prevents NSAID-induced intestinal bleeding in rodents. *J Pharm Pharmacol*, 57, 93–9.

DIAL EJ, ROMERO JJ, HEADON DR, and LICHTENBERGER LM, (2000). Recombinant human lactoferrin is effective in the treatment of Helicobacter felis-infected mice. *J Pharm Pharmacol*, 52, 1541–6.

DÍAZ M, and DECKER EA, (2004). Antioxidant mechanisms of caseinophosphopeptides and casein hydrolysates and their application in ground beef. *J Agric Food Chem*, 52, 8208–13.

DIEPVENS K, HÄBERER D, and WESTERTERP-PLANTENGA M, (2008). Different proteins and biopeptides differently affect satiety and anorexigenic/orexigenic hormones in healthy humans. *Int J Obes*, 32, 510–8.

DRAPEAU V, DESPRÉS JP, BOUCHARD C, ALLARD L, FOURNIER G, LEBLANC C, and TREMBLAY A, (2004). Modifications in food-group consumption are related to long-term body-weight changes. *Am J Clin Nutr*, 80, 29–37.

DRESCHER K, ROOS N, PFEUFFER M, SEYFERT HM, SCHREZENMEIR J, and HAGEMEISTER H, (1999). Recovery of 15N-lactoferrin is higher than that of 15N-casein in the small intestine of suckling, but not adult miniature pigs. *J Nutr*, 129, 1026–30.

DROUET L, BAL DIT SOLLIER C, CISSE M, PIGNAUD G, MAZOYER E, FIAT AM, JOLLES P, and CAEN JP, (1990). The antithrombotic effect of KRDS, a lactotransferrin peptide, compared with RGDS. *Nouv Rev Fr Hematol*, 32, 59–62.

ELIA D, STADLER K, HORVÁTH V, and JAKUS J, (2006). Effect of soy- and whey protein-isolate supplemented diet on the redox parameters of trained mice. *Eur J Nutr*, 45, 259–66.

ENGELI S, SCHLING P, GORZELNIAK K, BOSCHMANN M, JANKE J, AILHAUD G, TEBOUL M, MASSIÉRA F, and SHARMA AM, (2003). The adipose-tissue renin-angiotensin-aldosterone system: Role in the metabolic syndrome? *Int J Biochem Cell Biol*, 35, 807–25.

ERBA D, CIAPPELLANO S, and TESTOLIN G, (2002). Effect of the ratio of casein phosphopeptides to calcium (w/w) on passive calcium transport in the distal small intestine of rats. *Nutrition*, 18, 743–6.

ERMISCH A, BRUST P, and BRANDSCH M, (1989). beta-Casomorphins alter the intestinal accumulation of L-leucine. *Biochim Biophys Acta*, 982, 79–84.

FERRARETTO A, GRAVAGHI C, FIORILLI A, and TETTAMANTI G, (2003). Casein-derived bioactive phosphopeptides: Role of phosphorylation and primary structure in promoting calcium uptake by HT-29 tumor cells. *FEBS Lett* 551, 92–8.

FERRARETTO A, SIGNORILE A, GRAVAGHI C, FIORILLI A, and TETTAMANTI G, (2001). Casein phosphopeptides influence calcium uptake by cultured human intestinal HT-29 tumor cells. *J Nutr*, 131, 1655–61.

FIAT AM, MIGLIORE-SAMOUR D, JOLLÈS P, DROUET L, BAL DIT SOLLIER C, and CAEN J, (1993). Biologically active peptides from milk proteins with emphasis on two examples concerning antithrombotic and immunomodulating activities. *J Dairy Sci*, 76, 301–10.

GANAPATHY V, and MIYAUCHI S, (2005). Transport systems for opioid peptides in mammalian tissues. *AAPS*, 7, 852–6.

GREY V, MOHAMMED SR, SMOUNTAS AA, BAHLOOL R, and LANDS LC, (2003). Improved glutathione status in young adult patients with cystic fibrosis supplemented with whey protein. *J Cyst Fibros*, 2, 195–8.

GRILLENBERGER M, NEUMANN CG, MURPHY SP, BWIBO NO, VEER VAN'T P, HAUTVAST JG, and WEST CE, (2003). Food supplements have a positive impact on weight gain and the addition of animal source foods increase lean body mass of Kenyan schoolchildren. *J Nutr*, 133 (Suppl), 3957–64.

GÜLÇIN I, (2007). Comparison of in vitro antioxidant and antiradical activities of L-tyrosine and L-Dopa. *Amino Acids*, 32, 431–8.

GUNTHER CW, LEGOWSKI PA, LYLE RM, MCCABE GP, EAGAN MS, PEACOCK M, and TEEGARDEN D, (2005b). Dairy products do not lead to alterations in body weight or fat mass in young women in a 1-y intervention. *Am J Clin Nutr*, 81, 751–6.

GUNTHER CW, LYLE RM, LEGOWSKI PA, JAMES JM, MCCABE LD, MCCABE GP, PEACOCK M, and TEEGARDEN D, (2005a). Fat oxidation and its relation to serum parathyroid hormone in young women enrolled in a 1-y dairy calcium intervention. *Am J Clin Nutr*, 82, 1228–34.

GUSTAFSON DR, MCMAHON DJ, MORREY J, and NAN R, (2001). Appetite is not influenced by a unique milk peptide: Caseinomacropeptide (CMP). *Appetite*, 36, 157–63.

HA E, and ZEMEL MB, (2003). Functional properties of whey, whey components, and essential amino acids: Mechanisms underlying health benefits for active people. *J Nutr Biochem*, 14, 251–8.

HALL WL, MILLWARD DJ, LONG SJ, and MORGAN LM, (2003). Casein and whey exert different effects on plasma amino acid profiles, gastrointestinal hormone secretion and appetite. *Br J Nutr*, 89, 239–48.

HANSEN M, SANDSTRÖM B, and LÖNNERDAL B, (1996). The effect of casein phosphopeptides on zinc and calcium absorption from high phytate infant diets assessed in rat pups and Caco-2 cells. *Pediatr Res*, 40, 547–52.

HANSEN M, SANDSTRÖM B, JENSEN M, and SØRENSEN SS, (1997). Effect of casein phosphopeptides on zinc and calcium absorption from bread meals. *J Trace Elem Med Biol*, 11, 143–9.

HARA H, FUNABIKI R, IWATA M, and YAMAZAKI K, (1984). Portal absorption of small peptides in rats under unrestrained conditions. *J Nutr*, 114, 1122–9.

HARVEY-BERINO J, GOLD BC, LAUBER R, and STARINSKI A, (2005). The impact of calcium and dairy product consumption on weight loss. *Obes Res*, 13, 1720–6.

HATA Y, YAMAMOTO M, OHNI M, NAKAJIMA K, NAKAMURA Y, and TAKANO T, (1996). A placebo-controlled study of the effect of sour milk on blood pressure in hypertensive subjects. *Am J Clin Nutr*, 64, 767–71.

HAYES A, and CRIBB PJ, (2008). Effect of whey protein isolate on strength, body composition and muscle hypertrophy during resistance training. *Curr. Opin Clin Nutr Metab Care*, 11, 40–4.

HE F, TUOMOLA E, ARVILOMMI H, and SALMINEN S, (2001). Modulation of human humoral immune response through orally administered bovine colostrum. *FEMS Immunol Med Microbiol*, 31, 93–6.

HERNÁNDEZ-LEDESMA B, AMIGO L, RECIO I, and BARTOLOMÉ B, (2007). ACE-inhibitory and radical-scavenging activity of peptides derived from beta-lactoglobulin f(19–25). Interactions with ascorbic acid. *J Agric Food Chem*, 55, 3392–7.

HERNÁNDEZ-LEDESMA B, DÁVALOS A, BARTOLOMÉ B, and AMIGO L, (2005). Preparation of antioxidant enzymatic hydrolysates from alpha-lactalbumin and beta-lactoglobulin. Identification of active peptides by HPLC-MS/MS. *J Agric Food Chem*, 53, 588–93.

HIRA T, HARA H, TOMITA F, and AOYAMA Y, (2003). Casein binds to the cell membrane and induces intracellular calcium signals in the enteroendocrine cell: A brief communication. *Exp Biol Med*, 228, 850–4.

HIRAISHI H, TERANO A, OTA S, MUTOH H, SUGIMOTO T, HARADA T, RAZANDI M, and IVEY KJ, (1994). Protection of cultured rat gastric cells against oxidant-induced damage by exogenous glutathione. *Gastroenterology*, 106, 1199–207.

HOPPE C, MOLGAARD C, JUUL A, and MICHAELSEN KF, (2004c). High intakes of skimmed milk, but not meat, increase serum IGF-1 and IGFBP-3 in eight-year-old boys. *Eur J Clin Nutr*, 58, 1211–16.

HOPPE C, MOLGRAAD C, and MICHAELSON KF, (2006). Cow's milk and linear growth in industrilized and developing countries. *Annu Rev Nutr*, 26, 131–73.

HOPPE C, MOLGRAAD C, LYKKE THOMSEN B, JUUL A, and MICHAELSON KF, (2004a). Protein intake at 9 mo of age is associated with body size but not with body fat in 10-y-old Danish children. *Am J Clin Nutr*, 79, 494–501.

HOPPE C, UDAM TR, LAURIZEN L, MOLGAARD C, JUUL A, and MICHAELSEN KF, (2004b). Animal protein intake, serum insulin-like growth factor I, and growth in healthy 2,5-y-old Danish children. *Am J Clin Nutr*, 80, 447–52.

ISHII K, TAKAMURA N, SHINOHARA M, WAKUI N, SHIN H, SUMINO Y, OHMOTO Y, TERAGUCHI S, and YAMAUCHI K, (2003). Long-term follow-up of chronic hepatitis C patients treated with oral lactoferrin for 12 months. *Hepatol Res*, 25, 226–33.

JAHOVIC N, ÖGUNC AV, GRUZEL E, ARS D, ERCAN F, ERKANH G, YGEN BC, and YALCIN AS, (2005). Whey pretreatment ameliorates gastric and hepatic oxidative damage in ethanol-induced gastric ulcer via a neutrophil-dependent mechanism. *M Med J*, 18, 64–70.

JAUHIAINEN T, VAPAATALO H, POUSSA T, KYRÖNPALO S, RASMUSSEN M, and KORPELA R, (2005). Lactobacillus helveticus fermented milk lowers blood pressure in hypertensive subjects in 24-h ambulatory blood pressure measurement. *Am J Hypertens*, 18, 1600–5.

JOLLÈS P, LÉVY-TOLEDANO S, FIAT AM, SORIA C, GILLESSEN D, THOMAIDIS A, DUNN FW, and CAEN JP, (1986). Analogy between fibrinogen and casein. Effect of an undecapeptide isolated from kappa-casein on platelet function. *Eur J Biochem*, 158, 379–82.

JOLLÈS P, LOUCHEUX-LEFEBVRE MH, and HENSCHEN A, (1978). Structural relatedness of kappa-casein and fibrinogen gamma-chain. *J Mol Evol*, 11, 271–7.

KAMEMORI N, TAKEUCHI T, HAYASHIDA K, and HARADA E, (2004). Suppressive effects of milk-derived lactoferrin on psychological stress in adult rats. *Brain Res*, 1029, 34–40.

KATSANOS CS, KOBAYASHI H, SHEFFIELD-MOORE M, AARSLAND A, and WOLFE RR, (2006). A high proportion of leucine is required for optimal stimulation of the rate of muscle protein synthesis by essential amino acids in the elderly. *Am J Physiol Endocrinol Metab*, 291, 381–7.

KAWASE M, HASHIMOTO H, HOSODA M, MORITA H, and HOSONO A, (2000). Effect of administration of fermented milk containing whey protein concentrate to rats and healthy men on serum lipids and blood pressure. *J Dairy Sci*, 83, 255–63.

KAZLAUSKAITE J, BIZIULEVICIUS GA, ZUKAITE V, BIZIULEVICIENE G, MILIUKIENE V, and SIAURYS A, (2005). Oral tryptic casein hydrolysate enhances phagocytosis by mouse peritoneal and blood phagocytic cells but fails to prevent induced inflammation. *Int Immunopharmacol*, 5, 1936–44.

KEOGH JB, and CLIFTON P, (2008). The effect of meal replacements high in glycomacropeptide on weight loss and markers of cardiovascular disease risk. *Am J Clin Nutr*, 87, 1602–5.

KETTLE AJ, and WINTERBOURN CC, (1997). Myeloperoxidase: A key regulator of neutrophil oxidant production. *Redox Report*, 3, 3–15.

KONISHI M, IWASA M, YAMAUCHI K, SUGIMOTO R, FUJITA N, KOBAYASHI Y, WATANABE S, TERAGUCHI S, ADACHI Y, and KAITO M, (2006). Lactoferrin inhibits lipid peroxidation in patients with chronic hepatitis C. *Hepatol Res*, 36, 27–32.

KOO MW, (1994). The effects of milk and calcium on ethanol-induced gastric mucosal damage. *Pharmacol Res*, 29, 217–24.

KOOPMAN R, VERDIJK LB, BEELEN M, GORSELINK M, KRUSEMAN AN, WAGENMAKERS AJ, KUIPERS H, and VAN LOON LJ, (2008). Co-ingestion of leucine with protein does not further augment post-exercise muscle protein synthesis rates in elderly men. *Br J Nutr*, 99, 571–80.

KRUZEL ML, BACSI A, CHOUDHURY B, SUR S, and BOLDOGH I, (2006). Lactoferrin decreases pollen antigen-induced allergic airway inflammation in a murine model of asthma. *Immunology*, 119, 159–66.

KUHARA T, IIGO M, ITOH T, USHIDA Y, SEKINE K, TERADA N, OKAMURA H, and TSUDA H, (2000). Orally administered lactoferrin exerts an antimetastatic effect and enhances production of IL-18 in the intestinal epithelium. *Nutr Cancer*, 38, 192–9.

LAHOV E, and REGELSON W, (1996). Antibacterial and immunostimulating casein-derived substances from milk: Casecidin, isracidin peptides. *Food Chem Toxicol*, 34, 131–45.

LANDS LC, GREY VL, and SMOUNTAS AA, (1999). Effect of supplementation with a cysteine donor on muscular performance. *J Appl Physiol*, 87, 1381–5.

LEBLANC JG, MATAR C, VALDÉZ JC, LEBLANC J, and PERDIGON G, (2002). Immunomodulating effects of peptidic fractions issued from milk fermented with Lactobacillus helveticus. *J Dairy Sci*, 85, 2733–42.

LEJEUNE MP, KOVACS EM, and WESTERTERP-PLANTENGA MS, (2005). Additional protein intake limits weight regain after weight loss in humans. *Br J Nutr*, 93, 281–9.

LIN YC, LYLE RM, MCCABE LD, MCCABE GP, WEAVER CM, and TEEGARDEN D, (2000). Dairy calcium is related to changes in body composition during a two-year exercise intervention in young women. *J Am Coll Nutr*, 19, 754–60.

LIU HC, CHEN WL, and MAO SJ, (2007). Antioxidant nature of bovine milk beta-lactoglobulin. *J Dairy Sci*, 90, 547–55.

LIU S, SONG Y, FORD ES, MANSON JE, BURING JE, and RIDKER PM, (2005). Dietary calcium, vitamin D, and the prevalence of metabolic syndrome in middle-aged and older U.S. Women. *Diabetes Care*, 28, 2926–32.

LOTHIAN JB, GREY V, and LANDS LC, (2006). Effect of whey protein to modulate immune response in children with atopic asthma. *Int J Food Sci Nutr*, 57, 204–11.

LOW PP, RUTHERFURD KJ, GILL HS, and CROSS ML, (2003). Effect of dietary whey protein concentrate on primary and secondary antibody responses in immunized BALB/c mice. *Int Immunopharmacol*, 3, 393–401.

LUHOVYY BL, AKHAVAN T, and ANDERSON GH, (2007). Whey proteins in the regulation of food intake and satiety. *J Am Coll Nutr*, 26, 704S–12S.

MADANI S, PROST J, and BELLEVILLE J, (2000). Dietary protein level and origin (casein and highly purified soybean protein) affect hepatic storage, plasma lipid transport, and antioxidative defense status in the rat. *Nutrition* 16, 368–75.

MAENO M, YAMAMOTO N, and TAKANO T, (1996). Identification of an antihypertensive peptide from casein hydrolysate produced by a proteinase from Lactobacillus helveticus CP790. *J Dairy Sci*, 79, 1316–21.

MAES W, VAN CAMP J, VERMEIRSSEN V, HEMERYCK M, KETELSLEGERS JM, SCHREZENMEIR J, VAN OOSTVELDT P, and HUYGHEBAERT A, (2004). Influence of the lactokinin Ala-Leu-Pro-Met-His-Ile-Arg (ALPMHIR) on the release of endothelin-1 by endothelial cells. *Regul Pept*, 118, 105–9.

MALKOSKI M, DASHPER SG, O'BRIEN-SIMPSON NM, TALBO GH, MACRIS M, CROSS KJ, and REYNOLDS EC, (2001). Kappacin, a novel antibacterial peptide from bovine milk. *Antimicrob Agents Chemother*, 45, 2309–15.

MANSO MA, ESCUDERO C, ALIJO M, and LÓPEZ-FANDIÑO R, (2002). Platelet aggregation inhibitory activity of bovine, ovine, and caprine kappa-casein macropeptides and their tryptic hydrolysates. *J Food Prot*, 65, 1992–6.

MARIOTTI F, SIMBELIE KL, MAKARIOS-LAHHAM L, HUNEAU JF, LAPLAIZE B, TOMÉ D, and EVEN PC, (2004). Acute ingestion of dietary proteins improves post-exercise liver glutathione in rats in a dose-dependent relationship with their cysteine content. *J Nutr*, 134, 128–31.

MARKUS CR, PANHUYSEN GEM, TUITEN A, and KOPPERSCHAAR H, (2000). Effects of food on cortisol and mood in vulnerable subjects under controllable and uncontrollable stress. *Physiol Behav*, 70, 333–42.

MARQUES-VIDAL P, GONÇALVES A, and DIAS CM, (2006). Milk intake is inversely related to obesity in men and in young women: Data from the Portuguese Health Interview Survey 1998–1999. *Int J Obes*, 30, 88–93.

MARTIN GR, and WALLACE JL, (2006). Gastrointestinal inflammation: A central component of mucosal defense and repair. *Exp Biol Med*, 231, 130–37.

MASUDA O, NAKAMURA Y, and TAKANO T, (1996). Antihypertensive peptides are present in aorta after oral administration of sour milk containing these peptides to spontaneously hypertensive rats. *J Nutr*, 126, 3063–8.

MATAR C, VALDEZ JC, MEDINA M, RACHID M, and PERDIGON G, (2001). Immunomodulating effects of milks fermented by Lactobacillus helveticus and its non-proteolytic variant. *J Dairy Res*, 68, 601–9.

MATSUMOTO H, SHIMOKAWA Y, USHIDA Y, TOIDA T, and HAYASAWA H, (2001). New biological function of bovine alpha-lactalbumin: Protective effect against ethanol- and stress-induced gastric mucosal injury in rats. *Biosci Biotechnol Biochem*, 65, 1104–11.

MATTHIES H, STARK H, HARTRODT B, RUETHRICH HL, SPIELER HT, BARTH A, and NEUBERT K, (1984). Derivatives of beta-casomorphins with high analgesic potency. *Peptides*, 5, 463–70.

MAZOYER E, LÉVY-TOLEDANO S, RENDU F, HERMANT L, LU H, FIAT AM, JOLLÈS P, and CAEN J, (1990). KRDS, a new peptide derived from human lactotransferrin, inhibits platelet aggregation and release reaction. *Eur J Biochem*, 194, 43–9.

MEISEL H, (1997). Biochemical properties of regulatory peptides derived from milk proteins. *Biopolymers*, 43, 119–28.

MELANSON EL, DONAHOO WT, DONG F, IDA T, and ZEMEL MB, (2005). Effect of low- and high-calcium dairy-based diets on macronutrient oxidation in humans. *Obes Res*, 13, 2102–12.

MERRILEES MJ, SMART EJ, GILCHRIST NL, FRAMPTON C, TURNER JG, HOOKE E, MARCH ML, and MAGUIRE P, (2000). Effects of diary food supplements on bone mineral density in teenage girls. *Eur J Nutr*, 39, 256–62.

MICKE P, BEEH KM, and BUHL R, (2002). Effects of long-term supplementation with whey proteins on plasma glutathione levels of HIV-infected patients. *Eur J Nutr*, 41, 12–18.

MICKE P, BEEH KM, SCHLAAK JF, and BUHL R, (2001). Oral supplementation with whey proteins increases plasma glutathione levels of HIV-infected patients. *Eur J Clin Invest*, 31, 171–8.

MIDDLETON N, JELEN P, and BELL G, (2004). Whole blood and mononuclear cell glutathione response to dietary whey protein supplementation in sedentary and trained male human subjects. *Int J Food Sci Nutr*, 55, 131–41.

MINEHIRA K, INOUE S, NONAKA M, OSADA K, YAMADA K, and SUGANO M, (2000). Effects of dietary protein type on oxidized cholesterol-induced alteration in age-related modulation of lipid metabolism and indices of immune function in rats. *Biochim Biophys Acta*, 1483, 141–53.

MIRMIRAN P, ESMAILLZADEH A, and AZIZI F, (2005). Dairy consumption and body mass index: An inverse relationship. *Int J Obes*, 29, 115–21.

MIZUNO S, MENNEAR JH, MATSUURA K, and BERNARD BK, (2005). Studies of the toxicological potential of tripeptides (L-valyl-L-prolyl-L-proline and L-isoleucyl-L-prolyl-L-proline): V. A 13-week toxicity study of tripeptides containing casein hydrolysate in male and female rats, *Int J Toxicol*, 24 (Suppl), 41–59.

MÖLLER NP, SCHOLZ-AHRENS KE, ROOS N, and SCHREZENMEIR J, (2008). Bioactive peptides and proteins from foods: Indication for health effects. *Eur J Nutr*, 47, 171–82.

MORA-GUTIERREZ A, FARRELL HM, ATTAIE R, MCWHINNEY VJ, and WANG C, (2007). Influence of bovine and caprine casein phosphopeptides differing in alpha s1-casein content in determining the absorption of calcium from bovine and caprine calcium-fortified milks in rats. *J Dairy Res*, 74, 356–66.

MORENO YF, SGARBIERI VC, DA SILVA MN, TORO AA, and VILELA MM, (2006). Features of whey protein concentrate supplementation in children with rapidly progressive HIV infection. *J Trop Pediatr*, 52, 34–8.

MORIKAWA K, KONDO I, KANAMARU Y, and NAGAOKA S, (2007). A novel regulatory pathway for cholesterol degradation via lactostatin. *Biochem Biophys Res Commun*, 352, 697–702.

MULLALLY MM, MEISEL H, and FITZGERALD RJ, (1997). Identification of a novel angiotensin-I-converting enzyme inhibitory peptide corresponding to a tryptic fragment of bovine beta-lactoglobulin. *FEBS Lett*, 402, 99–101.

MURAKAMI K, OKUBO H, and SASAKI S, (2006). No relation between intakes of calcium and dairy products and body mass index in Japanese women aged 18 to 20 y. *Nutrition*, 22, 490–5.

NAGAOKA S, FUTAMURA Y, MIWA K, AWANO T, YAMAUCHI K, KANAMARU Y, TADASHI K, and KUWATA T, (2001). Identification of novel hypocholesterolemic peptides derived from bovine milk beta-lactoglobulin. *Biochem Biophys Res Commun*, 281, 11–7.

NAKAMURA Y, BANDO I, MENNEAR JH, and BERNARD BK, (2005). Studies of the toxicological potential of tripeptides (L-valyl-L-prolyl-L-proline and L-isoleucyl-L-prolyl-L-proline): IV. Assessment of the repeated-dose toxicological potential of synthesized L-valyl-L-prolyl-L-proline in male and female rats and dogs. *Int J Toxicol*, 24 (Suppl 4), 25–39.

NAKAMURA Y, YAMAMOTO N, SAKAI K, and TAKANO T, (1995). Antihypertensive effect of sour milk and peptides isolated from it that are inhibitors to angiotensin I-converting enzyme. *J Dairy Sci*, 78, 1253–7.

NARVA M, KÄRKKÄINEN M, POUSSA T, LAMBERG-ALLARDT C, and KORPELA R, (2003). Caseinphosphopeptides in milk and fermented milk do not affect calcium metabolism acutely in postmenopausal women. *J Am Coll Nutr*, 22, 88–93.

NASS N, SCHOEPS R, ULBRICH-HOFMANN R, SIMM A, HOHNDORF L, SCHMELZER C, RAITH K, NEUBERT RH, and EDER K, (2008). Screening for nutritive peptides that modify cholesterol 7alpha-hydroxylase expression. *J Agric Food Chem*, 56, 4987–94.

NEWBY PK, MULLER D, HALLFRISCH J, QIAO N, ANDRES R, and TUCKER KL, (2003). Dietary patterns and changes in body mass index and waist circumference in adults. *Am J Clin Nutr*, 77, 1417–25.

NILSSON M, STENBERG M, FRID AH, HOLST JJ, and BJORK IM, (2004). Glycemia and insulinemia in healthy subjects after lactose-equivalent meals of milk and other food proteins: The role of plasma amino acids and incretins. *Am J Clin Nutr*, 80, 1246–53.

NURMINEN ML, SIPOLA M, KAARTO H, PIHLANTO-LEPPÄLÄ A, PIILOLA K, KORPELA R, TOSSAVAINEN O, KORHONEN H, and VAPAATALO H, (2000). Alpha-lactorphin lowers blood pressure measured by radiotelemetry in normotensive and spontaneously hypertensive rats. *Life Sci*, 66, 1535–43.

OKADA T, (2004). Effect of cow milk consumption on longitudinal height gain in children. *Am J Clin Nutr*, 80, 1088–89.

ONER OZ, OĞÜNÇ AV, CINGI A, UYAR SB, YALÇIN AS, and AKTAN AO, (2006). Whey feeding suppresses the measurement of oxidative stress in experimental burn injury. *Surg Today*, 36, 376–81.

OTANI H, HORIMOTO Y, and MONNAI M, (1996). Suppression of interleukin-2 receptor expression on mouse CD4+ T cells by bovine kappa-caseinoglycopeptide. *Biosci Biotechnol Biochem*, 60, 1017–9.

OTANI H, NAKANO K, and KAWAHARA T, (2003). Stimulatory effect of a dietary casein phosphopeptide preparation on the mucosal IgA response of mice to orally ingested lipopolysaccharide from Salmonella typhimurium. *Biosci Biotechnol Biochem*, 67, 729–35.

PADDON-JONES D, SHEFFIELD-MOORE M, KATSANOS CS, ZHANG XJ, and WOLFE R, (2006). Differential stimulation of muscle protein synthesis in elderly humans following isocaloric ingestion of amino acids or whey protein. *Exp Gerontol*, 41, 215–9.

PANKSEPP J, NORMANSELL L, SIVIY S, ROSSI J 3RD, and ZOLOVICK AJ, (1984). Casomorphins reduce separation distress in chicks. *Peptides*, 5, 829–31.

PARK E, SHIN JI, PARK OJ, and KANG MH, (2005). Soy isoflavone supplementation alleviates oxidative stress and improves systolic blood pressure in male spontaneously hypertensive rats. *J Nutr Sci Vitaminol*, 51, 254–9.

PAZOS M, ANDERSEN ML, and SKIBSTED LH, (2006). Amino acid and protein scavenging of radicals generated by iron/hydroperoxide system: An electron spin resonance spin trapping study. *J Agric Food Chem*, 54, 10215–21.

PERDIGÓN G, RACHID M, DE BUDEGUER MV, and VALDEZ JC, (1994). Effect of yogurt feeding on the small and large intestine associated lymphoid cells in mice. *J Dairy Res*, 61, 553–62.

PEREIRA MA, JACOBS DR JR, VAN HORN L, SLATTERY ML, KARTASHOV AI, and LUDWIG DS, (2002). Dairy consumption, obesity, and the insulin resistance syndrome in young adults: The CARDIA Study. *JAMA*, 287, 2081–9.

PERPETUO E, JULIANO L, and LEBRUN I, (2003). Biochemical and pharmacological aspects of two bradykinin-potentiating peptides obtained from tryptic hydrolysis of casein. *J Protein Chem*, 22, 601–6.

PHILLIPS SM, HARTMAN JW, and WILKINSON SB, (2005). Dietary protein to support anabolism with resistance exercise in young men. *J Am Coll Nutr*, 24, 134S–139S.

PIHLANTO-LEPPÄLÄ A, KOSKINEN P, PIILOLA K, TUPASELA T, and KORHONEN H, (2000). Angiotensin I-converting enzyme inhibitory properties of whey protein digests: Concentration and characterization of active peptides. *J Dairy Res*, 67, 53–64.

PILVI TK, KORPELA R, HUTTUNEN M, VAPAATALO H, and MERVAALA EM, (2007). High-calcium diet with whey protein attenuates body-weight gain in high-fat-fed C57Bl/6J mice. *Br J Nutr*, 98, 900–7.

PINS JJ and KEENAN JM, (2006). Effects of whey peptides on cardiovascular disease risk factors. *J Clin Hypertens*, 8, 775–82.

RAGHUVEER TS, MCGUIRE EM, MARTIN SM, WAGNER BA, REBOUCHÉ CJ, BUETTNER GR, and WIDNESS JA, (2002). Lactoferrin in the preterm infants' diet attenuates iron-induced oxidation products. *Pediatr Res*, 52, 964–72.

RAHA S, DOSQUET C, ABGRALL JF, JOLLES P, FIAT AM, and CAEN JP, (1988). KRDS – a tetrapeptide derived from lactotransferrin–inhibits binding of monoclonal anti-body against glycoprotein IIb–IIIa on ADP-stimulated platelets and megakaryo-cytes. *Blood*, 72, 172–8.

RAJPATHAK SN, RIMM EB, ROSNER B, WILLETT WC, and HU FB, (2006). Calcium and dairy intakes in relation to long-term weight gain in US men. *Am J Clin Nutr*, 83, 559–66.

RANKIN JW, GOLDMAN LP, PUGLISI MJ, NICKOLS-RICHARDSON SM, EARTHMAN CP, and GWAZDAUSKAS FC, (2004). Effect of post-exercise supplement consumption on adaptations to resistance training. *J Am Coll Nutr*, 23, 322–30.

RANKIN JW, SHUTE M, HEFFRON SP, and SAKER KE, (2006). Energy restriction but not protein source affects antioxidant capacity in athletes. *Free Radic Biol Med*, 41, 1001–9.

REYNOLDS E, CAI F, SHEN P, and WALKER G, (2003). Retention in plaque and remineralization of enamel lesions by various forms of calcium in a mouthrinse or sugar-free chewing gum. *J Dent Res*, 82, 206–11.

REYNOLDS E, CAIN C, WEBBER F, BLACK C, RILEY P, JOHNSON I, and PERICH J, (1995). Anticariogenicity of calcium phosphate complexes of tryptic casein phosphopeptides in the rat. *J Dent Res*, 74, 1272–9.

RICH-EDWARDS JW, GANMAA D, POLLAK MN, NAKAMOTO EK, KLEINMAN K, TSERENDOLGOR U, WILLETT WC, and FRAZIER AL, (2007). Milk consumption and the prepubertal somatotropic axis. *Nutr J*, 6: 28.

RIEDEL WJ, KLAASEN T, and SCHMITT JAJ, (2002). Tryptophan, mood, and cognitive function. *Brain Behav Immunity*, 16, 581–9.

RIEU I, BALAGE M, SORNET C, DEBRAS E, RIPES S, ROCHON-BONHOMME C, POUYET C, GRIZARD J, and DARDEVET D, (2007). Increased availability of leucine with leucine-rich whey proteins improves postprandial muscle protein synthesis in aging rats. *Nutrition*, 23, 323–31.

RIVAL SG, BOERIU CG, and WICHERS HJ, (2001). Caseins and casein hydrolysates. 2. Antioxidative properties and relevance to lipoxygenase inhibition. *J Agric Food Chem*, 49, 295–302.

ROGERS I, EMMETT P, GUNNELL D, DUNGER D, HOLLY J, and ALSPAC STUDY TEAM, (2006). Milk as a food for growth? The insulin-like growth factors link. *Public Health Nutrition*, 9, 359–68.

ROOS N, MAHÉ S, BENAMOUZIG R, SICK H, RAUTUREAU J, and TOMÉ D, (1995). 15N-labeled immunoglobulins from bovine colostrum are partially resistant to digestion in human intestine. *J Nutr*, 125, 1238–44.

ROSANELI CF, BIGHETTI AE, ANTONIO MA, CARVALHO JE, and SGARBIERI VC, (2002). Efficacy of a whey protein concentrate on the inhibition of stomach ulcerative lesions caused by ethanol ingestion. *J Med Food*, 5, 221–8.

ROSANELI CF, BIGHETTI AE, ANTONIO MA, CARVALHO JE, and SGARBIERI VC, (2004). Protective effect of bovine milk whey protein concentrate on the ulcerative lesions caused by subcutaneous administration of indomethacin. *J Med Food*, 7, 309–14.

ROSELL M, HÅKANSSON NN, and WOLK A, (2006). Association between dairy food consumption and weight change over 9 y in 19352 perimenopausal women. *Am J Clin Nutr*, 84, 1481–8.

ROY BD, (2008). Milk: The new sports drink? A Review. *J Int Soc Sports Nutr*, 5, 15.

ROYLE PJ, MCINTOSH GH, and CLIFTON PM, (2007). Whey protein isolate and glycomacropeptide decrease weight gain and alter body composition in male Wistar rats. *Br J Nutr*, 100, 88–93.

SAITO Y, LEE YS, and KIMURA S, (1998). Minimum effective dose of casein phosphopeptides (CPP) for enhancement of calcium absorption in growing rats. *Int J Vitam Nutr Res*, 68, 335–40.

SAKAGUCHI M, KOSEKI M, WAKAMATSU M, and MATSUMURA E, (2003). Effects of beta-casomorphin-5 on passive avoidance response in mice. *Biosci Biotechnol Biochem*, 67, 2501–4.

SAKAGUCHI M, KOSEKI M, WAKAMATSU M, and MATSUMURA E, (2006). Effects of systemic administration of beta-casomorphin-5 on learning and memory in mice. *Eur J Pharmacol*, 530, 81–7.

SAKANAKA S, TACHIBANA Y, ISHIHARA N, and JUNEJA LR, (2005). Antioxidant properties of casein calcium peptides and their effects on lipid oxidation in beef homogenates. *J Agric Food Chem*, 53, 464–8.

SANDOMIRSKY BP, GALCHENKO SE, and GALCHENKO KS, (2003). Antioxidative properties of lactoferrin from bovine colostrum before and after its lyophilization. *Cryo Letters*, 24, 275–80.

SANGGAARD KM, HOLST JJ, REHFELD JF, SANDSTRÖM B, RABEN A, and THOLSTRUP T, (2004). Different effects of whole milk and a fermented milk with the same fat and lactose content on gastric emptying and postprandial lipaemia, but not on glycaemic response and appetite. *Br J Nutr*, 92, 447–59.

SATAKE M, ENJOH M, NAKAMURA Y, TAKANO T, KAWAMURA Y, ARAI S, and SHIMIZU M, (2002). Transepithelial transport of the bioactive tripeptide, Val-Pro-Pro, in human intestinal Caco-2 cell monolayers. *Biosci Biotechnol Biochem*, 66, 378–84.

SATO R, INANAMI O, TANAKA Y, TAKASE M, and NAITO Y, (1996). Oral administration of bovine lactoferrin for treatment of intractable stomatitis in feline immunodeficiency virus (FIV)-positive and FIV-negative cats. *Am J Vet Res*, 57, 1443–6.

SCHMITT JA, JORISSEN BL, DYE L, MARKUS CR, DEUTZ NE, and RIEDEL WJ, (2005). Memory function in women with premenstrual complaints and the effect of serotonergic stimulation by acute administration of an alpha-lactalbumin protein. *J Psychopharmacol*, 19, 375–84.

SCHOLZ-AHRENS KE, KOPRA N, and BARTH CA, (1990). Effect of casein phosphopeptides on utilization of calcium in minipigs and vitamin-D-deficient rats. *Z Ernahrungswiss*, 29, 295–8.

SCHULTE-FROHLINDE E, REINDL W, BIERLING D, LERSCH C, BRANTL V, TESCHEMACHER H, and SCHUSDZIARRA V, (2000). Effects of oral casokefamide on plasma levels, tolerance, and intestinal transit in man. *Peptides*, 21, 439–42.

SCHUSDZIARRA V, SCHICK A, DE LA FUENTE A, SPECHT J, KLIER M, BRANTL V, and PFEIFFER EF, (1983). Effect of beta-casomorphins and analogs on insulin release in dogs. *Endocrinology*, 112, 885–9.

SEPPO L, JAUHIAINEN T, POUSSA T, and KORPELA R, (2003). A fermented milk high in bioactive peptides has a blood pressure-lowering effect in hypertensive subjects. *Am J Clin Nutr*, 77, 326–30.

SFEIR RM, DUBARRY M, BOYAKA PN, RAUTUREAU M, and TOMÉ D, (2004). The mode of oral bovine lactoferrin administration influences mucosal and systemic immune responses in mice. *J Nutr*, 134. 403–9.

SHOJI H, OGUCHI S, SHINOHARA K, SHIMIZU T, and YAMASHIRO Y, (2007). Effects of iron-unsaturated human lactoferrin on hydrogen peroxide-induced oxidative damage in intestinal epithelial cells. *Pediatr Res*, 61, 89–92.

SIPOLA M, FINCKENBERG P, KORPELA R, VAPAATALO H, and NURMINEN ML, (2002). Effect of long-term intake of milk products on blood pressure in hypertensive rats. *J Dairy Res*, 69, 103–11.

SUETSUNA K, UKEDA H, and OCHI H, (2000). Isolation and characterization of free radical scavenging activities peptides derived from casein. *J Nutr Biochem*, 11, 128–31.

TAKEUCHI T, SHIMIZU H, ANDO K, and HARADA E, (2004). Bovine lactoferrin reduces plasma triacylglycerol and NEFA accompanied by decreased hepatic cholesterol and triacylglycerol contents in rodents. *Br J Nutr*, 91, 533–8.

TANG JE, and PHILLIPS SM, (2009). Maximizing muscle protein anabolism: The role of protein quality. *Curr Opin Clin Nutr Metab Care*, 12, 66–71.

TANIGUCHI M, YASUTAKE A, and TAKEDOMI K, (2000). Effects of dietary sulfur-containing amino acids on oxidative damage in rat liver caused by N-nitrosodimethylamine administration. *Br J Nutr*, 84, 211–7.

TEUCHER B, MAJSAK-NEWMAN G, DAINTY JR, MCDONAGH D, FITZGERALD RJ, and FAIRWEATHER-TAIT SJ, (2006). Calcium absorption is not increased by caseinophosphopeptides. *Am J Clin Nutr*, 84, 162–6.

THOMPSON WG, ROSTAD HOLDMAN N, JANZOW DJ, SLEZAK JM, MORRIS KL, and ZEMEL MB, (2005). Effect of energy-reduced diets high in dairy products and fiber on weight loss in obese adults. *Obes Res*, 13, 1344–53.

TOGAWA J, NAGASE H, TANAKA K, INAMORI M, NAKAJIMA A, UENO N, SAITO T, and SEKIHARA H, (2002). Oral administration of lactoferrin reduces colitis in rats via modulation of the immune system and correction of cytokine imbalance. *J Gastroenterol Hepatol*, 17, 1291–8.

TOZER RG, TAI P, FALCONER W, DUCRUET T, KARABADJIAN A, BOUNOUS G, MOLSON JH, and DRÖGE W, (2008). Cysteine-rich protein reverses weight loss in lung cancer patients receiving chemotherapy or radiotherapy. *Antioxid Redox Signal*, 10, 395–402.

TROMPETTE A, CLAUSTRE J, CAILLON F, JOURDAN G, CHAYVIALLE JA, and PLAISANCIE P, (2003). Milk bioactive peptides and beta-casomorphins induce mucus release in rat jejunum. *J Nutr*, 133, 3499–503.

TSUBOTA A, YOSHIKAWA T, NARIAI K, MITSUNAGA M, YUMOTO Y, FUKUSHIMA K, HOSHINA S, and FUJISE K, (2008). Bovine lactoferrin potently inhibits liver mitochondrial 8-OHdG levels and retrieves hepatic OGG1 activities in Long-Evans Cinnamon rats. *J Hepatol*, 48, 486–93.

TSUCHITA H, GOTO T, SHIMIZU T, YONEHARA Y, and KUWATA T, (1996). Dietary casein phosphopeptides prevent bone loss in aged ovariectomized rats. *J Nutr*, 126, 86–93.

TSUCHITA H, SEKIGUCHI I, KUWATA T, IGARASHI C, and EZAWA I, (1993). The effect of casein phosphopeptides on calcium utilization in young ovariectomized rats. *Z Ernahrungswiss*, 32, 121–30.

TSUCHITA H, SUZUKI T, and KUWATA T, (2001). The effect of casein phosphopeptides on calcium absorption from calcium-fortified milk in growing rats. *Br J Nutr*, 85, 5–10.

USHIDA Y, SHIMOKAWA Y, TOIDA T, MATSUI H, and TAKASE M, (2007). Bovine alpha-lactalbumin stimulates mucus metabolism in gastric mucosa. *J Dairy Sci*, 90, 541–6.

USHIDA Y, SHIMOKAWA Y, MATSUMOTO H, TOIDA T, and HAYASAWA H, (2003). Effects of bovine alpha-lactalbumin on gastric defense mechanisms in naive rats. *Biosci Biotechnol Biochem*, 67, 577–83.

VAN DER MEER R, DE VRIES HT, and VAN TINTELEN G, (1988). The phosphorylation state of casein and the species-dependency of its hypercholesterolaemic effect. *Br J Nutr*, 59, 467–73.

WAKABAYASHI H, TAKAKURA N, TERAGUCHI S, and TAMURA Y, (2003). Lactoferrin feeding augments peritoneal macrophage activities in mice intraperitoneally injected with inactivated Candida albicans. *Microbiol Immunol*, 47, 37–43.

WAKABAYASHI H, TAKAKURA N, YAMAUCHI K, and TAMURA Y, (2006). Modulation of immunity-related gene expression in small intestines of mice by oral administration of lactoferrin. *Clin Vaccine Immunol*, 13, 239–45.

WALSH DJ, BERNARD H, MURRAY BA, MACDONALD J, PENTZIEN AK, WRIGHT GA, WAL JM, STRUTHERS AD, MEISEL H, and FITZGERALD RJ, (2004). In vitro generation and stability of the lactokinin beta-lactoglobulin fragment (142–148). *J Dairy Sci*, 87, 3845–57.

WANG WP, IIGO M, SATO J, SEKINE K, ADACHI I, and TSUDA H, (2000). Activation of intestinal mucosal immunity in tumor-bearing mice by lactoferrin. *Jpn J Cancer Res*, 91, 1022–7.

WESTERTERP-PLANTENGA MS, LEJEUNE MP, NIJS I, VAN OOIJEN M, and KOVACS EMR, (2004). High protein intake sustains weight maintenance after body weight loss in humans. *Int J Obes*, 28, 57–64.

WILEY AS, (2005). Does milk make children grow? Relationships between milk consumption and height in NHANES 1999–2002. *Am J Hum Biol*, 17, 425–41.

WONG CW, and WATSON DL, (1995). Immunomodulatory effects of dietary whey proteins in mice. *J Dairy Res*, 62, 359–68.

WU G, FANG YZ, YANG S, LUPTON JR, and TURNER ND, (2004). Glutathione metabolism and its implications for health. *J Nutr*, 134, 489–92.

YAMAMOTO N, AKINO A, and TAKANO T, (1994). Antihypertensive effect of the peptides derived from casein by an extracellular proteinase from Lactobacillus helveticus CP790. *J Dairy Sci*, 77, 917–22.

YAMAUCHI K, WAKABAYASHI H, HASHIMOTO S, TERAGUCHI S, HAYASAWA H, and TOMITA M, (1998). Effects of orally administered bovine lactoferrin on the immune system of healthy volunteers. *Adv Exp Med Biol*, 443, 261–5.

YAMAUCHI R, OHINATA K, and YOSHIKAWA M, (2003). Beta-lactotensin and neurotensin rapidly reduce serum cholesterol via NT2 receptor. *Peptides*, 24, 1955–61.

ZAVORSKY GS, KUBOW S, GREY V, RIVERIN V, and LANDS LC, (2007). An open-label dose–response study of lymphocyte glutathione levels in healthy men and women receiving pressurized whey protein isolate supplements. *Int J Food Sci Nutr*, 58, 429–36.

ZEMEL MB, RICHARDS J, MATHIS S, MILSTEAD A, GEBHARDT L, and SILVA E, (2005a). Dairy augmentation of total and central fat loss in obese subjects. *Int J Obes*, 29, 391–7.

ZEMEL MB, RICHARDS J, MILSTEAD A, and CAMPBELL P, (2005b). Effects of calcium and dairy on body composition and weight loss in African-American adults. *Obes Res*, 13, 1218–25.

ZEMEL MB, TEEGARDEN D, VAN LOAN M, SCHOELLER DA, MATKOVIC V, LYLE RM, and CRAIG BA, (2004b). Role of dairy products in modulating weight and fat loss: A multi-center trial. Available from http://www.nationalcouncil.org. (Accessed May 10 2007).

ZEMEL MB, THOMPSON W, MILSTEAD A, MORRIS K, and CAMPBELL P, (2004a). Calcium and dairy acceleration of weight and fat loss during energy restriction in obese adults. *Obes Res*, 12, 582–90.

ZHANG X, and BEYNEN AC, (1993). Lowering effect of dietary milk-whey protein v. casein on plasma and liver cholesterol concentrations in rats. *Br J Nutr*, 70, 139–46.

ZOGHBI S, TROMPETTE A, CLAUSTRE J, EL HOMSI M, GARZON J, JOURDAN G, SCOAZEC J-Y, and PLAISANCIE P, (2006). beta-Casomorphin-7 regulates the secretion and expression of gastrointestinal mucins through a mu-opioid pathway. *Am J Physiol Gastrointest Liver Physiol*, 290, 1105–13.

ZOMMARA M, TOUBO H, SAKONO M, and IMAIZUMI K, (1998). Prevention of peroxidative stress in rats fed on a low vitamin E-containing diet by supplementing with a fermented bovine milk whey preparation: Effect of lactic acid and beta-lactoglobulin on the antiperoxidative action. *Biosci Biotechnol Biochem*, 62, 710–7.

ZOMMARA MA, TOUBO H, and IMAIZUMI K, (2002). Supplementing bovine milk immunoglobulin G prevents rats fed on a vitamin E-deficient diet from developing peroxidation stress. *Ann Nutr Metab* 46, 97–102.

ZUNQUIN G, ROULEAU V, BOUHALLAB S, BUREAU F, THEUNYNCK D, ROUSSELOT P, ARHAN P, and BOUGLE D, (2006). Iron and exercise induced alterations in antioxidant status. Protection by dietary milk proteins. *Free Radic Res*, 40, 535–42.

10

Bovine milk immunoglobulins against microbial human diseases

H. J. Korhonen and P. Marnila, MTT Agrifood Research Finland, Finland.

Abstract: Immunoglobulins (Igs), also called antibodies, are present in milk and colostrum of all lactating species. Igs are divided into classes having different physico-chemical structures and biological activities. The major Ig classes in bovine and human milk are IgA, IgG and IgM. Bovine colostral Igs provide the newborn calf with passive immune protection against microbial infections until the calf's own immune system matures. Colostral Ig preparations designed for farm animals have been commercially available for many years. Potential health benefits attributed to bovine colostrum have increased manufacture and marketing of colostral Ig-based dietary supplements also for human use. Furthermore, specific anti-microbial antibodies can be produced into colostrum by immunizing cows with vaccines made of pathogenic microorganisms. These antibodies can be concentrated and used to formulate so-called immune milk preparations. Such preparations have proven effective in prevention of animal and human infections caused, e.g. by rotavirus, *Shigella flexneri*, *Escherichia coli*, *Clostridium difficile*, *Streptococcus mutans*, *Cryptosporidium parvum* and *Helicobacter pylori*. Their therapeutic efficacy, however, seems limited. A few immune milk products have been commercialized and more can be expected in the future for use, e.g. as a supportive means in antibiotic treatments and for prevention of hospital infections.

Key words: immunoglobulin, antibody, colostrum, passive immunity, immune milk.

10.1 Introduction

It has been recognized for more than 100 years that maternal colostrum and milk offer passive protection to a newborn infant against enteric pathogens, primarily via the transfer of immunoglobulins (Igs) and associated bioactive factors (Ehrlich, 1892). Ruminant neonates are born virtually without Igs and therefore the colostral Igs are essential for survival. Probably due to this unique function, Igs represent the major protein fraction of colostrum, accounting for 70–80% of its total protein content (Butler, 1998; Marnila and Korhonen, 2002).

The Ig fraction can be concentrated from colostrum or cheese whey and is marketed commercially as a feed supplement and replacer of colostrum, mainly for neonate calves and pigs, to prevent gastrointestinal infections (Scammell, 2001; Mehra *et al.*, 2006).

The advent of functional foods has increased interest in the bioactive components of bovine colostrum and milk as potential ingredients for health-promoting foodstuffs and even biopharmaceuticals. To this end, scientific and commercial attention has been focused on the development of bovine-derived Ig products that contain specific antibodies targeted for prevention or treatment of microbial infections in humans (Korhonen *et al.*, 1998; Korhonen, 2002). Such preparations have been described as 'immune milk'. This concept dates back to the 1950s when Petersen and Campbell first suggested that orally administered bovine colostrum from hyperimmunized cows could provide passive immune protection for humans (Campbell and Petersen, 1963). Since then, a large number of animal and human studies have been carried out to demonstrate that these preparations can be effective in the prevention or treatment of human and animal diseases caused by various pathogenic microbes (for reviews see Weiner *et al.*, 1999; Korhonen *et al.*, 2000a; Lilius and Marnila, 2001; Hoerr and Bostwick, 2002; Korhonen and Marnila, 2006; Mehra *et al.*, 2006). This article reviews the current state of knowledge about the properties of bovine Igs, their utilization as ingredients for immune milk preparations, and the application of these preparations for the prevention and treatment of various microbial infections in humans.

10.2 Properties of immunoglobulins

10.2.1 Structure and physicochemical properties

Igs, which carry the biological function of antibodies, are present in colostrum and milk of all lactating species. In mammals, all five known classes of Igs have been characterized: IgG, IgM, IgA, IgD and IgE. The major Ig classes in both bovine and human milk are IgA, IgG and IgM. The basic chemical structure of all Igs is similar but their biological functions differ, although in principle they all contribute to the major defence mechanism against foreign materials recognized by the body's immune system. Igs account for up to 70–80% of the total protein content in colostrum, whereas in milk they account for only 1–2% of total protein. IgG1 is the predominant Ig class in bovine lacteal secretions as compared to IgA in human milk (Elfstrand *et al.*, 2002; Marnila and Korhonen, 2002). Table 10.1 provides the concentrations of different Ig classes in bovine colostrum and milk.

Ig molecules of all classes are symmetrical, multi-chain glycoproteins composed of two identical glycosylated heavy chains and two identical

Table 10.1 Immunoglobulin concentrations in bovine colostrum and milk

Immunoglobulin class	Molecular mass (kDa)	Concentration (gL^{-1})	
		Milk	Colostrum*
IgG_1	146–163	0.3–0.6	15–180
IgG_2	146–154	0.06–0.12	1–3
IgG total		0.15–0.8	20–200
SIgA	385–430	0.05–0.1	1–6
IgM	900	0.04–0.1	3–9

* = first milking.
Data compiled from Marnila and Korhonen (2002), Elfstrand *et al.* (2002) and Mehra *et al.* (2006).

non-glycosylated light chains. The basic structure of all monomeric Igs is similar (see Fig. 10.1). The IgG class can be considered as a general model of a monomeric Ig. The molecular weight of each light chain is around 23 kDa and of the heavy chains 53 kDa. The molecular weight of the complete Ig molecule varies around 160 kDa. Both the light and heavy chains contain domains referred as constant regions (C_L, C_H) and variable regions (V_L, V_H). The light chains are attached to the heavy chains by a disulphide bond, and also the two heavy chains are held together by two disulphide bonds near a hinge region which gives the molecule structural flexibility needed in antibody–antigen interactions (Nezlin, 1998a). The two identical antigen-binding sites needed in these interactions are formed by the N-terminal region of one heavy chain and the variable region of one light chain. The V_L-region determines the immunological specificity. Antigen binding occurs by the interactions between the antigen and these regions. The Ig classes and subclasses are determined by the genes encoding the constant regions of heavy chains (Butler, 1998). The bovine IgG molecule occurs dominantly in two sub-classes: IgG_1 and IgG_2.

Monomeric IgM and IgA have a similar basic structure to IgG except for the differences in heavy chain structures and the addition of a C-terminal octapeptide to the heavy chain of IgA (Butler, 1998; Nezlin, 1998b). Monomeric IgA occurs in serum, but in milk it is present as a dimer comprising two IgA molecules joined together by a polypeptide J-chain and an additional 75 kD secretory component. This secretory IgA (SIgA) has a molecular weight of about 380 kDa, and is more resistant to proteolysis and therefore more stable in the gastrointestinal tract than antibodies without the secretory component. IgM is a circular pentamer consisting of five subunits similar to those of monomeric IgA or IgG, which are linked together in a circular mode by disulphide bonds and a J chain: the molecular weight of IgM is approximately 900 kDa (see Fig. 10.2). The physicochemical properties of Igs are presented in Table 10.2.

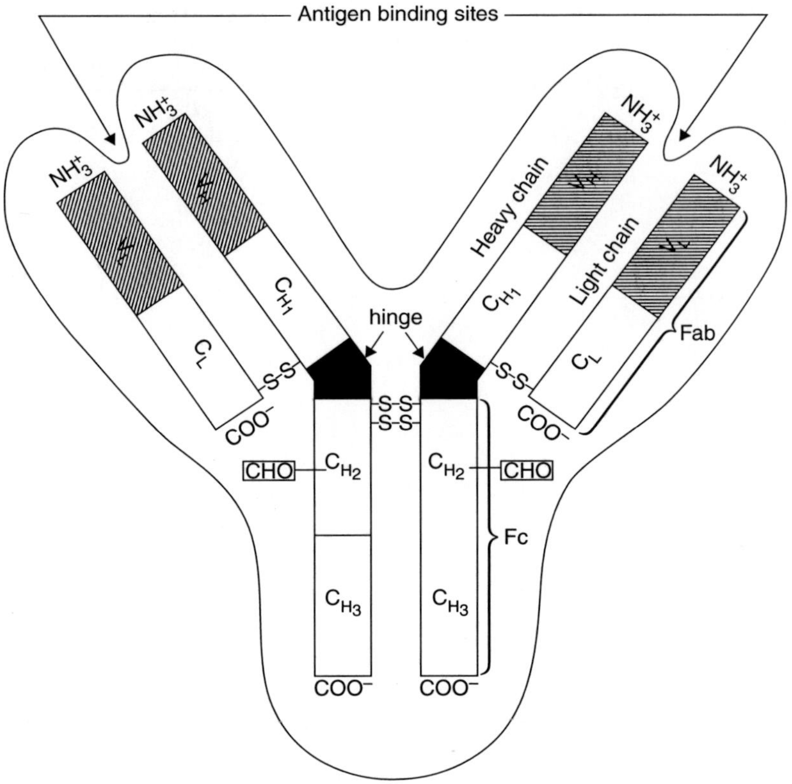

Fig. 10.1 Schematic diagram of a basic immunoglobulin. From Larson (1992).

Proteolysis of an Ig with papain cleaves it near the hinge region into two antigen binding-sites (Fab-fragments), which are formed by the N-terminal part of one heavy and one light chain, and to the third C-terminal fragment, termed the Fc-part. The differences between Ig classes impact upon the cleavage patterns in enzymatic proteolysis. For instance, cleavage of bovine IgG2a(A2) by pepsin results to two Fab and one Fc parts, whereas in IgG2a(A1) the Fab parts remain together by one disulphur bridge resulting in one $F(ab')_2$ and one Fc fragment which is cleaved further by pepsin (Heyermann and Butler, 1987).

10.2.2 Biological activities

Igs are produced by B lymphocytes and plasma cells diversified from B cells. The soluble Igs function as flexible adaptors between cellular and humoral immunity by binding to antigens and exhibiting one or more effector functions. The Fab parts attach the antigens while other parts (mostly the Fc region) interact with other elements of the immune system.

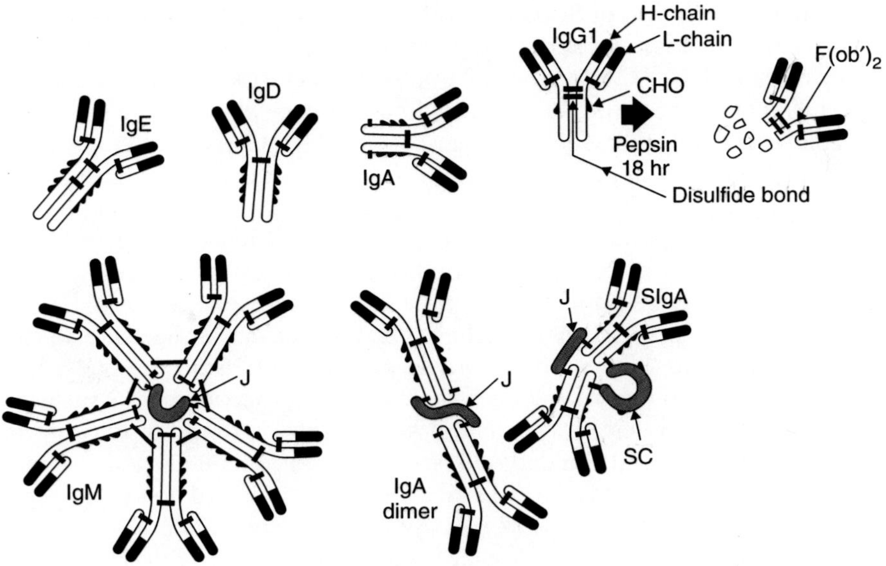

Fig. 10.2 Structure of five classes of immunoglobulins. Figure modified from Larson (1992).

Table 10.2 Properties of bovine immunoglobulins

	IgG_1	IgG_2	IgA	SIgA	IgM
Physico-chemical					
Molecular weight (kDa)	146–163	146–154	160	385–430	900
Heavy chain	56–59	54–59	61–63		62–76
Heavy chain type	γ1	γ2	α	α	μ
Number of H- and L-chains	2	2	2	4	10
Structure	Monomer	Monomer	Monomer	Dimer	Pentamer
Immunological					
Opsonization	+++	+	0	0	+++
Complement fixation	++	+	0	0	+++
Agglutination	+	+	++	++	+++

0 = no activity, + = low activity, ++ = moderate activity, +++ = strong activity.
Data modified from Marnila and Korhonen (2002).

For many pathogenic organisms, the attachment to the epithelial lining is an obligatory first step in the establishment of colonization and infection. Many pathogens have receptors, e.g. fimbriae for epithelial surfaces. Thus, the prevention of the microbial adhesion to epithelial linings is an important mechanism of SIgA antibodies in protecting the host (Woof and Kerr, 2006). Colostrum and human milk protect the neonate intestinal mucosa against EPEC infections by inhibiting bacterial adhesion to epithelial cells (Carbonare *et al.*, 1997). This effect has been shown to be mediated by colostrum and milk SIgA (Cámara *et al.*, 1994; Carbonare *et al.*, 1997, 2005; Fernandes *et al.*, 2001; Corrêa *et al.*, 2006), and by oligosaccharides present in human milk (Cravioto *et al.*, 1991). M-cells in intestinal epithelia have specific receptors for SIgA, and antigen-SIgA-receptor complex can be transcytosed by M-cells into lymphoid tissue for antigen presenting and induction of mucosal immune responses (Woof and Kerr, 2006).

The ability of Igs to form cross-links between surface antigens results in a network of cells that can be removed mechanically from the body. This agglutination of microbes reduces their capability to adhere to surfaces. Agglutinated microbes are usually not able to release toxins or to colonize the host. Normal colostrum and milk are known to contain natural antibodies which can agglutinate a large number of pathogenic and non-pathogenic micro-organisms (Bostwick *et al.*, 2000; Korhonen *et al.*, 2000b).

Igs can bind bacterial toxins which, thereafter, can be recognized more effectively by phagocytic leukocytes. Many bacterial toxins must first be actively transported via receptors inside the host cells to cause cell death. Blocking the toxin may prevent its internalization in host cells. Also, Igs can inhibit or reduce the production of toxins and other harmful components by inhibiting bacterial metabolism and by blocking enzymes and receptors. This blocking may also reduce the ability of pathogens to produce structures needed in adherence to epithelia. Specific Igs can protect against viral infections by binding viruses and preventing the virus replication by blocking the receptor mediated internalisation of viruses in the host cells. Specific Igs augment the recognition and phagocytosis of antigens by leukocytes (opsonisation). Divalent or polyvalent binding of an Ig to antigenic structure results in conformational change of the Ig molecule, which process again enables the Fc portion to bind to the corresponding Fc receptors on the leukocyte surface. This receptor binding leads to various immune cell effector functions depending on the cell Ig and leukocyte type. Leukocytes are an integral part of normal milk and colostrum and are of vital importance in defending the mammary gland against pathogens (Korhonen *et al.*, 2000b).

Specific Igs are rarely cidal to micro-organisms but may disturb cellular metabolism by blocking receptors and enzymes, and cause structural alterations leading to immobilization, increased membrane permeability and impaired cell growth. Igs may contribute to the killing of microbes by activating the classical pathway of complement. In blood and tissues, the activation of complement mediated bacteriolytic reactions may be the most important function of Igs, but the significance of this mechanism in milk remains obscure. Bovine and human colostrum contain an active complement system participating in the immune defense of the udder (Butler, 1998).

IgM antibodies, although produced in smaller amounts than IgG, are considerably more efficient than IgG with regard to most of the above activities, including complement fixation, opsonization and agglutination of bacteria. Bovine IgG_2B is more effective than IgG_2A in activating the classical pathway of complement (Bastida-Corquera, 1999). Specific IgM can prevent effectively the migration of bacteria by binding the flagellas used in the movement. Bovine IgA does not fix complement or opsonize bacteria, but it agglutinates antigens, neutralizes viruses and bacterial toxins. The SIgA dimer is found in mucosal secretions and its main function is to protect the mucosal barriers by binding microbes and preventing their attachment to the epithelium, and to contribute to the antigen presenting. The milk Igs have also been found to exert a synergistic effect on the activity of non-specific antimicrobial factors in milk, such as lactoferrin and lysozyme, as well as the lactoperoxidase-thiocyanate-hydrogen peroxide system (Loimaranta *et al.*, 1998a; Bostwick *et al.*, 2000).

10.3 Production and isolation of immunoglobulins

Traditionally, the globulin fraction was isolated from colostral whey by precipitation with either ammonium sulphate or ethanol. These chemical methods are, however, economically not suitable for large scale-production of Igs although they yield rather pure Ig fractions. Other challenges in isolation of active Igs from colostrum has been its complex composition and the sensitivity of Igs to heat treatments. With the rapid development of new chromatographic and membrane-separation technologies in the last thirty years, it is now possible to isolate individual milk proteins on a large-scale. Based on these techniques, a great number of pilot- or industrial-scale methods have been developed and patented for fractionation and isolation of Igs from colostral or cheese whey (for reviews see Korhonen, 2004; Korhonen and Pihlanto, 2007). With these methods, the recovery rate of Igs has varied from 40% to 70% of the level present in the starting material (Elfstrand *et al.*, 2002). Specific chromatographic techniques, such as immobilized metal chelate chromatography, immunoaffinity chromatography and cation-exchange chromatography have been applied to improve the yield and purity of immunoglobulin preparations further (Fukumoto *et al.*, 1994; Akita and Li-Chan, 1998). Microfiltration (MF) combined with ultrafiltration (UF) of bovine, equine and caprine colostrum has led to IgG/total solids purity of more than 90% (Piot *et al.*, 2004). Using cheese whey, a 0.1 μm membrane and the concept of selective membrane separation through pH manipulation, Mehra and Kelly (2004) produced an Ig-rich preparation with a protein composition similar to that produced from colostral whey. Korhonen *et al.* (1998) used various MF methods, such as UF, MF and reverse osmosis, and a cation-exchange resin as a molecular sieve, to concentrate Igs from colostral whey. The Ig level of the final freeze-dried concentrates varied from 45% to 75%. Expanded Bed Adsorption Chromatography (EBAC) is a new technology which has been applied to isolate Igs from cheese whey using an adsorbent with tailored ligand chemistry (Nielsen *et al.*, 2002). EBAC provides significant advantages over conventional packed bed column chromatography, and an Ig-purity from 50% to 70% can be achieved with this up-scalable method.

The technological properties of Igs have been studied extensively, as these proteins are known to affect many dairy processes (see review by Mehra *et al.*, 2006). Igs may inhibit or retard the activity of renneting enzymes of bacterial origin. In normal milk this is not noticed due to the relatively low Ig concentration but in case of mastitic milk, or if the Ig concentration of milk is increased, e.g. by adding colostrum, renneting may be retarded. The antimicrobial properties of Igs may adversely impact upon fermentation processes. A retarded fermentation by dairy starters is noted in colostrum and mastitic milk, which contain increased amounts of Igs. Also, high Ig concentrations may adversely affect the antibiotic residual tests based on microbial growth, causing false positive results (Korhonen, 2004). Igs contribute to cream formation by agglutinating fat globules, which process accelerates the ascent of cream to the surface. This phenomenon is attributed primarily to IgM, which have been termed cryoglobulins or 'cold agglutinins'. The agglutination property of Igs can, however, be inactivated by heat treatment (pasteurization) and mechanical agitation.

Among dairy processes, the properties of Igs are most affected by the thermal treatments. In high temperature/short time (HTST) pasteurization (72°C/15 sec) about 25–40% of the Ig activity is lost, whereas ultra high temperature (UHT) treatment (138°C/4 sec) and evaporation processing destroy most of the specific immune activity of milk due to Ig denaturation (Li-Chan *et al.*, 1995). In contrast, other studies (Mainer *et al.*, 1997, 1999) have reported that bovine milk Igs could resist the HTST pasteurization treatment without affecting their structure. Only 1% of IgG, 2% of IgA, and 14% of IgM concentrations were denatured in laboratory experiments. Also, the HTST pasteurization had little effect on the activity of bovine colostral IgG as the original rotavirus neutralizing activity was reduced by only 0.5%. In a recent study by McMartin *et al.* (2006), it was demonstrated that a rapid heat inactivation of IgG started at temperatures higher than 65°C, and at 81°C, as much as 90% of the virus neutralization activity of Igs was lost in less than two minutes. On the other hand, heating moderate or high quality colostrum at 60°C for at least 120 min had no effect on mean IgG concentration or titer of neutralizing antibodies against bovine viral diarrhea virus Type 1. In storage stability studies, it has been observed that bovine IgG added into UHT-treated milk retained its specific immune activity for over several months (Fukumoto *et al.*, 1994). Also, Ig molecules seem to retain their specific activity well in milk powder, irrespective of the storage temperature (Mehra *et al.*, 2006). Further, bacterial fermentation of milk by yoghurt or probiotic bacteria have not been found to reduce significantly the activity of colostral antibodies added in milk during a storage period of 50 days of at 4°C. (Wei *et al.*, 2002).

10.4 Applications of immunoglobulins

10.4.1 Immunoglobulin and immune milk preparations

The progress in understanding the mechanisms of Ig mediated immune functions and the rapid development of industrial fractionation technologies have raised interest in developing formulations supplemented with bovine colostral or cheese whey derived Igs. Most of the current commercial Ig products are prepared from colostrum of non-immunized cows by removing the fat, followed by microfiltration or pasteurization under conditions that retain the biological activity of Igs. These products are usually in the form of spray-dried and freeze-dried powders, but some are in the form of filtered colostral whey liquids or concentrates. Dried products include whole colostrum powder, skim powders, skim colostrum protein concentrate and colostrum whey concentrates. Some of above preparations have been tested clinically for certain physiological functions or prevention/ treatment of microbial infections. Accordingly, a few products boast specific health or nutrition function claims, such as boosting immunity against microbial infections or speeding recovery from physical endurance exercises. (Scammel, 2001; Kelly, 2003; Tripathi and Vashishtha, 2006). However, in most cases, the clinical evidence related to these products is very limited or not available.

The specificity of natural antibodies found in milk and colostrum of different cows reflects the wide spectrum of antigens the animals have encountered in the past in their environment and in ingested feedstuffs (Korhonen *et al.*, 2000a; Kelly, 2003). The antibody titre against certain antigenic pathogens or structure, e.g. virulence factor, can be raised up to several hundred times by immunizing the cow before parturition with vaccines containing the antigens (Korhonen *et al.*, 1995). The resulting immune colostra and products made thereof have fundamentally different antimicrobial properties and efficacies against pathogens than normal colostrum and these two concepts have, therefore, to be differentiated. This is emphasized by the fact that the normal colostrum and preparations made from it are, in most countries, regarded as food or dietary supplement, whereas the Ig containing preparations from immunized cows are often regarded as pharmaceuticals, e.g. in the EU and USA, or their regulatory status is not defined (Scammel, 2001; Hoerr and Bostwick, 2002; Mehra *et al.*, 2006). Table 10.3 lists commercial Ig and immune milk preparations developed over the last decade.

10.4.2 Proteolysis of immunoglobulins in the gastrointestinal tract

It is well known that the low pH of gastric acid reduces significantly the activities of ingested Igs. Furthermore, the ingested Igs are subjected to

Table 10.3 Commercial colostrum and immune milk products

Product	Company	Claimed health benefits
Intact™	Numico RA (Australia)	Immune enhancing, athletic performance
Gastrogard-R™ (from immunized cows)	Northfield Laboratories, (Australia)	Prevents diarrhea caused by rotavirus in infants and children <4 years
PRO-IMMUNE 99	GalaGen Inc., (USA)	Prevents scours caused by *E. coli* in calves.
Proventra™	GalaGen Inc., (USA)	Boosts immunity and enhances body's natural resistance
Lactimmunoglobulin Biotest	Biotest Pharm GmbH (Germany)	Product for treatment of diarrhea in AIDS patients
ColostrumGold™ liquid Colostrumune™ powder	Sterling Technology, Inc (USA)	Immune system booster
First Defence® (from immunized cows)	Immucell (USA)	Reduces mortality and morbidity from scours caused by *E.coli* K99+ and coronavirus in calves
Anti-CD WPC (from immunized cows)	MucoVax Ltd (Netherlands)	Prevents relapse of *C. difficile* diarrhea
Glycomax™ Immunoglobulin	Probiotec Nutritionals Ltd (Australia)	Improves immunity against microbes and improves health and wellbeing
ColoPlus™	ColoPlus Ltd. (Sweden)	Alleviates HIV-associated diarrhea

Data modified from Mehra *et al.* (2006).

degradation by intestinal proteases. In the stomach, the enzymatic hydrolysis by pepsin, fragments the IgG molecule to $F(ab')_2$, Fab/c, pFc and Fv fragments (Nezlin, 1998a). In the small intestine, trypsin, chymotrypsin, carboxypeptidase and elastase initially degrade the antibodies to $F(ab')_2$, Fab and Fc fragments. The secretory piece of milk SIgA protects this Ig form against digestion by proteolytic enzymes. Of the SIgA present in human colostrum, 20–80% passes undegraded through the gut of the human infant. Also, bovine IgG_1 makes an exception since it is rather resistant to trypsin. Bovine colostrum and mastitic milk contain a compound which inhibits trypsin activity. The resulting $F(ab')_2$ and Fab fragments retain at least part of the neutralizing and adhesion inhibiting activities in the intestine. Pacyna *et al.* (2001) supplemented whole milk with immune colostrum containing specific Igs against rotavirus. When 100 ml of this supplemented whole milk was administered to 105 children three times a day for six days, the anti-rotavirus Ig activity was detected by the ELISA method in 521 of 602 fecal specimens obtained during the study. Kelly *et al.* (1997)

demonstrated that 10%–30% of orally administered bovine anti-*Clostridium difficile* Ig concentrate could be recovered from the stools of human infants and adults in the form of $F(ab')_2$ and Fab fragments. The survival of IgG increased remarkably by encapsulation with gelatin.

10.4.3 Efficacy of immune milk preparations against microbial pathogens

The concept of 'immune milk', i.e. the transfer of passive immunity by milk Igs, dates back to the 1950s. Since then, a great number of clinical studies have been published on the efficacy of various immune and non-immune milk Ig preparations in the prevention and treatment of various gastrointestinal microbial diseases. Recently, these studies have been reviewed by Mehra *et al.* (2006), Korhonen and Marnila (2006) and Hammarström and Weiner (2008). Thus, the protocols and results of these numerous studies are not discussed in detail in this article. Briefly, the general tendency of these studies has been that orally administered bovine milk or colostral Igs have, in most cases, proved to be effective in the prevention of orally mediated infections. Colostral or milk Igs can effectively agglutinate bacteria (Xu *et al.*, 2006), and bovine colostrum and colostral preparations prevent effectively the attachment of a pathogen to the cells in epithelial lining. However, since bovine colostrum contain more than one adhesion inhibiting factor, the importance of specific Igs is often not known. Palmeira *et al.* (2001) observed that inhibition of enteropathogenic *E. coli* adherence to HEp-2 cultured cells by bovine colostrum was not caused by IgG fraction and was mediated by a high molecular weight fraction. However, in some studies the inhibition of pathogenic adhesion to epithelial cells has been attributed to presence of specific Igs. Casswall *et al.* (2002) showed that a specific bovine immune colostral preparation blocked almost 90% of *Helicobacter pylori* attachment to human gastric mucosal tissue *in vitro* and 95% of binding to Lewis b glycoconjugate, while a control colostral preparation from nonimmunized cows was inefficient. Doyle *et al.* (1993) reported that hyperimmune bovine colostral Ig inhibited *Cryptosporidium parvum* adherence to epithelial MDCK cells and infectivity in an *in vitro* assay, and this inhibition was correlated with the protective capacity of the bovine colostrum *in vivo*. No significant adhesion inhibition was observed for control preparation from colostra of sham immunized cows. Bojsen *et al.* (2007) studied the efficacy of bovine macromolecular whey protein fractions in prevention of infection of two human intestinal cell lines (Caco-2 and FHs 74 Int) by four different rotavirus strains (Wa, RRV, YM, RF). The major component of the protein fraction that inhibited effectively the rotavirus infectivity *in vitro* was shown to be bovine IgG. The same fraction was the only one affecting rotavirus shedding in an *in vivo* mouse model.

In the treatment of already established infections, promising therapeutic effects have been reported, mainly in such diseases where the infection is maintained through a reattachment and reinfection, e.g. inside the oral cavity or the gastrointestinal lumen, and where the secretion of toxins or other inflammatory compounds is involved, which can be neutralized by the specific colostral Igs. Diseases caused by enterotoxigenic *Escherichia coli* strains, *Cryptosporidium parvum*, *Candida spp*., *Shigella flexneri* and rotavirus are examples. As instances of clinical studies, a closer look is taken hereunder at recent progress in developing bovine immune milks targeted against *Streptoccus mutans* and *Clostridium difficile.* Table 10.4 provides data on recent clinical studies carried out with colostral Ig and immune milk preparations.

Streptoccus mutans *immune milk*

The pathogenesis of dental caries caused by *Streptococcus mutans* involves a series of attachment and binding events that lead to accumulation of cariogenic bacteria sufficient to cause the caries. Immunization strategies against streptococcal adhesins or glucosyltransferase enzymes (GTFs) have been shown to effectively interfere with the pathogenesis of mutans streptococci (Smith *et al.*, 2001). Blocking of the activity of *S. mutans* glucan binding proteins serves another effective strategy. Since humans cannot be immunized actively against caries bacteria due to the risk of side effects, bovine colostral antibodies have been studied in local passive immunization to prevent dental caries (for a review see Koga *et al.*, 2002).

An immune milk preparation from cows immunized with a *S. mutans* and *Streptococcus sobrinus* whole cell vaccine inhibited *in vitro* glucosyltransferase and fructosyltransferase enzyme activities of *S. mutans* (Loimaranta *et al.*, 1997), promoted aggregation of *S. mutans*, and inhibited adherence of the bacteria to saliva-coated hydroxyapatite particles (Loimaranta *et al.*, 1998b). The same immune preparation resulted in a higher resting pH in dental plaque of adult volunteers as compared to control groups, when used as a mouth rinse for three days (Loimaranta *et al.*, 1999). After the rinsing period with the immune product, the relative number of mutans streptococci had decreased significantly as compared to the controls.

Chicken egg yolk Igs obtained from immunized hens with *S. mutans* glucan binding protein had a clear protective effect against *S. mutans* infection and caries development in a rat model (Smith *et al.*, 2001). Mitoma *et al.* (2002) immunized cows with a fusion protein prepared by a fusion of a saliva-binding alanine-rich region of a cell surface protein antigen (Pac) and a glucan binding domain of the glucosyltransferase-I cell surface protein from *S. mutans*. The immune colostrum preparation effectively prevented dental caries development in a rat model when given as concentrate once a day for 55 days, together with cariogenic diet. Shimazaki *et al.* (2001) examined the effect of immune colostrum containing Igs against the same fusion protein on adult subjects after an antibiotic (cetylpyridinium chloride) treatment. The immune preparation inhibited significantly

Table 10.4 Recent *in vivo* studies on the efficacy of orally administered immunoglobulin preparations on health

Antigen used in immunization	Disease or condition	Treatment regimen	Treatment effect	Reference
S. mutans and *S. sobrinus* whole cell vaccine	Dental caries	Mouth rinse 3 times daily for 3 days (human volunteers)	Higher resting pH and smaller proportion of *S. mutans* in dental plaque	Loimaranta *et al.* (1999)
Virulence factors of *S. mutans*	Dental caries	Mouth rinse twice per day for 14 days by 10 ml of immune milk (human volunteers)	Inhibited recolonization of *S. mutans* after antibiotic treatment	Shimazaki *et al.* (2001)
Helicobacter felis whole cell vaccine	Gastritis	0.2 ml of immune whey (7.5% IgG) before infection or 3 times daily for 4 weeks in treatment of infected mice	Prevented infection in non-infected mice and decreased gastric inflammation and colonization in readily infected mice	Marnila *et al.* (2003)
Helicobacter pylori urease	Gastritis	150 ml of yogurt with 1% avian IgY and probiotic bacteria 3 times daily for 4 weeks	Decreased values in urea breath test indicating decrease in colonization	Horie *et al.* (2004)
No immunization	EHEC-colitis	Colostral preparation IMMULACT® to mice ad libitum (around 300 mg per day) for 3 weeks	Decreased rapidly EHEC colonization and decresed attachment to cecum walls and decreased mortality in mouse	Funatogawa *et al.* (2002)
Polyvalent or monovalent *E. coli* vaccine	Diarrhea	IgG supplemented baby formula, daily dose 0.5 g of IgG per kg of body weight	Lower incidence of diarrhea and shorter duration of diarrhea episodes in human infants during follow-up period for 6 months	Tawfeek *et al.* (2003)
Shigella dysenteriae antigen I	Shigellosis	100 ml orally 3 times per day for 3 days in combination with antibiotics	No significant difference in any clinical parameter	Ashraf *et al.* (2001)
No immunization	HIV-associated diarrhea	Colostral preparation ColoPlus® to HIV-associated diarrhea patients for 7 weeks	Substantial decrease in stool frequency and in fatigue and increase in body weight	Florén *et al.* (2006)

Table 10.4 *Cont'd*

Antigen used in immunization	Disease or condition	Treatment regimen	Treatment effect	Reference
Cocktail of 17 strains of pathogenic diarrhea bacteria	*E. coli* and *Salmonella* diarrhea	Orally 10 mg of specific Ig per day for 10 days starting 6 or 8 days before infection	Prevented enteroinvasive *E. coli* and *Salmonella typhi* diarrhea and normalized immunological parameters	Xu *et al.* (2006)
Cocktail of 17 strains of pathogenic diarrhea bacteria	*E. coli*, *Salmonella* and *Shigella* diarrhea	Orally once per day for 10 days starting 6 or 8 days before infection	Decreased the clinical signs of diarrhea and supported splenic NK-cell functions	Huang *et al.* (2008)
No immunization	Murine rotavirus infection in mouse	Milk IgG fraction orally at time of infection and 12 hours post infection	Decreases rotavirus shedding in stools	Bojsen *et al.* (2007)
Clostridium difficile toxin and *C. difficile* whole cells	*C. difficile* diarrhea	Orally for 2 weeks as supportive treatment after antibiotic treatment	*C. difficile* toxins eradicated from 15 of 16 patients and no relapses in any patient during 11-month follow-up period	Van Dissel *et al.* (2005)
No immunization	Mild hypercholesterolemia	Orally 5 g of blood derived IgG daily for 3 or 6 weeks	Both total cholesterol and LDL levels decreased from baselines	Earnest *et al.* (2005)
No immunization	Upper respiratory tract infections	60 g of colostral protein daily for 8 weeks	Reduced significantly incidence of self-estimated symptoms of respiratory infections but no difference in duration	Brinkworth and Buckley (2003)
No immunization	Endotoxemia due to abdominal surgery	Colostral product Lactobin® 52 g daily in 4 doses orally for 3 days before surgery	Lower levels of endotoxin and endotoxin neutralizing capacity in blood suggesting reduced endotoxemia due to surgery	Bölke *et al.* (2002a)
No immunization	Endotoxemia due to coronary surgery	Colostral product Lactobin® 42 g daily doses orally for 2 days before surgery	Lower levels of CRP but no effect on perioperative endotoxemia	Bölke *et al.* (2002b)

Parts of this table compiled and edited from Korhonen *et al.* (2000) and Mehra *et al.* (2006).

recolonization of *S. mutans* in the saliva and plaque as compared to the control group, and the ratios of *S. mutans* to total streptococci in saliva and plaque were lower than in the control group. Wei *et al.* (2002) studied the combined effect of specific colostral Ig preparation against caries streptococci with the probiotic bacterium *Lactobacillus rhamnosus* GG, ATCC 53103 (LGG). LGG added in milk was earlier shown to reduce the risk of caries in day-care children (Näse *et al.*, 2001). The LGG-bacteria and specific Igs in LGG-fermented milk synergistically inhibited adhesion of *S. mutans* to saliva-coated hydroxyapatite particles. The Igs remained active in UHT milk and in LGG fermented milk over the whole period of expected shelf-life of these products. Thus, the rinsing with bovine immune whey indicates favourable effects on human dental plaque by controlling *S. mutans* in the human oral cavity and it may be beneficial to combine specific bovine milk antibodies against mutans streptococci to probiotic LGG-containing milk products.

Clostridium difficile *immune milk*

Colonization of *Clostridium difficile* in the intestine may cause severe infectious diarrhea and colitis. The most important virulence factors are the toxins A and B, which are associated with the development of the disease. Outbreak of *C. difficile* infection often results from antibiotic treatments. In animal models, bovine immune colostrum preparations have been effective in the treatment of experimental *C. difficile* diarrhea. Van Dissel *et al.* (2005) used a preparation made of milk from cows immunized against *C. difficile* toxins and whole cell *C. difficile* as a supportive treatment for two weeks after a standard antibiotic treatment in an uncontrolled cohort study. Nine of 16 patients had a history of relapsing *C. difficile* diarrhea. After the regimen in 15 patients, the *C. difficile* toxins had disappeared from feces, and during the follow-up period of 11 months, none of the patients had another episode of *C. difficile* diarrhea. Numan *et al.* (2007) assessed the efficacy of *C. difficile* immune milk preparation in aiding the prevention of relapses in *C. difficile* patients. The immune milk was administered orally 5 g/day, divided into three equal doses for 14 days starting two weeks after standard treatment with oral metronidazole or vancomycin. In 109 disease episodes, 11 (about 10%) were followed by a relapse, whereas in contemporary controls the relapse rate was 20–25%. Young *et al.* (2007) evaluated the safety of the same immune milk administered to 77 *C. difficile* patients, similarly as in the previous study. Adverse effect monitoring, physical examinations and haematological and biochemical assessments showed no adverse effects in this group. Mattila *et al.* (2008) immunized cows with a whole-cell vaccine made of two toxigenic *C. difficile* strains and the immune whey preparation was made from the colostra of these cows. In a controlled double-blind randomized study, 18 patients with *C. difficile* associated diarrhea received orally 1.6 g of Ig in 200 ml volume twice per day for 2 weeks, and 20 patients received 400 mg of metronidazole twice per day. A

C. difficile culture and toxin test was made on days 0, 14 and 28, respectively. At day 14, 100% of the metronidazole group and 83% of the immune colostrum group, respectively, responded positively to the treatment. The authors concluded that the immune colostrum treatment is somewhat less effective than the standard treatment with antibiotics but, on the other hand, it does not cause antibiotic resistance problems and does not alter the normal colonic bacterial flora as antibiotics do. In future, by optimizing the dosage, the treatment time, and the antigens used in immunizing the cows, the efficacy of bovine milk Ig preparations could be improved and this treatment strategy could also be used as a supportive treatment together with conventional antibiotics.

10.5 Future trends

As a result of progress made in membrane separation and chromatographic techniques it is now possible to isolate Igs from bovine colostrum and cheese whey on a large scale. This progress has facilitated and boosted manufacture of Ig concentrates as dietary supplements or ingredients. Some of these preparations are being marketed with health or nutrition function claims attached, even though the clinical evidence is limited or not reported at all. More scientific research is therefore needed in this field to substantiate the claimed health benefits. As for immune milk preparations, there are already a few commercial products on the market with quite well established clinical evidence about the efficacy. The main limitation of the clinical use of bovine milk antibodies for humans is that, as proteins of another species, the Igs can be used only orally against gastrointestinal pathogens. An interesting attempt to overcome this limitation is a transchromosomic calf that has been cloned for producing humanized polyclonal Igs. As a result of five sequential genetic modifications and seven consecutive cloning events, this calf was reported to have full human IgG 10–20% of total serum IgG. After hyperimmunization of this calf with an anthrax protective antigen, both full human IgG and chimeric IgG were found to be effective in a toxin-neutralization assay (Kuroiwa *et al.*, 2009).

The regulatory status regarding products containing specific Igs may be considered as the main reason why products (for instance those containing specific Igs against cariogenic bacteria) are not on the market in most countries. The regulatory status of immune colostral preparations was undetermined for a long time and for some years these products have been classified as pharmaceuticals in the EU and USA. In the case of diseases that can be controlled also by conventional means, e.g. dental caries, this regulatory approach has slowed down the development of immune milk preparations in many countries. However, in the case of severe diseases caused by bacterial strains resistant to antibiotics, such as *C. difficile*, the regulatory status of a pharmaceutical has delayed but not hampered completely the progress of such product development (Hoerr and Bostwick,

2002; Mattila *et al.*, 2008). Currently, the US Food and Drug Administration (FDA) have accepted the safety of hyperimmune milks on the basis of clinical studies that show no adverse health effects from these products (Gingerich and McPhillips, 2005; Krissansen, 2007).

Incidences if diarrhea and mortality rates among children in many areas of developing countries are high. The prevention of diarrhea would improve the nutritional status and health in general. The emergence of antibiotic resistant pathogen strains puts emphasis on the need to develop alternative ways to prevent and treat gastrointestinal infections. In developed countries, the control of gastrointestinal microbial flora has become an integral part of health promotion. It is concluded that synergistic effects of Igs with probiotics and other milk bioactives, such as lactoferrin, may open new prospects for developing novel means to prevent microbial diseases by nutritional intervention.

10.6 References

AKITA E M, LI-CHAN E C Y (1998), 'Isolation of bovine immunoglobulin G subclasses from milk, colostrum, and whey using immobilised egg yolk antibodies.' *Journal of Dairy Science*, 81, 54–63.

ASHRAF H, MAHALANABIS D, MITRA A K, TZIPORI S, FUCHS G J (2001), 'Hyperimmune bovine colostrum in the treatment of shigellosis in children: A double-blind, randomized, controlled trial', *Acta Paediatrica*, 90, 1373–1378.

BASTIDA-CORCUERA F D, BUTLER J E, YAHIRO S, CORBEIL L B (1999), 'Differential complement activation by bovine IgG2 allotypes' *Veterinary Immunololy and Immunopathology*, 71(2), 115–123. DOI:10.1016/S0165-2427(99)00095-1

BOJSEN A, BUESA J, MONTAVA R, KVISTGAARD A S, KONGSBAK M B, PETERSEN T E, HEEGAARD C W, RASMUSSEN J T (2007), 'Inhibitory activities of bovine macromolecular whey proteins on rotavirus infections in vitro and in vivo', *Journal of Dairy Science* 90, 66–74.

BÖLKE E, JEHLE P M, HAUSMANN F, DÄUBLER A, WIEDECK H, STEINBACH G, STORCK M, ORTH K (2002a), 'Preoperative oral application of immunoglobulin-enriched colostrum milk and mediator response during abdominal surgery', *Shock*, 17, 9–12.

BÖLKE E, ORTH K, JEHLE P M, SCHWARZ A, STEINBACH G, SCHLEICH S, ULMER C, STORCK M, HANNEKUM A (2002b), 'Enteral application of an immunoglobulin-enriched colostrum milk preparation for reducing endotoxin translocation and acute phase response in patients undergoing coronary bypass surgery – a randomized placebo-controlled pilot trial', *Wiener Klinische Wochenschrift*, 114(21-22), 923–928.

BOSTWICK E F, STEJINS J, BRAUN S (2000), 'Lactoglobulins.' In: *Natural Food Antimicrobial Systems*. Ed. Naidu A S. pp. 133–158. Boca Raton, Florida, USA: CRC Press.

BRINKWORTH G D, BUCKLEY J D (2003), 'Concentrated bovine colostrum protein supplementation reduces the incidence of self-reported symptoms of upper respiratory tract infection in adult males', *European Journal of Nutrition*, 42(4), 228–232.

BUTLER J E (1998), 'Immunoglobulin diversity, B-cell and antibody repertoire development in large farm animals,' *Revue Scientifique et Technique*, 17(1), 43–70.

CÂMARA L M, CARBONARE S B, SILVA M L, CARNEIRO-SAMPAIO M M (1994), 'Inhibition of enteropathogenic Escherichia coli (EPEC) adhesion to HeLa cells by human colostrum: Detection of specific sIgA related to EPEC outer-membrane proteins', *International Archives of Allergy and Immunology*, 103(3), 307–310.

CAMPBELL B, PETERSEN W E (1963), 'Immune milk – a historical survey.' *Dairy Science Abstracts*, 25, 345–358.

CARBONARE S B, PALMEIRA P, SILVA M L M, CARNEIRO-SAMPAIO M M S (1997), 'Human colostrum IgA antibodies reacting to enteropathogenic Escherichia coli antigens and their persistence in the faeces of a breastfed infant', *Journal of Diarrhoeal Diseases Research*, 15, 53–58.

CARBONARE C B, CARBONARE S B, CARNEIRO-SAMPAIO M M (2005), 'Secretory immunoglobulin A obtained from pooled human colostrum and milk for oral passive immunization', *Pediatric Allergy and Immunology*, 16(7), 574–581.

CASSWALL T H, NILSSON H O, BJÖRCK L, SJÖSTEDT S, XU L, NORD C K, BORÉN T, WADSTRÖM T, HAMMARSTRÖM L (2002), 'Bovine anti-Helicobacter pylori antibodies for oral immunotherapy', *Scandinavian Journal of Gastroenterology*, 37(12), 1380–1385.

CORRÊA S, PALMEIRA P, CARNEIRO-SAMPAIO M M, SANAE NISHIMURA L, GUTH B E (2006), 'Human colostrum contains IgA antibodies reactive to colonization factors I and II of enterotoxigenic Escherichia coli. *FEMS Immunology and Medical Microbiology*, 47(2), 199–206.

CRAVIOTO A, TELLO A, VILLAFÁN H, RUIZ J, VEDOVO S, NEESER J R (1991), 'Inhibition of localized adhesion of enteropathogenic Escherichia coli to HEp-2 cells by immunoglobulin and oligosaccharide fractions of human colostrum and breast milk', *The Journal of Infectious Diseases*, 163, 1247–1255.

DOYLE P S, CRABB J, PETERSEN C (1993), 'Anti-Cryptosporidium parvum antibodies inhibit infectivity in vitro and in vivo,' *Infection and Immunity*, 61(10), 4079–4084.

EARNEST C P, JORDAN A N, SAFIR M, WEAVER E, CHURCH T S (2005), 'Cholesterol-lowering effects of bovine serum immunoglobulin in participants with mild hypercholesterolemia', *American Journal of Clinical Nutrition*, 81(4), 792–798.

EHRLICH P (1892), 'Über Immunität durch Verebung und Zeugung.' *Zeitschrift für Hygiene und Infektionskrankheiten*, 12, 183–203.

ELFSTRAND L, LINDMARK-MÅNSSON H, PAULSSON M, NYBERG L, ÅKESSON B (2002), 'Immunoglobulins, growth factors and growth hormone in bovine colostrums and the effects of processing.' *International Dairy Journal*, 12, 879–887.

FERNANDES R M, CARBONARE S B, CARNEIRO-SAMPAIO M M, TRABULSI L R (2001), 'Inhibition of enteroaggregative Escherichia coli adhesion to HEp-2 cells by secretory immunoglobulin A from human colostrum', *The Pediatric Infectious Disease Journal*, 20(7), 672–678.

FLORÉN C H, CHINENYE S, ELFSTRAND L, HAGMAN C, IHSE I (2006), 'ColoPlus, a new product based on bovine colostrum, alleviates HIV-associated diarrhoea', *Scandinavian Journal of Gastroenterology* 41(6), 682–686.

FUKUMOTO L R, LI-CHAN E, KWAN L, NAKAI S (1994), 'Isolation of immunoglobulins from cheese whey using ultrafiltration and immobilized metal affinity chromatography.' *Food Research International*, 27(4), 335–348.

FUNATOGAWA K, IDE T, KIRIKAE F, SARUTA K, NAKANO M, KIRIKAE T (2002), 'Use of immunoglobulin enriched bovine colostrum against oral challenge with enterohaemorrhagic Escherichia coli O157:H7 in mice', *Microbiology and Immunology*, 46(11), 761–766.

GINGERICH D A, MCPHILLIPS C A (2005), 'Analytical approach to determination of safety of milk ingredients from hyperimmunized cows', *Regulatory Toxicology and Pharmacology*, 41(2), 102–112.

HAMMARSTRÖM L, WEINER C K (2008), 'Targeted antibodies in dairy-based products', *Advances In Experimental Medicine And Biology*, 606, 321–343.

HEYERMANN H, BUTLER J E (1987), 'The heterogeneity of bovine IgG2-IV. Structural differencies between IgG2a molecules of the A1 and A2 allotypes', *Molecular Immunology*, 24, 1327–1334.

HOERR R A, BOSTWICK E F (2002), 'Commercializing colostrum based products: A case study of Galagen Inc.', *International Dairy Federation Bulletin*, 375, 33–46.

HORIE K, HORIE N, ABDOU A M, YANG J O, YUN S S, CHUN H N, PARK C K, KIM M, HATTA H (2004), 'Suppressive effect of functional drinking yogurt containing specific egg yolk immunoglobulin on Helicobacter pylori in humans', *Journal of Dairy Science*, 87(12), 4073–4079.

HUANG X H, CHEN L, GAO W, ZHANG W, CHEN S J, XU L B, ZHANG S Q (2008), 'Specific IgG activity of bovine immune milk against diarrhea bacteria and its protective effects on pathogen-infected intestinal damages' *Vaccine* 26, 5973–5980. DOI:10.1016/j.vaccine.2008.08.040

KELLY C P, CHETMAN S, KEATES S, BOSTWICK E, ROUSH A M, CASTAGLIOLO I, LAMONT J T, POTHOULAKIS C (1997), 'Survival of anti-*Clostridium difficile* bovine immunoglobulin concentrate in the human gastrointestinal tract', *Antimicrobial Agents and Chemotherapy*, 41, 236–241.

KELLY K S (2003), 'Bovine colostrums: A review of clinical uses', *Alternative Medicine Review*, 8(4), 378–394.

KOGA T, OHO T, SHIMAZAKI Y, NAKANO Y (2002), 'Immunization against dental caries', *Vaccine*, 20, 2027–2044.

KORHONEN H, SYVÄOJA E-L, AHOLA-LUTTILA H, SIVELÄ S, KOPOLA S, HUSU J, KOSUNEN T U (1995), 'Bactericidal effect of bovine normal and immune serum, colostrum and milk against Helicobacter pylori' *Journal of Applied Bacteriology*, 78(6), 655–662.

KORHONEN H, SYVÄOJA E-L, VASARA E, KOSUNEN T, MARNILA P (1998), 'Pharmaceutical composition, comprising complement proteins, for the treatment of *Helicobacter* infections and a method for the preparation of the composition.' PCT Patent Application No. WO98/00150.

KORHONEN H, MARNILA P, GILL H (2000a), 'Bovine milk antibodies for health: A review.' *British Journal of Nutrition*, 84, Suppl. 1, S1–S7.

KORHONEN H, MARNILA P, GILL H S (2000b), 'Milk immunoglobulins and complement factors.' *British Journal of Nutrition*, 84(Suppl. 1), S75-S80.

KORHONEN H (2002), 'Technology options for new nutritional concepts.' *International Journal of Dairy Technology*, 55, 79–88.

KORHONEN H (2004), 'Isolation of immunoglobulins from colostrum.' *International Dairy Federation Bulletin*, 389, 78–81.

KORHONEN H, MARNILA P (2006), 'Bovine milk antibodies for protection against microbial human diseases.' In: *Nutraceutical Proteins and Peptides in Health and Disease*. Eds. Mine Y, Shahidi F. Series: Nutraceutical Science and Technology 4. Series Editor Shahidi F. CRC Press, Taylor & Francis Group LLC, Boca Raton, USA, pp. 137–159.

KORHONEN H, PIHLANTO A (2007), 'Technological options for the production of health-promoting proteins and peptides derived from milk and colostrum.' *Current Pharmaceutical Design*, 13(8), 829–843.

KRISSANSEN G W (2007), 'Emerging health properties of whey proteins and their clinical implications.' *Journal of the American College of Nutrition*, 26(6), 713S–23S.

KUROIWA Y, KASINATHAN P, SATHIYASEELAN T, JIAO J A, MATSUSHITA H, SATHIYASEELAN J, WU H, MELLQUIST J, HAMMITT M, KOSTER J, KAMODA S, TACHIBANA K, ISHIDA I, ROBL J M (2009), 'Antigen-specific human polyclonal antibodies from hyperimmunized cattle', *National Biotechnology*, 2009 Jan 18. [Epub ahead of print]

LARSON B L (1992), 'Immunoglobulins of the mammary secretions.' In: Fox PF (ed.) *Advanced Dairy Chemistry 1: Proteins*, pp. 231–254. London: Elsevier Science Publishers.

LI-CHAN E, KUMMER A, LOSSO J N, KITS D D, MAKAI S (1995), 'Stability of bovine immunoglobulins to thermal treatment and processing.' *Food Research International*, 28(1), 9–16.

LILIUS E-M, MARNILA P (2001), 'The role of colostral antibodies in prevention of microbial infections', *Current Opinion in Infectious Diseases*, 14, 295–300.

LOIMARANTA V, TENOVUO J, VIRTANEN S, MARNILA P, SYVÄOJA E-L, TUPASELA T, KORHONEN H (1997), 'Generation of bovine immune colostrum against Streptococcus mutans and Streptococcus sobrinus and its effect on glucose uptake and extracellular polysaccharide formation by mutans streptococci.' *Vaccine*, 15(11), 1261–1268.

LOIMARANTA V, TENOVUO J, KORHONEN H (1998a), 'Combined inhibitory effect of bovine immune whey and peroxidase-generated hypothiocyanite against glucose uptake by Streptococcus mutans.' *Oral Microbiology and Immunology*, 13(6), 378–381.

LOIMARANTA V, CARLEN A, OLSSON J, TENOVUO J, SYVÄOJA E-L, KORHONEN H (1998b), 'Concentrated bovine colostral whey proteins from Streptococcus mutans/Strep. sobrinus immunized cows inhibit the adherence of Strep. mutans and promote the aggregation of mutans streptococci.' *Journal of Dairy Research*, 65(4), 599–607.

LOIMARANTA V, LAINE M, SÖDERLING E, VASARA E, ROKKA S, MARNILA P, KORHONEN H, TOSSAVAINEN O, TENOVUO J (1999), 'Effects of bovine immune and non-immune whey preparations on the composition and pH response of human dental plaque.' *European Journal of Oral Sciences*, 107(4), 244–250.

MAINER G, SÁNCHEZ L, ENA J M, CALVO M (1997), 'Kinetic and thermodynamic parameters for heat denaturation of bovine milk IgG, IgA and IgM.' *Journal of Food Science*, 62(5), 1034–1038.

MAINER G, DOMINGUEZ E, RANDRUP M, SANCHEZ L, CALVO M (1999), 'Effect of heat treatment on anti-rotavirus activity of bovine colostrum.' *Journal of Dairy Research*, 66, 131–137.

MARNILA P, KORHONEN H (2002), 'Immunoglobulins.' In: Eds. Roginski H, Fuquay J W, Fox P F. *Encyclopedia of Dairy Sciences*, pp. 1950–1956. Academic Press, London.

MARNILA P, ROKKA S, REHNBERG-LAIHO L, KÄRKKÄINEN P, KOSUNEN T U, RAUTELIN H, HÄNNINEN M-L, SYVÄOJA E-L, KORHONEN H (2003), 'Prevention and suppression of Helicobacter felis infection in mice using colostral preparation with specific antibodies,' *Helicobacter* 8(3), 192–201.

MATTILA E, ANTTILA V J, BROAS M, MARTTILA H, POUKKA P, KUUSISTO K, PUSA L, SAMMALKORPI K, DABEK J, KOIVUROVA O P, VÄHÄTALO M, MOILANEN V, WIDENIUS T (2008), 'A randomized, double-blind study comparing Clostridium difficile immune whey and metronidazole for recurrent Clostridium difficile-associated diarrhoea: Efficacy and safety data of a prematurely interrupted trial' *Scandinavian Journal of Infectious Diseases* 40(9), 702–708.

MCMARTIN S, GODDEN S, METZGER L, FEIRTAG J, BEY R, STABEL J, GOYAL S, FETROW J, WELLS S, CHESTER-JONES H (2006), 'Heat treatment of bovine colostrum. I: Effects of temperature on viscosity and immunoglobulin G level.' *Journal of Dairy Science*, 89(6), 2110–2118.

MEHRA R, KELLY P M (2004), 'Whey protein fractionation using cascade membrane filtration.' *International Dairy Federation Bulletin*, 389, 40–44.

MEHRA R, MARNILA P, KORHONEN H (2006), 'Milk immunoglobulins for health promotion: A review.' *International Dairy Journal*, 16(11), 1262–1271.

MITOMA M, OHO T, MICHIBATA N, OKANO K, NAKANO Y, FUKUYAMA M, KOGA T (2002), 'Passive immunization with bovine milk containing antibodies to a cell surface protein antigen-glucosyltransferase fusion protein protects rats against dental caries', *Infection and Immunity* 70(5), 2721–2724.

NÄSE L, HATAKKA K, SAVILAHTI E, SAXELIN M, PÖNKÄ A, POUSSA T, KORPELA R, MEURMAN J H (2001), 'Effect of long-term consumption of a probiotic bacterium, Lactobacillus rhamnosus GG, in milk on dental caries and caries risk in children.' *Caries Research*, 35(6), 412–420.

NEZLIN R (1998a), 'Chapter 1. General characteristics of immunoglobulin molecules.' In: *The Immunoglobulins*. Ed: Nezlin R. Academic Press: New York, USA.

NEZLIN R (1998b), 'Chapter 2. Animal and human immunoglobulins.' In: *The Immunoglobulins*. Ed: Nezlin R. Academic Press: New York, USA.

NIELSEN W K, OLANDER M A, LIHME A (2002), 'Expanding the frontiers in separation technology.' *Scandinavian Dairy Information*, Issue 2.

NUMAN S C, VELDKAMP P, KUIJPER E J, VAN DEN BERG R J, VAN DISSEL J T (2007), 'Clostridium difficile-associated diarrhoea: Bovine anti-Clostridium difficile whey protein to help aid the prevention of relapses', *Gut* 56(6), 888–889.

PACYNA J, SIWEK K, TERRY S J, ROBERTON E S, JOHNSON R B, DAVIDSON G P (2001), 'Survival of rotavirus antibody activity derived from bovine colostrum after passage through the human gastrointestinal tract', *Journal of Pediatric Gastroenterology and Nutrition* 32, 162–167.

PALMEIRA P, CARBONARE S B, SILVA M L, TRABULSI L R, CARNEIRO-SAMPAIO M M (2001), 'Inhibition of enteropathogenic Escherichia coli (EPEC) adherence to HEp-2 cells by bovine colostrum and milk', *Allergologia et Immunopathologia (Madr)*, 29(6), 229–237.

PIOT M, FAUQUANT J, MADEC M-N, MAUBOIS J-L (2004), 'Preparation of serocolostrum by membrane microfiltration.' *Lait*, 84, 333–341.

SCAMMELL A W (2001), 'Production and uses of colostrum.' *Australian Journal of Dairy Technology*, 56(2), 74–82.

SHIMAZAKI Y, MITOMA M, OHO T, NAKANO Y, YAMASHITA Y, OKANO K, NAKANO Y, FUKUYAMA M, FUJIHARA N, NADA Y, KOGA T (2001), 'Passive immunization with milk produced from an immunized cow prevents oral recolonization by *Streptococcus mutans*.' *Clinical and Diagnostic Laboratory Immunology*, 8, 1136–1139.

SMITH D J, KING W F, GODISKA R (2001), 'Passive transfer of immunoglobulin Y antibody to Streptococcus mutans glucan binding protein B can confer protection against experimental dental caries.' *Infection and Immunity*, 69(5), 3135–3142.

TAWFEEK H I, NAJIM N H, AL-MASHIKHI S (2003), 'Efficacy of an infant formula containing anti-*Escherichia coli* colostral antibodies from hyperimmunized cows in preventing diarrhea in infants and children: A field trial.' *International Journal of Infectious Diseases*, 7, 120–128.

TRIPATHI V, VASHISHTHA B (2006), 'Bioactive compounds of colostrum and its application.' *Food Review International*, 22, 225–244.

VAN DISSEL J T, DE GROOT N, HENSGENS C M, NUMAN S, KUIJPER E J, VELDKAMP P, VAN 'T WOUT J (2005), 'Bovine antibody-enriched whey to aid in the prevention of a relapse of *Clostridium difficile*-associated diarrhoea: Preclinical and preliminary clinical data', *Journal of Medical Microbiology*, 54(2), 197–205.

WEI H, LOIMARANTA V, TENOVUO J, ROKKA S, SYVÄOJA E-L, KORHONEN H, JOUTSJOKI V, MARNILA P (2002), 'Stability and activity of specific antibodies against *Streptococcus mutans* and *Streptococcus sobrinus* in bovine milk fermented with *Lactobacillus rhamnosus* strain GG or treated at ultra-high temperature', *Oral Microbiology and Immunology* 17, 9–15.

WEINER C, PAN Q, HURTIG M, BOREN T, BOSTWICK E, HAMMARSTRÖM L (1999), 'Passive immunity against human pathogens using bovine antibodies.' *Clinical and Experimental Immunology*, 116, 193–205.

WOOF J M, KERR M A (2006), 'The function of immunoglobulin A in immunity', *Journal of Pathology*, 208, 270–282.

XU L B, CHEN L, GAO W, DU K H (2006), 'Bovine immune colostrum against 17 strains of diarrhea bacteria and in vitro and in vivo effects of its specific IgG', *Vaccine* 24(12), 2131–2140. DOI:10.1016/j.vaccine.2005.11.009

YOUNG K W, MUNRO I C, TAYLOR S L, VELDKAMP P, VAN DISSEL J T (2007), 'The safety of whey protein concentrate derived from the milk of cows immunized against Clostridium difficile', *Regulatory Toxicology and Pharmacology* 47(3), 317–326.

11

Lactoferrin for human health

P. Marnila and H. J. Korhonen, MTT Agrifood Research Finland, Finland

Abstract: Lactoferrin (LF) is an iron-binding glycoprotein found in various biological fluids of mammals and in neutrophils. LF is considered to be an important host defense molecule and exhibits a diverse range of physiological functions such as antimicrobial, immunomodulatory and antioxidant activity. Experimental and clinical research carried out over the last thirty years has accumulated increasing evidence about the potential beneficial health effects of LF and its derivatives. These benefits may include anti-infective, anti-cancer and anti-inflammatory effects. The expanding knowledge about the multifunctional role of LF has enlarged its scope of applications from food preservation to health-promoting foods and supplements, pharmaceuticals, healthcare products, cosmetics and animal feeds. This chapter summarizes the current scientific knowledge about the functions and activities of LF, and highlights its present and possible future applications, also taking into consideration the emerging safety aspects.

Key words: lactoferrin, lactoferricin, antimicrobial, cancer prevention, inflammation.

11.1 Introduction

Lactoferrin (LF) is a multifunctional iron binding glycoprotein of the innate immune system. It is present in exocrine secretions such as milk, tears, nasal exudates, saliva, bronchial and cervico-vaginal mucus, seminal plasma and gastrointestinal fluid. LF is synthesized in various exocrine glands such as submucosal glands of conducting airways, epithelial cells in mammary gland and in neutrophils. LF is promptly delivered by circulating neutrophils to sites of microbial invasion. Thus, in the blood of septic patients, or in mastitic milk, the LF content increases many fold due to the high amount of activated and degranulating neutrophils, which is an early indication of host response to infection or invasion.

LF is composed of a single chain polypeptide sequence. Human LF (hLF) contains 691 and bovine LF (bLF) 689 amino acids, the degree of sequence homology being 69%. The molecular mass of hLF is between 77 and 82 kDa and that of bLF is 80–84 kDa, depending on the glycosylation degree. One LF molecule can bind two ferric (Fe^{3+}) ions, with the concomitant incorporation of bicarbonate ion (HCO_3^-). The affinity of LF to ferric ions is high (Kd~10^{-30}), which is about 300 times higher than in transferrin. In bovine milk, 20–30% of bLF is in iron saturated form (holo-LF) whereas in human milk 92–94% of hLF is in iron depleted form (apo-LF) and only 6–8% is holo-LF. A higher proportion of holo-LF gives to bovine milk Lf fraction the typical reddish color (Lönnerdal and Iyer, 1995; Pan *et al.*, 2006; Conesa *et al.*, 2008).

Although gene polymorphism occur in hLF, there is only one single functional LF gene located in chromosome 3 (Teng, 2002). The protein products expressed in human milk and neutrophils by LF gene are identical but there are some differences in glycosylation. Also, LF present in bovine milk and neutrophils appears to be a product of the same gene, possessing identical polypeptide chains. The differences between the milk and neutrophil LF are limited to the attached glycans. Indeed, bovine milk LF is heterogeneous, with two major glycosylation forms identified (Tsuji *et al.*, 1989). It is not known if these differences in glycosylation have an affect on protein functions, except that the fully glycosylated forms seem to be more resistant to intestinal proteases such as trypsin. The glycosylation degree does not, however, affect the iron binding affinity.

LF was first fractionated as an unknown ‘red fraction’ from cows’ milk by Sørensen and Sørensen in 1939, and the red protein from both human and bovine milk was later defined as a transferrin-like glycoprotein (Groves, 1960). Since then, a great amount of basic, applied and clinical research has been done with hLF, bLF and LF derived peptides. The well documented antimicrobial, anti-infective, cancer preventive and immune response regulating activities, as well as effects on iron metabolism, have led to the development of applications varying from clinical and nutraceutical uses to infant milk formulas, oral hygiene, cosmetic and various food supplements (Tomita *et al.*, 2002; Weinberg, 2007). Though LF is a molecule of our own immune system and is thus well tolerated, possible hazards have been documented and further research is needed to define the safe ways to exploit LF’s therapeutic properties (Weinberg, 2003).

11.2 Antimicrobial activities

11.2.1 Antibacterial effects and mechanisms

The first recognized antibacterial mechanism of LF was its ability to bind iron from medium, hence preventing bacterial growth. The majority of bacteria require iron and its availability correlates strongly with bacterial virulence. The LF found in human secretions and mucosal surfaces is almost entirely in iron free apo-form (Makino and Nishimura, 1992). In inflammatory sites containing a high amount of activated and degranulating neutrophils, LF scavenges iron effectively. Experimental studies have shown that iron sequestering by apo-LF can strongly inhibit the growth of many bacterial species (Weinberg, 2007). However, some bacterial pathogenic species counter this iron sequestering effect. Under conditions of iron deprivation, small iron-chelating molecules, siderophores are synthesized and secreted. Siderophores bind ferric ions with a high affinity and transport them into cells through specific receptors (Braun and Killmann, 1999; Andrews *et al.*, 2003). Some pathogens, such as *Neisseria, Haemophilus* and *Helicobacter* species, express specific outer cell membrane receptors capable of binding and removing iron directly from LF or transferrin (Andrews *et al.*, 2003). These bacteria may benefit from the host cellular inflammatory response. Bacteria capable of behaving as intracellular pathogens (e.g. *Listeria monocytogenes* and *Legionella pneumophila*), and multiplying in epithelial cells or in macrophages of the host, encounter problems in iron supply. Byrd and Horwitz (1991) showed that internalization of apo-LF into monocytes prevents the growth of *L. pneumophila* whereas holo-LF provide an iron source for the growth and inhibit the antimicrobial effect of interferon gamma (INF-γ).

Later *in vitro* studies have shown that bLF and its derivative peptide lactoferricin (LFcin) are able to induce apoptosis, programmed cell death, in *L. monocytogenes* infected THP-1 cells (Longhi *et al.*, 2005) and in Caco-2 cells (Valenti *et al.*, 1999). The activation of the apoptotic cell death accompanied with neutrophil chemoattractant release induced by *L. monocytogenes* infection of hepatocytes is a critical step in launching an innate immune defense action against this intracellular pathogen.

The addition of apo-LF to many other antimicrobial agents enhance their antimicrobial activity, since the efficacy of the majority of antimicrobial compounds is inhibited by iron. The inhibitory concentrations of vancomycin, nafcillin, amphotericin B, fluconazole, chloramphenicol, doxycycline and rifampicin are lowered *in vitro* in the presence of LF (Weinberg, 2003; Venkatesh and Rong, 2008). LF acts synergistically also with lysozyme. Both proteins are present in high concentrations in neutrophil granules and at sites of inflammation. Lysozyme hydrolyses beta-1,4 linkages between N-acetylmuramic acid and 2-acetyl-amino-2-deoxy-D-glucose residues in peptidoglycan of the bacterial cell wall. However, lysozyme alone is not considered to be very effective against Gram-negative bacteria due to the lipopolysaccharide (LPS) layer in its outer membrane. LF can bind to the LPS layer to enable lysozyme to gain access to the peptidoglycan in the bacterial cell wall. Combinations of LF and lysozyme are far more bactericidal than these factors alone. This action is inhibited if hLF is iron saturated (Ellison and Giehl, 1991). Similarly, LF and lysozyme are synergistic against Gram-positive bacteria but the efficacy is suppressed by saturating LF with iron (Leitch and Willcox, 1998).

In many cases, bacteria live in biofilms which are matrix-encased communities specialized for surface persistence. The bacterial transition from a free-living, independent life to a biofilm lifestyle can be devastating for the host, because biofilms notoriously resist killing by the host defence mechanisms and antibiotics. LF is able to block biofilm development of the opportunistic pathogen *Pseudomonas aeruginosa* at concentrations below those that kill or prevent growth. By chelating iron, LF stimulates surface motility, causing the bacteria to wander across the surface instead of forming cell clusters and biofilms (Singh *et al.*, 2002).

The proteolysis of LF by gastric pepsin generates an antimicrobial peptide, LFcin from the N-terminal region. Bovine LFcin is a 25 amino acid peptide consisting of residues 17–41, whereas the human LFc (hLFcin) represents a 47 amino acid peptide consisting of residues 1–47. These peptides are 2 to 12-fold more active against *Escherichia coli* compared with the corresponding undigested LF molecule. Bovine LFcin is a more potent bactericidal peptide than hLFcin and has antimicrobial activity against a wide range of microbes (for a review, see Wakabayashi *et al.*, 2003). The mechanisms of the cidal effects against Gram-positive bacteria are probably similar to those of other cationic and amphipathic antibacterial peptides. LFcin binds to Gram-positive bacteria through electrostatic interactions between the negatively charged cell membrane and cationic amino acid residues. Then amphipathic residues perturb the non-polar membrane interior by hydrophobic interactions. Bovine LFcin exerts slow inhibitory and bactericidal activity, causing depolarization of the cytoplasmic membrane, but does not lyse the bacteria and leaves the integrity of the membrane intact, indicating a possible intracellular target. Thus, it is probable that LFcin has multiple targets (Ulvatne *et al.*, 2004). The permeabilizing effect on the cytoplasmic membrane may disrupt the general energy metabolism in the cell, resulting in the inhibition of macromolecular biosynthesis. By means of incorporating radioactive precursors into DNA, RNA and proteins, Ulvatne *et al.* (2004) studied effects of bLFcin on the macromolecular synthesis in the Gram-negative *E. coli* and in the Gram-positive *Bacillus subtilis*. At sub-lethal concentrations, bLFcin inhibited the biosynthesis of macromolecules in both bacteria. On the other hand, a breakdown of macromolecular biosynthesis may also lead to permeabilization of the cytoplasmic membrane.

Recently, a new protease-like antimicrobial activity towards some bacterial virulence factors has been found in LF. For example, hLF degrades IgA1-protease and Hap-adhesin in *Haemophilus influenzae* (Plaut *et al.*, 2000), and degrades also invasion plasmid antigens of *Shigella flexneri* (Gomez *et al.*, 2003). Furthermore, hLF has been shown to decrease invasiveness of enteropathogenic *E. coli* by blocking the proteins that contribute to adherence and cause hemolysis (Ochoa *et al.*, 2003 and 2004).

11.2.2 Antiviral activities

LF inhibits the replication of a wide spectrum of RNA and DNA viruses. The list includes several enveloped viruses such as herpes simplex viruses, hepatitis B and C viruses, human cytomegalovirus, hantavirus, HIV and respiratory syncytial virus etc; also some naked viruses, e.g. rotavirus, poliovirus, adenovirus, human papillomavirus and enterovirus 71. The antiviral activity does not involve iron binding. LF is considered to inhibit viral replication by preventing the viral particles from infecting new host cells, either by binding to host cell receptors or to the virions. Bovine LF has been shown to be more active against viruses than hLF, and apo-LF is less effective than holo-LF (Valenti and Antonini, 2005). The observations that LF is strongly antiviral when present at the time of infection but far less active if given later, suggests that antiviral therapy with LF may, in many cases, not be practicable (Weinberg, 2003). However, in some viral infections the immune system regulating effect may be beneficial. Shin *et al.*, (2005) demonstrated in a mouse model that orally administered bLF did not exert any effect on influenza virus replication, but the amount of inflammatory macrophages and neutrophils in the lungs was reduced significantly. Thus, orally administered bLF suppressed substantially the strong inflammatory reaction of the murine immune system against influenza virus infection and therefore attenuated the severity of pneumonia.

11.2.3 Antifungal activities

Bovine LF, hLF as well as bLFcin have a well documented cidal activity *in vitro* against human pathogenic fungi, especially *Candida* species, including *C. albicans*. The cidal effect is related to the adsorption of LF to the cell surface rather than to iron deprivation of this fungal species. The candidacidal effect of hLF is reported to be dependent on the extracellular cation concentration and on the metabolic state of the cell (Viejo-Diaz *et al.*, 2004). Recent clinical and animal model studies suggest that the main antimycotic activity of bLF derived peptides is dependent on the N-terminal half of bLF and that bLF and bLFcin exert their antifungal effects by enhancing the host immune defense, especially upregulating the neutrophil function rather than by a direct antimycotic effect (Ueta *et al.*, 2001).

Yamauchi *et al.* (2000a) studied the efficacy of orally administered bLF (0.6, or 2 g, or placebo, daily for eight weeks) in the treatment of adults with mild or moderate tinea pedis. The treatment with bLF resulted in significantly decreased symptom scores in subjects with moderate vesicular or interdigital tinea pedis. Costantino and Guaraldi (2008) tested the clinical efficacy of a cream containing 4 % of LF in the treatment of an acute vulvovaginal candidiosis. The treatment with 5 g of cream in the inflamed area twice a day for seven days cured completely 27 of 34 patients, five showed good improvement and two patients suffered from a vaginitis after the treatment.

Human subjects with a dry mouth due to hyposalivation have a lowered level of salivary hLF. The occurrence of a *Candida* invasion in the mouth is associated with the dry mouth symptoms. Some patients with an advanced stage of AIDS also have lowered salivary LF levels and have repeated periods of oral candidiosis. Oral formulations containing LF and lysozyme have been found effective in the treatment of oral candidiosis, even in patients who fail to respond to the conventional antifungal drug therapy (Masci, 2000). Thus, the use of LF as an antifungal agent alone or in combination with antifungal drugs opens promising prospects for the therapy of opportunistic fungal infections.

11.3 Cancer prevention and immunological effects

In all mammalian species, LF and peptides derived from it are an essential part of innate immunity participating in the defense against neoplasia. Bovine LFcin is, *in vitro*, cytotoxic for a variety of human and rodent cancer cells. Relatively little is known about the mechanisms by which LF exerts its anticancer activity. Human holo-LF prevents the growth of breast carcinoma cells by inhibiting G1 cyclin dependent kinases (Damiens *et al.*, 1999). The exposure of murine Meth A fibrosarcoma cells to bLFcin leads to the loss of membrane integrity and to lysis *in vitro* (Eliassen *et al.*, 2002). Bovine LFcin is, *in vitro*, a potent inducer of apoptosis in the human monocytic leukaemia cell line THP-1, but the intact bLF does not trigger apoptosis even at higher concentrations (Yoo *et al.*, 1997). Mader *et al.* (2005) reported that bLFcin induces a rapid apoptosis in several different human leukaemia and carcinoma cell lines, including the Jurkat T leukemic T cell line. In contrast, bLFcin was not cytotoxic to primary cultures of resting or mitogen activated normal human T lymphocytes, to fibroblasts, or to endothelial cells. In Jurkat T cells, the bLFcin treatment triggered the production of reactive oxygen species (ROS) followed by a caspase-2 induced dissipation in the mitochondrial transmembrane potential and a subsequent activation of the caspase-9 and -3 mediated apoptosis. The authors conclude that bLFcin kills cancer cells *in vitro* by triggering the mitochondrial pathway of apoptosis, at least in part through generation of ROS (Mader *et al.*, 2005).

In animal model studies, bLF and bLFcin have exhibited a potent activity against cancer cells *in vivo*. Orally administered bLF reduced the development of neoplastic cells in the colon, oesophagus, and lungs of rats treated with carcinogens (Ushida *et al.*, 1999). Black tea polyphenols (Polyphenon-B) and bLF had a synergistic effect in the prevention of chemically induced buccal pouch carcinogenesis in a hamster model (Mohan *et al.*, 2008). Although the dietary bLF and Polyphenon-B alone significantly reduced the tumour incidence, a combined administration of these compounds was more effective in inhibiting the chemically-induced genotoxicity and development of carcinomas. Recently, patients with progressive advanced solid tumours, and those whose conventional therapy had failed, were given orally recombinant hLF (talactoferrin) in doses varying from 1.5 g to 9 g daily, using a two weeks on/two weeks off therapy. Among eight patients who could be evaluated radiologically, the tumour growth rate had decreased in seven cases (Hayes *et al.*, 2006).

Yoo *et al.* (1997) demonstrated that bLF suppresses tumour growth and metastasis in a mouse model and may, moreover, inhibit angiogenesis. The growth of new capillaries is essential for the progression and metastasis of solid tumours. Recent results suggest that the regulation of angiogenesis depends on the balance between stimulators and inhibitors. Cytokines such as INF-γ inducing protein 10, platelet factor 4, interleukin- (IL-) 4, IL-12 and IL-18 are reported to suppress tumour growth or metastasis due to their angiogenesis inhibiting effects. Kuhara *et al.* (2000) reported that the orally administered bLF induced INF-γ and IL-18 in the mouse intestinal epithelium and that this phenomenon might be connected to the anti-tumour activity of bLF. Soon after, Norrby *et al.* (2001) found that bLF inhibited a chemically-induced angiogenesis in rats. Using the dorsal air sacs assay, Shimamura *et al.* (2004) studied the influence of a bLF administration to mice on tumour-induced angiogenesis, and endothelial cell functions and angiogenesis related cytokine production. Both the orally and intraperitoneally given bLF suppressed, dose-dependently, the Lewis lung carcinoma (3LL) cell-induced angiogenesis. In the serum of mice given bLF orally, the concentration of IL-18, but not that of INF-γ, was markedly elevated. The authors concluded that bLF participates in the regulation of angiogenesis, possibly by blocking the endothelial function and inducing IL-18 production. The bLF antitumour effect *in vivo* may, therefore, be partly mediated by the inhibition of angiogenesis. LF has also been demonstrated to activate natural killer (NK) cells, polymorphonuclear cells and lymphokine activated killer cells which contribute to immune reactions against neoplasia (Kruzel *et al.*, 2007).

Hirotani *et al.* (2008) demonstrated in a Caco-2 cell model that bLF (0.4–1 mg/ml) attenuated rupturing and permeability changes induced by LPS from *E. coli* in the epithelial cell lining. In a mouse model, an intraperitoneal injection of bLF (5 mg per mouse) given 24 h before an intravenous injection of a sublethal *E. coli* LPS dose reduced tumour necrosis factor-alpha (TNF-α) serum levels and increased anti-inflammatory cytokine levels (Kruzel *et al.*, 2002). Orally-administered bLF is not completely degraded in the gastrointestinal tract, but it is retained to some degree in the form of bLFcin-containing peptides (Kuwata *et al.*, 2001). It is not likely that the orally administered bLF or peptides derived thereof are transported from the intestine to the blood circulation (Wakabayashi *et al.*, 2004).

Receptors for LF have been found in several cell types, including T and B lymphocytes, platelets and intestinal epithelial cells (Suzuki *et al.*, 2005) and it is likely that orally-administered bLF or its digestion products act on the intestinal immune system and then promote systemic effects. In the gastrointestinal tissue of a mouse, an orally-administered dose of bLF has been shown to trigger the production of IL-18 in epithelial cells, and IL-10 and INF-γ in lymphocytes; it also increases the number of CD4+ and CD8+ cells and NK cells in the intestinal mucosa. The systemic effects of bLF given orally to mice include increased cell numbers in lymph nodes and spleen, enhanced activity of peritoneal macrophages and NK cells and an enhanced production of Th-1 type cytokines IL-12 and INF-γ (Zimecki and Kruzel, 2007; Tomita *et al.*, 2008). In healthy human volunteers, a treatment with oral bLF of 50 mg/day increased significantly the proportion of neutrophilic precursors in the peripheral blood and reduced the spontaneous production of IL-6 and TNF-α by cultured peripheral blood mononuclear cells and enhanced chemotactic reactions by promoting the recruitment of leukocytes to the inflammatory site (Zimecki *et al.*, 1999). Spadaro *et al.* (2008) reported recently that human recombinant LF induced the maturation of dendritic cells from human monocytes *in vitro*. The authors concluded that a key immunomodulative function of LF may be to link the innate immune response to an adaptive one by promoting dendritic cell maturation. It is postulated that LF can control the cytokine-induced proinflammatory cascade during the development of a systemic inflammatory response (Kruzel *et al.*, 2007). The activation of the monocyte/macrophage system results in the production of IL-1β, IL-6 and TNF-α which in turn enhances the inflammatory reaction. Activated neutrophils release LF by degranulating at the site of inflammation. The released LF binds to specific receptors on monocytes and capillary endothelia and attenuates the priming reaction and onset of production of proinflammatory cytokines via the nuclear factor kappa B (NF-κB) and by stimulating the production of anti-inflammatory cytokines IL-4 and IL-10. LF also scavenges non-protein-bound iron in the inflamed area so as to suppress damage on host tissues due to ROS produced by neutrophil oxidative burst reaction. Further, in the case of an infection by Gram-negative bacteria, LF binds to LPS and inhibits the progressive inflammatory cascade caused by cell wall particles of destructed bacteria (Kruzel *et al.*, 2007).

Another anti-inflammatory role of LF is its capability to inhibit the classical pathway of complement reaction by blocking the formation of the C3 convertase, thus preventing the cleavage of C3. This inhibition is exerted by peptides derived from the N-terminal region of both hLF and bLF, and also from both hLFcin and bLFcin. No inhibitory effect by these peptides was observed on the alternative complement pathway in an AP50 assay (Samuelsen, 2004).

It seems that, in many cases, an inflammatory reaction by the host's immune system causes more harm than the invading pathogen or trauma that started the cascade. LF appears to pose an essential part of the regulatory feedback loop in the inflammatory process.

11.4 Other biological activities

Recently, oral bLF was reported to increase the haemoglobin and total iron concentration in human blood serum. Paesano *et al.*, (2006) administered pregnant women orally with 100 mg of 30% iron saturated bLF twice a day or ferrous sulphate 520 mg per day, respectively. After 30 days, treatment, both haemoglobin and serum iron were significantly higher in the bLF group than in the ferrous sulphate group or in the control group (haemoglobin 12.8 g/dL, 11.9 g/dL and 10.2 g/dL, respectively). Unlike ferrous sulphate, bLF did not result in any side effects. Koikawa *et al.* (2008) reported that bLF tablets taken (1.8 g of bLF/day, 6 mg of iron/day) daily for eight weeks prevented the lowering of haemoglobin level in female long-distance runners during the training period, whereas the same dosage of iron as ferric pyrophosphate did not prevent anaemia due to training. In another study with young US women, Lönnerdal and Bryant (2006) concluded that iron was equally well absorbed from purified recombinant hLF produced in rice as from ferrous sulphate. Thus, oral LF can exert influence on iron homeostasis directly or through other proteins of intestinal cells.

An acute ultraviolet (UV) exposure causes photokeratitis, and induces apoptosis in corneal cells of the eye. LF supplementation has been shown to suppress UV-B-induced oxidation in cultures of human corneal epithelial cells. Fujihara *et al.* (2000) investigated the protective effect of LF containing eyedrops against UV-B-induced corneal damage in a rat keratitis model by irradiating anesthetized rats and then scoring the corneal epithelial defect after 24 h from irradiation. The vehicle or LF containing eyedrops were applied either before or after the irradiation. The post-treatment with LF did not inhibit the extent of corneal damage and did not affect wound healing. However, the pre-treatment by topical application of LF suppressed the development of a corneal epithelial defect induced by UV-B irradiation. Therefore, the presence of LF in the human tear fluid may inhibit UV-induced corneal epithelial damage due to sunlight.

Cornish (2004) demonstrated that both bLF and hLF are anabolic factors for the bone *in vivo* in a mouse model. Local injections of bLF into adult mice resulted in an increased bone growth. In *in vitro* studies, bLF stimulated the proliferation and differentiation of osteoblasts and the apoptosis was inhibited at a bLF concentration of 100 μg/mL which level is within the limits of physiological concentrations. Also, the chondrocyte proliferation was stimulated (10 and 20 μg/mL of bLF) and the development of osteoclasts was inhibited at the bLF concentration of 10 μg/mL and arrested completely at 100 μg/mL. The osteogenic activity of bLF is maintained *in vitro* in deglycosylated, apo- and holo-forms and it can be detected in various fragments of bLF, suggesting that the regulatory activity is mediated via several distinct receptors and pathways (Cornish *et al.*, 2006).

11.5 Applications and safety aspects

Over the last thirty years, many methods have been developed for the isolation and purification of bLF from colostral or cheese whey. Most of these methods are based on gel filtration and chromatographic techniques (Law and Reiter, 1977; Tamura, 2004). Patented, large-scale production techniques based on cation-exchange resins followed by gel filtration or UF have been developed to isolate both LF and lactoperoxidase (LP) from cheese whey (Uchida *et al.*, 1996). A microporous membrane containing immobilized sulphonic acid has been applied to fractionate LF and LP from cheese whey (Chiu and Etzel, 1997): in this method, the average recovery rates were 73% for LP and 50% for LF. Zhang *et al.* (2002) developed a simple, rapid method for large-scale fractionation of LF from bovine colostrum based on high-performance liquid chromatography (HPLC), coupled with an SP-Sepharose cation exchanger. Other methods developed for the isolation of LF, in particular, include monoclonal antibody immunoaffinity, copper ions immobilized on Sepharose 6B (metal chelate interaction chromatography) and affinity chromatography using immobilized ferritin, heparin, DNA or textile dyes (Mulvihill and Ennis, 2003).

Industrial large-scale production of bovine LF began in Belgium in 1985. Commercial exploitation of bLF has, however, developed relatively slowly until recent years, perhaps reflecting the lack of break-through results from scientific research. On the other hand, Asian companies have applied bLF already for more than 20 years as an ingredient of commercial foodstuffs, most of them being milk-based products (Tomita *et al.*, 2008). In 1986, marketing of a bLF-containing infant formula, *BF-L Dry Milk*, was started by Morinaga Milk Industry in Japan. Currently, bLF-containing infant formulas are sold in Indonesia and Korea, as well as in Japan. Over the years, scientific evidence on the benefits with supplementation with bLF in infant formulas has been accumulating. Roberts *et al.*, (1992) reported that an infant formula containing 1 mg/ml of bLF established a *Bifidobacterium*-dominant faecal flora in normal infants, whereas infant formulas containing none or 0.1 mg/ml of bLF did not have a similar effect. In an intervention study with weaned Peruvian children, the test group (12–36 months of age, n = 146), which received orally 0.5 g of bLF daily for nine months, was demonstrated to have a significantly reduced rate of faecal colonization by *Giardia* species as compared to the control group (n = 174), which received maltodextrin as placebo (Ochoa *et al.*, 2008).

Other bLF-containing products include yoghurt, skim milk, milk-type drinks, supplemental tablets, pet food, oral care products (mouth wash, gel, toothpaste and chewing gum) and cosmetics which are mainly skin-care products (Wakabayashi *et al.*, 2006). The beneficial effects of these bLF-containing products on the health have been proved in clinical and animal studies. The favourable effect of yoghurt on rotavirus gastroenteritis and the similar effect of tablets on chronic hepatitis C, rotaviral gastroenteritis, and *Helicobacter pylori* gastric infection have been reported. The therapeutic effect of pet food on dermatitis in dogs and cats was also shown (see review by Tomita *et al.*, 2008).

The antimicrobial and antioxidative effects of bLF favour its use as a food preservative. The US Food and Drug Administration (FDA) has approved the use of bLF (at not more than 2% by weight) as a spray to reduce microbial contamination on the surface of raw beef carcasses. FDA has granted to bLF a 'Generally Recognized As Safe' (GRAS, GRN 67) status and this determination accounts for uses at defined levels in beef carcasses, sub-primals, and finished cuts (Taylor *et al.*, 2004). Another potential application area for bLF is the use as a component of edible coatings. Brown *et al.* (2008) tested the antimicrobial efficacy of bLF, alone and with lysozyme, against *E. coli* O157:H7 and *L. monocytogenes* in chitosan films. The incorporation of bLF alone into a chitosan film did not exhibit significant antimicrobial activity against these bacteria but the combination of bLF with lysozyme in the coating significantly decreased the growth of *E. coli* O157:H7.

Many microbes isolated from mastitic milk are sensitive to bLF or bLFcin, whereas many pathogenic strains of *E. coli* and *S. aureus* show resistance or can develop resistance to LF. However, bLF possesses a synergistic effect with penicillin against mastitic *S. aureus* (Diarra *et al.*, 2002). In an *in vitro* study, bLF increased the inhibitory activity of penicillin up to 4-fold in most penicillin susceptible mastitic *S. aureus* strains, whereas this increase was 4- to 16-fold in penicillin resistant strains (Lacasse *et al.*, 2008). Komine *et al.* (2006) reported that a combination of bLF and antibiotics were more effective than antibiotic or bLF treatments alone against clinical *S. aureus* mastitis in drying cows and the combination treatment also inhibited the production of TNF-α via the inhibition of NF-κB activation. In a study by Petitclerc *et al.* (2007), an experimental *S. aureus* mastitis was induced in 19 lactating cows. An intramammary treatment of cows with combinations of bLF and penicillin G resulted in 3–5 times higher cure rates than penicillin G treatment alone in cows with stable *S. aureus* infections showing high resistance to beta lactam antibiotics. These results would suggest that bLF added to penicillin may offer an effective combination for the treatment of chronic infections caused by beta lactam antibiotic resistant *S. aureus* strains.

Recently, a topical administration of talactoferrin (recombinant hLF expressed and produced in *Aspergillus niger*) was reported to promote wound healing in an experimental model with normal and diabetic mice (Engelmayer *et al.*, 2008). Talactoferrin promoted the closure rate in the period of 12 to 19 days after cutting the wounds, as well as in the 75% closure incidence and the time to 50% closure versus the control group treated with the vehicle. The improvement in wound healing was small but statistically significant. The wounds in talactoferrin-treated normal mice had also significantly higher levels of TNF-α, IL-6 and IL-8 than in mice treated with the vehicle at an early stage of the repair process, suggesting that the wound-healing promoting effect is associated with a regulated enhancement of the inflammatory phase of the wound repair (Engelmayer *et al.*, 2008). In diabetic patients suffering from neuropathic ulcers in feet, the topical application of talactoferrin-containing gel (1%, 2.5% or 8.5%) twice daily for 30 days was tolerated without any side effects.

A further topical treatment twice per day for twelve weeks with the gels containing 2.5% or 8.5% of talactoferrin promoted wound healing of the ulcers. The study groups receiving the 2.5% and 8.5% gels had about twice the incidence of ≥75% reduction in ulcer size as compared with the placebo group (47%, 53%, and 25%, respectively) (Lyons *et al.*, 2007). It may be concluded from these results that the recombinant hLF treatment appears to enhance the healing rate in the ulcers, which are a common pathological complication requiring hospitalization of diabetic people.

Bovine LF and recombinant hLF are now commercially available for development into preservatives, nutraceuticals and pharmaceutical products. Current research is focused on bone remodelling, prevention of infections, prevention of neoplasia and treatment of cancers, treatment of inflammations in intestine and joints, and promotion of wound healing as well as enhancement of antimicrobial and anti-neoplastic drugs, and prevention of iron induced oxidation of food products (Wakabayashi *et al.*, 2006, Tomita *et al.*, 2008). A safety evaluation of these potential application areas of LF are of critical importance before such products can be launched on the market.

Cerven *et al.* (2008a and 2008b) studied the safety of oral administration of purified recombinant human apo-LF and holo-LF, expressed and produced in rice grain, for 28 days in Wistar rats using 180 and 1800 mg/kg of body weight daily. No treatment-related changes in clinical parameters were observed in any of these preparations. Toxicological studies in rats for 13 weeks with a daily oral administration of bLF (Yamauchi *et al.*, 2000b) and of recombinant hLF produced in transgenic cows (Appel *et al.*, 2006) revealed that the levels of no observed adverse effect were at least 2 g/kg of body weight/day. Tamano (2008) studied the occurrence of potential toxic effects in rats fed a basal diet containing 0.2 % bLF for 40 weeks. No adverse findings were observed, and the serum triglyceride level was significantly decreased (72% of the control level), suggesting preventive effects against the metabolic syndrome. In a further experiment, male and female F344/DuCrj rats were fed a basal diet containing 0.02, 0.2, 2.0 and 5.0% bLF, 2.0% bLF hydrolysate or 0.1% LFcin for 60 weeks in males and 65 weeks in females. No toxicological effects, including carcinogenicity, were evident in either sex. The results of these studies can be regarded as supportive of the safety of clinical studies with bLF for use in food products and supplements (Tamano, 2008). Talactoferrin has received a fast-track designation for the treatment of non-small cell lung cancer and is entering Phase III clinical trials (Spadaro *et al.*, 2008). Generally, talactoferrin is very well tolerated and no toxicities have been reported.

Many bacteria, e.g. many *E. coli* and *S. aureus* strains, possess some degree of resistance against bLF or bLFcin. This resistance can be due to proteases on cell outer membranes or it may be due to positively charged groups on cell surfaces preventing the incorporation of cationic sequences of LF or LFcin (Ulvatne *et al.*, 2001 and 2002). Samuelsen *et al.* (2005) investigated inducible intrinsic resistance of *S. aureus* strains against bLFcin. A serial passage of seven *S. aureus* strains in medium with increasing concentrations of bLFcin resulted in an induced resistance at various levels in all strains. The resistance decreased rapidly during passages without bLFcin, but the minimum inhibitory concentrations remained elevated after 30 passages. Further, a cross resistance to penicillin G was observed. Thus, a widespread use of LF or LFcin in, for example, the food industry may lead to development of resistant pathogens. The risks due to possibly emerging LF-resistant microbes may be even higher than those of antibiotics because, in the case of LF or LFcin, potential pathogens are in practice educated to resist the molecules of our own innate immunity. Another type of risk may emerge owing to LF's ability to stimulate the replication of human T cell leukemia virus Type 1. Also, orally-ingested LF may be harmful to humans with infections by gastrointestinal bacteria that can use iron from LF (Weinberg, 2003). Immunological effects of orally administered LF are diverse and this far only partially known. Therefore, further experimental research concomitant with relevant safety evaluation appears necessary in those particular application areas.

11.6 References

ANDREWS S C, ROBINSON A K, RODRÍGUEZ-QUIÑONES F (2003), 'Bacterial iron homeostasis', *FEMS Microbiol Rev*, 27(2–3), 215–37. DOI: 10.1016/S0168-6445(03)00055-X

APPEL M J, VAN VEEN H A, VIETSCH H, SALAHEDDINE M, NUIJENS J H, ZIERE B, DE LOOS F (2006), 'Sub-chronic (13-week) oral toxicity study in rats with recombinant human lactoferrin produced in the milk of transgenic cows', *Food Chem Toxicol*, 44(7), 964–73. DOI: 10.1016/j.fct.2005.11.012

BRAUN V, KILLMANN H (1999), 'Bacterial solutions to the iron-supply problem', *Trends Biochem Sci*, 24(3), 104–9. DOI: 10.1016/S0968-0004(99)01359-6

BROWN C A, WANG B, OH J H (2008), 'Antimicrobial activity of lactoferrin against foodborne pathogenic bacteria incorporated into edible chitosan film', *J Food Prot*, 71(2), 319–24.

BYRD T F, HORWITZ M A (1991), 'Lactoferrin inhibits or promotes Legionella pneumophila intracellular multiplication in nonactivated and interferon gamma-activated human monocytes depending upon its degree of iron saturation. Iron-lactoferrin and nonphysiologic iron chelates reverse monocyte activation against Legionella pneumophila', *J Clin Invest*, 88(4), 1103–12. DOI: 10.1172/JCI115409.

CERVEN D, DEGEORGE G, BETHELL D (2008a), '28-Day repeated dose oral toxicity of recombinant human apo-lactoferrin or recombinant human lysozyme in rats', *Regul Toxicol Pharmacol*, 51(2), 162–7. DOI: 10.1016/j.yrtph.2008.03.007

CERVEN D, DEGEORGE G, BETHELL D (2008b), '28-Day repeated dose oral toxicity of recombinant human holo-lactoferrin in rats', *Regul Toxicol Pharmacol*, [Epub ahead of print] DOI: 10.1016/j.yrtph.2008.07.003

CHIU C K, ETZEL M R (1997), 'Fractionation of lactoperoxidase and lactoferrin from bovine whey using a cation exchange membrane'. *J Food Sci*, 62, 996–1000.

CONESA C, SÁNCHEZ L, ROTA C, PÉREZ M-D, CALVO M, FARNAUD S, EVANS R W (2008), 'Isolation of lactoferrin from milk of different species: Calorimetric and antimicrobial studies', *Comp. Biochem Physiol. B*, 150, 131–9. DOI: 10.1016/j.cbpb.2008.02.005

CORNISH J. (2004), 'Lactoferrin promotes bone growth', *Biometals*, 17(3), 331–5. DOI: 10.1023/B:BIOM.0000027713.18694.91

CORNISH J, PALMANO K, CALLON K E, WATSON M, LIN J M, VALENTI P, NAOT D, GREY A B, REID I R (2006), 'Lactoferrin and bone: Structure–activity relationships', *Biochem Cell Biol*, 84(3), 297–302. DOI: 10.1139/O06-057

CONSTANTINO D, GUARALDI C (2008), 'Preliminary evaluation of a vaginal cream containing lactoferrin in the treatment of vulvovaginal candidosis', *Minerva Ginecol*, 60(2), 121–5.

DAMIENS E, EL YAZIDI I, MAZURIER J, DUTHILLE I, SPIK G, BOILLY-MARER Y (1999), 'Lactoferrin inhibits G1 cyclin-dependent kinases during growth arrest of human breast carcinoma cells', *J Cell Biochem*, 74(3), 486–98.

DIARRA M S, PETITCLERC D, LACASSE P (2002), 'Effect of lactoferrin in combination with penicillin on the morphology and the physiology of Staphylococcus aureus isolated from bovine mastitis', *J Dairy Sci*, 85(5), 1141–9.

ELIASSEN L T, BERGE G, SVEINBJØRNSSON B, SVENDSEN J S, VORLAND L H, REKDAL Ø (2002), 'Evidence for a direct antitumor mechanism of action of bovine lactoferricin', *Anticancer Res*, 22(5), 2703–10.

ELLISON R T, GIEHL T J (1991), 'Killing of gram-negative bacteria by lactoferrin and lysozyme', *J Clin Invest*, 88(4), 1080–91. DOI: 10.1172/JCI115407

ENGELMAYER J, BLEZINGER P, VARADHACHARY A (2008), 'Talactoferrin stimulates wound healing with modulation of inflammation', *J Surg Res*, 149(2), 278–86. DOI: 10.1016/j.jss.2007.12.754

FUJIHARA T, NAGANO T, ENDO K, NAKAMURA M, NAKATA K (2000), 'Lactoferrin protects against UV-B irradiation-induced corneal epithelial damage in rats', *Cornea*, 19(2), 207–11.

GOMEZ H F, OCHOA T J, CARLIN L G, CLEARY T G (2003), 'Human lactoferrin impairs virulence of Shigella flexneri', *J Infect Dis*, 187(1), 87–95. DOI: 10.1086/345875

GROVES M L (1960), 'The isolation of a red protein from milk', *J Am Chem Soc*, 82, 3345–50.

HAYES T G, FALCHOOK G F, VARADHACHARY G R, SMITH D P, DAVIS L D, DHINGRA H M, HAYES B P, VARADHACHARY A (2006), 'Phase I trial of oral talactoferrin alfa in refractory solid tumors', *Invest New Drugs*, 24(3), 233–40. DOI: 10.1007/s10637-005-3690-6

HIROTANI Y, IKEDA K, KATO R, MYOTOKU M, UMEDA T, IJIRI Y, TANAKA K (2008), 'Protective effects of lactoferrin against intestinal mucosal damage induced by lipopolysaccharide in human intestinal Caco-2 cells', *Yakugaku Zasshi*, 128(9), 1363–8. DOI: 10.1248/yakushi.128.1363

KOIKAWA N, NAGAOKA I, YAMAGUCHI M, HAMANO H, YAMAUCHI K, SAWAKI K (2008), 'Preventive effect of lactoferrin intake on anemia in female long distance runners', *Biosci Biotechnol Biochem*, 72(4), 931–5. DOI: 10.1271/bbb.70383

KOMINE Y, KUROISHI T, KOBAYASHI J, ASO H, OBARA Y, KUMAGAI K, SUGAWARA S, KOMINE K (2006), 'Inflammatory effect of cleaved bovine lactoferrin by elastase on staphylococcal mastitis, *J Vet Med Sci*, 68(7), 715–23. DOI:10.1292/jvms.68.715

KRUZEL M L, HARARI Y, MAILMAN D, ACTOR J K, ZIMECKI M (2002), 'Differential effects of prophylactic, concurrent and therapeutic lactoferrin treatment on LPS-induced inflammatory responses in mice', *Clin Exp Immunol*, 130(1), 25–31. DOI: 10.1046/j.1365-2249.2002.01956.x.

KRUZEL M L, ACTOR J K, BOLDOGH I, ZIMECKI M (2007), 'Lactoferrin in health and disease', *Postepy Hig Med Dosw*, 61, 261–7.

KUHARA T, IIGO M, ITOH T, USHIDA Y, SEKINE K, TERADA N, OKAMURA H, TSUDA h (2000), 'Orally administered lactoferrin exerts an antimetastatic effect and enhances production of IL-18 in the intestinal epithelium', *Nutr Cancer*, 38(2), 192–9.

KUWATA H, YAMAUCHI K, TERAGUCHI S, USHIDA Y, SHIMOKAWA Y, TOIDA T, HAYASAWA H (2001), 'Functional fragments of ingested lactoferrin are resistant to proteolytic degradation in the gastrointestinal tract of adult rats', *J Nutr*, 131(8), 2121–7.

LACASSE P, LAUZON K, DIARRA M S, PETITCLERC D (2008), 'Utilization of lactoferrin to fight antibiotic-resistant mammary gland pathogens', *J Anim Sci*, 86(13 Suppl), 66–71. DOI:10.2527/jas.2007-0216

LAW B, REITER B (1977), 'The isolation and bacteriostatic properties oflactoferrin from bovine milk whey', *J Dairy Res*, 44, 595–61.

LEITCH E C, WILLCOX M D (1998), 'Synergic antistaphylococcal properties of lactoferrin and lysozyme', *J Med Microbiol*, 47(9), 837–42.

LONGHI C, CONTE M P, RANALDI S, PENTA M, VALENTI P, TINARI a, SUPERTI F, SEGANTI L (2005), 'Apoptotic death of Listeria monocytogenes-infected human macrophages induced by lactoferricin B, a bovine lactoferrin-derived peptide', *Int J Immunopathol Pharmacol*, 18(2), 317–25.

LÖNNERDAL B, IYER S (1995), 'Lactoferrin: A general review', *Haematologica* 80, 252–267.

LÖNNERDAL B, BRYANT A (2006), 'Absorption of iron from recombinant human lactoferrin in young US women', *Am J Clin Nutr*, 83(2), 305–9.

LYONS T E, MILLER M S, SERENA T, SHEEHAN P, LAVERY L, KIRSNER R S, ARMSTRONG D G, REESE A, YANKEE E W, VEVES A (2007), 'Talactoferrin alfa, a recombinant human lactoferrin promotes healing of diabetic neuropathic ulcers: A phase 1/2 clinical study', *Am J Surg*, 193(1), 49–54. DOI:10.1016/j.amjsurg.2006.07.010

MADER J S, SALSMAN J, CONRAD D M, HOSKIN D W (2005), 'Bovine lactoferricin selectively induces apoptosis in human leukemia and carcinoma cell lines', *Mol Cancer Ther*, 4(4), 612–24.

MAKINO Y, NISHIMURA S (1992), 'High-performance liquid chromatographic separation of human apolactoferrin and monoferric and diferric lactoferrins', *J Chromatogr*, 579(2), 346–9.

MASCI J R (2000), 'Complete response of severe, refractory oral candidiasis to mouthwash containing lactoferrin and lysozyme', *AIDS*, 14(15), 2403–4. DOI: 00000478-199706000-00008

MOHAN K V, LETCHOUMY P V, HARA Y, NAGINI S (2008), 'Combination chemoprevention of hamster buccal pouch carcinogenesis by bovine milk lactoferrin and black tea polyphenols', *Cancer Invest*, 26(2), 193–201. DOI: 10.1080/07357900701511961

MULVIHILL D M, ENNIS M P (2003), 'Functional milk proteins: Production and utilization'. In Fox P F, McSweeney P L H, Eds, *Adv Dairy Chem Proteins*, New York, Kluwer Academic/Plenum Publishers, Vol 1, 1175–228.

NORRBY K, MATTSBY-BALTZER I, INNOCENTI M, TUNEBERG S (2001), 'Orally administered bovine lactoferrin systemically inhibits VEGF(165)-mediated angiogenesis in the rat', *Int J Cancer*, 91(2), 236–40. DOI: 10.1002/1097-0215(200002)9999:9999 <::AID-IJC1024> 3.0.CO;2-T

OCHOA T J, NOGUERA-OBENZA M, EBEL F, GUZMAN C A, GOMEZ H F, CLEARY T G (2003), 'Lactoferrin impairs Type III secretory system function in enteropathogenic Escherichia coli', *Infect Immun*, 71(9), 5149–55. DOI: 10.1128/IAI.71.9.5149-5155.2003

OCHOA T J, CLEARLY T G (2004), 'Lactoferrin disruption of bacterial Type III secretion systems', *Biometals*, 17(3), 257–60. DOI: 10.1023/B:BIOM.0000027701.12965.d4

OCHOA T J , CHEA-WOO E, CAMPOS M, PECHO I, PRADA A, MCMAHON R J, CLEARY T G (2008), 'Impact of lactoferrin supplementation on growth and prevalence of Giardia colonization in children', *Clin Infect Dis*, 46(12), 1881–3. DOI: 10.1086/588476

PAESANO R, TORCIA F, BERLUTTI F, PACIFICI E, EBANO V, MOSCARINI M, VALENTI P (2006), 'Oral administration of lactoferrin increases hemoglobin and total serum iron in pregnant women', *Biochem Cell Biol*, 84(3), 377–80. DOI: 10.1139/O06-040

PAN Y, LEE A, WAN J, COVENTRY M J, MICHALSKI W P, SHIELL B, ROGINSKI H (2006), 'Antiviral properties of milk proteins and peptides', *Int Dairy J*, 16(11), 1252–1261.

PETITCLERC D, LAUZON K, COCHU A, STER C, DIARRA M S, LACASSE P (2007), 'Efficacy of a lactoferrin–penicillin combination to treat {beta}-lactam-resistant Staphylococcus aureus mastitis', *J Dairy Sci*, 90(6), 2778–87. DOI:10.3168/jds.2006-598

PLAUT A G, QIU J, ST GEME J W (2000), 'Human lactoferrin proteolytic activity: Analysis of the cleaved region in the IgA protease of Haemophilus influenzae', *Vaccine*, 19 Suppl 1, S148–52. DOI:10.1016/S0264-410X(00)00296-6

ROBERTS A K, CHIERICI R, SAWATZKI G, HILL M J, VOLPATO S, VIGI V (1992), 'Supplementation of an adapted formula with bovine lactoferrin: 1. Effect on the infant faecal flora', *Acta Paediatr*, 81(2), 119–24.

SAMUELSEN Ø, HAUKLAND H H, ULVATNE H, VORLAND L H (2004), 'Anti-complement effects of lactoferrin-derived peptides', *FEMS Immunol Med Microbiol*, 41(2), 141–8. DOI: 10.1016/j.femsim.2004.02.006

SAMUELSEN Ø, HAUKLAND H H, JENSSEN H, KRÄMER M, SANDVIK K, ULVATNE H, VORLAND L H (2005), 'Induced resistance to the antimicrobial peptide lactoferricin B in Staphylococcus aureus', *FEBS Lett*, 579(16), 3421–6. DOI: 10.1016/j.febslet.2005.05.017

SHIMAMURA M, YAMAMOTO Y, ASHINO H, OIKAWA T, HAZATO T, TSUDA H, IIGO M (2004), 'Bovine lactoferrin inhibits tumor-induced angiogenesis', *Int J Cancer*, 111(1), 111–6. DOI: 10.1002/ijc.20187

SHIN K, WAKABAYASHI H, YAMAUCHI K, TERAGUCHI S, TAMURA Y, KUROKAWA M, SHIRAKI K (2005), 'Effects of orally administered bovine lactoferrin and lactoperoxidase on influenza virus infection in mice', *J Med Microbiol*, 54(Pt 8), 717–23. DOI: 10.1099/jmm.0.46018-0

SINGH P K, PARSEK M R, GREENBERG E P, WELSH M J (2002), 'A component of innate immunity prevents bacterial biofilm development', *Nature*, 417(6888), 552–5.

SPADARO M, CAORSI C, CERUTI P, VARADHACHARY A, FORNI G, PERICLE F, GIOVARELLI M (2008), 'Lactoferrin, a major defense protein of innate immunity, is a novel maturation factor for human dendritic cells', *FASEB J*, 22(8), 2747–57.

SUZUKI Y A, LOPEZ V, LÖNNERDAL B (2005), 'Mammalian lactoferrin receptors: Structure and function', *Cell Mol Life Sci*, 62(22), 2560–75. DOI: 10.1007/s00018-005-5371-1

TAMANO S, SEKINE K, TAKASE M, YAMAUCHI K, IIGO M, TSUDA H (2008), 'Lack of Chronic oral toxicity of chemopreventive bovine lactoferrin in F344/DuCrj rats', *Asian Pac J Cancer Prev*, 9(2), 313–6.

TAMURA Y (2004), 'Production and application of bovine lactoferrin'. *IDF Bulletin*, 389, 64–8.

TAYLOR S, BROCK J, KRUGER C, BERNER T, MURPHY M (2004), 'Safety determination for the use of bovine milk-derived lactoferrin as a component of an antimicrobial beef carcass spray', *Regul Toxicol Pharmacol*, 39(1), 12–24. DOI: 10.1016/j.yrtph.2003.10.001

TENG C T (2002), 'Lactoferrin gene expression and regulation: An overview', *Biochem Cell Bio*l, 80(1), 7–16.

TOMITA M, WAKABAYASHI H, YAMAUCHI K, TERAGUCHI S, HAYASAWA H (2002), 'Bovine lactoferrin and lactoferricin derived from milk: Production and applications', *Biochem Cell Biol*, 80(1), 109–12.

TOMITA M, WAKABAYASHI H, SHIN K, YAMAUCHI K, YAESHIMA T, IWATSUKI K (2008), 'Twenty-five years of research on bovine lactoferrin appications', *Biochimie*, Epub ahead. DOI: 10.1016/j.biochi.2008.05.021

TSUJI S, HIRATA Y, MATSUOKA K (1989), 'Two apparent molecular forms of bovine lactoferrin', *J Dairy Sci*, 72(5), 1130–6.

UCHIDA T, SATO K, KAWASAKI Y, DOSAKO S (1996), *Separation of lactoperoxidase, secretory component and lactoferrin from milk or whey with a cation exchange resin*, US Patent No 5,516,675.

UETA E, TANIDA T, OSAKI T (2001), 'A novel bovine lactoferrin peptide, FKCRRWQWRM, suppresses Candida cell growth and activates neutrophils', *J Pept Res*, 57(3), 240–9. DOI: 10.1111/j.1399-3011.2001.00821

ULVATNE H, HAUKLAND H H, OLSVIK O, VORLAND L H (2001), 'Lactoferricin B causes depolarization of the cytoplasmic membrane of Escherichia coli ATCC 25922 and fusion of negatively charged liposomes', *FEBS Lett*, 492(1–2), 62–65. DOI: 10.1016/S0014-5793(01)02233-5

ULVATNE H, HAUKLAND H H, SAMUELSEN Ø, KRÄMER M, VORLAND L H (2002), 'Proteases in Escherichia coli and Staphylococcus aureus confer reduced susceptibility to lactoferricin B', *J Antimicrob Chemother*, 50(4), 461–7.

ULVATNE H, SAMUELSEN Ø, HAUKLAND H H, KRÄMER M, VORLAND L H (2004), 'Lactoferricin B inhibits bacterial macromolecular synthesis in Escherichia coli and Bacillus subtilis', *FEMS Microbiol Lett*, 237(2), 377–84. DOI: 10.1016/j.femsle.2004.07.001

USHIDA Y, SEKINE K, KUHARA T, TAKASUKA N, IIGO M, MAEDA M, TSUDA H (1999), 'Possible chemopreventive effects of bovine lactoferrin on esophagus and lung carcinogenesis in the rat', *Jpn J Cancer Res*, 90(3), 262–7.

VALENTI P, GRECO R, PITARI G, ROSSI P, AJELLO M, MELINO G, ANTONINI G (1999), 'Apoptosis of Caco-2 intestinal cells invaded by Listeria monocytogenes: Protective effect of lactoferrin', *Exp Cell Res*, 250(1), 197–202. DOI: 10.1006/excr.1999.4500

VALENTI P, ANTONINI G (2005), 'Lactoferrin: An important host defence against microbial and viral attack', *Cell Mol Life Sci*, 62(22), 2576–87. DOI: 10.1007/s00018-005-5372-0

VENKATESH M P, RONG L (2008), 'Human recombinant lactoferrin acts synergistically with antimicrobials commonly used in neonatal practice against coagulase-negative staphylococci and Candida albicans causing neonatal sepsis', *J Med Microbiol*, 57(Pt 9), 1113–21.

VIEJO-DÍAZ M, ANDRÉS M T, FIERRO J F (2004), 'Modulation of in vitro fungicidal activity of human lactoferrin against Candida albicans by extracellular cation concentration and target cell metabolic activity', *Antimicrob Agents Chemother*, 48(4), 1242–8. DOI: 10.1128/AAC.48.4.1242-1248.2004.

WAKABAYASHI H, TAKASE M, TOMITA M (2003), 'Lactoferricin derived from milk protein lactoferrin', *Curr Pharm Des*, 9(16), 1277–87.

WAKABAYASHI H, KUWATA H, YAMAUCHI K, TERAGUCHI S, TAMURA Y (2004), 'No detectable transfer of dietary lactoferrin or its functional fragments to portal blood in healthy adult rats', *Biosci Biotechnol Biochem*, 68(4), 853–60. DOI:10.1271/bbb.68.853

WAKABAYASHI H, YAMAUCHI K, TAKASE M (2006), 'Lactoferrin research, technology and applications', *Int Dairy J*, 16, 1241–51.

WEINBERG E D (2003), 'The therapeutic potential of lactoferrin', *Expert Opin Investig Drugs*, 12(5), 841–51. DOI: 10.1517/13543784.12.5.841

WEINBERG E D (2007), 'Antibiotic properties and applications of lactoferrin', *Curr Pharm Des*, 13(8), 801–11.

YAMAUCHI K, HIRUMA M, YAMAZAKI N, WAKABAYASHI H, KUWATA H, TERAGUCHI S, HAYASAWA H, SUEGARA N, YAMAGUCHI H (2000a), 'Oral administration of bovine lactoferrin for treatment of tinea pedis. A placebo-controlled, double-blind study', *Mycoses*, 43(5), 197–202. DOI: 10.1046/j.1439-0507.2000.00571.x

YAMAUCHI K, TOIDA T, NISHIMURA S, NAGANO E, KUSUOKA O, TERAGUCHI S, HAYASAWA H, SHIMAMURA S, TOMITA M (2000b), '13-Week oral repeated administration toxicity study of bovine lactoferrin in rats', *Food Chem Toxicol*, 38(6), 503–12. DOI: 10.1016/S0278-6915(00)00036-3

YOO Y C, WATANABE R, KOIKE Y, MITOBE M, SHIMAZAKI K, WATANABE S, AZUMA I (1997), 'Apoptosis in human leukemic cells induced by lactoferricin, a bovine milk protein-derived peptide: Involvement of reactive oxygen species', *Biochem Biophys Res Commun*, 237(3), 624–8. DOI:10.1006/bbrc.1997.7199

ZHANG N T, NAKANO T, OZIMEK L (2002), 'Isolation of lactoferrin from bovine colostrum by SP-Sepharose cation-exchange chromatography', *Milchwissenschaft*, 57: 614–7.

ZIMECKI M, SPIEGEL K, WŁASZCZYK A, KÜBLER A, KRUZEL M L (1999), 'Lactoferrin increases the output of neutrophil precursors and attenuates the spontaneous production of TNF-alpha and IL-6 by peripheral blood cells', *Arch Immunol Ther Exp (Warsz)*, 47(2), 113–8.

ZIMECKI M, KRUZEL M L (2007), 'Milk-derived proteins and peptides of potential therapeutic and nutritive value', *J Exp Ther Oncol*, 6(2), 89–106..

12

Harnessing milk oligosaccharides for nutraceutical applications

R. Hickey, Moorepark Food Research Centre, Teagasc, Ireland

Abstract: Given the vast quantities of oligosaccharides which are present in human milk, it is hardly surprising that during postnatal development these molecules play an important physiological role. There is increasing evidence of the local effects of human milk oligosaccharides within the gastrointestinal tract. Human milk oligosaccharides have been shown to modulate the intestinal flora, prevent infection by pathogenic bacteria and viruses and interact directly with immune cells. Therefore researchers are beginning to focus their attention on sources of oligosaccharides other than those from human milk which may have similar biological outcomes. Milk oligosaccharides from domestic animals have some structural similarity to human milk oligosaccharides but in many cases only trace amounts are present in milk, and further studies are required to validate their similarity of function. Despite these hurdles, the milk from dairy animals, in particular, remains an attractive source of potentially therapeutic oligosaccharides and warrants further investigation.

Key words: Oligosaccharides, prebiotics, anti-adhesion, anti-inflammatory.

12.1 Introduction

Oligosaccharides are the third largest solid component of human milk, following lactose and lipids, with concentrations reaching up to 50 g/L or more in colostrum to an average of 10–15 g/L in mature milk (Kunz *et al.*, 1999; Kunz and Rudloff, 2008). Indeed, when human milk is compared with milk of the most relevant domestic mammals, the oligosaccharide concentration is higher by a factor of 10 to 100 (Boehm and Stahl, 2003). Interestingly, although human milk contains such a large amount of this molecular class, infants are unable to digest these milk components (Gnoth *et al.*, 2000; Engfer *et al.*, 2000). This is remarkable, given the evolutionary selective

Table 12.1 Beneficial roles of milk oligosaccharides

Site of action	Possible bioactive role	References
Oral cavity, respiratory, gastrointestinal, urinary tracts	Decoy receptors	Wernersson *et al.* 2006; Thomas and Brooks 2004; Morrow *et al.* 2005; Coppa *et al.* 1990
Gastrointestinal system	Regulation of growth-related characteristics of intestinal cells	Deters *et al.* 2005; Kuntz *et al.* 2007
Gastrointestinal system	Prebiotic effect	Coppa *et al.* 2000
Gastrointestinal system	Glycome modification	Angeloni *et al.* 2005
Nervous system	Brain development	Wang *et al.* 2003
Immune system	Modulation of leukocyte extravasation	Bode *et al.* 2004a
Immune system	Formation of platelet–neutrophil complexes (PNC)	Bode *et al.* 2004b
Immune system	Maturation of the immune system	Eiwegger *et al.* 2004
Immune system	Interactions with DC-SIGN, Selectins, Galectins	Naarding *et al.* 2005; Bode, 2006

pressure of lactation, in which nourishment of the infant must constantly correspond to the metabolic and nutritional demands of the mother. Given their abundance, it became clear that human milk oligosaccharides are likely to have specific biological functions and these have been summarized in Table 12.1. Such functions may include prebiotic activity, antiadhesion effects, anti-inflammatory properties, glycome modification, brain development, growth-related characteristics of intestinal cells and other uncharacterized effects (for reviews, see Newburg *et al.*, 2005; Bode, 2006; Kunz and Rudloff, 2006). Many efforts are therefore, being made to replicate the effects of human milk oligosaccharides by searching for alternative compounds that may produce similar biological outcomes, particularly in relation to infant formula compositions. Given that little information exists regarding the biological activity associated with alternative compounds such as plant oligosaccharides and milk oligosaccharides from animal species, this review will focus on the beneficial effects that are linked to the oligosaccharide fraction of human milk, considering it as the gold standard for infant nutrition. In addition, the data which is known relating to milk oligosaccharides from animals, and in particular those used for dairy purposes, will be discussed given their structural similarity to human oligosaccharides and their commercial availability. Current methods for isolating and analyzing milk oligosaccharides and the path to commercialization are also summarized.

12.2 Composition of oligosaccharides in milk

As mentioned, the biological activity of human milk oligosaccharides is dependent on their individual biochemical structures and for this reason a brief overview is given on the composition of milk oligosaccharides. Indeed, in terms of human milk, oligosaccharide composition is very complex, generally ranging in size from three to eight sugars, although highly complex oligosaccharides may contain up to 32 sugars (Stahl *et al.*, 1994). It can be estimated that as many as 900 structures could exist in human milk, considering all of the monosaccharide combinations and possible linkages (Yarema, 2005). Currently, more than 90 structures have been elucidated and approximately 200 species have been identified (Kunz *et al.*, 2000; Finke *et al.*, 1999; Ninonuevo *et al.*, 2007). The monomers of milk oligosaccharides are D-glucose (Glc), D-galactose (Gal), N-acetylglucosamine (GlcNAc), L-fucose (Fuc) and sialic acid (NeuAc). The molecules are synthesized in the breast, starting with lactose at the reducing end (Kunz *et al.*, 2000). The core molecule is characterized by repetitive attachment of Gal and GlcNAc in β-glycosidic linkage to lactose (Boehm and Stahl, 2003). Thus, a large number of core structures can be formed, with further variations occurring due to the attachment of fucose and fucose and/or sialic acid to the respective core molecules (Egge, 1993; Stahl *et al.*, 1994).

Given that a number of acidic bovine milk oligosaccharides are structurally similar to those found in human milk, it is likely that they would also have similar biological functions. Table 12.2 summarizes the distribution of oligosaccharides in human milk, bovine colostrum and bovine mature milk. Gopal and Gill (2000) reported ten sialylated oligosaccharides in bovine milk and colostrum, while nine neutral oligosaccharides have been described (Urashima *et al.*, 2001). However, Lebrilla (2007) reasons that as many as 30 oligosaccharides can be found in bovine milk. Human milk is considered to be unique in its high content of complex fucosylated and sialylated oligosaccharides (Kunz *et al.*, 1999). In contrast, bovine milk contains mainly sialylated species (Veh *et al.*, 1981; Saito *et al.*, 1984). Quantitatively, when compared with human milk/colostrum, the levels of oligosaccharides in milk of domestic mammalian animals (cows, sheep, and goat) are much lower (Urashima *et al.*, 1997). In fact, oligosaccharides are present in mature bovine milk in only trace amounts, while considerably higher concentrations are present in bovine colostrum. As milk production matures post-parturition, the concentration of these oligosaccharides declines rapidly. This makes isolation and analysis technically difficult and has hindered their utilization in the healthcare and food sector as biologically active ingredients that promote health in infants and adults (Gopal and Gill, 2000). The most abundant acidic oligosaccharide reported in bovine colostrum is 3′-sialyllactose followed by sialyllactosamine, 6′-sialyllactose and disialyllactose. Together, 3′- and 6′-sialyllactose account for more than 50% of the total oligosaccharides present in bovine colostrum. Interestingly, 3′- and

Table 12.2 Distribution of oligosaccharides in human and bovine milk

Oligosaccharide	Human milk (g L^{-1})	Bovine milk (g L^{-1})	Bovine colostrums (g L^{-1})
Lactose	55–70	40–50	40–50
Lacto-*N*-tetraose	0.5–1.5	Trace	
Lacto-*N*-fucopentaose I	1.2–1.7		
Lacto-*N*-fucopentaose II	0.3–1.0		
Lacto-*N*-fucopentaose III	0.01–0.2		
Lacto-*N*-difucohexaose I	0.1–0.2		
Lacto-*N*-novopentaose			*
N-acetylgalactosaminyl glucose			*
N-acetylgalactosyl-lactose			*
3′-galactosyl-lactose			*
β-3′-galactosyl-lactose			*
6′-galactosyl-lactose			*
N-acetyl-lactoseamine			*
NeuAc(2–6)lactose	0.3–0.5	0.03–0.06 (combined)	0.019
NeuAc(2–3)lactose	0.1–0.3		0.095
NeuAc-lacto-*N*-tetraose a	0.03–0.2	Trace	
NeuAc-lacto-*N*-tetraose c	0.1–0.6	Trace	
NeuAc2-lacto-*N*-tetraose	0.2–0.6	Trace	
6-Sialyl-lactosamine			0.047
3-Sialyl-galactosyl-lactose			Trace (3 μmol L^{-1})
Disialyl-lactose			0.028
Sialyl-lactose-1-phosphate			Trace (3 μmol L^{-1})
Sialyl-lactose-6-phosphate			Trace (1 μmol L^{-1})
3-Glucolylneuraminyl-lactose			Trace (2 μmol L^{-1})
6-Glucolylneuraminyl-lactose			*

* Oligosaccharide detected and structurally characterized, but concentration not reported by authors.
Data compiled from Kunz and Rudloff (2002), Gopal and Gill (2000), Nakamura and Urashima (2004) and Mehra and Kelly (2006).

6′-sialyllactose are also major components of human milk and have been reported to reach concentrations of 1.0 g/L (McVeagh and Brand Miller, 1997). Analytical and structural data on milk oligosaccharides from animal sources is much needed, given the large amounts of human milk oligosaccharides that would be required for clinical trials. This has necessitated research directed towards development of methods/processes for large-scale separation and enrichment of bovine milk oligosaccharides, as well as for expression of human oligosaccharides in non-human milk. These approaches include (i) production of human milk oligosaccharides by fermentation of genetically engineered bacteria, (ii) concentration/fractionation technologies such as membrane filtration for bovine milk oligosaccharides, and (iii) expression of human milk oligosaccharides in transgenic animals (reviewed by Mehra and Kelly, 2006). Also of importance however, is the development of research initiatives that tackle compositional questions such as why the range and concentration of certain oligosaccharides change over the lactation period in both humans and animals, and how these changes affect the physiology and health of the host.

12.3 Nutraceutical potential of milk oligosaccharides

It is obvious that further studies are required in order to reveal the full repertoire of activities that may be associated with human milk oligosaccharides, not to mention oligosaccharides from the milk of animals. The probability of finding new biological functions should, in theory, be high, given the amount and types of mammalian cells exposed to milk oligosaccharides with each suckling. The following section describes the functional activity that has been assigned to human milk oligosaccharides to date and also presents examples of oligosaccharides from the milk of dairy animals where biological function has been established. It is only through an understanding of the structural and functional properties of these molecules that such compounds and their equivalents can reach their potential as food ingredients.

12.3.1 Prebiotic activity

In the weeks after birth, the type of feeding is thought to be the major factor influencing the intestinal flora of the human infant (Coppa *et al.*, 2004). Several studies suggest that bifidobacteria and lactobacilli (up to 90%) constitute the dominant microflora in breast-fed infants, while bottle-fed infants develop a more diverse microflora with a lower number of bifidobacteria (40–60%) and the presence of potential pathogens such as *Clostridium*, *Staphylococcus* and *Bacteroides* (Benno *et al.*, 1984; Moreau *et al.*, 1986; Balmer and Wharton, 1989; Millar *et al.*, 1996; Harmsen *et al.*, 2000; Favier *et al.*, 2002). Several beneficial properties have been associated with the presence of Bifidobacteria in breast-fed infants, such as inhibition of pathogenic organisms, modulation of mucosal physiology, barrier function and systemic immunologic and inflammatory responses (Sudo *et al.*, 1997; Agostoni *et al.*, 2004). In this respect, interest has arisen in the properties of human milk that evidently promote bacteria which are beneficial to the newborn.

György *et al.* (1954) found that a mixture of oligosaccharides containing N-acetylglucosamine (GlcNAc), which they referred to as gynolactose, was a growth-promoting so-called 'bifidus' factor for a 'bifidum mutant' called *Bifidobacterium bifidum* subspecies *pennsylvancium*, which had been isolated from the faeces of breast-fed infants. Subsequently, Kuhn (1958) found that GlcNAc-containing oligosaccharides were the most active 'bifidus' factor. More recent studies (Engfer *et al.*, 2000; Chaturvedi *et al.*, 2001; Coppa *et al.*, 2001) on the metabolic fate of human milk oligosaccharides have demonstrated that these molecules are capable of resisting digestion until they reach the colon, where they stimulate the growth of the bifidus-predominant flora. The remaining oligosaccharides that are not used in the colon are excreted in the faeces (Coppa *et al.*, 2000) and, to a lesser extent, in the urine (Rudloff *et al.*, 1996). In contrast, only traces of oligosaccharides are found in the faeces of formula-fed infants (Coppa *et al.*, 2004). Upon meeting the above criteria, it becomes apparent that human milk oligosaccharides can act as prebiotics. The definition of a prebiotic is a 'non-digestible food ingredients that beneficially affect the gut by selectively stimulating the growth and/or activity of one or a limited number of bacteria in the colon, that can improve host health' (Gibson and Roberfroid, 1995). In fact, human milk oligosaccharides are the first and most important prebiotics in humans. Indeed, they act as the model for the addition of prebiotics to infant formulae, even though their biological role appears far more complex than the roles of the simple oligosaccharides presently added to formulae.

Salminen *et al.* (2004) postulate that 'early colonization guides subsequent microbiota development which may later impact on health, to the extent of predisposing some infants towards specific diseases'. Rastall *et al.* (2005) also speculate on the feasability of programming humans at birth to receive a microbiota best able to maintain their life-long health. Therefore, future research on milk oligosaccharides may involve targeting the microflora of the most defenseless groups of society, notably the very young and the elderly, to promote the colonization of beneficial bacteria, thereby contributing to good health. It is tempting to suggest that, as more milk oligosaccharides are characterized, it may be possible to match these molecules with specific strains of beneficial bacteria that can selectively utilize the matching oligosaccharide(s), thereby producing specific positive physiological effects. Indeed, LoCascio *et al.* (2007) revealed, through glycoprofiling of human milk oligosaccharide consumption by bifidobacteria, that such scenerio may be possible in the future. This study revealed that one species, *Bifidobacterium longum* biovar. *infantis* ATCC 156697, an isolate from the infant gut, preferentially consumes small mass oligosaccharides, representing 63.9% of the total human milk oligosaccharides available. The study of commensal genomics is a recent development and in fact, only a relatively small number of bifidobacterial strains have been sequenced. Probably with the accumulation of more genomic data, researchers will be able to predict the ability of specific strains to utilize certain oligosaccharides. It is evident that bifidobacteria have a superior ability to flourish in the infant intestine, a feature that may be linked to their ability to utilize diverse carbohydrates. For instance, genome sequence analysis of *Bifidobacterium longum* NCC2705 revealed that more than 8.5% of the total predicted proteins were involved in the degradation of oligo- and polysaccharides (Schell *et al.*, 2002). This is 30% more than *E. coli*, *Enterococcus faecium*, *Lactococcus lactis*, *Bacillus halodurans* and *B. subtilis* and twice the number for *Mycobacterium leprae* and *Deinococcus radiodurans* (Schell *et al.*, 2002). More recently, German *et al.*, 2008, while studying the genome of *Bifidobacteria longum* biovar *infantis*; reported approximately 700 genes that are unique to infantis, including a variety of co-regulated glycosidases, relative to other Bifidobacteria, implying a co-evolution of human milk oligosaccharides and the genetic capability of certain intestinal bacteria to utilize them. The challenge therefore emerges to identify oligosaccharides other than those found in human milk that not only promote a bifidobacteria-dominant microflora in formula-fed infants but also select for specific flora that match the individual infant's health requirements.

12.3.2 Antiadhesion effects of milk oligosaccharides

It is well documented that adhesion of enteric, oral, and respiratory bacteria is required for colonization and subsequent development of disease. When in the adherent state, these bacteria are more likely to survive as their resistance to cleansing mechanisms, immune factors, bacteriolytic enzymes and antibiotics is higher (Ofek *et al.*, 2003). Therefore, prevention of adhesion at an early stage following exposure of the host to pathogens should prevent disease. It is now well accepted that milk oligosaccharides have a direct inhibitory effect on certain virulence-related abilities of pathogenic microorganisms by interfering with their adhesion to human cells. These molecules can withstand the low pH of the gut, and resist degradation through enzymes from the pancreas and brush border membrane (Bode, 2006). Indeed, with each suckling, the infant's orogastrointestinal tract is rinsed with 100 to 500 mg of oligosaccharides. This large amount makes it very likely that milk oligosaccharides have local effects on the mucosal surface, or even within the cell, as these molecules can effectively reach the sites where bacterial adhesion is known to occur. Today, many free oligosaccharides from milk, as well as glycoproteins, are considered to be soluble receptor analogs of epithelial cell surface carbohydrates (Kunz *et al.*, 2000). These glycans display structural homology to host cell receptors and thus function as receptor decoys to which pathogens can bind instead of the host, as presented in Fig. 12.1. Milk glycans can also inhibit pathogens by competitive binding with the host cell-surface receptors (Morrow *et al.*, 2005).

Oligosaccharide expression in human milk is determined genetically and their synthesis is linked to the presence of enzymes in the mammary acinar cells, whose expression varies among different populations. These enzymes or related enzymes also function in the production of glycoproteins that are present on the red blood cells and mucosal epithelial cells or which occur

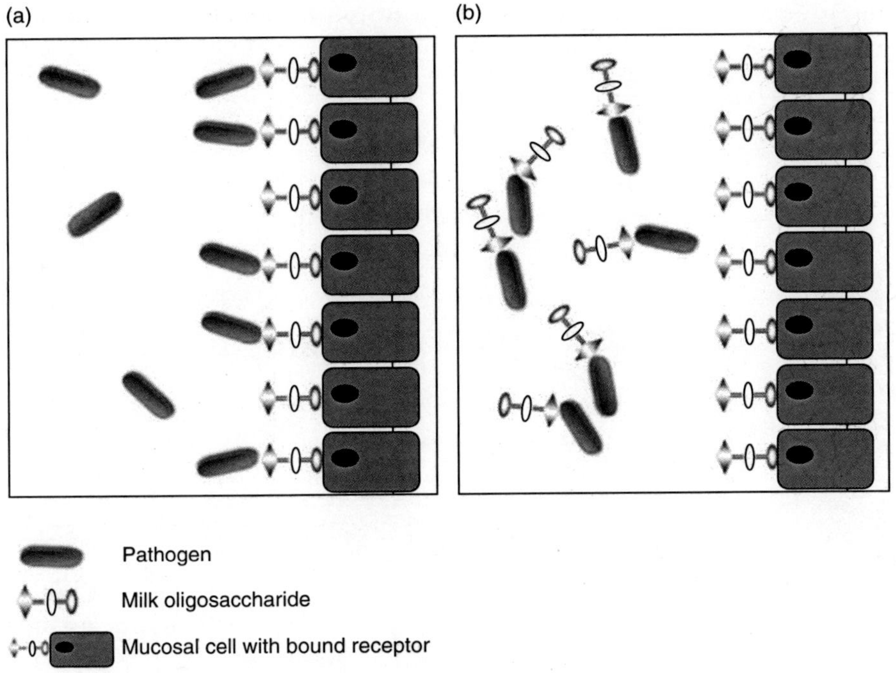

Fig. 12.1 Microbial adhesion and anti-adhesion by milk oligosaccharides. (a) Bacteria adhere via their adhesins to cognate receptors on epithelial cells of the mucosal surfaces to withstand normal cleansing mechanisms aimed at eradication of the invading pathogen. (b). In the presence of inhibitors of adhesion such as milk oligosaccharides, the bacteria are eradicated by the normal cleansing mechanism. (Adapted from Newburg, 1999; Ofek *et al.*, 2003.)

as free antigens in other biological fluids such as blood, saliva and intestinal contents. Polymorphisms of the secretor and Lewis genes are known to determine expression of the Lewis blood group type and the biosynthetic pathways involved are summarized in Fig. 12.2. For instance, oligosaccharides containing a 2-linked fucose are synthesized by-products of the secretor gene, and oligosaccharides containing a 3- or 4-linked fucose are synthesized by-products of the Lewis gene and related genes (Morrow *et al.*, 2004). Interestingly, numerous links have been made between blood group type and susceptibility to diseases, since these glycoconjugates can be used as receptors by various pathogens (Newburg *et al.*, 2004). For instance, O blood group individuals have increased susceptibility to cholera and to Norwalk virus (Glass *et al.*, 1985; Hutson *et al.*, 2002) while the P blood group type has been linked with susceptibility to hemolytic uremic syndrome (Newburg *et al.*, 1993). An association between Lewis and secretor histo-blood group genotypes and risk of infection with *Helicobacter*

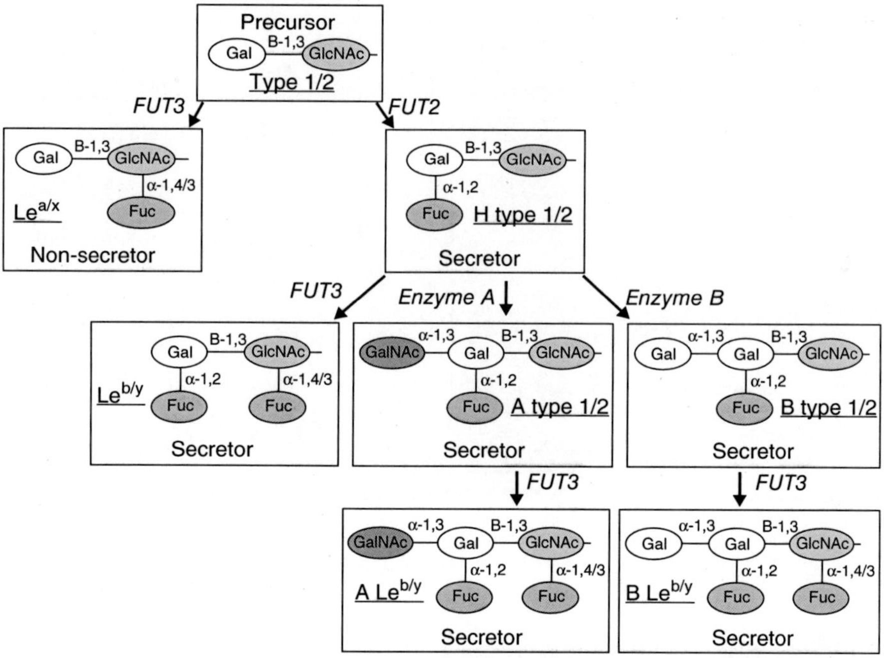

Fig. 12.2 The biosynthetic pathways of the Human ABO and Lewis histo-blood group antigens based on the Type 1 and Type 2 precursors. The names of the antigens are underlined (Adapted from Newburg *et al.*, 2005).

pylori and influenza virus has also been reported (Boat *et al.*, 1978; Ikehara *et al.*, 2001). This phenomenon may explain the frequency of specific blood group types in regions where certain pathogens are prevalent. For example, there a low incidence of nonsecretors among the Mexican mestizo population. This may result in a greater susceptibility of infants to the stable toxin of *E. coli*, campylobacter and some noroviruses, given that Mexican mestizo infants may receive milk containing less protective α1,2-linked glycans (Newburg *et al.*, 2004).

From a medical point of view, using milk oligosaccharides to prevent infection may be safer when compared with current approaches, as the bacteria are not killed but just made non-infective, thus less selection pressure is produced. In these circumstances, bacteria resistant to carbohydrates may occur through mutation but their infectivity should be significantly reduced as such mutations would result in the inability of the bacteria to bind the carbohydrates. This, in turn, implies they are unable to bind to the cell wall and infect their targets (Newburg *et al.*, 2005; Pieters, 2007). While, many human milk oligosaccharides are known to possess antiadhesive properties, as shown in Table 12.3, little research has focussed

Table 12.3 Free oligosaccharides and glycoconjugates from human milk that inhibit bacterial adherence

Microbe	Decoy	References
Escherichia coli	Fucosyloligosaccharide	Newburg *et al.* 2004
Camplyobacter jejuni	Fucosyl α1,2 oligosaccharide	Cervantes *et al.*, 1995
Listeria monocytogenes	Neutral oligosaccharides	Coppa *et al.* 2003
Streptococcus pneumoniae	Lacto-*N*-neotetraose (LNnT)	Andersson *et al.*, 1983; 1986
Haemophilus influenzae	Sialylated oligosaccharides	St Geme, 1994; Reddy *et al.* 1996
Helicobacter pylori	3′-sialyllactose	Mysore *et al.*, 1999
Streptococcus mutans	Glycomacroopeptide	Kawakami, 1997
S. suis	Sialylated oligosaccharides	Liukkonen *et al.*, 1992
S. sobrinus	Casein-derived components	Neeser *et al.*, 1994
S. sanguis	Glycomacropeptide (GMP)	Kawakami, 1997
Mycoplasma pneumoniae	3′-sialyllactose	Roberts *et al.*, 1989
Pseudomonas aeruginosa	3′ and 6′-sialyllactose	Thomas and Brooks, 2004
Neisseria meningitidis	Neutral human oligosaccharides	Hakkarainen *et al.*, 2005
Salmonella fyris	Neutral oligosaccharides	Coppa *et al.*, 2006
Salmonella typhimurium	Glycomacropeptide (GMP)	Brück *et al.*, 2006
Vibrio cholerae	Neutral oligosaccharides	Coppa *et al.*, 2006
Shigella	Globotriaosylceramide (Gb3)	Newburg *et al.*, 1992
Candida albicans	Fuc-α1-2Galβ	Brassart *et al.*, 1991
HIV	Lewis X	Naarding *et al.*, 2005
Noroviruses	Lacto-*N*-difucohexose (LDFH-I)	Morrow *et al.*, 2004
Rotavirus	Lactadherin	Newburg *et al.*, 1998
Influenza virus	3′ and 6′-sialyllactose	Kunz and Rudloff, 1993

on the antiadhesive potential of oligosaccharides from the milk of other species. Therefore, the following section of the review examines more closely the antiadhesive potential of milk oligosaccharides from domestic animals, using selected examples.

Protection against Helicobacter pylori

Helicobacter pylori infections are associated with peptic ulcer disease, mucosa-associated lymphoid-tissue (MALT) lymphoma and gastric adenocarcinoma. The bacterium colonizes the gastric mucosa or adheres to the epitheial cells lining the stomach. It is carried by 50% of the population in the developed world and there is an even higher incidence in developing countries (Blaser, 1996; Crespo and Suh, 2001). The binding of *H. pylori* is mediated by adhesions expressed on its surface, three of which have been characterized (BabA, SabA and HP0721) (Ilver *et al.*, 1998; Mahdavi *et al.*, 2002; Bennett and Roberts, 2005). BabA binds to Lewis^b while SabA and HP0721 bind sialic acid.

Based on hemagglutination and cell binding studies, Evans *et al.* (1988) proposed that *H. pylori* was able to recognize the oligosaccharide 3′-sialyllactose, which occurs naturally in human and bovine milk. Indeed, Simon *et al.* (1997) found that bovine 3′-sialyllactose was capable of inhibiting adherence of clinical isolates of *H. pylori* with a millimolar 50% inhibitory concentration (IC50). Subsequently, Mysore *et al.* (1999) investigated the effect of 3′-sialyllactose on colonization of *H. pylori* in rhesus monkeys. Twelve *H. pylori*-positive rhesus monkeys were given 3′-sialyllactose, either alone or in combination with omeprazole (a proton pump inhibitor), or bismuth subsalicylate (a stomach mucosa coating agent). Of the six monkeys given the milk oligosaccharide, two were cured permanently, and a third animal was transiently cleared. The three other animals remained persistently colonized. The authors concluded that antiadhesive therapy can cure or decrease *H. pylori* colonization in some rhesus monkeys, but the addition of a proton pump inhibitor or bismuth subsalicylate does not increase cure rate. Wang *et al.* (2001b) explored whether sialylated glycoconjugates from bovine milk could inhibit an experimental *H. pylori* infection in a mouse model. The researchers found that gastric colonization by *H. pylori* was remarkably decreased in all mice treated with bovine milk glycoconjugates, and the inflammation score was also significantly lower in treated mice than in infected control animals. More recently, Gustafsson *et al.* (2006) evaluated the capacity of pig milk to inhibit *H. pylori* binding to neoglycoproteins carrying Lewis^b and sialyl-di-Lewis^x. Transgenic mice, known to express Lewis^b and sialyl-Lewis^x in their gastric epithelium, were colonized by *H. pylori* and were subsequently treated with Lewis^b and sialyl-Lewis^x expressing or non-expressing porcine milk, or water (control) only. The expression of the Lewis^b and sialyl-Lewis^x carbohydrate epitopes on pig milk proteins was breed- and individual-specific and correlated with the ability of porcine milk to inhibit *H. pylori* adhesion *in vitro* and *H. pylori* colonisation *in vivo*. The authors concluded that milk from certain pig breeds may have a therapeutic and/or prophylactic effect on *H. pylori* infection. However, the challenge in preventing colonization of this bacterium lies in the fact that it can also bind to a range of other carbohydrate specificities, which makes treatment difficult (Bavington and Page, 2005; Pieters, 2007). In fact, Aspholm-Hurtig *et al.* (2004) have shown that the bacterium adapts its binding preferences in accordance with the host population. Therefore, one approach that may have potential in combating *H. pylori* is the use of multivalent inhibitors. This involves the simultaneous binding of multiple ligands on one biological entity to multiple receptors on another (Gustafsson *et al.*, 2006). Therefore, future studies involving *H. pylori* colonization should focus on developing milk oligosaccharide antiadhesives, which are both multifunctional in order to block more than one bacterial binding ligand and multivalent, in order to increase the affinity of the antiadhesive for its receptor.

Protection against Neisseria meningitidis

Neisseria meningitidis, or the meningococcus, is the cause of significant morbidity and mortality worldwide. Meningococci express Type IV pili, which mediate the colonization of their human-specific host by adherence, leading to life-threatening septicemia and meningitis in newborns (Nassif, 1999). Hakkarainen *et al.* (2005) tested the binding of *Neisseria meningitidis* to glycoproteins, with the aim of investigating the ability of milk oligosaccharides to inhibit this binding. The group used a pili binding assay, which resulted in *N. meningitidis* pili binding to bovine thyroglobulin and human salivary agglutinin glycoproteins. The binding to these glycoproteins was then subsequently inhibited effectively by neutral human milk oligosaccharides or acidic bovine milk oligosaccharides. The authors conclude that the elucidation of the individual milk oligosaccharides involved in inhibition may provide valuable information on the type of receptor active sequences present in *Neisseria meningitidis* pili.

Protection against Influenza virus

Influenza epidemics cause serious respiratory illness in 30–50 million people and kill 250000 to 500000 people worldwide every year; a pandemic could lead to millions of deaths (Suzuki, 2005). Influenza viruses are classified into three types (species), A, B and C, on the basis of the identity of the major internal protein antigens. Influenza A viruses are the most virulent of the three, and cause severe or fatal acute respiratory disease, which is epidemic and sometimes worldwide pandemic (reviewed by Suzuki, 1994, 2002). Influenza viruses bind to sialyl sugar chain receptors on the host cell membranes. The most dominant receptors of influenza A and B viruses were found to be the sialylglycoproteins and gangliosides containing monosialo-lactosamine Types I and II (Suzuki *et al.*, 1985a, b, 1986, 1992, 1997; Ito *et al.*, 1997).

A number of early studies investigated factors in human milk which could interfere with the activity of the influenza virus (Silver *et al.*, 1956; Schmidt, 1960). Indeed, Schmidt and Janssen (1967) discovered the protective effect that human and cow milk mucoids can exert in experimental influenza virus infection of the mouse. Later studies began to reveal the mechanisms involved in the antiviral activity. For instance, Matrosovich *et al.* (1993) investigated the receptor binding sites of the H1 and H3 influenza A and influenza B virus hemagglutinins by using bovine colostrum oligosaccharides as probes. Thirty human influenza A and B virus strains were studied in a competitive ligand binding assay. The following bovine colostrum oligosaccharides were included in the study: 3′-sialyllactose,

6′-sialyllactose, 6′-sialyl-*N*-acetyllactosamine and disialyllactose. The results obtained indicated that while the functional groups of sialic acid studied are recognized by the receptor binding sites of all influenza viruses, the magnitude of their contribution to the binding energy, as well as the contribution of the asialic portion of the receptor, may vary in dependence upon the virus type, subtype, and strain. Design of polyvalent synthetic sialic acid-containing inhibitors of virus attachment, modelled on milk oligosaccharide structures, may reveal a higher anti-influenza activity. Indeed, optimally designed synthetic inhibitors have already been shown to be more potent *in vitro* than known natural inhibitors and are showing promising results in first animal experiments (Matrosovich and Klenk, 2003).

12.3.3 Anti-inflammatory properties of milk oligosaccharides

There is accumulating evidence that milk oligosaccharides have direct effects on immune function. For instance, the composition of the intestinal flora plays an important role in the postnatal development of the immune system (Björkstén *et al.*, 2001; Ouwehand *et al.*, 2002) which, as described, is strongly influenced by milk oligosaccharides. Human milk oligosaccharides vary among individuals and over the course of lactation, which no doubt results in variations in gut microflora, which in turn may lead to differences in the immune system. For example, several studies have shown that allergic children are often less colonized with bifidobacteria and lactobacilli (Björkstén *et al.*, 1999, 2001; Kalliomäki *et al.*, 2001). In addition to altering intestinal flora, human milk oligosaccharides are now thought to have immunomodulatory properties at the systemic level, by altering protein–carbohydrate interactions. Indeed, as mentioned previously, their presence in the urine of breast-fed infants (Rudloff *et al.*, 1996) supports the hypothesis that milk oligosaccharides may be present in the infant's blood. Although, these molecules have not been directly detected in blood yet, it has been estimated that a concentration of 100–200 mg/L may be present, based on their concentration in milk, the daily intake, the infant's blood volume and the amount excreted in the urine over time (Obermeier *et al.*, 1999; Bode, 2006). The known immunomodulating effects that have been linked to human milk oligosaccharides are discussed in the following sections.

Modulation of leukocyte extravasation

Selectins belong to a subclass of carbohydrate-binding proteins that bind to fucosylated and sialylated glycoprotein ligands during inflammatory processes (Bevilacqua and Nelson, 1993). They are expressed on the surface of leukocytes and lymphocytes, as well as on endothelial cells, and are involved in the cell adhesion cascade leading to the extravasation of blood cells into the tissue during the course of inflammation (Springer, 1994; McEver, 1997). Selectins on activated endothelial cells help leukocytes to

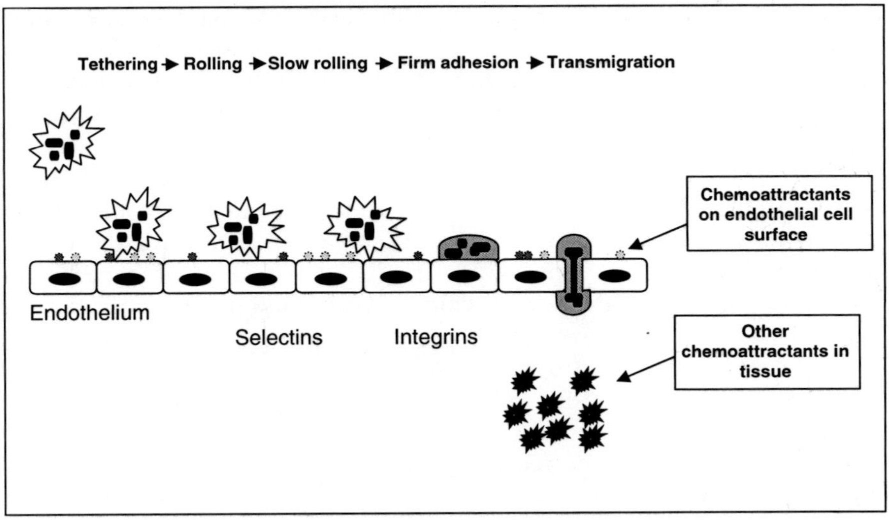

Fig. 12.3 Leukocyte–endothelial cell interactions are characterized by a casade of events in which selectins play an important role. They mediate the first cell–cell contact and initiate firmer adhesion (Adapted from Kunz *et al.*, 2003.)

engage the vessel wall in an initial interaction known as tethering, after which there is a subsequent rotational movement along the endothelium known as rolling (Reviewed by Kelly *et al.*, 2007). An overview of these interactions is presented in Fig. 12.3. There are three members of the selectin family, referred to as leuckocyte (L), endothelium (E) and platelet (P), and all three selectins recognize the same physiological-binding determinant, the sialyl-tetrasaccharide, sialyl Lewis^x (Phillips *et al.*, 1990; Erbe *et al.*, 1993; Varki, 1994). Accumulation of leukocytes in tissues is essential for an effective host defence, but excessive leukocyte infiltration can cause damage in healthy tissue, as observed in a variety of diseases (Carden and Granger, 2000; Laroux and Grisham, 2001; Schon *et al.*, 2002). Therefore, the interactions of selectins and their oligosaccharide ligands have been examined more closely in an effort to modulate leukocyte extravasation in cases of excessive immune responses (Weyrich *et al.*, 1993, Ma *et al.*, 1993). Interestingly, soluble selectin ligand analogs have been reported to reduce leuckocyte adhesion and excessive transmigration (Simanek *et al.*, 1998; Theoret *et al.*, 2001). In this respect, several glycoproteins and oligosaccharides that carry the the selectin ligand, sialyl Lewis^x, have been detected in human milk (Rudloff *et al.*, 2002; Schwertmann *et al.*, 1996).

Indeed, Klein *et al.* (2000) found that human milk oligosaccharides inhibited neutrophil adhesion to stimulated vascular endothelial cells in a dose-dependent fashion. In addition, Bode *et al.* (2004a) investigated the impact of human milk oligosaccharides on leukocyte rolling and adhesion. The researchers found that the acidic oligosaccharide fraction of milk reduced monocyte adhesion to endothelial cells in a concentration-dependent manner. At the highest concentration, monocyte, lymphocyte and neutrophil adhesion were reduced by approximately 50% while the physiological binding determinant, sialyl Lewis^x, had a less pronounced effect and lowered leukocyte adhesion by 36%. The authors concluded that the oligosaccharides within the acidic fraction of milk carry more than one binding determinant for selectins and they can be regarded as multivalent, unlike the monvalent sialyl Lewis^x. Interestingly, the acidic oligosaccharides involved included 3′ sialyllactose, which is also found in bovine milk.

Formation of platelet–neutrophil complexes (PNC)

Neutrophils associated with platelets or platelet–neutrophil complexes (PNC) are highly activiated and have a higher capacity for adhesion, phagocytosis and production of reactive oxygen species (Peters *et al.*, 1999). Furthermore, although PNCs exist in healthy individuals, they have been linked to tissue damage in certain diseases (Kubes *et al.*, 1991; Hsueh *et al.*, 2003). The initial step in PNC formation is mediated by P-selectin on activated platelets and P-selectin glycoprotein ligand 1 (PSGL-1) on neutrophils (Larson *et al.*, 1989; Bode *et al.*, 2004b). In addition, P-selectin binding to PSGL-1 also initiates signalling pathways which enhance β 2 integrin expression on the surface of neutrophils (Piccardoni *et al.*, 2001). Considering that human milk oligosaccharides may act as soluble selectin-ligand analogs, Bode *et al.* (2004b) investigated their effect on the formation of PNC and PNC-associated neutrophil activation. The group found that within physiological concentrations, the acidic oligosaccharide fraction of milk reduced PNC formation up to 20%, and β 2 integrin expression by 30%. Furthermore, as in the above study (Bode *et al.*, 2004a), the physiological binding determinant sialyl Lewis^x, had a less pronounced effect on PNC formation and neutrophil activation when compared with the acidic oligosaccharide fraction of milk. These differences may be explained by the occurrence of multivalent binding sites on complex milk oligosaccharides. Recently, Schumacher *et al.* (2006) strengthened the evidence that human milk oligsaccharides affect P-selectin binding capacities. The researchers investigated various human milk oligosaccharide fractions with regard to their ability to inhibit cell rolling in a dynamic flow chamber model that simulated blood flow conditions. The results also indicated that the acidic fraction interferes with specific P-selectin ligand binding.

An emerging question that has arisen from the above studies is whether the immuno-modulating effects of milk oligosaccharides can be regarded as an advantage or a disadvantage. Bode *et al.* (2004a) have postulated that the lower incidence of inflammatory diseases in breast-fed infants when compared with formula-fed infants suggests that the effects are beneficial. These benefits may become apparent in cases of excessive immune responses, such as necrotizing enterocolitis (NEC). Indeed, a prospective, multi-centred trial has reported that the incidence of NEC was six times higher in neonates fed formula compared with those fed breast milk (Lucas and Cole, 1990). NEC is the most common and potentially fatal gastrointestinal emergency in neonates. A cascade of inflammatory events take place during the development of NEC, including neutrophil activation, increase in vascular permeability, and release of reactive oxygen species that can eventually lead to breakdown of the mucosal barrier resulting in severe NEC, shock, sepsis, and sometimes death (Patole, 2007). Reactive oxygen species production is significantly higher in neutrophils associated with platelets. However, production can be decreased by interfering with P-selectin-mediated PNC formation. The above studies indicate that human milk oligosaccharides reduce PNC formation, which may, in turn, reduce the production of reactive oxygen species and therefore diminish the harmful effects of neutrophils in NEC.

Maturation of the immune system

Other studies examining the role of human milk oligosaccharides in immune function have focussed on maturation of the newborn immune system. At birth, although developed, the lymphoid system is not yet mature and the immune system is dominated by helper T cells (Th) of subtype-2. Maturation of the immune system subsequently occurs and helper T cells (Th) of subtype-1 become the more dominant cell type (Field *et al.*, 2001; Schack-Nielsen and Michaelsen, 2007). Th1 is proinflammatory and high levels are associated with Crohn's and autoimmune diseases while Th2 is immunoregulatory and high levels are associated with asthma and atrophic dermatitis (Stanner and Smith, 2005). Therefore, striking the right balance with regard to lymphoid cell development is a complex process involving the timely expression of growth factors (cytokines, chemokines), receptors and adhesion molecules (Niers *et al.*, 2007). The helper T cells have different cytokine profiles. For example, Th1 cells secrete more IL-2 and INF-γ, while Th2 cells secrete more IL-4, IL-10, and IL-13 (Schack-Nielsen and Michaelsen, 2007). It is now apparent that the oligosaccharide component of human milk may play a role in modulating the T-cell type. Velupillai and Harn (1994) examined whether neutral milk oligosaccharide structures such as lacto-*N*-fucopentose III (LNFPIII) and lacto-*N*-neotetraose (LNnT) were lymphostimulatory and/or able to induce factors known to down-regulate Th1 cells in a murine model. They found that LNFP-III and related sugars stimulate B cells to proliferate while LNFP-III induced the production of large amounts of interleukin 10 (IL-10), which is known to down-regulate Th1 cells. Modulation of T-cell type by the same oligosaccharides is further discussed by Terrazas *et al.* (2001) who examined the mechanisms that lead to Th2–biasing and immune suppression in mice. The researchers found that these oligosaccharides rapidly expand a suppressor cell population that mediates suppression via an IFN-γ- and a nitric oxide-dependent mechanism, with a discrete role played by IL-10. The authors concluded that these oligosaccharides in milk play a role in driving early immunological events towards Th2-type or anti-inflammatory responses and may have potential as therapeutic agents in the treatment of Th1-mediated autoimmune diseases. A further study by Eiwegger *et al.* (2004) investigated the influence of the neutral and acidic oligosaccharide fraction of human milk on cytokine production and activation of cord blood mononuclear cells (CBMC). The data showed that the acidic oligosaccharide fraction under *in vitro* conditions stimulated cord blood T cells towards enhanced ability of interferon (IFN)-γ as well as TH2-cytokine production. The authors suggest that absorbed acidic human milk oligosaccharides may play a direct role in the postnatal maturation of the immune system and may result in allergy prevention in breast-fed infants.

The above studies provide evidence that specific human milk oligosaccharides may serve as anti-inflammatory components. Whether animal milk oligosaccharides have similar properties warrants investigation. Interestingly, Otani and Monnai (1993) found that the sialic acid fraction of bovine glycomacropeptide may be involved in modulation of the immune system in a mouse model. Moreover, studies on goat milk oligosaccharides suggest they may have important roles in intestinal protection and repair, after damage caused by dextran sodium sulfate-induced colitis (Lara-Villoslada *et al.*, 2006) and hapten-induced colitis in rats (Daddaoua *et al.*, 2006).

12.3.4 Glycome modification by milk oligosaccharides

Mammalian cells are coated with a dense and complex array of glycans (Glycocalyx) which consists of various forms of glycoconjugates. Varki and Varki (2007) estimate that 1% of the genome of humans and other vertebrates is involved in producing the developmentally regulated and tissue-specific glycosylation of each cell type. It is these glycans that act as recognition sites, barriers and carriers at the cell surface. During postnatal development, there is a drastic modification in the surface glycosylation pattern of the intestinal epithelium (Dai *et al.*, 2000; Biol-N'garagba and Louisot, 2003). The terminal glycan chains produced by these cells result from the activity of the corresponding glycosyltransferases and glycosidases. Indeed, a number of studies have investigated the role of nutritional factors in the regulation of glycosyltransferases in the intestinal epithelium of rats (Lenoir *et al.*, 1995, 2000). These studies demonstrate that the intestinal galactotransferase and fucosyltransferase activities can be modified by dietary changes in the postnatal period (by milk or solid diet). Moreover, in the rat small intestine there is a shift from sialylation to fucosylation of the microvillous membrane with weaning (Torres-Pinedo and Mahmood, 1984). In this respect, the possibility that milk might influence the expression of these enzymes in the intestinal epithelium is highly likely, particularly since these cells are the first cells of the infant to encounter milk. Regulation of glycan expression may be useful in controlling bacterial adhesion, either by preventing unwanted pathogen adhesion or by promoting the binding of benefical microbes. In a recent review, Bode (2006) suggests that modification of the glycome of intestinal epithelial cells may occur upon exposure to human milk oligosaccharides. Indeed, a study performed by Angeloni *et al.* (2005) using micro-arrays, demonstrates the probability of this scenario. Using a method known as glycoprofiling, this group were able to monitor changing glycosylation patterns of Caco2 epithelial cells in response to treatment with 3′sialyllactose, the predominant sialylated human and bovine milk oligosaccharide. Gene expression levels of a number of glycosyltransferases were reduced in seven day cultured cells which were treated with 3′sialyllactose, when compared with control cells. The significance of the resultant glycan changes was evaluated by challenging treated and control cells with enteropathogenic *E. coli* (EPEC). Total bacterial adherence was monitored, and treated seven day cultured cells displayed a 50% reduction in adhesion when compared to the control cells. The reduced adherence correlated with reduced sialic acid and lactosamine epitopes, which are key glycans for EPEC adhesion (Vanmaele *et al.*, 1995, 1999). Deciphering surface glycosylation patterns in response to milk oligosaccharides may facilitate the design of strategies to protect against certain pathogens. No doubt, as the versatility of micro-array platforms expands, novel strategies for glycoprofiling will be developed. Indeed, recently it has become possible to monitor changes of the live mammalian cell surface glycome (Tateno *et al.*, 2007). Future studies using such technologies may reveal that cells from organs other than the intestine are subject to altered glycosyltransferase expression and, in turn, changes in their glycocalyx in the presence of milk oligosaccharides.

12.3.5 The role of milk oligosaccharides in brain development

Numerous studies have shown that brain structure and function in animals is permanently influenced by early nutrition (Morgan and Winick, 1980, 1981; Carlson and House, 1986). Among the nutritional factors involved in this process, sialic acid is thought be one of the most critical (reviewed by Sampathkumar *et al.*, 2006). Indeed, experimental evidence linking sialic acid intake with optimal brain development and cognitive abilities was initially reported by Morgan and Winick for rodents (1980). These researchers found that in the rat brain, malnutrition produces a reduction in the extent of dendritic arborization, a decreased concentration of ganglioside and glycoprotein sialic acid, and corresponding deficits in learning behaviour during the first few weeks of life. Interestingly, human milk oligosaccharides are a rich source of this 9-carbon sugar molecule (Wang *et al.*, 2001a) and it has been postulated that human milk sialic acid is used for the synthesis of infant brain gangliosides and sialyl glycoproteins (Wang *et al.*, 2003). Gangliosides are classified as neuronal membrane glycosphingolipids because they contain sialic acid linked to an oligoglycosyl backbone attached to a ceramide base (Ledeen and Yu, 1982). These molecules are important for differentiation (Rosner, 1998), synaptogenesis (Ledeen *et al.*, 1998) and neurotransmission (Rahmann *et al.*, 1990). The sialic acid moieties of gangliosides and glycoproteins in the frontal cortex play both a structural and functional role and may be involved in cell recognition, cell to cell contact formation, receptor binding and modulation, immunological properties and biosignal transduction (Wang and Brand-Miller, 2003). Indeed, the fact that neural cell membranes contain 20 times more sialic acid that other types of membrane clearly indicates a role in neural structure (Schauer, 1982).

Human milk sialyl-oligosaccharide concentration gradually decreases over the course of lactation and, by day 90, there is only 30–50% of that present in colostrum (Martín-Sosa *et al.*, 2003; Wang and Brand-Miller, 2003). Asakuma *et al.* (2007) have recently investigated changes in sialyl-oligosaccharides during the first three days of lactation and found that concentrations change, in addition to the presence of specific types. Such variations most likely occur in accordance with the physiological demands of the newborn infants. In this respect, Wang *et al.* (2001a, 2003) postulate that sialic acid might be a conditionally essential nutrient in infancy because of high demand coupled with a limited capacity for biosynthesis. The analysis of 25 human infants who died of sudden infant death syndrome provides evidence that dietary intake of human milk sialic acid does indeed promote development of the human brain (Wang *et al.*, 2003). Ganglioside-bound and protein-bound sialic acid concentrations were 32% and 22% higher, respectively, in the frontal cortex gray matter of breast-fed infants than that in formula-fed infants, suggesting increased synaptogenesis and differences in neurodevelopment.

Occasionally, sialic acids are linked to each other to form a polymerized structure, polysialic acid. In polymeric form, sialic acid makes the functional element of neural cell adhesion molecule (NCAM). NCAM is a prominent cell adhesion molecule in the nervous system, where it has been shown to participate in a number of developmental processes including neurite cell migration, axonal growth and path finding, synaptogenesis, and synaptic functions associated with learning, memory and circadian rhythm (Troy, 1996; Bruses and Rutishauser, 2001; Yabe *et al.*, 2003). An increase in NCAM sialylation has been associated with higher learning and memory (Cremer *et al.*, 1994), in addition to influencing the NCAM adhesive property (Rutishauser and Landmesser, 1996). Experiments involving NCAM carrying various polysialic acid levels has indicated that these polymers may play a role in neural development (Doherty *et al.*, 1990; Landmesser *et al.*, 1990; Rutishauser, 1996; reviewed by Lowe and Marth, 2003). Bork *et al.* (2005) have found that the concentration of sialic acid in cells regulates the amount of polysialic acid on NCAM. Moreover, two polysialyltransferases (ST8SiaII and ST8SiaIV) were found to be the key enzymes involved in sialic acid metabolism in the neonatal period and have been linked to learning behaviour (Muhlenhoff *et al.*, 2001). Wang *et al.* (2007) designed a series of studies to determine whether early sialic acid supplementation increased learning and memory performance in an appropriate animal model of the human infant. For this study, piglets were fed increasing amounts of sialic acid over 35 days. Brain ganglioside and sialoprotein concentrations and mRNA expression of two learning-associated genes (ST8SiaIV and GNE) were measured. In a dose-response manner, supplementary sialic acid was associated with faster learning, higher concentrations of protein-bound sialic acid in the frontal cortex, and 2 to 3-fold higher mRNA levels of the two learning-related genes. The researchers concluded that sialic acid supplementation can facilitate early brain development in piglets when fed in amounts up to and including the amount present in mature sow milk. It therefore seems that an exogenous source of sialic acid may be critical under conditions of extremely rapid brain growth, particularly during the first months after birth. Infant formulas have a relatively small amount of sialic acid (0–200 mg/ml) when compared with mature human milk (700 mg/ml), (Wang *et al.*, 2003, 2007). Future studies and further trials are required to test the effectiveness of supplementing infant formulas with sialic acid to ensure that formula-fed infants reach their full genetic potential, particularly in the case of premature infants.

12.3.6 Growth-related characteristics of intestinal cells

Regeneration of the intestinal villi occurs through a highly regulated process of proliferation, differientation and apoptosis (Karam, 1999; Studzinski and Harrison, 1999; Sreedharan and Mehta, 2004). A number of studies have shown that milk can influence intestinal development and functions (Ichiba *et al.*, 1992; Hirai *et al.*, 2002; Takeda *et al.*, 2004). This effect has been attributed mainly to the many growth factors that are present in milk, since they have been shown to stimulate proliferation of intestinal cells and the formation of the mucosal barrier (Ichiba *et al.*, 1992; Wagner *et al.*, 1998; Wagner and Forsythe, 2000; Donovan, 2006). Some studies have focussed on investigating the intracellular signalling pathways of many of these growth factors (Moghal and Sternberg, 1999; Takeda *et al.*, 2004). Indeed, Cai *et al.* (2006) has shown that certain oligosaccharides, known as advanced glycation end products (AGE), are involved in growth factor signalling. Kuntz *et al.* (2007) have recently discovered a role for human milk oligosaccharides in various stages of gastrointestinal development *in vitro*. In this study, three epithelial cell lines (Human intestinal cell lines, HT-29 and Caco-2) were exploited to investigate the effects of both neutral and acidic human milk oligosaccharides on proliferation, differentiation and apoptosis. These cell lines vary in their degree of differentiation and were chosen to represent different stages of intestinal cell development. For instance, human intestinal cell lines can be considered to be intestinal stem-like cells, HT-29 cells also represent a very low differentiated phenotype, while

Caco-2 cells reveal a moderate differentiation status. The researchers found that dose-dependent growth inhibition in all three types of epithelial cells occurred in the presence of both neutral and acidic oligosaccharide fractions. A stimulatory effect on cell differentiation was observed in human intestinal cell lines and HT-29 cells but not in Caco-2 cells in the presence of both oligosaccharide fractions. In addition, the neutral oligosaccharide fraction was shown to induce apoptosis in human intestinal cell lines and HT-29 cells but not in Caco-2 cells. The authors concluded that two different mechanisms exist for growth inhibition in intestinal cells by human milk oligosaccharides: (i) by suppressing cell cycle progression through induction of differentiation and/or (ii) by influencing apoptosis. The exact mechanisms by which the oligosaccharide cell surface interactions influence growth remains to be elucidated. However, it is worth noting that the strongest effect on cell differentiation was found in response to sialyllactose, which is also the major acidic oligosaccharide in bovine milk.

12.3.7 Uncharacterized effects of milk oligosaccharides

Bode (2006) postulates that milk oligosaccharides may target a number of protein–carbohydrate interactions. Due to structural similarities, he suggests that human milk oligosaccharides may interfere with interactions involving galectins. Galectins are a family of endogenous lectins with affinity for β-galactoside- and poly-N-acetyllactosamine containing oligosaccharides. The backbone of human milk oligosaccharides is formed by these poly-N-acetyllactosamines (Kunz *et al.*, 2000). Several roles have been assigned to galectins, which range from cell adhesion, regulation of cell growth, and embryonic development, to immune processes such as inflammation. Even metastasis and apoptosis are modulated by the interactions of these molecules. The underlying principle of all these functions is carbohydrate recognition (Hasan *et al.*, 2007). Suffice to say, the possibility of discovering milk oligosaccharides that can interact with these molecules is of major interest.

According to Bode (2006), milk oligosaccharides may also interfere with protein–carbohydrate interactions involving sialic-acid-binding immunoglobulin-like lectins (Siglecs). Siglecs are type I transmembrane proteins belonging to the immunoglobulin superfamily, which bind to terminal sialic acid (α2-3 or/and α 2-6-bound) (McMillan and Crocker, 2008); as described, they are also present on certain milk oligosaccharides. Neutrophils and monocytes/macrophages are important elements of the innate immune defense against invasive bacterial pathogens. These cells express different members of a family of siglecs called the CD33rSiglecs (neutrophils, Siglec-5 and Siglec-9; monocytes, Siglec-5, Siglec-7, and Siglec-9; and macrophages, Siglec-5 and Siglec-11) that are capable of sending inhibitory signals to the cells to modulate inflammatory responses. It has been proposed that the recognition of 'self' sialoglycoconjugates by siglecs could regulate cellular activation during inflammatory and immune responses (Crocker and Varki, 2001), and support for this model has been obtained from studies of CD22/Siglec-2 on B cells (Lanoue *et al.*, 2002) and Siglec-7 on Natural Killer cells (Nicoll *et al.*, 2003). However, given that certain pathogens can also express sialic acids that are important for pathogenicity, it is possible that the interaction between a pathogen's sialoglycoconjugates and siglecs could modulate the immune response, possibly to the advantage of the pathogen (Avril *et al.*, 2006). The question therefore arises: can milk oligosaccharides intervene before pathogens begin to interact with siglecs?

Another potential milk oligosaccharide target could be the DC-specific ICAM3-grabbing non-integrin (DC-SIGN) receptor (Bode, 2006). Dentric cells (DCs) can capture an array of infectious agents (including HIV-1, hepatitis C, Ebola, cytomegalovirus, Dengue virus, *Mycobacterium*, and *Candida albicans*), (Van Kooyk and Geijtenbeek, 2003; Bode, 2006) and present their antigens to T lymphocytes. DCs express, among other receptors, the DC-specific ICAM3-grabbing non-integrin (DC-SIGN) receptor, a C-type lectin, which contains an external calcium-dependent mannose-binding lectin domain (Mitchell *et al.*, 2001; Weis *et al.*, 1998). Interestingly, Naarding *et al.* (2005) revealed that the Lewis^X motif present in human milk can bind to DC-SIGN and inhibit DC-SIGN-mediated transfer of HIV-1 to $CD4^+$ T lymphocytes. Interaction of Lewis^X-containing human milk compounds with DC-SIGN may influence the immune responses summoned in infants to incoming pathogens by preventing their interaction with the DCs and thereby prevent the presentation of pathogen-specific antigens and the subsequent activation of $CD4^+$ T lymphocytes.

12.4 Isolation of milk oligosaccharides for use as food additives

Numerous publications and reviews are available which describe the various extraction and separation techniques used in milk oligosaccharide analysis (Boehm and Stahl, 2003; Mehra and Kelly, 2006; Niñonuevo *et al.*, 2006, 2008). To date, the majority of methods that have been established relate to human milk oligosaccharides analysis; however, such techniques can readily be applied to the investigation of milk oligosaccharides from other species. Separation of oligosaccharides generally begins with removal of fat by centrifugation followed by the protein, via precipitation (Egge *et al.*, 1983). To separate milk oligosaccharides from lactose, charcoal column chromatography (Brand-Miller *et al.*, 1998) or ultrafiltration can be used (Sarney *et al.*, 2000). Kunz *et al.* (1996) used high-pH anion-exchange chromatography and pulsed amperometric detection (HPAEC-PAD) to separate and characterize neutral and acidic lactose-derived oligosaccharides without prior derivatization or reduction. Another common method for quantification of milk oligosaccharides is capillary electrophoresis (CE)

(Song *et al.*, 2002). Indeed, Nakajima *et al.* (2006) have successfully developed a capillary affinity electrophoresis (CAE) method for analyzing milk oligosaccharides based on their affinities for lectins.

For structural analyses of distinct oligosaccharides, nuclear magnetic resonance (NMR) spectrometry proves a powerful method (reviewed by Boehm and Stahl, 2003) but, if only minute amounts of sample are available, mass spectrometry (MS) plays an important role in sequencing milk oligosaccharides. The introduction of matrix-assisted laser desorption/ionization MS (MALDI-MS) and carbohydrate compatible matrices greatly expanded the utility of MS analysis of milk oligosaccharides (Niňonuevo *et al.*, 2006; reviewed by Boehm and Stahl, 2003; Mehra and Kelly, 2006). Of great importance to the field was the introduction of the HPLC chip developed by Agilent Technologies and a team of researchers at UC Davis. An analytical strategy was devised by this group that allows the routine profiling of milk oligosaccharides using HPLC chip/TOF (time of flight) technology (Niňonuevo *et al.*, 2005, 2006, 2008). This technique uses an integrated microfluidic chip, coupled with a high mass accuracy time of flight mass analyzer. The chip technologies allow the determination and identification of new and known oligosaccharides in essentially any milk sample. With a few microliters of milk corresponding to pico- to nanomoles of oligosaccharides, complete profiling can be readily performed with high reproducibility. New structural elucidation technology, including the use of infrared lasers to degrade oligosaccharide ions, yield structural information on as little as a few femtomoles (one quadrillionth mole) of material (Lebrilla, 2007).

12.5 Commercial products containing oligosaccharides

As previously mentioned, the biochemical composition of human milk oligosaccharides appears unique. Additionally, these molecules cannot be synthesized in large amounts and to do so is costly. Therefore, much focus has been given to the production of other carbohydrates which, while structurally different, have similar biological function. In particular, prebiotic oligosaccharides that promote the growth of bifidobacteria and lactobacilli in the colon, thereby reproducing the effects of human oligosaccharides, have been of major interest. The best characterized prebiotic is a mixture of galacto-oligosaccharides and fructo-oligosaccharides, well known as GOS/FOS mixture (Bruzzese *et al.*, 2006). Recently, it has been shown that the supplementation of infant formula with a mixture of GOS/FOS is effective in modifying the composition of the intestinal flora in both term and preterm infants (Moro *et al.*, 2002; Boehm *et al.*, 2002). Indeed, many recent studies involving non-digestable oligosaccharides other than those found in human milk, suggest that many beneficial properties can be associated with their presence. For instance, these oligosaccharides and their breakdown products in the form of short chain fatty acids (SCFAs) have been linked with alleviation of constipation, improved mineral absorption, regulation of lipid metabolism, cancer prevention, treatment of hepatic encephalopathy, modulation of glycemia/insulinemia and modulation of the immune system (reviewed by Swennen *et al.*, 2006). Whether many of these functions occur in the human milk-fed infant has yet to be proven. Researchers are also focusing their attention on milk oligosaccharides from other mammals, particularly farm animals, as a source for novel prebiotics. One alternative may be to use bovine colostrum, given its higher content of oligosaccharides when compared with mature bovine milk (Davis *et al.*, 1983; Martinez-Ferez *et al.*, 2006). Infant formulas designed for nutrition during the first weeks after birth, are currently produced from different bovine milk derivatives. Therefore, isolation, fractionation or enrichment of fractions containing bovine milk oligosaccharides may have potential applications if used in infant formulas.

12.6 Perspectives and future trends

Validating novel bioactivities associated with milk oligosaccharides will differentiate dairy from other ingredients in the food industry. Milk oligosaccharides that can be associated with clearly defined physiological outcomes and that can be measured in a quantitative fashion will lead to a strong marketing position. By understanding milk oligosaccharide structure and functions, it is tempting to suggest that in the future many novel foods and beverages, having undergone regulatory approval, will be marketed with the aim of promoting health of the consumer. Before this scenario can be realized, economical and rapid enrichment and purification procedures must be put in place, in addition to the various animal studies and human clinical trials which will be required to validate the different bioactivities associated with these molecules.

12.7 References

AGOSTONI C, AXELSSON I, BRAEGGER C, GOULET O, KOLETZKO B, MICHAELSEN K F, RIGO J, SHAMIR R, SZAJEWSKA H, TURCK D and WEAVER L T (2004), 'Probiotic bacteria in dietetic products for infants: A commentary by the ESPGHAN Committee on Nutrition', *J. Pediatr. Gastroenterol. Nutr.*, 38, 365–74.

ANDERSSON B, DAHMEN J, FREJD T, LEFFLER H, MAGNUSSON G, NOORI G and EDEN C S (1983), 'Identification of an active disaccharide unit of a glycoconjugate receptor for pneumococci attaching to human pharyngeal epithelial cells', *J. Exp. Med.*, 158, 559–70.

ANDERSSON B, PORRAS O, HANSON L A, LAGERGARD T and SVANBORG-EDEN C (1986), 'Inhibition of attachment of *Streptococcus pneumoniae* and *Haemophilus influenzae* by human milk and receptor oligosaccharides', *J. Infect. Dis.*, 153, 232–7.

ANGELONI S, RIDET J L, KUSY N, GAO H, CREVOISIER F, GUINCHARD S, KOCHHAR S, SIGRIST H and SPRENGER N (2005), 'Glycoprofiling with micro-arrays of glycoconjugates and lectins', *Glycobiology*, 15, 31–41.

ASAKUMA S, AKAHORI M, KIMURA K, WATANABE Y, NAKAMURA T, TSUNEMI M, ARAI I, SANAI Y and URASHIMA T (2007), 'Sialyl oligosaccharides of human colostrum: Changes in concentration during the first three days of lactation', *Biosci. Biotechnol. Biochem.*, 71, 1447–51.

ASPHOLM-HURTIG M, DAILIDE G, LAHMANN M, KALIA A, ILVER D, ROCHE N, VIKSTROM S, SJOSTROM R, LINDEN S, BACKSTROM A, LUNDBERG C, ARNQVIST A, MAHDAVI J, NILSSON U J, VELAPATINO B, GILMAN R H, GERHARD M, ALARCON T, LOPEZ-BREA M, NAKAZAWA T, FOX J G, CORREA P, DOMINGUEZ-BELLO M G, PEREZ-PEREZ G I, BLASER M J, NORMARK S, CARLSTEDT I, OSCARSON S, TENEBERG S, BERG D E and BOREN T (2004) 'Functional adaptation of BabA, and *H. pylori* ABO blood group antigen binding adhesion', *Science*, 305, 519–22.

AVRIL T, WAGNER E R, WILLISON H J and CROCKER P R (2006), 'Sialic acid-binding immunoglobulin-like lectin 7 mediates selective recognition of sialylated glycans expressed on *Campylobacter jejuni* lipooligosaccharides', *Infect. Immun.*, 74, 4133–41.

BALMER S E and WHARTON B A (1989), 'Diet and faecal flora in the newborn: Breast milk and infant formula', *Arch. Dis. Child.*, 64, 1672–7.

BAVINGTON C and PAGE C (2005), 'Stopping bacterial adhesion: A novel approach to treating infections', *Respiration*, 72, 335–44.

BENNETT H J and ROBERTS I S (2005), 'Identification of a new sialic acid-binding protein in *Helicobacter pylori*', *FEMS Immunol. Med. Microbiol.*, 44, 163–9.

BENNO Y, SAWADA K and MITSUOKA T (1984), 'The intestinal microflora of infants: Composition of fecal flora in breast-fed and bottle-fed infants', *Microbiol. Immunol.*, 28, 975–86.

BEVILACQUA M P and NELSON R M (1993), 'Selectins', *J. Clin. Invest.*, 91, 379–87.

BIOL-N'GARAGBA M C and LOUISOT P (2003), 'Regulation of the intestinal glycoprotein glycosylation during postnatal development: Role of hormonal and nutritional factors', *Biochimie*, 85, 331–52.

BJÖRKSTÉN B, NAABER P, SEPP E and MIKELSAAR M (1999), 'The intestinal microflora in allergic Estonian and Swedish 2-year-old children', *Clin. Exp. Allergy*, 29, 342–6.

BJÖRKSTÉN B, SEPP E, JULGE K, VOOR T and MIKELSAAR M (2001), 'Allergy development and the intestinal microflora during the first year of life', *J. Allergy Clin. Immunol.*, 108, 516–20.

BLASER M J (1996), 'The bacteria behind ulcers', *Sci. Am.*, 274, 104–7.

BOAT T F, DAVIS J, STERN R C and CHENG P W (1978), 'Effect of blood group determinants on binding of human salivary mucous glycoproteins to influenza virus', *Biochim. Biophys. Acta*, 540, 127–33.

BODE L (2006), 'Recent advances on structure, metabolism, and function of human milk oligosaccharides', *J. Nutr.*, 136, 2127–30.

BODE L, KUNZ C, MUHLY-REINHOLZ M, MAYER K, SEEGER W and RUDLOFF S (2004a), 'Inhibition of monocyte, lymphocyte, and neutrophil adhesion to endothelial cells by human milk oligosaccharides', *Thromb. Haemost.*, 92, 1402–10.

BODE L, RUDLOFF S, KUNZ C, STROBEL S and KLEIN N (2004b), 'Human milk oligosaccharides reduce platelet–neutrophil complex formation leading to a decrease in neutrophil beta 2 integrin expression', *J. Leukoc. Biol.*, 76, 820–6.

BOEHM G, LIDESTRI M, CASETTA P, JELINEK J, NEGRETTI F, STAHL B and MARINI A (2002), 'Supplementation of a bovine milk formula with an oligosaccharide mixture increases counts of faecal bifidobacteria in preterm infants', *Arch. Dis. Child. Fetal Neonatal Ed.*, 86, F178–81.

BOEHM G and STAHL B (2003), 'Oligosaccharides'. In: T. Mattila and M. Saarela, Editors, *Functional Dairy Products*, Woodhead Publishing Limited, Cambridge, England, 203–243.

BORK K, REUTTER W, GERARDY-SCHAHN R and HORSTKORTE R (2005), 'The intracellular concentration of sialic acid regulates the polysialylation of the neural cell adhesion molecule', *FEBS Lett.*, 579, 5079–83.

BRAND-MILLER J C, MCVEAGH P, MCNEIL Y and MESSER M (1998), 'Digestion of human milk oligosaccharides by healthy infants evaluated by the lactulose hydrogen breath test', *The Journal of Pediatrics*, 133, 95–8.

BRASSART D, WOLTZ A, GOLLIARD M and NEESER J R (1991), '*In vitro* inhibition of adhesion of *Candida albicans* clinical isolates to human buccal epithelial cells by Fuc alpha 1-2Gal beta-bearing complex carbohydrates', *Infect. Immun.*, 59, 1605–13.

BRÜCK W M, KELLEHER S L, GIBSON G R, GRAVERHOLT G and LONNERDAL B L (2006), 'The effects of alpha-lactalbumin and glycomacropeptide on the association of Caco-2 cells by enteropathogenic *Escherichia coli*, *Salmonella typhimurium* and *Shigella flexneri*', *FEMS Microbiol. Lett.*, 259, 158–62.

BRUSES J L and RUTISHAUSER U (2001), 'Roles, regulation, and mechanism of polysialic acid function during neural development', *Biochimie*, 83, 635–43.

BRUZZESE E, VOLPICELLI M, SQUAGLIA M, TARTAGLIONE A and GUARINO A (2006), 'Impact of prebiotics on human health', *Dig. Liver. Dis.*, 38 Suppl 2, S283–7.

CAI W, HE J C, ZHU L, LU C and VLASSARA H (2006), 'Advanced glycation end product (AGE) receptor 1 suppresses cell oxidant stress and activation signaling via EGF receptor', *Proc. Natl. Acad. Sci. U S A*, 103, 13801–6.

CARDEN D L and GRANGER D N (2000), 'Pathophysiology of ischaemia-reperfusion injury', *J. Pathol.*, 190, 255–66.

CARLSON S E and HOUSE S G (1986), 'Oral and intraperitoneal administration of N-acetylneuraminic acid: Effect on rat cerebral and cerebellar N-acetylneuraminic acid', *J. Nutr.*, 116, 881–6.

CERVANTES L E, NEWBURG D S and RUIZ-PALACIOS G M (1995), 'α1-2 Fucosylated chains (H-2 and Lewisb) are the main human milk receptor analogs for *Campylobacter* (abstract).' *Pediatr. Res.*, 37, 171A.

CHATURVEDI P, WARREN C D, BUESCHER C R, PICKERING L K and NEWBURG D S (2001), 'Survival of human milk oligosaccharides in the intestine of infants', *Adv. Exp. Med. Biol.*, 501, 315–23.

COPPA G V, BRUNI S, MORELLI L, SOLDI S and GABRIELLI O (2004), 'The first prebiotics in humans: Human milk oligosaccharides', *J. Clin. Gastroenterol.*, 38, S80–3.

COPPA G V, BRUNI S, ZAMPINI L *et al.* (2000), 'Oligosaccharides of human milk'. In: Muzzarelli RAA, ed. *Chitosan per os: From Dietary Supplement to Drug Carrier*, Grottammare: Atec, 313–330.

COPPA G V, BRUNI S, ZAMPINI L, GALEAZZI T and GABRIELLI O (2003), Oligosaccharides of human milk inhibit the adhesion of *Listeria monocytogenes* to Caco-2 cells', *Ital. J. Pediatr.*, 29, 61–68.

COPPA G V, GABRIELLI O, GIORGI P, CATASSI C, MONTANARI M P, VARALDO P E and NICHOLS B L (1990), 'Preliminary study of breastfeeding and bacterial adhesion to uroepithelial cells', *Lancet*, 335, 569–71.

COPPA G V, PIERANI P, ZAMPINI L, BRUNI S, CARLONI I and GABRIELLI O (2001), 'Characterization of oligosaccharides in milk and feces of breast-fed infants by high-performance anion-exchange chromatography', *Adv. Exp. Med. Biol.*, 501, 307–14.

COPPA G V, ZAMPINI L, GALEAZZI T, FACINELLI B, FERRANTE L, CAPRETTI R and ORAZIO G (2006), 'Human milk oligosaccharides inhibit the adhesion to Caco-2 cells of diarrheal pathogens: *Escherichia coli*, *Vibrio cholerae*, and *Salmonella fyris*', *Pediatr. Res.*, 59, 377–82.

CREMER H, LANGE R, CHRISTOPH A, PLOMANN M, VOPPER G, ROES J, BROWN R, BALDWIN S, KRAEMER P, SCHEFF S and *et al.* (1994), 'Inactivation of the N-CAM gene in mice results in size reduction of the olfactory bulb and deficits in spatial learning', *Nature*, 367, 455–9.

CRESPO A and SUH B (2001), '*Helicobacter pylori* infection: Epidemiology, pathophysiology, and therapy', *Arch. Pharm. Res.*, 24, 485–98.

CROCKER P R and VARKI A (2001), 'Siglecs, sialic acids and innate immunity', *Trends Immunol*, 22, 337–42.

DADDAOUA A, PUERTA V, REQUENA P, MARTINEZ-FEREZ A, GUADIX E, DE MEDINA F S, ZARZUELO A, SUAREZ M D, BOZA J J and MARTINEZ-AUGUSTIN O (2006), 'Goat milk oligosaccharides are anti-inflammatory in rats with hapten-induced colitis', *The Journal of Nutrition*, 136, 672–6.

DAI D, NANTHKUMAR N N, NEWBURG D S and WALKER W A (2000), 'Role of oligosaccharides and glycoconjugates in intestinal host defense', *J. Pediatr. Gastroenterol. Nutr.*, 30 Suppl 2, S23–33.

DAVIS D T, HOLT C and CHRISTIE W W (1983), 'Biochemistry of lactation', 71–117.

DETERS A M, LENGSFELD C and HENSEL A (2005), 'Oligo- and polysaccharides exhibit a structure-dependent bioactivity on human keratinocytes *in vitro*', *J. Ethnopharmacol.*, 102, 391–399.

DOHERTY P, FRUNS M, SEATON P, DICKSON G, BARTON C H, SEARS T A and WALSH F S (1990), 'A threshold effect of the major isoforms of NCAM on neurite outgrowth', *Nature*, 343, 464–6.

DONOVAN S (2006), 'Role of human milk components in gastrointestinal development: Current knowledge and future needs.' *J. Pediatr.*, 149, S49-S61.

EGGE H (1993), 'The diversity of oligosaccharides in human milk'. In *New Perspectives in Infant Nutrition*, ed. B Renner, G Sawatzki, Thieme, Stuttgart, New York, 12–26.

EGGE H, DELL A and VON NICOLAI H (1983), 'Fucose containing oligosaccharides from human milk. I. Separation and identification of new constituents', *Arch. Biochem. Biophys.*, 224, 235–53.

EIWEGGER T, STAHL B, SCHMITT J, BOEHM G, GERSTMAYR M, PICHLER J, DEHLINK E, LOIBICHLER C, URBANEK R and SZEPFALUSI Z (2004), 'Human milk-derived oligosaccharides and plant-derived oligosaccharides stimulate cytokine production of cord blood T-cells in vitro', *Pediatr. Res.*, 56, 536–40.

ENGFER M B, STAHL B, FINKE B, SAWATZKI G and DANIEL H (2000), 'Human milk oligosaccharides are resistant to enzymatic hydrolysis in the upper gastrointestinal tract', *Am. J. Clin. Nutr.*, 71, 1589–96.

ERBE D V, WATSON S R, PRESTA L G, WOLITZKY B A, FOXALL C, BRANDLEY B K and LASKY L A (1993), 'P- and E-selectin use common sites for carbohydrate ligand recognition and cell adhesion', *The Journal of Cell Biology*, 120, 1227–35.

EVANS D G, EVANS D J, JR., MOULDS J J and GRAHAM D Y (1988), 'N-acetylneuraminyllactose-binding fibrillar hemagglutinin of *Campylobacter pylori*: A putative colonization factor antigen', *Infect Immun*, 56, 2896–906.

FAVIER C F, VAUGHAN E E, DE VOS W M and AKKERMANS A D (2002), 'Molecular monitoring of succession of bacterial communities in human neonates', *Appl. Environ. Microbiol.*, 68, 219–26.

FIELD C J, CLANDININ M T and VAN AERDE J E (2001), 'Polyunsaturated fatty acids and T-cell function: Implications for the neonate', *Lipids*, 36, 1025–32.

FINKE B, STAHL B, PFENNINGER A, KARAS M, DANIEL H and SAWATZKI G (1999), 'Analysis of high-molecular-weight oligosaccharides from human milk by liquid chromatography and MALDI-MS', *Anal. Chem.*, 71 3755–3762.

GERMAN J B, FREEMAN S L, LEBRILLA C B and MILLS D A (2008), 'Human milk oligosaccharides: Evolution, structures and bioselectivity as substrates for intestinal bacteria', *Nestle Nutr Workshop Ser Pediatr Program*, 62, 205–18; discussion 218–22.

GIBSON G R and ROBERFROID M B (1995), 'Dietary modulation of the human colonic microbiota: Introducing the concept of prebiotics', *J. Nutr.*, 125, 1401–12.

GLASS R I, HOLMGREN J, HALEY C E, KHAN M R, SVENNERHOLM A M, STOLL B J, BELAYET HOSSAIN K M, BLACK R E, YUNUS M and BARUA D (1985), 'Predisposition for cholera of individuals with O blood group. Possible evolutionary significance', *Am. J. Epidemiol.*, 121, 791–6.

GNOTH M J, KUNZ C, KINNE-SAFFRAN E and RUDLOFF S (2000), 'Human milk oligosaccharides are minimally digested in vitro', *J. Nutr.*, 130, 3014–20.

GOPAL P K, GILL, H S (2000), 'Oligosaccharides and glycoconjugates in bovine milk and colostrum', *British Journal of Nutrition* 84 (suppl. 1), S69–S74.

GUSTAFSSON A, HULTBERG A, SJOSTROM R, KACSKOVICS I, BREIMER M E, BOREN T, HAMMARSTROM L and HOLGERSSON J (2006), 'Carbohydrate-dependent inhibition of *Helicobacter pylori* colonization using porcine milk', *Glycobiology*, 16, 1–10.

GYÖRGY P, NORRIS R F and ROSE C S (1954), 'Bifidus factor. I. A variant of *Lactobacillus bifidus* requiring a special growth factor', *Arch. Biochem. Biophys.*, 48, 193–201.

HAKKARAINEN J, TOIVANEN M, LEINONEN A, FRANGSMYR L, STROMBERG N, LAPINJOKI S, NASSIF X and TIKKANEN-KAUKANEN C (2005), 'Human and bovine milk oligosaccharides inhibit *Neisseria meningitidis* pili attachment *in vitro*', *J. Nutr.*, 135, 2445–8.

HARMSEN H J, WILDEBOER-VELOO A C, RAANGS G C, WAGENDORP A A, KLIJN N, BINDELS J G and WELLING G W (2000), 'Analysis of intestinal flora development in breast-fed and formula-fed infants by using molecular identification and detection methods', *J. Pediatr. Gastroenterol. Nutr.*, 30, 61–7.

HASAN S S, ASHRAF G M and BANU N (2007), 'Galectins–potential targets for cancer therapy', *Cancer Lett.*, 253, 25–33.

HIRAI C, ICHIBA H, SAITO M, SHINTAKU H, YAMANO T and KUSUDA S (2002), 'Trophic effect of multiple growth factors in amniotic fluid or human milk on cultured human fetal small intestinal cells', *J. Pediatr. Gastroenterol. Nutr.*, 34, 524–8.

HSUEH W, CAPLAN M S, QU X W, TAN X D, DE PLAEN I G and GONZALEZ-CRUSSI F (2003), 'Neonatal necrotizing enterocolitis: Clinical considerations and pathogenetic concepts', *Pediatr. Dev. Pathol.*, 6, 6–23.

HUTSON A M, ATMAR R L, GRAHAM D Y and ESTES M K (2002), 'Norwalk virus infection and disease is associated with ABO histo-blood group type', *J. Infect. Dis.*, 185, 1335–7.

ICHIBA H, KUSUDA S, HAGANE Y, FUJITA K and ISSIKI C (1992), 'Measurement of growth promoting activity in human milk using a fetal small intestinal cell line', *Biol. Neonate*, 61, 47–53.

IKEHARA Y, NISHIHARA S, YASUTOMI H, KITAMURA T, MATSUO K, SHIMIZU N, INADA K, KODERA Y, YAMAMURA Y, NARIMATSU H, HAMAJIMA N and TATEMATSU M (2001), 'Polymorphisms of two fucosyltransferase genes (Lewis and Secretor genes) involving type I Lewis antigens are associated with the presence of anti-*Helicobacter pylori* IgG antibody', *Cancer Epidemiol. Biomarkers Prev.*, 10, 971–7.

ILVER D, ARNQVIST A, OGREN J, FRICK I M, KERSULYTE D, INCECIK E T, BERG D E, COVACCI A, ENGSTRAND L and BOREN T (1998), '*Helicobacter pylori* adhesin binding fucosylated histo-blood group antigens revealed by retagging', *Science*, 279, 373–7.

ITO T, SUZUKI Y, TAKADA A, KAWAMOTO A, OTSUKI K, MASUDA H, YAMADA M, SUZUKI T, KIDA H and KAWAOKA Y (1997), 'Differences in sialic acid–galactose linkages in the chicken egg amnion and allantois influence human influenza virus receptor specificity and variant selection', *J. Virol.*, 71, 3357–62.

KALLIOMÄKI M, KIRJAVAINEN P, EEROLA E, KERO P, SALMINEN S and ISOLAURI E (2001), 'Distinct patterns of neonatal gut microflora in infants in whom atopy was and was not developing', *J. Allergy Clin. Immunol.*, 107, 129–34.

KARAM S M (1999), 'Lineage commitment and maturation of epithelial cells in the gut', *Front. Biosci.*, 4, D286–98.

KAWAKAMI H (1997), 'Biological significance of sialic acid-containing substances in milk and their application', *Rec. Res. Dev. Agric. Biol. Chem.*, 1, 193–208.

KELLY M, HWANG J M and KUBES P (2007), 'Modulating leukocyte recruitment in inflammation', *J. Allergy Clin. Immunol.*, 120, 3–10.

KLEIN N, SCHWERTMANN A, PETERS M, KUNZ C and STROBEL S (2000), 'Immunomodulatory effects of breast milk oligosaccharides', *Adv. Exp. Med. Biol.*, 478, 251–9.

KUBES P, ARFORS K E and GRANGER D N (1991), 'Platelet-activating factor-induced mucosal dysfunction: Role of oxidants and granulocytes', *Am. J. Physiol.*, 260, G965–71.

KUHN R (1958), 'Oligosaccharides of milk.' *Bull. Soc. Chim. Biol.*, 40, 297–314.

KUNTZ S, RUDLOFF S and KUNZ C (2007), 'Oligosaccharides from human milk influence growth-related characteristics of intestinally transformed and non-transformed intestinal cells', *Br. J. Nutr.*, 1–10.

KUNZ C, BODE L and RUDLOFF S (2003), 'Genetic Variability of Human Milk Oligosaccharides: Are there Biologic Consequences?' In: *Genetic Expression and Nutrition*. Eds. Bachmann, C. and Koletzko, B. Nestle Nutrition Workshop Series, Paediatric Program Vol. 50 137–152.

KUNZ C and RUDLOFF S (1993), 'Biological functions of oligosaccharides in human milk', *Acta paediatrica* Oslo, Norway, 82, 903–12.

KUNZ C and RUDLOFF S (2002), 'Health benefits of milk-derived carbohydrates', *Bull. Int. Dairy Fed.*, 375, 72–79.

KUNZ C and RUDLOFF S (2006), 'Health promoting aspects of milk oligosaccharides', *Int. Dairy J.*, 16, 1341–1346.

KUNZ C and RUDLOFF S (2008), 'Potential anti-inflammatory and anti-infectious effects of human milk oligosaccharides', *Adv. Exp. Med. Biol.*, 606, 455–65.

KUNZ C, RUDLOFF S, BAIER W, KLEIN N and STROBEL S (2000), 'Oligosaccharides in human milk: Structural, functional, and metabolic aspects', *Annu. Rev. Nutr.*, 20, 699–722.

KUNZ C, RUDLOFF S, HINTELMANN A, POHLENTZ G and EGGE H (1996), 'High-pH anion-exchange chromatography with pulsed amperometric detection and molar response factors of human milk oligosaccharides', *Journal of Chromatography. B, Biomedical Applications*, 685, 211–21.

KUNZ C, RUDLOFF S, SCHAD W and BRAUN D (1999), 'Lactose-derived oligosaccharides in the milk of elephants: Comparison with human milk', *The British Journal of Nutrition*, 82, 391–9.

LANDMESSER L, DAHM L, TANG J C and RUTISHAUSER U (1990), 'Polysialic acid as a regulator of intramuscular nerve branching during embryonic development', *Neuron*, 4, 655–67.

LANOUE A, BATISTA F D, STEWART M and NEUBERGER M S (2002), 'Interaction of CD22 with alpha 2,6-linked sialoglycoconjugates: Innate recognition of self to dampen B cell autoreactivity?' *Eur. J. Immunol.*, 32, 348–55.

LARA-VILLOSLADA F, DEBRAS E, NIETO A, CONCHA A, GALVEZ J, LOPEZ-HUERTAS E, BOZA J, OBLED C and XAUS J (2006), 'Oligosaccharides isolated from goat milk reduce intestinal inflammation in a rat model of dextran sodium sulfate-induced colitis', *Clinical Nutrition (Edinburgh, Scotland)*, 25, 477–88.

LAROUX F S and GRISHAM M B (2001), 'Immunological basis of inflammatory bowel disease: Role of the microcirculation', *Microcirculation*, 8, 283–301.

LARSEN E, CELI A, GILBERT G E, FURIE B C, ERBAN J K, BONFANTI R, WAGNER D D and FURIE B (1989), 'PADGEM protein: A receptor that mediates the interaction of activated platelets with neutrophils and monocytes', *Cell*, 59, 305–12.

LEBRILLA C B (2007), 'Nutrition and Health – Milk bioactives', *California Dairy Research Foundation*, Annual Report.

LEDEEN R W, WU G, LU Z H, KOZIRESKI-CHUBACK D and FANG Y (1998), 'The role of GM1 and other gangliosides in neuronal differentiation. Overview and new finding', *Ann. N. Y. Acad. Sci.*, 845, 161–75.

LEDEEN R W and YU R K (1982), 'Gangliosides: Structure, isolation, and analysis', *Methods Enzymol.*, 83, 139–91.

LENOIR D, GRECO S, LOUISOT P and BIOL M C (2000), 'Implication of insulin and nutritional factors in the regulation of intestinal galactosyltransferase activity during postnatal development', *Metabolism*, 49, 526–31.

LENOIR D, RUGGIERO-LOPEZ D, LOUISOT P and BIOL M C (1995), 'Developmental changes in intestinal glycosylation: Nutrition-dependent multi-factor regulation of the fucosylation pathway at weaning time', *Biochim. Biophys. Acta.*, 1234, 29–36.

LIUKKONEN J, HAATAJA S, TIKKANEN K, KELM S and FINNE J (1992), 'Identification of N-acetylneuraminyl alpha 2-3 poly-N-acetyllactosamine glycans as the receptors of sialic acid-binding *Streptococcus suis* strains', *J. Biol. Chem.*, 267, 21105–11.

LOCASCIO R G, NINONUEVO M R, FREEMAN S L, SELA D A, GRIMM R, LEBRILLA C B, MILLS D A and GERMAN J B (2007), 'Glycoprofiling of bifidobacterial consumption of human milk oligosaccharides demonstrates strain specific, preferential consumption of small chain glycans secreted in early human lactation', *J. Agric. Food Chem.*, 55, 8914–9.

LOWE J B and MARTH J D (2003), 'A genetic approach to Mammalian glycan function', *Annu. Rev. Biochem.*, 72, 643–91.

LUCAS A and COLE T J (1990), 'Breast milk and neonatal necrotising enterocolitis', *Lancet*, 336, 1519–23.

MA X L, WEYRICH A S, LEFER D J, BUERKE M, ALBERTINE K H, KISHIMOTO T K and LEFER A M (1993), 'Monoclonal antibody to L-selectin attenuates neutrophil accumulation and protects ischemic reperfused cat myocardium', *Circulation*, 88, 649–58.

MAHDAVI J, SONDEN B, HURTIG M, OLFAT F O, FORSBERG L, ROCHE N, ANGSTROM J, LARSSON T, TENEBERG S, KARLSSON K A, ALTRAJA S, WADSTROM T, KERSULYTE D, BERG D E, DUBOIS A, PETERSSON C, MAGNUSSON K E, NORBERG T, LINDH F, LUNDSKOG B B, ARNQVIST A, HAMMARSTROM L and BOREN T (2002), '*Helicobacter pylori* SabA adhesin in persistent infection and chronic inflammation', *Science*, 297, 573–8.

MARTÍN-SOSA S, MARTIN M J, GARCIA-PARDO L A and HUESO P (2003), 'Sialyloligosaccharides in human and bovine milk and in infant formulas: Variations with the progression of lactation', *J. Dairy Sci.*, 86, 52–9.

MARTINEZ-FEREZ A, RUDLOFF S, GUADIX A, HENKEL C A, POHLENTZ G and BOZA J J *et al.* (2006) 'Goats' milk as a natural source of lactose-derived oligosaccharides: Isolation by membrane technology', *Int. Dairy J.*, 16, 173–181.

MATROSOVICH M and KLENK H D (2003), 'Natural and synthetic sialic acid containing inhibitors of influenza virus receptor binding', *Reviews in Medical Virology*, 13, 85–97.

MATROSOVICH M N, GAMBARYAN A S, TUZIKOV A B, BYRAMOVA N E, MOCHALOVA L V, GOLBRAIKH A A, SHENDEROVICH M D, FINNE J and BOVIN N V (1993), 'Probing of the receptor-binding sites of the H1 and H3 influenza A and influenza B virus hemagglutinins by synthetic and natural sialosides', *Virology*, 196, 111–21.

MCEVER R P (1997), 'Selectin–carbohydrate interactions during inflammation and metastasis', *Glycoconj. J.*, 14, 585–91.

MCMILLAN S J and CROCKER P R (2008), 'CD33-related sialic-acid-binding immunoglobulin-like lectins in health and disease', *Carbohydr. Res.*, 343, 2050–6.

MCVEAGH P and BRAND MILLER J B (1997), 'Human milk oligosaccharides: Only the breast', *J. Paediatr. Child Health*, 33, 281–6.

MEHRA R and KELLY P (2006), 'Milk Oligosaccharides: Structural and technological aspects', *International Dairy Journal*, 16, 1334–1340.

MILLAR M R, LINTON C J, CADE A, GLANCY D, HALL M and JALAL H (1996), 'Application of 16S rRNA gene PCR to study bowel flora of preterm infants with and without necrotizing enterocolitis', *J. Clin. Microbiol.*, 34, 2506–10.

MITCHELL D A, FADDEN A J and DRICKAMER K (2001), 'A novel mechanism of carbohydrate recognition by the C-type lectins DC-SIGN and DC-SIGNR. Subunit organization and binding to multivalent ligands', *The Journal of Biological Chemistry*, 276, 28939–45.

MOGHAL N and STERNBERG P W (1999), 'Multiple positive and negative regulators of signaling by the EGF-receptor', *Curr. Opin. Cell Biol.*, 11, 190–6.

MOREAU M C, THOMASSON M, DUCLUZEAU R and RAIBAUD P (1986), 'Kinetics of establishment of digestive microflora in the human newborn infant as a function of the kind of milk', *Reprod. Nutr. Dev.*, 26, 745–53.

MORGAN B L and WINICK M (1980), 'Effects of administration of N-acetylneuraminic acid (NANA) on brain NANA content and behavior', *J. Nutr.*, 110, 416–24.

MORGAN B L and WINICK M (1981), 'The subcellular localization of administered N-acetylneuraminic acid in the brains of well-fed and protein restricted rats', *Br. J. Nutr.*, 46, 231–8.

MORO G, MINOLI I, MOSCA M, FANARO S, JELINEK J, STAHL B and BOEHM G (2002), 'Dosage-related bifidogenic effects of galacto- and fructooligosaccharides in formula-fed term infants', *J. Pediatr. Gastroenterol. Nutr.*, 34, 291–5.

MORROW A L, RUIZ-PALACIOS G M, ALTAYE M, JIANG X, GUERRERO M L, MEINZEN-DERR J K, FARKAS T, CHATURVEDI P, PICKERING L K and NEWBURG D S (2004), 'Human milk oligosaccharides are associated with protection against diarrhea in breast-fed infants', *J. Pediatr.*, 145, 297–303.

MORROW A L, RUIZ-PALACIOS G M, JIANG X and NEWBURG D S (2005), 'Human-milk glycans that inhibit pathogen binding protect breast-feeding infants against infectious diarrhea', *J. Nutr.*, 135, 1304–7.

MUHLENHOFF M, MANEGOLD A, WINDFUHR M, GOTZA B and GERARDY-SCHAHN R (2001), 'The impact of N-glycosylation on the functions of polysialyltransferases', *J. Biol. Chem.*, 276, 34066–73.

MYSORE J V, WIGGINTON T, SIMON P M, ZOPF D, HEMAN-ACKAH L M and DUBOIS A (1999), 'Treatment of *Helicobacter pylori* infection in rhesus monkeys using a novel antiadhesion compound', *Gastroenterology*, 117, 1316–25.

NAARDING M A, LUDWIG I S, GROOT F, BERKHOUT B, GEIJTENBEEK T B, POLLAKIS G and PAXTON W A (2005), 'Lewis X component in human milk binds DC-SIGN and inhibits HIV-1 transfer to CD4+ T lymphocytes', *The Journal of Clinical Investigation*, 115, 3256–64.

NAKAJIMA K, KINOSHITA M, MATSUSHITA N, URASHIMA T, SUZUKI M, SUZUKI A and KAKEHI K (2006), 'Capillary affinity electrophoresis using lectins for the analysis of milk oligosaccharide structure and its application to bovine colostrum oligosaccharides', *Anal. Biochem.*, 348, 105–14.

NAKAMURA T and URASHIMA T (2004), 'The milk oligosaccharides of domestic farm animals', *Trends Glycosci. Glycotechnol.*, 16, 135–142.

NASSIF X (1999), 'Interactions between encapsulated *Neisseria meningitidis* and host cells', *Int. Microbiol.*, 2, 133–6.

NEESER J R, GOLLIARD M, WOLTZ A, ROUVET M, DILLMANN M L and GUGGENHEIM B (1994), '*In vitro* modulation of oral bacterial adhesion to saliva-coated hydroxyapatite beads by milk casein derivatives', *Oral Microbiol. Immunol.*, 9, 193–201.

NEWBURG D S, ASHKENAZI S and CLEARY T G (1992), 'Human milk contains the Shiga toxin and Shiga-like toxin receptor glycolipid Gb3', *J. Infect. Dis.*, 166, 832–6.

NEWBURG D S, CHATURVEDI P, LOPEZ E L, DEVOTO S, FAYAD A and CLEARY T G (1993), 'Susceptibility to hemolytic-uremic syndrome relates to erythrocyte glycosphingolipid patterns', *J. Infect. Dis.*, 168, 476–9.

NEWBURG D S, PETERSON J A, RUIZ-PALACIOS G M, MATSON D O, MORROW A L, SHULTS J, GUERRERO M L, CHATURVEDI P, NEWBURG S O, SCALLAN C D, TAYLOR M R, CERIANI R L and PICKERING L K (1998), 'Role of human-milk lactadherin in protection against symptomatic rotavirus infection', *Lancet*, 351, 1160–4.

NEWBURG D S (1999) 'Human milk glycoconjugates that inhibit pathogens' *Current Med. Chem.* 6(2), 117–27.

NEWBURG D S, RUIZ-PALACIOS G M, ALTAYE M, CHATURVEDI P, MEINZEN-DERR J, GUERRERO MDE L and MORROW A L (2004), 'Innate protection conferred by fucosylated oligosaccharides of human milk against diarrhea in breastfed infants', *Glycobiology*, 14, 253–63.

NEWBURG D S, RUIZ-PALACIOS G M and MORROW A L (2005), 'Human milk glycans protect infants against enteric pathogens', *Annu. Rev. Nutr.*, 25, 37–58.

NICOLL G, AVRIL T, LOCK K, FURUKAWA K, BOVIN N and CROCKER P R (2003), 'Ganglioside GD3 expression on target cells can modulate NK cell cytotoxicity via siglec-7-dependent and -independent mechanisms', *Eur. J. Immunol.*, 33, 1642–8.

NIERS L, STASSE-WOLTHUIS M, ROMBOUTS F M and RIJKERS G T (2007), 'Nutritional support for the infant's immune system', *Nutr. Rev.*, 65, 347–60.

NINONUEVO M, AN H, YIN H, KILLEEN K, GRIMM R, WARD R, GERMAN B and LEBRILLA C (2005), 'Nanoliquid chromatography–mass spectrometry of oligosaccharides employing graphitized carbon chromatography on microchip with a high-accuracy mass analyzer', *Electrophoresis*, 26, 3641–9.

NINONUEVO M R, PARK Y, YIN H, ZHANG J, WARD R E, CLOWERS B H, GERMAN J B, FREEMAN S L, KILLEEN K, GRIMM R and LEBRILLA C B (2006), 'A strategy for annotating the human milk glycome', *J. Agric. Food Chem.*, 54, 7471–80.

NINONUEVO M R, PERKINS P D, FRANCIS J, LAMOTTE L M, LOCASCIO R G, FREEMAN S L, MILLS D A, GERMAN J B, GRIMM R and LEBRILLA C B (2008), 'Daily variations in oligosaccharides of human milk determined by microfluidic chips and mass spectrometry', *J. Agric. Food Chem.*, 56, 618–26.

NINONUEVO M R, WARD R E, LOCASCIO R G, GERMAN J B, FREEMAN S L, BARBOZA M, MILLS D A and LEBRILLA C B (2007), 'Methods for the quantitation of human milk oligosaccharides in bacterial fermentation by mass spectrometry', *Anal. Biochem.*, 361, 15–23.

OBERMEIER S, RUDLOFF S, POHLENTZ G, LENTZE M J and KUNZ C (1999), 'Secretion of 13C-labelled oligosaccharides into human milk and infant's urine after an oral [13C]galactose load', *Isotopes Environ. Health Stud.*, 35, 119–25.

OFEK I, HASTY D L and SHARON N (2003), 'Anti-adhesion therapy of bacterial diseases: Prospects and problems', *FEMS Immunol. Med. Microbiol.*, 38, 181–91.

OTANI H and MONNAI M (1993), 'Inhibition of proliferative responses of mouse spleen lymphocytes by bovine milk kappa-casein digests', *Food Agri. Immunol.*, 5, 219–229.

OUWEHAND A, ISOLAURI E and SALMINEN S (2002), 'The role of the intestinal microflora for the development of the immune system in early childhood', *Eur. J. Nutr.*, 41 Suppl 1, I32–7.

PATOLE S (2007), 'Prevention and treatment of necrotising enterocolitis in preterm neonates', *Early Hum. Dev.*, 83, 635–42.

PETERS M J, DIXON G, KOTOWICZ K T, HATCH D J, HEYDERMAN R S and KLEIN N J (1999), 'Circulating platelet–neutrophil complexes represent a subpopulation of activated neutrophils primed for adhesion, phagocytosis and intracellular killing', *Br. J. Haematol.*, 106, 391–9.

PHILLIPS M L, NUDELMAN E, GAETA F C, PEREZ M, SINGHAL A K, HAKOMORI S and PAULSON J C (1990), 'ELAM-1 mediates cell adhesion by recognition of a carbohydrate ligand, sialyl-Lex', *Science*, 250, 1130–2.

PICCARDONI P, SIDERI R, MANARINI S, PICCOLI A, MARTELLI N, DE GAETANO G, CERLETTI C and EVANGELISTA V (2001), 'Platelet/polymorphonuclear leukocyte adhesion: A new role for SRC kinases in Mac-1 adhesive function triggered by P-selectin', *Blood*, 98, 108–16.

PIETERS R J (2007), 'Intervention with bacterial adhesion by multivalent carbohydrates', *Med. Res. Rev.*, 27, 796–816.

RAHMANN H, KORTJE K H, SEYBOLD V and ROSNER H (1990), 'Ultrastructural localization of gangliosides, calcium and a high-affinity Ca(2+)-ATPase in nerve terminals: A contribution to the possible functional role of gangliosides', *Indian J. Biochem. Biophys.*, 27, 420–4.

RASTALL R A, GIBSON G R, GILL H S, GUARNER F, KLAENHAMMER T R, POT B, REID G, ROWLAND I R and SANDERS M E (2005), 'Modulation of the microbial ecology of the human colon by probiotics, prebiotics and synbiotics to enhance human health: An overview of enabling science and potential applications', *FEMS Microbiol. Ecol.*, 52, 145–52.

REDDY M S, BERNSTEIN J M, MURPHY T F and FADEN H S (1996), 'Binding between outer membrane proteins of nontypeable *Haemophilus influenzae* and human nasopharyngeal mucin', *Infect. Immun.*, 64, 1477–9.

ROBERTS D D, OLSON L D, BARILE M F, GINSBURG V and KRIVAN H C (1989), 'Sialic acid-dependent adhesion of *Mycoplasma pneumoniae* to purified glycoproteins', *J. Biol. Chem.*, 264, 9289–93.

ROSNER H (1998), 'Significance of gangliosides in neuronal differentiation of neuroblastoma cells and neurite growth in tissue culture', *Ann. N. Y. Acad. Sci.*, 845, 200–14.

RUDLOFF S, POHLENTZ G, DIEKMANN L, EGGE H and KUNZ C (1996), 'Urinary excretion of lactose and oligosaccharides in preterm infants fed human milk or infant formula', *Acta Paediatr.*, 85, 598–603.

RUDLOFF S, STEFAN C, POHLENTZ G and KUNZ C (2002), 'Detection of ligands for selectins in the oligosaccharide fraction of human milk', *Eur. J. Nutr.*, 41, 85–92.

RUTISHAUSER U (1996), 'Polysialic acid and the regulation of cell interactions', *Curr. Opin. Cell Biol.*, 8, 679–84.

RUTISHAUSER U and LANDMESSER L (1996), 'Polysialic acid in the vertebrate nervous system: A promoter of plasticity in cell–cell interactions', *Trends Neurosci.*, 19, 422–7.

SAITO T, ITOH T and ADACHI S (1984), 'Presence of two neutral disaccharides containing N-acetylhexosamine in bovine colostrum as free forms', *Biochim. Biophys. Acta*, 801, 147–50.

SALMINEN S, GIBSON G R, MCCARTNEY A L and ISOLAURI E (2004), 'Influence of mode of delivery on gut microbiota composition in seven year old children', *Gut*, 53, 1388–9.

SAMPATHKUMAR S G, LI A and YAREMA K J (2006), 'Sialic acid and the central nervous system: Perspectives on biological functions, detection, imaging methods and manipulation', *CNS Neurol. Disord. Drug Targets*, 5, 425–40.

SARNEY D B, HALE C, FRANKEL G and VULFSON E N (2000), 'A novel approach to the recovery of biologically active oligosaccharides from milk using a combination of enzymatic treatment and nanofiltration', *Biotechnol. Bioeng.*, 69, 461–7.

SCHACK-NIELSEN L and MICHAELSEN K F (2007), 'Advances in our understanding of the biology of human milk and its effects on the offspring', *J. Nutr.*, 137, 503S-510S.

SCHAUER R (1982), 'Chemistry, metabolism, and biological functions of sialic acids', *Adv. Carbohydr. Chem. Biochem.*, 40, 131–234.

SCHELL M A, KARMIRANTZOU M, SNEL B, VILANOVA D, BERGER B, PESSI G, ZWAHLEN M C, DESIERE F, BORK P, DELLEY M, PRIDMORE R D and ARIGONI F (2002), 'The genome sequence of *Bifidobacterium longum* reflects its adaptation to the human gastrointestinal tract', *Proc. Natl. Acad. Sci. U S A*, 99, 14422–7.

SCHMIDT E (1960), 'Influenza virus inhibitor in human milk and cow's milk. Part 1. Comparative studies on neuraminic acid content and the level of the inhibitor titer in the hemagglutination test', *Z. Kinderheilkd.*, 84, 339–48.

SCHMIDT E and JANSSEN M (1967), 'The protective effect of human and cow milk mucoids in experimental influenza virus infection of the mouse', *Z. Kinderheilkd.*, 98, 307–14.

SCHON M P, KRAHN T, SCHON M, RODRIGUEZ M L, ANTONICEK H, SCHULTZ J E, LUDWIG R J, ZOLLNER T M, BISCHOFF E, BREMM K D, SCHRAMM M, HENNINGER K, KAUFMANN R, GOLLNICK H P, PARKER C M and BOEHNCKE W H (2002), 'Efomycine M, a new specific inhibitor of selectin, impairs leukocyte adhesion and alleviates cutaneous inflammation', *Nat. Med.*, 8, 366–72.

SCHUMACHER G, BENDAS G, STAHL B and BEERMANN C (2006), 'Human milk oligosaccharides affect P-selectin binding capacities: *In vitro* investigation', *Nutrition*, 22, 620–7.

SCHWERTMANN A, RUDLOFF S and KUNZ C (1996), 'Potential ligands for cell adhesion molecules in human milk', *Ann. Nutr. Metab.*, 40, 252–62.

SILVER R K, BRAUN G, ZILLIKEN F, WERNER G H and GYORGY P (1956), 'Factors in human milk interfering with influenza-virus activities', *Science*, 123, 932–3.

SIMANEK E E, MCGARVEY G J, JABLONOWSKI J A and WONG C H (1998), 'Selectin–carbohydrate interactions: From natural ligands to designed mimics', *Chem. Rev.*, 98, 833–862.

SIMON P M, GOODE P L, MOBASSERI A and ZOPF D (1997), 'Inhibition of *Helicobacter pylori* binding to gastrointestinal epithelial cells by sialic acid-containing oligosaccharides', *Infect. Immun.*, 65, 750–7.

SONG J F, WENG M Q, WU S M and XIA Q C (2002), 'Analysis of neutral saccharides in human milk derivatized with 2-aminoacridone by capillary electrophoresis with laser-induced fluorescence detection', *Anal. Biochem.*, 304, 126–9.

SPRINGER T A (1994), 'Traffic signals for lymphocyte recirculation and leukocyte emigration: The multistep paradigm', *Cell*, 76, 301–14.

SREEDHARAN R and MEHTA D I (2004), 'Gastrointestinal tract', *Pediatrics*, 113, 1044–50.

ST GEME J W, 3rd (1994), 'The HMW1 adhesin of nontypeable *Haemophilus influenzae* recognizes sialylated glycoprotein receptors on cultured human epithelial cells', *Infect. Immun.*, 62, 3881–9.

STAHL B, THURL S, ZENG J, KARAS M, HILLENKAMP F, STEUP M and SAWATZKI G (1994), 'Oligosaccharides from human milk as revealed by matrix-assisted laser desorption/ionization mass spectrometry', *Anal. Biochem.*, 223, 218–26.

STANNER S and SMITH E (2005), 'Breastfeeding: Early influences on later health', *Nutr. Bull.*, 30, 94–102.

STUDZINSKI G P and HARRISON L E (1999), 'Differentiation-related changes in the cell cycle traverse', *Int. Rev. Cytol.*, 189, 1–58.

SUDO N, SAWAMURA S, TANAKA K, AIBA Y, KUBO C and KOGA Y (1997), 'The requirement of intestinal bacterial flora for the development of an IgE production system fully susceptible to oral tolerance induction', *J. Immunol.*, 159, 1739–45.

SUZUKI T, HORIIKE G, YAMAZAKI Y, KAWABE K, MASUDA H, MIYAMOTO D, MATSUDA M, NISHIMURA S I, YAMAGATA T, ITO T, KIDA H, KAWAOKA Y and SUZUKI Y (1997), 'Swine influenza virus strains recognize sialylsugar chains containing the molecular species of sialic acid predominantly present in the swine tracheal epithelium', *FEBS Lett.*, 404, 192–6.

SUZUKI Y (1994), 'Gangliosides as influenza virus receptors. Variation of influenza viruses and their recognition of the receptor sialo-sugar chains', *Prog. Lipid Res.*, 33, 429–57.

SUZUKI Y (2002), 'Recent Advances in Influenza Virus Research', *Research Signpost*, India, 161–168.

SUZUKI Y (2005), 'Sialobiology of influenza: Molecular mechanism of host range variation of influenza viruses', *Biol. Pharma. Bull.*, 28, 399–408.

SUZUKI Y, MATSUNAGA M and MATSUMOTO M (1985a), 'N-Acetylneuraminyllactosyl ceramide, GM3-NeuAc, a new influenza A virus receptor which mediates the adsorption–fusion process of viral infection. Binding specificity of influenza virus

A/Aichi/2/68 (H3N2) to membrane-associated GM3 with different molecular species of sialic acid', *J. Biol. Chem.*, 260, 1362–5.

SUZUKI Y, MATSUNAGA M, NAGAO Y, TAKI T, HIRABAYASHI Y and MATSUMOTO M (1985b), 'Ganglioside GM1b as an influenza virus receptor', *Vaccine*, 3, 201–3.

SUZUKI Y, NAGAO Y, KATO H, MATSUMOTO M, NEROME K, NAKAJIMA K and NOBUSAWA E (1986), 'Human influenza A virus hemagglutinin distinguishes sialyloligosaccharides in membrane-associated gangliosides as its receptor which mediates the adsorption and fusion processes of virus infection. Specificity for oligosaccharides and sialic acids and the sequence to which sialic acid is attached', *J. Biol. Chem.*, 261, 17057–61.

SUZUKI Y, NAKAO T, ITO T, WATANABE N, TODA Y, XU G, SUZUKI T, KOBAYASHI T, KIMURA Y, YAMADA A and *et al.* (1992), 'Structural determination of gangliosides that bind to influenza A, B, and C viruses by an improved binding assay: Strain-specific receptor epitopes in sialo-sugar chains', *Virology*, 189, 121–31.

SWENNEN K, COURTIN C M and DELCOUR J A (2006), 'Non-digestible oligosaccharides with prebiotic properties', *Crit. Rev. Food Sci. Nutr.*, 46, 459–71.

TAKEDA T, SAKATA M, MINEKAWA R, YAMAMOTO T, HAYASHI M, TASAKA K and MURATA Y (2004), 'Human milk induces fetal small intestinal cell proliferation–involvement of a different tyrosine kinase signaling pathway from epidermal growth factor receptor', *J. Endocrinol.*, 181, 449–57.

TATENO H, UCHIYAMA N, KUNO A, TOGAYACHI A, SATO T, NARIMATSU H and HIRABAYASHI J (2007), 'A novel strategy for mammalian cell surface glycome profiling using lectin microarray', *Glycobiology*, 17, 1138–46.

TERRAZAS L I, WALSH K L, PISKORSKA D, MCGUIRE E and HARN D A, Jr. (2001), 'The schistosome oligosaccharide lacto-N-neotetraose expands Gr1(+) cells that secrete anti-inflammatory cytokines and inhibit proliferation of naive CD4(+) cells: A potential mechanism for immune polarization in helminth infections', *J. Immunol.*, 167, 5294–303.

THEORET J F, BIENVENU J G, KUMAR A and MERHI Y (2001), 'P-selectin antagonism with recombinant p-selectin glycoprotein ligand-1 (rPSGL-Ig) inhibits circulating activated platelet binding to neutrophils induced by damaged arterial surfaces', *J. Pharmacol. Exp. Ther.*, 298, 658–64.

THOMAS R and BROOKS T (2004), 'Common oligosaccharide moieties inhibit the adherence of typical and atypical respiratory pathogens', *J. Med. Microbiol.*, 53, 833–40.

TORRES-PINEDO R and MAHMOOD A (1984), 'Postnatal changes in biosynthesis of microvillus membrane glycans of rat small intestine: I. Evidence of a developmental shift from terminal sialylation to fucosylation', *Biochem. Biophys. Res. Commun.*, 125, 546–53.

TROY F A I (1996), 'Sialobiology and the polysialic acid glycotope'. In 'Biology of the Sialic Acids', ed. Rosenberg, A. Plenum Press, New York and London, 95–144.

URASHIMA T, MURATA S and NAKAMURA T (1997), 'Structural determination of monosialyl trisaccharides obtained from caprine colostrum', *Comparative Biochemistry and Physiology. Part B, Biochem. Mol. Biol.*, 116, 431–5.

URASHIMA T, SAITO T, NAKAMURA T and MESSER M (2001), 'Oligosaccharides of milk and colostrum in non-human mammals', *Glycoconj. J.*, 18, 357–71.

VAN KOOYK Y and GEIJTENBEEK T B (2003), 'DC-SIGN: Escape mechanism for pathogens', *Nat. Rev. Immunol.*, 3, 697–709.

VANMAELE R P, FINLAYSON M C and ARMSTRONG G D (1995), 'Effect of enteropathogenic *Escherichia coli* on adherent properties of Chinese hamster ovary cells', *Infect. Immun.*, 63, 191–8.

VANMAELE R P, HEERZE L D and ARMSTRONG G D (1999), 'Role of lactosyl glycan sequences in inhibiting enteropathogenic *Escherichia coli* attachment', *Infect. Immun.*, 67, 3302–7.

VARKI A (1994), 'Selectin ligands', *Proc. Natl. Acad. Sci. U. S. A.*, 91, 7390–7.

VARKI N M and VARKI A (2007), 'Diversity in cell surface sialic acid presentations: Implications for biology and disease', *Lab. Invest.*, 87, 851–7.

VEH R W, MICHALSKI J C, CORFIELD A P, SANDER-WEWER M, GIES D and SCHAUER R (1981), 'New chromatographic system for the rapid analysis and preparation of colostrum sialyloligosaccharides', *J. Chromatogr.*, 212, 313–22.

VELUPILLAI P and HARN D A (1994), 'Oligosaccharide-specific induction of interleukin 10 production by B220+ cells from schistosome-infected mice: A mechanism for regulation of CD4+ T-cell subsets', *Proc. Natl. Acad. Sci. U S A*, 91, 18–22.

WAGNER C L and FORSYTHE D W (2000), 'Effect of human milk and recombinant EGF, TGFalpha, and IGF-1 on small intestinal cell proliferation', *Adv. Exp. Med. Biol.*, 478, 373–4.

WAGNER C L, FORSYTHE D W and WAGNER M T (1998), 'The effect of recombinant TGFalpha, human milk, and human milk macrophage media on gut epithelial proliferation is decreased in the presence of a neutralizing TGFalpha antibody', *Biol. Neonate*, 74, 363–71.

WANG B and BRAND-MILLER J (2003), 'The role and potential of sialic acid in human nutrition', *Eur. J. Clin. Nutr.*, 57, 1351–69.

WANG B, BRAND-MILLER J, MCVEAGH P and PETOCZ P (2001a), 'Concentration and distribution of sialic acid in human milk and infant formulas', *Am. J. Clin. Nutr.*, 74, 510–5.

WANG B, MCVEAGH P, PETOCZ P and BRAND-MILLER J (2003), 'Brain ganglioside and glycoprotein sialic acid in breastfed compared with formula-fed infants', *Am. J. Clin. Nutr.*, 78, 1024–9.

WANG B, YU B, KARIM M, HU H, SUN Y, MCGREEVY P, PETOCZ P, HELD S and BRAND-MILLER J (2007), 'Dietary sialic acid supplementation improves learning and memory in piglets', *Am. J. Clin. Nutr.*, 85, 561–9.

WANG X, HIRMO S, WILLEN R and WADSTROM T (2001b), 'Inhibition of *Helicobacter pylori* infection by bovine milk glycoconjugates in a BAlb/cA mouse model', *J. Med. Microbiol.*, 50, 430–5.

WEIS W I, TAYLOR M E and DRICKAMER K (1998), 'The C-type lectin superfamily in the immune system', *Immunol. Rev.*, 163, 19–34.

WERNERSSON J, DANIELSSON NIEMI L, EINARSON S, HERNELL O and JOHANSSON I (2006), 'Effects of human milk on adhesion of *Streptococcus mutans* to saliva-coated hydroxyapatite *in vitro*', *Caries Res.*, 40, 412–7.

WEYRICH A S, MA X Y, LEFER D J, ALBERTINE K H and LEFER A M (1993), '*In vivo* neutralization of P-selectin protects feline heart and endothelium in myocardial ischemia and reperfusion injury', *J. Clin. Invest.*, 91, 2620–9.

YABE U, SATO C, MATSUDA T and KITAJIMA K (2003), 'Polysialic acid in human milk. CD36 is a new member of mammalian polysialic acid-containing glycoprotein', *J. Biol. Chem.*, 278, 13875–80.

YAREMA K J (2005), *Handbook of Carbohydrate Engineering*, Taylor & Francis, New York.

13

Lipids from the milk fat globule membrane as a health ingredient: Composition, properties and technological aspects

S. Mezouari, and Y. Pouliot, Laval University, Canada; M. Britten, Agriculture and Agri-Food Canada, Canada

Abstract: Recently, the milk fat globule membrane (MFGM) has attracted great attention as a potential source of functional ingredients with respect to its essential fatty acids, liposoluble vitamins and phospholipids. Progress in the utilization of MFGM as an ingredient in functional food formulations is dependent upon obtaining a thorough knowledge of the characteristics of its components, the changes induced by processing, and the specific requirements of the individual prepared-food product. This chapter gives a global insight into the current knowledge of milk lipids in a globule membrane material with regard to composition, separation, and potentially beneficial health effects.

Key words: milk fat globule membrane, phospholipids, liposoluble vitamins, sterols, fatty acids.

13.1 Introduction

In the past, food quality was mainly defined by organoleptic parameters such as taste, odor and color. Today, however, more attention is paid to nutritional aspects, because of increased consumer awareness that food contains microcomponents that have beneficial effects for disease prevention and health maintenance. In order to comply with these new quality standards, scientists are interested in searching for novel natural sources of bioactive components to enhance the nutritional profile of food products (Milner, 1999; Hillbrick and Augustin, 2007). Milk is recognized as an important source of nutrients in the human diet, providing a source of energy, high-quality proteins, and essential minerals and vitamins. In addition, milk fat is a determining factor in the cost, nutrition, and physical and sensory characteristics imparted to dairy products (Kaylegain *et al.*, 1995). Milk fat exists as spherical droplets, referred to as milk fat globules (Heid and Keenan, 2005), with natural diameters ranging from 0.2 to 10 μm. About 20% of milk fat globules (MFG) have a diameter larger than 1 μm and this fraction contains more than 90% of the total fat (Keenan and Dylewski, 1995; Deeth, 1997). The MFG is surrounded by a biological membrane called the milk fat globule membrane (MFGM), which represents from about 2 to more than 6% of the mass of fat globules (Keenan and Mather, 2002). Characterization of the MFGM has been the subject of many studies over a number of years (Jenness and Palmer, 1945; Thomson *et al.*, 1959, 1961; Bracco *et al.*, 1972; Patton and Keenan, 1975; Ray and Singh, 1974; Kuchroo and Narayanan, 1981; Hamzawi and Shahin, 1986; Sharma *et al.*, 1987; Mather and Keenan, 1988). Although neutral (triglycerides, diglycerides, monoglycerides, cholesterol and its esters) and polar lipids (phospholipids) were identified in these studies, there were generally large differences in the results. The available data on the overall composition of MFGM clearly show this uncertainty. Therefore, it is difficult to show their specific composition since different studies with different preparation, analytical methods, and milk sample histories were used. Recently, MFGM has attracted great attention as a potential source of functional ingredients with respect to its essential fatty acids, liposoluble vitamins and phospholipids. Progress in the utilization of MFGM as an ingredient in functional food formulations is dependent upon a thorough knowledge of the characteristics of its components, the changes induced by processing, and the specific requirements of the individual prepared-food product. Physical, chemical, and performance analysis tools are available for the development of such dairy products and the evaluation of their nutritional quality.

This chapter gives a global insight into the current knowledge of milk lipids in a globule membrane material with regard to composition, separation, and potentially beneficial health effects.

13.2 Composition of milk fat globule membrane lipids

13.2.1 Triglycerides

Over 400 different fatty acids (FAs) have been identified in milk fat, but only 12 individual fatty acids account for more than 1% of the total fat (Jensen, 2002). The vast range and wide variety of fatty acids lead to a considerable number of possible molecular species of triglycerides (TAGs). TAGs are synthesized in the mammary gland by enzyme mechanisms that exert some selectivity over the esterification of different fatty acids at each position of the sn-glycerol moiety (Moore and Christie, 1979). Determination of the stereo-specific structure has revealed that the fatty acids display a high selectivity in their arrangement and distribution on the triglycerides.

It has been found that for cows fed a normal diet, the fatty acids C4:0 and C6:0 are esterified at position sn-3, while C12:0 and C14:0 accumulate at position sn-2, and C16:0 tends preferentially to occupy the positions sn-1 and sn-2. C18:0 is esterified at position sn-1, and C18:1 and C18:2 tend to be esterified at positions sn-1 and sn-3 (Taylor and Hawke, 1975; Parodi, 1979; MacGibbon and Taylor, 2006). The positional distribution of fatty acids in TAG molecular species is not only a determining factor for physico-chemical properties (texture and thermal behavior) but also an important indicator for organoleptic and nutritional aspects. It is well known that long-chain fatty acids in external position sn-1 and sn-3 affect fat absorption and consequently the physiological response to dietary fat. Bracco *et al.* (1972) found that in human milk, 69% of C14:0 and 67% of C16:0 are located at the sn-2 position, which accounts for the better absorption of human milk compared to cow milk. TAGs are the major fraction of neutral lipids in MFGM. The levels reported in the literature range from 20% to 80% and are highly influenced by the method of isolation (Kitchen, 1977; McPherson and Kitchen, 1983; Kanno and Kim, 1990; Keenan and Patton, 1995; Fong *et al.*, 2007). The fatty-acid compositions associated with MFGM-TAGs are characterized by their high proportions of long-chain fatty acids C16:0, C18:0 and C18:1 and low contents of fatty acids such as C14:1, C16:1, C18:2 (Kitchen, 1977; McPherson and Kitchen, 1983; Fong *et al.*, 2007). The isolation of high-melting triacylglycerol from the fat globule membrane prepared from washed cream was first reported by Palmer and Wiese in 1933. Wolf and Dugan (1964) confirmed the occurrence of high-melting triacylglycerol and reported that this fraction consisted of 71% trisaturated, 26% disaturated, and 0.0% triunsaturated. Newman and Harrison (1973) deduced from the microelectrophoretic characteristics of milk fat globules that the high-melting triglycerides are located predominately along the inner face of the membrane, because the membrane surface is strongly ionogenic and contains small amounts of neutral lipids. However, it is often reported that the significant amount of high-melting triglycerides results from contamination (from the core of the MFG) during isolation of the membrane (Walstra, 1974, 1985). Furthermore, it has been observed that when fat globules are destabilized at temperatures higher than 37°C, the MFGM triglycerides present the same fatty-acid distribution as that of bulk milk fat. It is also important to note that the technological parameters used during cooling and churning have a substantial effect on the composition of MFGM TAGs. The triglycerides' molecular species and their fatty acids have different crystallization and melting temperatures ranging from –40 to 40°C. Based on their carbon number and chain length, the triglycerides are classified into short- (C26–C36), medium- (C38–C44) and long-chain triglycerides (C46–C56), and these factors partly determine milk fat's rheological and thermal properties.

The MFGM contains a high level of saturated fatty acids (Kitchen, 1977; Fong *et al.*, 2007) because of extensive biohydrogenation in the rumen.

Linolenic and linoleic acid can be hydrogenated by rumen flora to stearic acid by a series of reactions. However, its relatively high content of monounsaturated fatty acids is due to the presence of a desaturase in mammary and adipose tissues. Heptadecanoic (C17:0) and pentadecanoic acid (C15:0) are also present in MFGM (Kitchen, 1977; Fong *et al.*, 2007). These fatty acids are specific to ruminant fat and are synthesized by bacterial flora in the rumen (Wu and Palmiquist, 1991). Their presence in adipose tissue or serum is used as a biomarker for long-term milk fat intake, since the human body is not able to synthesize them (Wolk *et al.*, 2001). Although several investigations have been conducted to find out to what extent milk fat globule size could influence the fatty-acid composition of fat globules (Brunner, 1965; Timmen and Patton, 1988; Briard *et al.*, 2003; Wiking *et al.*, 2004; Fauquant *et al.*, 2005), their effects on the fatty-acid composition of MFGM triglycerides are not widely documented. Information on whether there are differences in fatty-acid composition between individual bovine milk fat globule membranes is of practical importance. Compositional variations may offer the possibility of selecting MFGMs according to their composition and properties to meet the needs of different applications.

Due to its high content of lauric acid and palmitic acid, milk fat is considered hypercholesterolemic compared to unsaturated vegetable oils. Therefore, changes in the composition of milk fat by dietary influences have been extensively investigated, because of the repercussions from such changes on the structure and function of milk fat (Wong *et al.*, 1973; Anderson, 1974; Wood *et al.*, 1975; Smith *et al.*, 1977; Bitman and Wood, 1990; Palmquist and Schanbacher, 1991; Bitman *et al.*, 1995; Stockdale *et al.*, 2003). Lopez *et al.* (2008) found that, compared with the total lipids extracted from whole milk, the MFGM in milk from cows fed a diet rich in polyunsaturated fatty acids contained a higher concentration of unsaturated fatty acids (C18:1, C18:2, and C18:3) and very long-chain fatty acids (C22:0, C23:0, C24:0, C20:5 n-3, C22:6 n-3). Palmquist and Schanbacher, (1991) characterized the fatty-acid composition of MFGM in lactating dairy cows and reported that the ratio of polyunsaturated/saturated fatty acids was increased in the MFGM by feeding the cows a high corn diet or by infusing soy oil; an increase from 0.31 to 1 of the 18:2/18:1 ratio has also been observed after infusing soy oil for only 4 days, which confirms the high turnover of mammary cell membranes involved in milk fat secretion (Mather and Keenan, 1983). The MFGM is directly derived from the apical plasma membrane of mammary epithelial cells, so consequently these changes in the FA composition of the MFGM mean that FA synthesis in the mammary gland is closely dependent on the cows' diet, which eventually influences the composition of the mammary cell membrane. On one hand, the MFGM from cows fed on a diet rich in polyunsaturated fatty acids could be a good source of essential fatty acids, which have an important biological function (Sacks and Katan, 2002). Unlike saturated and monounsaturated fatty acids, polyunsaturated fatty acids, including linoleic acid of the n-6 series (C18:2 n-6) and α-linolenic acid of the n-3 series (C18:3 n-3), cannot be synthesized by mammals and must be supplied in the diet. On the other hand, MFGM rich in unsaturated fatty acids is more susceptible to oxidation, photoxidation and thermal treatments, automatically leading to its degradation and consequently resulting in the deterioration of whole milk, dairy products, and MFGM-enriched foods. Therefore more precautions are required to protect MFGM against oxidative deterioration (O'Connor and O'Brien, 2006).

In the fat globule membrane, the lipids coexist with the proteins and glycoproteins, but little is known on the types of interaction between these molecules and their impact on nutritional and technological aspects. In this context, an interesting report by Keenan and Heid (1982) found that, in MFGM prepared from bovine and caprine milk, palmitic (16:0), stearic (18:0) and oleic (18:1) acids are so tightly and strongly attached to a number of membrane proteins that the release of these bound fatty acids could be achieved only through an alkali treatment, suggesting a covalent attachment. Whether such an attachment exists with the other MFGM constituents has yet to be confirmed.

13.2.2 Phospholipids

Phospholipids, commonly referred to as 'lecithins', are amphiphilic molecules with a hydrophobic tail and a hydrophilic head group. Their physicochemical properties are mostly governed by the degree of unsaturation of their fatty acids and the type of head group present. Known for their properties as emulsifying, wetting and dispersing agents, they are widely used in the food industry (margarine, chocolate, bakery goods) and also in pharmaceuticals. Soybeans and eggs are the richest lecithin sources used in the food industry. The phospholipids in milk are important for the structure of MFGM and emulsion stability. The concentration of phospholipids ranges from 0.5 to 1% of the total milk lipids (Jenssen *et al.*, 1991; Patton and Jenssen, 1976). A decrease in the concentration of phospholipids through lactation has been observed by Bitman and Wood (1990). They are synthesized *de novo* in the mammary gland and in milk they are mostly associated with MFGM (60–65%), making buttermilk a rich source of phospholipids. The rest are located in the membrane material of skimmed milk and may also be found as lipoprotein complexes (Patton and Keenan 1975; Patton and Jenssen, 1976; Moore and Christie, 1979; Keenan, 2001). Glycerophospholipids consist of a glycerol backbone on which two fatty acids are esterified on positions sn-1 and sn-2; a phosphate residue may be linked with different organic groups such as choline, serine, ethanolamine, and linositol. Generally, the fatty-acid chain on the sn-1 position is more saturated compared with the one on the sn-2 position (Kuchroo and Narayanan, 1981; Newburg and Chaturvedi, 1992; Pfeuffer and Schrezenmeir, 2001; Christie, 2003; Rombaut and Dewettinck, 2006; Fong *et al.*, 2007). The phospholipids are asymmetrically located on the MFGM; phosphatidylethanolamine (PE), phosphatidylserine (PS) and phosphatidylinositol (PI) are located on the inner membrane, while phosphatidylcholine (PC) and sphyngomyelin (SM) are on the outside (Deeth, 1997). According to the studies on phospholipid classes from various animal species, zwitterionic PE, PC and SM are the major phospholipids; the anionic forms PI and PS are present at lower concentrations (Morrison, 1968; Kataoka and Nakae, 1973; Keenan and Dylewski, 1995). Morrison (1968) has explained the constant amount of cholin-containing phospholipids, PC and SM, by the probability that these phospholipids perform the same function in all milks. However, the composition of phospholipids is variable and strongly dependent on the conditions and method of isolation (Bracco *et al.*, 1972; Keenan and Dylewski; 1995; Fauquant *et al.*, 2007). Compared to bulk milk fat, MFGM phospholipids are richer in unsaturated fatty acids. Oleic acid (C18:1) and linoleic acid (C18:2) are present in appreciable concentrations, whereas the short- and medium-chain fatty acids (C4–C10) specific to milk fat are present at very low levels (Bracco *et al.*, 1972; Kitchen, 1977; McPherson and Kitchen, 1983). Further published data (Fauquant *et al.*, 2005) have confirmed the high level of unsaturated fatty acids in MFGM phospholipids (C18:1: 26–30%; C18:2: 6%), and shown no significant difference in the fatty-acid composition of MFGM phospholipids from small and large milk fat globules selected by microfiltration. Phosphatidylcholine and phosphatidyethanolamine contain 40–60% unsaturated fatty acids, of which about one third are polyunsaturated fatty acids (Christie, 1995). It is important to mention that MFGM phospholipids also contain a relatively high proportion of saturated fatty acids, such as palmitic C16:0 and stearic C18:0 acids. Moreover, the sphingomyelin is characterized by its high saturated fatty-acid content with an appreciable amount of C22, C23 and C24 (Keenan *et al.*, 1970; Keenan and Huang, 1972; McPherson and Kitchen, 1983). Polyunsaturated fatty acids are esterified preferentially at the sn-2 position, while the monounsaturates and the saturates (14:0, 16:0) are evenly distributed between the sn-2 and sn-1 positions. Stearic acid (18:0) is located at sn-1 position. However, the phospholipids in human milk show high selectivity for C18:0 at the sn-1 and C18:2 at the sn-2 positions (MacGibbon and Taylor, 2006). The composition of fatty acids in phospholipids is dependent on feeding regimes. Sleigh and Burley (1974) reported an increase of unsaturated fatty acids in MFGM phospholipids when cows were fed supplements of protected polyunsaturated vegetable oil. Smith *et al.* (1977) found that feeding a supplement rich in linoleic acid increased the unsaturation of the phospholipids in the inner and outer milk fat globule membranes. However, this unsaturation was less than that of the core lipids. Lopez *et al.* (2008) found that when cows were fed a diet rich in polyunsaturated fatty acids, the content of phospholipids and sphingomyelin in the MFGM increased by about 18% and 30%, respectively: additionally, a higher proportion of stearic acid and unsaturated fatty acids was observed.

Huang and Kuksis (1967) found little quantitative difference in phospholipids and neutral lipids in MFGM prepared from summer and winter milk. Additional factors such as temperature, age, bacteriological quality, and stage of lactation can also influence these changes (Evers, 2004a).

The molecular characterization of phospholipid assemblies and the mechanisms affecting their behavior and affinity with the surrounding molecules in the MFGM is important in order to better understand the colloidal stability of MFGM in milk, and of foods to which they might be added. In this context, phospholipid–protein interactions have been an attractive subject for several research works. The reported data show that the phospholipids do not exist in a free form in milk, but are strongly associated with proteins (Deeth, 1997). Freundenstein *et al.* (1979) observed that butyrophilin retains a small amount of plasma membrane phospholipids. In addition, when added to food products, the phospholipids of the MFGM can interact with other surface active molecules, such as β-lactoglobulin, to form complexes (Leaver *et al.*, 1999). Electrostatic interactions between phospholipids and β-lactoglubilin have been observed using phospholipid monolayers and/or bilayers as model membrane systems (Lefèvre and Subirade, 2000). The tendency of β-lactoglobulin to interact preferentially with charged phospholipids rather than with neutral lipids has also been reported. Better knowledge of the interaction mechanisms between milk phospholipids and other components should enhance their utilization and optimize their functions in food matrices.

13.2.3 Sterols

Sterols are polycyclic alcohols having a secondary –OH group at Position 3, which makes them more polar than triglycerides. Sterols exist in a free form, esterified with fatty acids, or as steryl glycosides. The determination of sterols in dairy products is usually limited to measuring the cholesterol content in order to estimate the nutritional value and detect any form of adulteration in butter and cream from adding vegetable fats (since the latter are rich sources of phytosterols, of which β-sitosterol is the most abundant constituent) (Contarini *et al.*, 2002). A survey of the literature shows that the occurrence of sterols in MFGM has not been widely investigated and they are generally taken as total cholesterol. Controversial reports about their qualitative and quantitative composition have been reported. The inconsistent data seem to be related to the preparative methods, which induce changes in their distribution between the milk fat globule and MFGM (Keenan *et al.*, 1988). Jenssen *et al.* (1991) found that about 0.2 to 0.5% of total lipids are sterols, and these are mostly located in the MFGM. Cholesterol, the major sterol in milk, ranged between 0.25–0.5 wt % of the total lipids. In most cases, it is estimated to be about 0.4 wt % (Hung and Kuksis, 1967; Walstra and Jenness, 1984; Frede *et al.*, 1990; Jensen *et al.*, 1991). Cholesterol accounts for more than 94% of the total sterols, of which

10% or fewer are present in esterified form (Jenssen, 1995). Using electron microscopy, cholesterol has been located in the MFGM and the globule core (Martin, 1989). About 17 different sterol molecules have been detected in cow milk (Jenssen, 1995, p22), but to date they have not been fully identified. Brewington *et al.* (1970) reported the presence of lanosterol, β-sitosterol and dihydrolanosterol in the unsaponifiable fraction of milk lipids. Braco *et al.* (1972) analyzed the unsaponifiable fraction of MFGM and detected cholesterol and 7-dihydrocholesterol (provitamin D3). In addition, other sterols, namely β-sitosterol, stigmasterol and campesterol, have been identified (Mincione *et al.*, 1977). The presence of such sterols has also been detected in milk lipid samples from animal species other than cows. Desmosterol, lathosterol and lanosterol have been identified in buffalo milk (Addeo *et al.*, 1981) and dihydrolanosterol in goat milk (Fraga *et al.*, 2000). The most recent study, where many more bioactive sterols were detected in the MFGM than had previously been reported in the literature, is that of Fauquant *et al.* (2007), who used a patented cross-flow microfiltration system and separated the MFGMs according to their size. The most important hydrocarbon identified in the MFGM was squalen, which displayed an even distribution among lipid fractions (Bracco *et al.*, 1972). Thomson *et al.* (1961) found that squalen averaged 0.6% of the total membrane lipids. This highly unsaturated compound is an acyclic terpene ($C_{30}H_{50}$) and is an important intermediary in the biosynthesis of cholesterol.

13.2.4 Vitamin E

Vitamin E exhibits a complex chemistry as there are eight different homologs, four tocopherols and four tocotrienols (α, β, γ and δ) which reflect its activity. Cereals, legumes, nuts, oil seeds and vegetable oils are rich sources of tocopherols and their isomers. In animal and human fat, tocopherols are present in trace quantities. Despite its relatively low content, α-tocopherol seems to be the most important natural antioxidant present in the MFGM. Milk fat contains approximately 20 μg α-tocopherol/g (Erickson *et al.*, 1964; Kanno *et al.*, 1968; Bruhn and Franke, 1971; Jenssen, 1995; O'Connor and O'Brien, 2006), and its concentration in MFGM lipids is at least three times higher than in the core of the fat globule (Erickson *et al.*, 1963). The content of vitamin E is highly dependent on the amount consumed by the cows; the highest levels are normally found during spring and summer, when cows are fed on fresh, vitamin-rich pasture. According to the data of Yeargan *et al.* (1979), about 45 μg α-tocopherol/g milk fat is the upper physiological limit for the secretion of vitamin E in milk; however, the mechanism involved in the transfer of fat-soluble vitamins from feed to milk is poorly understood. Only a few reports on the presence of tocopherols other than the α-form in the MFGM are available. Thomson *et al.* (1964) detected the presence of δ-tocopherol in some milk samples from cows fed with brewers' grain. However, Kanno *et al.* (1968) found that in addition to α-tocopherol, γ-tocopherol was naturally present in milk fat and its content varied from 1.0 to 2.5 in summer and from 0.3 to 2.9 μg per gram of fat in winter, but no information on its presence in the MFGM was given. The demands on milk oxidative stability and vitamin E level increase when cows are fed an unsaturated fatty-acid-rich diet to modify the milk's fatty-acid composition. Therefore, using feed management to control the concentration of fat-soluble components, such as vitamin E, in milk and dairy products has been a point of interest in recent years. However, no information is available on the influence of feed management on the distribution of tocopherols in the MFGM.

13.2.5 Carotenoids and vitamin A

Carotenoids are polyisoprenoids which contain 40 carbon atoms, with one or two ring structures at the end of their extensive system of conjugated double bonds. β-carotene is the most widely investigated, due to its protective effect in retarding the photooxidation of lipids by an energy transfer mechanism, and its capacity to react with the triplet state of the excited sensitizer. In bovine milk, between 75 and 95% of the carotenoids are β-carotene, which is an important form of vitamin A (provitamin A) and a natural colorant responsible for the yellow color of butter. In the spring, cows produce milk with a higher carotene content. It is notable that the milk fat of many ruminant species is colorless, e.g. that of goats. Patton and Fowkes (1967) suggested that the milk fat droplets acquire their carotene when enveloped by plasma membranes at the secretion stage. Patton *et al.* (1980a) found that the intracellular membranes of the tissue, particularly mitochondria, are richer in carotenoids than the secreted fat globules, which implies that the milk fat globules acquire β-carotene during their formation in the cell. Despite several biochemical studies on milk lipid fractions, the exact location of carotene has not been completely identified. A high level of carotene in MFGM was reported by Thomson *et al.* (1961). White and Eaton (1954) found that smaller fat globules have a higher concentration of carotenoids than larger ones. Mulder and Walstra (1974) stated that the approximate distributions of carotene and vitamin A content between the MFGM and the core was 5 and 95%, respectively. This was confirmed by Walstra and Jenness (1984). The presence of vitamin A in the MFGM and the fat globule was also shown by Zahar and Smith (1995). However, Smith *et al.* (1977) and Patton *et al.* (1980a) observed no carotene in the MFGM and this was supported by Jensen and Nielson (1996), who reported that α-tocopherol was the only fat-soluble vitamin detected in the MFGM, whereas γ-tocopherol, β-carotene and vitamin A were only present in the inner core of the milk fat globule. Patton *et al.* (1980a) attributed the absence of carotene in the MFGM to its destruction by oxidases or selective removal at the time of fat droplet secretion. In their recent study, Calderon *et al.* (2007) suggested that, under a high carotenoid diet, the secretion of β-carotene in milk is not limited by its concentration at the mammary gland but by the mechanism of its transfer from plasma to milk. These conflicting results require more investigation to confirm or to refute the presence of carotene and vitamin A in the MFGM.

13.3 Separation of milk fat globule membrane lipids

The recovery of MFGM lipids from whey and buttermilk without impairing their composition and their nutritional value is still a challenging problem for food chemists and the dairy industry. In spite of the excellent recovery possible from the use of organic solvents, we still do not use this method to extract milk lipids. Such solvents are directly responsible for serious problems such as solvent escape, fire, explosion and air pollution; furthermore, when inhaled by humans, they are known to dissolve in neural lipids, thus affecting the nervous system. All these concerns have stimulated interest in research into alternative processes for polar MFGM-lipid isolation. No optimal procedure has been adopted as yet, since many difficulties have arisen in relation to the yield and degree of purity of the final fraction, which could be attributed, among other factors, to technological parameters such as pH and temperature (Daufin *et al.*, 1991; Cheryan, 1998; Rao, 2002). The isolation of the MFGM has been reviewed by Ward *et al.* (2006). Despite these technological difficulties, it should be emphasized that the phospholipids fraction could be concentrated to some extent using membrane separation technologies. Sachdeva and Buchheim (1997) obtained a variable recovery of purified phospholipids from buttermilk by using ultrafiltration combined with microfiltration and diafiltration. The microfiltration of whey from cheddar cheese using a 0.2 μm ceramic filter led to the recovery of a phopholipids fraction with a high proportion of phosphatidylcholine and phosphatidylethanolamine, which means that the separation was very selective with respect to the yield of individual phospholipids (Boyd *et al.*, 1999). Morin *et al.* (2004) succeed in doubling the polar lipid concentration using a 0.1 μm ceramic membrane. However, they concluded that the optimal fractionation of MFGM lipids could not be achieved using only microfiltration. The final fraction of phospholipids in the extract is largely dependent on the quality of the butter milk. Rombaut *et al.* (2007) also used microfiltration with a 0.15 μm cellulose acetate membrane, and obtained an increase in polar lipid retention from 64 to 98% by increasing the pH from 4.6 to 7.5 at 50°C. Although microfiltration seems to be a promising method for obtaining an MFGM-rich fraction, several attempts have been made to optimize its use by using different pretreatments to reduce the presence of casein, which is concentrated together with the MFGM during filtration, reducing the purity of the extract. The microfiltration of buttermilk or butter serum with minimum retention of casein after filtration can be achieved by adding a dissociation agent such as urea, ethanol, citrate, or sodium citrate (McGann and Fox, 1974; Zadow, 1993; Udabage *et al.*, 2000; O'Connell *et al.*, 2001; Corredig *et al.*, 2003). A cream washing treatment was shown to improve the purity of phospholipids extracted from buttermilk (Britten *et al.*, 2008; Lamothe *et al.*, 2008). Morin *et al.* (2007a) found 74% less protein in buttermilk when cream was washed prior to churning.

A survey of the literature shows that various separation methods have been tested on different samples with different starting compositions and histories. As previously mentioned, the composition of milk fat and the MFGM is variable, and therefore it is difficult to put forward general statements on the efficiency of the separation techniques. A systematic study under the same conditions, using samples from the same origin, is needed for a better comparison. The structure, the composition and the nature of the interactions between the membrane components are not fully understood as yet. The MFGM presents a complex mixture of lipids, proteins, lipoproteins, glycolipids and enzymes with different polarity, physicochemical properties and affinities. In this context, the difficulties encountered in isolating the MFGM and its polar fractions in their intact and pure forms are not surprising. MFGM characteristics should be taken into consideration when selecting the method and the technological parameters to be used for their recovery. Thus, each separation technique adopted should be adjusted to the characteristics of the starting material and to the fraction to be purified.

13.4 Effect of processing on milk fat globule membrane lipids

Besides environmental factors, including the presence of bacteria in milk or the presence of mastitis pathogens in mastitic milk (Erwin and Randolph, 1975; McPherson and Kitchen, 1983), it is well known that processing conditions markedly affect the composition and stability of the MFGM. So far, the effects of various operations used in the processing (agitation, homogenization, heat treatment, concentration, drying and freezing) of MFGM proteins have been extensively investigated (Jackson and Brunner, 1960; Anderson *et al.*, 1977; Oortwijn and Walstra, 1979; Walstra and Oortwijn, 1982; McPherson and Kitchen, 1983; Dalgleish and Banks, 1991; Sharma and Dalgleish, 1994; Kim and Jimenez Flores, 1995; van Boekel and Walstra, 1995; Corredig and Dalgleish, 1996; Walstra *et al.*, 1999; Ye *et al.*, 2002, 2004, 2007; Lee and Sherbon, 2002; Evers, 2004a). However, the effects of such technological treatments on the composition and stability of the MFGM lipids fraction have not been widely studied. The effects of technological processes on MFGM lipids have mostly been reported in the form of observations on the quantitative changes, without describing the mechanism of the degradation that may have occurred. In the MFGM, the proteins coexist with lipids and any degradation of the former should affect the integrity of the whole MFGM, including the lipid fraction. Thus, in addition to the binding of caseins and whey proteins into the MFGM, heat treatment and homogenization induce a loss of MFGM-lipid constituents such as triglycerides and phospholipids (Greenbank and Pallansch, 1961; Darling and Butcher, 1978; Keenan *et al.*, 1983; McPherson *et al.*, 1984; Houlihan *et al.*, 1992). Migration of phospholipids from the MFGM to the aqueous phase has been observed (Radema, 1956; Koops and Tarassuk, 1959; Houlihan *et al.*, 1992) and probably results from the detachment of membrane fragments from the surface of the fat globule (Mulder and Walstra, 1974). Heat treatment could induce hydrolysis of the phospholipids (Nakanishi and Kaya, 1970). A decrease of triglyceride in the MFGM has been observed as a result of heating at 80°C (Houlihan *et al.*, 1992). Lee and Sherbon (2002) reported a loss of 20% of lipids when MFGM was heated. Spray-drying induced a decrease in the content of MFGM phospholipids (Morin *et al.*, 2007b). Furthermore, it should be emphasized that the artificial membrane formed after homogenization does not isolate the fat from the lipase and oxidative stress. As a consequence, the MFGM and milk fat are more vulnerable to hydrolytic rancidity and photo-oxidation (light-activated flavor), and to oiling off of the fat after homogenization. On the other hand, homogenization has been shown to slightly decrease metal-catalyzed oxidation, which was attributed to the rearrangement of the phospholipids rich in polyunsaturated fatty acids. To avoid lipid degradation, pasteurization of the milk is recommended prior to or immediately after homogenization. However, according to Mulder and Walstra (1974), heating the milk induces the migration of copper from the milk plasma to the MFGM, which may catalyze and contribute to the oxidative degradation of the unsaturated fatty acids of membrane phospholipids (McPherson and Kitchen, 1983; van Boekel and Walstra, 1995). Additionally, severe heating results in the removal of the lipids from the membrane and increases the tendency to coalescence. Freezing milk or cream produces ice crystals, which disrupt the lipoprotein complex and damage the MFGM.

Generally, the degree of MFGM damage is directly related to the severity of the technological treatments to which the milk is subjected. With regard to the MFGM-lipid fraction, the impact of such treatments is estimated according to the changes that occur in the distribution of the phospholipids between the cream, milk plasma and fat phase (Koops and Tarassuk, 1959; Greenbank and Pallansch, 1961; Patton *et al.*, 1980b; Evers, 2004a). Another indicator that has been proposed for evaluating MFGM damage is the concentration of free fatty acids (FFAs). FFAs are formed by the hydrolysis of triglycerides, and mono- and diglycerides due to the enzymatic activity of lipase. While free fatty acids indicate hydrolytic degradation of fats, they are, however, quite inadequate to characterize the quality of the MFGM, because the determination of FFAs by titration does not differentiate between free fatty acids formed by enzymatic hydrolysis and those formed as a consequence of MFGM damage. The methodologies for the measurement of MFGM damage have been reviewed in detail by Evers (2004b).

13.5 Beneficial effects of milk-fat globule membrane lipids

Fat is a carrier of many lipophilic micronutrients, such as vitamins. There is a high correlation between the absorption of fat-soluble nutrients and fat content (Cooper *et al.*, 1997). Triglycerides supply metabolic energy by providing 38 KJ/g, mainly through β-oxidation of the constituent fatty acids. The useful energy available to the body depends on the digestibility of the fat, which in turns depends on the chain length of the constituent fatty acids. The MFGM has a relatively high content of monounsaturated and saturated fatty acids. The efficiency of oleic acid (C18:1) in lowering serum cholesterol is now recognized, with the important advantage of being resistant to oxidation. Oleic acid-enriched diets may reduce LDL (low-density lipoprotein) oxidation and slow down the progress of atherosclerosis. Adversely, lauric (12:0), myristic (14:0) and palmitic acid (16:0) are known to raise plasma LDL cholesterol. However, stearic acid (18:0) has a neutral effect on lipoprotein cholesterol levels. In the MFGM, the polyunsaturated fatty acids are present in relatively small quantities compared to the saturated ones; however, their physiological activity is of utmost importance. Linoleic acid (18:2 n-6) and linolenic acid (18:3 n-3) are essential fatty acids and play a key role as precursors of arachidonic acid (20:4 n-6), which is converted to eicosanoids (C20:5 n-3, C22:6 n-3), known for their potent biological activity at extremely low concentrations. Therefore, these essential fatty acids are absolutely necessary for growth and reproduction. Their activity in lowering plasma LDL cholesterol is well demonstrated. However, their high susceptibility to oxidation, producing cytotoxin in the LDLs, may cause atherosclerotic changes in the blood vessels. Their consumption should therefore be limited and controlled to some extent (Hegsted *et al.*, 1965; Keys *et al.*, 1965; Dupont *et al.*, 1991; Bonanome and Grundy, 1988; Grundy and Denkey, 1990; Katan *et al.*, 1994; Parthasarathy and Santanam, 1994; Zock, 1995; Frankel, 1998).

Because the addition of antioxidants to dairy products is not legally allowed in most countries, the antioxidative potential in milk is provided primarily by its endogenous components. Lipids are protected against oxidative degradation by antioxidants naturally present in the raw milk (Lindmark-Mansson and Akesson, 2000; Molkentin, 2000). A lipophilic antioxidant such as α-tocopherol is an effective scavenger of the lipid peroxy radical and is efficient in protecting unsaturated fatty acids against lipid peroxidation by donating a hydrogen atom to the lipid radical. The resulting tocopheryl radical is relatively stable and unreactive (Lea and Ward, 1959; Frankel, 1991; Pongracz *et al.*, 1995; Kamal-eldin and

Appelqvist, 1996; Fox and McSweeney, 1998). A high correlation between oxidative stability and the concentration of tocopherols in the MFGM lipids and fat fraction has been found (Krukovsky *et al.*, 1950; Erickson *et al.*, 1963, 1964). Furthermore, vitamin E has attracted greater interest because it reduces oxidation problems in milk (Nicholson and St-Laurent, 1991; Charmley and Nicholson, 1994). Vitamin E is also of interest in relation to health problems such as mastitis and the immune defence of cows (Johnston and Chew, 1984; Batra *et al.*, 1992; Politis *et al.*, 1995). Several investigations have looked at the impact of a diet enriched with vitamin E on the concentration of α-tocopherol in milk, but the published data are still controversial. It has been widely reported (Dunkley and Ronning, 1966; King *et al.*, 1966; St. Laurent *et al.*, 1990; Barrefors *et al.*, 1995; Focant *et al.*, 1998; Granelli *et al.*, 1998) that the increase of α-tocopherol content in the milk decreases spontaneous and copper-induced oxidation. However, no beneficial impact from supplementing the diet with vitamin E has been observed (Schingoethe *et al.*, 1979; Charmley and Nicholson, 1994). Besides its protective effect, there is now accumulating evidence that vitamin E may also play other biological roles that do not necessarily involve its antioxidant function. A possible structural role in the maintenance of cell membrane integrity has already been mentioned. α-Tocopherol inhibits LDL oxidation *in vivo* and *in vitro*. Based on animal studies, the role of α-tocopherols in retarding the progression of atherosclerosis has been shown. In addition, the intake of vitamin E correlates with reduced coronary diseases (Stampfer, 1993; Jialal, 1995).

In addition to their emulsifying properties, phospholipids are generally considered to have an antioxidant activity. Phospholipids seem to have synergistic activity in mixtures with tocopherols. This synergistic activity is often related to the metal-scavenging ability of phospholipids and thus they act as a secondary antioxidant; but it seems that the acidic phospholipids such as PE and PI are more effective than PC (Hudson and Mahgoub, 1981; Nath and Rama Murthy, 1983; Hildebrand *et al.*, 1984; Frede *et al.*, 1990; Chen and Nawar, 1991). In addition, recent nutritional studies suggest that the consumption of phospholipids may bring health-related benefits, including antiviral and antimicrobial effects (Van Hooijdonk *et al.*, 2000; Kivinen *et al.*, 1992). A protective effect against gastric ulceration and a high gastroprotective effect by milk phospholipids have been demonstrated in humans (Kivinen *et al.*, 1992; Vesper *et al.*, 1999; Pfeuffer and Schrezenmeir, 2001). Furthermore, a positive effect from phosphatidylserin on Alzheimer's disease (Gindin *et al.*, 1998; Hashioka *et al.*, 2004) has been reported. Sphyngomyelin and its metabolites, ceramides, appear to have tumor-suppressing properties (Parodi, 1996, 1997); and it is believed that neurotransmitters and their synthesis, which are important for memory, are promoted by SM and PC (Blusztajn, 1998; Szuhaj and Nieuwenhuyzen, 2003). The beneficial effects of the MFGM's bioactive components have been recently reviewed by Spitsberg (2005) and German *et al.* (2006).

Atherosclerosis and coronary heart disease are related to a high level of plasma cholesterol. However, cholesterol is important for fat resorption and in the adrenals is a precursor of steroid hormones. In the brain and other nervous tissues, cholesterol is an important component of myelin. Although their composition in the MFGM is not fully identified, it is worth noting that sterols are also important bioactive components. One example of a successful functional food is the incorporation of plant sterols/phytosterols into vegetable-oil spreads. This type of product is now available in the marketplace and has been proven to lower blood LDL-cholesterol by around 10–15% as part of a healthy diet (Mortan *et al.*, 1995; Ling and Jones, 1995).

13.6 Conclusion and perspectives

The lipid fraction of the MFGM could be considered as a valuable alternative source of bioactive components. However, more studies are required for an exhaustive identification and quantification of these bioactive microcomponents, as the functional and nutritional relevance of the MFGM is closely related to its specific chemical composition, which is controlled by numerous factors including season, region of production, stage of lactation, breeding, and diet composition. An understanding of the nutritional effects, physical properties and technological factors affecting the composition and stability of MFGM lipids is necessary for developing innovative strategies that allow optimization of their use as functional ingredients. Processing conditions should be selected to ensure minimum changes in MFGM composition to prevent any reduction in its stability and nutritional value. Because of the limited knowledge in this area, more investigations are necessary to comprehensively explain the effects of changes in the MFGM during processing on the properties of the end-products. Therefore it is necessary to understand the dynamic interactions that occur among the various components of the MFGM.

13.7 References

ADDEO F, KUZDZAL-SAVOIE S, CHIANESE L, MALORNI A and SEPE C (1981), 'Liposoluble compounds in buffalo milk', *Lait*, 61, 187–212.

ANDERSON A (1974), 'Milk fat globule membrane composition and dietary change: Supplements of coconut oil fed in two physical forms', *J Dairy Sci*, 57, 399–404.

ANDERSON M, BROOKER B E, CAWSTON T E and CHEESEMAN G C (1977), 'Changes during storage in stability and composition of ultra-heat-treated aseptically-packed cream of 18% fat content', *J Dairy Res*, 44, 111–124.

BARREFORS P, GRANELLI K, APPELQVIST L A and BJOERCK L (1995), 'Chemical characterization of raw milk samples with and without oxidative off-flavor', *J Dairy Sci*, 78, 2691–2699.

BATRA T R, SINGH K, HO S K and HIDIROGLOU M (1992), 'Concentrations of plasma and milk vitamin E and plasma β-carotene of mastitic and healthy cows', *Int J Vitam Nutr Res*, 62, 233–237.

BITMAN J and WOOD D L (1990), 'Changes in milk fat phospholipids during lactation', *J Dairy Sci*, 1990, 73, 1208–1216.

BITMAN J, WOOD D L, MILLER R H, WILK J C and MOORE E D (1995), 'Comparison of lipid composition of milk from half-Danish jersey cows and United States jersey cows', *J Dairy Sci*, 78, 655–658.

BLUSZTAJN J K (1998), 'Developmental neuroscience – Choline', a vital amine, Science, 281, 794–795.

BONANOME A and GRUNDY S M (1988), 'Effect of dietary stearic acid on plasma cholesterol and lipoprotein levels', *New England Journal of Medicine*, 318, 1244–1248.

BOYD L C, DRYE N C and HANSEN A P (1999), Isolation and characterization of whey phospholipids, *J Dairy Sci*, 82, 2550–2557.

BRACCO U, HIDALGO J and BOHREN H (1972), 'Lipid composition of the fat globule membrane of human and bovine milk', *J Dairy Sci*, 55, 165–172.

BREWINGTON C R, CARESS E A and SCHWARTZ D P (1970), 'Isolation and identification of new constituents in milk fat', *J Lipid Res*, 11, 355–361.

BRIARD V, LECONTE N, MICHEL F and MICHALSKI M C (2003), 'The fatty acid composition of small and large naturally occurring milk fat globules', *Eur J Lipid Sci Technol*, 105, 677–682.

BRITTEN M, LAMOTHE S and ROBITAILLE G (2008), 'Effect of cream treatment on phospholipids and protein recovery in butter making process', *Int J Food Sci Technol*, 43, 651–657.

BRUHN J C and FRANKE A A (1971), 'Influence of copper and tocopherol on the susceptibility of herd milk to spontanous oxidized flavor', *J Dairy Sci*, 54, 761–762.

BRUNNER J R (1965), 'Physical equilibria in milk: The lipid phase', in Webb B H, Johnson A H, *Fundamentals of Dairy Chemistry*, Westport (USA), AVI Publ. Co. Inc., pp. 403–505.

CALDERON F, CHAUVEAU-DURIOT B, PRADEL P, MARTIN B, GRAULET B, DOREAU M and NOZIÈRE P (2007), 'Variations in carotenoids, vitamin A and E, and color in cow's plasma and milk following a shift from hay diet to diets containing increasing levels of carotenoids and vitamin E', *J Dairy Sci*, 90, 5651 5664.

CHARMLEY E and NICHOLSON J W G (1994), 'Influence of dietary fat source on oxidative stability and fatty acid composition of milk from cows receiving a low or high level of dietary vitamin E', *Ca J Anim Sci*, 74, 657–664.

CHEN Z Y and NAWAR W W (1991), 'Prooxidative and antioxidative effects of phospholipids on milk fat', *J Am Oil Chem Sci*, 938–940.

CHERYAN M (1998), *Ultrafiltration and Microfiltration Handbook*, 2nd ed, Boca Raton, FL, CRC Press.

CHRISTIE W W (1995), 'The composition and structure of milk lipids', in Fox P F, *Advanced Dairy Chemistry. 2. Lipids*, London (UK), Chapman & Hall, 1–36.

CHRISTIE W W (2003), *Lipid Analysis: Isolation, Separation, Identification and Structural Analysis of Lipids* (3rd ed), Bridgwater, Oily Press.

CONTARINI G, POVOLO M, BONFITTO E and BERARDI S (2002), 'Quantitative analysis of sterols in dairy products: Experiences and remarks', *Int Dairy J*, 12, 573–578.

COOPER D A, BERRY D A, SPENDEL V A, JONES M B, KIORPES A L, PETERS J C (1997), 'Nutritional status of pigs fed olestra with and without increased dietary levels of vitamin A and E in long-term studies', *J Nutr*, 127, 169–1635.

CORREDIG M and DALGLEISH D G (1996), 'Effect of different heat treatments on the strong binding interactions between whey proteins and milk fat globules in whole milk', *J Dairy Res*, 63, 441–449.

CORREDIG M, ROESCH R R and DALGLEISH D G (2003), 'Production of a novel ingredient from buttermilk', *J Dairy Sci*, 86, 2744–2750.

DALGLEISH D G and BANK J M (1991), 'The formation of complexes between serum proteins and fat globules during heating of whole milk', *Milchwissenschaft*, 46, 75–78.

DARLING D F and BUTCHER D W (1978), 'Milk fat globule membrane in homogenized cream', *J Dairy Res*, 45, 197–208.

DAUFIN G, LABBE J P, QUEMERAIS A and MICHEL F (1991), 'Fouling of an inorganic membrane during ultrafiltration of defatted whey-protein concentrates', *Neth Milk Dairy J*, 45, 259–272.

DEETH H C (1997), 'The role of phospholipids in the stability of milk fat globules', *Aust J Dairy Tech*, 52, 44–46.

DUNKLEY W L and RONNING M (1966), Influence of supplemental tocopherol and carotene on oxidative stability of milk and milk fat, *Proc. 17th Int Dairy Congr (Munich)* A, 223–227.

DUPONT J, WHITE P J and FELDMAN E B (1991), 'Saturated and hydrogenated fats in food in relation to health', *J Am College Nutr*, 10, 577–592.

ERICKSON D R, DUNKLEY W L and RONNING M (1963), 'Effect of intravenously injected tocopherol on oxidized flavor in milk', *J Dairy Sci*, 46, 911–915.

ERICKSON D R, DUNKLEY W L and SMITH L M (1964), 'Tocopherol distribution in milk fraction and its relation to antioxidant activity', *J Food Sci*, 29, 269–275.

ERWIN R E and RANDOLPH H E (1975), 'Influence of mastitis on properties of milk. XI. Fat globule membrane'. *J Dairy Sci*, 58, 9–12.

EVERS J M (2004a), 'The milkfat globule membrane – compositional and structural changes post secretion by the mammary secretory cell', *Int Dairy J*, 14, 661–674.

EVERS J M (2004b), 'The milkfat globule membrane – methodologies for measuring milkfat globule (membrane) damage', *Int Dairy J*, 14, 747–760.

FAUQUANT C, BRIAR V, LECONTE N and MICHALSKI M C (2005), 'Differently sized native milk fat globules separated by microfiltration: Fatty acid composition of the milk fat globule membrane and triglyceride core', *Eur J Lipid Sci Technol*, 107, 80–86.

FAUQUANT C, BRIAR-BION V, LECONTE N, GUICHARDANT M and MICHALSKI M C (2007), 'Membrane phospholipids and sterols in microfiltered milk fat globules', *Eur J Lipid Sci Technol*, 109, 1167–1173.

FOCANT M, MIGNOLET E, MARIQUE M, CLABOTS F, BREYNE T, DALEMANS D and LARODELLE Y (1998), 'The effect of vitamin E supplementation of cow diets containing rapeseed and linseed on the prevention of milk fat oxidation', *J Dairy Sci*, 81, 1095–1101.

FONG B Y, CARMEN S N and MACGIBBON A K H (2007), 'Protein and lipid composition of bovine milk-fat-globule membrane', *Int Dairy J*, 17, 275–288.

FOX P F and MCSWEENEY P L H (1998), *Dairy Chemistry and Biochemistry*, London, UK: Blackie Academic & Professional.

FRAGA B M J, FONTECHA J, LOZADA L, MARTNEZ-CASTRO I and JUAREZ M (2000), 'Composition of the sterol fraction of caprine milk fat by gas chromatography and mass spectrometry', *J Dairy Res*, 67, 437–441.

FRANKEL N E (1991), 'Recent advances in lipids oxidation', *J Sci Food Agric*, 54, 495–510.

FRANKEL, N E (1998), *Lipid Oxidation*, U K, The Oily Press.

FREDE E, PRECHT D and TIMMEN H (1990), 'Lipide: Fettsäuren, Fette und fettbegleitstoffe einschliesslich fettlöslicher Vitamine', in Schlimme E, *Kompendium zur Milchwirtschaftlichen Chemie*, München, Volkswirtschaftlicher Verlag, 57–78.

FREUDENSTEIN C T W, KEENAN W N, EIGEL M, SASAKI J and STADLER W W F (1979), 'Preparation and characterization of the inner coat material associated with fat globule membranes from bovine and human milk', *Ex Cell Res*, 118, 277–294.

GERMAN J B and DILLARD C J (2006), 'Composition, structure and absorption of milk lipids: A source of energy, fat-soluble nutrients and bioactive molecules', *Crit Rev Food Sci Nutr*, 46, 57–92.

GINDIN J, NAOR S, STOVICHECK Y, DUSHENAT M and KONSTANTIN N (1998), 'Phosphatidylserine effect on functioning and general condition of Alzheimer's disease patients', *J Am Geriatr Soc*, 46, S100.

GRANELLI K, BARREFORS P, BJÖRCK L and APPELQVIST L-Å (1998), 'Further studies on lipid composition of bovine milk in relation to spontanous oxidized flavor', *J Sci Food Agric*, 77, 161–171.

GREENBANK G R and PALLANSCH M J (1961), 'Migration of phosphatides in processing dairy products', *J Dairy Sci*, 44, 1597–1602.

GRUNDY S M and DENKE M A (1990), 'Dietary influences on serum lipids and lipoproteins', *J Lipid Res*, 31, 1149–1172.

HAMZAWI L F and SHAHIN Y (1986), 'Fatty acid composition of milk fat globule membrane phospholipids from various species', *Milchwissenschaft*, 41, 754–756.

HASHIOKA S, MONJI A, HAN Y H, NAKANISHI H and KANBA S (2004), 'Potentiality of phosphatidylserine (PS)-liposomes for treatment of Alzheimer's disease', *Journal of Neuroimmunology*, 154, 119.

HEGSTED D M, MCGANDY R B, MYERS M L and STARE F J (1965), 'Quantitative effects of dietary fat on serum cholesterol in man', *Am J Clin Nutr*, 17, 281–295.

HEID H W and KEENAN T W (2005), 'Intracellular origin and secretion of fat globules,' *Eur J Cell Biol*, 84, 245–258.

HILDEBRAND D H, TERAO J and KITO M (1984), 'Phospholipids plus tocopherols increase soybean oil stability', *J Am Oil Chem Sci*, 61, 552–555.

HILLBRICK G and AUGUSTIN M A (2007), 'Milkfat characteristics and functionality: Opportunities for improvement', *Aust J Dairy Technol*, 57, 45–51.

HOULIHAN A V, GODDARD P A, NOTTINGHAM S M, KITCHEN B J and MASTER C J (1992), 'Interactions between the bovine milk membrane and skim milk components on heating whole milk', *J Dairy Res*, 59, 187–195.

HUANG T C and KUKSIS A (1967), 'A comparative study of the lipids of globule membrane and fat core and of the milk serum of cows', *Lipids*, 2, 453–460.

HUDSON B J F and MAHGOUB S E O (1981), 'Synergism between phospholipids and naturally-occuring antioxidants in leaf lipids', *J Sci Food Agric*, 32, 208–210.

JACKSON RH and BRUNNER J R (1960), 'Characteristics of protein fractions isolated from the fat/plasma interface of homogenized milk', *J Dairy Sci*, 43, 912–919.

JENNESS R and PALMER L S (1945), 'Substances adsorbed on the fat globules in cream and their relation to churning. V. Composition of the "membrane" and distribution of the adsorbed substances in churning', *J Dairy Sci*, 28, 611–623.

JENSEN R G, FERRIS A M and LAMMI-KEEFE C J (1991), 'The composition of milk fat', *J Dairy Sci*, 74, 3228–3243.

JENSEN R G (1995), *Handbook of Milk Composition*, New York, Academic Press.

JENSEN R G (2002), 'The composition of bovine milk lipid: January 1995 to December 2000', *J Dairy Sci*, 85, 295–350.

JENSEN S K and NIELSEN K N (1996), 'Tocopherols, retinol, β-carotene and fatty acids in fat globule membrane and fat globule core in cows' milk', *J Dairy Res*, 63, 565–574.

JIALAL I, FULLER C J and HUET B A (1995), 'The effect of alpha-tocopherol supplementation on LDL oxidation: A dose response study', *Arterioscler Thromb*, 15, 192–198.

JOHNSTON L A and CHEW B P (1984), 'Peripartum changes of plasma and milk vitamin A and β-carotene among dairy cows with and without mastitis,' *J Dairy Sci*, 67, 1832–1840.

KAMAL-ELDIN A and APPELQVIST L Å (1996), 'The chemistry and antioxidant properties of tocopherols and tocotrienols', *Lipids*, 31, 671–701.

KANNO C, YAMAUCHI K and TSUGO T (1968), 'Occurence of γ-tocopherol and variation of α- and β-tocopherol in bovine milk fat', *J Dairy Sci*, 51, 1713–1719.

KANNO C and KIM D H (1990), 'A simple procedure for the preparation of bovine milk fat globule membrane and a comparison of its composition, enzymatic activities, and electrophoretic properties with those prepared by other methods', *Agric Biol Chem*, 54, 2845–2854.

KATAN M B, ZOCK P L and MENSINK R P (1994), 'Effects of fat and fatty acids on blood lipids in humans: An overview', *Am J Clin Nutr*, 60, 1017–1022.

KATAOKA K and NAKAE T (1973), 'Comparative studies of milk constituents of various mammals in Japan. VIII. Comparison in phospholipids composition of the milk from various mammals', *Japan J Dairy Sci*, 22, 137–142.

KAYLEGAIN K E and LINDSAY R C (1995), *Handbook of Milk Fractionation Technology and Applications*, Champaign, IL, AOCS Press.

KEENAN T W, MORRE D J, OLSON D E, YUNGHANS W N and PATTON S (1970), 'Biochemical and morphological comparison of plasma membrane and milk fat globule membrane from bovine mammary gland', *J Cell Bio*, 44, 80–93.

KEENAN T W and HUANG C M (1972), 'Membranes of mammary gland. VI. Lipid and protein composition of golgie apparatus and rough endoplasmic reticulum from bovine mammary gland', *J Dairy Sci*, 55, 1586–1596.

KEENAN T W and HEID H W (1982), 'Tight attachment of fatty acids to proteins associated with milk lipid globule membrane', *Eur J Cell Bio*, 26, 270–276.

KEENAN T W, MOON T W and DYLEWSKI D P (1983), 'Lipid globules retain globule membrane material after homogenization', *J Dairy Sci*, 66, 196–203.

KEENAN T W, MATHER I H and DYLEWSKI D P (1988), 'Physical equilibria: Lipid phase'. In Wong N P, Jenness R, Keeney M, and Marth E H, *Fundamentals of Dairy Chemistry*, New York, Van Nostrand-Reinhold, 511–582.

KEENAN T W and DYLEWSKI D P (1995), 'Intracellular origin of milk lipid globules and the nature of structure of milk fat globule membrane'. In Fox P F, *Advanced Dairy Chemistry Lipids, Vol. 2*, London, Chapman and Hall, 89–130.

KEENAN T W and PATTON S (1995), 'The structure of milk: Implication for sampling and storage. The milk fat globule membrane'. In Jensen R G, *Handbook of Milk Composition*, New York, Academic Press, 5–50.

KEENAN T W (2001), 'Milk lipid globules and their surrounding membrane: A brief history and perspectives for future research', *J Mammary Gland Biol Neoplasia*, 6, 365–371.

KEENAN T W and MATHER I H (2002), 'Milk fat globule membrane'. In Roginski H, Fuquay J, Fox P F, *Encyclopedia of Dairy Science*, London, Academic Press, 1568–1576.

KEYS A, ANDERSON J T and GRANDE F (1965), 'Serum cholesterol reponse to changes in the diet. IV. Particular saturated fatty acids in the diet', *Metabolism*, 14, 776–787.

KIM H H Y and JIMENEZ-FLORES R (1995), 'Heat–induced interactions between the proteins of milk fat globule membrane and skim milk', *J Dairy Sci*, 78, 24–35.

KING R L, TIKRITI H H and OSKARSSON M (1966), 'Natural and supplemented tocopherol in the dairy ration and oxidized flavor', *J Dairy Sci*, 49, 157 (abstr).

KITCHEN B J (1977), 'Fractionation and characterization of the membranes from bovine milk fat globules', *J Dairy Res*, 44, 469–482.

KIVINEN A, TARPILA S, SALMINEN S and VAPAATALO H (1992), 'Gastroprotection with milk phospholipids – A 1st human study', *Milchwissenschaft*, 47, 694–696.

KOOPS J and TARASSUK N P (1959), 'The effect of various processing treatments on the partition of phosphatids between the fat phase and the milk plasma', *Neth Milk Dairy J*, 13, 180–189.

KRUKOVSKI V N, WHITING F and LOOSLI J K (1950), 'Tocopherol, caratenoid, and vitamin A content of the milk fat and the resistance of milk to the development of oxidized flavor as influenced by the breeds and season', *J Dairy Sci*, 33, 791–796.

KUCHROO T K and NARAYANAN K M (1981), 'Composition of fat globule membrane phospholipids', *Ind J Dairy Sci*, 34, 88–90.

LAMOTHE S, ROBITAILLE G, ST-GELAIS D and BRITTEN M (2008), 'Butter making from caprine creams: Effect of washing treatment on phospholipids and milk fat globule membrane proteins distribution', *J Dairy Res*, 75, 439–443.

LEA C H and WARD R J (1959), 'Relative antioxidant activities of the seven tocopherols', *J Sci Food Agric*, 10, 537–547.

LEAVER J, LAW A J R and BRECHANY E Y (1999), 'Covalent modification of emulsified β-casein resulting from lipid peroxidation', *J Colloid Interface Sci*, 210, 207–214.

LEE S J and SHERBON J W (2002), 'Chemical changes in bovine milk fat globule membrane caused by heat treatment and homogenization of whole milk', *J Dairy Res*, 69, 555–567.

LEFÈVRE T and SUBIRADE M (2000), 'Interaction of β-lactglobulin with phospholipid bilayers: A molecular level elucidation as revealed by infrared spectroscopy', *Int J Biol Macromol*, 28, 59–67.

LINDMARK-MANSSON H and AKESSON B (2000), 'Antioxidative factors in milk', *Br J Nutr*, 84, 103–110.

LING W and JONES P (1995), 'Dietary phytosterols: A review of metabolism, benefits and side effects', *Life Science*, 57, 195–206.

LOPEZ C, BION V B, MENARD O, ROUSSEAU F, PRADEL P and BESLE J M (2008), 'Phospholipid, sphingolipid, and fatty acid composition of the milk fat globule membrane are modified by diet', *J Agric Food Chem*, 56, 5226–5236.

MACGIBBON A K H and TAYLOR M W (2006), 'Composition and structure of bovine milk lipids'. In Fox P F and McSweeney P L H, *Advanced Dairy Chemistry, Volume 2: Lipids*, New York, Springer, 1–35.

MARTIN R W JR (1989), 'Electron microscopic localization of cholesterol in bovine milk fat globules', *Food Microstructure*, 8, 3–10.

MATHER I H and KEENAN T W (1983), 'Function of endomembranes and the cell surface in the secretion of organic milk constituents'. In Mepham T B, *Biochemistry of Lactation*, Amsterdam, Elsevier, 231–283.

MATHER I H and KEENAN T W (1988), 'Origin and secretion of milk lipids', *J Mammary Gland Biol Neoplasia*, 3, 259–273.

MCGANN T C A and FOX P F (1974), 'Physicochemical properties of casein micelles reformed from urea-treated milk', *J Dairy Res*, 41, 45–53.

MCPHERSON A V and KITCHEN B J (1983), 'The bovine milk fat globule membrane – its formation, composition, structure and behaviour in milk and dairy products', *J Dairy Res*, 50, 107–133.

MCPHERSON A V, DASH M C and KITCHEN B J (1984), 'Isolation and composition of milk fat globule membrane material. I. From pasteurized milks and creams', *J Dairy Res*, 51, 279–287.

MILNER J A (1999), 'Functional food and health promotion', *J Nutri*, 129, 1395–1397.

MINCIONE B, SPAGNA MUSSO S and DE FRANCISCUS G (1977), 'Studies on the different species. Sterol content of cows' milk', *Milchwissenschaft*, 32, 599–603.

MOLKENTIN J (2000), 'Occurence and biochemical characteristics of natural bioactive substances in bovine milk lipids', *Br J Nutr*, 84, S47-S53.

MOORE J H and CHRISTIE W W (1979), 'Lipid metabolism in the mammary gland of ruminant animals', *Prog Lipid Res*, 17, 347–395.

MORIN P, JIMÉNEZ-FLORES R and POULIOT Y (2004), 'Effect of temperature and pore size on the fractionation of fresh and reconstituted buttermilk by microfiltration', *J Dairy Sci*, 87, 267–273.

MORIN P, BRITTEN M, JIMÉNEZ-FLORES R and POULIOT Y (2007a), 'Microfiltration of buttermilk and washed cream buttermilk for concentration of milk fat globule membrane components', *J Dairy Science*, 90, 2132–2140.

MORIN P, JIMÉNEZ-FLORES R and POULIOT Y (2007b), 'Effect of processing on the composition and microstructure of buttermilk and its milk fat globule membranes', *Int Dairy J*, 17, 1179–1187.

MORRISON W R (1968), 'The distribution of phospholipids in some mammalian milks', *Lipids*, 3, 101–103.

MORTAN G, LEE S, BUSS D and LAUWRENCE P (1995), 'Intakes and major dietary sources of cholesterol and phytosterols in the British diet', *J Hum Nutr Diet*, 8, 429–440.

MULDER H and WALSTRA P (1974), *The Milk Fat Globule: Emulsion Science as Applied to Milk Products and Comparable Foods.* Pudoc, Wageningen, Farnham Royal, Commonwealth Agricultural Bureau.

NAKANISHI T and KAYA K (1970), Changes of phospholipids in cows' milk by heating, *Proc. XVIII Int Dairy Congr*, Sydney, 1E, 75.

NATH S B and RAMA MURTHY M K (1983), 'Autoxidative stability of cow and buffalo milk fat triglycerides as influenced by other minor lipid constituents', *Ind J Dairy Sci*, 36, 151–156.

NEWBURG D S and CHATURVEDI P (1992), 'Neutral glycolipids of human and bovine milk', *Lipids*, 27, 923–927.

NEWMAN R A and HARRISON R (1973), 'Characterisation of the surface of bovine milk fat globule membrane using microelectrophoresis', *Biochim Biophys Acta*, 298, 798–809.

NICHOLSON J W G and ST-LAURENT A M (1991), 'Effect of forage type and suplemental dietary vitamin E on milk oxidative stability', *Ca J Anim Sci*, 71, 1181–1186.

O'CONNELL J E, KELLY A L, AUTY M A E, FOX P F and DE KRUIF K G (2001), 'Ethanol-dependent heat-induced dissociation of casein micelles', *J Agric Food Chem*, 49, 4420–4423.

O'CONNOR T P and O'BRIEN N M (2006), 'Lipid oxidation'. In Fox P F and McSweeney P L H, *Advanced Dairy Chemistry Lipids*, Vol. 2, New York, Springer, 557–600.

OORTWIJN H and WALSTRA P (1979), 'The membranes of recombined fat globules. 2. Composition', *Neth Milk Dairy J*, 33, 134–154.

PALMER L S and WIESE H F (1933), 'Substances adsorbed on the fat globule in cream and their relation to churning. II. The isolation and identification of adsorbed substances', *J Dairy Sci*, 16, 41–57.

PALMQUIST D L and SCHANBACHER F L (1991), 'Dietary fat composition influences fatty acid composition of milk fat globule membrane in lactating cows', *Lipids*, 26, 718–722.

PARODI P W (1979), 'Stereospecific distribution of fatty acids in bovine milk fat triglycerides', *J Dairy Res*, 46, 75–81.

PARODI P W (1996), 'Milk fat components: Possible chemopreventive agents for cancer and other dieses', *Aust J Dairy Technol*, 51, 24–32.

PARODI P W (1997), 'Cows' milk fat components as potential anticarcinogenic agents', *J Nutr*, 127, 1055–1060.

PARTHASARATHY S and SANTANAM N (1994), 'Antioxidants and atherosclerosis', *Current Opinion in Lipidology*, 5, 371–375.

PATTON S and FOWKES F M (1967), 'The role of the plasma membrane in the secretion of milk fat', *J Theoret Biol*, 15, 274–281.

PATTON S and KEENAN T W (1975), 'The milk fat globule membrane', *Biochim Biophys Acta*, 415, 273–309.

PATTON S and JENSEN R G (1976), *Biomedical Aspects of Lactation*, Oxford, UK, Pergamon Press.

PATTON S, KELLY J J and KEENAN T W (1980a), 'Carotene in bovine milk fat globules: Observations on origin and high content in tissue mitochondri', *Lipids*, 15, 33–38.

PATTON S, LONG C and SOKKA T (1980b), 'Effect of storing milk on cholesterol and phospholipid of skim milk', *J Dairy Sci*, 63, 697–700.

PFEUFFER M and SCHREZENMEIR J (2001), 'Dietary sphingolipids: Metabolism and potential health implications', *Kieler Milchwirtschaftliche Forschungsberichte*, 53, 31–42.

POLITIS I, HIDIROGLOU M, BATRA T R, GILMORE J A, GOREWIT R C and SCHERF H (1995), 'Effects of vitamin E on immune function of dairy cows', *Am J Vet Re*, 56, 179–184.

PONGRACZ G, WEISER H and MATZINGER D (1995), 'Tocopherols – antioxidants in nature', *Fett Wissenschaft Technologie*, 97, 90–104.

RADEMA L (1956), The effect of pasteurization on the partition of phosphatides between the fat and aqueous phase of cream, *Proc XIV, Int Dairy Congr 1*, Part 2, 403–413.

RAO H G R (2002), 'Mechanisms of flux decline during ultrafiltration of dairy products and influence of pH on flux rates of whey and buttermilk', *Desalination*, 144, 319–324.

RAY T K and SINGH G P (1974), Lipid composition of milk fat globule membrane fractions, in *Proc XIX Int Dairy Congr*, Vol. IE, I D F, 211–212.

ROMBAUT R and DEWETTINCK K (2006), 'Properties, analysis and purification of milk polar lipids', *Int Dairy J*, 16, 1362–1373.

ROMBAUT R, DEJONCKHEERE V and DEWETTINCK K (2007), 'Filtration of milk fat globule membrane fragments from acid buttermilk cheese whey', *J Dairy Sci*, 90, 1662–1673.

SACHDEVA S and BUCHHEIM W (1997), 'Recovery of phospholipids from buttermilk using membrane processing', *Kieler Milchwirtschaftliche Forschungsberichte*, 49, 47–68.

SACKS F M and KATAN M (2002), 'Randomized clinical trials on the effects on dietary fat and carbohydrate on plasma lipoproteins and cardiovascular disease', *Am J Med*, 113, 13–24.

SCHINGOETHE D J, PARSONS J G, LUDENS F C, SCHAFFER L V and SHAVE H J (1979), 'Response of lactating cows to 300 mg of supplemental vitamin E daily', *J Dairy Sci*, 62, 333–338.

SHARMA S and DALGLEISH D G (1994), 'Effect of heat treatments on the incorporation of milk serum proteins into the fat globule membrane of homogenized milk', *J Dairy Res*, 61, 375–384.

SHARMA K C, KUMARI A, SAREEN V K and SINGH S (1987), 'Lipid composition of fat globule membrane from buttermilk and butter serum of buffalo and cow milk', *Milchwissenschaft*, 42, 439–441.

SLEIGH R W and BURLEY R W (1974), *Proc. of the Austral Biochem Soc*, 7, 60 (cited in Smith *et al.*, 1977).

SMITH L M, BIANCO D H and DUNKLEY W L (1977), 'Composition of milk fat globules with increased linoleic acid', *J Am Oil Chem Soc*, 54, 132–137.

SPITSBERG V L (2005), 'Bovine milk fat globule membrane as a potential nutraceutical', *J Dairy Sci*, 88, 2289–2294.

STAMPFER M J, MANSON J E, COLDITZ G A, SPEOZER F E, WILLET W C and HENNEKENS C H (1993), 'A prospective study of vitamin E supplementation and risk of coronary disease in women', *N Engl J Med*, 328, 1444–1449.

ST-LAURENT A M, HIDIROGLOU M, SNODDON M and NICHOLSON J W G (1990), 'Effect of α-tocopherol supplementation to dairy cows on milk and plasma α-tocopherol concentrations and on spontaneous oxidized flavor in milk', *Can J Anim Sci*, 70, 561–570.

STOCKDALE C R, WALKER G P, WALES W J, DALLEY D E, BIRKETT A, SHEN Z and DOYLE P T (2003), 'Influence of pasture and concentrates in the diet of grazing dairy cows on the fatty acid composition of milk', *J Dairy Res*, 70, 267–276.

SZUHAJ B F and NIEUWENHUYZEN W V (2003), *Nutrition and Biochemistry of Phospholipids*, Champaign, IL, AOCS Press.

TAYLOR M W and HAWKE J C (1975), 'Structural analysis of the triacylglycerols of bovine milkfats', *N Z J Dairy Sci Technol*, 10, 49–57.

THOMSON M P, BRUNNER J R and STINE C M (1959), 'Characteristics of high-melting triglyceride fractions from the fat-globule membrane and butter oil of bovine milk', *J Dairy Sci*, 42, 1651–1658.

THOMSON M P, BRUNNER J R, LINQUIST K and STINE C M (1961), 'Lipid components of the fat globule membrane', *J Dairy Sci*, 44, 1589–1596.

THOMSON S Y, HENRY K M and KON S K (1964), 'Factors affecting the concentration of vitamins in milk. I. Effect of breed, season and geographical location on fat-soluble vitamins', *J Dairy Res*, 31, 1–25.

TIMMEN H and PATTON S (1988), 'Milk fat globules: Fatty acid composition, size and in vivo regulation of fat liquidity', *Lipids*, 23, 685–689.

UDABAGE P, MCKINNON I R and AUGUSTIN M A (2000), 'Mineral and casein equilibria in milk: Effect of added salts and calcium chelating agents', *J Dairy Res*, 67, 361–370.

VAN BOEKEL M A J S and WALSTRA P (1995), 'Effect of heat treatment on chemical and physical changes to milk fat globules'. In Fox P F, *Heat-induced Changes in Milk*, Brussels IDF, 51–61.

VAN HOOIJDONK A C M, KUSSENDRAGER K D and STEIJNS J M (2000), 'In vivo antimicrobial and antiviral activity of components in bovine milk and colostrums involved in non-specific defense', *Br J Nutr*, 84, 127–134.

VESPER H, SCHMELZ E M, NIKOLOVA-KARAKASHIAN M N, DILLEHAY D L, LYNCH D V and MERRILL A H JR (1999), 'Sphingolipids in food and the emerging importance of sphingolipids to nutrition', *J Nutr*, 129, 1239–1250.

WALSTRA P (1974), 'High-melting triglycerides in fat globule membrane-artifact', *Neth Milk Dairy Res J*, 28, 3–9.

WALSTRA P and OORTWIJN H (1982), 'The membranes of recombined fat globules. 3. Mode of formation', *Neth Milk Dairy Res J*, 36, 103–113.

WALSTRA P and JENNESS R (1984), *Dairy Chemistry and Physics*, New York, John Wiley & Sons.

WALSTRA P (1985), 'Some comments on the isolation of fat globule membrane material', *J Dairy Res*, 52, 309–312.

WALSTRA P, GEURTS T J, NOOMEN A, JELLEMA A and VAN BOEKEL M A J S (1999), *Dairy Technology: Principles of Milk Properties and Milk Process*, New York, Marcell Dekker.

WARD R E, GERMAN J B and CORREDIG M (2006), 'Composition, applications, fractional significance of milk fat globule membrane material'. In Fox P F and McSweeney P L H, *Advanced Dairy Chemistry, Volume 2: Lipids*, New York, Springer, 213–244.

WHITE R F and EATON H D (1954), 'Relationship between fat globule surface area and carotenoids and vitamin A content of milk in successive portions of a milking', *J Dairy Sci*, 37, 147–155.

WIKING L, STAGSTED J, BJÖRCK L and NIELSEN J H (2004), 'Milk fat globule size is affected by fat production in dairy cows', *Int Dairy J*, 14, 909–913.

WOLF D P and DUGAN L R (1964), 'Structure of high melting glycrides from the milk fat globule membrane', *J Am Oil Chem Soc*, 41, 139–141.

WOLK A, FURUHEIM M and VESSBY B (2001), 'Fatty acid composition of fatty acid of adipose tissue and serum lipids are valid biological markers of dairy fat intake in men', *J Nutr*, 131, 828–833.

WONG N P, WALTER H E, VESTAL J H, LACROIX D E and ALFORD J A (1973), 'Cheddar cheese with increased polyunsaturated fatty acids', *J Dairy Sci*, 56, 1271–1275.

WOOD F W, MURPHY M F and DUNKLEY W L (1975), 'Influence of elevated polyunsaturated fatty acids on processing and physical properties of butter', *J Dairy Sci*, 58, 839–845.

WU Z, PALMQUIST L (1991), 'Synthesis and biohydrogenation of fatty acids by ruminal microorganisms in vitro', *J Dairy Sci*, 74, 3035–3036.

YE A, SINGH H, TAYLOR M W and ANEMA S (2002), 'Characterization of protein components of natural and heat treated milk fat globule membranes', *Int Dairy J*, 12, 393–402.

YE A, SINGH H, JAMES D O and ANEMA S (2004), 'Kinetics of heat-induced association of beta-lactoglobulin and alpha-lactalbumin with milk fat globule membrane in whole milk', *Int Dairy J*, 14, 389–398.

YE A, ANEMA S G and SINGH H (2007), 'Behaviour of homogenized fat globules during the spray drying of whole milk', *Int Dairy J*, 17, 374–382.

YEARGAN M R, OSHIDARI S, MITCHELL G E, TUCKER R E, SCHELLING G T and HEMKEN R W (1979), 'Mammary transfer of vitamin E in cows treated with vitamin A or linoleic acid', *J Dairy Sci*, 62, 1734–1738.

ZADOW J G (1993), 'Alcohol-mediated temperature-induced reversible dissociation of the casein micelle in milk', *Aust J Dairy Tech*, 48, 78–81.

ZAHAR M and SMITH D E (1995), 'Vitamin A distribution among fat core, fat globule membrane, and serum fraction in milk', *J Dairy Sci*, 78, 498–505.

ZOCK P L (1995), *Dietary fatty acids and risk factors for coronary heart disease: Controlled studies in healthy volunteers*, PhD Thesis, University of Wageningen.

Part III

Technological functionality of dairy components, and food and non-food applications

14

Molecular understanding of the interaction of dairy proteins with other food biopolymers

M. Corredig, University of Guelph, Canada

Abstract: When mixing dairy proteins with a biopolymer, two possible scenario may occur: the mixture either remains stable or after a certain amount of time shows phase separation. In this chapter, we attempt to describe the principles driving the interactions between the milk proteins and biopolymers, with focus on the applications in dairy products. Various types of associative and segregative behaviours are discussed, as often these mechanisms of destabilization lead to changes in structure and texture in food products.

Key words: proteins, whey proteins, caseins, emulsions, polysaccharides, phase separation, coacervates.

14.1 Introduction

The ability to predict the behaviour of biopolymers in foods is critical in achieving optimal quality as most macroscopic properties, such as texture, mouthfeel and shelf life, depend on the interactions occurring between the biopolymers during processing and storage. Polysaccharides are often included in dairy formulations because of their viscosifying, texturing and stabilizing effects. Polysaccharides are also increasingly included in formulations because their consumption has been associated with wellness and prevention of chronic diseases, and in this case they are added at levels not only nutritionally significant but also that impart changes in the overall quality, texture and stability. Although the chapter deals mainly with the polymers commonly used in dairy products, the fundamental aspects discussed herein are most likely applicable to many other biopolymers when present in protein matrices.

In foods and consumer products, the addition of biopolymers to a formulation often leads to changes in structure and texture, such as the formation of weak networks or solid gels, or separated phases. The control of texture is of critical importance to the final quality of the product and, for this reason, the interactions between biopolymers in food have been widely researched in the past. In dairy systems, the interactions between polysaccharides and milk proteins occur at different levels, as proteins are present either in molecular form or as supramolecular assemblies and colloidal particles. Whey proteins constitute about 1% of the total volume fraction of milk and are present in milk as monomers or oligomers, a few nanometer in size. With heating, whey proteins denature and form aggregates much larger in size, and they can be present either in the soluble phase or adsorbed on casein particles. The casein proteins in milk are self associated in colloidal particles polydispersed in size (ranging between 80 and 400 nm). These particles, held together by calcium phosphate bridges and sterically stabilized by κ-casein, occupy about 10% of the volume fraction of milk. When caseinates are employed as ingredients, they are no longer structured as casein micelles, but they are a mix between monomeric proteins and small particles of about 30 nm of diameter. The interactions between biopolymers and caseins will depend on their state of aggregation and processing history. In addition, in dairy products, often proteins are adsorbed on the surface of lipid droplets in emulsion systems, and the behaviour of the emulsion droplets will depend on the type of protein present at the interface. It is therefore often necessary to distinguish between the different types of interactions that may occur between biopolymers and dairy proteins, depending on the proteins involved and the processing history of the milk matrix. The present chapter will attempt to illustrate these differences. The first part of the chapter will deal with biopolymer interactions with monomeric proteins, the second part will focus on casein micelles interactions with polysaccharides, and the third part of the chapter will discuss the interactions between polymers and proteins when the proteins are adsorbed at an oil–water interface.

When mixing proteins with a biopolymer, at very dilute regimes the system will often appear stable and homogeneous, as the entropy of mixing dominates; however, at high enough concentrations, destabilization will occur because of either associative or segregative behaviour (Tolstoguzov, 1991) (Fig. 14.1). When two or more polymers with different structures are mixed in solution, they tend to show a segregative behaviour. At close approach, the molecules will lose conformational entropy and there will be a tendency to reduce the concentration of one polymer in close proximity of the other polymer. In a protein–polysaccharide system such as, for example, milk with a neutral polysaccharide, at high enough concentrations there will be separation of the mix into two phases, one protein-rich and the other polysaccharide-rich. In systems where the biopolymer is mixed with large protein particles or oil droplets, the excluded volume effect, which is entropy driven, plays an important role in causing phase separation (Vrij, 1976; Poon, 1998; de Kruif and Tuiner, 2001). The phase behaviour

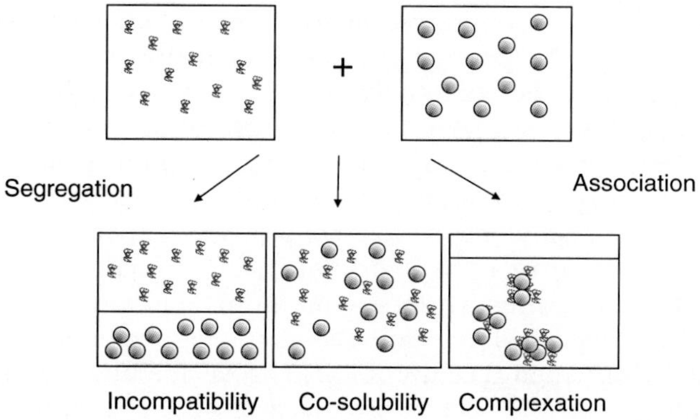

Fig. 14.1 When two polymers are present in a mixture, co-solubility occurs at low concentrations, while at high enough concentrations, associative or segregative interactions will cause phase separation or the formation of complexes (coacervates).

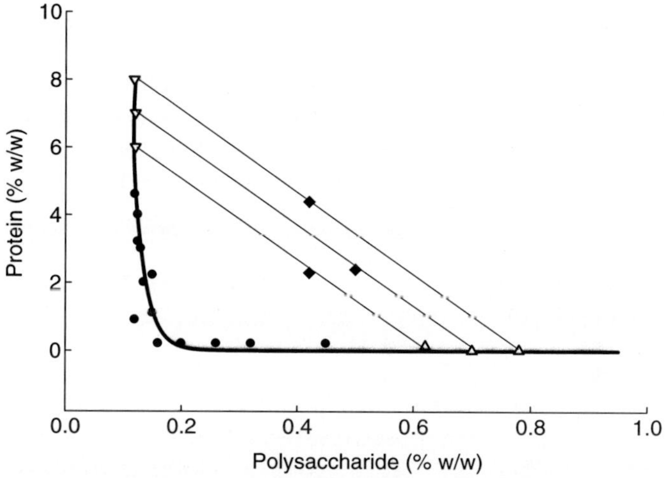

Fig. 14.2 Phase separation diagram for a binary mixture containing protein and polysaccharide. Binodal (solid line, ●), initial mixtures (◆), upper phase (△), lower phase (▽).

of many food biopolymers has been extensively described by various phase diagrams for binary systems (Grinberg and Tolstoguzov, 1997). In binary mixtures there is a small region of compatibility, below the tie line (Fig. 14.2). The closer the tie line to the axes, the higher the polymer incompatibility.

If the two biopolymers present in solution interact with one another, instability will occur because of associative interactions (Fig. 14.1). Many of the polysaccharides added to milk matrices carry a charge, because they have carbohydrates with carboxylic, phosphate or sulphate groups. These polysaccharides will associate with the proteins, creating soluble or insoluble complexes, bridging between two or more protein particles. This process is called coacervation and the biopolymer complexes are called coacervates. The interactions between the biopolymers are usually electrostatic in nature, and therefore strongly dependent on pH and ionic strength. These aggregates have been widely studied as they are of practical importance. For example, it has been recently suggested that the interactions between salivary proteins and the pectins present in wine are relevant to a better taste perception, as they decrease the tannins–protein interactions that cause the highly astringent taste in the mouth (Carvalho *et al.*, 2006). The control of coacervates formation is not only of great importance to the design of microstructure and appearance of foods, but also in the engineering of systems for the controlled release of bioactives in the gastrointestinal tract. In addition, it has been shown that the efficacy of digestive enzymes is different when molecules are present in a coacervate system from when they are in isolation, and this can have profound implications for allergenicity reduction (Peyron *et al.*, 2006).

In this chapter, we attempt to describe the principles driving the interactions between milk proteins and biopolymers, and their applications in milk products. This overview is by no means exhaustive, but brings a series of examples from the widespread use of polysaccharides in dairy products.

14.2 Interpolymer complexes: Interactions of polysaccharides with whey proteins or sodium caseinate

Whey proteins and caseins are dairy ingredients commonly used in foods. Whey protein isolates are composed mostly of β-lactoglobulin, α-lactalbumin and bovine serum albumin. These are compact globular proteins, which will modify their structure with heating. Above their denaturation temperature, they unfold and form aggregates. The protein that dominates the behaviour of whey proteins is β-lactoglobulin, and for this reason β-lactoglobulin is the most studied of the whey proteins. Native β-lactoglobulin has a molecular mass of about 18 kDa, is positively charged below its isoelectric point (pH 5.2), and it has been shown to form complexes with charged polysaccharides (Swaisgood, 1992).

The caseins are phosphoproteins with specific areas showing high charge density. For example, α_{s1}-casein has an acidic peptide region of about forty amino acids that contains seven phosphate groups and twelve carboxyl groups. The κ-casein has a hydrophilic region containing glycosilated moieties and fourteen carboxylic groups (Holt, 1992). For this reason, these highly charged moieties may interact with charged polysaccharides under specific conditions. It is important to note that, while great effort has been made in the past to research the interactions of polysaccharides with micellar caseins (see below), very little has been reported on the interactions between polysaccharides and non-micellar casein ingredients (such as sodium caseinate).

In the past few years there has been considerable progress in understanding the associative interactions of whey proteins (especially β-lactoglobulin in isolation or whey protein isolates) with polysaccharides in binary mixtures (Turgeon *et al.*, 2007). Complexes form between proteins and polysaccharide molecules when opposite net charges (or charge patches) are present. With the exception of chytosan (α(1,4)2-amino-2deoxy-d-glucose polymer), which is positively charged, most of the charged polysaccharides used in dairy systems are negatively charged. For this reason, most complexes form with the milk proteins at a pH below their isoelectric point (for example, at pH < 5 for β-lactoglobulin). These complexes are non-covalent and are affected by pH, ionic strength, temperature, and changes in processing conditions (e.g. rate of acidification or changes in shear). At a pH near the isoelectric point, the proteins will start interacting with the polymers, increasing the ionization of both polyelectrolytes at the expense of the portion of the polymer in solution. To be attractive, the decrease of electrostatic free energy caused by the formation of the coacervate needs to be larger than the increase in free energy caused by the mixing term and the conformational changes needed to form the complex (Biesheuvel and Cohen Stuart, 2004).

The formation of coacervates in β-lactoglobulin and gum arabic mixtures is a clear example of the associative behaviour between charged polysaccharides and whey proteins. Gum Arabic is a negatively charged polysaccharide composed of D-galactose, L-arabinose, L-rhamnose and D-glucuronic acid. In addition, it has a polypeptide moiety which is responsible for its emulsifying properties (Williams *et al.*, 1990). Under particular conditions, the mixture forms two separate phases, with one rich in the coacervates. By using slow acidification (achieved by the addition of glucono-δ-lactone to the binary mixture), it has been possible to accurately describe the effect of pH on the complex formation between the two polymers (Mekhloufi *et al.*, 2005). During acidification, at a critical pH, primary soluble complexes form. During complexing, there is a clear change in the structure of the β-lactoglobulin (Mekhloufi *et al.*, 2005). A further decrease in pH results in visible phase separation, with a continuous coarsening of the aggregates (Sanchez *et al.*, 2006). The whey proteins and the gum arabic polymers present in the coacervate phase show independent translational movement. This has been interpreted as electrostatic interactions in the coacervates as they continue to break and reform, showing a continuous rearrangement of the coacervates (Weinbreck *et al.*, 2004a). The structure of the coacervates is pH dependent: the structure density is directly proportional to the extent of the electrostatic interactions (Weinbreck *et al.*, 2004a). It has been shown that the coacervates have different functional properties compared to the two biopolymers in isolation. In mixes containing whey protein isolate and gum acacia, the complexes cause an increase in viscosity; the weaker the electrostatic interactions, the lower the viscosity of the coacervate phase (Weinbreck *et al.*, 2004b).

Mixing whey proteins with negatively charged polysaccharide molecules will cause the formation of different types of aggregates, depending on the stiffness of the polysaccharide and its charge density, as well as the pH and the ionic strength of the mixture. An optimal ratio of the two biopolymers needs also to be achieved, as this will correspond to an electrostatic equivalent number of protein and polysaccharide molecules. Soluble complexes have been obtained mixing whey proteins with xanthan gum or pectin (Kazmierski *et al.*, 2003; Girard *et al.*, 2004; Laneuville *et al.*, 2005). Pectin is a cell-wall polysaccharide, generally extracted from citrus, apple or beet. Pectin is an anionic polysaccharide with D-galacturonic acid residues esterified with methoxyl groups, the extent of methylesterification influencing the overall charge and its distribution on the chain (Voragen *et al.*, 1995). Xanthan polymers are also negatively charged. Xanthan, which is produced by Xhanthomonas sp., has a pentasaccharide repeating unit with acetyl and pyruvate substitutions, with carboxylic groups (Laneuville *et al.*, 2005). Xanthan has a low charge density when compared to other negatively charged polysaccharides used in dairy systems. When β-lactoglobulin is mixed with xanthan, a coupled gel network forms having clusters of associated protein–xanthan complexes, with junction zones at very low concentrations of biopolymer (0.1%) (Laneuville *et al.*, 2006). The rheological properties of the gel network can be controlled by modulating the ratio of protein and xanthan polymer, or by varying the polymer molecular mass (Laneuville *et al.*, 2006).

The interfacial properties of the coacervates are also often distinct from those of the biopolymers in isolation (Ganzevles *et al.*, 2006; Guzey and McClements, 2006): For example, the foam stabilizing properties of the electrostatic complexes can be superior to those of the protein alone, mostly because they provide a better barrier to gas diffusion and oxidation (Schmitt *et al.*, 2005).

Compared to whey proteins, much less work has been carried out on the interactions of non-micellar caseins with charged polysaccharides. Recently, it has been shown that sodium caseinate forms nanoparticles between 250 and 350 nm when mixed with chitosan (Anal *et al.*, 2008). Chitosan is composed of N-acetylglucosamine and N-glucosamine repeating units and it is cationic (with a pKa of about 6.5). Its composition and charge make it unique, as most of the commercially available polysaccharides employed as functional ingredients in dairy products are either neutral or negatively charged. Chitosan forms electrostatic soluble or insoluble complexes with proteins (Laplante *et al.*, 2005; Hong and McClements, 2007). Chitosan molecules form electrostatic complexes with casein proteins, in the pH range 4.8 to 6.0. It has been hypothesized that the particular structural features of the casein proteins, with well defined charged moieties, are very desirable in the formation of nanoparticles (Anal *et al.*, 2008). These complexes may result in unique functional ingredients used to design foods with novel textures, or nanoparticles with unique release properties at a particular pH range (between 3.5 and 5.5).

A specific associative behaviour can be noted when mixing casein proteins with κ-carrageenan. Carrageenans, derived from red algae, are polysaccharides containing sulphate moieties, with repeating galactose and 3,6 anhydrogalactose joined by alternating α-(1,3) and β-(1,4) links (Rochas and Landry, 1987). These polysaccharides are classified depending on the number and position of their sulphate groups: one, two or three, for κ-, ι-, and λ-, respectively. When dispersed in solution, κ- and ι- carrageenan undergo a temperature-dependent, reversible coil-to-helix transition. At neutral pH, κ-carrageenan is negatively charged; however, it has been shown to interact specifically with κ-casein. This interaction has been attributed to the positively charged region of κ-casein between amino acids 97 to 112 (Snoeren *et al.*, 1975). Size exclusion chromatography of mixtures containing pure caseins (β, α_s and κ) as well as sodium caseinate, in the absence and the presence of κ-carrageenan, clearly show that κ-casein is the protein involved in this specific interaction, as only in the presence of κ-casein do κ-carrageenan molecules form complexes with the proteins. This specific interaction is of great importance in the stabilization of dairy products as κ-carrageenan adsorbs at neutral pH to the surface of the casein micelles, providing additional steric stabilization in milk and ice-cream mixes (see below).

Segregative interactions often occur between negatively charged polymers and milk proteins at pH near neutral, because of thermodynamic incompatibility; that is to say, the interactions between different molecules are energetically less favourable than the interactions between like-molecules. Thermodynamic incompatibility leads to phase separation. A recent example of thermodynamic incompatibility has been shown in the study of mixtures of whey proteins and β-glucans (Kontogiorgos *et al.*, 2009). This group of neutral, linear homopolysaccharides containing 1-4-β-glucosyl residues, have received considerable attention because of the physiological benefits related to their consumption (Lazaridou and Biliaderis, 2007). Phase diagrams of binary mixes of whey protein and β-glucan at neutral pH show close proximity of the bimodal line to the axes, indicating high incompatibility in a wide range of concentrations (Kontogiorgos *et al.*, 2008).

Segregative interactions are of particular importance during heating of whey proteins. With the presence of charged polysaccharides, different types of heat-induced aggregates form. Heat treatment causes structural changes in whey proteins, with disulphide interchange reactions leading to the formation of protein aggregates. At low ionic strength and at neutral pH, denatured β-lactoglobulin monomers form well-defined clusters of about 30 nm (Aymard *et al.*, 1996), but in the presence of κ-carrageenan or pectin, the aggregation is affected (Capron *et al.*, 1999; Beaulieu *et al.*, 2005), and it depends on the charge density of the polymer, as shown in a study using pectin molecules with similar charges but different charge distributions (Kazmierski *et al.*, 2003). It has been reported that if mixtures are heated at 70°C with κ-carrageenan at levels below phase separating concentrations, neither the rate nor the structure of the aggregates is affected (Croguennoc *et al.*, 2001). When heating at 90°C, the presence of κ-carrageenan causes the formation of larger and more polydisperse intermediate aggregates, modifying the make-up of the final building blocks of the gel networks (Flett and Corredig, 2009). At high enough concentrations, these mixed systems show a faster heat-induced aggregation and gelation (Ould Eleya and Turgeon, 2000).

Only recently, the influence of shear on the phase behaviour during heating of whey protein isolate solutions containing κ-carrageenan has been considered, although it is known that shear can be used to modulate bulk properties such as texture, stability, sensorial perception and microstructure of the mixtures. The phase behaviour depends on the shear applied during heating and cooling, with a shift of the bimodal tie lines towards lower concentrations of whey proteins and a change in the microstructure and the distribution of the polymers' concentrations in the two phases (Gaaloul *et al.*, 2008).

14.3 Interactions of casein micelles with polysaccharides

Casein proteins are present in milk as supramolecular aggregates, the casein micelles. These protein particles, polydisperse in size (with a diameter between 80 and 400 nm) are an association of four individual molecules, α_{s1}, α_{s2}, β and κ-casein, with κ-casein present primarily on the surface. The hydrophilic C-terminal part of κ-casein protrudes in solution, imparting a polyelectrolyte brush character to the surface of the micelles, causing steric stabilization. When polysaccharides are added to milk, different effects will occur depending on the presence of associative or segregative interactions of these biopolymers with the casein micelles, which, at the natural pH of milk, are negatively charged.

When non-interacting polysaccharides are added to milk, phase separation often occurs by a depletion mechanism. The polymer molecules are in fact excluded from the surface, creating a depletion layer around the particle, and the casein micelles are then driven together (Asakura and Oosawa, 1958; Vrij, 1976). The thickness of the depletion layer is not affected by the concentration of the polymer (Vrij, 1976), until concentrations in the

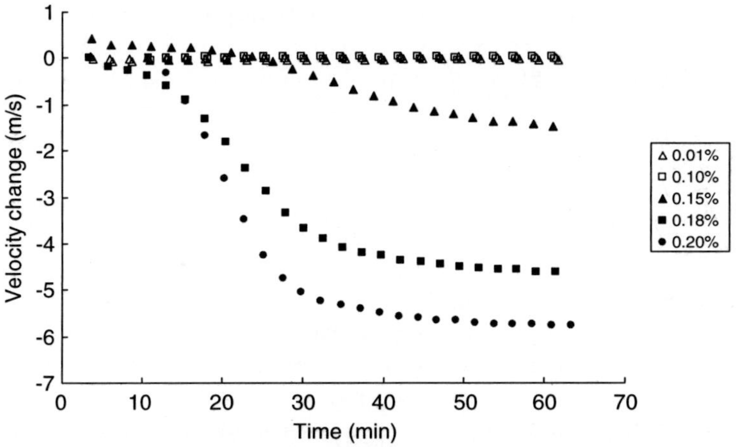

Fig. 14.3 Effect of the addition of different concentrations of high methoxyl pectin in skim milk on the velocity of ultrasonic waves (8 MHz) traveling through the sample over time. Velocity is measured relative to the initial value right after mixing. Above a critical concentration, the ultrasonic velocity shows a clear decrease until reaching a plateau, indicating an increase in the overall compressibility of the mixture because of the formation of casein micelle clusters.

semi-dilute regime, where there are significant increases in the osmotic pressure. The thickness of the layer decreases with the polymer concentration (De Gennes, 1981).

The phase separation occurring in milk systems in the presence of the non-interacting polysaccharide is affected by the molecular mass of the biopolymer. This has been well demonstrated in a study of milk and guar gum mixtures (Tuiner *et al.*, 2000). Guar gum is a galactomannan, consisting of β (1-6)-linked mannose repeating units and side groups of β (1-4)-linked galactose in a 2:1 ratio. It has been shown that the critical polymer concentration to cause phase separation increases with decreasing molecular weight. Charged polysaccharides also cause phase separation at neutral pH, if the molecules do not interact with the casein micelles. Figure 14.3 illustrates the effect of increasing concentrations of high methoxyl pectin (negatively charged) on the stability of milk as indicated by changes in the velocity of ultrasonic waves travelling through the sample. This parameter indicates the energy losses to which the ultrasonic wave will be subjected while travelling through the sample. Above a critical amount of high methoxyl pectin, the ultrasonic attenuation changes dramatically with time, indicating that the non-interacting polymer is structuring and causing the casein micelles to cluster themselves, eventually leading to phase separation (Acero-Lopez *et al.*, 2009). It has been recently demonstrated, using diffusing wave spectroscopy and ultrasonic spectroscopy, that clustering of the casein micelles occurs even at concentrations of polysaccharide lower than those needed to cause visible phase separation (Acero-Lopez *et al.*, 2009). To be able to control phase separation in complex mixtures is critical not only to control stability but, more importantly, because in processing phase-separating systems, controlling the extent of the phase separation has potential for the creation of new textures and structures.

Scattering techniques, combined with microscopy at appropriate length scales, as well as rheological measurements, have provided a large amount of information on the overall mechanisms involved in polysaccharide-induced phase separation in milk (Doublier *et al.*, 2000; de Kruif and Tuiner, 2001). In acid-induced casein gels, increasing the amount of non-interacting polysaccharide (e.g. locus bean gum or xanthan) leads to the formation of a more compact casein network, with an increased porosity of the gels, as shown by electron and phase contrast microscopy (Sanchez *et al.*, 2000). Different structures can also form by controlling the extent of phase separation in milk during rennet-induced aggregation (Acero-Lopez *et al.*, 2009). The microstructure, in turn, affects the fracture behaviour of the food system, and this is one of the most important parameters in sensory perception (Van den Berg *et al.*, 2007).

The specific associative interactions occurring between κ-casein and κ-carrageenan are of critical importance to the stability of many dairy products. At pH near neutral, κ-carrageenan (as well as ι-carrageenan) molecules adsorb on the surface of casein micelles, imparting additional stabilization to these protein particles. Dynamic light scattering measurements demonstrate that with addition of κ-carrageenan there is an increase in the size of the micelles (Dalgleish and Morris, 1988). Below the coil to helix transition of κ-carrageenan, the charge density of the polysaccharide chain increases and interactions occur (Syrbe *et al.*, 1998). At sufficient concentrations, helix–helix associations between polysaccharides form a three-dimensional network (Rochas and Landry, 1987). It is important to note that at temperatures above the coil–helix transition, at a sufficient concentration of κ-carrageenan, phase separation by depletion flocculation occurs (Langendorff *et al.*, 1997).

Images from field emission scanning electron microscopy have clearly illustrated that κ-carrageenan, when present at concentrations below gelling, interacts with the casein micelles forming bridges between the protein particles (Martin *et al.*, 2006). At these low concentrations, κ-carrageenan inhibits macroscopic phase separation when other non-interacting polysaccharides are present. Figure 14.4 illustrates the difference in phase separation behaviour for a milk mixture containing flaxseed gum with and without κ-carrageenan (below gelling concentration). Flaxseed gum is a water extract from the seed and it is composed of an acidic and a neutral fraction, containing four major sugars, L-galactose, D-xylose, L-rhamnose and D-galacturonic acid (Oomah *et al.*, 1995). Flaxseed gum in milk causes phase separation by a depletion mechanism,

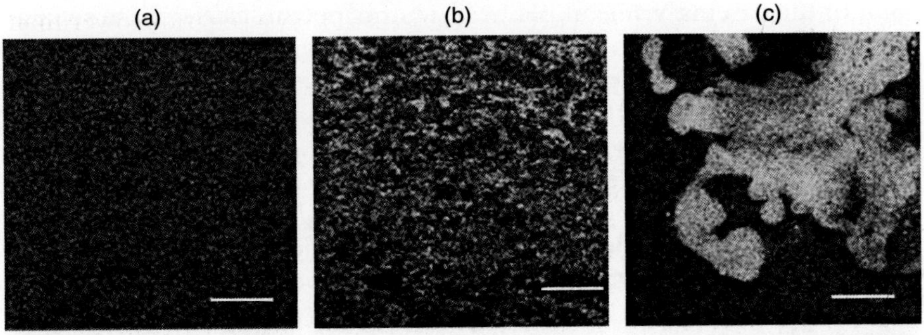

Fig. 14.4 Confocal micrographs of skim milk mixtures observed one hour after mixing. Bar 50 μm. Protein stained with FTIC appears brighter. (a) Skim milk, (b) Skim milk containing 0.30% flaxseed gum, (c) skim milk containing 0.30% flaxseed and 0.03% k-carrageenan.

and the presence of κ-carrageenan inhibits visible phase separation. As is clearly shown in Fig. 14.4, and confirming previous literature results, the mixtures containing κ-carrageenan and phase separating polysaccharides are still microscopically organized in clusters, with discrete micelle-rich domains and a continuous phase rich in the non-interacting polysaccharide (Schorsch *et al.*, 2000; Thaiudom and Goff, 2003). Bridging of the polysaccharide between casein micelles and the formation of a loose network seem to be the key to inhibiting macroscopic phase separation (Bourriot *et al.*, 1999; Spagnuolo *et al.*, 2005). It has been in fact proven that when helix–helix association is hindered, by the use of NaI or comparing ι-carrageenan to κ-carrageenan, without inhibiting the adsorption of the polymer to the micelles, the ability of κ-carrageenans to stabilize mixtures against macroscopic phase separation is significantly reduced (Spagnuolo *et al.*, 2005).

An interesting feature of the association and bridging between κ-carrageenan and κ-casein is that the mixtures do not show a significant decrease in the ability of the micelles to freely diffuse. Diffusing wave spectroscopy data have recently shown that the casein micelles themselves have considerable mobility even when associated with κ-carrageenan (Alexander and Dalgleish, 2007). This unique interaction mechanism between κ-casein and κ-carrageenan has not yet been fully exploited. Recently it has been shown that mixing κ-carrageenan with milk at appropriate ratios, under shear, causes the formation of large aggregates which show a very different functionality to that of the original biopolymers (Ji *et al.*, 2008). This implies that value-added ingredients containing aggregates of caseins and κ-carrageenan may be obtained by modulating the interactions between the biopolymers.

Other negatively charged polysaccharides (such as pectin) form electrostatic complexes with casein micelles at pH values close to the isoelectric point of the caseins. While at pH near neutral, pectins cause phase separation, at low-pH, high methoxyl pectins (i.e. with lower charges on the backbone) are often employed to stabilize low-pH milk products, such as acidified milk beverages (Maroziene and de Kruif, 2000). The adsorption of pectin around the casein particles seems to occur at a pH between 5.0 and 4.5, right when the casein micelles show the collapse of the κ-casein polyelectrolyte brush. The newly adsorbed layer of pectin will increase the negative charge on the surface of the protein particles and will also impart more steric stability. To stabilize an acid milk beverage, after acidification of the milk, high methoxyl pectin is added and the mix is homogenized. At a pH close enough to the isoelectric point of the caseins, the high methoxyl pectin will adhere to the particles and also form a loose network, which is at least in part responsible for the stabilization of the acid drinks as well as to the improvement of the mouthfeel (Boulenguer and Laurent, 2003; Tromp *et al.*, 2004). Indeed, only a fraction of the biopolymer is actually associated with the micelles, and researchers have shown that the casein particles are still stable when the unadsorbed part of the pectin is removed (Tromp *et al.*, 2004). The dispersed particles are stabilized by an electro–steric mechanism, and differences in the charge of the polymer strongly affect the stabilizing efficiency. Although low methoxyl pectins are more charged at a particular pH than high methoxyl pectins, they do not stabilize the acid dispersions, and amongst the high methoxyl pectins, the least charged are the most effective (Liu *et al.*, 2006).

Other polysaccharides are also of interest in low pH casein gels. Soybean soluble polysaccharide (SSPS), a polysaccharide derived from soy cotyledons, has also been employed to stabilize milk beverages. SSPS contains a rhamnogalacturonan backbone branched with galactose and arabinose chains. In addition, SSPS contains a small protein portion, which plays an important role in the adsorption of this polysaccharides at oil–water interfaces (Nakamura *et al.*, 2004). SSPS has been shown to be more efficient than pectin in imparting stability to acid milk at pH < 4.0 (Liu *et al.*, 2006), forming electrostatic complexes with the casein particles. At these pH values, high methoxyl pectins lose their effectiveness in binding and stabilizing acid milk dispersions. It has been hypothesized that the neutral sugar size chains of SSPS protrude into solution, causing steric stabilization.

Perhaps one of the most complex examples of the interaction dynamics between a biopolymer and milk proteins (especially casein micelles) can be found in the effect of exopolysaccharides (EPS), produced by lactic acid bacteria, in texture formation during acid-induced gelation of milk. Lactic acid bacteria that produce EPS are often used in starter cultures, to enhance the viscosity or the water holding capacity of the acid gels. Recently, EPS production by lactic acid bacteria has been a focus of attention for the dairy industry. Although there may be a direct relation between the viscosity and structure formation, and the amount of EPS produced by the bacterial strain, other factors such as the molecular structure and characteristics of the EPS as well as the interactions occurring with the milk proteins play an important role (Tuinier *et al.*, 2001; Ayala-Hernandez *et al.*, 2008a). EPS can be attached either as capsules to the bacterial cells or excreted by the bacteria as unattached material (Hassan, 2007). The 'ropy' characteristics of some fermented milks are associated with particular types of unattached EPS, and could be related to the ability of the polymer to interact with milk proteins (Ayala-Hernandez *et al.*, 2008b). The dynamics of the interactions are complex, as EPS is gradually produced during fermentation, while the environmental conditions as well as the physico-chemical characteristics of the casein micelles are also changing. For this reason, the higher pH of gelation induced by cultures producing EPS has been attributed to phase separation because of a lack of interactions between EPS and the caseins at pH > gelation (Girard and Shaffer-Lequart, 2007a). On the other hand, there is also evidence of the interactions between EPS and proteins at acid pH (Girard and Shaffer-Lequart, 2007a, 2007b; Ayala-Hernandez *et al.*, 2008a). A microstructural study of milk fermented with a strain producing EPS showed gels with large interstitial cavities filled with bacterial colonies surrounded by polysaccharide (Hassan *et al.*, 2003). This was attributed to phase separation and incompatibility between the protein network and the EPS produced, which remains in close proximity to the bacterial cells. It was also reported that the formation of EPS-rich domains in the gels contributes to the texture and mouthfeel of acid milk gels (Hassan *et al.*, 2003). This is an area still under debate, most likely because of the complications due to lack of accurate methods to measure the amounts of polysaccharide present in the gels and the differences in EPS composition and molecular characteristics depending on the bacterial strain. However, the technological benefits, such as water binding and improved mouthfeel in reduced fat cheeses, as well as structuring characteristics of the EPS, continue to emerge and more research will be carried out to understand their mechanisms of interaction with the milk proteins in fermented dairy matrices.

14.4 Polysaccharides interactions with milk proteins adsorbed at oil–water interfaces

In dairy emulsions, the primary emulsifying agent is the proteins (i.e. caseins, in either monomeric or micellar form, or whey proteins). The proteins adsorb at the oil/water or air/water interface, protecting the interfacial membranes from coalescence. Milk proteins function as surface active ingredients in emulsions because of their amphiphilic structure, and they contribute to the stability of the emulsion droplets by a combination of electrostatic and steric stabilization mechanisms (Dalgleish, 1997). The behaviour of the oil droplets will depend on the type of protein adsorbed at the interface and the surface load. In aqueous conditions, at pH near neutral and at low ionic strength, casein proteins form loops and tails that protrude into solution, and under these conditions the oil droplets covered by sodium caseinate do not behave as hard spheres, because of the strong steric repulsion between the droplets (Gaygadzhiev *et al.*, 2008). Globular whey proteins form thinner and denser layers, and stabilization occurs mostly by electrostatic repulsion; at high ionic strength they are present as flocculated systems (Gaygadzhiev *et al.*, 2008).

Emulsions stabilized by dairy proteins often show poor stability to changes in pH, ionic strength and processing conditions such as heating, cooling or shearing. For this reason, polysaccharides are often added, mostly to control the rheological properties of the continuous phase. The addition of these biopolymers will improve stability and affect the texture and sensory characteristics of the emulsions. Most polysaccharides added do not show interfacial activity, with the exception of a few, such as gum arabic or SSPS. These polysaccharides contain a polypeptide fraction associated to the carbohydrate backbone. This polypeptide is the surface active fraction that anchors to the surface of the oil droplet, with the hydrophilic carbohydrate portion extending in solution and providing the steric and electrostatic stabilization (Williams *et al.*, 1990; Nakamura *et al.*, 2004).

The addition of polysaccharides to emulsions needs to be carefully controlled, because of the interactions occurring between the polysaccharides and the proteins adsorbed at the interface. These interactions can be segregative or associative in nature, and they will have implications for stability and the formation of particular textures and structures.

When non-interacting, non-adsorbing polysaccharide is added to emulsions, depletion flocculation may occur. This has been reported for dairy stabilized emulsions at pH near neutral with added pectin, κ-, ι-, λ-carrageenan, xanthan gum or flaxseed gum, amongst others (Surh *et al.*, 2006; Sun *et al.*, 2007). Phase separation is of great importance as it can help in the formation of structure; for example, the development of multiple emulsions of the oil-in-water-in-water type. These are mixtures of oil droplets and water droplets dispersed in a continuous water phase, with this latter phase in a gel-like state to improve stability (Khalloufi *et al.*, 2008, 2009).

The presence of unadsorbing polysaccharide can also influence the heat stability of emulsions, as already discussed for the heat-induced aggregation of whey protein solutions (Capron *et al.*, 1999; Flett and Corredig, 2009). It has been demonstrated that the presence of xanthan or carrageenan molecules increases the rate of aggregation of emulsions stabilized by whey protein concentrates. As the polysaccharide concentration increases, the heat stability of the emulsions is significantly reduced (Euston *et al.*, 2002).

The association of polysaccharide molecules with proteins can occur via covalent bonds or physical interactions (for example, electrostatic interactions and formation of coacervates). Protein–polysaccharide conjugates can be prepared via non-enzymatic browning (Maillard-type reactions) using dry heating, and these conjugates have shown improved surface activity and good emulsifying and stabilizing properties, especially if the protein present contains a number of hydrophobic moieties (Dickinson and Semenova, 1992; Wooster and Augustin, 2006; Akhtar and Dickinson, 2007). These complexes, formed via heat processing, mimic the structural characteristics of surface active polysaccharides such as gum arabic and SSPS. This area will surely see developments in the future. In addition to dry heating as a method to prepare these protein–polysaccharide conjugates, it is possible to form covalent protein–polysaccharide complexes at the interface using enzymes such as laccase (Littoz and McClements, 2008).

When charged polysaccharides are added to emulsions, electrostatic interactions play an important role in the interactions occurring at the interface. In dairy emulsions, typically, attractive interactions occur at pH values close to or below the isoelectric point of the proteins, although this is not the case for chytosan, a positively charged polysaccharide soluble at pH < 6 (Hong and McClements, 2007). The addition of a polysaccharide that interacts with the proteins adsorbed at the interface can often cause bridging between oil droplets, followed by destabilization and phase separation. The destabilization is clearly shown using optical microscopy. Figure 14.5 illustrates the effect of the addition of various concentrations of flaxseed gum to a 10% oil-in-water emulsion stabilized by whey protein at pH 3.5. At this pH, flaxseed gum forms complexes with the protein adsorbed at the interface, and bridging between oil droplets occurs (Khalloufi *et al.*, 2009). Bridging flocculation leads to emulsions with very different bulk properties from those of emulsions destabilized by depletion interactions; for example, the flocs can withstand shear stresses and show higher water holding capacity than depleted emulsions (Blijdenstein *et al.*, 2004).

Multilayered interfaces (containing protein and polysaccharide interacting with the protein) can be designed to improve oil-in-water emulsion stability to environmental stresses and processing conditions (Guzey and McClements, 2006). The charge density of the polysaccharide is important in the adsorption of the polymer to protein-stabilized emulsions; for example, in the case of β-lactoglobulin covered oil droplets, ι-carrageenan, the most charged helical structure amongst the carrageenan molecules, is the most effective in creating charged interfacial membranes (Gu *et al.*, 2005).

Emulsions containing electrostatic complexes of proteins and polysaccharide at the interface can be obtained by either emulsifying the oil in an aqueous dispersion containing the coacervate, or by a layer-by-layer deposition technique. In this case, the emulsion droplets are first emulsified with milk proteins (i.e. whey proteins, caseins) and then the polysaccharide is added to the soluble phase. The molecules form electrostatic complexes at the oil–water interface. The sequence of adsorption of the biopolymers at the interface affects the structure and stabilizing properties of the interfacial layer, by creating differences in the gradients and heterogeneities in

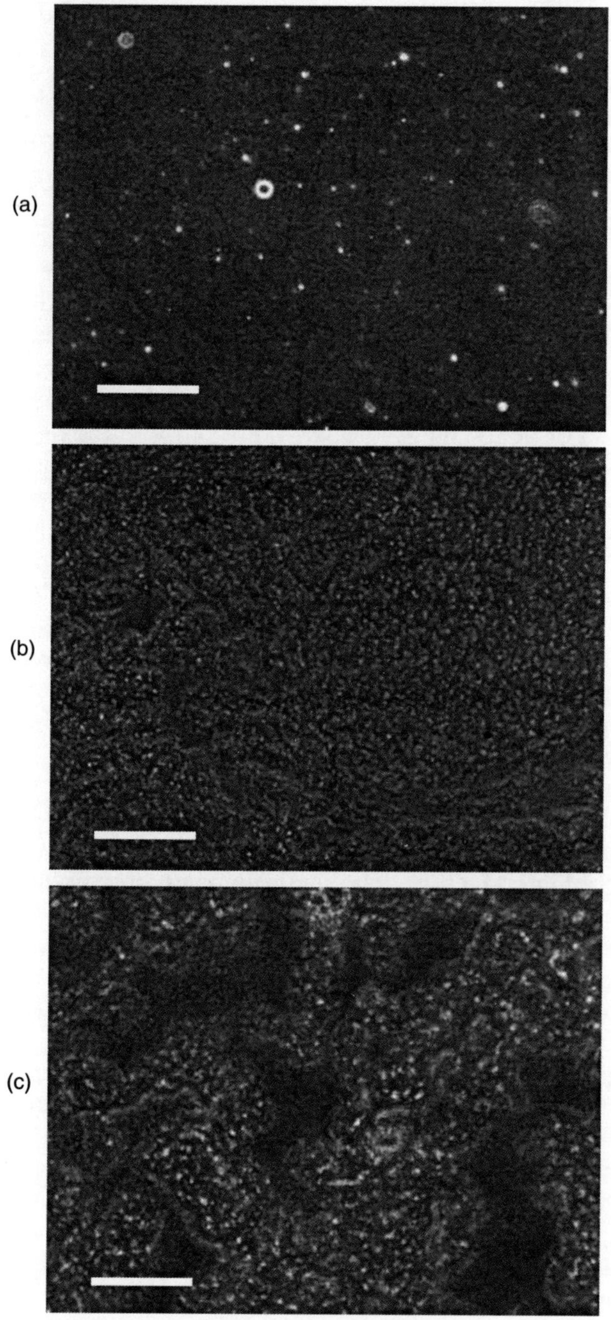

Fig. 14.5 Optical images of emulsions prepared with whey protein isolate and adjusted to pH 3.5. (a) Control, (b) emulsions with 0.02% and (c) 0.1% flaxseed gum. Bar is 20 μm.

the complexes present on the surface. This has been shown clearly, using spectroscopic and reflectivity experiments (Guzey and McClements, 2006; Ganzevles *et al.*, 2008). The formation of electrostatic complexes at the interface is of considerable importance when considering its potential to design new delivery systems that incorporate sensitive bioactives in the coacervates. The protection of the nutrients against environmental and processing stresses can be enhanced by protecting the nutrients within the interfacial layers.

Layer by layer deposition can also occur dynamically during processing; however, very few studies have been reported on this topic. For example, in a mixed system containing emulsion and high methoxyl pectin (negatively charged) at a pH of 6.8, we have no interactions between the polysaccharide and the protein adsorbed at the interface. In fact, at high enough concentrations of pectin, it is possible to notice the occurrence of phase separation by a depletion mechanism. However, if the emulsion droplets undergo a change in pH (as for example, during direct acidification or fermentation), at a pH low enough, electrostatic adsorption of the high methoxyl pectin on sodium caseinate oil droplets will occur (Fig. 14.6) (Bonnet *et al.*, 2005; Liu *et al.*, 2007). Without polysaccharide, the emulsions undergo a sol–gel transition, but with a sufficient amount of high methoxyl pectin, the droplets are stable and, depending on the amount of

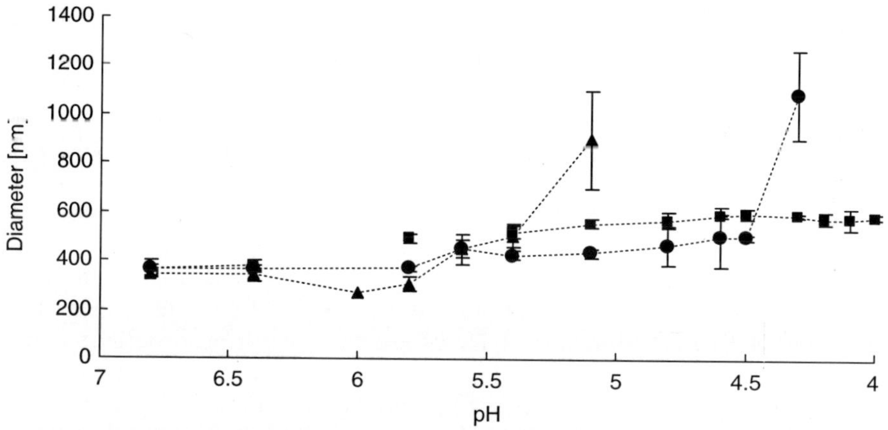

Fig. 14.6 Changes in the average hydrodynamic diameter measured by dynamic light scattering during acidification of emulsions stabilized with sodium caseinate and with added high methoxyl pectin. The emulsions were diluted in their soluble phase extracted using ultrafiltration. While control emulsions (▲) clearly show aggregation, and the addition of a low amount of high methoxyl pectin delays aggregation (0.05%, ●), when enough pectin is present (0.2%, ■) aggregation is inhibited and there is a clear increase in size, caused by the adsorption of the high methoxyl pectin at the interface at pH 5.5.

polysaccharide present, it is possible to modify the pH of gelation of the emulsions (Fig. 14.6). Only recently, using non-disruptive techniques such as ultrasonic and diffusing wave spectroscopy, it has been possible to show that the dynamic changes occurring during processing can be optimized to obtain different structures and textures.

14.5 Conclusions

An understanding of the dynamic changes occurring during mixing of polymers with milk proteins is paramount not only for imparting stability to the food products, but more importantly to optimize delivery of bioactives and design particular structures and textures tailored to the ever increasingly sophisticated consumer. Much work still needs to be carried out to understand the complexity of mixtures evolving over time and changing during the various stages of processing, storage and digestion.

The creation of encapsulated complexes as used for delivery systems of bioactives offers great potential. Much effort is currently focused on interfacial engineering of encapsulated structures tailored to specific functions, such as limited fat intake, controlled release of bioactives, or creation of triggers for specific reactions at a particular stage of the consumption and digestion. While information on the combined physical and biochemical changes occurring in these complex systems is still sparse, some studies are beginning to emerge; however, novel analytical methodologies need to be developed to be able to follow the dynamics of the interactions and the coacervates or in the water-in-water or oil-in-water-in-water emulsion systems generated from segregative interactions.

Although many biopolymers are commonly added to dairy products, surprisingly little is understood on the interactions of some polysaccharides commonly found in products (such as starch, inulin or β-glucan). The formation of new biopolymers using enzymes to create covalent crosslinks is also not yet fully exploited, as well as much less studied are complexes between milk proteins and neutral polymers. Once we clearly understand how to control the dynamics of change of the interactions and to fine tune segregative and associative behaviour, we will be able to create novel structures and delivery systems. In dairy products, structuring processes commonly employed, such as renneting, acidification and heating, should also be carefully controlled when other ingredients are present. A better knowledge of these interactions will lead to the manipulation of food sensory properties and digestive behaviour of the complexes.

14.6 References

ACERO-LOPEZ A, CORREDIG M, ALEXANDER M (2009), 'Diffusing wave and ultrasonic spectroscopy of rennet-induced gelation of milk in the presence of high methoxyl pectin'. *Food Biophysics*. DOI:10.1007/s11483-009-9122y

AKHTAR M, DICKINSON E (2007), 'Whey protein–maltodextrin conjugates as emulsifying agents: An alternative to gum arabic'. *Food Hydrocoll.* 21, 607–616 DOI:10.1016/j.foodhyd.2005.07.014

ALEXANDER M, DALGLEISH D G (2007), 'The interaction of casein micelles with κ-carrageenan studied by diffusing wave spectroscopy'. *Food Hydrocoll.* 21, 128–136. DOI: 10.1016/j.foodhyd.2006.03.003

ANAL A K, TOBIASSEN A, FLANAGAN J, SINGH H (2008), 'Preparation and characterization of nanoparticles formed by chitosan–caseinate interactions'. *Coll. Surf. B. Biointerfaces.* 64, 104–110. DOI:10.1016/j.colsurfb.2008.01.010

ASAKURA S, OOSAWA F (1958), 'Interaction between particles suspended in a solution of macromolecules'. *J. Polym. Sci.* 33, 183–192.

AYALA-HERNANDEZ I, HASSAN A, GOFF H D, MIRA DE ORDUNA R, CORREDIG M (2008a), 'Production, isolation and characterization of exopolysaccharides produced by *Lactococcus Lactis* subsp. cremoris JFR1 and their interaction with milk proteins: Effect of pH and media composition'. *Int. Dairy J.* 18, 1109–1118. DOI:10.1016/j.idairyj.2008.06.008

AYALA-HERNANDEZ I, GOFF H D, CORREDIG M (2008b), 'Interactions between milk proteins and exopolysaccharides produced by *Lactococcus Lactis* observed by scanning electron microscopy'. *J. Dairy Sci.* 91, 2583–2590. DOI:10.3168/jds.2007-0876

AYMARD P, DURAND D, NICOLAI T (1996), 'The effect of temperature and ionic strength on the dimerization of β-lactoglobulin'. *Int. J. Biol. Macromol.* 19, 213–221.

BEAULIEU M, CORREDIG M, TURGEON S L, WICKER L, DOUBLIER J (2005), 'The formation of heat-induced protein aggregates in whey protein/pectin mixtures studied by size exclusion chromatography coupled with multi-angle laser light scattering detection'. *Food Hydrocoll.* 19, 803–812. DOI:10.1016/j.foodhyd.2004.10.025

BIESHEUVEL P M, COHEN STUART M A (2004), 'Electrostatic free energy of weakly charged macromolecules in solution and intermacromolecular complexes consisting of oppositely charged polymers. *Langmuir.* 20, 2785–2791. DOI: 10.1021/la036204l

BLIJDENSTEIN T B J, VAN WINDEN A J M, VAN VLIET T, VAN DER LINDEN E, VAN AKEN G A (2004), 'Serum separation and structure of depletion and bridging flocculated emulsions, a comparison. *Coll. Surf. A: Physicochem. Eng. Aspects.* 245, 41–48. DOI:10.1016/j.colsurfa.2004.07.002

BONNET C, CORREDIG M, ALEXANDER M (2005), 'Stabilization of caseinate-covered oil droplets during acidification with high methoxyl pectin'. *J. Agric. Food Chem.* 53, 8600–8606. DOI:10.1021/jf0511998

BOULENGUER P, LAURENT M A (2003), 'Comparison of the stabilization mechanism of acid dairy drinks induced by pectin and soluble soybean polysaccharide (SSP)'. In Voragen A G J, *Advances in Pectin and Pectinase Research*, Dordrecht, Kluver Academic Publishers, 467–480.

BOURRIOT S, GARNIER C, DOUBLIER J L (1999), 'Micellar casein-κ-carrageenan mixtures. I. Phase separation and ultrastructure'. *Carb. Polym.* 40, 145–157.

CAPRON I, NICOLAI T, DURAND D (1999), 'Heat induced aggregation and gelation of β-lactoglobulin in the presence of κ-carrageenan'. *Food Hydrocoll.* 13, 1–5.

CARVALHO E, MATEUS N, PLET B, PIANET I, DUFOURC E, DE FREITAS V (2006), 'Influence of wine pectic polysaccharides on the interactions between tannins and salivary proteins'. *J. Agric. Food Chem.* 54, 8936–8944. DOI:10.1021/jf061835h

CROGUENNOC P, NICOLAI T, DURAND D, CLARK A (2001), 'Phase separation and association of globular protein aggregates in the presence of polysaccharides: 2. Heated mixtures of native β-lactoglobulin and κ-carrageenan'. *Langmuir.* 17, 4380–4385. DOI: 10.1021/la001675i

DALGLEISH D G (1997), 'Adsorption of protein and stability of emulsions' *Trends Food Sci. Techn.* 8, 1–6.

DALGLEISH D G, MORRIS E R (1988), 'Interactions between carrageenans and casein micelles: Electrophoretic and hydrodynamic properties of particles'. *Food Hydrocoll.* 2, 311–320.

DE GENNES P G (1981), 'Polymer solutions near an interface. Adsorption and depletion layers'. *Macromol.* 14, 1637–1644.

DE KRUIF C G, TUINER R (2001), 'Polysaccharide protein interactions'. *Food Hydrocoll.* 15, 555–563.

DICKINSON E, SEMENOVA M G (1992), 'Emulsifying properties of covalent protein–dextran hybrids'. *Coll. Surf.* 64, 299–310.

DOUBLIER J L, GARNIER C, RENARDA D, SANCHEZ C (2000), 'Protein–polysaccharide interactions'. *Curr. Opin. Coll. Interf. Sci.* 5, 202–214.

EUSTON S R, FINNIGAN S R, HIRST R L (2002), 'Kinetics of droplet aggregation in heated whey protein-stabilized emulsions: Effect of polysaccharide'. *Food Hydrocoll.* 16, 499–505.

FLETT K, CORREDIG M (2009), 'Whey protein aggregate formation during heating in the presence of κ-carrageenan'. *Food Chem.* 115, 1479–1485. DOI: 10/1016/j.foodhyd.2009.01.077

GAALOUL S, TURGEON S L, CORREDIG M (2008), 'Influence of shearing on the physical characteristics and rheological behaviour of an aqueous whey protein isolate–κ-carrageenan mixture'. *Food Hydrocoll.* 23, 1243–1252. DOI: 10.1016/j.foodhyd.2008.09.011

GANZEVLES R A, ZINOVIADOU K, VAN VLIET T, COHEN STUART M A, DE JONGH H H J (2006), 'Modulating surface rheology by electrostatic protein/polysaccharide interactions'. *Langmuir.* 22, 10089–10096. DOI: 10.1021/la061537e

GANZEVLES R A, FOKKINK R, VAN VLIET T, COHEN STUART M A, DE JONGH H H J (2008), 'Structure of mixed β-lactoglobulin/pectin adsorbed layers at air/water interfaces; A spectroscopy study. *J. Colloid Interf. Sci.* 317, 137–147. DOI:10.1016/j.jcis.2007.09.030

GAYGADZHIEV Z, CORREDIG M, ALEXANDER M (2008), 'Diffusing wave spectroscopy study of the colloidal interactions occurring between casein micelles and emulsion droplets: Comparison to hard-sphere behaviour'. *Langmuir.* 24, 3794–3800. DOI: 10.1021/la703265c

GIRARD M, SCHAFFER-LEQUART C (2007a), 'Gelation and resistance to shearing of fermented milk: Role of exopolysaccharides'. *Int. Dairy J.* 17, 666–673. DOI:10.1016/j.idairyj.2006.08.007

GIRARD M, SCHAFFER-LEQUART C (2007b), 'Gelation of skim milk containing anionic polysaccharides and recovery of texture after shearing'. *Food Hydrocoll.* 21, 1031–1040. DOI:10.1016/j.foodhyd.2006.07.012

GIRARD M, SANCHEZ C, LANEUVILLE S, TURGEON S L, GAUTHIER S F (2004), 'Associative phase separation of β-lactoglobulin/pectins solutions: A kinetic study by small angle static light scattering'. *Coll. Surf. B Biointerf.* 35, 15–22. DOI: 10.1016/j.colsurfb.2004.02.002

GRINBERG V Y, TOLSTOGUZOV V B (1997), 'Thermodynamic incompatibility of proteins and polysaccharides in solution'. *Food Hydrocoll.* 11, 145–158.

GU Y S, DECKER E A, MCCLEMENTS D J (2005), 'Influence of pH and carrageenan type on properties of β-lactoglobulin stabilized oil-in-water emulsions'. *Food Hydrocoll.* 19:83–91. DOI:10.1016/j.foodhyd.2004.04.016

GUZEY D, MCCLEMENTS D J (2006), 'Formation, stability and properties of multilayer emulsions for application in the food industry'. *Adv. Coll. Interf. Sci.* 128, 227–248. DOI:10.1016/j.cis.2006.11.021

HASSAN A N, IPSEN R, JANZEN T, QVIST K B (2003), 'Microstructure and rheology of yogurt made with cultures differing only in their ability to produce exopolysaccharides'. *J. Dairy Sci.* 86, 1632–1638.

HASSAN A N (2007), 'ADSA Foundation Scholar Award: Possibilities and challenges of exopolysaccharide-producing lactic cultures in dairy foods'. *J. Dairy Sci.* 91, 1282–1298. DOI:10.3168/jds.2007-0558

HOLT C (1992), 'Structure and stability of bovine casein micelles'. *Adv. Prot. Chem.* 43, 63–151.

HONG Y H, MCCLEMENTS D J (2007), 'Modulation of pH sensitivity of surface charge and aggregation stability of protein-coated lipid droplets by chitosan addition'. *Food Biophys.* 2, 46–55. DOI: 10.1007/s11483-007-9028-5

JI S, CORREDIG M, GOFF H D (2008), 'Production and functional properties of micellar casein/κ-carrageenan aggregates'. *Int. Dairy J.* 18, 64–71. DOI:10.1016/j.idairyj.2007.07.001

KAZMIERSKI M, WICKER L M, CORREDIG M (2003), 'Interactions of β-lactoglobulin and pectins in acidified systems'. *J. Food Sci.* 68, 1673–1679.

KHALLOUFI S, ALEXANDER M, GOFF H D, CORREDIG M (2008), 'Physicochemical properties of whey protein isolate stabilized oil-in-water emulsions when mixed with flaxseed gum at neutral pH'. *Food Res. Int.* 41, 964–972. DOI:10.1016/j.foodres.2008.07.019

KHALLOUFI S, CORREDIG M, GOFF, H D, ALEXANDER M (2009), ' Flaxseed gums and their adsorption on whey protein stabilized oil-in-water emulsions'. *Food Hydrocoll.* 23, 611–618. DOI:10.1016/j.foodhyd.2008.04.004

KONTOGIORGOS V, TOSH S M, WOOD P J (2009), 'Phase behaviour of high molecular weight oat β-glucan/whey protein isolate binary mixtures'. *Food Hydrocoll.* DOI:10.1016/j.foodhyd.2008.07.005

LANEUVILLE S I, SANCHEZ C, TURGEON S L, HARDY J, PAQUIN P (2005), 'Small angle static light scattering study of associative phase separation kinetics in β-lactoglobulin+xanthan gum mixtures under shear'. in Dickinson E, *Food Colloids – Interactions, Microstructure and Processing*, Cambridge, Royal Society of Chemistry, 443–465.

LANEUVILLE S I, TURGEON S L, SANCHEZ C, PAQUIN P (2006), 'Gelation of native β-lactoglobulin induced by electrostatic attractive interaction with xanthan gum'. *Langmuir.* 22, 7351–7357. DOI: 10.1021/la060149+

LANGENDORFF V, CUVELIER G, LAUNAY C, PARKER A (1997), 'Gelation and flocculation of casein micelle/carrageenan mixtures'. *Food Hydrocoll.* 11, 35–42.

LAPLANTE S, TURGEON S L, PAQUIN P (2005), 'Emulsion stabilizing properties of various chitosans in the presence of whey protein isolate'. *Carb. Polym.* 59, 425–434. DOI:10.1016/j.carbpol.2004.08.031

LAZARIDOU A, BILIADERIS C G (2007), 'Molecular aspects of cereal β-glucan functionality: Physical properties, technological applications and physiological effects'. *J. Cereal Sci.* 46, 101–118. DOI:10.1016/j.jcs.2007.05.003

LITTOZ F, MCCLEMENTS D J (2008), 'Bio-mimetic approach to improving emulsion stability: Cross-linking adsorbed beet pectin layers using laccase'. *Food Hydrocoll.* 22, 1203–1211. DOI:10.1016/j.foodhyd.2007.06.009

LIU J, NAKAMURA A, CORREDIG M (2006), 'Addition of pectin and soy soluble polysaccharide affects the particle size distribution of casein suspensions prepared from acidified skim milk'. *J. Agric. Food Chem.* 54, 6241–6246. DOI:10.1021/jf060113n

LIU J, CORREDIG M, ALEXANDER M (2007), 'A diffusing wave spectroscopy study of the dynamic of interactions between high methoxyl pectin and sodium caseinate emulsions during acidification'. *Coll. Surf. B: Biointerf.* 59, 164–170. DOI:10.1016/j.colsurfb.2007.05.004

MAROZIENE A, DE KRUIF C G (2000), 'Interaction of pectin and casein micelles'. *Food Hydrocoll.* 14, 391–394. DOI:10.1016/S0268-005X(00)00019-9

MARTIN A H, GOFF H D, SMITH A, DALGLEISH D G (2006), 'Immobilization of casein micelles for probing their structure and interactions with polysaccharides using scanning electron microscopy (SEM)'. *Food Hydrocoll.* 20, 817–824. DOI:10.1016/j.foodhyd.2005.08.004

MEKHLOUFI G, SANCHEZ C, RENARD D, GUILLEMIN S, HARDY J (2005), 'pH-induced structural transitions during complexation and coacervation of β-lactoglobulin and acacia gum'. *Langmuir.* 21, 386–394. DOI: 10.1021/la0486786

NAKAMURA A, YOSHIDA R, MAEDA H, FURUTA H, CORREDIG M (2004), 'Study of the role of the carbohydrate and protein moieties of soy soluble polysaccharides in their emulsifying properties'. *J. Agric. Food Chem.* 52, 5506–5512. DOI:10.1021/jf049728f

OOMAH B D, KENASCHUK E O, CUI W W, MAZZA G (1995), 'Variation in the composition of water soluble polysaccharides in flaxseed'. *J. Agric. Food Chem.* 43, 1484–1488.

OULD ELEYA M M, TURGEON S L (2000), 'Rheology of κ-carrageenan and β-lactoglobulin mixed gels'. *Food Hydrocoll.* 14, 29–40.

PEYRON S, MOUÉCOUCOU J, FRÉMONT S, SANCHEZ C, GONTARD N (2006), 'Effect of heat treatment and pectin addition on β-lactoglobulin allergenicity'. *J. Agric. Food Chem.* 54, 5643–5650.

POON W C K (1998), 'Phase separation, aggregation and gelation in colloid–polymer mixtures and related systems'. *Curr. Opin. Coll. Interf. Sci.* 3, 593–602.

ROCHAS C, LANDRY S (1987), 'Molecular organization of kappa carrageenan in aqueous solution. *Carb. Polym.* 7, 435–447.

SANCHEZ C, ZUNIGA-LOPEZ R, SCHMITT C, DESPOND S, HARDY J (2000), 'Microstructure of acid-induced skim milk – locus bean – gum xanthan gels'. *Int. Dairy J.* 10, 199–212.

SANCHEZ C, MEKHLOUFI G, RENARD D (2006), 'Complex coacervation between β-lactoglobulin and acacia gum: A nucleation and growth mechanism'. *J. Coll. Interf. Sci.* 299, 867–873. DOI:10.1016/j.jcis.2006.02.031

SCHMITT C, PALMA DA SILVA T, BOVAY C, SHOJAEI-RAMI S, FROSSARD P, KOLODZIEJCZYK E, LESER M E (2005), 'Effect of time on the interfacial and foaming properties of β-lactoglobulin–acacia gum electrostatic complexes and coacervates at pH 4.2'. *Langmuir.* 21, 7786–7795. DOI: 10.1021/la0510984

SCHORSCH C, JONES M, NORTON I T (1999), 'Thermodynamic incompatibility and microstructure of milk protein/locus bean gum/sucrose systems'. *Food Hydrocoll.* 13, 89–99.

SNOEREN T H M, PAYENS T A J, JEUNINK J, BOTH P (1975), 'Electrostatic interaction between κ-carrageenan and κ-casein'. *Milchwiss.* 30, 393–396.

SPAGNUOLO P A, DALGLEISH D G, GOFF H D, MORRIS E R (2005), 'Kappa-carrageenan interactions in systems containing casein micelles and polysaccharide stabilizers'. *Food Hydrocoll.* 19, 371–377. DOI:10.1016/j.foodhyd.2004.10.003

SUN C H, GUNASEKARAN S, RICHARDS M P (2007), 'Effect of xanthan gum on physicochemical properties of whey protein isolate stabilized oil-in-water emulsions'. *Food Hydrocoll.* 21, 555–564. DOI:10.1016/j.foodhyd.2006.06.003

SURH J, DECKER E A, MCCLEMENTS D J (2006), 'Influence of pH and pectin type on properties and stability of sodium-caseinate stabilized oil-in-water emulsions'. *Food Hydrocoll.* 20, 607–618. DOI:10.1016/j.foodhyd.2005.07.004

SWAISGOOD H E (1992), 'Advanced dairy chemistry', in: Fox P F, *Proteins*, London, Elsevier Applied Science, 63–70.

SYRBE A, BAUER W K, KLOSTERMEYER H (1998), 'Polymer science concepts in dairy systems'. *Int. Dairy J.* 8, 179–198.

THAIUDOM S, GOFF H D (2003), 'Effect of κ-carrageenan on milk protein polysaccharide mixtures'. *Int. Dairy J.* 13, 763–771. DOI:10.1016/S0958-6946(03)00097-9

TOLSTOGUZOV V B (1991), 'Functional properties of food proteins and role of protein–polysaccharide interaction'. *Food Hydrocoll.* 4, 429–435.

TROMP R H, DE KRUIF C G, VAN EIJK M, ROLIN C (2004), 'On the mechanism of stabilization of acid milk drinks by pectin'. *Food Hydrocoll.* 18, 565–572. DOI:10.1016/j.foodhyd.2003.09.005

TUINIER R, TEN GROTENHUIS E, DE KRUIF C G (2000), 'The effect of depolymerized guar gum on the stability of skim milk'. *Food Hydrocoll.* 14, 1–7.

TUINIER R, VAN CASTEREN W H M, LOOIJESTEIJN P J, SCHOLS H A, VORAGEN A G J, ZOON P (2001), 'Effects of structural modifications on some physical characteristics of exo-polysaccharides from Lactococcus lactis'. *Biopolym.* 59, 160–166.

TURGEON S L, SCHMITT C, SANCHEZ C (2007), 'Protein–polysaccharide complexes and coacervates'. *Curr. Opin. Coll. Interf. Sci.* 12, 166–178. DOI:10.1016/j.cocis.2007.07.007

VAN DEN BERG L, VAN VLIET T, VAN DER LINDEN E, VAN BOEKEL M A J S, VAN DE VELDE F (2007), 'Breakdown properties and sensory perception of whey proteins polysaccharide mixed gels as a function of microstructure'. *Food Hydrocoll.* 21, 961–976. DOI:10.1016/j.foodhyd.2006.08.017

VORAGEN A G J, PILNIK W, THIBAULT J F, AXELOS M A V, RENARD C M G C (1995), 'Pectins', in: Stephen A M, *Food Polysaccharides and Their Applications*. New York, Marcel Dekker, 287–339.

VRIJ A (1976), 'Polymers at interfaces and the interactions in colloidal dispersions'. *Appl. Chem.* 48, 471–483.

WEINBRECK F, ROLLEMA H S, TROMP R H, DE KRUIF C G (2004a), 'Diffusivity of whey protein and gum arabic in their coacervates'. *Langmuir.* 20, 6389–6395. DOI:10.1021/la049908j

WEINBRECK F, WIENTJES R H W, NIEUWENHUIJSE H, ROBIJN G W, DE KRUIF C G (2004b), 'Rheological properties of whey protein/gum arabic coacervates'. *J. Rheol.* 48, 1215–1228.

WILLIAMS P A, PHILLIPS G O, RANDALL R C (1990), 'Structure function relationships of gum arabic', in Phillips G O, Wedlock D J, Williams P A, *Gums and Stabilizers for the Food Industry*, Oxford, IRL Press, 25–36.

WOOSTER T J, AUGUSTIN M A (2006), 'β-lactoglobulin–dextran Maillard conjugates: Their effect on interfacial thickness and emulsion stability'. *J. Coll. Interf. Sci.* 303, 564–572. DOI:10.1016/j.jcis.2006.07.081

15

Novel applications of enzymes in the dairy sector: Optimizing functional properties of milk proteins by enzymatic cross-linking

T. Huppertz, NIZO food research, The Netherlands

Abstract: Milk proteins are probably the most-used protein source world-wide, which is no doubt related to their excellent functional properties. In particular, their gelation and emulsification properties are utilized in a wide variety of popular food products. Functional properties can be modified through enzymatic cross-linking, which is commonly performed with the enzyme transglutaminase (TGase). Treatment of milk proteins with TGase improves their acid gelation properties and thus it has considerable scope for the improvement of texture in acid-coagulated dairy products. Enzymatic coagulation of milk is hindered by TGase treatment, which limits options for its application in cheese production, but may offer opportunities in preventing undesirable enzymatic coagulation of milk in other products. The stability of emulsion droplets can also be increased by cross-linking the proteins on their surfaces. By targeted application of TGase or cross-linking enzymes, many other valuable applications for enzymatic cross-linking can undoubtedly be developed.

Key words: transglutaminase, cross-link, casein, whey protein, yoghurt, stability.

15.1 Introduction

In the current food market, an overwhelming proportion of products contain one or more sources of proteins. While these proteins clearly fulfill a nutritional role in the products, they more often than not also confer other desirable properties with respect to product texture, flavour or stability. Protein functionality can be described as the desirable, or undesirable, attributes conferred by the proteins through interactions with other components of the system to which they are applied. Functional properties can include, e.g. the formation and stabilization of emulsions and foams, the creation and stabilization of textures, the binding of water, and the building of viscosity.

Due to their abundance and relative ease of isolation, milk proteins are probably the best-characterized group of food proteins to date. As a result, the functionality of milk proteins has long been recognized and has been exploited in the utilization of milk as a base material or ingredient in the preparation of a wide variety of popular food products, such as cheese, yoghurt and ice cream. Additionally, milk proteins are also isolated from milk or other dairy streams and applied as crude, semi-purified or highly purified fractions in a wide variety of food and non-food products. Major milk protein ingredients include sodium and calcium caseinate, milk protein concentrate, whey protein concentrate and isolate, and rennet and acid casein. A review of the preparation and uses of milk protein ingredients is provided by Mulvihill and Ennis (2003) and Rollema (2003). Applications of caseins and caseinates in non-food products are reviewed by De Kruif (2003).

Although functionality of milk proteins is already very high, studies over the past two decades have focused on further improving their functionality, by enzymatic cross-linking, which creates covalent linkages between proteins. In this Chapter, the use of enzymatic cross-linking of milk proteins to tailor the functional properties of milk is described. Focus is upon two of the most-utilized properties, namely gelation properties and emulsification properties. These properties are outlined in Section 15.2, and Section 15.3 deals with the enzymatic cross-linking of milk proteins. The physico-chemical properties of cross-linked milk proteins are outlined in Section 15.4. Sections 15.5 and 15.6 deal with the gelation and emulsification properties of cross-linked milk proteins, respectively.

15.2 Functional properties of milk proteins

15.2.1 Gelation properties of milk proteins

The gelation properties of milk proteins, in particular the caseins, form the basis for a wide variety of desirable and undesirable phenomena in milk and dairy products. Prime examples of desirable gelation phenomena in milk are the enzymatic gelation of milk as the primary step in cheese-making, as well as the acid-induced gelation of milk in the manufacture of yoghurt or other acid-coagulated dairy products. On the other hand, examples of undesirable gelation properties of milk are also in abundance, prime examples being the heat-induced gelation of milk and the enzyme-induced age-gelation of UHT-treated milk products. The gelation properties of caseins are closely related to the degree of the colloidal stability of the association colloids in which they naturally occur, i.e. the casein micelles. To induce gelation, the colloidal stability of the micelles is reduced, initially leading to aggregation and flocculation of the destabilized micelles, eventually followed by the formation of a gel.

Casein micelles are the crucial features in the aforementioned gelation properties of milk. Casein micelles are highly hydrated association colloids with an average radius of ~100 nm. The micellar dry matter consists of ~94% protein and 6% inorganic material, referred to as micellar calcium phosphate (MCP). MCP is crucial in maintaining the structural integrity of the micelles (De Kruif and Holt, 2003). The substructure of the casein micelle has been an issue of debate for the past five decades, and will undoubtedly remain so in the future. However, a consensus is now being reached on MCP, in that it is present in amorphous nanoclusters consisting of predominantly calcium phosphate, in which phosphorylated serine residues of the caseins participate (De Kruif and Holt, 2003). However, in terms of gelation properties, the composition and properties of the micellar surface are of far greater importance. The micellar surface consists predominantly of κ-casein, although at low temperature, significant amounts β-casein are also found on the micelle surface (De Kruif and Roefs, 1996). The hydrophilic and partially glycosylated C-terminus of κ-casein extends from the surface, and has been referred to as a 'hairy layer', but is better termed a polyelectrolyte brush. The polyelectrolyte brush stabilizes the micelles through providing both steric and electrostatic repulsion (De Kruif, 1999), although the former probably predominates since ionic strength in milk is already quite high and can be increased considerably further without inducing micellar aggregation.

As with any colloidal particle stabilized by a polyelectrolyte brush, the colloidal stability of casein micelles can be effectively reduced by either one, or a combination, of two routes: (i) removal of the polyelectrolyte brush, or (ii) collapse of the polyelectrolyte brush. Enzymatic coagulation is a typical example where colloidal stability is reduced by removal of the brush, whereas acid-induced coagulation is brought about by a collapse of the brush, as will be outlined in the next two subsections. Heat-induced coagulation is, most likely, the result of a combined removal of some of the brush and collapse of the residual brush, and is described in third subsection.

Enzymatic gelation of milk

Enzymatic coagulation of milk forms the first step in the production of most varieties of cheese. During this process, chymosin hydrolyzes the Phe105–Met106 bond in κ-casein, as a result of which its C-terminus, consisting of residues 106–169 and termed the caseinomacropeptide, is released into the serum phase of the milk. The residual N-terminus of κ-casein, termed para-κ-casein, remains attached to the micelles, but is unable to provide the para-casein micelle with sufficient steric and electrostatic repulsion for the micelles to remain as separate entities. When sufficient κ-casein has been hydrolyzed by chymosin, i.e. >90%, the residual κ-casein is unable to stabilize the micelles and flocculation commences, following which the para-casein micelles form a self-supporting gel. In subsequent stages of the cheese-making process, the gel is cut, which induces syneresis and the expulsion of whey. Rennet-induced gelation of milk is reviewed in detail by Dalgleish (1992), Hyslop (2003), Horne and Banks (2004) and Dejmek and Walstra (2004).

Acid-induced gelation of milk

Acid-induced gelation of milk is based on the destabilization of the casein micelles as a result of a collapse of the stabilizing polyelectrolyte brush. The protruding C-terminus of κ-casein, which predominates in the brush, is soluble primarily as a result of its considerable net-negative charge at neutral pH. This negative charge is derived largely from a high number of glutamic acid residues, as well as N-acetylneuraminic acid groups of the glycosylated fraction of κ-casein. When pH is reduced, the carboxyl-groups of the glutamic acid residues (pKa = 4.1) gradually become protonated and net-negative charge is thus reduced. Significant protonation of N-acetylneuraminic acid residues (pKa = 2.6) does not occur until pH < 4, but the residual negative charge provided therefrom is insufficient to retain solvency of the polyelectrolyte brush and the brush collapses (De Kruif and Zhulina, 1996). As a result, residual inter-micellar repulsion is insufficient to maintain the micelles as individual entities. Acid-induced flocculation of casein micelles thus commences, ultimately leading into the formation of a self-supporting gel network. Acid-induced gelation of milk is reviewed in detail by Lucey and Singh (1997, 2003).

Heat-induced gelation of milk

Heat-induced coagulation of milk is an undesirable phenomenon which occurs on heating milk for a prolonged time at high temperature, e.g. 140°C for unconcentrated milk or 110–120°C for concentrated milk. The heat stability of milk is influenced by a myriad of factors, as reviewed extensively by O'Connell and Fox (2003) and Singh (2004). At 140°C, the heat coagulation time (HCT) of bulked unconcentrated milk samples increases with increasing pH, up to a maximum at pH ~6.7, followed by a decrease in HCT to a minimum at pH ~6.9, on the alkaline side of which HCT increases with increasing pH (O'Connell and Fox, 2003; Singh, 2004). For concentrated milk, HCT at 120°C increases with increasing pH, up to a maximum at pH ~6.6, above which HCT progressively decreases with increasing pH (O'Connell and Fox, 2003). The influence of pH on HCT of unconcentrated and concentrated milk is thought to derive from its effects on calcium activity, heat-induced acidification, heat-induced dissociation of κ-casein and the denaturation of whey protein as well as the denatured whey proteins' subsequent association with micellar and non-micellar caseins (O'Connell and Fox, 2003; Singh, 2004). Since the pH-dependency of the aforementioned contributors to stability and instability varies, typical HCT profiles are obtained. For instance, the concentration of ionic calcium and heat-induced acidification of milk are predominantly important in the acidic region, whereas heat-induced dissociation of κ-casein occurs extensively only at pH values in the alkaline region of the pH–HCT profile.

15.2.2 Emulsification properties

Because most proteins have significant amounts of both polar and apolar areas, they tend to be surface active and thus accumulate at the oil–water, or air–water, interface. The conformation that a protein adopts at the oil–water interface depends on its molecular structure and interactions. Natively-unfolded proteins, e.g. caseins, adopt a conformation at the interface whereby the polar segments protrude into the aqueous phase, the non-polar segments protrude into the oil phase and neutral legions lie flat on the interface. As a result, relatively open and thick membranes of low visco-elasticity are formed. In contrast, globular proteins, such as the whey proteins, have a defined orientation at the interface, whereby the predominantly apolar regions of the surface face the oil phase and the predominantly polar regions face the aqueous phase. Following adsorption, globular proteins can undergo further interactions, as a result of which a highly visco-elastic membrane forms (McClements, 1999). Both caseins and whey proteins are excellent emulsifiers, which is of importance, for instance, in the homogenization of milk, the preparation of recombined milk, and the use of milk powders, caseinates and whey protein isolates and concentrates in a wide variety of products, including clinical and infant foods, processed cheese and cream liqueur. Emulsification properties of milk proteins are reviewed in detail by Dickinson (2003).

15.3 Enzymatic cross-linking of milk proteins

15.3.1 Transglutaminase-induced cross-linking of milk proteins

Cross-linking of milk proteins with transglutaminase (TGase) in an attempt to further improve their functionality has been of great academic and industrial interest in since the 1980s. TGase catalyzes the acyl-transfer reaction, where the γ-carboximide group of glutamine acts as an acyl donor. Acyl-acceptors can, as outlined in Fig. 15.1, be (A) primary amines, including (B) the ε-amino-group of lysine. In the latter case, this leads to the formation of an intermolecular protein cross-link. In the absence of a primary amine, water can act as an acyl-acceptor, thereby leading to deamidation of the glutamine residue (Fig. 15.1C). The use of TGase in food systems is reviewed by Motoki and Seguro (1998), Nielsen (1995), Zhu *et al.* (1995), Kuraishi *et al.* (2001) and De Jong and Koppelman (2002). The use of TGase in dairy applications is reviewed by Jaros *et al.* (2006a).

Transglutaminases cover a group of enzymes that occur widely in nature, having been discovered in plants and most bodily tissues and fluids. Molecular properties of transglutaminases are described by Folk (1980) and

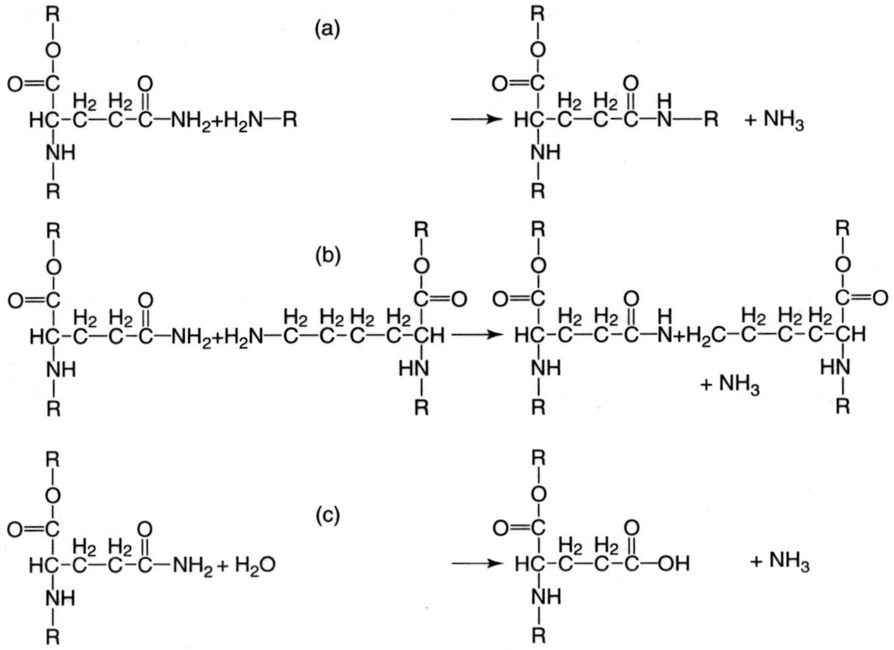

Fig. 15.1 General reactions catalyzed by transglutaminase; (a) acyl-transfer reaction; (b) protein-cross-linking reaction; (c) deamidation.

Folk and Finlayson (1977). In the human body, TGase, also known as Factor XIII, is involved in the blood coagulation system and cross-links fibrin. Applications for TGase in food production became commercially relevant following the discovery of extracellular TGase production by *Streptoverticillium* species, as reported by Ando *et al.* (1989). Since then, large-scale food-grade production of TGase by *Streptoverticillium mobaraence* S-8112 has been commercialized by the Japanese company Ajinomoto. The production of TGase by microbial fermentation is reviewed by Zhu *et al.* (1995). In contrast to most other types of TGase, microbially-produced TGase does not require the presence of Ca^{2+} for the expression of activity. Activity of microbial TGase increases ~10-fold in near-linear fashion between 5 and 50°C, with activity rapidly decreasing at temperatures >60°C due to inactivation of the enzyme. Optimal pH for TGase is in the range 5–8, with little or no activity observed at pH < 4 or >9 (Seguro *et al.*, 1996).

When milk is incubated with TGase, the caseins are rapidly cross-linked, whereas the whey proteins remain unaffected (Hinz *et al.*, 2007). This is undoubtedly related to the fact that glutamine- and lysine-residues in the natively-unfolded caseins are readily accessible, whereas these residues are largely buried inside the globular whey proteins. Accessibility of the proteins to TGase can be enhanced by treatment with dithiolthreitol (DTT), which unfolds the protein through cleavage of disulphide bonds (Coussons *et al.* 1992; De Jong and Koppelman, 2002; Lee *et al.*, 2002) or by increasing pH to 8.5–9.0, where β-lactoglobulin is partially-unfolded (Faergemand *et al.*, 1997a) or by denaturation of β-lactoglobulin as a result of heat treatment (O'Sullivan *et al.*, 2002a) or high pressure treatment (Lauber *et al.*, 2003). In addition, TGase-induced cross-linking of α-lactalbumin can be improved by removing the calcium-ion bound in the native structure (Sharma *et al.*, 2002; Nieuwenhuizen *et al.*, 2003). In whole milk, buttermilk, and cream, the proteins of the milk fat globule membrane are also susceptible to TGase-induced cross-linking (Hinz *et al.*, 2007).

Compared to in milk, cross-linking of caseins occurs much more readily in sodium caseinate (Lorenzen, 2000a), which is undoubtedly related to the more accessible structure of the protein aggregates of sodium caseinate, which are approximately an order of magnitude smaller than the casein micelles and hence have a considerably higher accessible surface area.

In milk at 30–40°C and natural pH, the rate of TGase-induced cross-linking of caseins is in the order κ-casein > β-casein > α_s-caseins (Sharma *et al.*, 2001; Smiddy *et al.*, 2006; Huppertz and De Kruif, 2007a,b; Hinz *et al.*, 2007). This order of cross-linking probably largely reflects the accessibility of the various caseins for TGase, with κ-casein being found predominantly on the readily-accessible micellar surface, whereas the α_s-caseins are largely found in the poorly-accessible micellar core (De Kruif and Holt, 2003). Altering micelle structure influences the susceptibility of the caseins towards cross-linking, as is exemplified by studies by Moon *et al.* (2009) who observed that, particularly, susceptibility of κ-casein to TGase-induced cross-linking strongly depends on milk pH, with considerably less β-casein being cross-linked at high pH (8–9) than at low pH (5.5–6.5). In sodium caseinate, the susceptibility of caseins to cross-linking is κ-casein > α_s-caseins > β-casein (Tang *et al.*, 2005).

TGase-induced cross-linking of milk proteins is by no means optimal in milk. This has been related to the potential presence of a low molecular mass TGase-inhibitor in milk (De Jong *et al.*, 2003). Cross-linking of proteins in milk is further improved by pre-heating the milk (Bonisch *et al.*, 2004, 2006; Rodriguez-Nogales, 2006), but this cannot be unambiguously attributed to heat-induced inactivation of a potential enzyme inhibitor, as suggested by Bonisch *et al.* (2004, 2006), since heat treatment may also improve accessibility of the caseins for TGase through changes in micelle structure. The addition of glutathione to milk also improves TGase-induced cross-linking of milk (Bonisch *et al.*, 2007a), which may again be due to improved substrate accessibility. Glutathione induces sulphydryl–disulphide interchange reactions, which reduce the size of disulphide-linked protein aggregates (Dong and Hoseney, 1995), and can thus break down the κ-casein aggregates present at the micellar surface.

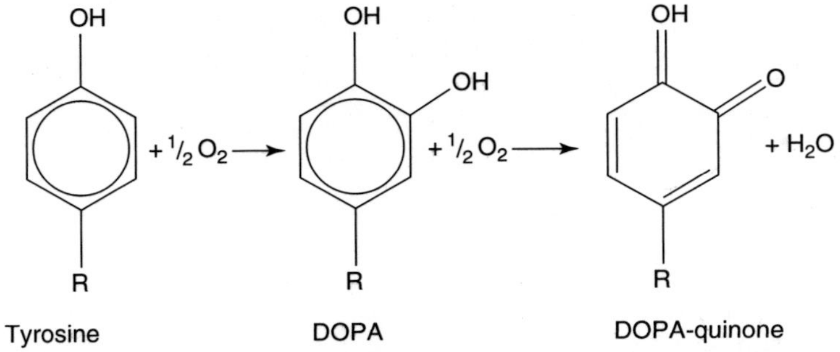

Fig. 15.2 Tyrosinase-catalyzed hydroxylation and subsequent oxidation of the phenol-group of tyrosine.

15.3.2 Tyrosinase and laccase-induced cross-linking of milk proteins

Tyrosinase (EC 1.14.18.1; *monophenol monooxygenase*) belongs to the group of phenol oxidases and is widespread in nature. The enzyme is found in animals, plants, fungi and bacteria, where it is involved in the biosynthesis of melanin pigments from the amino acid tyrosine, as well as from other phenolic compounds (Claus and Decker, 2006). Tyrosinase catalyzes the oxidation of various phenolic compounds to their corresponding quinones, using oxygen as an electron acceptor. These quinones are highly reactive and may undergo further non-enzymatic reactions (Selinheimo *et al.*, 2006). In tyrosinase-induced protein cross-linking, tyrosinase catalyzes two distinct subsequent oxidative reactions: the first reaction involves a monophenolase activity, whereby the phenol groups of tyrosine is hydroxylated; the second reaction involves a diphenolase activity, whereby the substrate is further oxidized to form a quinone. Both reactions are outlined in Fig. 15.2. The quinone may subsequently react non-enzymatically with another quinone, the free sulphydryl group of a cysteine residue, or the ε-amino-group of a lysine residue, forming inter-molecular cross-links, as outlined in Fig. 15.3.

When pasteurized skim milk is incubated at 30°C with *Trichoderma reesei* tyrosinase, cross-linking of caseins occurs readily at 30°C in the presence oxygen, whereas the whey proteins α-lactalbumin and β-lactoglobulin are not cross-linked under these conditions (Lille *et al.*, 2007). All caseins are susceptible to tyrosinase-induced cross-linking under the aforementioned conditions, with κ-casein and β-casein apparently being most susceptible to tyrosinase-induced cross-linking. Pre-heating pasteurized milk at 90°C for 5 min prior to incubation with tyrosinase does not yield significant additional tyrosinase-induced cross-linking of caseins and does not improve the susceptibility of whey proteins to tyrosinase-induced cross-linking (Lille *et al.*, 2007).

Like tyrosinase, some laccases can also induce cross-linking of proteins. Laccases are mainly found in fungi and plants, but have also been found in

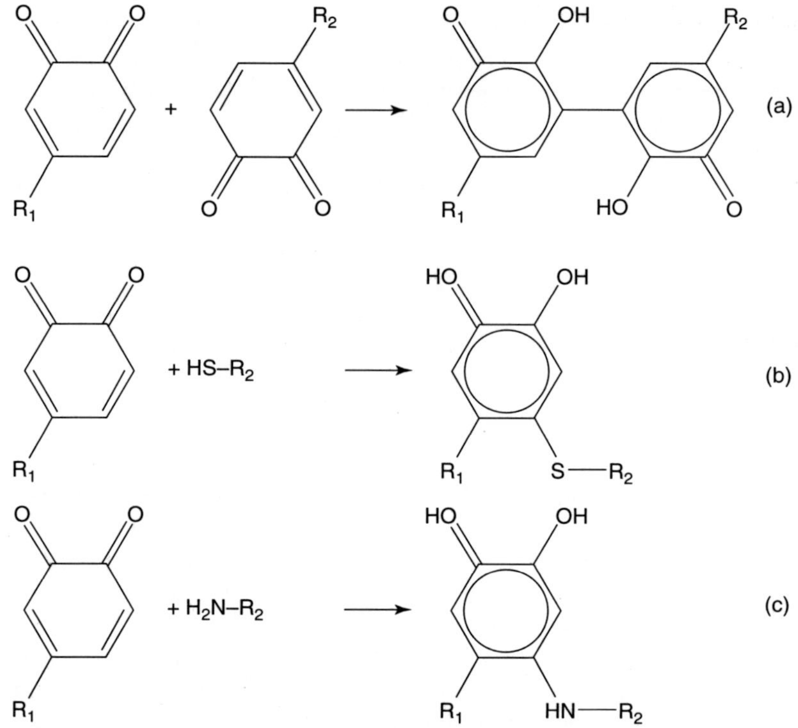

Fig. 15.3 Cross-link formation between tyrosinase-induced quinone of tyrosine with (a) another tyrosine-quinone, (b) the free sulphydryl group of a cysteine residue or (c) the ε-amino group of a lysine-residue.

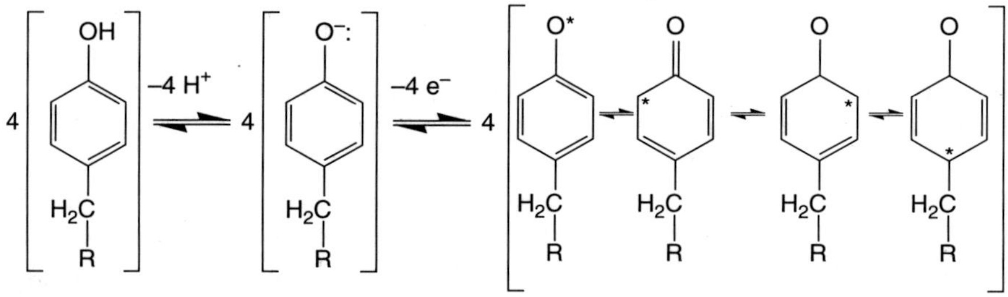

Fig. 15.4 Laccase-induced oxidation of phenol-groups.

insects and bacteria. Laccases are able to oxidize various phenolic compounds and amines. Laccase-induced oxidation occurs through the removal of one electron, which yields the formation of free radicals, as outlined in Fig. 15.4. In proteins, tyrosine, cysteine and tryptophan residues can be oxidized by laccase (Mattinen *et al.*, 2006). Studies on laccase-induced cross-linking of proteins in milk have not been reported to date. Such studies may be of interest in considering alternatives for TGase to induced cross-linking of milk proteins, but the low pH optimum (4–5) for most laccases should be taken into account and may induce limitations in applicability.

15.4 Physicochemical properties of cross-linked milk proteins

15.4.1 Casein micelles

Data on the physicochemical properties of casein micelles cross-linked by tyrosinase or laccase have not been published to date, but over the last decade or so a wealth of information has become available on the physicochemical properties of caseins cross-linked by TGase. Following incubation with TGase, casein micelle size in unconcentrated milk remains unaffected (O'Sullivan *et al.*, 2002a; Vasbinder *et al.*, 2003; Mounsey *et al.*, 2005; Huppertz and De Kruif, 2007b, 2008; Huppertz *et al.*, 2007), suggesting that cross-linking in this medium is exclusively intra-micellar and not inter-micellar. However, when concentrated milk systems are incubated with TGase, gelation can occur, suggesting that when inter-micellar distance is sufficiently small, inter-micellar cross-linking can occur (Nonaka *et al.*, 1992). The zeta-potential of cross-linked casein micelles is also comparable to that of native casein micelles (Huppertz and De Kruif, 2008), which is not surprising considering that it is, to a large extent, determined by the negative charge on glutamic acid residues of κ-casein as well as the N-acetylneuraminic acid of the glycosylated κ-casein, neither of which is involved in the cross-linking reaction. Microstructure of cross-linked casein micelles is comparable to that of native casein micelles, as evidenced by similarities in scattering spectra obtained by small-angle neutron scattering (SANS) and small-angle X-ray scattering (SAXS) (Huppertz and De Kruif, 2008). Improved water-binding of TGase-treated casein micelles has been reported by Imm *et al.* (2000).

The ability of the micelles to retain their structural integrity under unfavourable conditions is improved significantly as a result of TGase-induced cross-linking. Native casein micelles are association colloids, whose structure is determined by cooperative weak interactions between the proteins and ionic interactions between the phosphorylated serine residues of the caseins with the calcium phosphate nanoclusters (De Kruif and Holt, 2003). As a result, casein micelles are disrupted when the protein–protein or protein–nanocluster interactions are disturbed. In contrast, when all micellar casein is cross-linked, micelles become completely stable against disruption and start behaving more like micro- or nanogel particles, which are fully stabilized against disruption but shrink or swell as a result of altered solvent conditions (Huppertz *et al.*, 2007).

The increased stability of cross-linked casein micelles is exemplified in their increased stability on addition of the chaotropic agent urea (O'Sullivan *et al.*, 2002a; Smiddy *et al.*, 2006), the detergent sodium dodecyl sulphate (Smiddy *et al.*, 2006), or the calcium chelator trisodium citrate (O'Sullivan *et al.*, 2002a; Smiddy *et al.*, 2006), or on an improvement in solvent quality on heating milk in the presence of ethanol (O'Sullivan *et al.*, 2002a; Smiddy *et al.*, 2006) or on the treatment of milk at high hydrostatic pressure (O'Sullivan *et al.*, 2002a; Smiddy *et al.*, 2006; Huppertz and Smiddy, 2008). When all caseins are cross-linked, the light scattering behaviour of the micelles in the presence of urea, citrate or a combination thereof, fits that predicted when taking into account the influence of such additives on refractive index and micellar mass, hence suggesting complete stability of the particles (Huppertz *et al.*, 2007). Cross-linked micelles swell in the presence of urea or on chelation of micellar calcium (Huppertz *et al.*, 2007).

15.4.2 Sodium caseinate

Properties of TGase-treated sodium caseinate have been widely reported. After incubation of sodium caseinate with TGase, aggregates have a molecular mass up to ~2000 kDa, i.e. up to ~100 casein monomers (Lorenzen *et al.*, 1998), which suggests largely intra-particle cross-linking. As a result of TGase-induced cross-linking, gelation of concentrated caseinate suspensions or caseinate-stabilized emulsions can occur, provided the protein content is sufficiently high. A sodium caseinate suspension containing 2.7% protein did not gel on incubation with TGase, but incubation of a caseinate suspension containing 7.2% protein yielded the formation of a transparent gel which did not exhibit syneresis on cutting (Partanen *et al.*, 2008). A more concentrated suspension of sodium caseinate (12.5% protein), as well as an emulsion prepared therefrom, also gelled on incubation with TGase (Nonaka *et al.*, 1992). TGase-induced gelation of sodium caseinate suspensions is probably aided by the fact that the water-binding capacity of the product is improved compared with its untreated variant (Lorenzen, 2000a, 2007). Acid-induced gelation of sodium caseinate is also affected by TGase treatment, with gels formed from TGase-treated sodium caseinate showing considerably higher firmness and finer gel structure (Myllarinen *et al.*, 2007; Partanen *et al.*, 2008).

Solubility of sodium caseinate at pH < 4 and at pH ~5 is improved considerably on treatment with TGase (Flanagan *et al.*, 2003) and can be improved further by combining TGase treatment with partial enzymatic hydrolysis by a *Bacillus* proteinase, either before or after TGase treatment (Flanagan and Fitzgerald, 2002). Furthermore, TGase-treatment of sodium caseinate suspension reduced the development of turbidity and pH 4.6-soluble nitrogen during heat treatment at 140°C, suggesting that treatment with TGase improves the stability of sodium caseinate to heat-induced coagulation and heat-induced hydrolysis (Flanagan *et al.*, 2003). Treatment of sodium caseinate with TGase offers little potential for improving foaming properties (Flanagan *et al.*, 2003) and some minor improvements in emulsifying properties (Flanagan *et al.*, 2003; Tang *et al.*, 2005) that are outlined further in Section 15.6.

15.4.3 Whey proteins

Compared to casein, as described above, studies on the properties of TGase-treated whey proteins are far less in abundance. TGase-induced gelation of whey protein solutions has not been observed, but Eissa and Kahn (2005) showed that the storage modulus of acid-induced gels from preheated whey protein were approximately 2-fold higher for TGase-treated variants than for untreated variants. In addition, Gauche *et al.* (2008) showed that incubation of reconstituted whey powder (50%, m/m) with TGase changed the flow behaviour of the sample towards less Newtonian and more pseudoplastic behaviour. TGase-treatment also increased the heat stability of a whey protein concentrate solution, but did not influence its solubility or viscosity (Lorenzen, 2007).

15.5 Gelation properties of milk systems subjected to enzymatic cross-linking

15.5.1 Acid-induced gelation

Acid-coagulation properties of casein micelles are strongly affected by TGase treatment, with acid-induced flocculation of cross-linked casein micelles commencing at a considerably higher pH than that for native casein micelles (Schorsch *et al.*, 2000; Anema *et al.*, 2005; Huppertz and De Kruif, 2008). In addition, improved storage modulus was observed for acid gels prepared from TGase-treated pasteurized skim milk (Anema *et al.*, 2005) or phosphocasein suspensions (Schorsch *et al.*, 2000). However, these results could not be confirmed for acid gels from 1.5% fat UHT milk, for which Jaros *et al.* (2006b) observed improved large-deformation properties but no effect of small deformation properties. The reason for these differences is unclear at present; it could be related to the presence of fat in this system, which, as discussed later, reduces the positive effects of TGase, or due to the use of UHT-treated milk, which notably produces rather poor acid milk gels. Acid-gels prepared from TGase-treated phosphocasein suspensions were also shown to be less susceptible to syneresis and temperature-induced changes in microstructure (Schorsch *et al.*, 2000). Acid gelation properties of TGase-treated milk could be further enhanced by performing the incubation with TGase at 400 MPa rather than at atmospheric pressure (Anema *et al.*, 2005). Acid coagulation of completely cross-linked casein micelles can further be manipulated by adjustment of their micellar calcium phosphate content; the pH at which acid induced flocculation of cross-linked casein micelles commenced increases in near-linear fashion with increasing micellar calcium phosphate content (Huppertz and De Kruif, 2008).

Perhaps quite surprisingly, the mechanism for the TGase-induced reduction in acid stability of casein micelles has received little attention thus far. Huppertz and De Kruif (2008) reasoned that TGase-induced reductions in the stability of casein micelles to acid-induced coagulation are probably related to a reduced degree of steric repulsion provided by the casein brush on the micelle surface, due to reduced mobility of the cross-linked proteins. Steric stabilization of casein micelles contains a large entropic term (Tuinier and De Kruif, 2002), which is likely to be reduced considerably when the casein brush is cross-linked.

The improved acid-induced coagulation properties of cross-linked casein micelles offer great potential for improving the texture of yoghurt and other acid-coagulation dairy products. Particularly, the application of TGase in set-yoghurt manufacture has been studied extensively. Both the pre-treatment of milk with TGase and TGase treatment during acidification have been studied, as outlined in Tables 15.1 and 15.2 respectively. From these tables, it is apparent that yoghurt prepared from TGase-treated milk has a higher firmness and lower syneresis than equivalents produced from untreated milk. Particularly in set-yoghurt prepared from TGase-treated milk, it is apparent that the effects of TGase are greatest in skim milk yoghurt and gradually diminish with increasing fat content (Table 15.1). In addition, the sensory perception of the yoghurt is also influenced. Sensory evaluation by Lorenzen *et al.* (2002) revealed that set-yoghurt prepared from TGase-treated milk was perceived as less acidic, less intense in odour and taste, and firmer, compared with yoghurt prepared from untreated skim milk. In contrast, set-style yoghurt produced by the simultaneous TGase-treatment and acidification of milk was perceived as more flaky and gritty (Faegemand *et al.*, 1999a).

Stirred yoghurt prepared from TGase-treated milk also shows improved viscosity, when TGase treatment is carried out prior to or during acidification (Table 15.3). Care should be taken, however, with the combined acidification and TGase treatment, although presented by Bonisch *et al.* (2007b,c) as a rapid method of improving yoghurt properties, compared to the more lengthy pre-treatment of milk with TGase. The results of Bonisch *et al.* (2007b) clearly indicate that TGase remains active in the final product. Since residual activity is detectable, this would necessitate the inclusion of TGase on product labelling as a 'novel food ingredient' in most jurisdictions, whereas TGase can be classified as a 'processing aid' once fully inactivated.

15.5.2 Rennet-induced gelation

As a result of the TGase-induced cross-linking of caseins, the rennet gelation of milk is impaired, as highlighted by increases in the time required to

Table 15.1 Properties of set-style yoghurt prepared from TGase-treated milk

Reference	Substrate type	Major effects of TGase
Faergemand and Qvist (1997)	Reconstituted skim milk	• Increased storage modulus (~5-fold) • Increased yield stress (~3-fold) • Reduced permeability (~2-fold) • Finer protein network in yoghurt
Lorenzen and Schlimme (1998)	Whole milk (3.5% fat)	• Increase fermentation time • Increased gel strength • Reduced syneresis • No effect on acidity
Lorenzen *et al.* (1999)	Milk (3.2% protein, 0–10% fat)	• Increased fermentation time (~15%) • Increased gel strength for skim milk (~100%), 1.5% fat milk (~50%) and 3.5% fat milk (~30%) • Reduced gel strength for 10% fat milk (~10%) • Reduced syneresis (~2-fold) • Reduced acidity (~10%)
Lauber *et al.* (2000)	Skim milk	• Increased breaking strength (~40%) (breaking strength correlates directly with degree of cross-linking.
Neve *et al.* (2001)	Pre-heated (5 min 92°C) skim milk	• Little effect on growth rate of bacteria • Increase in fermentation time (~10%) • Reduced acidity • Increase gel strength (effect increases with storage time if TGase is not inactivated) • Reduced syneresis (effect increases with storage time if TGase is not inactivated)
Lorenzen *et al.* (2002)	Pre-heated (5 min 92°C) skim milk, low-fat milk (1.5% fat) or full-fat milk (3.5% fat)	• Reduced acidity (~10% in skim milk and full-fat milk) • Increased gel strength (effect greatest in skim milk and smallest in full-fat milk; effect enhanced during storage if TGase is not inactivated) • Reduced syneresis (effect greatest in skim milk and smallest in full-fat milk; effect enhanced during storage if TGase is not inactivated) • Less-intense flavour and odour (sensory assessment) • Increased firmness, creaminess and graininess (sensory assessment)

Table 15.2 Properties of set-style yoghurt prepared using the simultaneous acidification and TGase-treatment of milk

Reference	Substrate type	Major effects of TGase
Faergemand *et al.* (1999a)	Recombined milk (3.7,4.3 or 4.7% protein, 1.5 or 3.4% fat)	• Increased gel firmness (effect of greater at higher protein and higher fat content and higher enzyme concentration) • Decreased gel permeability (effect greater at lower protein and lower fat content and higher enzyme concentration) • Texture properties (sensorial assessment) of low-fat TGase variant comparable to high fat variant without TGase
Lorenzen and Schlimme (1998)	Whole milk (3.5% fat)	• Increase fermentation time • Increased gel strength • Reduced syneresis • No effect on acidity

Table 15.3 Properties of stirred-style yoghurt prepared using TGase

Reference	TGase treatment	Substrate type	Major effects of TGase
Jaros *et al.* (2007)	Prior to fermentation	Pre-heated (30 min 90°C) reconstituted skim milk	• Increased viscosity (effect decreases at high shear rate) • Increased thixotropic behaviour • Increased complex modulus (G*)
Bonisch *et al.* (2007b)	During fermentation	Pre-heated (3 min 95°C) reconstituted skim milk	• Increased viscosity (effect increases during storage) • Coarser product appearance
Bonisch *et al.* (2007c)		Pasteurized skim milk	• Increased viscosity (effect increases with enzyme concentration and over storage)

induce visual coagulation, i.e. the rennet coagulation time (RCT) (Lorenzen, 2000b; O'Sullivan *et al.*, 2002a; Huppertz and de Kruif, 2007a) and reduction in the rate of development of firmness of the coagulation (Huppertz and De Kruif, 2007a; Bonisch *et al.*, 2008). These effects are unlikely to be related to cross-linking of proteins in the micellar core, since TGase treatment of para-casein micelles prepared by cold-renneting did not impair milk's coagulation properties (O'Sullivan *et al.*, 2002a). More likely, impaired renneting of TGase-treated milk is due to the cross-linking of caseins, particularly κ-casein, on the micellar surface. According to Lorenzen (2000b), O'Sullivan *et al.* (2002b) and Bonisch *et al.* (2008), impaired renneting of TGase-treated casein micelles is due to the fact that the enzymatic hydrolysis of κ-casein by chymosin is inhibited. They supported this claim with experimental data showing a reduced rennet-induced release of caseinomacropeptide into the milk serum from TGase-treated milks. If reduced enzymatic hydrolysis of κ-casein was the cause of TGase-induced impaired renneting of milk, one would logically expect the time required for the onset of flocculation , which is identifiable as the minimum in time-profiles determined by light scattering or rheological measurements, to increase significantly. This expected increase in the time required for the onset of gelation to occur, however, was not observed in the diffusing wave spectroscopy studies of Huppertz and De Kruif (2007a). These authors observed only a small TGase-induced increase in the flocculation time, and observed that it is primarily the rate of aggregation of the renneted casein micelles that is affected by treatment with TGase. The rate of rennet-induced aggregation of TGase-treated casein micelles showed a positive linear correlation with the degree of non-cross-linked κ-casein and a positive linear correlation with the reciprocal of RCT; the degree of non-cross-linked κ-casein also showed a linear correlation with the reciprocal of RCT (Huppertz and De Kruif, 2007a). These findings strongly suggest that the cross-linking of κ-casein on the micelle surface results in a reduced rate of aggregation of renneted micelles, as a result of which a weaker gel, or no gel at all, will form. The impaired renneting of cross-linked casein micelles limits the applicability of TGase-treatment as a pretreatment in cheese-making. However, studies on the inclusion of TGase in the cheese-matrix at a later stage may still yield interesting options for the modification and tailoring of textural properties of cheese.

15.5.3 Heat-induced gelation of milk

TGase treatment of milk has also been studied as a pre-treatment with the aim of extending the stability of milk against heat-induced gelation. As outlined in Section 15.2.1 (*Heat induced gelation of milk*), heat-induced dissociation of κ-casein from the micelle has been attributed a major role in promoting the heat-induced coagulation of unconcentrated and concentrated milk products, so prevention thereof can thus contribute to an improved stability of milk against heat-induced coagulation. TGase treatment of milk transforms the heat coagulation time–pH profile of unconcentrated milk from one containing a maximum and minimum to a sigmoidal one, where heat coagulation time increases with increasing pH. TGase-induced improvements in the heat stability of unconcentrated milk are indeed observed, but only at pH-values >~6.6 (O'Sullivan *et al.*, 2001, 2002b; Mounsey *et al.*, 2005; Huppertz and De Kruif, 2008). At lower pH, TGase-treatment actually reduces the heat stability of unconcentrated milk (Mounsey *et al.*, 2005; Huppertz and De Kruif, 2008). The improved heat stability at pH > 6.7 is probably due to the fact that heat-induced dissociation of κ-casein is prevented by TGase-induced cross-linking. The TGase-induced reduction in the heat stability of unconcentrated milk at pH < 6.7 is probably due to the fact that TGase-treatment improves the susceptibility of casein micelles to acid-induced coagulation, which plays a major role in the heat-induced coagulation at low pH (Huppertz and De Kruif, 2008). The heat stability of cross-linked casein micelles can be further modified by adjusting their MCP content; the sigmoidal HCT–pH profile of TGase-treated milk progressively shifts to more alkaline values with decreasing MCP content (Huppertz and De Kruif, 2008).

The heat stability of concentrated milk is also influenced considerably by treatment with TGase. The original profile with a maximum at pH 6.5–6.6 is changed into one where, as for unconcentrated milk, heat coagulation time increases with increasing pH (O'Sullivan *et al.*, 2001). Heat stability is improved at pH > ~6.6 but reduced at lower pH. It is likely that, as for unconcentrated milk, the reduced dissociation of κ-casein and the reduced stability of the cross-linked micelles to acid-induced coagulation are responsible for the effects observed, although further study may be required to confirm this. Overall, TGase-induced changes in the heat stability of milk offer relatively few commercial opportunities, since the improvements in heat stability that are observed occur at pH values far above the natural pH of such milk systems.

15.6 Emulsification properties of cross-linked milk proteins

Hinz *et al.* (2007) studied the emulsification properties of milk proteins cross-linked by TGase-treatment in raw whole milk, and by preparing recombined milk from skim milk treated with transglutaminase. In both systems, pre-treatment of the milk proteins by TGase prior to homogenization impaired the emulsification properties of the milk proteins, as evidenced by the formation of clusters of fat globules, particularly at high homogenization pressures. This is caused by the fact that the cross-linked casein micelles are not able to spread over the oil–water interface like native casein micelles. As a result, the cross-linked casein micelles are not able to cover the full fat globule surface in homogenized milk and homogenization clusters are formed (Hinz *et al.*, 2007). These results indicate that care should be taken in the use of cross-linked casein micelles as an emulsifier as they are able to cover a far smaller fat globule surface than their native counterparts. However, there may be some advantages as well, as the use of cross-linked casein micelles on the droplet surface is likely to increase protein-surface load considerably, and hence potentially increase stability of the droplet and make them less susceptible to creaming.

In addition, TGase-induced cross-linking of the proteins of the milk fat globule membrane also occurs and, when performed in cream, such treatment stabilizes the fat globules against coalescence (Hinz *et al.*, 2007). Although further study of these phenomena is required, the findings may be exploited for improving the stability of cream against rebodying or plasticization which occurs on temperature cycling of the cream.

The influence of cross-linking of caseinates, whey proteins and individual whey proteins on their emulsifying properties was studied by Faergemand and co-workers. Treatment of emulsions stabilized by sodium caseinate, β-casein or α_{s1}-casein with TGase resulted in a rapid increase in the surface shear viscosity, the extent of the rate of increase being >5-fold higher for the first two emulsions than for the last (Faergemand *et al.*, 1997b, 1999b). Such increases in surface shear viscosity indicate that TGase treatment can be used to improve the stability of casein-stabilized emulsion droplets, which was indeed observed by Faergemand *et al.* (1998). Treatment of caseinate-stabilized emulsions with TGase was shown to improve the stability of the emulsion towards creaming, Ostwald ripening and ethanol-induced flocculation (Faergemand *et al.*, 1998). Surface shear viscosity of β-lactoglobulin-stabilized emulsion droplet, which is naturally higher than that of casein-stabilized emulsion droplets, also increased as a result of cross-linking, but to a far lesser extent (Faergemenad *et al.*, 1997b). In addition, the stability of β-lactoglobulin-stabilized emulsion droplets to Ostwald-ripening, creaming and ethanol-induced flocculation was also improved by treatment with TGase (Faergemand *et al.*, 1998).

15.7 Conclusions

The enzymatic cross-linking of milk proteins alters milk's properties considerably, as a result of which major changes in functionality are observed. One of the most obvious positive examples is the improved acid-induced gelation properties of milk treated with TGase or tyrosinase; the former is now commercially applied. Other applications are arguably a few steps further away from commercial application but are nonetheless equally promising. In particular, findings on the ability of TGase to improve the stability of emulsions are worthy of further exploration. Likewise, although TGase treatment impairs rennet coagulation of milk, it may be applicable in products where enzymatic coagulation is actually an undesirable phenomenon, e.g. UHT-treated milk products. The field of utilization of TGase or other cross-linkers in modifying the functionality of milk protein ingredients is also still an area that shows considerable promise. Finally, further study on the use of cross-linking enzymes other than TGase should highlight whether these can be utilized to achieve desirable effects not readily achievable with TGase.

15.8 References

ANDO H, ADACHI M, UMEDA K, MATSUURA A, NONAKA M, UCHIO R, TANAKA H and MOTOKI M (1989), 'Purification and characteristics of a novel transglutaminase derived from microorganisms', *Agric Biol Chem* 53, 2613–2617.

ANEMA S G, LAUBER S, LEE S K, HENLE T and KLOSTERMEYER H (2005), 'Rheological properties of acid gels prepared from pressure- and transglutaminase-treated skim milk', *Food Hydrocolloid* 19, 879–887.

BONISCH M P, LAUBER S and KULOZIK U (2004), 'Effect of ultra-high temperature treatment on the enzymatic cross-linking of micellar casein and sodium caseinate by transglutaminase', *J Food Sci* 69, E398–E404.

BONISCH M P, TOLKACH A and KULOZIK U (2006), 'Inactivation of an indigenous transglutaminase inhibitor in milk serum by means of UHT-treatment and membrane separation techniques', *Int Dairy J* 16, 699–678.

BONISCH M P, LAUBER S and KULOZIK U (2007a), 'Improvement of enzymatic cross-linking of casein micelles with transglutaminase by glutathione addition', *Int Dairy J* 17, 3–11.

BONISCH M P, HUSS M, LAUBER S and KULOZIK U (2007b), 'Yoghurt gel formation by means of enzymatic cross-linking during fermentation', *Food Hydrocolloid* 21, 585–595.

BONISCH M P, HUSS M, WEITL K and KULOZIK U (2007c), 'Transglutaminase cross-linking of milk proteins and impact on yoghurt gel properties', *Int Dairy J* 17, 1360–1371.

BONISCH M P, HEIDEBACH T C and KULOZIK U (2008), 'Influence of transglutaminase protein cross-linking on the rennet coagulation of casein', *Food Hydrocolloid* 22, 288–297.

CLAUS H and DECKER H (2006), 'Bacterial tyrosinases', *Syst Appl Microbiol* 29, 3–14.

COUSSONS P J, PRINCE N C, KELLY S M, SMITH B and SAWYER L (1992), 'Transglutaminase catalyzes the modification of glutamine side chains in the C-terminal region of bovine β-lactoglobulin. *Biochem J* 283, 803–806.

DALGLEISH D G (1992), 'The enzymatic coagulation of milk', in Fox P F, *Advanced Dairy Chemistry Volume 1: Proteins*, London, Elsevier, 579–619.

DEJMEK P and WALSTRA P (2004), 'The syneresis of rennet-coagulated curd', in Fox P F, McSweeney P L H, Cogan T M and Guinee T P, *Cheese: Chemistry, Physics and Microbiology*, Amsterdam, Elsevier, 71–103.

DE JONG G A H and KOPPELMAN S J (2002), 'Transglutaminase catalyzed reactions: Impact on food applications', *J Food Sci* 67, 2798–2806.

DE JONG G A H, WIJNGAARDS S and KOPPELMAN S J (2003), 'Transglutaminase inhibitor from milk', *J Food Sci* 68, 820–825.

DE KRUIF C G (1999), 'Casein micelle interactions', *Int Dairy J* 9, 183–188.

DE KRUIF C G (2003), 'Non-food applications of caseins', *Progr Biotechnol* 23, 259–263, 269.

DE KRUIF C G and ROEFS S P F M (1996), 'Skim milk acidification at low temperatures: A model for the stability of casein micelles', *Neth Milk Dairy J* 50, 113–120.

DE KRUIF C G and ZHULINA E B (1996), 'Kappa-casein as a polyelectrolyte brush on the surface of casein micelles', *Colloid Surface A* 117, 151–159.

DE KRUIF C G and HOLT C (2003), 'Casein micelle structure, functions and interactions', in Fox P F and McSweeney P L H, *Advanced Dairy Chemistry Volume 1: Proteins*, New York, Kluwer Academic/Plenum Publishers, 233–276.

DICKINSON E (2003), 'Interfacial, emulsifying and foaming properties of milk proteins', in Fox P F and McSweeney P L H, *Advanced Dairy Chemistry Volume 1: Proteins*, New York, Kluwer Academic/Plenum Publishers, 1229–1260.

DONG W and HOSENEY R C (1995), 'Effects of certain breadmaking oxidants and reducing agents on dough rheological properties', *Cereal Chem* 72, 58–64.

EISSA A and KAHN S A (2005), 'Acid-induced gelation of enzymatically modified, preheated whey proteins', *J Agric Food Chem* 53, 5010–5017.

FAERGEMAND M and QVIST K B (1997), 'Transglutaminase: effect on rheological properties, microstructure and permeability of set-style acid skim milk gel', *Food Hydrocolloid* 11, 287–292.

FAERGEMAND M, OTTE J and QVIST K B (1997a), 'Enzymatic crosslinking of whey proteins by a Ca2+ independent microbial transglutaminase from Streptococcus lydicus', *Food Hydrocolloid* 11, 19–25.

FAERGEMAND M, MURRAY B S and DICKINSON E (1997b), 'Crosslinking of milk proteins with transglutaminase at the oil–water interface', *J Agric Food Chem* 45, 2514–2519.

FAERGEMAND M, OTTE J and QVIST K B (1998), 'Emulsifying properties of milk proteins cross-linked with microbial transglutaminase', *Int Dairy J* 8, 715–723.

FAERGEMAND M, SORENSEN M V, JORGENSEN U, BUDOLFSEN G and QVIST K (1999a), 'Transglutaminase: Effect on instrumental and sensory texture of set-style yoghurt', *Milchwissenschaft* 54, 563–566.

FAERGEMAND M, MURRAY B S, DICKINSON E and QVIST K B (1999b), 'Cross-linking of adsorbed casein films with transglutaminase', *Int Dairy J* 9, 343–346.

FLANAGAN J, GUNNING Y and FITZGERALD R J (2003), 'Effect of cross-linking with transglutaminase on the heat stability and some functional characteristics of sodium caseinate', *Food Res Int* 36, 267–274.

FLANAGAN J and FITZGERALD R J (2002), 'Physicochemical and nitrogen solubility properties of Bacillus proteinase hydrolysates of sodium caseinate incubated with transglutaminase pre- and post-hydrolysis', *J Agric Food Chem* 50, 5429–5436.

FOLK J E (1980), 'Transglutaminases', *Ann Rev Biochem* 49, 517–531.

FOLK J E and FINLAYSON J S (1977), 'The ε-(γ-glutamyl)lysine crosslink and the catalytic role of transglutaminases', *Adv Protein Chem* 31, 1–133.

GAUCHE C, VIERA J T C, OGLIARI P J and BORDIGNON-LUIZ M T (2008), 'Crosslinking of whey proteins by transglutaminase', *Proc Biochem* 43, 788–794.

HINZ K, HUPPERTZ T, KULOZIK U and KELLY A L (2007), 'Influence of enzymatic cross-linking on milk fat globules and emulsifying properties of milk proteins', *Int Dairy J* 17, 289–293.

HORNE D S and BANKS J M (2004), 'Rennet-induced coagulation of milk', in Fox P F, McSweeney P L H, Cogan T M and Guinee T P, *Cheese: Chemistry, Physics and Microbiology*, Amsterdam, Elsevier, 47–70.

HUPPERTZ T and DE KRUIF C G (2007a), 'Rennet-induced coagulation of enzymatically cross-linked casein micelles', *Int Dairy J* 17, 442–447.

HUPPERTZ T and DE KRUIF C G (2007b), 'Ethanol stability of casein micelles cross-linked by transglutaminase', *Int Dairy J* 17, 436–441.

HUPPERTZ T and DE KRUIF C G (2008), 'Structure and stability of nanogel particles prepared by internal cross-linking of casein micelles', *Int Dairy J* 18, 556–565.

HUPPERTZ T and SMIDDY M A (2008), 'Behaviour of partially cross-linked casein micelles under high pressure', *Int J Dairy Technol* 61, 51–55.

HUPPERTZ T, SMIDDY M A and DE KRUIF C G (2007), 'Biocompatible protein micro-gel particles from cross-linked casein micelles', *Biomacromol* 8, 1300–1305.

HYSLOP D B (2003), 'Enzymatic coagulation of milk', in Fox P F and McSweeney P L H, *Advanced Dairy Chemistry Volume 1: Proteins*, New York, Kluwer Academic/Plenum Publishers, 839–878.

IMM J Y, LIAN P and LEE C M (2000), 'Gelation and water binding properties of transglutaminase-treated skim milk powder', *J Food Sci* 65, 200–205.

JAROS D, PARTSCHEFELD C, HENLE T and ROHM H (2006a), 'Transglutaminase in dairy products: Chemistry, physics and applications', *Journal Texture Stud* 37, 113–155.

JAROS D, PATZOLD J, SCHWARZENBOLZ U and ROHM H (2006b), 'Small and large deformation rheology of acid gels from transglutaminase treated milks. *Food Biophys* 1, 124–132.

JAROS D, HEIDIG C and ROHM H (2007), 'Enzymatic modification through microbial transglutaminase ennhances the viscosity of stirred yohurt', *J Texture Stud* 38, 179–198.

KURAISHI C, YAMAZAKI K and SUSA Y (2001), 'Transglutaminase: Its utilization in the food industry', *Food Rev Int* 17, 221–246.

LAUBER S, HENLE T and KLOSTERMEYER H (2000), 'Relationship between the crosslinking of caseins by transglutaminase and the gel strength of yoghurt', *Eur Food Res Technol* 210, 305–309.

LAUBER S, KRAUSE, I, KLOSTERMEYER H and HENLE T (2003), 'Microbial transglutaminase crosslinks β-casein and β-lactoglobulin to heterologous oligomers under high pressure', *Eur Food Res Technol* 216, 15–17.

LEE D S, MATSUMOTO S, MATSUMURA Y and MORI T (2002), 'Identification of the ε-(γ-glutamyl)lysine cross-linking sites in α-lactalbumin polymerized by mammalian and microbial transglutaminases. *J Agric Food Chem* 50, 7412–7419.

LILLE M, AUTIO K and LANTTO R (2007), 'Tyrosinase-induced crosslinking of milk proteins', Poster presented at *World Dairy Summit*, Dublin, Ireland, October 2007.

LORENZEN P C (2000a), 'Techno-functional properties of transglutaminase-treated milk', *Milchwissenschaft* 55, 667–670.

LORENZEN P C (2000b), 'Renneting properties of transglutaminase-treated milk', *Milchwissenschaft* 55, 433–437.

LORENZEN P C (2007), 'Effects of varying time/temperature conditions of pre-heating and enzymatic cross-linking on techno-functional properties of reconstituted dairy ingredients', *Food Res Int* 40, 700–708.

LORENZEN P C and SCHLIMME E (1997), 'Pilot studies on the effect of enzymatic crosslinking in dairying', *Kieler Milchwirtschaftliche Forschungsberichte* 49, 221–227.

LORENZEN P C, SCHLIMME E and ROOS N (1998), 'Crosslinking of sodium caseinate by a microbial transglutaminase', *Nahrung* 42, 151–154.

LORENZEN P C, MAUTNER A and SCHLIMME E (1999), Auswirkung der enzymatischen Quervernetzung von Milchproteinen auf die resultierended Eigenschaften van Joghurt-erzeugnissen', *Kieler Milchwirtschaftliche Forschungsberichte* 51, 89–97.

LORENZEN P C, NEVE H, MAUTNER A and SCHLIMME E (2002), 'Effect of enzymatic cross-linking of milk proteins on functional properties of set-style yogurt', *Int J Dairy Technol* 55, 152–157.

LUCEY J A and SINGH H (1997), 'Formation and physical properties of acid milk gels: A review', *Food Res Int* 30, 529–542.

LUCEY J A and SINGH H (2003), 'Acid coagulation of milk', in Fox P F and McSweeney P L H, *Advanced Dairy Chemistry Volume 1: Proteins*, New York, Kluwer Academic/Plenum Publishers, 1001–1025.

MATTINEN M-L, HELLMAN M, PERMI P, AUTIO K, KALKKINEN N and BUCHERT J (2006), 'Effect of protein structure on laccase-catalyzed protein oligomerization', *J Agric Food Chem* 54, 8883–8890.

MCCLEMENTS D J (1999), *Food Emulsions: Principles, Practice and Techniques*, CRC Press. Boca Raton, FL.

MOON J-H, HONG Y-H, HUPPERTZ T, FOX P F and KELLY A L (2009), 'Properties of casein micelles cross-linked by transglutaminase', *Int J Dairy Techn* 62, 27–32.

MOTOKI M and SEGURO K (1998), 'Transglutaminase and its use for food processing', *Trend Food Sci Technol* 9, 204–210.

MOUNSEY J S, O'KENNEDY B T and KELLY P M (2005), 'Influence of transglutaminase treatment on properties of micellar casein and products made therefrom', *Lait* 85, 405–418.

MULVIHILL D M and ENNIS M P (2003), 'Functional milk proteins: Production and utilization', in Fox P F and McSweeney P L H, *Advanced Dairy Chemistry Volume 1: Proteins*, New York, Kluwer Academic/Plenum Publishers, 1175–1228.

MYLLARINEN P, BUCHERT J and AUTIO K (2007), 'Effect of transglutaminase on rheological properties and microstructure of chemically acidified sodium caseinate gels', *Int Dairy J* 17, 800–807.

NEVE H, LORENZEN P C, MAUTNER A, SCHLIMME E and HELLER K J (2001), 'Effects of transglutaminase treatment on the production of set skim milk yoghurt: Microbiological aspects', *Kieler Milchwirtschaftliche Forschungsberichte* 53, 347–361.

NIELSEN P M (1995), 'Reactions and potential industrial applications of transglutaminase: Review of literature and patents', *Food Biotechnol* 9, 119–156.

NIEUWENHUIZEN W F, DEKKER H L, DE KONING L J, GRONEVELD T, DE KOSTER C G and DE JONG G A H (2003), 'Modification of glutamine and lysine residues in holo and apo α-lactalbumin with microbial transglutaminase', *J Agric Food Chem* 51, 7132–7139.

NONAKA M, SAKAMOTO H, TOIGUCHI S, KAWAJIRI H, SOEDA T and MOTOKI M (1992), 'Sodium caseinate and skim milk gels formed by incubation with microbial transglutaminase', *J Food Sci* 57, 1214–1218, 1241.

O'CONNELL J E and FOX P F (2003), 'Heat-induced coagulation of milk', in Fox P F and McSweeney P L H, *Advanced Dairy Chemistry Volume 1: Proteins* New York, Kluwer Academic/Plenum Publishers, 879–945.

O'SULLIVAN M M, LORENZEN P C, O'CONNELL J E, KELLY A L, SCHLIMME E and FOX P F (2001), 'Influence of transglutaminase on the heat stability of milk', *J Dairy Sci* 84, 1331–1334.

O'SULLIVAN M M, KELLY A L and FOX P F (2002a), 'Influence of transglutaminase treatment on some physico-chemical properties of milk', *J Dairy Res* 69, 433–442.

O'SULLIVAN M M, KELLY A L and FOX P F (2002b), 'Effect of transglutaminase on the heat stability of milk: A possible mechanism', *J Dairy Sci* 85, 1–7.

PARTANEN R, AUTIO K, MYLLARINEN P, LILLE M, BUCHERT J and FORSELL P (2008), 'Effect of transglutaminase on structure and syneresis of neutral and acidic sodium caseinate gels', *Int Dairy J* 18, 414–421.

RODRIGUEZ-NOGALES J M (2006), 'Enhancement of transglutaminase-induced protein cross-linking by preheat treatment of cows' milk: A statistical approach', *Int Dairy J* 16, 26–32.

ROLLEMA H S (2003), 'Caseinates', *Progr Biotechnol* 23, 255–259, 269.

SCHORSCH C, CARRIE H and NORTON I T (2000), 'Cross-lining casein micelles by a microbial transglutaminase: Influence of cross-links in acid-induced gelation', *Int Dairy J* 10, 529–539.

SEGURO K, NIO N and MOTOKI M (1996), 'Some characteristics of a microbial protein cross-linking enzyme: Transglutaminase', in Parris N, Kato A, Creamer L K and Pearce J (Eds.), *Macromolecular Interactions in Food Technology*, ACS Symposium Series, Vol. 650 (pp. 271–280). Washington, DC, USA: American Chemical Society.

SELINHEIMO E, SALOHEIMO M, AHOLA E, WESTERHOLM-PARVINEN A, KALKKINE N, BUCHERT J and KRUUS K (2006), 'Production and characterization of a secreted, C-terminally processed tyrosinase from the fimanentous fungus *Trichoderma reesei*', *FEBS J* 273, 4322–4335.

SHARMA R, LORENZEN P C and QVIST K B (2001), 'Influence of transglutaminase treatment of skim milk on the formulation of ε-(γ-glutamyl)lysine and susceptibility of individual proteins towards crosslinking', *Int Dairy J* 11, 785–793.

SHARMA R, ZAKORA M and QVIST K B (2002), 'Susceptibility of an industrial α-lactalbumin concentrate to cross-linking by microbial transglutaminase. *Int Dairy J* 12, 1005–1012.

SINGH H (2004), 'Heat stability of milk', *Int J Dairy Techn* 57, 111–119.

SMIDDY M A, MARTIN J E G H, KELLY A L, DE KRUIF C G and HUPPERTZ T (2006), 'Stability of casein micelles cross-linked by transglutaminase. *J Dairy Sci* 89, 1906–1914.

TANG C, YANG X Q, CHEN Z, WU H and PENG Z Y (2005), 'Physicochemical and structural characteristics of sodium caseinate biopolymers induced by microbial transglutaminase', *J Food Biochem* 29, 402–421.

TUINIER R and DE KRUIF C G (2002), 'Stability of casein micelles', *J Chem Phys* 117, 1290–1295.

VASBINDER A J, ROLLEMA H S, BOT A and DE KRUIF C G (2003), Gelation mechanism of milk as influenced by temperature and pH; studied by the use of transglutaminase cross-linked casein micelles', *J Dairy Sci* 86, 1556–1563.

ZHU Y, RINZEMA A, TRAMPER J and BOL J (1995), Microbial transglutaminase – a review of its production and application in food processing. *Appl Microbiol and Biotech* 44, 277–282.

16

Improving technological and functional properties of milk by high-pressure processing

A. L. Kelly, K. I. Kothari, D. D. Voigt, University College Cork, Ireland; T. Huppertz, NIZO food research, The Netherlands

Abstract: High-pressure processing has become an established technology for treating meat, fruit and fruit juices, shellfish and ready meals over the last 15 years. In the same period, the changes in milk constituents have been studied in detail, and found to be often unique: for example, changes in casein micelle properties and denaturation of whey protein affect the gel assembly process in cheese and yoghurt. Today, the understanding of the effects of high pressure on milk is quite advanced, and is reviewed in this chapter, along with current and possible commercial implications of these changes.

Key words: high-pressure processing; milk; cheese; yoghurt; ice cream.

16.1 Introduction to high-pressure processing in the food industry

The principal manner of extending the shelf-life and ensuring the safety of food products involves the application of heat treatment; the scientific basis for this principle was established by Louis Pasteur over 100 years ago. Interestingly, however, not long after Pasteur's studies on the preservation of a range of foods by heating, an American agricultural chemist, Bert Hite, demonstrated that applying an entirely different physical force, pressure, could also achieve inactivation of undesirable micro-organisms. A number of subsequent studies in the early years of the 20th century showed that high-pressure (HP) treatment could also induce changes in food properties, such as coagulation of egg proteins. However, HP processing remained a curious footnote in the annals of food research for almost another eighty years, due to a lack of available equipment to extend this proven concept to practical commercial processing. Only after engineering developments in materials science allowed the production of large-scale vessels capable of reaching and maintaining extremely high pressures did the first HP-treated food products reach the market, with the launch of HP-treated jams, jellies and other products in Japan around 1990.

Today, HP processing has become established as a successful niche application in several countries, in all cases for products where the unique benefits of the technology outweigh the very significant capital investment in equipment required for commercial application. Table 16.1 summarises

Table 16.1 Some commercially available HP-processed food products. (Adapted from Hogan *et al.*, 2005 and based on personal communication from Dr Carole Tonello, NC Hyperbaric, Burgos, Spain.)

Product	Manufacturer	Country
Guacamole, salsa dips, ready-meals and fruit juices	Fressurized Foods	USA
Hummus	Hannah International	USA
Processed poultry products	Purdue Farms	USA
Oysters	Motivatit Seafoods	USA
	Goose Point Oysters	USA
	Joey Oysters	USA
Lobsters	Clearwater	Canada
	Ocean Choice	Canada
	Seafood 2000	Canada
Crab	Phillip's Seafoods	USA
Desalted cod	Ghezzi	Italy
Fruit and vegetable juices	Ultifruit	France
	Beskyd Frycovice	Czech Republic
	Danny Boy	Australia
	Ata	Italy
	Leahy Orchards	Canada
	Frubaca	Portugal
Tomato sauces	SimplyFresco	USA
Avocado products	San Lorenzo	Mexico
	Calavo	USA
Soya products	Toby's	USA
Sliced ham and tapas	Espuña	Spain
Sliced, cured and marinated meat	Campofrio	Spain
RTE sliced meats	Hormel	USA
	Kraft Foods	USA
Cured smoked ham	Abraham	Germany
Preservative-free chicken strips	Purdue Farms	USA
Sliced and diced poultry products	Tyson Foods	USA
RTE meals and salads	Maple Leaf	Canada
Sliced harm and cured meats	Santa Maria Foods	Canada
Chicken sausages	Lou famous	USA
Cheese sandwich filling	Rodilla	Spain

the key current areas of commercial application of HP processing. The relative under-representation of dairy products within the commodities listed is noteworthy, and this chapter will outline some of the reasons for this, and explore potential future applications for HP processing in the dairy area, particularly for functional products.

16.2 Effects of high-pressure treatment on constituents and properties of milk

16.2.1 Effect of high pressure on physico-chemical equilibria

The physicochemical properties of milk under any given conditions are determined by a myriad of equilibria. For instance, the intra- and intermolecular associations of proteins are dependent on a balance between association of parts of the protein and their solvation by water. Cooperative attraction may result from van der Waals attractions, hydrogen bonding, hydrophobic interactions and electrostatic interactions, whereas repulsion will be predominantly electrostatic in nature. Hydration of proteins (either favourable or unfavourable) is determined by the physicochemical properties of both the solvent and the protein.

In milk, the stability of casein micelles is also largely dependent on a further equilibrium, viz. the solubility of calcium phosphate in the serum phase of the product. When considering the effects of HP on the constituents and properties of milk, it is therefore critical to establish how the aforementioned equilibria are affected by pressure.

When milk, or any other system at equilibrium, is placed under pressure, the system reacts according to Le Chatelier's principle, which states that, when a stress is applied to a system at equilibrium, the system will react in such a way to counteract the applied stress. When pressure is applied to a system, the overall volume of the system is reduced, and thus reactions that reduce the volume of the system are promoted, whereas those that involve an increase in the volume of the system are suppressed (Atkins, 1995). For aqueous systems, such as milk, it is important to consider that the volume of its main constituent, water, has by far the largest influence on the overall volume of the system, and that the specific promotion or suppression of certain reactions under HP are primarily related to the degree to which they affect the spatial arrangement of water molecules. This importance of the spatial arrangement of water is exemplified by the influence of HP on the ionisation of salts. Because of the electrostrictive effect, i.e. the fact that the arrangement of water around charged ions is more compact than that around uncharged salts, ionisation of salts is promoted under pressure. As a result, the solubility of salts such as calcium phosphate, also increases under pressure.

The effects of high pressure on proteins are more complex and less unambiguous. As previously suggested by Hvidt (1975), HP-induced changes in protein structure and interactions are predominantly related to their effect on the spatial arrangement of the solvent. It is, however, difficult, if not impossible, to attribute such changes to particular types of bonding involved in protein structure or interaction, e.g. hydrogen bonding, hydrophobic interactions or electrostatic interactions, particularly considering that the exact nature of hydrophobic bonding is still unclear and that hydrophobic bonding appears to exist only by virtue of the existence of hydrogen bonds. Furthermore, it appears important to distinguish between intra- and intermolecular interactions.

In terms of milk proteins, the globular whey proteins unfold under pressure, indicating that solvation of the amino acid residues is favoured under pressure. In contrast, the rheomorphic caseins, with little globular folding, show association behaviour; in the case of β-casein, micellisation is suppressed with increasing pressure up to ~150 MPa, but is strongly promoted at higher pressures, thus indicating that solvation of particular regions of the protein becomes unfavourable (Payens and Heremans, 1969).

16.2.2 Effect of high pressure on milk salts

Because milk contains far more calcium phosphate than is soluble under physiological conditions, approximately two thirds of all calcium and half of all inorganic phosphate is 'deposited' in the casein micelles in the form of nanometer-sized clusters and referred to as micellar calcium phosphate (MCP) nanoclusters. MCP also contains magnesium and citrate and smaller amounts of other elements (Walstra *et al.*, 2006). When pressure is increased, mineral solubility increases because of the aforementioned electrostrictive effect. As a result, MCP is solubilised, as is highlighted by *in-situ* NMR (Hubbard *et al.*, 2002) and light-scattering measurements (Huppertz and De Kruif, 2007a). Solubilisation of MCP proceeds in a near-linear fashion with increasing pressure up to ~400 MPa, at which pressure all MCP is solubilised (Huppertz and De Kruif, 2007a). On release of pressure, solubilization of MCP is rapidly and fully reversible (Hubbard *et al.*, 2002). Reversibility of HP-induced solubilisation of MCP is also indicated by the fact that the concentration of calcium in ultrafiltration permeate of HP-treated milk is comparable to that of untreated milk (Regnault *et al.*, 2006). HP-induced increases in the concentrations of calcium and inorganic phosphate in the ultracentrifugal supernatant of milk have been observed (Schrader and Buchheim, 1998; Lopez-Fandino *et al.*, 1998) but, in the context of the aforementioned findings, these should probably be interpreted as HP-induced reductions in the level of sedimentable protein-bound calcium and phosphorus, rather than increases in the level of serum calcium and phosphate in HP-treated milk. Small HP-induced increases in the concentration of ionic calcium have also been observed (Lopez-Fandino *et al.*, 1998; Zobrist *et al.*, 2005), but these reverse rapidly on storage at temperatures at or above ambient (Zobrist *et al.*, 2005).

16.2.3 Effect of high pressure on caseins and casein micelles

When skim milk, or a suspension of casein micelles, is placed under pressure, its ability to scatter light is reduced, as shown by decreases in turbidity and increases in light transmission, which have been observed using *in-situ* measurements (Kromkamp *et al.*, 1996; Gebhart *et al.*, 2005; Huppertz *et al.*, 2006a, b; Huppertz and De Kruif, 2006, 2007b; Orlien *et al.*, 2006; Huppertz and Smiddy, 2008). Since casein micelles are the predominant particles which scatter light in skim milk, HP-induced reductions in turbidity indicate disruption of the casein micelles. Micellar disruption under pressure increases with increasing pressure, up to ~400 MPa (Kromkamp *et al.*, 1996; Huppertz *et al.*, 2006b; Orlien *et al.*, 2006), where full micellar disruption is achieved and the turbidity of the system is comparable to that of the serum phase of milk (Huppertz *et al.*, 2006b). Micellar disruption under HP is a time-dependent process, which proceeds considerably faster at higher pressures (Kromkamp *et al.*, 1996; Huppertz *et al.*, 2006b; Orlien *et al.*, 2006). In addition, reducing temperature (Orlien *et al.*, 2006) or pH (Huppertz and De Kruif, 2006) also increase the extent of micellar disruption, whereas increasing the aforementioned parameters has the opposite effect.

All these observations appear to be related to HP-induced solubilisation of MCP, which has been suggested as the main factor causing micellar disruption under HP (Huppertz and De Kruif, 2006). Solubility of MCP increases under HP, and reaches completion in unconcentrated milk at natural pH at ~400 MPa (Huppertz and De Kruif, 2007a). A suspension of casein micelles at 400 MPa or higher is likely to resemble a sodium caseinate suspension, since all MCP is solubilised but protein–protein interactions are favoured. The presence of whey proteins does not affect the HP-induced disruption of casein micelles (Huppertz and De Kruif, 2007b).

When casein micelles are held at 'intermediate' pressures, e.g. 200–300 MPa, for prolonged times, the initial disruption of casein micelles is followed by a reformation process, during which casein particles of colloidal dimensions are formed (Huppertz *et al.*, 2006b, c; Orlien *et al.*, 2006; Huppertz and De Kruif, 2006, 2007b; Huppertz and Smiddy, 2008). This reformation process does not occur when micellar disruption is complete (Huppertz *et al.*, 2006b; Orlien *et al.*, 2006), and is probably related to the association of micellar fragments. Reformation is more extensive at higher temperatures (Orlien *et al.*, 2006) and higher pH values (Huppertz and De Kruif, 2006), but is not influenced by the presence of whey proteins (Huppertz and De Kruif, 2007b). Reformation of casein particles is further promoted on release of pressure, as highlighted by increases in turbidity and reductions in light transmission (Kromkamp *et al.*, 1996; Huppertz *et al.*, 2006b). Such reformation is probably mediated by calcium phosphate, because solubility of calcium phosphate rapidly decreases on reducing pressure (Hubbard *et al.*, 2002).

As a result of the aforementioned changes in casein micelles during HP treatment and subsequent decompression, the casein micelles in HP-treated milk differ considerably from those in untreated milk. Such differences are highlighted by extensive changes in a number of physicochemical parameters, most notably average casein micelle size, sedimentation behaviour of the caseins, and the turbidity or lightness of the milk. Casein micelle size is affected little by treatment at pressures <200 MPa, but treatment at 250–300 MPa for >15 min can result in considerable increases in micelle size (Gaucheron *et al.*, 1997; Needs *et al.*, 2000; Huppertz *et al.*, 2004a,b; Regnault *et al.*, 2004; Anema *et al.*, 2005a, 2008). When unconcentrated milk at natural pH is treated at 250 MPa for 30 min at ambient temperature, micelle size is increased by up to 30% (Huppertz *et al.*, 2004a,b; Regnault *et al.*, 2004), but this increase in micelle size is far greater when treatment temperature is increased (Gaucheron *et al.*, 1997; Huppertz *et al.*, 2004a; Anema *et al.*, 2005a) or milk pH is increased prior to HP treatment (Huppertz *et al.*, 2004a). Likewise, reducing treatment temperature or pH has the opposite effects, and may even result in reductions in micelle size on treatment at 250 MPa (Huppertz *et al.*, 2004a; Regnault *et al.*, 2004). HP-induced increases in casein micelle size are, at least partially, reversible on subsequent storage of the milk (Huppertz *et al.*, 2004a).

Treatment of milk at 300–800 MPa generally results in considerable reductions in casein micelle size, to approximately half of that in untreated milk (Gaucheron *et al.*, 1997; Needs *et al.*, 2000; Huppertz *et al.*, 2004a,b; Regnault *et al.*, 2004; Anema *et al.*, 2005a, 2008). However, when treatment at 300–400 MPa is performed at high temperature, e.g. >40°C, micelle size is considerably higher (Huppertz *et al.*, 2004a; Anema *et al.*, 2005a; Garcia-Risco *et al.*, 2000). HP-induced reductions in casein micelle size are generally irreversible on subsequent storage (Huppertz *et al.*, 2004a).

HP-induced changes in casein micelle size are also accompanied by changes in the turbidity or lightness of skim milk. These optical parameters are affected little by treatment at 100–200 MPa but decrease strongly on treatment at 200–400 MPa; at higher pressures, little further effect is observed (Needs *et al.*, 2000; Huppertz *et al.*, 2004b; Reganult *et al.*, 2004). While decreases in lightness or turbidity of skim milk are invariably associated with changes in micelle size, the decrease therein observed at 250 MPa, despite aforementioned increases in micelle size at this pressure, may at first seem contradictory. However, it should be noted that turbidity and lightness are dependent on both micelle size as well as the number of micelles. Hence, the decrease in turbidity or lightness observed in milk treated at 250 MPa, despite increases in micelle size, probably suggests the presence of a very large number of particles in such milk which are too small to scatter light extensively, and that the fraction of large micelles resulting in the increased micelle size is relatively small. HP-induced decreases in turbidity and lightness are generally, at least partially, reversible on subsequent storage, particularly at temperatures at or above ambient (Huppertz *et al.*, 2004b).

16.2.4 Effect of high pressure on whey proteins

Treatment of milk at up to 100 MPa does not denature β-lactoglobulin (β-lg), but treatment at higher pressures results in considerable denaturation of this protein, up to ~90% of total β-lg in milk treated at 400 MPa (Lopez-Fandino *et al.*, 1996; Gaucheron *et al.*, 1997; Scollard *et al.*, 2000; Huppertz *et al.*, 2004a,c). Denaturation of α-lactalbumin (α-la) only occurs at >400 MPa, and reaches ~70% after treatment at 800 MPa (Huppertz *et al.*, 2004a,c). The higher barostability of α-la than β-lg may be due to the higher number of intra-molecular disulphide bonds (Hinrichs *et al.*, 1996; Gaucheron *et al.*, 1997) and the absence of a free sulphydryl group (Lopez-Fandino *et al.*, 1996) in α-la. The extent of HP-induced denaturation of α-la and β-lg increases with treatment time (Scollard *et al.*, 2000; Hinrichs and Rademacher, 2004; Huppertz *et al.*, 2004a), treatment temperature (Gaucheron *et al.*, 1997; Lopez-Fandino and Olano, 1998; Garcia-Risco *et al.*, 2000; Huppertz *et al.*, 2004a), milk pH (Arias *et al.*, 2000; Huppertz *et al.*, 2004a) and the level of micellar calcium phosphate in the milk (Huppertz *et al.*, 2004c). HP-induced denaturation of whey proteins is largely prevented by adding a sulphydryl-blocking agent to milk prior to HP treatment (Huppertz *et al.*, 2004c).

Denatured β-lg in HP-treated milk is found in various forms; in a manner analogous to heat treatment, HP-denatured β-lg associates with the casein micelles, as shown by transmission electron microscopy coupled with immunogold labelling (Scollard *et al.*, 2000), as well as by a reduced proportion of total β-lg in the ultracentrifugal supernatant from HP-treated milk (Gaucheron *et al.*, 1997; Garcia-Risco *et al.*, 2003). Quantitative studies have shown that the majority of denatured β-lg in HP-treated skim milk is associated with the casein micelles, with a small proportion remaining non-sedimentable, either in the form of whey protein aggregates or in association with casein particles too small to be sedimented (Gaucheron *et al.*, 1997; Huppertz *et al.*, 2004a; Zobrist *et al.*, 2005). However, recent studies by Anema (2008) showed that most whey protein remains in the serum phase of concentrated HP-treated milk. In HP-treated whole milk, denatured α-la and β-lg are also found associated with the milk fat globule membrane (Ye *et al.*, 2004).

Huppertz *et al.* (2004c) suggested the following mechanism for HP-induced denaturation of α-la and β-lg in milk and whey: under HP, β-lg unfolds (Kuwata *et al.*, 2001), which results in the exposure of its free sulphydryl group (Tanaka *et al.*, 1996; Moller *et al.*, 1998; Stapelfeldt *et al.*, 1999); this free sulphydryl group can interact with κ-casein, α-la or β-lg, and possibly α_{s2}-casein, through sulphydryl-disulphide interchange reactions. On release of pressure, unfolded α-la and β-lg molecules that have not interacted with another protein may refold to a state closely related to that of native β-lg, as no difference in ^{1}H nuclear magnetic resonance (NMR) profile are observed between untreated and HP-treated β-lg

(Ikeuchi *et al.*, 2001; Beloque *et al.*, 2000); the close structural similarity between monomeric untreated and HP-treated β-lg strongly suggests that the sulphydryl–disulphide interchange reactions, which result in association of whey proteins, occur during HP treatment and not post-HP treatment, since the free sulphydryl group of β-lg is not available for interaction post-HP treatment. Calcium may facilitate close approach of unfolded whey proteins to other proteins through charge shielding. Thus, the lower extent of HP-induced denaturation of α-la and β-lg in rennet whey than in milk appears to be a result of the absence of κ-casein from, and a considerably lower calcium concentration in, the former (Huppertz *et al.*, 2004c).

Compared to α-la and β-lg, HP-induced denaturation of other whey proteins has received little attention to date. Lopez-Fandino *et al.* (1996) reported that no denaturation of bovine serum albumin (BSA) occurred in bovine milk at 100–400 MPa; the high barostability of BSA is probably related to the fact that this molecule, through its 17 intra-molecular disulphide bonds, has an extremely rigid molecular structure which remains largely unaffected under HP (Lopez-Fandino *et al.*, 1996). The immunoglobulins are also relatively stable to HP processing, with ~90% of colostral IgG surviving treatment at 500 MPa for 5 minutes (Indyk *et al.*, 2008).

16.2.5 Effects of high pressure on milk fat

Compared to milk proteins, very few studies have considered HP-induced changes in milk fat. HP treatment of bovine milk at up to 600 MPa does not affect fat globule size significantly (Huppertz *et al.*, 2003; Ye *et al.*, 2004). Similarly, HP treatment (500 MPa) of ewes' milk had no effect on the fat globule size, although an increase in the number of small milk fat globules (1–2 μm) was observed following HP treatment at 200 MPa at 25°C (Gervilla *et al.*, 2001).

However, HP treatment of cream (30% fat) at 800 MPa for 10 min significantly increased fat globule size (Kanno *et al.*, 1998), with treatment at lower pressures (100–600 MPa at <40°C) not affecting the fat globule size (Dumay *et al.*, 1996; Kanno *et al.*, 1998). HP treatment of bovine milk also results in the association of β-lg (>100 MPa), α-la (≥700 MPa), and κ-CN (500 MPa) with the milk fat globule membrane (MFGM) (Ye *et al.*, 2004).

Creaming of fat globules on cold storage is greater in milk treated at 100–250 MPa than in untreated milk, whereas treatment at 400 or 600 MPa significantly reduces creaming (Huppertz *et al.*, 2003). Improved whipping properties of cream, with reduced whipping time and loss of serum, was observed following HP treatment (500–600 MPa for 1–2 min) of cream (26–29% fat), probably due to enhanced crystallisation of milk fat (Eberhard *et al.*, 1999). Furthermore, crystallisation of milk fat in HP-treated cream occurs at a temperature higher than in cream retained at atmospheric pressure, which may have potential applications in ageing of ice cream mix and physical ripening of cream for butter making (Buchheim and Abou El-Nour,

1992; Dumay *et al.*, 1996; Frede and Buchheim, 2000). HP treatment at up to 800 MPa does not rupture the MFGM; however, aggregation of MFGM is observed following HP treatment (Kanno *et al.*, 1998).

No increase in the level of free fatty acids or lipolysis is observed following HP treatment of ewes' milk. Kanno *et al.* (1998) suggested that HP treatment of cream could even repair a ruptured or damaged MFGM, through aggregation of protein in the MFGM. It has also been reported that HP treatment (400 MPa for 4 h at 45°C) of the hard (high-melting) fraction of milk fat, commonly used in baking and confectionary applications, significantly suppressed thermal deterioration (Abe *et al.*, 1999).

16.3 Applications of high-pressure treatment in processing liquid milk

Probably one of the most studied and successful food processing operations in the world is the commercial pasteurisation of milk for consumption, typically achieved by heating milk to 72–74°C for 15–30 s; such treatment renders milk safe for consumption by killing all vegetative pathogens, and reducing levels of spoilage micro-organisms to a level such that the shelf-life of milk at refrigeration temperatures is significantly extended relative to that of raw milk, and with minimal alterations in nutritional or sensory properties.

To be an interesting alternative to such an efficient process, HP treatment must offer a significant added-value element, such as a significant extension of shelf-life; however, studies undertaken for such applications have generally concluded that extensions of shelf-life are relatively minimal and unlikely to be commercially applicable. Huppertz *et al.* (2006c) reviewed the studies published to date on inactivation of spoilage and pathogenic bacteria in milk, and concluded that treatment at very high pressures (>600 MPa) is required to induce significant levels of inactivation, and that considerable strain-to-strain variation in bacterial resistance may present an additional complication. Interestingly, many bacteria are more barotolerant in milk than in simpler systems such as buffers, which has been mechanistically linked to the protective role of MCP solubilised by HP (Black *et al.*, 2007). While extension of the shelf-life of milk by HP treatment has been demonstrated (Garcia-Risco *et al.*, 1998), such a benefit must be weighed against the capital equipment costs, particularly compared to other technological solutions such as microfiltration. In addition, while psychotropic bacteria are very sensitive to pressure, their extracellular proteases are very pressure-resistant and may subsequently negatively influence dairy product quality, while the extreme pressure-resistance of spore-forming bacteria presents a significant challenge for neutral-pH food products such as milk. Nonetheless, there has been some recent commercial success in HP treatment of colostrum, as will be discussed later in this chapter.

Of course, any consideration for use of HP for processing of milk must also consider effects such as colour changes for skim milk and altered creaming of whole milk, as discussed earlier, which may represent either advantages or disadvantages for certain product applications.

16.4 High-pressure treatment of milk for the manufacture of acid-coagulated milk products

Due to consumer preference for thick and viscous yoghurt (Tamime and Robinson, 1999; Sodini *et al.*, 2004), pre-heat treatment of milk (85–90°C for 5–30 min) is commercially practiced in yoghurt production, as it markedly improves the texture of yoghurt, mainly through denaturation of whey proteins and their subsequent structural role in the gel (Dannenberg and Kessler, 1988; Mulvihill and Grufferty, 1995; Lucey *et al.*, 1997, 1998; Anema *et al.*, 2004). However, with increasing interest in alternatives to conventional heat treatment and growing demand for natural food products, the application of HP processing to yoghurt milk has received considerable attention. As HP treatment of milk alters casein micelles and denatures whey proteins, this may potentially yield different acid-milk gel structures from those obtained using heated milk.

Two different approaches involving HP processing for the production of acidified/cultured milks have been studied:

(i) HP treatment of milk or similar base material prior to acidification and gel formation, to enhance the textural properties of yoghurt;
(ii) HP treatment of the final yoghurt to extend shelf-life by microbial inactivation, which will not be discussed further here.

HP treatment of milk or similar base material at 100–700 MPa for up to 30 min has been generally studied either as an alternative to conventional heat treatment or in sequential combination with heat treatment in order to produce set or stirred yoghurt with improved textural and water-holding properties.

16.4.1 Impact of high-pressure treatment on yoghurt microstructure and texture

Casein micelles aggregate as milk pH decreases towards the isoelectric point of casein (4.6) and form a three-dimensional matrix of micellar chains and clusters, with the fat globules, if present, becoming part of the gel network, particularly when homogenisation has been applied. Several studies have compared the microstructure of yoghurt prepared from HP-treated milk to that made from heat-treated milk. As expected from studies on milk, smaller casein micelles were present in set yoghurt made from HP-treated milk than those made from heat-treated milk (Needs *et al.*, 2000; Harte *et al.*, 2002; Penna *et al.*, 2007a). Also, casein micelles in yoghurt from HP-treated milk appeared as smooth-surfaced particles with bound denatured whey protein, while denatured whey protein was also found in amorphous clumps; the gel consisted of densely packed strands, and micelles did not appear to be fused (Needs *et al.*, 2000). However, in yoghurt from heat-treated milk, the micelles were bridged by dense filamentous projections, probably of denatured whey protein, at their surfaces, and formed a denser and more continuous network than in the HP yoghurt (Needs *et al.*, 2000). Harte *et al.* (2002) reported that HP treatment of milk at 200 MPa resulted in gels having large aggregates of fused micelles, similar to those observed in yoghurt from raw milk. Combined heat and HP treatment of milk led to compact yoghurt gels with increasingly larger casein micelle clusters interspaced by void spaces, and exhibiting a high degree of cross-linking (Penna *et al.*, 2007a).

Knudsen *et al.* (2006) reported that any difference in the gel network structure arising from heat or HP treatment of the milk before acidification disappeared in stirred acid-gels; HP-induced dissociation of casein micelles in milk was thus of minor importance for the rheological properties of subsequently stirred acidified gels.

A small number of studies have reported effects of HP treatment of milk on yoghurt texture or rheological properties. Desobry-Banon *et al.* (1994) and Needs *et al.* (2000) reported that yoghurt made from HP-treated milk exhibited higher elastic modulus than yoghurt made from heat-treated milk, suggesting that many close micelle–micelle bonds or interactions were established in the former yoghurt. HP treatment of full-fat milk (676 MPa for 30 min) resulted in yoghurt gels with equivalent rheological properties to those obtained from heat-treated milk (85°C for 35 min) (Harte *et al.*, 2002). Yoghurt made from ultrafiltrated skim milk (4.85% protein) HP-treated at 600 MPa for 15 min had higher gel firmness, fracture stress and fracture strain values than yoghurt made from heat-treated (85°C for 20 min) concentrated skim-milk (Capellas and Needs, 2003). The firmness of ewes' milk yoghurt was also significantly higher when made from HP-treated milk (350 or 500 MPa for 15 min at 10–55°C) compared to yoghurt made from heat-treated milk (Ferragut *et al.*, 2000). When yoghurt made from HP-treated milk was compared with conventional heat-treated milk yoghurt in a consumer test, lower fracture stress values in HP yoghurt were correlated with a higher score for creaminess, and a mixture of 90% heat-treated and 10% HP-treated milk gave a creamy product that had a similar taste to conventional yoghurt (Capellas *et al.*, 2002).

In contrast, Walsh-O'Grady *et al.* (2001) reported that heat treatment of milk was more efficient than HP treatment at inducing casein/whey protein interactions that were favourable for the formation of acid gels; HP-induced denaturation of whey proteins did not result in the same degree of interaction between the casein and whey protein as that caused by heat treatment. Also, Knudsen *et al.* (2006) reported that rheological properties of stirred acid-gels made from milk either subjected to heat treatment or HP treatment were similar.

Combining heat and HP treatment of milk may offer a promising route for achieving desirable rheological and textural properties in yoghurt without addition of stabilisers (Harte *et al.*, 2003; Penna *et al.*, 2007a). Also, it has been reported that combining treatment of milk with HP and cross-linking by transglutaminase (TG) prior to acidification markedly reduced the gelation time, increased the gelation pH, and resulted in gels with higher elastic modulus than gels that were made from either TG or HP-treated milk alone, indicating possible synergistic effects of HP and enzymatic cross-linking (Anema *et al.*, 2005b).

16.4.2 Impact of HP treatment on syneresis of yoghurt

Separation of the liquid phase from acid-milk gels, i.e. syneresis or whey separation, occurs when the protein network is unable to firmly hold water. In general, the more open the gel structure, the lower the water-holding capacity. Some studies have reported that yoghurt made from HP-treated milk showed increased resistance to syneresis and better water-retention capacity than yoghurt produced from high-heat-treated milk (Johnston *et al.*, 1993; Capellas and Needs, 2003). The improved water-binding characteristics of the protein gels from HP-treated milk may reflect HP-induced changes in protein association, and interactions, and subsequent microstructural alterations to the gel. Some studies have reported that HP treatment of milk increased milk-protein hydration (Gaucheron *et al.*, 1997; Huppertz *et al.*, 2004f). HP treatment of ewes' milk improved water retention in yoghurt (Ferragut *et al.*, 2000). Harte *et al.* (2003) reported that yoghurt made from HP-treated milk exhibited the highest whey retention properties when determined by centrifugation; however, when whey retention was evaluated as whey separated from yoghurt on a filter paper, yoghurts made from milk subjected to heat treatment and then HP-treated at >400 MPa exhibited higher whey retention. Penna *et al.* (2007b) observed no difference in water-holding capacity between yoghurts prepared from either heat-treated milk or milk subjected to heat treatment followed by HP treatment.

16.5 High-pressure treatment of milk for cheese manufacture

Cheese is probably one of the most complex dairy systems; its manufacture involves a combination of enzymes and microorganisms interacting with each other in a finely tuned manner to concentrate and metabolise proteins and fat from milk and then yield the desired aroma, flavour and texture of cheese. In addition, cheeses are ripened for periods from ten days (e.g. Camembert) to over two years (e.g. Parmesan) and continuously change during this time; only acid-coagulated fresh cheeses are produced for direct consumption.

Table 16.2 Inactivation of food-borne pathogens in cheese

Pathogen	HP conditions	Log reduction	References
L. monocytogenes NCTC 11994 and Scott A	500 MPa, 10 min, 5–20°C	5	Lopez-Pedemonte *et al.* (2007)
Salmonella enteritis	600 MPa,10 min, 21.5°C	7	Chen (2007)
E.coli K-12	500 MPa, 20 min, 20°C	6	O'Reilly *et al.* (2000)
E. coli 0157:H7	600 MPa,10 min, 21.5°C	5.5	Chen *et al.* (2007)
S. aureus ATCC 6538	600 MPa, 20 min, 20°C	4	O'Reilly *et al.* (2000)
S. aureus ATCC 12600	600 MPa, 10 min, 21°C	7	Guan *et al.* (2006)

The effects of HP processing on milk constituents, rates of enzymatic reactions, and, most importantly, inactivation of spoilage and pathogenic bacteria have made its application to cheese or its manufacture a subject of much research interest (O'Reilly *et al.*, 2001; Rastogi *et al.*, 2007) (Table 16.2). Studies have addressed the options of either HP-treating milk and making cheese therefrom, or making cheese from normal milk (raw or pasteurised) and subsequently HP-treating the cheese itself; only the former case will be considered here.

In the context of safety concerns associated with cheese, *Listeria monocytogenes*, sometimes associated with soft cheese, still has a high rate of mortality (Makino *et al.*, 2005; Gianfranchesi *et al.*, 2006); a key challenge to deal with is the ability of the pathogen to proliferate at refrigeration temperatures (Vasscur *et al.*, 1999). Other pathogenic bacteria in cheese, such as *Salmonella enteritis (*Keene, 1999; Ratnam, 1999), *E.coli O157:H7* (Espié *et al.*, 2006) and *Clostridium botulinum* (Townes, 1996) can cause gastoenteritis, kidney damage, or even death. It has been proposed that HP may offer a good alternative for pre-treatment of milk for cheese manufacture, since pathogens can be destroyed but flavours should be preserved (San Martin Gonzales *et al.*, 2006; Trujillo *et al.*, 2002b; Guamis 2000).

To date, five different cheese types have been made from milk treated at HP under different conditions (Table 16.3). Buffa *et al.* (2001a,b,c) produced cheese from goat milk treated at 500 MPa (15 min, 20°C) and Sandra *et al.* (2004) produced Queso-fresco from milk treated at 400 MPa (20 min, 20°C), and both found an overall reduction in the total bacterial microflora of the manufactured cheese. In later studies, when milk was spiked with two or four log cycles of *L.monocytogenes* and HP-treated at 500 MPa (10 min, room temperature), *L. monocytogenes* was not detectable in Camembert cheese produced from this milk and ripened for three weeks (Linton *et al.*, 2008).

Table 16.3 Cheese types manufactured from HP-treated milk

Cheese type	HP treatment conditions	References
Camembert	500 MPa, 10 min, 20°C	Linton *et al.* (2008)
Cheddar	483, 676 MPa, 5 min, 10–40°C	San Martin-Gonzales *et al.* (2007)
	300,400 MPa 32–90 min	Pandey *et al.* (2003)
	345, 586 MPa, 1-min cycles, 15 min	Drake *et al.* (1997)
Goat milk cheese	500 MPa, 15 min, 20°C	Buffa *et al.* (2005, 2004, 2003, 2001a,b,c, 2000)
		Trujillo *et al.* (2002a, 2000, 1999)
Queso fresco	400 MPa, 20 min, 20°C	Sandra *et al.* (2004)
Reduced-fat cheese	400 MPa, 15 min, 22°C	Molina *et al.* (2000)

As well as changes in microbiology, HP-induced changes in milk proteins can lead to changes in cheese manufacture and ripening. In the manufacturing process of most types of cheese, fermentation of lactose by the lactic acid bacteria plays a major role and determines the final pH. HP treatment at 300–400 MPa delays the pH changes during Cheddar cheese manufacture (Pandey *et al.*, 2003), possibly due to changed buffering capacity after HP treatment (Sala n *et al.*, 2005). In contrast, an increased rate of acidification by lactic acid bacteria after HP treatment at 600 MPa for 30 min was reported by Huppertz *et al.* (2004d) and was attributed to increased availability of proteinaceous material for bacterial growth.

The first key step in cheese manufacture is the coagulation of milk, which occurs in two steps; firstly, chymosin or another suitable enzyme cleaves the Phe_{105}-Met_{106} bond of κ-CN, which separates the stabilising layer of the casein micelle into two parts: *para*-κ-CN (f1–105), which remains associated with the casein micelle, and caseinomacropeptide (CMP); in the second step, the *para*-CN micelles aggregate into a gel. From a number of studies of the effect of HP-treatment of milk on rennet coagulation, it seems that HP treatment at ≤ 300 MPa reduces rennet coagulation time (RCT) but, above this pressure, RCT increases (Desobry-Banon *et al.*, 1994; López-Fandiño *et al.*, 1997; Drake *et al.*, 1997). It has been proposed that the binding of denatured β-lactoglobulin to the surface of the casein micelle cause delays in rennet coagulation, reduces syneresis, in turn, increases moisture content and yield of the final cheese (Zobrist *et al.*, 2005). Yield increases of up to 24% were found in model curd after applying a pressure of 600 MPa on the milk, which was largely due to increased moisture content and incorporation of whey proteins into the curd (Huppertz *et al.*, 2004e). In pilot trials, Cheddar cheese manufactured from milk HP-treated at 483 MPa showed an 11.54% increase in yield in comparison to cheese

Table 16.4 Effect of HP on yield and moisture increases of model curd and cheese

Cheese type	HP conditions	Increase in yield	Increase in moisture*	References
Cheddar	483 MPa, 5 min, 30°C	11.54%	11.6%	San Martin-Gonzales *et al.* (2007)
	586 MPa, 1-min cycles	11.3%	5.4%	Drake *et al.* (1997)
Model curd	400 MPa, 30 min, 20°C	10%	2.4%	Huppertz *et al.* (2004e)
	600 MPa, 30 min, 20°C	24%	4.5%	
Reduced-fat	400 MPa, 15 min, 22°C	15.6%	8.9%	Molina *et al.* (2000)

* Moisture increase in comparison to pasteurised milk cheese.

from pasteurised milk and 11.6% higher moisture content (San Martin-Gonzales *et al.*, 2007; Table 16.4). This increase in moisture content was also observed in goat milk cheese (Truijillo *et al.*, 2002a) and Queso-fresco cheese (Sandra *et al.*, 2004).

The texture of cheese depends mainly on its moisture content, and higher moisture results in a softer cheese texture. Cheddar cheese manufactured from HP-treated milk showed a pasty and weak texture (Drake *et al.*, 1997), and rheological analysis of Cheddar cheese manufactured from milk treated at 483 MPa at 10°C showed a more solid-like behaviour, which was related to the higher moisture and protein content of the cheeses compared to cheese manufactured from raw or pasteurised milk (San Martin-Gonzales *et al.*, 2007). Low-fat cheese made from milk HP-treated at 400 MPa (22°C, 15 min) and ripened for 60 days had much higher texture scores than cheese manufactured from pasteurised milk (Molina *et al.*, 2000).

During ripening of cheese, a number of biochemical events take place, namely, proteolysis, lipolysis and glycolysis, which are all responsible for the aroma, texture and flavour of the final cheese. When caprine milk cheese was manufactured from HP-treated milk (500 MPa, 15 min, 20°C), the resulting cheese had higher levels of β-lactoglobulin, faster proteolysis of α_{s1}-CN and β-CN, and higher levels of arginine during ripening, in comparison with pasteurised milk cheese that was produced with a similar level of moisture in non-fat substances. Nevertheless, indicators for secondary proteolysis, such as levels of water-soluble or trichloroacetic-acid-soluble nitrogen, showed no significant differences between control and HP-milk cheese, although levels of hydrophobic peptides increased in the HP-milk cheeses after 30 days of ripening (Buffa *et al.*, 2005). Studies of lipolysis showed higher levels of short-chain free fatty acids (C_4–C_8) and an overall development of total free fatty acids similar to that in raw milk cheese.

Sensory evaluation showed higher scores for the overall aroma and taste of cheese manufactured from HP-treated milk; nevertheless, differences were not significant (Buffa *et al.*, 2001c). Sensory analysis carried out on Cheddar cheese made from HP-treated milk (Drake *et al.*, 1997) showed a bitter, high acid, whey taint and weak flavour; however, control cheeses, made from pasteurised milk, also showed bitter flavours and high acid defects.

16.6 High-pressure treatment of milk for ice-cream manufacture

Compared to the widely examined potential applications of HP in the manufacture of cheese or yoghurt as discussed above, applications in other product categories have received far less attention. However, HP treatment of ice cream mix may have some interesting applications with respect to the properties of the final product, and has been subject to a patent application (Keenan *et al.*, 2000). The mechanism behind this application is that HP treatment of mixtures of micellar casein and sugar induces gelation (Keenan *et al.*, 2001), as a result of which desirable rheological properties of the mix can be obtained at far lower content of sugar, protein and/or fat, and possibly also in the absence of stabilisers (Huppertz *et al.*, unpublished data). Ice cream prepared from HP-treated mix has been shown to melt considerably slower than ice cream prepared from a comparable untreated mix (Keenan *et al.*, 2000).

16.7 Commercial prospects for high-pressure processing in the dairy industry

The discussions above clearly indicate that the effects of HP on milk and dairy products are complex, frequently unique, and, in some cases, demanding of closer study before use in commercial processes. The slowness of adoption of HP by the dairy sector may be interpreted as a sign of lack of potential commercial usefulness, but we argue that this is not the case, but rather that lack of commercialisation instead stems from two attributes of HP processing as it relates to dairy products, as follows:

(i) *Complexity*: The effects of pressure on the milk system are clearly more complex than those for most food products that have reached commercialisation stage, and arguably the need for a solid base of under-pinning scientific research has delayed the evaluation of possible applications. The number of papers appearing in the scientific literature regarding studies of milk is substantially higher than that for any other product type, confirming the enormous research effort in the dairy field and the number of scientific findings of sufficient novelty to warrant publication in peer-reviewed journals. It seems likely, however, that the exploratory research phase for HP and milk has reached a peak, and that the exploitation phase should, in theory, follow accordingly

(ii) *Scale/cost*: The scale of manufacture of many dairy products in developed countries is very large, and the ensuing cost of incorporating HP treatment into production lines is a major consideration and probable barrier to commercial application. The ratio of cost (investment, operation and maintenance) to benefits must be sufficient to make a strong business case for the use of HP equipment. In other words, the unique benefits of HP treatment, relative to cheaper available technologies, must be clear enough (either to manufacturers, or consumers, or both) to convince companies that HP is a realistic and viable processing option.

The issues relating to commercialisation of HP processing for dairy products have been considered in detail by Patel *et al.* (2008). There are a large number of international patents held by big dairy companies relating to applications of HP to dairy products, demonstrating both that companies have investigated this possibility, and that results have indicated outcomes worth legal protection (Table 16.5).

The recently proposed commercialisation of applications for fermented dairy products and colostrum (by Fonterra in New Zealand) and cheese for convenience food products (in Spain) have marked a tipping point where HP has broken out of the laboratory into the dairy processing plant for the first time, and it can only be postulated that future developments are likely at an accelerated pace.

In the case of colostrum, the key interest relates to the fact that the biofunctional properties of immunoglobulins are readily destroyed by the heat treatment which must inevitably be applied to ensure a commercially-useful shelf-life for distribution. The Fonterra process involves production of an acidified colostrum-based drink in which these goals are achieved by application of high-pressure treatment under conditions which are microbiologically efficacious but retain a very high percentage of the biological activity of the key immunoglobulins (Indyk *et al.*, 2008). In the case of yoghurt, to achieve a longer shelf-life of cultured milk products without compromising their quality, Fonterra has patented the use of HP treatment to extend the shelf-life of yoghurt for up to 90 days when stored at refrigeration temperature (Carroll *et al.*, 2004). This patent claims that the product obtained following HP treatment (~300–400 MPa at <20°C) is almost free from spoilage by yeast and mould during its shelf-life and still has at least 10^7 cfu/ml of beneficial live bacteria in it; however, careful selection of selective baro-resistant bacterial strains is of foremost importance.

Table 16.5 A selection of patents relating to high-pressure processing and dairy products

Year	Patent number	Title	Patent holder
1992	JP4262749	Preparation of milk protein	Snow Brand Milk Product Co Ltd
1992	EP0469857	Method for accelerating cheese ripening	Fuji Oil Co Ltd
1994	JP6046754	Method for controlling ageing of semisoft cheese	Yotsuba Nyugyo KK
1994	DE4226255	High-pressure treatment of fat mixes, e.g. milk products, ice cream, to alter crystal size, crystal shape, and/or crystalline arrangement in inner phase of emulsion	APV Gaulin GMBH
1998	WO9818350	Method for the preparation of a foodstuff	Unilever PLC
2002	WO0245528	Method for high-pressure preservation	Unilever PLC
2003	WO03007724	Yogurt manufacture using high hydrostatic pressure and thermal treatment	University of Washington
2003	WO03003845	Process for pasteurising cheese products, and cheese products formed thereby	SPA Egidio Galbani
2004	WO2004/045295	Dairy product and process	New Zealand Dairy Board
2004	WO2004/032655	Pressure treating foods to reduce spoilage	Fonterra Co-operative Group
2004	WO2004/091309	Enhancing clarity and/or stability properties of protein-containing liquids and gels	New Zealand Dairy Board
2006	WO2006/096074	High-pressure processing of bioactive compositions	Fonterra Co-operative Group
2006	UA14335U	Method of high-pressure milk processing	Tuhan Baranovsky Donetsk State University
2008	WO2006/096074	High-pressure processing of metal ion lactoferrin	Fonterra Co-operative Group

16.8 Conclusion

Subjecting a complex and physicochemically fragile system such as milk to the extreme stresses encountered at very high hydrostatic pressures, perhaps not surprisingly, has very significant implications for its constituents and properties at both microscopic and macroscopic levels. Significant research has addressed these effects over the last 15 years or so, and much is now known about pressure-induced changes in milk characteristics. The next phase, whereby companies exploit these changes in commercial product and process applications, is probably just in its infancy, and it remains to be seen for which niches high-pressure processing will ultimately become the technology of preference.

16.9 References

ABE S, KAWASHIMA A, MASUDA T, SUZUKI K and SUZUKI K (1999), 'Evaluation of heat deterioration of hard milk fat as a function of pressurization time by chemiluminescence analysis', *Food Sci Tech Res*, 5, 381–383.

ANEMA S G (2008), 'Effect of milk solids concentration on whey protein denaturation, particle size changes and solubilization of casein in high-pressure-treated skim milk', *Int Dairy J*, 18, 228–235.

ANEMA S G, LOWE E K and STOCKMAN G (2005a), 'Particle size changes and casein solubilisation in high-pressure-treated skim milk', *Food Hydrocolloid*, 19, 257–267.

ANEMA S G, LAUBER S, LEE S K, HENLE T and KLOSTERMEYER H (2005b), 'Rheological properties of acid gels prepared from pressure- and transglutaminase-treated skim milk', *Food Hydrocolloid*, 19, 879–887.

ANEMA S G, LEE S K, LOWE E K and KLOSTERMEYER H (2004), 'Rheological properties of acid gels prepared from heated pH-adjusted skim milk', *J Agric Food Chem*, 52, 337–343.

ARIAS M, LOPEZ-FANDINO R and OLANO A (2000), 'Influence of pH on the effects of high pressure on milk proteins', *Milchwissenschaft*, 55, 191–194.

ATKINS P W (1995), *Concepts in Physical Chemistry*, Oxford, Oxford University Press.

BELLOQUE J, LOPEZ-FANDINO R and SMITH G M (2000), 'A ^{1}H-NMR study on the effect of high pressures on β-lactoglobulin'. *J Agric Food Chem*, 48, 3906–3912.

BLACK E P, HUPPERTZ T, KELLY A L and FITZGERALD G F (2007), 'Baroprotection of vegetative bacteria by milk constituents: A study of *Listeria innocua*', *Int Dairy J*, 17, 104–110.

BUCHHEIM W and ABOU EL-NOUR A M (1992), 'Induction of milk fat crystallization in the emulsified state by high hydrostatic pressure', *Fat Sci Tech*, 94, 369–373.

BUFFA M N, TRUJILLO A J, PAVIA M and GUAMIS B (2001a), 'Changes in textural, microstructural, and colour characteristics during ripening of cheeses made from raw, pasteurized or high-pressure-treated goats' milk', *Int Dairy J*, 11–12, 927–934.

BUFFA M, TRUJILLO A J, ROYO C and GUAMIS B (2000), 'Changes in chemical and microbiological characteristics of goat cheese made from raw, pasteurised and high-pressure treated milk', *High Pressure Res*, 19, 27–32.

BUFFA M, GUAMIS B, ROYO C and TRUJILLO A J (2001b), 'Microbiological changes throughout ripening of goat cheese made from raw, pasteurized and high-pressure-treated milk', *Food Microbiol*, 18, 45–51.

BUFFA M, GUAMIS B, SALDO J and TRUJILLO A J (2004), 'Changes in organic acids during ripening of cheeses made from raw, pasteurized or high-pressure-treated goats' milk', *Lebensm-Wiss Technol*, 37, 247–253.

BUFFA M, GUAMIS B, PAVIA M and TRUJILLO A J (2001c), 'Lipolysis in cheese made from raw, pasteurized or high-pressure-treated goats' milk', *Int Dairy J*, 11, 175–179.

BUFFA M, GUAMIS B and TRUJILLO A J (2005), 'Specific effect of high-pressure treatment of milk on cheese proteolysis', *J Dairy Res*, 72, 385–392.

BUFFA M, GUAMIS B, SALDO J and TRUJILLO A J (2003), 'Changes in water binding during ripening of cheese made from raw, pasteursied or high-pressure-treated goat milk', *Lait*, 83, 89–96.

CAPELLAS M and NEEDS E (2003), 'Physical properties of yoghurt prepared from pressure-treated concentrated or fortified milks', *Milchwissenschaft*, 58 , 46–48.

CAPELLAS M, NORONHA R, MOR-MUH M and NEEDS E (2002), 'Effect of high pressure on the consumer liking and preference of yoghurt', *High Pressure Res*, 22, 701–704.

CARROLL T, CHEN P, HARNETT M and HARNETT J (2004), *Pressure treating food to reduce spoilage*, International Patent No. WO 2004/032655.

CHEN H (2007), 'Use of linear, Weibull, and log-logistic functions to model pressure inactivation of seven foodborne pathogens in milk', *Food Microbiol*, 24, 3, 197–204.

DANNENBERG F and KESSLER H G (1988), 'Effect of denaturation of β-lactoglobulin on texture properties of set-style nonfat yoghurt. 2. Firmness and flow properties', *Milchwissenschaft*, 43, 700–704.

DESOBRY-BANON S, RICHARD F and HARDY J (1994), 'Study of acid and rennet coagulation of high pressurized milk', *J Dairy Sci*, 77, 3267–3274.

DRAKE M A, HARRISON S L, ASPLUND M, BARBOSA-CANOVAS G and SWANSON B G (1997), 'High Pressure Treatment of Milk and Effects on Microbiological and Sensory Quality of Cheddar Cheese', *J Food Sci*, 64, 843–860.

DUMAY E, LAMBERT C, FUNTENBERGER S and CHEFTEL J C (1996), 'Effects of high pressure on the physico-chemical characteristics of dairy creams and model oil/water emulsions', *Lebensm-Wiss Technol*, 29, 606–625.

EBERHARD P, STRAHM W and EYER H (1999), 'High pressure treatment of whipped cream', *AgrarForschung*, 6, 352–354.

ESPIÉ E, VAILLANT V, MARIANI-KURKDJIAN P, GRIMONT F, MARTIN-SCHALLER R, DE VALK HAND VERNOZY-ROZAND C (2006), 'Short Report *Escherichia coli* O157 outbreak associated with fresh unpasteurised goats' cheese', *Epidemil Infect*, 134, 143–146.

FERRAGUT V, MARTINEZ V M, TRUJILLO A J and GUAMIS B (2000), 'Properties of yoghurts made from whole ewes' milk treated by high hydrostatic pressure', *Milchwissenschaft*, 55, 267–269.

FREDE E and BUCHHEIM W (2000), 'The influence of high pressure upon the phase behaviour of milk-fat and milk-fat fractions', *Milchwissenschaft*, 55, 683–686.

GARCIA-RISCO M R, CORTES E, CARRACOSA AV and LOPEZ-FANDINO R (1998), 'Microbiological and chemical changes in high-pressure-treated milk during refrigerated storage', *J Food Prot*, 61, 735–737.

GARCIA-RISCO M R, OLANO A, RAMOS M and LOPEZ-FANDINO R (2000), 'Micellar changes induced by high pressure. Influence in the proteolytic activity and organoleptic properties of milk', *J Dairy Sci*, 83, 2184–2189.

GARCIA-RISCO M R, RECIO I, MOLINA E and LOPEZ-FANDINO R (2003), 'Plasmin activity in pressurized milk', *J Dairy Sci*, 86, 728–734.

GAUCHERON F, FAMELART M H, MARIETTE F, RAULOT K, MICHEL F and LE GRAET Y (1997), 'Combined effects of temperature and high-pressure treatments on physicochemical characteristics of skim milk', *Food Chem*, 59, 439–447.

GEBHART R, DOSTER W and KULOZIK U (2005), 'Pressure-induced dissociation of casein micelles: Size distribution and effects of temperature', *Braz J Med Biol Res*, 38, 1209–1214.

GERVILLA R, FERRAGUT V and GUAMIS B (2001), 'High hydrostatic pressure effects on color and milk-fat globule of ewes' milk', *J Food Sci*, 66, 880–885.

GIANFRANCHESI M, D'OTTAVIO M C, GATTUSO A, POURSHABAN M, BERTOLETTI I, BIGNAZZI R, MANZONI P, MARCHETTI M and AURELI P (2006), 'Listeriosis Associated with Gorgonzola (Italien Blue-veined Cheese)', *Foodborne Pathog Dis*, 3, 190–195.

GUAMIS B (2000), 'High pressure applications in milk treatment for cheese-making and accelerated ripening', *High Pressure Res*, 19, 19–25.

GUAN D, CHEN H, TING E Y and HOOVER D G (2006), 'Inactivation of Staphylococcus aureus and Escherichia coli O157:H7 under isothermal-endpoint pressure conditions', *J Food Eng*, 77, 620–627.

HARTE F, AMONTE M, LUEDECKE L, SWANSON B G and BARBOSA-CANOVAS G V (2002), 'Yield stress and microstructure of set yoghurt made from high hydrostatic pressure-treated full fat milk', *J Food Sci*, 67, 2245–2250.

HARTE F, LUEDECKE L, SWANSON B and BARBOSA-CANOVAS G V (2003), 'Low-fat set yoghurt made from milk subjected to combinations of high hydrostatic pressure and thermal processing', *J Dairy Sci*, 86, 1074–1082.

HINRICHS J and RADEMACHER B (2004), 'Kinetics of combined thermal and pressure-induced whey protein denaturation in bovine skim milk', *Int Dairy J*, 14, 315–323.

HINRICHS J, RADEMACHER B and KESSLER H G (1996), 'Reaction kinetics of pressure-induced denaturation of whey proteins', *Milchwissenschaft*, 51, 504–509.

HOGAN E, KELLY A L and SUN D-W (2005), 'High pressure processing of foods: An overview', in Sun D-W (ed.) *Emerging Technologies for Food Processing*, London, Elsevier, 3–32.

HUBBARD C D, CASWELL D, LÜDEMANN H D and ARNOLD M (2002), 'Characterisation of pressure-treated skimmed milk powder dispersions: Application of NMR spectroscopy', *J Sci Food Agr*, 82, 1107–1114.

HUPPERTZ T and DE KRUIF C G (2006), 'Disruption and reassociation of casein micelles under high pressure: Influence of milk serum composition and casein micelle concentration', *J Agric Food Chem*, 54, 5903–5909.

HUPPERTZ T and DE KRUIF C G (2007a), 'High pressure-induced solubilisation of micellar calcium phosphate from cross-linked casein micelles', *Colloid Surface A*, 295, 1–3, 264–268.

HUPPERTZ T and DE KRUIF C G (2007b), 'Disruption and reassociation of casein micelles during high pressure treatment: Influence of whey proteins', *J Dairy Res*, 74, 194–197.

HUPPERTZ T and SMIDDY M A (2008), 'Behaviour of partially cross linked casein micelles under high pressure', *Int J Dairy Technol*, 61, 51–55.

HUPPERTZ T, FOX P F and KELLY A L (2003), 'High pressure-induced changes in the creaming properties of bovine milk', *Innov Food Sci Emerg Technol*, 4, 349–359.

HUPPERTZ T, FOX P F and KELLY A L (2004a), 'High pressure treatment of bovine milk: Effects on casein micelles and whey proteins', *J Dairy Res*, 71, 97–106.

HUPPERTZ T, FOX P F and KELLY A L (2004b), 'Properties of casein micelles in high pressure-treated bovine milk', *Food Chem*, 87, 103–110.

HUPPERTZ T, FOX P F and KELLY A L (2004c), 'High pressure-induced denaturation of α-lactalbumin and β-lactoglobulin in bovine milk and whey: A possible mechanism', *J Dairy Res*, 71, 489–495.

HUPPERTZ T, FOX P F, KELLY A L (2004d), 'Influence of high pressure treatment on the acidification of bovine milk by lactic acid bacteria', *Milchwissneschaft*, 59 (5/6), 246–249.

HUPPERTZ T, FOX P F and KELLY A L (2004e), 'Effects of high pressure treatment on the yield of cheese curd from bovine milk', *Innov Food Sci Emerg Technol*, 5, 1–8.

HUPPERTZ T, FOX P F and KELLY A L (2004f), 'Dissociation of caseins in high pressure-treated bovine milk', *Int Dairy J*, 14, 675–680.

HUPPERTZ T, FOX P F, DE KRUIF K G and KELLY A L (2006a), 'High pressure-induced changes in bovine milk proteins: A review', *Biochimica et Biophysica Acta – Proteins and Proteomics*, 1764, 593–598.

HUPPERTZ T, KELLY A L and DE KRUIF C G (2006b), 'Disruption and reassociation of casein micelles under high pressure', *J Dairy Res*, 73, 294–298.

HUPPERTZ T, SMIDDY M A, UPADHYAY V K and KELLY A L (2006c), 'High pressure-induced changes in bovine milk: A review', *Int J Dairy Technol*, 59, 58–66.

HVIDT A A (1975), 'Discussion of pressure–volume effects in aqueous protein solutions', *J Theor Biol*, 50, 245–252.

IKEUCHI Y, NAKAGAWA K, ENDO T, SUZUKI A, HAYASHI T and ITO T (2001), 'Pressure-induced denaturation of monomer β-lactoglobulin is partially reversible: Comparison of monomer form (highly acidic pH) with dimer form (neutral pH)', *J Agric Food Chem*, 49, 4052–4059.

INDYK H E, WILLIAMS J W and PATEL H A (2008), 'Analysis of denaturation of bovine IgG by heat and high pressure using an optical biosensor', *Int Dairy J*, 18, 359–366.

JOHNSTON D E, AUSTIN B A and MURPHY R J (1993), 'Properties of acid-set gels prepared from high pressure treated skim milk', *Milchwissenschaft*, 48, 206–209.

KANNO C, UCHIMURA T, HAGIWARA T, AMETANI M and AZUMA N (1998), 'Effect of hydrostatic pressure on the physicochemical properties of bovine milk fat globules and the milk fat globule membrane'. In Isaacs N S, *High Pressure Food Science, Bioscience and Chemistry*, The Royal Society of Chemistry, Cambridge, 182–192.

KEENAN R D, WIX L ANDYOUNG D (2000), *Method for the preparation of a foodstuff*, US Patent 6, 156, 367.

KEENAN R D, YOUNG D J, TIER C M, JONES A D and UNDERDOWN J (2001), 'Mechanism of pressure-induced gelation of milk'. *J Agric Food Chem*, 49, 3394–3402.

KEENE W E (1999), 'Lessons from investigations of foodborne disease outbreaks', *J Am Med Assoc*, 281, 1845–1847.

KNUDSEN J C, KARLSSON A O, IPSEN R and SKIBSTED L H (2006), 'Rheology of stirred acidified skim milk gels with different particle interactions', *Colloid Surface A*, 274, 56–61.

KROMKAMP J, MOREIRA R M, LANGEVELD L P M and VAN MIL P J J M (1996), Microorganisms in milk and yoghurt: Selective inactivation by high hydrostatic pressure. In *Heat Treatments and Alternative Methods*, pp. 266–271. Special Issue No 9602. Brussels: International Dairy Federation.

KUWATA K, LI H, YAMADA H, BATT C A, GOTO Y and AKASAKA K (2001), 'High pressure NMR reveals a variety of flucutating confomers in β-lactoglobulin', *J Mol Biol*, 305, 1073–1083.

LINTON M, MACKLE A B, UPADHYAY V K, KELLY A L and PATTERSON M F (2008), 'The fate of Listeria monocytogenes during the manufacture of Camembert-type cheese: A comparison between raw milk and milk treated with high hydrostatic pressure', *Innov Food Sci Emerg Tech*, 9, 4, 423–428, doi:10.1016/j.ifset.2008.01.001.

LÓPEZ-FANDIÑO R, CARRASCOSA A V and OLANO A (1996), 'The effects of high pressure on whey protein denaturation and cheese-making properties of raw milk', *J Dairy Sci*, 79, 929–936.

LÓPEZ-FANDIÑO R, CARRASCOSA A V and OLANO A (1997), 'Rennet coagulation of milk subjected to high pressures', *J Agric Food Chem*, 45, 3233–3237.

LÓPEZ-FANDIÑO R, DE LA FUENTE M A, RAMOS M and OLANO A (1998), 'Distribution of minerals and proteins between the soluble and colloidal phases of pressurized milks from different species', *J Dairy Res*, 65, 69–78.

LOPEZ-PEDEMONTE T, ROIG-SAGUES A, DE LAMO S, HERNANDEZ-HERRERO M and GUAMIS B (2007), 'Reduction of counts of *Listeria monocytogenes* in cheese by means of high hydrostatic pressure', *Food Microbiology*, 24, 1, 59–66.

LUCEY J A, MUNRO P A and SINGH H (1998), 'Rheological properties and microstructure of acid milk gels as affected by fat content and heat treatment', *J Food Sci*, 63, 660–664.

LÓPEZ-FANDIÑO R and OLANO A (1998), Effects of high pressures combined with moderate temperatures on the rennet coagulation properties of milk. *International Dairy Journal*, 8, 623–627.

LUCEY J A, TEO C T, MUNRO P A and SINGH H (1997), 'Rheological properties at small (dynamic) and large (yield) deformations of acid gels made from heated milk', *J Dairy Res*, 64, 591–600.

MAKINO S-I, KAWAMOTO K, TAKESHI K, OKADA Y, YAMASAKI M, YAMAMOTO M and IGIMI S (2005), 'An outbreak of food-borne listeriosis due to cheese in Japen, during 2001', *Int J Food Microbiol*, 104, 189–196.

MOLINA E, DOLORES ALVAREZ M, RAMOS M, OLANO A and LÓPEZ-FANDIÑO R (2000), 'Use of high-pressure-treated milk for the production of reduced-fat cheese', *Int Dairy J*, 10, 467–475.

MOLLER, R E, STAPELFELDT H and SKIBSTED, L H (1998), 'Thiol reactivity in pressure-unfolded β-lactoglobulin. Antioxidative properties and thermal refolding, *J Agric Food Chem*, 46, 425–430.

MULVIHILL D M and GRUFFERTY M B (1995), 'Effect of thermal processing on the coagulability of milk by acid'. In *Heat-induced Changes in Milk*, Brussels: International Dairy Federation – sp. Issue No. 9501 2nd edition, 188–205.

NEEDS E C, CAPELLAS M, PATRICIA BLAND A, MANOJ P, MACDOUGAL D and PAUL G (2000), 'Comparison of heat and pressure treatments of skim milk, fortified with whey protein concentrate, for set yoghurt preparation: Effects on milk proteins and gel structure', *J of Dairy Res*, 67, 329–348.

NEEDS E C, STENNING R A, GILL A L, FERRAGUT V and RICH G T (2000), 'High-pressure treatment of milk: Effects on casein micelle structure and on enzymic coagulation', *J Dairy Res*, 67, 31–42.

O'REILLY C E, O'CONNOR P M, KELLY A L, BERESFORD T P and MURPHY P M (2000), 'Use of hydrostatic pressure for inactivation of microbial contaminants in cheese', *Appl Environ Microbiol*, 66, 11, 4890–4896.

O'REILLY C E, KELLY A L, MURPHY P M and BERESFORD T P (2001), 'High pressure treatment: Applications in cheese manufacture and ripening', *Trends Food Sci Techn*, 12, 51–59.

ORLIEN V, KNUDSEN J C, COLON M and SKIBSTED L H (2006), 'Dynamics of casein micelles in skim milk during and after high pressure treatment', *Food Chem*, 98, 513–521.

PANDEY P K, RAMASWAMY H S and ST-GELAIS D (2003), 'Evaluation of pH change kinetics during various stages of Cheddar cheese-making from raw, pasteurized, micro-filtered and high-pressure-treated milk', *Lebensmittel-Wissenschaft und-Technologie*, 36, 5, 497–506.

PATEL H, CARROLL T and KELLY A L (2008), 'New dairy processing technologies'. In *Dairy Processing Technology and Quality Assurance*, Chandan, R.C., Kilara, A. and Shah, N.P. (eds.) London: Elsevier, pp. 465–482.

PAYENS T A J and HEREMANS K (1969), 'Effect of pressure on the temperature-dependent association of β-casein', *Biopolymers*, 8, 335–345.

PENNA A L B, GURRAM S and BARBOSA-CANOVAS G V (2006), 'Effect of high hydrostatic pressure processing on rheological and textural properties of probiotic low-fat yoghurt fermented by different starter cultures', *J Food Proc Eng*, 29, 447–461.

PENNA A L B, GURRAM S and BARBOSA-CANOVAS G V (2007a), 'Effect of milk treatment on acidification, physicochemical characteristics, and probiotic cell counts in low fat yoghurt', *Milchwissenschaft*, 62, 48–52.

PENNA A L B, GURRAM S and BARBOSA-CANOVAS G V (2007b), 'High hydrostatic pressure processing on microstructure of probiotic low-fat yoghurt', *Food Res Int*, 40, 510–519.

RASTOGI N K, RAGHAVARAO K S M S, BALASUBRAMANIAM V M, NIRANJAN K and KNORR D (2007), 'Opportunities and Challenges in High Pressure Processing of Foods', *Crit Rev Food Sci Nutr*, 47, 69–112.

RATNAM, S (1999), 'Salmonella Enteritis outbreak due to contaminated cheese-Newfoundland', *Canada Communicable Disease Report*, 25–03.

REGNAULT S, DUMAY E and CHEFTEL J C (2006), 'Pressurisation of raw skim milk and of a dispersion of phosphocaseinate at 9°C or 20°C: Effects on the distribution of minerals and proteins between colloidal and soluble phases', *J Dairy Res*, 73, 91–100.

REGNAULT S, THIEBAUD M, DUMAY E and CHEFTEL J C (2004), 'Pressurisation of raw skim milk and of a dispersion of phosphocaseinate at 9°C or 20°C: Effects on casein micelle size distribution', *Int Dairy J*, 14, 55–68.

SALAUN F, MIETTON B and GAUCHERON F (2005), 'Buffering capacity of dairy products', *Int Dairy J*, 15, 95–109.

SAN MARTIN-GONZALEZ M F, RODRIGUEZ J J, GURRAM S, CLARK S, SWANSON B G and BARBOSA-CANOVAS G V (2007), 'Yield, composition and rheological characteristics of cheddar cheese made with high pressure processed milk', *LWT – Food Sci Technol*, 40, 697–705.

SAN MARTIN-GONZALEZ M F, WELTI-CHANES J and BARBOSA-CANOVAS G V (2006), 'Cheese Manufacture Assisted by High Pressure', *Food Rev Int*, 22, 275–289.

SANDRA S, STANFORD M A and MEUNIER GODDIK L (2004), 'The use of high-pressure processing in the production of Queso fresco cheese', *J Food Sci*, 69, 153–158.

SCHRADER K and BUCHHEIM W (1998), 'High pressure effects on the colloidal calcium phosphate and the structural integrity of micellar casein in milk. II. Kinetics of the casein micelle disintegration and protein interactions in milk', *Kieler Milchw Forsch*, 50, 79–88.

SCOLLARD P G, BERESFORD T P, NEEDS E C, MURPHY P M and KELLY A L (2000), 'Plasmin activity, β-lactoglobulin denaturation and proteolysis in high pressure treated milk', *Int Dairy J*, 10, 835–841.

SODINI I, REMEUF F, HADDAD S and CORRIEU G (2004), 'The relative effect of milk base, starter, and process on yoghurt texture: A review', *Crit Rev Food Sci Nutr*, 44, 113–137.

STAPELFELDT H, OLSEN C E and SKIBSTED L H (1999), 'Spectrofluorometric characterization of β-lactoglobulin B covalently labeled with 2-(4'-maleimidylanilino) naphthalene-6-sulfonate', *J Agric Food Chem*, 47, 3986–3990.

TAMIME A Y and ROBINSON R K (1999), *Yoghurt Science and Technology*- 2nd edition, Cambridge, UK, Woodhead Publishing Ltd.

TANAKA N, TSURUI Y, KOBAYASHI I and KUNUGI S (1996), 'Modification of the single unpaired sulphydryl group of β-lactoglobulin under high pressure and the role of intermolecular S–S exchange in the pressure denaturation', *Int J Biol Macromol*, 19, 63–68.

TOWNES J M (1996), 'An outbreak of Type A botulismus associated with a commercial cheese sauce', *Ann Intern Med*, 125, 558–563.

TRUJILLO A J, CAPELLAS M, SALDO J, GERVILLA R and GUAMIS B (2002b), 'Applications of high-hydrostatic pressure on milk and dairy products: A review', *Innov Food Sci Emerg Tech*, 3, 295–307.

TRUJILLO A J, GUAMIS B and CARRETERO C (2000), 'A procedure for the manufacture of goat milk cheese with controlled-microflora by means of high hydrostatic pressure', *Food Chem*, 69, 73–79.

TRUJILLO A J, ROYO C, FERRAGUT V and GUAMIS B (1999), 'Ripening profiles of goat cheese produced from milk treated with high pressure', *J Food Sci*, 64, 833–837.

TRUJILLO A J, BUFFA M, CASALS I, FERNANDEZ P and GUAMIS B (2002a), 'Proteolysis in goat cheese made from raw, pasteurized or pressure-treated milk', *Innov Food Sci Emerg Technol*, 3, 309–319.

TRUJILLO A J, CAPELLAS M, SALDO J, GERVILLA R and GUAMIS B (2002), 'Applications of high-hydrostatic pressure on milk and dairy products: A review', *Innov Food Sci Emerg Technol*, 3, 295–307.

TRUJILLO A J, CAPELLAS M, BUFFA M, ROYO C, GERVILLA R, FELIPE X, SENDRA E, SALDO J, FERRAGUT V and GUAMIS B (2000), 'Application of high pressure treatment for cheese production', *Food Research Int*, 33, 311–316.

VASSEUR C, BAVEREL L, HÉBRAUD M, LABADIE J (1999), 'Effect of osmotic, alkaline, acid or thermal stresses on the growth and inhibition of *Listeria monocytogens*', *J Appl Microbiol*, 86, 469–476.

WALSH-O'GRADY C D, O'KENNEDY B T, FITZGERLAD R J and LANE C N (2001), 'A rheological study of acid-set "simulated yoghurt milk" gels prepared from heat-or pressure-treated milk proteins', *Lait*, 81, 637–650.

WALSTRA P, WOUTERS J T M and GEURTS T J (2006), *Dairy Science and Technology*, Boca Raton: CRC Press.

YE A, ANEMA S G and SINGH H (2004), 'High-pressure-induced interactions between milk fat globule membrane proteins and skim milk proteins in whole milk', *J Dairy Sci*, 87, 4013–4022.

ZOBRIST M R, HUPPERTZ T, UNIACKE T, FOX P F and KELLY A L (2005), 'High pressure-induced changes in rennet-coagulation properties of bovine milk', *Int Dairy J*, 15, 655–662.

17

Impact of dairy ingredients on the flavor profiles of foods

M. A. Drake, J. Wright, M. Whitson and M. Lloyd, North Carolina State University, USA

Abstract: Dairy ingredients encompass a wide range of highly functional and nutritious ingredients. Flavor of these products can have a significant impact on the quality and shelf-life of ingredient applications. The flavor variability and shelf stability of these products and the influence of processing steps and storage are the topics of this chapter.

Key words: flavor, dairy ingredients, whole milk powder, skim milk powder, whey protein.

17.1 Defining flavor of dried ingredients

Dairy foods have traditionally enjoyed a positive flavor image from consumers. A consistent, desirable and long-lasting (shelf-life) flavor is a key aspect of quality (Drake, 2004). While basic capitalism was likely the initial source of flavor quality criteria (the best product demanded the best price), the dairy industry recognized the role of flavor quality and flavor consistency nearly 100 years ago, when commodity grading and judging evolved (Singh *et al.*, 2003; Drake *et al.*, 2009a). The concept of product grades and judging were to establish basic uniform criteria for flavor and, where applicable, texture of specific dairy commodities. The goal was to ensure a minimal level of product consistency. Lists of common product defects were developed and point or quality deductions for the presence of these defects established. Many of these grades are still in use today (www.usda.ams.gov) and dairy products judging remains a popular activity for college students (Bodyfelt *et al.*, 2008). Grades and dairy products judging represent the first documented attempt to define flavor of dairy products. The pros and cons of these approaches have been compared elsewhere (Singh *et al.*, 2003; Delahunty and Drake, 2004; Drake, 2007; Bodyfelt *et al.*, 2008).

Grades and product judging criteria remain a useful tool to establish minimal quality criteria and to train students, but were not designed to specifically document flavor (or texture) properties. Instead, modern sensory analysis techniques have evolved to address these needs (Drake and Civille, 2003; Drake, 2007, 2009). A defined sensory language or lexicon is simply a language to document specific sensory attributes of a product (Drake and Civille, 2003). The development of defined sensory languages for dairy commodities and foods has dramatically changed and improved the application of sensory analysis as a scientific tool to document product flavors. These languages have definitions and food and/or chemical references for each attribute. As such, the language can be widely applied to train sensory panels at multiple locations (Drake *et al.*, 2002). A trained sensory panel requires a significant and long-term time and financial investment (Drake, 2007); however, the potential benefits are numerous. These languages can also serve as platforms to relate instrumental and consumer perception to defined trained panel attributes from the language.

An established language for whey and dried dairy ingredients is listed in Table 17.1. The dried dairy ingredient language has been adapted for use with liquid whey, dried whey proteins, caseins, fluid milk, skim and whole milk powders and dairy ingredient applications (Carunchia Whestine *et al.*, 2003, 2005; Karagul Yuceer *et al.*, 2003a; Beucler *et al.*, 2005; Russell *et al.*, 2006; Wright *et al.*, 2006; Drake, 2006; Drake *et al.*, 2006; Croissant *et al.*, 2007; Childs *et al.*, 2007; Lloyd *et al.*, 2009a,b). The impact of processing parameters, starter culture rotation, pasture feeding, between-facility variability, and storage have been specifically documented and, in many cases, related to volatile component measurements or consumer perception. Such information is invaluable to effectively and strategically market dairy ingredients in today's competitive and global environment.

17.2 Sources of flavors in dried dairy ingredients

In order to provide dried ingredients with a consistent minimal flavor, we must first identify and characterize flavors that are present in these products. The development of the dried dairy ingredient sensory language (Table 17.1) provides the tool to accomplish this goal. The next step is to identify sources of these flavors. For all dried dairy ingredients, the raw product stream, processing, and storage influence flavor, flavor variability and storage stability. In the case of whey proteins, the influence of the cheesemake procedure (e.g. starter culture) also has a direct influence on final product flavor. All of these factors will influence the flavor of the final product, which also will subsequently influence the impact it has on a final ingredient application.

Before the milk arrives at the processing plant, many variables can affect its flavor quality, including the diet of the herd, season of the year, and

Table 17.1 Sensory language for whey and dried whey ingredients

Term	Definition	References
Overall aroma intensity	The overall orthonasal aroma impact of the rehydrated sample	
Flavors (evaluated in the mouth)		
Sweet aromatic	The sweet aromatics associated with dairy products (diacetyl is one example)	Diacetyl (2,3-butanedione), mild Cheddar or Colby-jack shreds
Sour aromatic	Sour aromatic associated with fermented dairy products	Aroma of plain yogurt and sour cream (not diacetyl)
Cardboard	Aromatics associated with wet cardboard	Pentanal, cardboard in water
Brothy	Aromatics associated with vegetable stock and boiled potatoes	Methional, broth from canned potatoes
Fatty	Stale aromatics associated with old frier oil	2, 4-Decadienal
Serumy/metallic	Aromatics associated with raw steak and hamburger	Raw steak juice
Soapy	Aromatics associated with medium chain fatty acids and soaps	Decanoic acid, unscented plain bar soap in water
Cabbage	Sulfurous aromatic associated with cooked cruciferous vegetables	Dimethyl trisulfide, boiled fresh cut cabbage
Cereal/oatmeal	A sweet aromatic reminiscent of cooked cereal grains such as oatmeal	Freshly cooked oat flakes
Animal	Aromatic associated with animals and animal hair	Knox unflavored gelatin, dissolve one bag of gelatin (28 g) in two cups of distilled water
Cucumber/green	Sweet aromatic associated with freshly cut cucumber	2, 6-Nonadienal, 20 ppm on filter paper in sniff jar
Raisin	Sweet aromatic reminiscent of cooked raisins	Boil 3 tsp raisins in 1 cup water, use water as aroma and flavor reference
Musty/earthy	Aromatic reminiscent of damp basements and/or damp soil	Potting soil
Barny/phenolic	Aromatics reminiscent of ruminant sweat and urine	p-Cresol
Grassy	Aromatics reminiscent of cut grasses and hay	Hexanal
Bitter	Fundamental taste sensation elicited by caffeine, quinine	Caffeine (0.08% in water)
Sweet taste	Fundamental taste sensation elicited by sugars	Sucrose (3% in water)
Astringent	Drying sensation on the tongue and oral cavity surfaces	Black tea, alum

Adapted from Drake *et al.*, 2003; Russell *et al.*, 2006; Wright *et al.*, 2006, 2009.

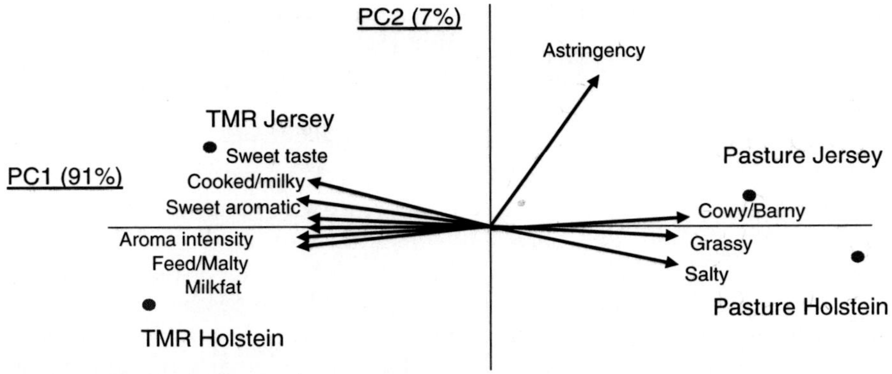

Fig. 17.1 Principal component biplot of sensory profiles of pasteurized whole milk from cows fed a total mixed ration (TMR) or pasture-based (pasture) diet.

microbiological quality. The diet that a cow eats will affect the flavor of the milk it produces (Al-Mabruk *et al.*, 2004; Bendall, 2001; Coulon and Priolo, 2002; Croissant *et al.*, 2007; Fearon *et al.*, 1998; Mounchili *et al.*, 2004; Palmquist *et al.*, 1993; Randby *et al.*, 1999; Toso *et al.*, 2002; Urbach, 1990). Croissant *et al.* (2007) recently demonstrated that pasteurized fluid milk from cows fed on pasture was distinct in flavor profile from cows fed a total mixed ration (Fig. 17.1). Several researchers have found that the season of the year affects whole milk powder (WMP) quality and stability (Baldwin and Ackland, 1991; Baldwin *et al.*, 1991; Biolatto *et al.*, 2007; Negri *et al.*, 2004; Steen, 1977; Van Mil and Jans, 1991). Steen (1977) collected WMP from two manufacturing facilities in Denmark over a year period to observe the effect of season on a number of quality attributes, including physical properties, flavor, and storage stability up to 24 months. The thiobarbituric acid (TBA) value, a measure of lipid oxidation, was higher in the summer, which was consistent with Biolatto and others (2007) who found that straight chain aldehydes were higher in WMP produced during the spring and summer compared to fall and winter. They explained that the variation was likely due to the feed of the cows, consistent with Toso and others (2002), who reported higher levels of straight chain aldehydes in raw milk from cows fed grass silage, compared to hay or maize silage. Van Mil and Jans (1991) also found that WMP produced in the spring was less stable and decreased in flavor score more rapidly than WMP produced in other seasons. The microbiological profile of raw milk can also affect the quality of WMP made from it (Celestino *et al.*, 1997; de Oliveira *et al.*, 2000). Ipsen and Hansen (1988) found that high coliform counts in raw milk were detrimental for the flavor of WMP stored for a year. Grassy flavors have been documented in fluid milk, butter, and whole milk powder from cows that were pasture-fed (Carunchia Whetstine and Drake, 2007; Croissant *et al.*, 2007; Krause *et al.*, 2007; Lloyd *et al.*, 2009b), and pasture-feeding has been attributed to international differences in Cheddar cheese flavor (Drake *et al.*, 2005). Such differences would also be expected in skim milk powders and dried whey ingredients.

Protein content will also influence flavor profile. Drake *et al.* (2009b) recently documented that flavor profile of milk protein concentrates changed as protein content increased. Lower protein content products displayed sweet aromatic and cooked flavors, similar to rehydrated skim milk powder, while higher protein content products (>70%) displayed tortilla and animal flavors (Drake *et al.*, 2009b). Carunchia Whestine *et al.* (2005) and Russell *et al.* (2006) noted similar flavor and volatile compound differences between WPC80 (80% protein) and WPI (at least 90% protein). Such differences in flavor reflect the different sources for flavor in those products. Products with less protein have more available lactose to generate volatile compounds responsible for sweet, milky flavors (caramelization, Maillard products) while proteins and amino acids would be the major reservoir for volatile compound generation in higher protein content products.

Volatile compound sources of flavor for skim and whole milk powders include an array of heat-generated compounds (caramelization, Maillard products, lactones) as well as lipid oxidation products and free fatty acids (Karagul Yuceer *et al.*, 2001, 2002, 2003a; Carunchia Whetstine and Drake, 2007). Both heat-generated and lipid oxidation products increase in concentration and variety as the fluid product progresses through concentration and spray drying (Drake *et al.*, 2006). Heat treatment during processing, which designates whether the finished spray dried powder is low, medium, or high heat based on undenatured whey protein nitrogen (WPN), (≥6, 1.51–5.99, and ≤1.5 mg undenatured whey protein/g powder, respectively, USDEC, 2005) will also influence flavor profile and volatile compounds. Medium and high heat milk powders are characterized by more intense cooked/sulfurous flavors and lower intensities of sweet aromatic flavors. High and medium heat skim milk powders (SMP) may even have a distinct burnt or scorched-like flavor. Volatile sulfur compounds are more prevalent in these milk powders (Drake *et al.*, 2006). The spray drying process itself contributes flavors. Drake *et al.* (2006) documented flavor development in medium heat SMP across processing and across a 40 h production run. Higher concentrations and a wider variety of Maillard and caramelization compounds were formed as the product progressed through processing, and these compounds increased with production time. That is, spray dried product collected after continuous processing for 40 h had more compounds and higher concentrations of compounds than spray dried product collected from the beginning of a processing run. However, sensory changes were not documented, and the role that these volatile compound differences played in physical properties or shelf stability was not investigated.

Table 17.2 Published values for shelf-life of whole milk powder

Shelf-life	Conditions	Source
3–4 months	21°C and 50% relative humidity	Carunchia Whetstine and Drake, 2007; Lloyd *et al.*, 2009a,b
6–9 months	<27°C and <65% relative humidity	USDEC, 2005
9–12 months	35°C in air, moderate heat treat (>95°C, 20 s)	Baldwin *et al.*, 1991
3 years	Optimal conditions: (i) moderate heat treatment (>95°C, 2 min) (ii) moisture content at least 2.8% (iii) storage temperature of 20°C or less (iv) air and moisture proof reduced O_2 packaging	Van Mil and Jans, 1991

Some factors that influence shelf-life of WMP include initial milk quality, processing variables, moisture content, packaging, oxygen exposure, exposure to light, and storage temperature (Hough *et al.*, 2002; Lloyd *et al.*, 2004; Nielsen *et al.*, 1997; Ruckold *et al.*, 2003). The shelf-life of WMP is closely tied to the development of off-flavors, especially those attributed to lipid oxidation. The shelf-life of US WMP is reported to be six to nine months when stored at temperatures less than 27°C and relative humidity less than 65% (USDEC, 2005), while international WMP claims a shelf-life of one year (S. Hess, Hershey Foods, personal communication, 2008). Values in the literature for shelf-life of WMP range from three to four months (Carunchia Whetstine and Drake, 2007; Lloyd *et al.*, 2009) to three years (Van Mil and Jans, 1991) (Table 17.2). Baldwin *et al.* (1991) detected oxidized flavor in WMP stored in air at 35°C after nine to twelve months. Van Mil and Jans (1991) reported that WMP processed and stored under optimal conditions could be kept for three years without quality loss. Optimal conditions included: (i) moderate heat treatment, (ii) moisture content at least 2.8%, (iii) storage temperature of 20°C or less, and (iv) air and moisture proof oxygen reduced packaging. Since WMP has a high fat content, it is very susceptible to lipid oxidation. Common storage-related off-flavors include grassy and painty/fatty flavors. Grassy flavor may also be associated with milk from pasture-fed animals and, in that case, is not a storage-related flavor. However, the flavor when detected in milk from total mixed ration (TMR)-fed animals, generally signifies the initiation of loss of product quality (Carunchia Whetstine and Drake, 2007). Painty/fatty flavors are definitive indicators of lipid oxidation and a sign that the product is beyond shelf-life (Carunchia Whetstine and Drake, 2007; Lloyd *et al.*, 2009a,b). The shelf-life of WMP can be extended by oxygen-free packaging, cool and dry storage conditions, or the use of antioxidants (Lloyd *et al.*, 2009a).

Numerous storage studies have shown that WMP and cream powder flavor is better preserved under low oxygen conditions (Andersson and Lingnert, 1998; Hall and Andersson, 1985; Steen, 1977; Van Mil and Jans, 1991). Van Mil and Jans (1991) concluded that WMP should be made with a moderate heat treatment and a moisture content of 2.8–5%, then stored at less than 20°C with lowered oxygen levels. Some research has shown slight benefits from added antioxidants in WMP (Abbot and Waite, 1962, 1965; Hall and Andersson, 1985), but this is not a common commercial practice (Dan Meyer, American Dairy Products Institute, personal communication, 2008). Lloyd *et al.* (2009a) recently demonstrated that application of nitrogen flushing prevented development of painty off flavors in WMP for up to one year at 21°C storage compared to three months for WMP packaged with no nitrogen flushing.

As with WMP, the same parameters influence final flavor profile and shelf stability of skim milk powder (SMP), but in a different manner since the composition of SMP is distinct from WMP; namely, a lot less fat is present. SMP contains not more than 1.25% (Grade A and USDA/ADPI standards) or 1.5% fat (Codex standard 1999) compared to 26 to 28% fat for WMP. As such, the shelf-life for SMP under ambient temperature storage is considerably longer than shelf-life of WMP. Similarly to WMP, a wide projected range of shelf-life has been reported for SMP with estimates as high as five years. Carunchia Whetstine and Drake (2007) documented changes in the sensory profile of rehydrated skim milk powders within twelve months storage at 21°C, 50% relative humidity, but concluded that the quality of ingredient applications would likely not be impacted until a later storage date. Changes in flavor with storage were characterized first by a decrease in fresh flavor intensities (cooked, sweet aromatic) and then the evolution of storage flavors (cardboard, fatty) which were associated with lipid oxidation. They concluded, in conjunction with previous studies (Karagul-Yuceer *et al.*, 2001, 2002) that lipid oxidation was the primary source of loss of fresh flavor in SMP stored under ambient storage conditions. These results were confirmed in a subsequent larger study with SMP manufactured at multiple facilities, collected within one month of manufacture, and then stored under identical conditions for 24 months (Drake *et al.*, 2006). However, variability in flavor stability was observed among SMP from different facilities and was not related to moisture content or transition mineral content. Differences were attributed to differences specific to processing facilities (e.g. different equipment, minor differences in process steps). A conservative shelf-life for low heat SMP stored at 21°C was 18 to 24 months.

Flavor stability and shelf-life of SMP were distinct at different storage temperatures. Figure 17.2 demonstrates the differences in flavor develop-

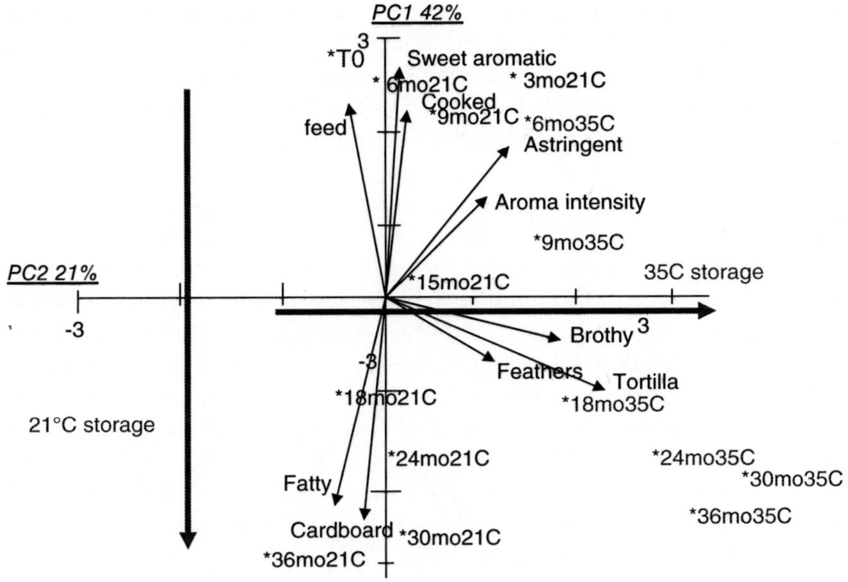

Fig. 17.2 Principal component biplot of sensory profiles of rehydrated skim milk powder stored for various times at 21 or 35°C.

ment in skim milk powder with different storage temperatures. In contrast to cardboard and fatty flavors, which were indicative of loss of freshness at 21°C storage, products stored at higher temperature developed distinct off flavors of burnt feathers, brothy, and tortilla, most likely due to a difference in type of reactions predominating. Lipid oxidation is the primary source of volatile compound changes in SMP stored at 21°C (Karagul Yuceer *et al.*, 2002; Drake *et al.*, 2006). Maillard reactions play a smaller role. It is likely that Maillard reactions play a larger role at higher storage temperature and thus influence the prevalence and type of volatile compounds produced and the sensory flavors. These differences also demonstrate that caution should be taken when evaluating products under accelerated storage conditions and relating those changes to traditional storage time and temperature. The shelf-life for SMP stored at 35°C was deemed to be 12–15 months (versus 18 to 24 months at 21°C).

In the case of whey proteins, the influence of the cheesemake procedure (e.g. starter culture) also has a direct influence on final product flavor. Whey flavor is directly affected by the type of cheese produced as well as the quality of milk and the method of whey handling after draining, including the time elapsed (Bodyfelt *et al.*, 1988; Tomaino *et al.*, 2004; Gallardo Escamilla *et al.*, 2005; Rankin, 2009). Fresh liquid whey is expected to possess a delicate and bland flavor that should relate into powdered whey when reconstituted (Drake *et al.*, 2003; Drake, 2006). Bodyfelt *et al.* (1988) described liquid whey as exhibiting a slightly dirty-sweet/acidic taste and odor by qualitative dairy judging/grading criteria. More recent modern descriptive sensory analysis has documented a host of specific flavors in liquid and dried whey ingredients (Karagul-Yuceer *et al.*, 2003b; Drake *et al.*, 2003, 2009b; Carunchia Whetstine *et al.*, 2003, 2005; Tomaino *et al.*, 2004; Gallardo-Escamilla *et al.*, 2005; Russell *et al.*, 2006; Wright *et al.*, 2006, 2009; Drake, 2006; Table 17.1). Fresh liquid Cheddar whey has a distinct sweet aromatic and cooked/milky flavor that rapidly diminishes with cold storage to cardboard and metallic/serumy flavors which were likely what Bodyfelt *et al.* (1988) sought to describe (Carunchia Whetstine *et al.*, 2003; Tomaino *et al.*, 2004). These flavors, as well as additional flavors not documented in liquid whey, have been reported in dried whey ingredients (Table 17.1). Volatile compound analysis has also been conducted to pinpoint specific compounds responsible for liquid and dried whey protein flavors. Table 17.3 outlines volatile compounds that have been identified in the literature as being present in liquid whey, WPC80 and/or WPI (Karagul-Yuceer *et al.*, 2003b; Drake *et al.*, 2003, 2009b; Carunchia Whetstine *et al.*, 2003, 2005; Mahajan *et al.*, 2004; Tomaino *et al.*, 2004; Gallardo-Escamilla *et al.*, 2005; Russell *et al.*, 2006; Wright *et al.*, 2006, 2009; Drake, 2006; Javidipour and Qian, 2008).

Processing will also influence flavor of dried whey ingredients. As with milk powders, the spray drying process has an effect, but whey protein powders are also subjected to membrane fractionation procedures, primarily UF (ultrafiltration) and NF (nanofiltration) and MF (microfiltration) to a lesser degree. Residual detergents from recently cleaned membranes or flavor-contributing compounds from fouled membranes can all contribute to flavor and flavor variability. Processing parameters also vary widely between facilities and within a facility. Ultimately, identical whey ingredients from different suppliers will have distinctive flavors (Drake *et al.*, 2009b). Figures 17.3 and 17.4 demonstrate basic processing steps for a commercial WPC80 and WPI manufacturer, respectively. The flavor of the whey changes as it progresses through processing from liquid whey to spray dried WPC80 or WPI (Tables 17.4 and 17.5). Sweet and sour aromatics and cooked/milky flavors (hallmarks of fresh whey flavor) and basic tastes diminish as the whey is processed and concentrated. Concurrently, volatile compound changes can also be observed (Tables 17.6 and 17.7). These differences reflect the changing composition of the whey as well as the unit operations to which it is subjected. For example, diacetyl (2, 3 butandione) decreases as the liquid product is concentrated and then increases after spray drying. It is possible that this compound is destroyed or breaks down during processing but then is formed again during the spray drying step (Table 17.6). Lactones, present initially in the fresh whey prior to fat removal, probably partition into fat and are removed during processing (Table 17.7). In contrast, dimethyl trisulfide has been documented as an

Table 17.3 Volatile compounds identified in liquid whey and dried whey products

Volatile	Whey type[ab]
(E)-2-hexenal	WPC80, agglom. WPC80, WPI, agglom. WPI
(E)-2-nonenal	liquid, SWP, WPC80, WPI
(E)-2-octenal	SWP, WPC80
(E,E)-2,4-decadienal	liquid, SWP, WPC80, WPI
(E,E)-2,4-nonadienal	SWP, WPI
(E,E)-2,4-octadienal	SWP
(E,Z)-2,4-decadienal	SWP
(E,Z)-2,4-nonadienal	SWP
(E-Z)-2,6-nonadienal	liquid, SWP, WPC80, WPI
(Z)-2-nonenal	liquid, SWP, WPI
(Z)-4-heptanal	WPC80
(Z)-4-heptenal	SWP
1,2-propadiene; alkenyl	liquid
1,5-octadienone	WPI
1-dodecane	liquid
1-octen-3-ol	WPC80, agglom. WPC80, WPI, agglom. WPI
1-octen-3-one	liquid, SWP, WPC80, inst. WPC80, WPI
1-propanol	liquid
2,3 methyl butanol	liquid
2,3,5-trimethyl pyrazine	SWP
2,3-dimethyl pyrazine	SWP
2,3-methylbutanoic acid	liquid
2,5-dimethyl pyrazine	SWP
2,5-dimethyl-4-hydroxy-3-(2H) furanone (Furaneol)	SWP, WPC80, WPI
2,6-dimethyl pyrazine	SWP
2-acetyl-1-pyrroline	liquid, SWP, WPC80, WPI
2-acetylpyrrole	SWP
2-acetylthiazole	SWP
2-butanol	liquid
2-butanone	liquid
2-ethyl pyrazine	liquid, SWP
2-ethyl-1-hexanol	WPC80, agglom. WPC80, WPI, agglom. WPI
2-furfural	liquid
2-heptanol	WPC80,inst.WPC80
2-heptanone	liquid, WPC80, agglom. WPC80, WPI, agglom. WPI
2-isobutyl-3-methoxypyrazine	liquid, WPC80, WPI
2-methoxy phenol (guiacol)	WPC80
2-methoxy-3-isopropylpyrazine	liquid
2-methyl propanoic acid	SWP
2-methyl-3-furanthiol	liquid, WPI
2-nonanol	WPC80,inst.WPC80
2-nonanone	liquid, WPC80, agglom. WPC80, WPI, agglom. WPI

Table 17.3 *Cont'd*

Volatile	Whey type[ab]
2-octanone	WPC80,inst.WPC80
2-pentyl furan	WPC80, agglom. WPC80, WPI, agglom. WPI
2-phenethanol	WPC80, WPI
2-propanol	liquid
2-propionyl-1-pyrroline	SWP
2-undecanone	WPC80, agglom. WPC80, WPI, agglom. WPI
3-hydroxy-4,5-dimethyl-2-(5H)-furanone (Sotolon)	liquid, SWP, WPC80, WPI
3-methoxy-4-hydroxy benzaldehyde (vanillin)	WPC80
3-methyl butanoic acid	SWP
3-methyl furan	liquid
4-methyl octanoic acid	WPC80, WPI
9-decanoic acid	SWP
acetaldehyde	liquid
acetic acid	liquid, SWP, WPC80, WPI
acetoin	liquid
acetone	liquid
benzaldehyde	WPC80, agglom. WPC80, WPI, agglom. WPI
butadiene	liquid, WPC80, WPI
butanoic acid	liquid, SWP, WPC80, WPI
butanol	liquid
decanal	WPC80, WPI
decanoic acid	SWP, WPC, WPI
delta-decalactone	liquid, SWP, WPC80, WPI
delta-dodecalactone	SWP, WPC80, WPI
delta-octalactone	SWP
delta-undecalactone	SWP
diacetyl	liquid, SWP, WPC80, agglom. WPC80, WPI, agglom. WPI
dimethyl disulfide	SWP, WPC80, inst.WPC80, agglom. WPC80, WPI, agglom. WPI
dimethyl sulfide	liquid, WPC80, WPI
dimethyl trisulfide	liquid, SWP, WPC80, agglom. WCP80, WPI, agglom. WPI
dimethylamine allyl	liquid
dodecanoic acid	SWP
ethanol	liquid
ethyl acetate	liquid
formic acid	SWP
furfuryl alcohol	SWP
gamma-decalactone	SWP, WPC80
gamma-dodecalactone	SWP
gamma-hexalactone	SWP
gamma-nonalactone	WPC80
heptanal	liquid, SWP, WPC80, inst. WPC80, agglom. WPC80, WPI, agglom. WPI

Table 17.3 *Cont'd*

Volatile	Whey type[ab]
heptanoic acid	SWP, WPC80, WPI
heptanone	liquid
hexanal	liquid, SWP, WPC80, inst. WPC80, agglom. WPC80, WPI, agglom. WPI
hexanoic acid	liquid, SWP, WPC80, WPI
hydrocarboxyl	liquid
isobutyric acid	liquid
maltol	liquid, SWP
methional	liquid, SWP, WPC80, WPI
methyl propanoic acid	liquid
nonanal	liquid, SWP, WPC80, agglom. WPC80, WPI, agglom. WPI
nonanoic acid	WPC80, WPI
nonanol	WPC80,inst.WPC80
nonanone	liquid
o-aminoacetophenone	WPC80, WPI
octanal	liquid, WPC80, agglom. WPC80, WPI, agglom. WPI
octanoic acid	liquid, SWP, WPC80, WPI
octanol	WPC80,inst.WPC80
p-Cresol	SWP
pentanal	liquid
pentanoic acid	liquid, SWP, WPC80, WPI
phenol	liquid
phenyl ethyl acetate	WPC80
phenylacetaldehyde	SWP, WPI
propan-1-ol; alkyl	liquid
propanoic acid	liquid, SWP
skatol	SWP
toluene	WPC80, agglom. WPC80, WPI, agglom. WPI
fatty acids	
caproic	liquid
caprylic	liquid
capric	liquid
lauric	liquid
myristic	liquid
palmitic	liquid
palmitoleic	liquid
stearic	liquid
oleic	liquid
linoleic	liquid

[a] SWP – Sweet Whey Powder; WPC80 – Whey Protein Concentrate 80% protein; inst. WPC80 – instantized WPC80; agglom. WPC80 – agglomerated WPC80; WPI – Whey Protein Isolate >90% protein; agglom. WPI – agglomerated WPI.

[b] Cheese source: *liquid* – Cheddar, Gouda, Mozzeralla, Paneer, Quarg, rennet casein, acid casien, lactic acid casein; *SWP* – Cheddar; *WPC80* – Cheddar, Mozzerella, Monterey Jack; *inst. WPC80* – Cheddar, Monterey Jack; *agglom. WPC80* – Cheddar, Mozzerella; *WPI* – Cheddar; *agglom. WPI* – Cheddar.

Karagul-Yuceer *et al.*, 2003b; Drake *et al.*, 2003, 2009b; Mahajan *et al.*, 2004; Tomaino *et al.*, 2004; Carunchia Whetstine *et al.*, 2003, 2005; Gallardo-Escamilla *et al.*, 2005; Russell *et al.*, 2006; Drake, 2006; Wright *et al.*, 2006, 2009; Javidipour and Qian, 2008.

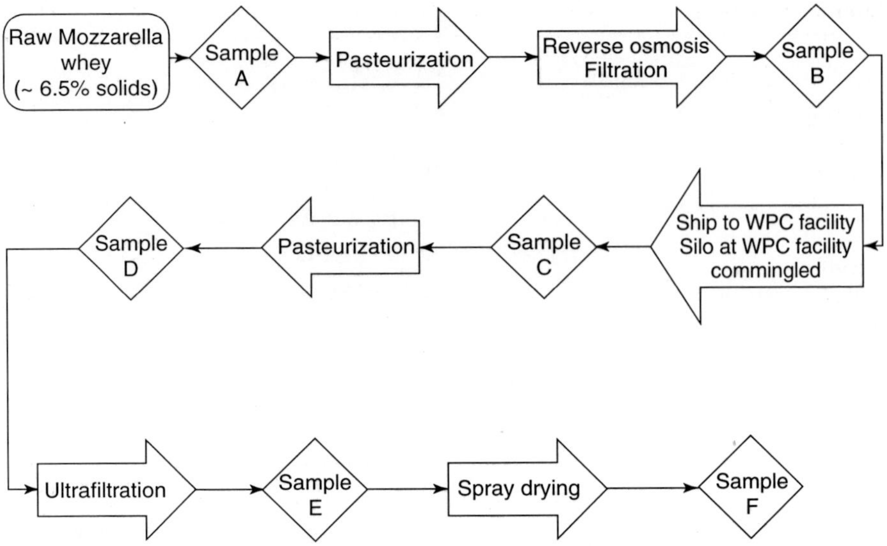

Fig. 17.3 General WPC processing flow chart with sampling points (Mozzarella).

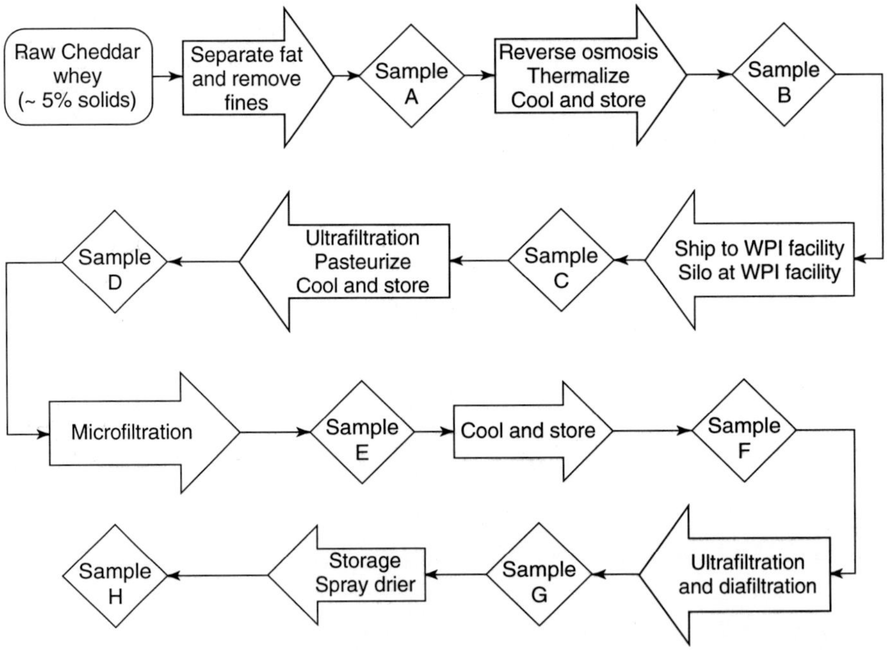

Fig. 17.4 General WPI processing flow chart with sampling points (Cheddar).

Table 17.4 Descriptive sensory profiles (mean +/– standard deviation) of commercial WPC80 throughout production

Sample	A	B	C	D	E	F
Aroma intensity	2.4 +/– 0.3[a]	1.0 +/– 0.2[c]	1.6 +/– 0.2[b]	2.0 +/– 0.1[b]	1.5 +/– 0.4[bc]	1.0 +/– 0.2[c]
Cooked/ milky	1.8 +/– 0.2[a]	1.8 +/– 0.2[a]	1.4 +/– 0.6[a]	1.0 +/– 0.5[b]	ND	1.5 +/– 0.6[a]
Sweet aromatic/ diacetyl	1.6 +/– 0.2[a]	1.4 +/– 0.5[a]	0.8 +/– 0.3[a]	1.2 +/– 0.2[a]	0.9 +/– 0.2[a]	1.3 +/– 0.3[a]
Cardboard	ND	ND	0.8 +/– 0.2[b]	1.2 +/– 0.2[b]	1.9 +/– 0.3[a]	0.9 +/– 0.2[b]
Salty taste	1.6 +/– 0.2[a]	0.5 +/– 0.1[b]	0.6 +/– 0.1[b]	0.8 +/– 0.1[b]	ND	ND[b]
Sweet taste	1.6 +/– 0.1[b]	3.0 +/– 0.2[a]	2.4 +/– 0.1[a]	2.8 +/– 0.2[a]	ND	ND[c]
Sour taste	1.3 +/– 0.2[a]	1.5 +/– 0.2[a]	1.3 +/– 0.3[a]	1.4 +/– 0.3[a]	ND	ND
Astringency	1.2 +/– 0.2[b]	1.5 +/– 0.2[b]	1.7 +/– 0.2[b]	1.5 +/– 0.3[b]	2.7 +/– 0.3[a]	2.6 +/– 0.2[a]

[a–c] Means in a row followed by different letters are different ($P < 0.05$).
All sensory analyses conducted at 10% solids (w/w) except for raw whey (6.5%).
Attributes were scored on a 0 to 15-point universal intensity scale (Meilgaard *et al.*, 1999) where most dairy flavors fall between 1 and 7 (Drake, 2004).
Capital letters refer to samples (Fig. 17.3).

off flavor source in WPI (Wright *et al.*, 2006) and appears to form during processing, likely as a by-product of sulfur-containing amino acid breakdown (Tables 17.5 and 17.7). Removal of many whey flavor compounds occurs during the processing of liquid whey into whey protein, especially during filtration steps. However, flavor precursors are still present and new flavors are formed or increase to detectable levels during the spray-drying step.

A detailed analysis of whey protein processing suggests that fat removal, heating, membrane filtration and spray drying all influence flavor and flavor formation. Current and ongoing work with serum protein processing (removal of whey proteins directly from fluid milk prior to the cheesemaking procedure) may also provide additional clarification of the role of these individual process steps (Nelson and Barbano, 2005a,b; Drake *et al.*, 2009b). Serum protein processing is similar to whey protein processing without the influence of starter culture fermentation and the addition of annatto. Production of serum proteins allows a direct comparison of the influence of the cheesemaking procedure on whey proteins. Few studies to date have directly compared the two products, but preliminary work suggests that significant flavor and functional differences exist between them (Drake *et al.*, 2009b).

In addition to the role of specific unit operation processes, time to process whey from fluid product to spray dried product also may influence

Table 17.5 Descriptive sensory profiles (mean +/– standard deviation) of commercial WPI throughout production

Sample	A	B	C	D	E	F	G	H
Aroma intensity	2.0 +/– 0.2^{b}	ND	ND	3.0 +/– 0.2^{a}	1.7 +/– 0.2^{b}	1.5 +/– 0.2^{c}	1.0 +/– 0.3^{c}	1.5 +/– 0.2^{c}
Cooked/milky	1.1 +/– 0.2^{b}	1.7 +/– 0.2^{a}	1.5 +/– 0.2^{ab}	1.4 +/– 0.3^{ab}	ND	ND	ND	ND
Sweet aromatic/diacetyl	2.0 +/– 0.2^{a}	1.5 +/– 0.3^{b}	1.4 +/– 0.2^{b}	1.3 +/– 0.2^{b}	0.5 +/– 0.1^{c}	0.5 +/– 0.2^{c}	ND	ND
Metallic/meat serum	ND	1.5 +/– 0.1^{a}	1.1 +/– 0.2^{a}	ND	ND	ND	ND	ND
Cardboard	ND	1.5 +/– 0.1^{b}	1.0 +/– 0.2^{c}	1.5 +/– 0.2^{b}	1.3 +/– 0.2^{bc}	0.6 +/– 0.1^{d}	ND	2.0 +/– 0.2^{a}
Sour aromatic	1.0 +/– 0.1^{a}	ND	ND	ND	ND	ND	ND	ND
Cabbage/brothy	ND	ND	ND	ND	ND	ND	ND	2.2 +/– 0.2
Salty taste	1.7 +/– 0.4^{b}	2.9 +/– 0.2^{a}	1.7 +/– 0.3^{b}	1.4 +/– 0.2^{b}	ND	ND	ND	ND
Sweet taste	1.6 +/– 0.4^{b}	3.0 +/– 0.2^{a}	3.1 +/– 0.1^{a}	2.0 +/– 0.3^{b}	ND	ND	ND	ND
Sour taste	1.4 +/– 0.3^{a}	1.2 +/– 0.2^{ab}	1.0 +/– 0.2^{b}	0.5 +/– 0.1^{c}	ND	ND	ND	ND
Astringency	1.1 +/– 0.4^{a}	1.5 +/– 0.3^{a}	1.6 +/– 0.5^{a}	1.5 +/– 0.3^{a}	1.7 +/– 0.4^{a}	1.1 +/– 0.2^{a}	1.2 +/– 0.2^{a}	1.6 +/– 0.6^{a}

$^{a–e}$ Means in a row followed by different letters are different ($P < 0.05$).
All sensory analyses conducted at 10% solids (w/w) except for raw whey (6.5%).
Attributes were scored on a 0 to 15-point universal intensity scale (Meilgaard *et al.*, 1999) where most dairy flavors fall between 1 and 7 (Drake, 2004).
Capital letters refer to samples (Fig. 17.4).

Table 17.6 Flavor dilution factors of selected aroma-active volatile compounds identified throughout the WPC 80 production process

Compound	FR[1]	Log_3FD[2] factors: Fresh whey	After RO & pasteurize	After pasteurize & UF	Finished WPC 80	RI DB-5	RI DB-Wax	ID[3]
1-octen-3-ol	NB	4	4	3	2	984	1280	odor, RI, MS
2-acetyl-1-pyrroline	NB	3	5	3	4	923	1320	odor, RI
2-isobutyl-3-methoxypyrazine	NB	ND[4]	ND	<1	<1	1186	1510	odor, RI
2-phenethanol	NB	3	5	1	2	1162		odor, RI
2,3 butanedione	NB	3	1	2	5	681	890	odor, RI
2,6-nonadienal	NB	1	4	2	2	1171	1596	odor, RI
acetic acid	AC	3	1	<1	4	656	1394	odor, RI, MS
butyric acid	AC	6	4	3	4	850		odor, RI, MS
hexanoic acid	AC	ND	ND	ND	3	1050	1757	odor, RI, MS
delta-decalactone	NB	4	<1	ND	<1	1500	2013	odor, RI
dimethyl trisulfide	NB	1	4	3	<1	978	1360	odor, RI, MS
hexanal	NB	2	1	1	2	792	1066	odor, RI, MS
methional	NB	5	<1	2	2	922	1428	odor, RI
nonanal	NB	1	<1	3	2	1108		odor, RI, MS
nonanoic acid	NB	3	6	4	4	1256		odor, RI, MS
octanal	NB	2	1	1	2	1010		odor, RI, MS

[1] Fraction (NB = neutral/basic, AC = acid).
[2] Flavor dilution factors were determined on a DB-5 column for NB compounds, and on a DB-Wax column for AC compounds.
[3] Method of identification. (RI = retention index, MS = mass spectra)
[4] Not detected.

Table 17.7 Flavor dilution factors of selected aroma-active volatile compounds identified throughout the WPI production process

Compound	FR[1]	RI DB-5	Log_3FD^2 factors		RI DB-5	RI DB-Wax	ID[3]
		After RO/Ship	After MF, UF, DF	Finished WPI			
1-octen-3-ol	NB	4	2	4	984	1280	odor, RI, MS
2-acetyl-1-pyrroline	NB	2	1	3	923	1320	odor, RI
2-isobutyl-3-methoxypyrazine	NB	2	2	1	1186	1510	odor, RI
2-phenethanol	NB	3	4	3	1162		odor, RI
2,3 butanedione	NB	2	<1	<1	681	890	odor, RI
2,6-nonadienal	NB	4	<1	2	1171	1596	odor, RI
acetic acid	AC	3	2	2	656	1394	odor, RI, MS
acetophenoneee	NB	3	<1	ND[4]	1094		odor, RI, MS
butyric acid	AC	2	ND	2	850		odor, RI, MS
hexanoic acid	AC	<1	<1	<1	1050	1757	odor, RI, MS
delta-decalactone	NB	1	<1	<1	1500	2013	odor, RI
dimethyl disulfide	NB	ND	ND	<1	773		odor, RI
dimethyl trisulfide	NB	<1	2	7	978	1360	odor, RI, MS
hexanal	NB	2	<1	1	792	1066	odor, RI, MS
maltol	AC	1	ND	4	1141		odor, RI, MS
methional	NB	5	3	3	922	1428	odor, RI
nonanal	NB	6	3	3	1108		odor, RI, MS
nonanoic acid	NB	3	1	3	1256		odor, RI, MS
octanal	NB	3	1	1	1010		odor, RI, MS

[1] Fraction (NB = neutral/basic, AC = acid).
[2] Flavor dilution factors were determined on a DB-5 column for NB compounds, and on a DB-Wax column for AC compounds.
[3] Method of identification (RI = retention index, MS = mass spectra).
[4] Not detected.

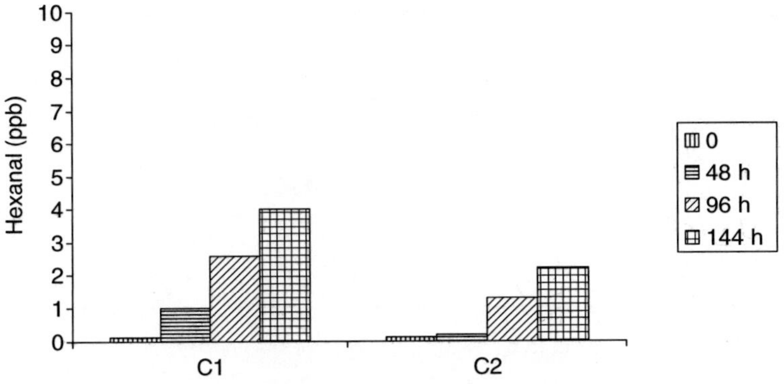

Fig. 17.5 Hexanal increases in liquid pasteurized Cheddar cheese whey with (C2) and without (C1) fat removal across 144 hours storage at 5°C. Hexanal was extracted and quantified by SPME GC-MS. Changes in hexanal concentration are significant ($p < 0.05$) between treatments at each timepoint.

flavor. The time to process liquid whey to a finished spray dried whey protein concentrate (WPC) may vary from 6 h to as much as 48 h depending on the volume of whey incoming to the processing facility. When volumes are low, liquid pasteurized whey or liquid pasteurized concentrated whey may be held at a low temperature for 6 to 36 h prior to subsequent processing. Tomaino *et al.* (2004) first documented that refrigerated storage of liquid pasteurized whey resulted in increases in lipid oxidation compounds. More recent work also confirms this observation; in Fig. 17.5, fresh Cheddar cheese whey was pasteurized, the fat removed, and then the whey was stored at 4°C for 96 h. Increases in lipid oxidation compounds occurred with cold storage and descriptive sensory analysis likewise revealed evolution of cardboard flavors and decreases in fresh cooked/milky and sweet aromatic flavors, characteristic of fresh whey. Examples with spray dried WPC80 also suggest that increases in process time (e.g. storage of liquid product at any step during processing) negatively impact upon final spray dried product flavor. Figure 17.6 demonstrates the sensory profiles of Mozzarella WPC80 spray dried following increasing storage time at 3°C of liquid protein retentate.

Bleaching is another step that occurs in some whey protein processing facilities that may influence flavor. A consistent cream colored powder that, when rehydrated, has little to no visible color is the desirable end product for the dried whey protein powders. Many cheeses (namely Cheddar) have annatto extract added to provide an orange color in the final finished cheese. Approximately 10 to 20% of the added annatto carries over into the fluid whey where it imparts an undesirable color to the finished spray dried whey protein. The whey is bleached to minimize color and produce

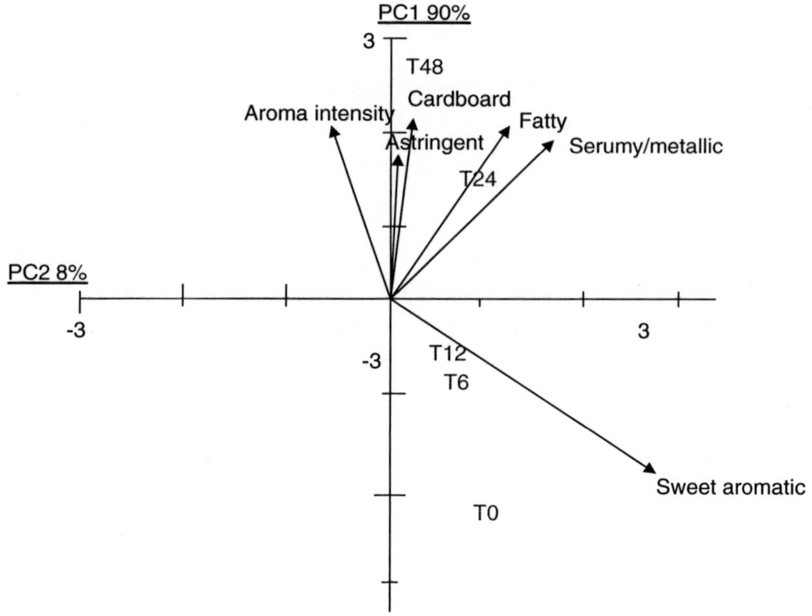

Fig. 17.6 Sensory changes in spray-dried Mozzarella WPC80 following 3°C storage of concentrated retentate for various timepoints. T0 – no storage prior to spray drying, T6 – 6 h storage prior to spray drying, T12 – 12 h storage prior to spray drying, T24 – 24 h storage prior to spray drying, T48 – 48 h storage prior to spray drying.

a consistent colored product for ingredients applications. Hydrogen peroxide and benzoyl peroxide are the only legally permitted bleaching agents for whey in the United States, with several restrictions to their use. Hydrogen peroxide decomposes to oxygen and water during bleaching (Smith, 2004). The residual hydrogen peroxide is removed from the whey by either heat and/or catalase. As a GRAS (Generally Recognized as Safe) substance, the maximum treatment level for bleaching whey using hydrogen peroxide is 0.05% (<500 ppm) (21CFR184.1366). Along with hydrogen peroxide, benzoyl peroxide is also a GRAS substance, and is used in food with no limitation other than current good manufacturing practices (21CFR184.1157). Benzoyl peroxide is typically used at a rate no more than 0.002%. The most effective conditions are 60°C for 15 minutes at pH 6 to 7. Longer holding times are required if lower temperatures are used. Benzoyl peroxide is effective at lower usage levels than hydrogen peroxide, is less corrosive on stainless steel and does not require catalase addition. However, application of benzoyl peroxide results in elevated levels of benzoic acid in dried WPC. Benzoic acid is not a desirable component to many Asian export markets and this represents a significant barrier to benzoyl peroxide usage.

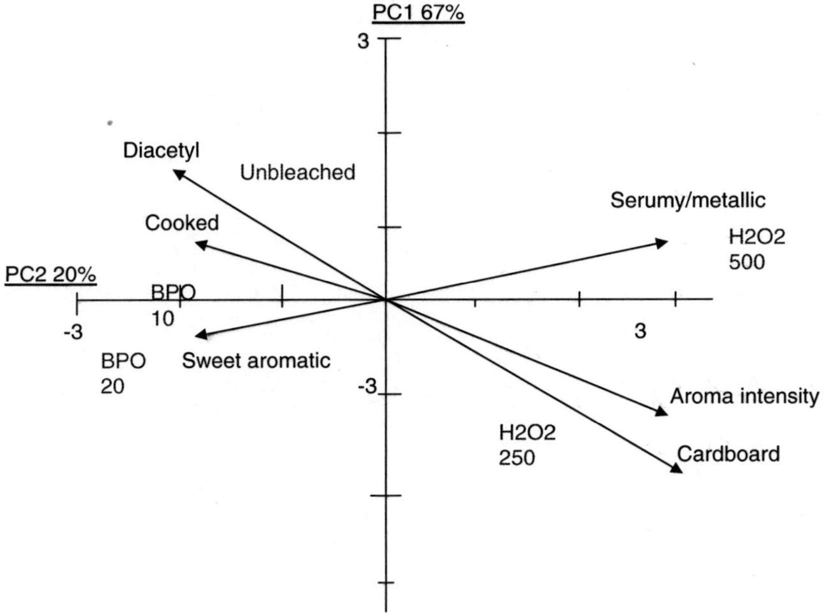

Fig. 17.7 Principal component biplot of the sensory profiles of fresh fluid colored Cheddar cheese whey with and without hydrogen peroxide (H_2O_2) or benzoyl peroxide (BPO) bleaching. Unbleached – colored whey with heating but no bleach; BPO10 – 10 ppm; BPO20 – 20 ppm; H_2O_2 250 – 250 ppm; H_2O_2 500 – 500 ppm.

Published quantitative studies on the impact of bleaching on whey protein functionality and flavor are sparse and dated. The functionality of whey protein may change due to bleaching with hydrogen peroxide (Cooney and Morr, 1972; Munyua, 1975; Chang *et al.*, 1977). Bleaching can impart oxidized tallowy flavors to milk and to whey (Munyua, 1975; Washam *et al.*, 1974). Oxidized flavors in whey that were apparent right after bleaching purportedly disappeared after evaporation and drying (McDonough *et al.*, 1968). A more complete and up-to-date scientific understanding of the impact of bleaching on whey protein flavor and functionality is needed. Figure 17.7 demonstrates recent studies on the differences in unbleached Cheddar whey compared with whey from the same cheese vat treated with different concentrations of benzoyl or hydrogen peroxide. The bleaching agent concentrations represent the range of concentrations that might be expected in current industrial practices, as does the time/temperature treatment. Clearly, flavor and volatile compound profiles are impacted upon by bleaching and by different bleaching agents (Figs 17.7 and 17.8). As might be expected, higher concentrations of lipid oxidation products are present in the bleached wheys compared to unbleached wheys. What

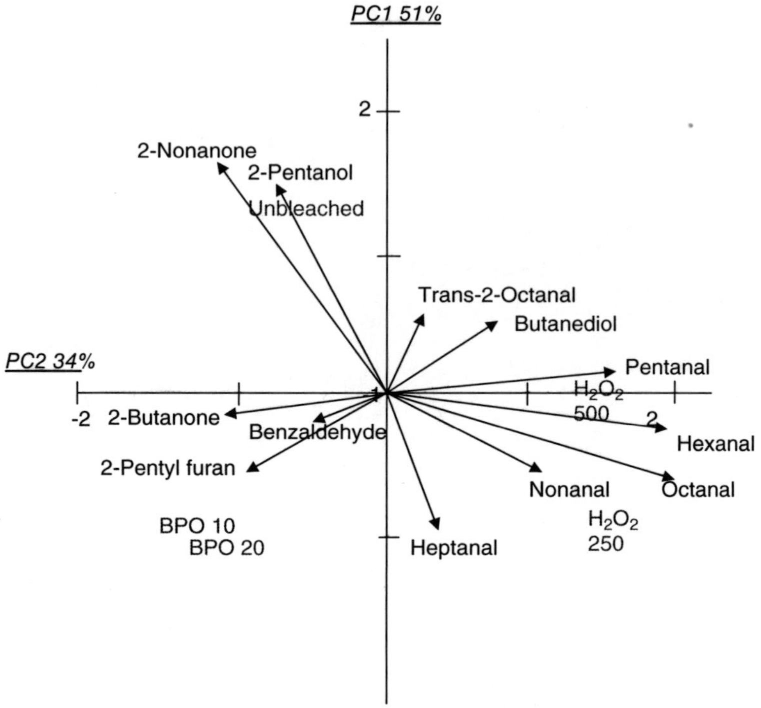

Fig. 17.8 Principal component biplot of the instrumental volatile profiles of fresh fluid colored Cheddar cheese whey with and without hydrogen peroxide (H_2O_2) or benzoyl peroxide (BPO) bleaching. Unbleached – colored whey with heating but no bleach; BPO10 – 10 ppm; BPO20 – 20 ppm; H_2O_2 250 – 250 ppm; H_2O_2 500 – 500 ppm.

remains to be seen is if these differences carry-through to finished spray dried products.

Finally, agglomeration and storage time can also impact dried ingredient flavor and flavor stability. Agglomeration is the process of producing agglomerates, small clumps of several particles. Agglomeration increases dispersability and decreases dispersion time because of the increase in the particle size and porosity of the powder (Turchiuli *et al.*, 2005; Peitsch, 2005). This process is often applied to SMP, WMP and WPC80 and WPI for beverage applications, as the product will mix and rehydrate more readily when agglomerated. Industrially, the most common way to produce agglomerated powders is by spray drying, followed by the rewetting and agglomeration of powders in fluidized beds (Pietsch, 2005). The product may have been spray dried at an earlier time, hence the name re-wet agglomeration is applied to this process. A solvent, called a binder, is sprayed onto the powder from above or inside of the bed (Turchiuli *et al.*, 2005). The binder can be water/steam or usually lecithin solution. Lecithin is used in this process to further increase dispersion properties (wettability) of the powder. The term 'instantized' generally refer to products agglomerated with lecithin, whereas the term 'agglomerated' may or may not imply that lecithin has been added. More recently, agglomeration has been conducted as a continuous process as the milk or whey protein is spray-dried, eliminating the need for the re-wet agglomeration process. This newer process is called single pass agglomeration (Henning *et al.*, 2006).

If lecithin is applied, careful attention should be paid to the source and age of the lecithin used. Lecithin is generally sourced from soy and has a distinct flavor. The product is itself prone to oxidation. A recent study demonstrated that the application of lecithin decreased the storage stability of WPC80 compared to product that had been agglomerated without lecithin or not agglomerated (Wright *et al.*, 2009). As with skim milk powder (SMP) and whole milk powder (WMP), whey proteins develop storage associated flavors (Drake *et al.*, 2009b; Wright *et al.*, 2009). Storage-associated flavors documented across 18 months' storage under ambient temperature included cardboard, raisin/brothy, cucumber and fatty. These flavors increased more rapidly in instantized or agglomerated WPC80 or WPI compared to non-agglomerated products (Wright *et al.*, 2009). Although WPC80 has a higher fat content that WPI, the same types of flavor changes were observed. Volatile compound analysis confirmed that volatile lipid oxidation products increased with storage time concurrently with flavor changes. Previous studies, focused solely on instrumental changes, have likewise confirmed the prevalence of lipid oxidation products in storage of WPC80 (Javidipour and Qian, 2008). Wright *et al.* (2009) suggested that the optimum shelf-life for non-agglomerated WPC80 and WPI stored at 21°C was 12–15 months, and 8–12 months for steam agglomerated or lecithin agglomerated WPC80 and WPI. Since lipid degradation plays a large role with WPC80 and WPI loss of shelf-life, nitrogen flushed packaging, as with WMP, would likely improve shelf stability. Mortenson *et al.* (2008) recently reported that processing effects (bleaching, agglomeration, cheese source) had few effects on sensory properties and volatile compounds of WPC34 and WPI. However, they used liquid products from multiple facilities which were then processed under pilot scale conditions. Multiple sources of process variability and lack of extensive training by the sensory panel may have confounded their ability to document specific differences. Certainly, the wealth of published and unpublished information suggests that their results were not representative.

17.3 Flavor carry-through with dried dairy ingredients

Dried dairy ingredients are highly versatile and value-added ingredients. In most cases, these products are not consumed directly. Instead they are used as ingredients in other products. For example, in the US, over 75% of the

WMP produced goes into confectionary products such as chocolate, with the remainder used in baked goods, dry sauce and soup mixes, dairy beverages, ice cream, recombined milk, or other products (USDEC, 2005). A rich dairy flavor is one of the reasons why this ingredient is used in confections and sauces. In other cases, such as most examples with whey proteins, the dairy ingredient is used specifically for its nutrition and/or functionality. Little or no flavor is the desired sensory profile in such cases and in either case, non-dairy or atypical flavors are certainly unacceptable and undesirable. Painty flavor in a dry soup mix from beyond-shelf life WMP, or potato flavor in a fruit-flavored whey protein beverage (from the whey protein), are extreme but realistic examples of flavor carry-through. Regardless of the application, flavor carry-through into ingredient applications is critical. Less attention has been paid to this issue in previous years when dried ingredients were treated as commodities, but with increased national and global competition, flavor quality of dried ingredients and flavor carry-through are now critical issues.

Flavor carry-through is influenced by the nature of the flavor or off-flavor in the ingredient and the specific ingredient application. The numerous sources of whey protein and dried ingredient variability have been discussed in this chapter. As such, whey proteins and other dried dairy ingredients from different suppliers generally have different and distinct flavor profiles (Drake *et al.*, 2009b). Other atypical flavors can also occur due to unexpected changes in raw product stream, processing and storage. Different products also have different tolerances for flavor variability. Caudle *et al.* (2005) demonstrated that SMP with off-flavors negatively impacted upon consumer acceptance of vanilla ice cream, yogurt, hot cocoa, and white chocolate to differing degrees. Some off-flavors were masked in certain ingredient applications but not in others. An animal off-flavor negatively affected acceptance of vanilla ice cream, strawberry yogurt and white chocolate but not hot cocoa. In contrast, fatty off-flavor negatively impacted on acceptance of vanilla ice cream and hot cocoa but not strawberry yogurt or white chocolate. Lloyd *et al.* (2009a) demonstrated similar effects with fresh and stored WMP in white and milk chocolate. White chocolate was more transparent to off-flavor carry-through than milk chocolate.

Several studies have demonstrated that variability in whey protein flavor (WPC80, WPI) impacts consumer acceptance of meal replacement beverages, bars and protein beverages (Drake, 2006; Childs *et al.*, 2007; Drake *et al.*, 2009b; Wright *et al.*, 2009). The issue of optimum whey protein shelf-life is demonstrated with flavor carry-through into 6% protein beverages and consumer acceptance of these beverages (Fig. 17.9). Commercial whey proteins (WPC80 and WPI, respectively) from the same facility were compared to effectively evaluate shelf-life and flavor stability rather than flavor differences characteristic of different facilities (Drake *et al.*, 2009b). Stored whey proteins developed distinct flavor differences, readily apparent to trained panelists when the whey proteins were rehydrated at 10% solids

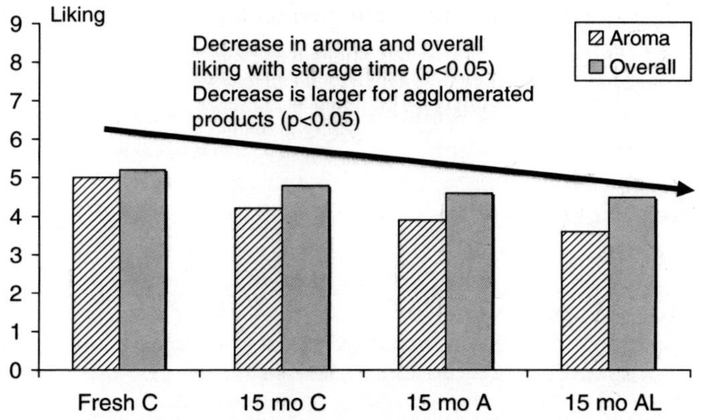

Fig. 17.9 Consumer overall acceptance and aroma liking (n = 75 consumers) of peach flavored beverages with 6% WPC80. Beverages made with fresh control (C) WPC80 received higher ($p < 0.05$) acceptance scores than beverages made with WPC80 stored for 15 months at 21°C (15 mo C – control; 15 mo A – agglomerated; 15 mo AL – agglomerated with lecithin).

and evaluated as-is (a standard practice for trained panelists) (Wright *et al.*, 2009). When these proteins were incorporated into 6% flavored protein beverages for consumers, distinct differences in consumer acceptance were observed between freshly manufactured product and stored products (Fig. 17.9). These results clearly demonstrated that off-flavors do carry-through into ingredient applications and can negatively affect consumer acceptance. These results also, in conjunction with previous studies, collectively suggest that trained panel profiles of dried dairy ingredients do correlate with consumer acceptance and relationships can be established between trained panel profiles of 'raw' ingredients and the potential for carry-through into ingredient applications. The take-home message is to know and understand the characteristic flavor profile of ingredients and to know and understand the characteristic flavor profile of the application(s). Shelf stability of the application must also be accounted for in this situation and emphasizes the need to be conservative with shelf-life estimates of dried ingredients. The shelf-life of many ingredient applications is six months or longer. Clearly, the use of dried ingredients that are close to the end of an established or recommended shelf-life, or that are already showing sensory signs of loss of quality, is not wise to achieve a high quality and competitive ingredient application.

17.4 Future trends

The demand for dried dairy ingredients continues to increase. However, competition from other ingredient sources is increasing concurrently.

Meeting the demand for a consistent product that has the desired target flavor profile and a long shelf-life requires a continued interest and investment in research. A full understanding of the impact of process steps on flavor and flavor stability will enable the industry to meet these demands.

17.5 References

ABBOT, J. and R. WAITE. 1962. Effect of antioxidants on keeping quality of whole milk powder .1. Flavones, gallates, butylhydroxyanisole and nordihydroguaiaretic acid. *J. Dairy Res.* 29:55–58.

ABBOT, J. and R. WAITE. 1965. Effect of antioxidants on keeping quality of whole milk powder 2. Tocopherols. *J. Dairy Res.* 32:143–147.

AL-MABRUK, R.M., N.F.G. BECK, and R.J. DEWHURST. 2004. Effects of silage species and supplemental vitamin E on the oxidative stability of milk. *J. Dairy Sci.* 87:406–412.

ANDERSSON, K. and H. LINGNERT. 1998. Influence of oxygen concentration on the flavor and chemical stability of cream powder. *Food Sci. Tech. Lebens. Wiss. Technol.* 31:245–251.

BALDWIN, A.J. and J.D. ACKLAND. 1991. Effect of preheat treatment and storage on the properties of whole milk powder – changes in physical and chemical properties. *Neth. Milk Dairy J.* 45:169–181.

BALDWIN, A.J., H.R. COOPER, and K.C. PALMER. 1991. Effect of preheat treatment and storage on the properties of whole milk powder – changes in sensory properties. *Neth. Milk Dairy J.* 45:97–116.

BENDALL, J.G. 2001. Aroma compounds of fresh milk from New Zealand cows fed different diets. *J. Agric. Food Chem.* 49:4825–4832.

BEUCLER, J.L., M.A. DRAKE, and E.A. FOEGEDING. 2005. Design of a beverage from whey permeate. *J. Food Sci.* 70:S277–285.

BIOLATTO, A., G. GRIGIONI, M. IRURUETA, A.M. SANCHO, M. TAVERNA, and N. PENSEL. 2007. Seasonal variation in the odour characteristics of whole milk powder. *Food Chem.* 103(3):960–967.

BODYFELT, F.W., J. TOBIAS, and G.M. TROUT. 1988. *The Sensory Evaluation of Dairy Products*. Van Nostrand Reinhold, New York 476 p.

BODYFELT, F.W., M.A. DRAKE, and S.A. RANKIN. 2008. Developments in dairy foods sensory science and education – from student contests to impact on product quality. *Int. Dairy J.* 18:729–734.

CARUNCHIA, M., J.D. PARKER, M.A. DRAKE, and D.K. LARICK. 2003. Impact of starter culture rotation on flavor of liquid Cheddar whey. *J. Dairy Science.* 86:439–448.

CARUNCHIA-WHETSTINE, M.E., A.E. CROISSANT, and M.A. DRAKE. 2005. Characterization of WPC80 and WPI flavor. *J. Dairy Science*. 88:3826–3829.

CARUNCHIA WHETSTINE, M.E. and M.A. DRAKE. 2007. 'The flavor and flavor stability of skim and whole milk powders.' Ch. 13 In *Flavor Chemistry of Dairy Products*. K.R. Cadwallader, M.A. Drake, R. McGorrin, Eds. ACS Publishing, Washington, DC. pp 217–252.

CAUDLE, A.D., Y. YOON, and M.A. DRAKE. 2005. Influence of flavor variability in skim milk powder on consumer acceptability of ingredient applications. *J. Food Sci.* 70:S427–S431.

CELESTINO, E.L., M. IYER, and H. ROGINSKI. 1997. The effects of refrigerated storage of raw milk on the quality of whole milk powder stored for different periods. *Int. Dairy J.* 7:119–127.

CHANG, J.E., E.G. HAMMOND, and G.W. REINBOLD. 1977. Reactions of benzoyl peroxide with whey. *J. Dairy Sci.* 60:40–44.

CHILDS, J.L., M.D. YATES, and M.A. DRAKE. 2007. Sensory properties of meal replacement bars and beverages made from whey or soy proteins. *J. Food Sci.* 72:S425–434.

COONEY, C.M. and C.V. MORR. 1972. Hydrogen peroxide alteration of whey proteins in whey and concentrated whey systems. *J. Dairy Sci.* 55:567–573.

COULON, J.B. and A. PRIOLO. 2002. Sensory properties of meat and dairy products are affected by the forages consumed by the animals. *Productions Animales* 15:333–342.

CROISSANT, A.E., S. WASHBURN, and M.A. DRAKE. 2007. Evaluation of chemical properties and consumer perception of fluid milk from conventional and pasture-based production systems. *J. Dairy Sci.* 90:4942–4953.

DELAHUNTY, C.M. and M.A. DRAKE. 2004. Sensory character of cheese and its evaluation. Ch 19. In: *Cheese: Chemistry, Physics and Microbiology, Volume 1 General Aspects, 3rd Edition* (P.F. Fox, P.L.H. McSweeney, T.M. Cogan and T.P. Guinee, Eds), Elsevier, London. pp 455–487.

DE OLIVEIRA, C.A.F., L. MESTIERI, M.V. SANTOS, J.F.G. MORENO, A. SPERS, and P.M.L. GERMANO. 2000. Effect of microbiological characteristics of raw milk on the quality of whole milk powder. *Brazilian J. Microbiol.* 31:95–98.

DRAKE, M.A., P.D. GERARD, S. WRIGHT, K.R. CADWALLADER, and G.V. CIVILLE. 2002. Cross validation of a sensory language for Cheddar cheese. *J. Sensory Stud.* 17:215–229.

DRAKE, M.A. and G.V. CIVILLE. 2003. Flavor Lexicons. *Compr. Rev. Food Sci.* 2:33–40.

DRAKE, M.A. Y. KARAGUL-YUCEER, K.R. CADWALLADER, G.V. CIVILLE, and P.S. TONG. 2003. Determination of the sensory attributes of dried milk powders and dairy ingredients. *J. Sensory Stud.* 18:199–216.

DRAKE, M.A. 2004. Defining dairy flavors. *J. Dairy Sci.* 87:777–784.

DRAKE, M.A., M.D. KEZIAH, P.D. GERARD, C.M. DELAHUNTY, C. SHEEHAN, R.P. TURNBULL, and T.M. DODDS. 2005. Comparison of cross-cultural differences between lexicons for descriptive analysis of Cheddar cheese flavor in Ireland, New Zealand, and the United States. *International Dairy Journal.* 15:473–483.

DRAKE, M.A. 2006. Flavor and flavor carry-through of whey proteins in beverages. pp 292–300. In *The Wonders of Whey ... Catch the Power. Proceedings of the 4th International Whey Conference.* American Dairy Products Institute, Elmhurst, IL. 440 pages.

DRAKE, M.A., R.E. MIRACLE, A.D. CAUDLE, and K.R. CADWALLADER. 2006. Relating Sensory and Instrumental Analyses. Ch. 2 in *Sensory-directed Flavor Analysis.* R. Marsili, Editor, CRC Press, Taylor and Francis Publishing, Boca Raton, FL. pp 23–55.

DRAKE, M.A. 2007. Sensory analysis of dairy foods. *J. Dairy Sci.* 90:4925–4937.

DRAKE, M.A., R.E. MIRACLE, and J.M. WRIGHT. 2009b. 'Sensory properties of dairy proteins.' Ch. 15 In *Milk Proteins: From Expression to Food.* Thompson, A., Boland, M., Singh, H. (Eds). Elsevier, Amsterdam, The Netherlands. pp 429–448.

DRAKE, M.A. 'History of sensory analysis'. Ch. 1 In *The Sensory Evaluation of Dairy Products.* S. Clark, M. Costello, M.A. Drake, F. Bodyfelt (Eds). 2nd Edition, Springer, New York, NY pp 1–6.

DRAKE, M.A., S.L. DRAKE, F. BODYFELT, S. CLARKE, and M. COSTELLO. 2009a. 'Modern sensory practices.' Ch. 17 In *The Sensory Evaluation of Dairy Products.* S. Clark, M. Costello, M.A. Drake, F. Bodyfelt (Eds). 2nd Edition, Springer, New York, NY pp 490–505.

FEARON, A.M., C.S. MAYNE, and C.T. CHARLTON. 1998. Effect of naked oats in the dairy cow's diet on the oxidative stability of the milk fat. *J. Science Food Agric.* 76:546–552.

GALLARDO-ESCAMILLA, F.J., A.L. KELLY, and C.M. DELAHUNTY. 2005. Sensory characteristic and related volatile flavor compound profile of different types of whey. *J. Dairy Sci.* 88:2689–2699.

HALL, G. and J. ANDERSSON. 1985. Flavor changes in whole milk powder during storage. 3. Relationships between flavor properties and volatile compounds. *J. Food Quality*. 7:237–253.

HENNING, D.R., R.J. BAER, A.N. HASSAN, and R. DAVE. 2006. Major advances in concentrated and dry milk products, cheese, and milk fat based spreads. *J. Dairy Sci.* 89:1179–1188.

HOUGH, G., R.H. SANCHEZ, G.G. DE PABLO, R.G. SANCHEZ, S.C. VILLAPLANA, A.M. GIMENEZ, and A. GAMBARO. 2002. Consumer acceptability versus trained sensory panel scores of powdered milk shelf-life defects. *J. Dairy Sci.* 85:2075–2080.

IPSEN, R. and P.S. HANSEN. 1988. 'Factors affecting the storage stability of whole milk powder.' In *Report No. 274*. Statens Mejeriforsøg, Hillerød, Denmark.

JAVIDIPOUR, I. and M.C. QIAN. 2008. Volatile component change in whey protein concentrate during storage investigated by headspace solid-phase microextraction gas chromatography. *Dairy Sci. Technol.* 88:95–104.

KARAGUL-YUCEER, Y., M.A. DRAKE, and K.R. CADWALLADER. 2001. Aroma active components of nonfat dried milk. *J. Agric. Food Chem.* 49:2948–2953.

KARAGUL-YUCEER, Y., K.R. CADWALLADER, and M.A. DRAKE. 2002. Identification of volatile components of stored nonfat dried milk. *J. Agric. Food Chem.* 50:305–312.

KARAGUL-YUCEER, Y.K., K. VLAHOVICH, M.A. DRAKE, and K.R. CADWALLADER. 2003a. Characteristic Aroma Components of Rennet Casein. *J. Agric. Food Chem.* 51:6797–6801.

KARAGUL-YUCEER, Y., M.A. DRAKE, and K.R. CADWALLADER. 2003b. Aroma active components of liquid Cheddar whey. *J. Food Sci.* 68:1215–1219.

KRAUSE, A.J., K. LOPETCHARAT, and M.A. DRAKE. 2007. Identification of the characteristics that drive consumer liking of butter flavor. *J. Dairy Sci.* 90:2091–2102.

LLOYD, M.A., J. ZOU, H. FARNSWORTH, L.V. OGDEN, and O.A. PIKE. 2004. Quality at time of purchase of dried milk products commercially packaged in reduced oxygen atmosphere. *J. Dairy Sci.* 87:2337–2343.

LLOYD, M.A., S. HESS, and M.A. DRAKE. 2009a. Effect of package and storage temperature on flavor and flavor stability of whole milk powder. *J. Dairy Sci.* 92:2409–2422.

LLOYD, M.A., M.A. DRAKE, and P.D. GERARD. 2009b. Flavor Variability and Flavor Stability of US-Produced Whole Milk Powder. *J. Food Sci.* In Press.

MAHAJAN, S.S., L. GODDIK, and M.C. QIAN. 2004. Aroma compounds in sweet whey powder. *J. Dairy Sci.* 87:4057–4063.

MCDONOUGH, F.E., R.E. HARGROVE, and R.P. TITTSLER. 1968. Decolorization of annatto in Cheddar cheese whey. *J. Dairy Sci.* 51:471–472.

MEILGAARD, M.M., G.V. CIVILLE, and T. CARR. 1999. *Sensory Evaluation Techniques*. 3rd Ed. CRC Press, New York, NY. 387 pp.

MORTENSON, M.A., Z.M. VICKERS, and G.A. REINECCIUS. 2008. Flavor of whey protein concentrates and isolates. *Int. Dairy J.* 18:649–657.

MOUNCHILI, A., J.J. WICHTEL, I.R. DOHOO, G.P. KEEFE, and L.J. HALLIDAY. 2004. Risk factors for milk off-flavors in dairy herds from Prince Edward Island, Canada. *Preventive Vet. Med.* 64:133–145.

MUNYUA, J.K. 1975. Hydrogen peroxide alteration of whey protein in skim milk and whey protein solutions. *Milch.*, 30:730–734.

NEGRI, L.M., M.S. CHAVEZ, M.A. TAVERNA, A.L. CUATRIN, and A.C. RUBIOLO. 2004. Note: The effect of silo milk composition parameters on heat stability of whole milk powder. *Food Sci. Tech. Int.* 10:415–420.

NELSON, B.K. and D.M. BARBANO. 2005a. A microfiltration process to maximize removal of serum proteins from skim milk before cheesemaking. *J. Dairy Sci.* 88:1891–1900

NELSON, B.K. and D.M. BARBANO. 2005b. Yield and aging of Cheddar cheeses manufactured from milks with different milk serum protein contents. *J. Dairy Sci.* 88:4183–4194.

NIELSEN, B.R., H. STAPELFELDT, and L.H. SKIBSTED. 1997. Early prediction of the shelf-life of medium-heat whole milk powders using stepwise multiple regression and principal component analysis. *Int. Dairy J.* 7:341–348.

PALMQUIST, D.L., A.D. BEAULIEU, and D.M. BARBANO. 1993. Feed and animal factors influencing milk-fat composition. *J. Dairy Sci.* 76:1753–1771.

PIETSCH, W. 2005. *Agglomeration in Industry: Occurrence and Applications*. VCH Verlag GmbH & Co. KGaA., Weinheim, Germany. pp 207–246.

RANDBY, A.T., I. SELMER-OLSEN, and L. BAEVRE. 1999. Effect of ethanol in feed on milk flavor and chemical composition. *J. Dairy Sci.* 82:420–428.

RANKIN, S.A. 2009. 'Concentrated and dried milk products.' Ch. 11 In *The Sensory Evaluation of Dairy Products*. S. Clark, M. Costello, M.A. Drake, F. Bodyfelt (Eds). 2nd Edition, Springer, New York, NY pp 333–386.

RUCKOLD, S., H.D. ISENGARD, J. HANSS, and K.H. GROBECKER. 2003. The energy of interaction between water and surfaces of biological reference materials. *Food Chem.* 82:51–59.

RUSSELL, T., M.A. DRAKE, and P.D. GERARD. 2006. Sensory properties of whey and soy proteins. *J. Food Sci.* 71:S447–S455.

SINGH, T., M.A. DRAKE, and K.R. CADWALLADER. 2003. Flavor of Cheddar cheese: A chemical and sensory perspective. *Compr. Rev. Food Sci.* 2:166–189.

SMITH, K. 2004. *Whey Processing CDR Technical Review: Bleaching*. Wisconsin Center for Dairy Research, Madison, WI

STEEN, K. 1977. Seasonal variations in the composition of dry whole milk and its keeping qualities under storage. *Maelkeritidende* 90:126–140, 152–168.

TOMAINO, R.M., L.G. TURNER, and D.K. LARICK. 2004. The effect of *Lactococcus lactis* starter cultures on the oxidative stability of liquid whey. *J. Dairy Sci.* 87:300–307.

TOSO, B., G. PROCIDA, and B. STEFANON. 2002. Determination of volatile compounds in cows' milk using headspace GC-MS. *J. Dairy Res.* 69:569–577.

TURCHIULI, C., Z. ELOUALIA, N.E. MANSOURI, and E. DUMOULIN. 2005. Fluidised bed agglomeration: Agglomerates shape and end-use properties. *Powder Technol.* 157:168–175.

URBACH, G. 1990. Effect of feed on flavor in dairy foods. *J. Dairy Sci.* 73(12):3639–3650.

[USDEC]. UNITED STATES DAIRY EXPORT COUNCIL. 2005. *Reference Manual for US Milk Powders: 2005 Revised Edition*. Available from: http://www.usdec.org/publications/PubDetail.cfm?ItemNumber=213 Accessed July 22, 2008.

VAN MIL, P. and J.A. JANS. 1991. Storage stability of whole milk powder – effects of process and storage conditions on product properties. *Neth. Milk Dairy J.* 45:145–167.

WASHAM, C.J., G.W. REINBOLD, E.R. VEDAMUTH, and R. JORGENSE. 1974. Changes in milk, whey, and blue cheese as induced by benzoyl peroxide. *J. Milk and Food Technol.* 37:244–249.

WRIGHT, J.M., M.E. CARUNCHIA WHETSTINE, R.E. MIRACLE, and M.A. DRAKE. 2006. Characterization of a cabbage off-flavor in whey protein isolate. *J. Food Sci.* 71:C91–96.

WRIGHT, B.J., S.E. ZEVCHAK, J.M. WRIGHT, and M.A. DRAKE. 2009. Impact of agglomeration on flavor and flavor stability of whey proteins. *J. Food Sci.* 74:S17–S29.

18

Production of dairy aromas and flavors: New directions

M. Sibeijn, Dishman Netherlands B.V., The Netherlands and J. A. Wouters, NIZO food research, The Netherlands

Abstract: Dairy-derived aromas and flavors are widely used in the food industry. Nowadays, the traditional milk-derived products, such as cheese and butter, are more and more replaced by novel ingredients that are obtained by fermentation or enzymic modification. These ingredients offer more intense flavors and lower costs. In the past, the emphasis has been on the development of enzymes, production strains and substrates for flavor and aroma production. Downstream processing was usually limited to concentration of the fermented liquid, optionally followed by drying. New developments in distillation and membrane technology offer new opportunities for the production of highly concentrated dairy-derived flavors and aromas.

Key words: dairy aroma, enzyme, fermentation, downstream processing, distillation, membrane.

18.1 Introduction

In the world of food ingredients, flavors play an important role. Dairy-derived flavors are widely used in various food applications, such as snacks, cheese sauces, dressings, baking products, confectionary, and many others. The flavors are usually divided into four main groups, according to the traditional products: milky, yoghurt, cheesy and creamy flavors.

In the past, flavoring of food ingredients with dairy aroma was done either by using the traditional dairy products as such or by using extracts or steam-distillates of these products. However, these flavors were usually weak and had high production costs (Eaton, 1994). As dairy-derived flavors are very complex and comprise a wide variety of components, it is difficult to mimic these flavors by a synthetic mix of the pure components. Usually some 'background' flavors are missing in synthetic mixtures. The development of a high quality artificial dairy-flavor would therefore be rather costly. In combination with the trend that consumers prefer natural ingredients instead of artificial ones, there was little incentive for the development of artificial dairy-flavors (Eaton, 1994). Moreover, natural dairy-derived flavors offer the advantage that all flavor components are already present and in the right concentrations. However, due to the increasing costs of dairy ingredients over the past years, the development of artificial flavors becomes more attractive; for instance, Shirakawa (2007) recently filed a patent on the preparation of a heat-stable, artificial milk flavoring.

The demand for natural dairy aroma and flavors is expected to increase. Major drivers are the increasing demand of healthier foods with reduced fat (and cholesterol) levels (Holland *et al.*, 2006) and the desire to use clean label, natural ingredients. Fat reduction leads to a change in flavor perception by the consumer, which is not always appreciated. Adding flavors to the food product may help to improve the final flavor. Flavors produced from milk and milk-derived products are considered as natural, and are widely used (Eaton, 1994). Consequently, there is a strong incentive to develop better natural flavors for applications in snack and other foods, especially for stronger and more characteristic cheese flavors.

Another trend is the demand for more intense flavors at a lower production cost. This especially applies to cheese flavors, because the production of natural cheese takes a relatively long time and is thus expensive.

So far, the downstream processing of fermented milk-derived flavors has not obtained much attention, apart from solving issues in flavor analysis. For analytical purposes, distillation and extraction procedures have been developed in order to be able to analyze the various components of the fermentation medium by chromatography.

The application of downstream processing techniques may open new opportunities for the production of novel type milk-derived flavors. Since, from an economic point of view, it is unattractive to isolate all individual components, it would be possible to isolate flavor blocks. The advantage of having different flavor blocks is that flavorists can combine these flavor blocks to create new compositions. Krammer *et al.* (2002) claimed the synthesis of a cheese flavoring by combining up to 14 flavor blocks of both volatile and non-volatile components.

18.2 Production methods for dairy aromas and flavors

18.2.1 Fermentation

For a long time, fermentation has been used for the production of various dairy products, such as cheese and yoghurt. For flavoring purposes, these products were applied as such, or extracts of steam-distillates of these products were used. Also, the starting material, milk, is an expensive raw material. Over the past years, milk prices have increased considerably, which makes the use of milk as a starting material for the production of a flavor less attractive. A lot of research efforts have therefore been directed towards the selection and development of production strains and the selection of alternative substrates for the production of dairy-like flavors (Smit *et al.*, 2005), and these are still going on. For instance, vegetable resources can be used as raw materials (Crow *et al.*, 2003; Ortega-Luis *et al.*, 2005; Pratt, 1987). It is therefore possible to select a cheap source for the production of dairy-like flavors.

18.2.2 Enzymes

Another development in cost reduction in the production of cheese flavors is the use of enzymes instead of fermentation. The application of enzymes leads to a significant reduction in processing time compared to the traditional processes, while also an increase in flavor concentration of 5 to 30 times is achieved (Wilkinson and Kilcawley, 2002). It also offers an opportunity to use other raw materials. Not only milk-ingredients, but vegetable resources can be applied as starting material (Tomasini *et al.*, 1995).

Especially for enzyme-modified cheese flavors (EMCs), a wide range of products is already available, including Cheddar, Mozzarella, Romano, Parmesan, Blue, Gouda, and many others. The flavor profile of these EMCs may be quite different from that of the natural cheese but, depending on the interaction with the applied base, the desired cheesy note may be obtained (Kilcawley *et al.*, 1998).

In an EMC process, selected enzymes (both proteases and lipases) are added to a cheese curd that is typically slurried with water (Kilcawley *et al.*, 2006). The applied enzymes amplify particular groups of flavorful compounds, such as short chain fatty acids, and peptides of a particular molecular weight. As a consequence, the products of these EMC processes lack the balanced flavor profile of a cheese that has been ripened naturally. An improvement of the flavor profile can be obtained by addition of a culture to an EMC (Crow *et al.*, 2003).

Enzymatic processes are also applied for the creation of enzyme modified butter (EMB) flavor, which uses lipase to produce free fatty acids. Volatile compounds are often lost, due to the processing conditions, leading to a lack of top notes in the produced flavor. Sometimes the flavors obtained by this type of process need an additional flavoring, especially when volatile components have a significant impact on the flavor. For example, a butter flavor is made from a combination of a starter distillate and enzyme-treated butterfat. Starter distillates contain >89% of water and the remainder is a mixture of volatile, butter-like flavor compounds, the so-called top notes. This starter distillate is prepared by (steam)distilling a fermentation medium prepared from milk ingredients in combination with a lipase (Rutloff, 1982).

18.3 Downstream processing of dairy aromas and flavors

18.3.1 Traditional methods

Apart from applying the dairy products, such as milk, cream, cheese or yoghurt, directly or in a concentrated form, also their extracts or

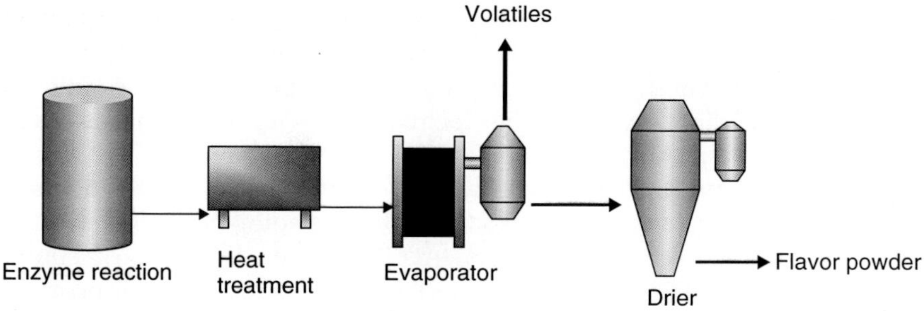

Fig. 18.1 Typical process scheme for the production of milk-derived flavors.

steam-distillates were used as a flavor. Both steam-distillation and extraction may lead to loss of flavors that are not removed from the original dairy product, leading to an incomplete flavor composition. As a result, a weak flavor is obtained with relatively high production costs.

The use of enzymes for the production of EMC and EMB lead to a strong cost reduction for these types of flavor. These flavors are usually manufactured as a powder. A typical process for the production of an EMC flavor is shown in Fig. 18.1. A slurry system of cheese or curd is incubated with enzymes under controlled conditions until the required level of flavor is obtained. Subsequently, the slurry is concentrated and spray-dried. Instead of concentration and spray-drying, freeze-drying may be used, especially in cases when heat-sensitive components are required in the flavor product.

Concentration and drying of flavors may lead to a loss of volatile compounds, and thus decrease the quality of the resulting flavor. A method to retain volatile components in the flavoring product is encapsulation. Whey powder is often used as the encapsulating agent. It is added, usually in combination with maltodextrin, prior to spray-drying (Anon., 1996; Bakal and Eisenstadt, 1988). Rapid cooling of the spray-dried material results in the formation of fat and sugar crystals in the matrix, which preserves the unique dairy flavor profiles (Anon., 1996). The use of ultrasound during preparation of the emulsion prior to spray-drying has a positive effect on the aroma retention and flavor quality (Mongenot *et al.*, 2000).

18.3.2 Distillation techniques

Steam-distillation has been used for a long time to produce dairy aromas from dairy products. However, high temperatures may affect the composition of the aroma obtained, because of degradation of heat-sensitive flavor components. Another disadvantage is that hydrophobic flavor components may be retained by the matrix of the original product, leading to a different composition of the aroma. As steam distillates contain a considerable amount of water (up to 90%), a weak flavor is obtained.

A relatively novel distillation technology is the Spinning Cone Column (SCC), developed and owned by FT Technologies. This technology is based on traditional steam-distillation, but in an advanced way. The SCC combines relatively mild process conditions in combination with a long path length and short residence time, which makes this equipment very suitable for processing heat-sensitive components. Moreover, this type of equipment can handle very viscous slurries, and is therefore suitable for processing fermentation media (Pyle, 1994). In the dairy industry in New Zealand, this technology is used for the removal of pasture-related flavors from cream in order to maintain a constant product quality throughout the year [www.ft-tech.net]. It also allows specific flavor profiles to develop, by adjustment of the stripping rate, the operating temperature and the off-set temperature. The technology is also used for aroma recovery in the (partly) de-alcoholization of wine, the recovery of coffee aromas, and the production of fruit aromas. Up to now, the production of dairy aromas by using SCC has not been reported.

18.3.3 Membrane technologies

Micro-, ultra- and nanofiltration (MF, UF or NF)

Membrane technologies are widely used to separate biomass from liquid. The membrane type (MF, UF, NF) is chosen depending on the desired permeate composition. In the case of NF, only small molecules will be present, whereas in the case of micro filtration, larger molecules, such as proteins, may also be present in the permeate. These types of filtrations are suitable for the preparation of aqueous solutions containing water-soluble aroma compounds.

Reversed osmosis (RO)

Reversed osmosis may be used as an alternative to concentration by evaporation in flavor production. Most of the flavor components will be retained, but loss of volatiles may occur as small molecules will pass through the RO membrane. A drawback of concentration by RO is the relative low concentration factor that can be obtained, due to the osmotic pressure of the concentrate. A new development in this respect is high pressure RO (up to 150–200 bars), allowing higher osmotic pressures of the concentrate. Tubular RO membranes with a large diameter, of about 14 mm, allow for high solids contents. This is especially attractive for fermentation media containing volatile components, as most of these components will disappear during evaporation. A separate removal of these volatiles is not required when applying high pressure RO, because most of the volatiles will be retained in the concentrate. As this technology is still under development, there are no turn-key solutions for implementation. High pressure RO can be carried out at ambient temperatures, which makes it very attractive for concentration of heat-sensitive materials.

Other osmotic techniques

Other concentration technologies based on osmotic principles are still under development. For direct osmosis, a semi-permeable, water-selective membrane is used. An aqueous solution with a higher osmotic pressure than the liquid feed is present on the other side of the membrane. The liquid feed is then concentrated by a difference in osmotic pressure as the driving force (Jariel *et al.*, 1996; Herron *et al.*, 1992).

Another development is osmotic evaporation. In this case, a macroporous hydrophobic membrane is used. Moreover, an aqueous solution with a high osmotic pressure is present at the permeate side. The driving force for the removal of water from a liquid stream is the difference in water vapour pressure of the liquid feed and the aqueous solution. In this case, very high concentration can be obtained. For instance, a fruit juice has been concentrated up to 65 ° Brix, while retaining most of the volatile components (Jariel *et al.*, 1996; Hogan *et al.*, 1998). Membrane distillation is based on the same driving force as osmotic evaporation, but instead of an aqueous solution, a gas phase is applied on the other side. However, more losses of volatiles are observed for membrane distillation, and therefore this technology will be less attractive for the concentration of flavors (Jariel *et al.*, 1996).

It should also be noted that the new technologies have been applied for the concentration of clarified juices, and that no information is available on their suitability for processing fermentation media.

Pervaporation

Another method for the recovery of volatile components is pervaporation. Pervaporation is a membrane-assisted separation process, in which a liquid feed mixture is separated by partial vaporization through a non-porous perm-selective membrane. The process results in a vapor permeate and a liquid retentate. One or more components in a liquid mixture absorb on one side of a dense polymeric membrane, diffuse through the membrane thickness, and desorb and evaporate at the opposite membrane surface, with the driving force being a difference in vapor pressure in the concentrate and permeate stream. The flux through the membrane is therefore dependent on the processing temperature, the concentration of the component and the applied vacuum at the permeate side. For removal of volatile organics from aqueous streams, hydrophobic membranes are used. Processing can be done at moderate temperatures, so this technology is also suitable for heat-sensitive materials.

Baudot and Marin (1996) described the recovery of methylthiobutanoate (hydrophobic molecule with cheese fragrance) and the recovery of diacetyl (hydrophilic, butter aroma) from aqueous model solutions using pervaporation. The highest enrichment (factor 290) was obtained at the lowest permeate pressure, as in this case the largest difference in vapor pressure is obtained. The optimal membrane is depending on the nature (hydrophobicity) of the volatile component. For the recovery of volatile components from an aqueous solution or medium, the use of a hydrophobic membrane is most suitable. Sibeijn *et al.* (2004) did not find any differences in two types of hydrophobic membranes for sulphur containing volatiles (dimethyl sulphide and dimethyl disulfide), whereas large differences were observed for aldehydes, ketones and esters.

Baudot and Marin (1996) assumed that the matrix of the feed may affect the vapor pressure of the aroma components and thus the flux of the aroma compounds over the membrane. For instance, lipid components present in the feed may specifically sorb hydrophobic components, which lowers the vapor pressure of these components in the feed. Also proteins, such as caseins, may interact with volatile compounds.

Sibeijn *et al.* (2004) showed that the feed matrix affects the recovery of volatile components by pervaporation. The composition of the permeate of an aqueous model solution containing key-components of a yoghurt flavor such as acetaldehyde, diacetyl, 2-butanone, 3-methyl butanal, 2-petanone, 2,3-pentadione and hexanal was compared with the permeate composition of a yoghurt fermentation. The enrichment factors of the various key-compounds in the permeate differed from each other, both in the model solution and in the yoghurt fermentation, which could only partly be ascribed to different concentrations in the feed. For 3-methyl butanal an enrichment factor of more than 1000 could be obtained, whereas for acetaldehyde an enrichment factor of 300 was found. Overington *et al.* (2008) also determined enrichment factors for various flavor compounds, such as esters, ketones and acids. In these experiments, different enrichment factors were also obtained. Differences can be ascribed to differences in vapor pressure (concentration in the feed), hydrophobic nature and chain length of the molecules, but also to interactions with the feed matrix.

The use of pervaporation for the recovery of dairy aromas will not result in a concentrate with the same volatile composition as the feed, because of the different permeation rates of the various components. Pervaporation may be used for the preparation of concentrated flavor blocks from (diluted) feeds, and can be used in combination with other flavor blocks to prepare a semi-natural flavor as described by Krammer *et al.* (2002).

18.4 Future trends

Up to now, milk-derived flavor production consisted of fermentation or enzymatic treatment of milk raw materials, followed by a heat-treatment to inactivate the micro-organisms or enzymes and concentration of the medium, optionally followed by drying. Other downstream technologies, either available or under development, can now be used for the production of a novel generation of milk-derived flavors.

In Table 18.1, a summary is given for the various available technologies that may be used in the production of milk-derived flavors. From this table it can be concluded that most direct concentration methods suitable for fermentation media containing volatiles or heat-sensitive materials, viz. high pressure RO, direct osmosis and osmotic evaporation, are still under development. High pressure RO is closest to realization on an industrial scale.

For the production of dairy flavors by using mature technologies, a combination of technologies will be required to produce highly concentrated flavors; for instance, by first removing the volatile or heat-sensitive components, concentration of the remaining process stream and subsequently addition of the volatiles to the concentrated process stream, optionally followed by drying. In Fig. 18.2 this principle is shown for the production of buttermilk powder. A buttermilk powder with a more intense flavor can be obtained when the volatiles are isolated using SSC prior to evaporation and are added to the concentrate prior to spray-drying. The same technology can be applied to a fermentation broth in order to obtain a powder with a more intense flavor. The processing will comprise removal of the volatile components; further concentration of the fermentation broth; addition of the isolated volatile components; and drying.

Alternatively, membrane filtration technologies may be used in the production of concentrated fermented flavors. For instance, the water-soluble, small flavor components can be separated from the fermentation medium by ultrafiltration or nanofiltration, depending on the desired composition of the permeate. Further concentration can be carried out with high pressure reverse osmosis (up to 150–200 bars), allowing higher osmotic pressures of the concentrate. The demand for more concentrated flavors can be met by applying a combination of technologies which separate the flavor components from the other ingredients of the fermentation medium by membrane filtration, and then by further concentrating using a mild concentration technology that retains the volatile components.

The use of various flavor blocks for composing a specific flavor may be an attractive way to produce flavors by fermentation of raw material containing proteins and/or fats, because it is possible to compose, for example, different (cheese) flavors by using a selection of the various flavor blocks (Krammer *et al.*, 2002). This offers an opportunity to produce naturally-derived flavors in a cost-effective way. The flavor blocks can be produced by concentrating the volatiles produced during the fermentation. For instance, distillation (SCC) or pervaporation of a fermented medium will yield a concentrate of top-notes. The composition will be dependent on the concentrations of volatiles present in the fermentation broth.

To conclude, development of processing routes for the isolation of milk-derived flavors should be done in combination with a good strain selection for fermentation and the required application for the specific flavoring, as it appears that the matrix in which the flavoring is applied affects the final

Table 18.1 Overview of various technologies for the processing of milk-derived flavors

Technology	Driving force	Advantage	Disadvantage	Challenge	Opportunity
Evaporation	Δ Vapor pressure	Mature technology High concentration factors	Loss of volatiles Degradation of heat-sensitive compounds High energy use		In combination with SCC or membrane filtrations in order to preserve volatiles or heat sensitive components
Spinning Cone Column	Δ Vapor pressure	Mature technology Heat-sensitive products Viscous and/or solids-containing feed High concentration factors	Isolation of top notes only Flavor composition different from source		Removal of volatiles prior to further processing and addition of volatiles to final product. Production of highly concentrated liquid flavors
Pervaporation	Δ Vapor pressure	Heat sensitive materials Selectivity High concentration factors	Isolation of volatiles Flavor composition different from source	Application in organic removal from aqueous streams Development of more selective membranes	Further concentration of top-notes Production of sulphur-containing flavor blocks
Membrane filtration (MF/UF/NF)	Δ Pressure	Heat sensitive materials Separation of solids and large molecules Size selectivity	No concentration method Flavor composition different from source		Production of concentrated liquid flavors in combination with RO. Separate processing of retentate and permeate, and recombination thereof

Table 18.1 *Cont'd*

Technology	Driving force	Advantage	Disadvantage	Challenge	Opportunity
Reverse osmosis	Δ Pressure	Identical flavor composition as source material Heat-sensitive materials	Relatively low concentration factor High-pressure equipment under development on industrial scale	High pressure RO to obtain higher concentration factors Application of viscous and/or solids-containing feed	Concentration of heat-sensitive streams. Production of concentrated flavors with identical profile as original source
Direct osmosis	Δ Osmotic Pressure	Heat-sensitive materials High concentration factors Viscous and solids-containing feed	Novel technology, still under development Low flux	Application to feeds other than fruit juices Scale up	Concentration of heat-sensitive streams. Production of concentrated flavors with identical flavor profiles original source
Osmotic evaporation	Δ Water vapor pressure	Heat-sensitive materials Very high concentration factors	Novel technology, still under development Low flux	Application to feeds other than fruit juices Scale up	Concentration of heat-sensitive streams Production of concentrated flavors with identical profile as original source
Membrane distillation	Δ Osmotic pressure	High concentration factors	Degradation of heat sensitive components, loss of volatiles Novel technology, still under development Low flux	Application to feeds other than fruit juices Scale up	Concentration of heat sensitive streams

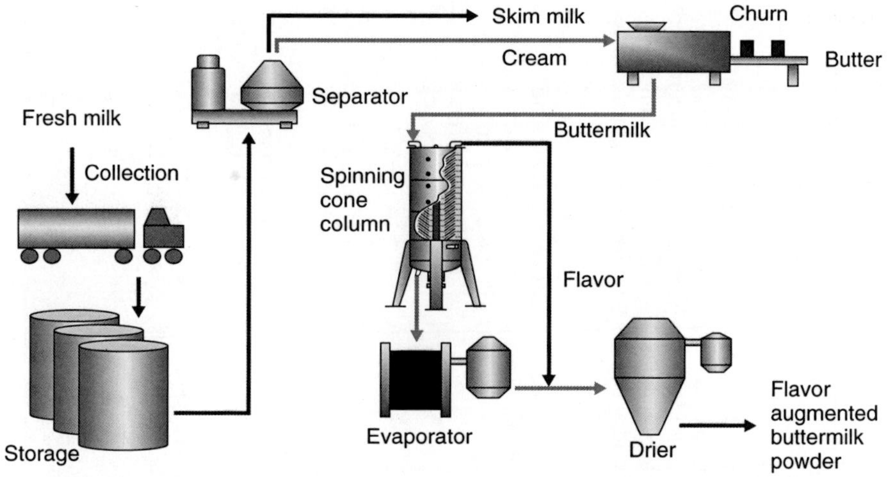

Fig. 18.2 Production of an improved buttermilk flavor by applying SSC-technology for isolation of the volatile components (courtesy of Flavourtech Europe Ltd).

flavor. Therefore, a close cooperation of microbiologists, process developers and application specialists is required for the successful development of novel strategies for flavor production.

18.5 References

ANONYMOUS (1996), Concentrate on Flavour, *Dairy Industries International* 61 (12), 33, 35.

BAKAL A.I. and M.E. EISENSTADT (1988), *Granulated butter-flavored products*, European Patent Application EP 0305042 to Cumberland Packing Corporation.

BAUDOT A. and M. MARIN (1996), Dairy aroma compounds recovery by pervaporation, *Journal of Membrane Science* 120, 207–220.

CROW V.L., M. HAYES, B.J. CURRY, P.K. SAMAL, R.H. SCHOFIELD and R. HOLLAND (2003), *Cheese flavor ingredient and method for its production*, WO03/047358.

EATON D.C. (1994), 'Dairy Flavors'. In: A. Gabelman (editor), *Bioprocess Production of Flavor, Fragrance and Color Ingredients*. John Wiley & Sons, New York 169–203.

HERRON J.R., E.G. BEAUDRY, C.E. JOCHUMS and L.E. MEDINA (1992), *Osmotic concentration apparatus and method for direct osmotic concentration of fruit juices*, US Patent 5281430.

HOGAN P.A., R.P. CANNING, P.A. PETERSON, R.A. JOHNSON and A.S. MICHAELS (1998), A new option: Osmotic distillation, *Chem. Eng. Progr.* July 49–61.

HOLLAND R., V. CROW, G. NORRIS, M. BENNETT and T. COOLBEAR (2006), Cheese flavor for the future – using traditional tools to deliver new flavors in new ways, *Australian Journal of Dairy Technology* 61, 97–104.

JARIEL O., M. REYNES, M. COUREL, N. DURAND, M. DORNIER and P. DEBLAY (1996), Comparison of some fruit juice concentration techniques, *Fruits* 51, 437–450.

KILCAWLEY K.N., M.G. WILKINSON and P.F. FOX (1998), Enzyme modified cheese, *International Dairy Journal* 8, 1–10.

KILCAWLEY K.N., M.G. WILKINSON and P.F. FOX (2006), A novel two-stage process for the production of enzyme modified cheese, *Food Research International* 39, 619–627.

KRAMMER G., P. WERKHOFF, G. HANSMANN and S. HIMMER (2002), *Cheese flavoring*, US patent 20020034570, EP 1166650.

MONGENOT N., S. CHARRIER and P. CHALIER (2000), Effect of ultrasound emulsification on cheese aroma encapsulation by carbohydrates, *Journal of Agricultural and Food Chemistry* 48, 861–867.

ORTEGA-LUIS A.G., I. MONTELONGO, D. MADRIGAL and E. ROGERT (2005), Cheese flavor for imitated fused cheese, *Ciencia y Technologia de Alimentos* 15, 53–55.

OVERINGTON A., M. WONG, J. HARRISON and L. FERREIRA (2008), Concentration of dairy flavor compounds using pervaporation, *International Dairy Journal* 18, 835–848.

PRATT N.G. (1987), *Preparation of a flavor*, EP 0312746.

PYLE L. (1994), Processed foods with natural flavor: The use of novel recovery technology, *Nutrition and Food Science* 94, 12–14.

RUTLOFF H. (1982), Biotechnology and aroma production, *Nahrung* 26, 575–589.

SHIRAKAWA T. (2007), *Perfume composition and process for producing the same.* US 7252849.

SIBEIJN M., H.C. VAN DER HORST, C. DE JONG and G. SMIT (2004), Isolation of fermented dairy flavors, *Bulletin of the International Dairy Federation* 389, 69–73.

SMIT G., B.A. SMIT and W. ENGELS (2005), Flavour formation by lactic acid bacteria and biochemical flavor profiling of cheese products. *FEMS Microbiology Reviews* 29, 591–610.

TOMASINI A., G. BUSTILLO and J.-M. LEBEAULT (1995), Production of blue cheese flavor concentrates from different substrates supplemented with lipolyzed cream, *International Dairy Journal* 5, 247–257.

WILKINSON M.G. and K.N. KILCAWLEY (2002), Technology of enzyme modified cheese and natural cheese, *Bulletin of the International Dairy Federation* 371, 10–15.

19

Dairy ingredients in non-dairy food systems

B. T. O'Kennedy, Moorepark Food Research Centre, Teagasc, Ireland

Abstract: This chapter includes a brief introduction to the dairy ingredients available for incorporation into a wide range of food products, their functionality in isolation and their resultant functionality in food systems.

Key words: dairy ingredients, processed meats, emulsions, chocolate, bakery.

19.1 Introduction to dairy ingredients and their potential use in other food systems

The manufacture of many dairy ingredients involves the fractionation of milk into its various components and subsequent dehydration. The dairy industry has the advantage of dealing with a food stream naturally in the liquid state, which makes large scale fractionation of the various components more efficiently realisable. This essentially uses the inate physicochemical properties of the milk components (solubility, enzymatic susceptibility, pH dependency, etc.) to separate, concentrate and usually dehydrate, to produce ingredients suitable for further exploitation. The use of dairy-based ingredients in other food systems is often driven by the same inate physicochemical properties that make separation possible. The ability of dairy ingredients to deal with water and other components present in food systems makes them invaluable as a tool for improving, stabilising or changing the structural, textural or organoleptic properties of a diverse range of food systems. This chapter will attempt to logically approach the introduction of dairy ingredients into such diverse food products as processed meat, chocolate, edible emulsions and bakery products. The main dairy ingredients obviously centre around the composition of the milk itself (protein, fat and lactose), thus giving a vast range of ingredients, often tailored for specific purposes. While bovine milk is specifically designed for nutritional purposes, getting the greatest nutritional packet into the most efficient volume, while keeping the inate viscosity of this nutritional 'soup' within limits for biological excretion, these co-operative properties can be further utilised for functional expression in other food systems. This requires a major co-operation between heterogeneous proteins and mineral salts, while still maintaining a unique stability to overaggregation which would be biologically unacceptable.

Milk is a dynamic material and its properties are subject to much variation due to, among other things, cattle breed and age, diet, health, stage of lactation and the environment where the animal lives. Milk contains ~13% solids, which are mainly composed of fat, protein and the milk sugar, lactose. As a mixed colloidal dispersion, milk is extremely stable and can be heated, frozen and dried without any outward sign of destabilisation.

In its unprocessed state, milk contains fat globules in the range 1–10 μm in diameter, which, on standing, form a typical cream layer that may be dispersed by gentle stirring. The capacity to cream at low centrifugal force makes milk fat separation relatively easy. The fat globules are stabilised by the natural milk fat globule membrane, which is composed of a mixture of proteins, phospholipids, lipoproteins, cerebrocides and sterols, some of which may have biological activity. Phospholipids are the most surface active components (surfactants) and can therefore form a barrier between the fat, and the water in which the fat globules are suspended. Sterols are highly insoluble in water and associate with phospholipids, the most well known being cholesterol.

The main proteins in bovine milk are casein and whey proteins. Casein exists as discrete spherical particles of 200 nm diameter, generally referred to as casein micelles, constituting 2.5–2.8% (w/w) of the milk. The structure of these casein micelles is determined by the interaction between the protein and calcium phosphate. This group of milk proteins are insoluble at their point of net neutral charge (pH 4.6) and this property is of major importance to the structure of some foods.

Whey proteins constitute about 0.5% (w/w) of milk and are classified as the proteins that remain in solution after coagulation of the casein through rennet action, or by acidification. This group of proteins are heat labile and, in certain food applications, are intentionally heat-denatured to promote functionality development.

Lactose is the major sugar in milk (4.6%, w/w) and is in true solution. It is a reducing disaccharide composed of glucose and galactose, and forms a major source of nutrition for most of the micro-organisms that grow in milk, to either ferment it or cause spoilage by acidification or gas production.

The main soluble salts comprise potassium, sodium, calcium, magnesium, chloride and phosphate. The organic anion, citrate, is also an integral part of this soluble salt solution. The importance of this soluble salt system in maintaining the integrity and stability of the complex colloidal dispersion cannot be over-emphasised. The general composition of milk is outlined in Table 19.1. It is important to note that these are average values and can

Table 19.1 Typical levels of the major colloidal and soluble components in bovine whole milk

Component	Colloidal	Soluble
Fat (%)	4.0–4.4	
Casein (%)	2.5–2.8	
Whey protein (%)		0.5–0.6
Lactose (%)		4.4–4.6
Calcium (mg/100 ml)	80	40
Magnesium (mg/100 ml)	3	8
Potassium (mg/100 ml)		150
Sodium (mg/100 ml)		50
Chloride (mg/100 ml)		100
Phosphorous (mg/100 ml)	45	45
Citrate (mg/100 ml)	15	170

vary, hence affecting the functionality of the milk. However, it is more important to visualise the various components and their place or function in the complex colloidal dispersion we call milk, and the food ingredients that can be derived from it.

19.1.1 Whole and skimmed milk powder

A major family of products in the dairy ingredients portfolio is dried milk, either as whole milk powder (WMP) or skimmed milk powder (SMP). While WMP is generally standardised to 26% milk fat with separated cream, SMP is produced from milk which has had the cream removed. The skimmed milk/whole milk is concentrated by vacuum evaporation to >45% solids prior to spray drying. The physical properties of the powders (bulk density, particle size, flow properties) are determined both by the solids content of the feed to the dryer and subsequent agglomeration of the powder particles. A pre-heat treatment of the milk is central to the final functionality of the rehydrated powders, which are generally available as low, medium and high heat powder. The preheat treatment can range from 72°C for 15 sec to 120°C for 2 min. Note, that high-heat treatments are performed on the unconcentrated milk prior to evaporation and drying. The major effect of the preheating temperature and duration is to alter the level of native whey proteins as they are denatured under severe heating conditions. This results in the formation of a new colloidal particle which is now composed of casein, whey protein and calcium phosphate. While this family of dairy ingredients is the simplest, in that the ingredients require little separation prior to concentration and drying, they form a historical source of 'milk' which is utilised in vast quantities in both dairy-related and other food systems. These types of dairy products are key ingredients because of their functional, nutritional and organoleptic characteristics in food applications, as well as their storage and economic versatility.

19.1.2 Casein-based ingredients

Casein represents a heterogeneous group of proteins composed of varying levels of α_{s1}, α_{s2}, β and κ-caseins. The interactions of these proteins under the conditions present in milk give casein its distinctive properties. Due to the liquid nature of milk and the physico-chemical differences between the various components, the dairy industry actively fractionates these components in a range of different forms. The solubility differences between casein and the whey proteins, and the inherent pH-dependent solubility characteristics of calcium phosphate, make the separation of casein and the production of a range of casein-based ingredients readily achievable.

Acid casein

The commercial manufacture of acid casein from skim milk has been reviewed by Mulvihill (1989). The production of this ingredient depends on the inate ability of casein to aggregate at a pH where the net charge on the protein tends towards zero (a property utilised in the production of fermented milk products). It involves acidification of pasteurised (72°C for 15 s) skim milk at 30°C with sufficient acid to change the pH to 4.4–4.7 (the isoelectric point of the casein). This can be achieved with mineral acids (HCl) or through fermentation of the lactose in skim milk, producing the more natural lactic acid. It is important that the initial temperature treatment causes minimal denaturation of whey proteins (heat denatured whey proteins will coagulate at the acidic pH, either separate from or in association with the casein). Casein, as it exists in fresh skim milk, is in micellar form (particles containing calcium phosphate with diameters in the range 60–300 nm). At acid pH values, however, colloidal calcium solubilises and remains with the acid whey following separation from the casein. The separated casein curd is washed to remove residual lactose, soluble salts and native whey proteins, before being dried. While dried acid casein itself is functionless (cannot be redispersed easily in water to its initial hydrated state prior to drying), it is used in the production of the highly functional caseinate family. These include sodium, potassium and calcium caseinate, depending on the alkali type used to increase the pH of the protein dispersion to a neutral pH (6.5–7.0). These are high protein ingredients (up to 90% protein in dry matter), and have large applications in the food industry as emulsifiers, whiteners, etc.

Rennet casein

Micellar casein as it exists in milk is susceptible to the action of the enzyme chymosin (often called rennet), resulting in the formation of a gel or curd. This enzyme cleaves a crucial fragment of κ-casein, which destabilises the casein micelle, forming a curd. The curd is washed free of lactose, whey proteins and soluble salts, subsequently dried and is generally referred to as rennet casein. Both rennet and acid casein use the same basic separation technologies (separation of coagulated casein from the soluble serum phase, resulting in a high protein concentrate). The major differences between acid and rennet casein are (i) κ-casein intact in acid casein; (ii) high calcium phosphate content of rennet casein and (iii) acid casein can solubilise on pH adjustment. Rennet casein requires the use of emulsifying salts (citrates, polyphosphates) to initiate functionality, hence their usage in analogue cheese. Both acid and rennet casein are from a practical point of view 'functionless' (extremely difficult to rehydrate in water to a level which existed prior to separation, concentration and dehydration of the original acid or rennet-induced gel).

Phosphocasein

This is essentially micellar casein as it occurs in milk and is generally produced using membrane technology, therefore not requiring a precipitation step as seen with both acid and rennet casein production (Schuck *et al.*, 1994). Membrane technology utilises retention and permeation principles, the membrane pore size determining whether any particular entity in the milk remains in the retentate or is eliminated in the permeate. Microfiltration results in the elimination of the native whey protein, lactose and soluble salt fractions, retaining the micellar casein with its associated colloidal calcium phosphate. Following microfiltration, diafiltration, evaporation and drying, a 75–80% protein powder is produced. Due to the inherent low ionic strength conditions prevailing in phosphocasein dispersions, stability to flocculation at processing temperatures (60–80°C) at $pH < 6.7$, is limited (Le Ray *et al.*, 1998). However, re-introduction of the soluble milk salts, in the form of milk ultrafiltrate, can restore the inate heat stability of the micellar casein. Phosphocasein exhibits physico-chemical and micellar behaviour similar to milk in terms of particle size, rennet gelation (Pires *et al.*, 1999) and acid gelation (Famelart *et al.*, 1996).

19.1.3 Whey protein based ingredients

The whey proteins present in milk are mainly comprised of β-lactoglobulin, α-lactalbumin, bovine serum albumin and immunoglobulins, in order of concentration. These proteins are particularly heat-labile and care needs to be exercised during processing to preserve their native structure. Ironically their potential functionality is generally observed when these proteins are heat-denatured. When whey proteins are heated above a critical temperature (the denaturation temperature), they change their conformational state and are now susceptible to aggregation (the extent of which is dependent on the environmental conditions, pH, ionic strength, etc.). In milk systems, this heat-induced aggregation of whey proteins can result in the association of the aggregated whey proteins with the casein micelle (Vasbinder and de Kruif, 2003). Such heat treatment of milk can have a negative or positive functional effect, depending on what the milk is used for. For yoghurt production, it is a necessary pre-requisite, improving the water holding capacity and strength of the acid gel. Conversely, for cheese production, a negative effect on rennet gel formation and subsequent syneresis is observed resulting in high moisture curd. Heat-induced denaturation and subsequent aggregation of concentrated whey protein dispersions can result in the formation of functional water binding gels.

Sweet whey is the liquid by-product of chymosin-coagulated milk, either from cheese production or rennet casein manufacture. It is termed sweet because it has a typical pH in the region of 6.0–6.4. Acid whey is generally a by-product of cottage cheese or acid casein production and has a pH of about 4.6. Typical compositions of both sweet and acid whey are outlined in Table 19.2. The major compositional differences between the different whey streams are the mineral content, lactate concentration, and the presence of glycomacropeptide. Glycomacropeptide is the κ-casein fragment cleaved through the action of chymosin. Acid whey from casein manufacture contains significantly higher levels of calcium, phosphate and chloride (HCl addition). Cheese whey contains some lactate from the action of the starter cultures.

Non-hygroscopic whey powder can be produced through evaporation of the whey to >60% solids followed by lactose crystallisation, giving a powder protein concentration of ~12% (w/w). The whey protein content of the powders can be increased by using ultrafiltration and diafiltration, prior to evaporation and drying. These are termed whey protein concentrates

Table 19.2 Typical composition of sweet and acid whey

Component	Rennet casein whey	Mineral acid whey	Cheddar cheese whey
Solids (g/L)	66	63	67
pH	6.4	4.7	5.9
Protein (g/L)	6.2	5.8	6.2
Non-protein nitrogen (g/L)	0.37	0.3	0.27
Lactose (g/L)	52.3	46.9	52.4
Lactate (g/L)	–	–	2.0
Ash (g/L)	5.0	7.9	5.2
Fat (g/L)	0.2	0.3	0.2
Calcium (g/L)	0.5	1.2	0.47
Phosphorous (g/L)	0.46	0.63	0.46
Magnesium (g/L)	0.07	0.11	0.08
Sodium (g/L)	0.58	0.5	0.58
Potassium (g/L)	1.45	1.4	1.5
Chloride (g/L)	1.02	2.25	1.0

(WPC) and can range in protein content from 35–80% (w/w). Demineralised whey powders can be made by ion-exchange and electrodialysis, and are often used in infant formulations. Whey protein isolates (WPI) are generally produced by ion-exchange and can be up to 92% protein (w/w). The application of these products in foods will depend on the functionality required. Whey powders are used as a source of lactose but where the more functional whey protein effect is required, the use of WPC may be the choice.

19.1.4 Milk protein concentrates/isolates

Milk protein concentrate (MPC) is usually produced using membrane technology and, as its name suggests, contains both casein and whey proteins in the ratio that exists in milk. MPC is available in a range of protein levels from 42 to 85%, the lactose level falling as the protein level increases. MPC is produced from skim milk by use of ultrafiltration and/or diafiltration, evaporation and drying technology. The proteins, along with the lactose and minerals that did not pass through the membrane (permeate), remain in the retentate. Diafiltration or washing of the retentate is necessary to give protein concentrations higher than 65% in the final powder. The casein fraction is in the native micellar form and therefore carries significant quantities of calcium and phosphate. MPC can be used for its nutritional and functional properties. The high protein, low lactose ratio makes MPC suitable for protein-fortified beverages and low-carbohydrate foods. The whey protein is usually still in the native form but can easily be denatured prior to processing if required. Milk protein isolate is defined as having at least 90% protein and is the top of the MPC range of ingredients. MPC can also be produced by precipitating the proteins out of milk or by dry-blending the milk proteins with other milk components. The selection of MPC ingredients available ranging from skim milk powder (35% protein), MPC 42, MPC 70, MPC 75, MPC 80, MPC 85 and finally MPI 90, give a selection of compositional and functional attributes as a result of the changing mineral to protein ratio.

19.1.5 Lactose

Lactose is a disaccharide composed of D-glucose and D-galactose, and is significantly less sweet than sucrose. It is a reducing sugar and is used to provide a Maillard reaction-linked browning in certain foods. It is also routinely added to soup and sauce formulations. It is the major carbohydrate substrate for Lactobacillus spp. during fermentation of milk. Industrially, it is produced from whey which has been concentrated to >60% solids. This forms a supersaturated lactose solution at temperatures >90°C. On cooling slowly, α-lactose monohydrate crystals are formed, which are separated using decanters, washed, and dried in fluidised bed driers.

19.1.6 Milk fat

Milk fat, in the form of butter or cream, is used as an ingredient in a number of food systems (confectionary, cream liqueur, etc.). Cream can have application in food products such as soups, sauces and toppings, and can be produced at various fat contents, depending on the function required. Cream is an oil-in-water emulsion. The milkfat globules in unhomogenised cream have a mean diameter of 3–4 μm. They are stabilised by their natural membrane comprised of phospholipids, lipoproteins, cerebrocides and proteins. The fat itself is composed of a range of triacylglycerols, the distinctive flavour being conveyed by the relatively high concentration of butyric acid. Cream is separated from milk by centrifugal separators which are capable of producing variable cream products from 12–70% fat.

The shelf-life of these creams is limited by microbiological contamination and by the separation of the fat phase during chilled storage to form a cream layer. Homogenisation reduces separation during storage and also varies the viscosity of the product. Cream for whipping, in general should not be homogenised, even at low pressures, as this will affect the functionality. The functionality (whipping ability) of whipping cream is dependent on fat concentration, temperature and the presence of the natural fat globule membrane. Whipping causes partial destabilisation of the cream emulsion. Gradually during whipping, networks of coalesced fat globules build up to surround the air bubbles and form a three-dimensional matrix which provides the foam's rigidity.

The production of sweet-cream butter or salted butter utilises the phase inversion of the oil-in-water emulsion (cream) to a water-in-oil emulsion (butter). Cooling the cream prior to churning is an important pre-requisite in the production of butter. Generally, cream is cooled to 4°C to allow seeding of the fat crystals to occur, reheated to 18–20°C for two hours, when slow crystallisation of the higher melting fat components can occur, and finally slowly cooled to the churning temperature (16°C). Salt is incorporated into butter primarily to enhance flavour, but also for microbiological stability.

19.2 Functionality of dairy ingredients

19.2.1 Effect of pH, ionic strength, ionic species, temperature

Varying ratios of proteins, fat, sugars, salts and water make up the compositional differences we observe as being different foods. The application of various processing conditions also affects the texture, structure and general organoleptic appreciation of foods in general. They may also contain permitted additives, such as emulsifiers, stabilisers, colours and flavours. The structure, texture and stability of any food that contains milk proteins will be affected by pH, ionic strength, and temperature. This will also be applicable to emulsions stabilised by milk proteins. To put this approach into perspective, casein micelles are unstable to heat and pH reduction when the ionic strength of the aqueous phase is too low (Auty *et al.*, 2005). This suggests that texture of foods from dairy sources or containing dairy protein ingredients is to a large extent dependant on the soluble salt content. Whey protein denaturation and subsequent aggregation is central to texture and structure development in dairy desserts, mousses and yoghurt. In these types of foods, heat-induced whey protein denaturation and aggregation may occur at neutral pH (6.7) prior to fermentation or acidification to their final pH values. In general, the structure of dairy products is determined by the pH and the ionic environment. It has been recognised for a long time that the presence of divalent cations (calcium) enhances the aggregation of casein at all temperatures and of whey proteins on heat-induced denaturation. Heat-induced denaturation/aggregation of β-lactoglobulin is highly pH-dependant. At low ionic strength, maximal denaturation/aggregation is achieved on heating at pH 5.0, with minima occurring at pH < 3.5 and 6.0 (O'Kennedy and Mounsey, 2008).

Protein hydration is generally considered to be the water that is more-or-less immobilised by a protein (Damodaran, 1997). Creamer (1985) showed that casein micelle hydration was affected by pH, NaCl and rennet treatment. The addition of NaCl was shown to increase casein micelle hydration at all pH values above 4.5, either by displacing calcium or calcium phosphate from the protein matrix (with a concomitant increase in the number of ionic groups and a consequent increase in the volume of the matrix) or by affecting the ability of the matrix to aggregate. The water binding of whey proteins is highly influenced by the heat-induced unfolding and aggregation of the individual whey proteins. The effect of pH and specific ions on whey protein aggregation is an important area for consideration when attempting to optimise the water holding ability (O'Kennedy and Mounsey, 2008).

19.2.2 Emulsification

Foods can include emulsions with a variable fat content. They can be water continuous such as yoghurt, or oil-continuous such as table spreads. The strongly amphipathic nature of proteins, particularly dairy proteins, resulting from their mixture of polar and nonpolar side chains, causes them to be concentrated at interfaces. At fluid/fluid interfaces, it is well established that globular proteins (whey proteins) in general, lose their tertiary structure, existing in extended configurations with hydrophobic side chains orientated towards the non-aqueous phase and hydrophilic side chains directed towards the aqueous phase. The effect of oil droplet size and emulsifier type on the rheology of emulsion gels was studied by McClements *et al.* (1993). The modulus increased as the oil droplet size decreased for droplets stabilised by protein, but decreased when the oil droplets were stabilised by low molecular weight surfactants (Tween). Oil droplets coated with protein could be incorporated into the protein network, reinforcing the structure and so increasing the modulus, whereas droplets coated with Tween (ethoxylated sorbitan), which could not be incorporated into the protein network and therefore disrupted the three-dimensional structure of the gel, decreased the modulus. This is a simplistic but intuitive view of the inherent complexity of multicomponent composite food systems. A more elaborate view of the problem is provided by Gaygadzhiev *et al.* (2009), where, using whey protein-stabilised fat globules in a dispersion of casein micelles, they highlight the importance of the flocculation state of the emulsion droplets in affecting the structure formation of the gel formed on renneting. An earlier study (Sala *et al.*, 2007) examined the modulus of emulsion-filled gels, which was varied by changing the interactions between oil droplets and the gel matrix. Where the oil droplets were bound to the gel matrix, the modulus increased, while unbound oil droplets decreased the modulus. However, they stressed that the degree of aggregation of oil droplets in the matrix affected the modulus to an extent related to the size and to the stiffness of the aggregates.

The source and conformation of the dairy-based protein emulsifier used to stabilise emulsions in foods can have great effects on texture. As the casein-based emulsifier changes from sodium caseinate to calcium caseinate to calcium phosphate caseinate, the properties of the resulting emulsions markedly change, due to the differing functionalities of the emulsifiers themselves.

19.2.3 Gelation

The formation of three-dimensional matrices during the processing of food systems is closely related to the functional attribute that is termed 'heat-gelation'. Whey proteins are globular proteins and hence are heat denaturable. They are widely used to enhance the nutritional value of formulated foods and are often cited as being highly functional through their ability to gel on heating (De Wit, 1989; Hermansson, 1975). Whey proteins denature when the temperature exceeds ~70°C and subsequently aggregate into complexes which, depending on the protein concentration, pH and general environmental conditions, may coagulate, precipitate, or gel. This is closely aligned with the water-holding capacity of the protein (see Section 19.2.1). Thermal induction of gelation is normally the functional property of interest, where large quantities of water are immobilised. Characteristics of whey protein gels may vary from elastic to curd-like, depending on the preparation technique. Physical properties of thermally induced whey protein gels depend on the level and type of soluble salt present in solution. Maximum gel firmness for whey protein concentrate was observed at 11 mM calcium or 200 mM sodium, with higher levels reducing the firmness of the gels (Mulvihill and Kinsella, 1988). As different WPC ingredients have different inate salt levels, WPC functionality can often be variable.

19.2.4 Heat stability

Foods formulated from added dairy ingredients generally go through a heating step in their preparation or formulation. The heating may be a pasteurisation step or have the purpose of initiating functional development. This can be viewed as a negative or a positive attribute, depending on whether a liquid or solid end result is required. The ability of complex colloidal mixtures of milk proteins, fat and minerals to be stable to high heating regimes is an important property of dairy-based ingredients. Generally, a complex sequence of events occurs prior to dehydration or post-reconstitution to achieve subsequent stability to heating at elevated temperatures (120 or 140°C). This may involve pre-heating to elevated temperatures (120°C × 2 min) prior to concentration of the milk and/or addition of phosphates/pH adjustment. The main milk components that make up these colloidal structures are casein, whey proteins, calcium and phosphorous. Since some or all of these components are markedly unstable to high heating regimes in isolation, the ability of the components to self-protect one another is central to heat stability. At low concentrations of milk solids (10% solids) most milk samples are stable to high heating regimes, provided the pH is in the normal range (pH 6.55–6.8). However, when milk is concentrated (>20% solids), it has reduced heat stability across the pH range (6.2–7.2).

19.2.5 Ethanol stability

The ability of milk to withstand the addition of significant quantities of ethanol without destabilisation has often been used as an indicator of milk quality. The relationship between milk constituents and ethanol stability was studied by Davies and White (1958) and the importance of pH and the composition of the milk aqueous phase was shown by Horne and Parker (1980, 1981a, 1981b). The main effects of increasing the ethanol concentration in aqueous ethanolic mixtures is a major shift in the dielectric constant, which decreases the solvent quality (makes it more hydrophobic), thus affecting the solubility of ionic species, and introduces the concept of apparent pH.

A study by O'Connell *et al.* (2001) emphasised the importance of temperature on the structure and behaviour of casein micelles in ethanolic solutions. They showed that the repulsive forces between caseins increase and solvent quality is enhanced with increasing temperature, which results in swelling of the micelle and eventual dissociation. They ascribed this behaviour of a dramatic decrease in cohesive interactions between casein molecules on heating in the presence of alcohols to be a consequence of a reduction in phosphoseryl-mediated cross-linking and an increase in protein hydrophilicity. They also concluded that the contribution of calcium phosphate to the cohesive interactions responsible for the micellarisation may be related to its effect on hydrophobic interactions.

These parameters impinge in the study of cream liqueur systems and the remarkable ability of sodium caseinate to stabilise the emulsion in ethanolic solutions.

19.3 Dairy ingredients in food systems

19.3.1 Processed meat products

The chemistry and functional behaviour of meat as a raw material for further processing is derived from the properties of muscle. The major components of muscle are water, protein and fat. The protein is classified into three major groups (i) myofibrillar, which are composed mainly of actin and myosin and are soluble in relatively high concentrations of salt (NaCl); (ii) sarcoplasmic, which are water soluble and (iii) connective tissue which is not soluble. Potential water-holding capacity is one of the major functional attributes of meat in the production of processed meat products. The ability of the myofibrillar meat proteins to go into solution in the presence of high salt concentrations, thus absorbing a lot of water, makes them very valuable in boneless and formed hams where subsequent heating gels these solubilised meat proteins, creating a solid matrix with good water-binding ability.

The reasons for using non-meat proteins in processed meat products could be cost-reduction or yield enhancement, or to improve some functional attribute of the product (reduction in shrink loss, cook yield, juiciness, fat mimetic or reduced packet purge). When using dairy-derived ingredients to replace meat or to improve the functionality of the processed meat product, it must be remembered that the functional expression of the dairy ingredient will be dictated by the same factors that control the functionality of the meat proteins. A notional example of a formulation for the production of a processed meat product is outlined in Table 19.3. It can be observed that 2.2% NaCl and 0.3% sodium tripolyphosphate are an integral

Table 19.3 Typical levels of the major ingredients in a Frankfurter-type formulation incorporating whey protein

Ingredient	Amount (%)
Lean beef	11.5
Pork trim	52
Whey protein isolate	1.73
Salt	2.2
Sodium tripolyphosphate	0.3
Cure (Nitrite)	0.14
Ice	32.13

part of the formulation together with a significant quantity of water and a quantity of non-meat protein (which is subject to legal limits that may be variable, depending on the country of production). Both casein-based and whey protein-based dairy ingredients are used in processed meat products and the effects of elevated NaCl and polyphosphate concentrations are very different upon these. Casein-based ingredient (sodium caseinate) solubility and hydration behaviour is totally dependant on the pH and ionic strength. Sodium caseinate is only sparingly soluble at pH 6 or lower under low ionic strength conditions; however, high concentrations of sodium caseinate can be solubilised at lower pH values when NaCl and/or sodium tripolyphosphate is present in high concentrations. Whey proteins, whether as WPC or WPI, are generally soluble over a wide pH range and do not exhibit their potential functionality until they are denatured, usually heat-induced. However, the concentration of whey protein, the pH at the denaturation temperature, the high ionic strength and the presence of divalent cations determine the degree and type of subsequent aggregation of the denatured proteins. Myosin, the salt soluble protein from chicken breast, has been shown to have three endothermic peaks on heating, namely 48.5, 53 and 58°C (Vittayanont *et al.*, 2001) while β-lactoglobulin was shown to have one endothermic peak at 73–76°C. Endothermic peaks are associated with unfolding of the protein (denaturation). The same authors concluded that the additive effect of heating on the gel strength of myosin/β-lactoglobulin mixtures could be achieved only if the temperature reached was greater than the gelling temperature of β-lactoglobulin (73–79°C). They were attempting to explain the variation observed in meat products that had added whey protein where the heating temperature did not exceed 71°C. A large number of additional studies have been carried out in related systems where meat proteins and whey proteins are mixed (Beuschel *et al.*, 1992; Hung and Smith, 1993; Smith and Rose, 1995; McCord *et al.*, 1998).

The use of cold setting whey protein might be seen as a method of overcoming this variation. Cold-setting whey protein dispersions are pre-denatured using heating to temperatures above the denaturation temperature for periods of time where all the whey protein is denatured and aggregated into soluble aggregates with diameters in the range 30–50 nm (Ju and Kilara, 1998). These particles have a high concentration of free sulphydryl groups and are highly reactive, especially when the ionic strength is high. Gelation of such preparations can now occur at lower temperatures. Where previously the myosin/β-lactoglobulin composite system was seen as two separate gelation occurrences because they occurred at different temperatures, it is now quite possible that an integrated gel system could be operative due to the effect of high ionic strength on both the myosin and the denatured whey protein, and the presence of effective concentrations of free sulphydryl groups.

It has already been noted (Section 19.2.2) that dairy protein ingredients are good emulsifiers and can therefore be utilised in the formation of high fat emulsions prior to addition to processed meat systems. Sodium caseinate and whey protein have been used for this purpose. Emulsion formation prior to introduction to the meat system not only introduces fat to the system but increases the effective concentration (as interfacial protein) of the dairy protein of interest which can be assimilated into the heated meat protein matrix on subsequent heating. It is most important that interactions between all available interfaces occur in processed meat products as adhesiveness is an important property of these types of food products. In this respect, the protein at the emulsion interface must adhere to the general meat protein matrix to achieve the required final structure and avoid fracture following processing. It must be realised that meat processing is often a very cost sensitive enterprise and high cost ingredients need to give a justifiable return.

19.3.2 Food emulsions

Food emulsions are generally of the oil-in-water or water-in-oil type but occasionally multiple emulsions are utilised. The oil-in-water emulsion would be typified by cream liqueur, ice-cream or mayonnaise while the water-in-oil type, although less prevalent, would be represented by table spreads. A suitable emulsifier is a prerequisite for efficient emulsion formation and these can include proteins and low molecular weight surfactants. The most widely used proteinaceous emulsifier for oil-in-water emulsions in the food industry is probably sodium caseinate. However, other casein-based (calcium caseinate, skimmed milk powder) and whey protein based ingredients (whey protein concentrate, whey protein isolate) are also used for this purpose. The manufacture of oil-in-water food emulsions utilises a lot of energy in creating the oil surface which is stabilised by the adsorbed protein. This is generally performed using high pressure homogenisers (valve-type), although crude emulsions can be produced using mixers of various types and nano-emulsions can be produced using microfluidisers.

Edible oil-in-water emulsions

In food systems, the chemical complexity of the oil phase can make emulsion behaviour or functionality hard to predict. Many food emulsions contain fat that is partially solid at ambient temperature, which can lead to crystalline inclusions which, in turn, may lead to instability due to coalescence (van Boekel and Walstra, 1981). Long-term stability of emulsions depends in part on the thickness and strength of the adsorbed films at the oil–water interface (Dickinson and Stainsby, 1982). Dairy-based primary emulsifiers (proteins) are available in a number of different states, which can have a dramatic effect on both the efficiency of emulsification and the subsequent functionality of the emulsion. These can range from the apparently simple sodium caseinate, which is the most disaggregated of the casein-type emulsifiers, to the more aggregated calcium and calcium phosphate caseinates, which can increase the protein load on the oil–water interface. Whey proteins are also used for their emulsification abilities and can be used in both the native or denatured state (provided the denatured whey protein aggregate is not too large). Heat denaturation of the whey proteins can improve emulsifying capacity, possibly by enhancing macromolecular flexibility and surface hydrophobicity (Dickinson and Stainsby, 1987). While this is a very general statement, it is probably based on a number of different sub-processes that can occur during the process of heat treatment and subsequent emulsification with oil on cooling. (i) The process of heat-induced denaturation and subsequent aggregation is dependent on pH, ionic strength, temperature and time. If the aggregation step is rate limiting, the final denaturation level will be low and any soluble aggregates will be small. Denatured whey protein therefore exists as aggregates of variable size. (ii) Emulsification of oil in water requires the movement of the whey protein aggregates to the oil–water interface as new surfaces are created through energy input. It is possible that when denatured whey protein aggregates stabilise the newly created oil–water interface, the protein load on the interface is increased compared to emulsification with native whey proteins, resulting in an emulsion with different properties. Whether this comprises an improvement or not depends on the application. It must be borne in mind that the properties of emulsions are largely dependent on the properties of the interfacial layer (the fat type will play a part), which are dependent on the environmental conditions pertaining in the aqueous phase. Ice-cream is a frozen, aerated, oil-in-water emulsion, where micellar casein and whey proteins can act as emulsifiers; coffee-whiteners are oil-in-water emulsions where caseinate may be the preferred emulsifier. Acidic milk-based emulsions may rely on the emulsified oil droplets being an integrated part of the gel structure through interaction of the interfacial layer with the bulk casein/whey protein complexes.

Polysaccharide-based biopolymers (e.g. guar gum, carrageenan, alginate, chitosan) are also incorporated into emulsion products for a variety of reasons. These might include viscosity control, water control or general stabilisation of the emulsion from creaming. It has been shown that interactions between polysaccharides and dairy proteins can have detrimental or beneficial effects on food product quality, depending on their nature and the required properties of the food in question (Tolstogusov, 2003). The use of hydrocolloids can positively influence emulsion creaming kinetics or a negative flocculation/serum separation. Mixtures of biopolymers are often observed to separate into distinct phases, which can be the result of thermodynamic incompatabilty (polymers do not like one another), complex coacervation (positive charges like negative charges, opposites attract) or depletion flocculation (localised concentration of polymer).

Edible water-in-oil emulsions

Variable-fat spreads, whether of the high, reduced or low fat type, are usually water-in-oil emulsions and appear as plastic solids. Butter and margarine must, by law, contain a minimum of 80% fat, but spreads conventionally contain either 72–80% fat (full fat), 55–60% fat (reduced fat), 39–41% fat (low fat) or 20–30% fat (very low fat), but all are water-in-oil emulsions. The principle ingredients of fat spreads are fat (dairy or vegetable), fat-based emulsifier, milk protein, stabiliser, sodium chloride and water, and each of these will affect the emulsion, processing and consumer behaviour of the final product. The level of sodium chloride in the aqueous phase can vary but is usually in the region of 1.5%, w/w. The water-in-oil pre-emulsions of fat spreads are always stabilised by high shear working of the emulsion at low temperatures to a plastic consistency. Before this solidification step, emulsions can become unstable due to either phase separation or phase inversion (Mulder and Walstra, 1974). It is intuitively evident that the likelihood of phase inversion increases as the fraction of added disperse phase is increased. The processing of low-fat spreads comprises two steps, namely preparation of an aqueous-phase-in-oil emulsion while stirring, followed by pumping the emulsion through one or two scraped surface coolers in series at a defined agitation rate and at a defined refrigerant temperature (shear at low temperatures, as mentioned above). Patent literature has suggested that the higher the aqueous phase viscosity, the greater the stability to inversion (Platt, 1988). Sodium caseinate is often the protein of choice to aid in the stabilisation of the water-in-oil emulsion. While the introduction of NaCl into the aqueous phase was initially for organoleptic reasons, the interaction between NaCl and caseinate also has a significant effect on the stability of the emulsion prior to solidification. The viscosity of a caseinate solution is an indicator of the degree of bound water absorbed by the hydrophilic groups, as well as the water trapped inside the aggregated molecules (Korolczuk, 1982). Sodium caseinate contributes to the stability of the water-in-oil emulsion through steric and water binding effects (Keogh, 1992). The same author concluded that NaCl made a significant contribution to the aqueous phase viscosity. While final emulsion stability and its stability to inversion (e.g. becoming an oil-in-water emulsion) may be related to the viscosity of the caseinate-based aqueous phase, the interaction between the level of fat-soluble emulsifier and the aqueous caseinate may also be significant (Barfod *et al.*, 1989). Large reductions in the level of NaCl in the aqueous phase can lead to inversion problems during processing, and alternative methods of increasing the aqueous phase viscosity may have to be approached, e.g. hydrocolloids, if low salt spreads are required.

Cream Liqueur

Large quantities of dairy ingredients are utilised in the production of a specialised oil-in-water emulsion, namely, cream liqueurs. These would include the cream itself and the primary emulsifier of choice, sodium caseinate. As mentioned above, sodium caseinate is the sodium salt of acid casein and is a very efficient emulsifier. A typical composition of a cream liqueur

Table 19.4 Typical levels of the major ingredients in a cream liqueur formulation (12% fat)

Ingredient	Amount (%)
Cream (40% fat)	30
Sugar	20
Sodium caseinate	2.5
Ethanol	14
Trisodium citrate	1.0–1.5
Colours and flavours	~0.1
Water	31.9–32.4

is shown in Table 19.4. Essentially it is a dairy-fat based emulsion with an aqueous ethanolic/sugar/caseinate continuous phase. The fat content can vary from 5–16% (w/w), depending on the market, and these emulsions can be stable to separation problems for up to two years. Most of the problems observed in commercial cream liqueurs revolve around the fat fraction. Creaming, cream plug formation, flecking and gelation are all manifestations of fat globule destabilisation. However, the fat globules are emulsified by a surface layer of caseinate (Banks *et al.*, 1981a) which confers the sought after stability in the first place, so destabilisation may be a caseinate problem. Calcium-induced aggregation was considered to be the dominant factor controlling the shelf-life of cream liqueurs at high ambient temperatures (Banks *et al.*, 1981b), but other protein interactions might also be important when calcium chelators are present. The addition of trisodium citrate at levels normally present in milk (10 mM) reduced the incidence of instability due to increases in viscosity at high ambient temperature by reducing the ionic calcium level coming from the cream (Banks *et al.*, 1981b). While sodium caseinate is a relatively expensive ingredient, attempts to substitute it for a cheaper alternative have not had much success. This is mainly due to its high emulsification potential and stabilising power in aqueous alcoholic solutions. However, sodium caseinates from different sources have been shown to have different alcohol stabilities, which were attributed to different production protocols and different levels of damage to the protein during processing (Muir and Dalgleish, 1987). Pierre (1989) advocated that the final alcohol stability of both milk and sodium caseinate dispersions was determined by alcohol induced shifts in the pK values of the protein constituent amino acids, and on soluble calcium levels. While the alcohol stability of milk has limited practical applications, the stability of sodium caseinate to ethanol is an important requirement in cream liqueur production. Cream liqueurs (17% ethanol v/v, 40% solids, pH 6.8–7.0) contain up to 3% sodium caseinate to aid in the emulsification of cream and provide body to the resultant emulsion. Stability of the liqueurs to time/

temperature dependent viscosity increases is a necessary requirement for long-term storage of the product.

Cream liqueurs can be manufactured in a number of different ways but generally they follow similar basic steps. Sodium caseinate, trisodium citrate and sugars dissolved in water, 40–48% fat cream mixed in to give the required fat level, potable alcohol added to give 17% (v/v), colours and flavours as required, followed by homogenisation. A variation of this process would entail a two-step process whereby the emulsion is formed in the absence of ethanol, which is added later. As the presence of alcohol reduces the surface tension of the mixture during the homogenisation step, fat globule sizes after the one-step process are generally smaller than those using the two-step process.

Creaming can often be found in low fat liqueurs or in liqueurs made using a two-step production process, and is generally acceptable provided the cream layer disperses on shaking. It can be a problem where sufficient fat rises to eventually to form a cream plug. Gel formation is where the caseinate and/or the caseinate stabilised fat globules aggregate to form a three-dimensional structure. Gel formation can often follow the initial observed viscosity increase. Slightly higher concentrations of most of the essential ingredients will aid in gel formation, i.e. fat (18–20%), caseinate (4%), alcohol (20–22%, v/v). This would suggest that at 16% fat, 3% caseinate and 17% alcohol, the cream liqueur is on the edge, with good body, good stability. The caseinate stabilised fat globule/bulk solution caseinate is constrained and has a periodic structure resulting in a stable emulsion.

It has been established that the presence of calcium ions or higher concentrations of sodium ions will predispose the cream liqueur system towards gelation. The presence of calcium is due to the skim milk component of the cream and the quality of the added water. The requirement for the addition of tri-sodium citrate to chelate the calcium originating from the cream may paradoxically induce gelation through its sodium counterions, especially at high citrate additions. Temperature of storage is most important for the stability of liqueurs. Cream liqueurs, and even the aqueous phase from liqueur systems, are subject to increased viscosity if the temperature of storage is high. The life-time of the system is inversely related to the temperature at which it is stored or transported. If more than one of the above scenarios co-exist, the life span of the liqueur will be limited.

O'Kennedy *et al.* (2001) showed that pH, ionic strength and ethanol content play a significant part in determining the stability of sodium caseinate to aggregation and eventual precipitation. They also concluded that about one-third of the casein protein was susceptible to ethanol-induced aggregation, the balance remaining in a non-aggregated state under the conditions of pH, ionic strength and ethanol concentration used (pH 7, 25°C). It is well accepted that sodium caseinate is composed of a heterogeneous group of proteins, including the α_{s1}, β, κ and α_{s2} casein fractions. The main proteins susceptible to ethanol induced aggregation were the sulphydryl containing α_{s2} and κ casein fractions. The non-aggregated casein fraction showed no time-dependent increase in viscosity of cream liqueur analogues at 45°C over a 63-day period. An opportunity therefore exists for the development of a caseinate ingredient for the cream liqueur industry based on that fraction which does not aggregate in the presence of ethanol. Lynch and Mulvihill (1997) have shown that the apparent viscosity of cream liqueurs on storage at 45°C was dependant on the sodium caseinate source and the authors suggested that electrostatic and sulphydryl interactions were involved.

19.3.3 Chocolate and confectionary

Milk in its various forms (evaporated milk, sweetened condensed milk, milk powders and cream) is a valuable ingredient in many confectionary recipes, including chocolate, caramels, toffees and fudge. Butter or butteroil, whey powder, lactose and specialised milk powders can also be used for specific purposes. Milk components affect the consistency, colour, flavour, storage stability and nutritional status of chocolate (Campbell and Pavlasek, 1987), and the role of milk fat as an anti-blooming agent in chocolate is well established (Versteeg *et al.*, 1994).

Milk chocolate is an integral part of the confectionary area and utilises large quantities of dried dairy-based ingredients. These are mainly composed of WMP and chocolate crumb, but SMP and whey powders are also used.

Visualising milk chocolate requires a mental shift as we are used to dealing with milk protein functionality or sugar behaviour in an aqueous environment. Chocolate is a fat continuous system containing very little water (0.1%), composed of cocoa butter (mainly) with a dispersed phase of solid particles. These include milk powder, and sugar and cocoa liquor particles which are refined to variable sizes, usually small enough to be undetectable in the mouth. However, chocolate with a very small powder particle size distribution has a totally different mouthfeel than a chocolate with a larger size distribution. A typical composition of milk chocolate is outlined in Table 19.5. The ratio of these components in the initial mixture yields a

Table 19.5 Typical levels of ingredients in milk chocolate

Ingredient	Amount (%)
Cocoa butter	20
Whole milk powder	24.4
Sugar	43
Cocoa liquor	12.4
Lecithin	0.2

composite which is ideal in texture for refining using multiple rollers. It is important that the fat in the full cream milk powder migrates to the continuous cocoa butter phase during the initial mixing and refining stages, as this will increase the volume of oil available to disperse the insoluble powder particles. This is particularly important during the refining step as the surface area of the refined insoluble particles increases enormously. Full cream milk powders should ideally have a high free or readily extractable fat level, for incorporation into chocolate Whole milk powders with specific physical properties such as high free-fat content, large particle size and low vacuole volume have been shown to have beneficial effects on the properties of chocolate (Verhey, 1986; Dewettinck *et al.*, 1996; Twomey *et al.*, 2002). For these reasons, chocolate manufacturers prefer roller-dried to spray-dried powders, which typically do not have any of these attributes. The refiners reduce the particle size of the sugar, yielding a flake-like confection. While the gross composition of this refined flake is not unlike the finished chocolate, the refined flake has no propensity to flow when heated above the melting point of the cocoa butter. Following the refining step for particle size reduction, the flaked mixture is conched to remove a certain amount of residual water and a significant amount of volatiles which affect the flavour. As the particle size of chocolate decreases, the specific surface area increases and therefore the fat in the continuous phase is increasingly used to coat the dispersed surfaces. The conched mass becomes liquid when lecithin is added at temperatures above the melting point of the fat. This phenomenon is central to the rheology of milk chocolate, allowing the insoluble solid particles to flow over one another when dispersed in triglyceride oil.

Caramels, toffees and fudge

These confections owe their taste and appearance to the presence of milk proteins and butter when heated in the presence of an aqueous sugar solution. They develop a characteristic flavour due to the Maillard reaction between proteins and reducing sugars. The butter also gives an attractive flavour due to decomposition on heating, which cannot be duplicated using vegetable fats. Evaporated or concentrated milk can be used, but sweetened condensed milk is ideal. Reconstitution of full cream and/or skimmed milk powders to high solids concentrations can also be employed as a substitute for sweetened condensed milk but the heat stability of the milk protein has to be adequate. Evaporated or sweetened condensed milk have been pre-heated to high temperatures prior to concentration, hence their stability to further heating in the concentrated form, either in the canning process (retort) and/or in the caramel production itself, is adequate. It should be noted that in the production of caramel, the final boil temperatures can reach between 118°C and 131°C, depending on the hardness required. Caramel is basically a fat emulsion in a vitreous syrup phase with the milk protein dispersed, while fudge has a solid sugar crystal phase dispersed with the fat and milk protein in a syrup phase.

19.4 Bakery products

Other than wheat flour, the major sources of protein found in baked products are eggs, milk and soy. Gluten is the main structure-forming protein in wheat flour, and is responsible for the elastic properties of dough. Gluten removal from bakery fomulations results in major problems for bakers and currently many gluten-free products available on the market are of poor quality, showing poor mouthfeel and flavour. Gluten contains the protein fractions glutenin and gliadin. The former is a rough rubbery mass when fully hydrated, while gliadin produces a viscous, fluid mass on hydration. Gluten, therefore has both an elastic and a viscous component which together give gluten the important properties of extensibility, gas holding ability and resistance to stretch, which determine the properties of dough (Gallagher *et al.*, 2004).

19.4.1 Bread

Dairy ingredients can be used in bread for nutritional benefits, including increasing calcium content and protein efficiency ratio, and functional benefits including flavour, texture enhancement, and storage improvement. Dairy ingredients that enhance water absorption can also improve dough-handling properties (Cocup and Sanderson, 1987).

The gluten proteins of flour contain sulphydryl groups due to the presence of the amino acid cysteine in the primary structure. The sulfhydryl group in cysteine can react (oxidise) with another cysteine molecule to form a disulphide bond to make one molecule of cystine. When gluten molecules react with one another (become oxidised) during bread manufacture, dough strength increases, but extensibility decreases. Sulphydryl–disulphide interchange reactions have been deemed to be responsible for the correct development of dough structure.

The native whey protein seemingly interferes with gluten development and therefore has negative effects in breadmaking (Zadow *et al.*, 1983). Heat treatment of whey proteins changes their structure and this denaturation step eliminates the negative effect previously observed (Harper and Zadow, 1984). This suggested that heat-induced denaturation and, more importantly, subsequent aggregation into defined particulate structures prevented the negative interaction with gluten (presumably by disulphide–sulphydryl interchange reactions) which occurs when the native monomeric whey protein is heat-denatured in the presence of gluten.

Current research aims to completely substitute gluten with a functional casein-based ingredient (Stathopoulos and O'Kennedy, 2008). The principle behind this approach is that by increasing the calcium concentration to an optimum level in the casein/caseinate ingredient, it will be possible, under the correct pH and ionic strength conditions, to replace the highly functional (covalent) S–S bonds in a gluten-based dough with calcium-induced casein–casein complexes.

19.5 Future trends

Application of dairy-ingredients into food systems in the future will centre on their ability to not only have the ability to be functional during processing but also to have a nutritional edge. Much emphasis in recent years has been to eliminate or reduce certain traditional inclusions in food systems (gluten, fat, sugar, sodium), thus causing problems for producers with a loss in perceived product quality. A lot of advances have been made in understanding the factors that govern dairy ingredient functional expressional and their ability to be processed into a structural or liquid food system. Whey protein ingredients, for instance, have been widely used in recent years for supplying good nutritional protein to athletes of various disciplines, as well as to the general population. There will be a need to move this type of ingredient usage back into the general types of food, both to upgrade the nutritional effect of the food and to still maintain the quality aspect required.

It is the basis of research and subsequent development of dairy ingredients, that given the existing knowledge, efforts to further improve the functionalities of milk proteins will be undertaken. This could include the use of transglutaminase to enzymatically produce a covalent bond between the ε-amino group of lysine and the γ-carboximide group of glutamine. Caseins are readily susceptible to transglutaminase-induced cross-linking (Huppertz and de Kruif, 2007), while the whey protein are only susceptible when in the denatured form. Sodium caseinate is also markedly susceptible to the enzyme, where inter-molecular bond formation leads to gelation (Lorenzen, 2000). The ability to create new structures or change existing ones may aid in the general development of the rapidly changing nutritional requirements from our food.

While dairy proteins are efficient emulsifiers, there is always room for improvement and one area where application is ongoing is the formation of carbohydrate–protein conjugates. Covalent attachment of polysaccharides/sugars to a dairy protein by non-chemical means (Maillard reaction) can produce high molecular weight biopolymers (glycoconjugates). The Maillard reaction occurs when a protein and carbohydrate moity are processed at high temperatures. Many of the positive benefits of producing Maillard generated conjugates include improvements in emulsifying activity, foaming properties, calcium complexing, solubility and heat stability (Oliver *et al.*, 2006). Akhtar and Dickinson (2007) found that glycoconjugates made with maltodextrin showed enhanced emulsifying properties as the molecular weight of the polysaccharide was reduced. This could possibly prevent or control flocculation and coalescence which might improve food quality.

While the future production of dairy ingredients may involve some application of the above technologies, significant advances in the understanding of the factors that control their functional expression through binding of minerals, protein–protein interactions, protein–polysaccharide interactions, and protein–lipid interactions need to be further exploited to produce nutritional functional ingredients for the food industry.

19.6 References

AKHTAR, M. and DICKINSON, E. 2007. Whey protein–maltodextrin conjugates as emulsifying agents: An alternative to gum arabic. *Food Hydrocolloids* 21, 607–616.

AUTY, M.A.E., O'KENNEDY, B.T., ALLAN-WOJTAS, P. and MULVIHILL, D.M. 2005. The application of microscopy and rheology to study the effect of milk salt concentration on the structure of acidified micellar casein systems. *Food Hydrocolloids*, 19, 101–109.

BANKS, W., MUIR, D.D. and WILSON, A.G. 1981a. The formulation of cream-based liqueurs. *Milk Industry*, 83, 16–18.

BANKS, W., MUIR, D.D. and WILSON, A.G. 1981b. Extension of the shelf life of cream-based liqueurs at high ambient temperatures. *Journal of Food Technology*, 16, 587–595.

BARFOD, N.M., KROG, N. and BUCHEIM, W. 1989. 'Lipid–protein–emulsifier–water interactions in whippable emulsions'. In: Kinsella J E and Soucie W G, *Food Proteins*, Champaign, Illinois, AOCS, 144–158.

BEUSCHEL, B.C., PARTRIDGE, J.A. and SMITH, D.M. 1992. Insolubilised whey protein concentrate and/or chicken salt soluble protein gel properties. *J. Food Sci.* 57, 852–855.

CAMPBELL, L.B. and PAVLASEK, S.J. 1987. Dairy products as ingredients in chocolate and confections. *Food Tech.* 41, 78–85.

COCUP, R.O. and SANDERSON, W.B. 1987. Functionality of dairy ingredients in bakery products. *Food Tech.* 41, 102–104.

CREAMER, L.K. 1985. Water absorption by renneted casein micelles. *Milchwissenschaft* 40, 589–591.

DAMODARAN, S. 1997. 'Food proteins: An overview'. In: Damodaran S and Paraf A, *Food Proteins and their Applications*, New York, Marcel Dekker, Inc., 1–24.

DAVIES, D.T. and WHITE, J.C.D. 1958. The relation between the chemical composition of milk and the stability of the caseinate complex. II Coagulation by ethanol. *J. Dairy Res.* 25, 256–266.

DEWETTINCK, K., DE MOOR, H. and HUYGHEBAERT, A. 1996. The free-fat content of dried milk products and flow properties of milk chocolate. *Milchwissenschaft* 51, 25–28.

DE WIT, J.N. 1989. 'The use of whey protein products'. In: Fox P F, *Developments of Dairy Chemistry-4*, Elsevier Applied Science, London, 323–345.

DICKINSON, E. and STAINSBY, G. 1982. *Colloids in Food*. Applied Science, London.

DICKINSON, E. and STAINSBY, G. 1987. Progress in the formulation of food emulsions and foams. *Food Technology*. 41, 74–81.

FAMELART, M.H., LEPESANT, F., GAUCHERON, F., LE GRAET, Y. and SCHUCK, P. 1996. pH-Induced physiochemical modifications of native phosphocaseinate suspensions: Influence of aqueous phase. *Le Lait* 76, 445–460.

GALLAGHER, E., GORMLEY, T.R. and ARENDT, E.K. 2004. Recent advances in the formulation of gluten-free cereal-based products. *Trends Food Sci. Tech.* 15, 143–152.

GAYGADZHIEV, Z., CORREDIG, M. and ALEXANDER, M. 2009. The impact of the concentration of casein micelles and whey protein-stabilised fat globules on the rennet-induced gelation of milk. *Colloids and Surfaces B: Biointerfaces* 68, 154–162.

HARPER, W.J. and ZADOW, J.G. 1984. Heat induced changes in whey protein concentrates as related to bread manufacture. *NZ J. Dairy Sci. Technol.* 19, 229–237.

HERMANSSON, A.M. 1975. Functional properties of proteins for foods-flow properties. *Journal of Texture Studies* 5, 425–439.

HORNE, D.S. and PARKER, T.G. 1980. The pH sensitivity of individual cow milks. *Neth. Milk Dairy J.* 34, 126–130.

HORNE, D.S. and PARKER, T.G. 1981a. Factors affecting the ethanol stability of bovine milk. I. Effect of serum phase components. *J. Dairy Res.* 48, 273–284.

HORNE, D.S. and PARKER, T.G. 1981b. Factors affecting the ethanol stability of bovine milk. III. Substitution of ethanol by other organic solvents. *Int. J. Biol. Macro.* 3, 399–402.

HUNG, T.-Y. and SMITH, D.M. 1993. Dynamic rheological properties and microstructure of partially insoluble whey protein concentrate and chicken breast salt-soluble protein gels. *J. Agr. Food Chem.* 41, 1372–1378.

HUPPERTZ, T. and DE KRUIF, C.G. 2007. Rennet-induced coagulation of enzymatically cross-linked casein micelles. *Int. Dairy J.* 17, 442–447.

JU, Z.Y. and KILARA, A. 1998. Effect of preheating on properties of aggregates and of cold-set gels of whey protein isolate. *J. Agric. Food Chem.* 46, 3604–3608.

KEOGH, M.K. 1992. *The stability to inversion of a concentrated water-in-oil emulsion* Ph D Thesis, National University of Ireland.

KOROLCZUK, J. 1982. Hydration and viscosity of casein solutions. *Milchwissenschaft* 37, 274–276.

LE RAY, C., MAUBOIS, J.L., GAUCHERON, F., BRULE, G., PRONNIER, P. and GARNIER, F. 1998. Heat stability of reconstituted casein micelle dispersions: Changes induced by salt addition. *Lait* 78, 373–390.

LORENZEN, P.C. 2000. Techno-functional properties of transglutaminase-treated milk. *Milchwissenschaft* 55, 667–670.

LYNCH, A.G. and MULVIHILL, D.M. 1997. Efect of sodium caseinate on the stability of cream liqueurs. *Int. J. Dairy Tech.* 50, 1–7.

MCCLEMENTS, J., MONAHAN, F.J. and KINSELLA, J.E. 1993. Effect of emulsion droplets on the rheology of whey protein isolate gels. *Journal of Texture Studies* 24, 411.

MCCORD, A., SMYTH, A.B. and O'NEILL, E.E. 1998. Heat-induced gelation properties of salt-soluble muscle proteins as affected by non-meat proteins. *J. Food Sci.* 63, 580–583.

MUIR, D.D. and DALGLEISH, D.G. 1987. Differences in behaviour of sodium caseinate in alcoholic media. *Milchwissenschaft* 42, 770–772.

MULDER, H. and WALSTRA, P. 1974. 'Isolation of milk fat'. In: Mulder H and Walstra P, *The Milk Fat Globule*, Wageningen, Pudoc, 228–243.

MULVIHILL, D.M. and KINSELLA, J.E. 1988. Gelation of β-lactoglobulin: Effects of sodium chloride and calcium chloride on the rheological and structural properties of gels. *Journal of Food Science* 53, 231–236.

MULVIHILL, D.M. 1989. 'Caseins and caseinates: Manufacture'. In: Fox P F, *Developments in Dairy Chemistry-4*, Elsevier Applied Science Publishers, London, UK, 97–130.

O'CONNELL, J.E., KELLY, A.L., FOX, P.F. and DE KRUIF, C.G. 2001. Mechanism for the ethanol-dependent heat-induced dissociation of casein micelles. *J. Agr. Food Chem.* 49, 4424–4428.

O'KENNEDY, B.T., CRIBBEN, M. and KELLY, P.M. 2001. Stability of sodium caseinate to ethanol. *Milchwissenschaft* 56, 680–684.

O'KENNEDY, B.T. and MOUNSEY, J.S. 2008. The dominating effect of ionic strength on the heat-induced denaturation and aggregation of β-lactoglobulin in simulated milk ultrafiltrate. *Int. Dairy J.* 19, 123–128.

OLIVER, C.M., MELTO, L.D. and STANLEY, R.A. 2006. Creating proteins with novel functionality via the Maillard reaction. A review. *Critical Reviews in Food Sci. and Nutrition* 46, 337–356.

PIERRE, A. 1989. Milk stability in ethanolic solutions. *J. Dairy Res.* 56, 521–527.

PIRES, M.S., ORELLANA, G.A. and GATTI, C.A. 1999. Rennet coagulation of casein micelles and heated casein micelles: Action of Ca^{2+} and pH. *Food Hydrocolloids* 13, 235–238.

PLATT, B.L. 1988. *Low fat spread*, European Patent No 0 256 712.

SALA, G., VAN AKEN, G.A., COHEN STUART, M.A. and VAN DE VELDE, F. 2007. Effect of droplet–matrix interactions on large deformation properties of emulsion-filled gels. *J. Texture Stud.* 38, 511–535.

SCHUCK, P., PIOT, M., MEJEAN, S., FAUQUANT, J., BRULE, G. and MAUBOIS, J.L. 1994. Deshydratation des laits enrichis en caseine micellaire par microfiltration; Comparaison des proprietes des poudres obtenues avec celles d'une poudre de lait ultra-propre. *Lait* 74, 47–63.

SMITH, D.M. and ROSE, A.J. 1995. Properties of chicken salt soluble protein and whey protein concentrate gels as influenced by sodium tripolyphosphate. *Poultry Sci.* 74, 169–175.

STATHOPOULOS, C. and O'KENNEDY, B.T. 2008. A rheological evaluation of concentrated casein systems as replacement for gluten: Calcium effects. *Int. J. Dairy Tech.* 61, 397–402.

TOLSTOGUZOV, V.B. 2003. Some thermodynamic considerations to food formulation. *Food Hydrocolloids* 17, 1–23.

TWOMEY, M., KEOGH, M.K., O'KENNEDY, B.T. and MULVIHILL, D.M. 2002. Seasonal effects of some milk powder characteristics on the rheology of milk chocolate. *Irish J. Agric. Food Res.* 41, 105–116.

VAN BOEKEL, M.A.J.S. and WALSTRA, P. 1981. Stability of oil-in-water emulsions with crystals in the disperse phase. *Colloids Surf.* 3, 109.

VASBINDER, A.J. and DE KRUIF, C.G. 2003. Casein-whey protein interactions in heated milk: The influence of pH. *Int. Dairy J.* 13, 669–677.

VERHEY, J.G. 1986. Physical properties of dried milk in relation to chocolate manufacture. *Neth. Milk Dairy J.* 40, 261–268.

VERSTEEG, C., THOMAS, L.N., YEP, Y.L., PAPALOIS, M. and DIMICK, P.S. 1994. New fractionated milk fat products. *Aust. J. Dairy Tech.* 49, 57–61.

VITTAYANONT, M., VEGA-WARNER, V., STEFFE, J.F. and SMITH, D.M. 2001. Heat-induced gelation of chicken pectoralis major myosin and β-lactoglobulin. *J. Agric. Food Chem.* 49, 1587–1594.

ZADOW, J.G., HARDHAM, J.F. and MARSHALL, S.C. 1983. Sulphydryl residues in whey protein concentrates and their effect on bread baking characteristics in a model system. *Aust. J. Dairy Technol.* 38, 27–28.

20

The role of dairy ingredients in processed cheese products

T. P. Guinee, Moorepark Food Research Centre, Teagasc, Ireland

Abstract: The chapter reviews the manufacture of processed cheese products (PCPs) and how their rheological and cooking characteristics are affected by the addition of dairy ingredients, especially milk proteins. It discusses how casein and/or paracasein is the primary stabilisation agent in PCPs, and how the addition of emulsifying salts such as sodium phosphates/citrates mediate the demineralisation of the insoluble paracasein and its conversion to a functional protein that binds water and emulsifies free fat released during processing. Finally, it considers how differences in initial solubility, pH and mineral composition of cheese and milk proteins affect the characteristics of the resultant PCPs.

Key words: processed cheese products (PCPs), manufacture, dairy ingredients, cheese, cheese base, milk proteins, caseins, whey proteins.

20.1 Introduction

Pasteurised process cheese products (PCPs) are cheese-based foods prepared by comminuting, blending and melting one or more natural cheeses, water (usually), emulsifying salts (ES), and optional ingredients into a smooth homogeneous blend with the aid of heat and mechanical shear. Optional ingredients, which are determined by the type of PCP, include dairy ingredients (for example butter oil, casein, caseinates, whey proteins, whey powder), water, condiments or embellishments (for example preserved meat, fish, vegetable, salt, pepper preparations), flavours, colours and preservatives (Table 20.1). Following processing, the hot molten product can be filled/moulded into a large variety of shapes and sizes of pack, common formats for use in the industrial and food service sectors including firm sliceable PCPs in the form of blocks, slabs or sausage (chub) wrapped in various types of plastic film, and dips/sauces or pastes filled into drums or buckets. Retail PCPs are available in a multitude of formats including foil-covered portions (blocks, triangles), slices stacked or individually

Table 20.1 Ingredients, other than cheese, used in pasteurised processed cheese products[ab]

Ingredient type	Main function/effect	Examples
Dairy ingredients		
Milk fat	Gives desired composition, texture and meltability characteristics	Cream, butter, anhydrous milk fat
Milk proteins	Give desired compositional specification, texture and meltability characteristics; assist the production of a physico-chemically stable product	Milk protein isolates and concentrates, micellar casein powder, skim milk powder, acid casein, rennet casein, casein, sodium or calcium caseinates, whey protein isolates and concentrates, milk protein hydrolysates, ultrafiltered milk
Lactose	Low cost filler; may affect texture (e.g. fluidity) and taste (sweetness)	Whey powder, lactose powder, skim milk powder, evaporated milk, liquid whey
Stabilisers	Assist in formation a physico-chemically stable product; give desired texture and meltability characteristics	Emulsifying salts: sodium phosphates and sodium citrates; Hydrocolloids and gums: sodium alginate, κ-carrageenan, locus bean gum, guar gum, xanthan gum
Acid regulators/ pH controlling agents	Assist control of the pH of final product	Food-grade organic acids, e.g. lactic, acetic, citric, phosphoric acid
Flavours	Impart flavour, especially where much young cheese is used	Enzyme modified cheeses, hydrolysed butter oil, hydrolysed milk proteins, autolysed yeast extracts, paprika, starter culture distillate, smoke extracts
Flavour enhancers	Accentuates flavour	Salt (NaCl), yeast extract
Sweetening agents	Increase sweetness, especially in products targeted to young children	Sucrose, dextrose, corn syrup, hydrolysed lactose
Colours	Impart desired colour	Natural colours: Annatto extracts, β-carotene, paprika, curcumin (turmeric) Riboflavin, chlorophyll preparations
Preservatives	Retard mould growth; prolong shelf-life	Nisin, sodium- and/or potassium-salts of sorbic acid or propionic acid
Condiments/ embellishments	Impart variety to appearance, aroma and taste; product differentiation	Sterile preparations of meat, fish, vegetables, nuts and/or fruits

[a] Compiled using: state-of-the-art knowledge of commercial processed cheese operations, and information on processed cheese standards from Codex Alimentarius Commission and from Code of Federal Regulations (Federal government of the United States).
[b] The ingredients permitted are subject to the prevailing regulations in the region of manufacture.

Table 20.2 Functional properties of unheated and heated processed cheese products (PCP)

Unheated PCP	Heated/toasted/melted PCP[a]
Sliceability (e.g. to enable use in a variety of food service/domestic applications: sandwiches, cheese burgers, lunchbox) **Spreadability** (e.g. for use on crackers) **Shreddability** (e.g. to enable distribution in a range of applications such as sandwiches, lasagne) **Pliability/Mouldability** (e.g. for use as an ingredient in combination with other foods) **Crumbliness** (e.g. for use in tossed salads) **Convenience of use**, as affected by shape, size, packaging (e.g. portion size suitable for lunchbox)	**Softness-meltability** (for most applications) **Flowability/Spreadability (melt)** to a controlled level (for many applications such as pizza, toasted sandwiches, lasagne, gratins) Ability to **brown in colour** (e.g. in sandwiches) **Surface sheen (limited oil-release)** (e.g. in pizza and most heated applications) **Flow-resistance** (e.g. for applications where visible appearance of cheese piece following cooking is desired: ready prepared meals, cheese insets in burgers, deep-fried cheese sticks, savoury desserts)

[a] Melt refers to the ability of the cheese to soften on heating, even though the cheese may or may not flow. However, melt or melt properties are frequently used in a general way to describe the all the characteristics of heated cheese, and in particular the degree of flow.

wrapped in plastic film, sausage-shapes (chubs), fingers, tubes and tubs. Following packing, PCPs are stored, typically at temperatures <8°C.

Global production of PCPs is estimated to be ~1.6 million tonnes per annum, (Sørensen and Pedersen, 2005). Production in the EU 20 was ~0.56 million tonnes in 2004, with France, Germany, Ireland and Poland being the largest producers. Processed cheese production as a percentage of natural cheese is ~8.5%, and its volume exceeds that of the largest varieties of natural cheese, Cheddar (~1.3 M tonnes) and Mozzarella (~1.4 M tonnes). The continued growth and popularity of PCPs can be attributed primarily to their diverse range of flavours, textures (as perceived by touch and mouthfeel during handling, eating: e.g. firm, soft, adhesive, chewiness, cohesiveness, succulence), and functional properties of the unheated and heated products (Table 20.2). These properties of PCPS ensure a wide variety of end-use applications and overall consumer appeal, made possible by differences in formulation and processing conditions, condiment addition, and packaging technology.

20.2 Processed cheese types and permitted ingredients

There are various types of PCPs, the category/standard of which depend on national legislation. Examples include the code of Federal Regulations

Table 20.3 Ingredient specifications for processed cheese products as defined by Codex Alimentarius[ab]

Product category	Permitted ingredients
Named variety processed cheese and spreadable processed cheese	One or more varieties of cheese, with added cheese accounting for ≥51% (w/w) of the final product; milk fat [cream, butter oil] for standardisation of milk fat in final product; water; salt, vinegar, spices, seasonings, flavouring condiments at dry matter levels ≤16.7% (w/w) of dry matter of finished processed cheese, starter culture bacteria, and enzymes, emulsifying salts [sodium, potassium, calcium salts of citric and phosphoric acid at levels ≤0.4% (w/w)] ; pH regulators [food grade organic acids], colours at ≤0.06% (w/w); mould inhibitors [sorbic acid, potassium/sodium sorbate, and/or sodium propionates at levels ≤0.3% (w/w) or nisin at ≤0.001% (w/w)].
Processed cheese and spreadable processed cheese	As for Named variety processed cheese and spreadable processed cheese, except that there are prescribed levels for any one cheese variety in the cheese blend.
Processed cheese preparations – processed cheese foods and spreads	As for processed cheese and spreadable processed cheese, but with the following extra optional ingredients:
	Other dairy ingredients (milk, skim milk, buttermilk, cheese whey, whey proteins, caseins – in wet or dehydrated forms); Selected hydrocolloids and gums
	Taste intensifiers – sodium glutamate

[a] Summarised from Codex Alimentarius Standards, CODEX-STAN A-8(a), A-8(b) and A-8(c) (FAO/WHO, 2008).
[b] The composition of the various product categories are detailed in these standards, and relate to minimum contents of dry matter and fat-in-dry matter.

in the USA, which defines three standards based on permitted ingredients and composition: pasteurised process cheese, pasteurised process cheese food and pasteurised process cheese spread. International standards for processed cheese are defined by the Codex Alimentarius Commission (FAO/WHO, 2008); these standards, summarised in Table 20.3, include:

- Named variety of process(ed) cheese and spreadable process(ed) cheese: (Codex-Stan A-8(a)-1978);
- Process(ed) cheese and spreadable process(ed) cheese: (Codex-Stan A-8(b)-1978); and
- Process(ed) cheese preparations process(ed) cheese food and process(ed) cheese spread: (Codex-Stan A-8(c)-1978).

According to these standards (Table 20.3), the minimum content of natural cheese must be such that it contributes greater than or equal to 51% of the dry matter of the final PCP in the case of process cheese foods and spreads, and about 82%–96% of the dry matter in process cheese or spreadable process cheese, depending on the blend of the cheeses, the amount of milk fat required to standardise to the minimum fat-in-dry matter content, and the levels of added emulsifying and product flavourings.

20.3 Manufacture of processed cheese products

20.3.1 Overview of manufacture

The manufacturing protocol of processed cheese products has been extensively reviewed (Zehren and Nusbaum, 1992; Carić and Kaláb, 1993; Guinee *et al.*, 2004; Kapoor and Metzger, 2008), and is summarised below. It involves the following steps:

- *Formulation*, selecting the different types and levels of ingredients to give the desired end product characteristics;
- *Size reduction of added cheese* by shredding, grating or mincing to maximise the surface area of the cheese and facilitate heat transfer to the blend during subsequent processing;
- *Blending of formulation ingredients* to ensure homogeneity of all materials and uniform end product quality;
- *Processing of the blend*, heating by direct or indirect steam injection typically to ~75–85°C for 1–5 min, while constantly agitating/shearing, the functions being:
    - to kill any potential pathogenic and spoilage micro-organisms, and thereby extend the shelf-life of the final processed cheese product;
    - to induce the interaction between cheese and other ingredients (e.g. emulsifying salts);
    - to facilitate the physico-chemical and microstructural changes which transform the blend to an end product with the desired characteristics and physico-chemical stability;
- *Homogenization of the hot molten product* (typically at first and second stage pressures of 15 and 5 MPa, respectively), an optional step to:
    - assist further mixing, and size reduction of any coarse particles or undissolved particles (e.g. emulsifying salt, dry ingredient, cheese rind), and to
    - promote a finer dispersion of fat droplets, which leads to a smoother and creamier hot blend and thicker and firmer consistency in the final processed cheese;
- *Cooling and storage of the hot molten processed cheese*, to promote:
    - fat crystallisation and regulate the type and degree of protein interactions between the paracaseinate coated fat globules and the dispersed paracaseinate molecules,
    - the desired degree of setting and formation of the end-product texture characteristics to a degree regulated by the cooling rate.

20.3.2 Principles of manufacture

Natural cheese: Protein-mineral complex and structure

Protein in natural rennet-curd cheeses (e.g. Gouda, Cheddar) occurs as paracasein aggregates, rendered insoluble by inter-protein linkages mediated by calcium (attached to acidic amino acid residues such as glutamate and aspartate), colloidal calcium-phosphate (attached to serine phosphate groups), and hydrophobic interactions between uncharged amino acid residues. The depressing effect of calcium phosphate on paracasein hydration has been corroborated by studies in model (dilute) casein systems (Sood *et al.*, 1979; Green, 1982; Ruegg and Moor, 1984; Snoeren *et al.*, 1984) and cheese (Guinee *et al.*, 2000a). The calcium content of natural cheeses exhibits marked inter- (e.g. from ~15.2 $mg \cdot g^{-1}$ paracasein in Bavarian Blue cheese to 35.6 in Emmental) and intra- (e.g. 26.5 to 30 $mg \cdot g^{-1}$ paracasein in Cheddar cheese) variety difference. Noteworthy, is the fact that most (~50 to 60% of total at pH ~5.2) of calcium and phosphate in natural rennet-curd cheeses is insoluble, and this increases as the pH is increased between 5.0 and 6.0 (Guinee *et al.*, 2000a; Ge *et al.*, 2002) but decreases on ageing (Hassan *et al.*, 2004).

In acid curd cheeses, (e.g. Quark, Cottage cheese), protein occurs in the form insoluble casein/casein-whey protein aggregates. In contrast to natural rennet curd cheeses, the levels of calcium and inorganic phosphate are comparatively quite low (e.g. ~100 mg ca $\cdot$ 100 g^{-1}, ~5.8 $mg \cdot g^{-1}$ protein), due to the low pH of the gel prior to whey separation (~4.7 compared to >6.4 for most rennet-curd varieties) and to the ensuing solubilisation of colloidal calcium phosphate. Moreover, most of the remaining calcium in the product is soluble. The integrity of the aggregates is maintained by hydrophobic, electrostatic and covalent disulphide bonds, where the cheesemilk has been subjected to high heat treatment.

In both cheese types, the aggregates are fused into a network (matrix) of entangled strands of aggregates. The matrix encloses moisture and fat. Fat in most rennet-curd cheeses and some acid-curd cheeses occurs as globules enclosed by the native fat globule membrane that prevents the leakage of free fat, 'sweating' and/or greasiness of the cheese. Depending on the milk treatment (e.g. homogenisation), cheese type, and fat-in-dry matter content, the fat globules may be clumped to varying degrees, or partly-coalesced in the form of pools between the paracasein strands (McMahon *et al.*, 1993; Tunick *et al.*, 1993; Kaláb, 1995; Auty *et al.*, 2001). In some cheeses however, such as those made from recombined milk (skim milk powder, anhydrous butter oil and water) or acid-curd cheeses, with high fat-in-dry matter content (e.g. cream cheese), the milk is homogenised, typically at first and second stage pressures of ~5–25 MPa and 3–5 MPa, respectively, to facilitate recombination and/or prevent creaming of native fat globules during subsequent fermentation or acidification. The reformed fat globules become coated with a protein layer consisting of casein micelles, sub-micelles, and to a lesser extent whey protein, referred to as recombined fat globule membrane. The membrane enables the fat globules to behave as fat-filled protein particles, which can become an integral part of the gel network during acid gelation. Their participation in the formation of a composite gel thereby leads to a more uniform, stiffer gel at the end of fermentation (van Vliet and Dentener-Kikkert, 1982; Ortega-Fleitas *et al.*, 2000). These reformed fat globules are very stable, even when the cheese is subsequently heated to temperatures of ~100°C during melting (Guinee *et al.*, 2000b).

Processing natural cheese in the absence of emulsifying salt

Application of heat (70 to 90°C) and mechanical shear to natural cheese in the absence of stabilisers usually results in the formation of a heterogeneous, gummy, pudding-like, heavy mass that undergoes extensive oiling-off and moisture exudation during manufacture and on cooling. Such defects may be attributable to:

- the coalescence of non-globular fat due to the shearing of fat globule membranes, and
- partial dehydration/aggregation and shrinkage of the paracasein/casein matrix as affected by:
  - increased hydrophobic interactions as induced by the relatively low pH of cheese (for most cheeses, ~4.6–5.6) and high temperature applied during processing;
  - precipitation of soluble (serum) calcium and phosphate, leading to further calcium phosphate mediated interactions between the paracasein molecules (especially, in rennet-curd cheeses), and
  - the consequential decline in pH and negative charge.

The above effects are also evident when cooking natural cheeses quiescently at temperatures ~90–100°C, where exudation of some free oil contributes to desired flow, surface sheen and succulence of the hot molten product on dishes such as toasted sandwiches and pasta. However, examination after cooling shows the cheese to be congealed and greasy.

Yet, exceptions to the above may occur, especially when heating natural cheeses with a low level of intact casein (e.g. <70% of total), relatively low calcium phosphate content (e.g. <15 mg ca·g^{-1}) and high pH (e.g. >5.7). Such cheeses provide the basis for manufacture of pasteurised blended cheese (Code of Federal Regulations, 2008). Analogously, heating and kneading of curds with a high intact casein level (e.g. Mozzarella, Cagliata; intact casein ~96–98% of total) during the plasticisation process is possible but at much lower temperatures (e.g. <65°C) and shear rates than those used in processing. The pH (~5.2–5.3) at which these curds are normally plasticised coincides with the pH where solubilisation of colloidal calcium phosphate is maximum and the protein has net negative charge. Plasticisation at lower pH (close to isoelectric pH) results in excessive protein aggregation, as manifested by a heterogeneous, rough-textured, grainy, curdy mass.

At higher pH values and typical calcium (~27 $mg \cdot g^{-1}$ casein) and phosphate (~21 $mg \cdot g^{-1}$ casein) levels, the curd is typically dull, lumpy and tough, indicating a low degree of hydration. However, reducing the calcium phosphate level (e.g. calcium ~19 $mg \cdot g^{-1}$ casein), and hence, the degree of calcium phosphate mediated linkages between the paracasein molecules, enables the curd to be plasticised successfully at higher pH values (~5.8–5.9) (Guinee *et al.*, 2002).

The role of protein and emulsifying salts during processing

The addition of the sodium (or potassium) salts of citric acid and/or phosphoric acid, at levels 1.0–3.0 $g \cdot 100\ g^{-1}$ during processing, eliminates the defects discussed above and promotes the formation of smooth, homogeneous, stable processed cheese products. These salts, referred to as emulsifying salts (e.g. citrates, orthophosphates, pyrophosphates, polyphosphates), generally consist of a monovalent cation (e.g. sodium) and a polyvalent anion (e.g. phosphate, citrate). While these salts are not emulsifiers *per se*, together with heat and shear, they essentially convert the insoluble cheese protein into a more hydrated, soluble sodium paracaseinate that binds free water and emulsifies the free fat released from the natural cheese during processing. Hence, the re-solubilised cheese protein is the primary stabilising agent in the formation of pasteurised processed cheese products, even though other materials such as hydrocolloids and starches may have texture-modifying functionalities.

The emulsifying salts promote hydration and dispersion of the cheese protein by triggering a number of concurrent physico-chemical changes in the natural cheese. These changes include:

- upward adjustment of the pH and stabilisation (buffering);
- calcium sequestration;
- demineralisation and hydration/dispersal of the cheese protein;
- emulsification of free fat; and
- structure formation.

The use of the correct blend of emulsifying salts usually shifts the pH of cheese upwards (typically from ~5.0–5.5 in the natural cheese to 5.6–5.9 in the PCP) and stabilises it by virtue of their high buffering capacity. This change contributes to an enhanced dissociation and calcium-sequestering ability of the emulsifying salts, and an increased negative charge on the paracaseinate.

At the typical pH of processed cheese (pH 6.0), the emulsifying salts are sufficiently dissociated to sequester a large portion of the calcium attached to the casein. For example, typically cited pKa values for phosphoric acid are 2.15, 7.2 and 12.4, while those for citric acid are 3.1, 4.8 and 6.4 (http://www.chembuddy.com; http://www.zirchrom.com/organic.htm). At this pH, the added sodium phosphates and citrates may compete successfully with the casein (carboxyl and phosphoserine groups), not only for caseinate-bound calcium but also the native colloidal calcium phosphate, initiating its transfer from the paracasein and/or casein and subsequent deposition as insoluble calcium phosphate and/or calcium citrate inclusions (Gaucher *et al.*, 2007; Guinee and O'Kennedy, 2009). The demineralisation of casein at the elevated pH leads to a more open reactive paracaseinate/caseinate conformation with superior water-binding capacity compared to that in native cheeses, especially in rennet-curd cheeses where the pH is generally much lower and the level of casein bound calcium much higher. The protein matrix of the natural cheese is thereby transformed to a sodium caseinate or paracaseinate dispersion (sol) in acid-curd and rennet-curd cheeses, respectively (Nakajima *et al.*, 1975; Fox *et al.*, 1965; Lee *et al.*, 1981, 1986; Marchesseau *et al.*, 1997; Panouillé *et al.*, 2004; Gaucher *et al.*, 2007). The 'reactivated casein' then binds the free water and emulsifies free fat created during processing, and is thereby central to the formation of a stabilised processed cheese. These changes are confirmed by: (i) the large increase in the level of water-soluble protein during processing (from ~5 to 20% in the natural cheese to ~60 to 80% in processed cheese) (Fig. 20.1a); (ii) the high levels of water-insoluble calcium (~60 to 80% of total, Fig. 20.1b) and phosphate in processed cheese products for a range of levels and type of emulsifying salts used; and (iii) the reduction in fat globule size during processing. The interactive effects of the type (degree of polymerisation, dissociation constants, molecular mass) and level of emulsifying salt, pH, and processing conditions (time, temperature, and shear) determine the degrees of casein demineralisation and hydration during processing. Hence, the type/level of emulsifying salts and processing conditions have a major influence on the textural and melting attributes of PCPs and ACPs (analogue cheese product) (Rayan *et al.*, 1980; Thomas *et al.*, 1980; Lee *et al.*, 1981; Gupta *et al.*, 1984; Cavalier-Salou and Cheftel, 1991; Marchesseau *et al.*, 1997; Lu *et al.*, 2008; Brickley *et al.*, 2008).

On cooling, the homogeneous molten viscous mass sets to form a characteristic body, which, depending on blend formulation, processing conditions and cooling rate, may vary from firm and sliceable to soft and spreadable (Piska and Štětina, 2004). Micro-structural studies on PCPs indicate that the structure is an emulsion of discrete, rounded fat droplets of varying size in a continuous protein matrix (Kimura *et al.*, 1979; Taneya *et al.*, 1980; Rayan *et al.*, 1980; Heertje *et al.*, 1981; Lee *et al.*, 1981; Tamime *et al.*, 1990; Savello *et al.*, 1989; Marchesseau and Cuq, 1995; Marchesseau *et al.*, 1997). The matrix consists of strands that are finer than those of natural cheese, with the degree of fineness increasing as the pH is raised from 5.2 to 6.1, an effect attributed to a change in the proportions of different types of protein interactions: hydrophobic, electrostatic, hydrogen bonds, and residual calcium crosslinks (Marchesseau *et al.*, 1997). In contrast to natural cheese, the fat and paracasein are more uniformly distributed, the matrix is less compact and fused, there is less clumping/coalescence of fat globules, and the mean fat globule size is generally

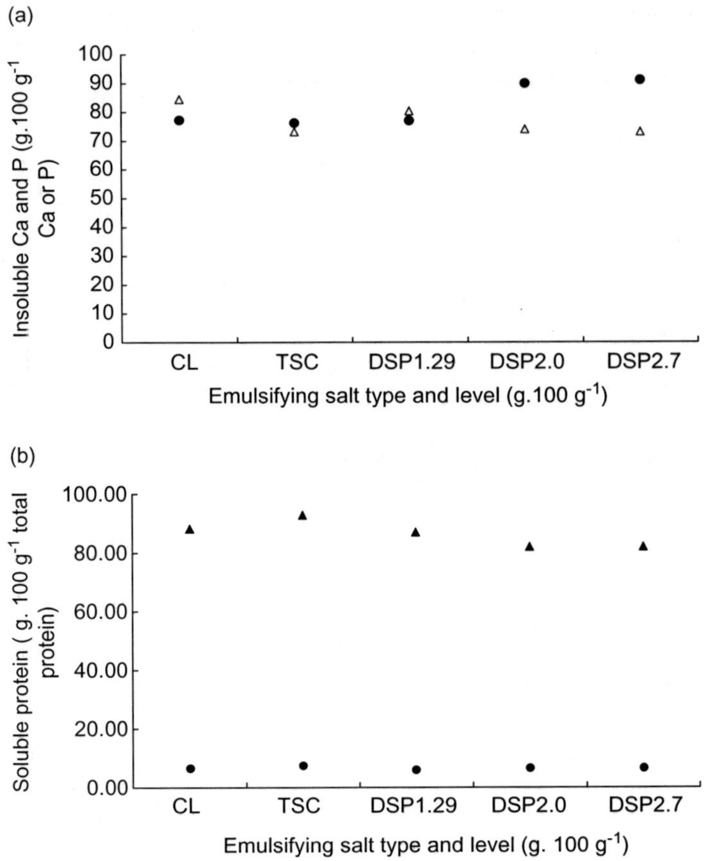

Fig. 20.1 Effect of emulsifying salt type and level on: a) levels of insoluble calcium (●) and phosphorous (△) and b) levels of water-soluble (▲) and pH 4.6 soluble protein (■) in experimental processed cheese. Emulsifying salt type: TSC, trisodium citrate dihydrate; DSP, disodium hydrogen orthophosphate anhydrous. The level of emulsifying salt ($g \cdot 100\ g^{-1}$) is shown by the number after the type; CL (control) consisted of a blend of 1.29 g TSC$\cdot 100\ g^{-1}$ and plus 0.71 g DSP$\cdot 100\ g^{-1}$; presented data are the means of four replicate trials.

much smaller than that of natural cheese (but depends on formulation and processing conditions, e.g. emulsifying salts, milk protein additions, processing time and extent of shear). The incorporation of emulsified paracaseinate-coated fat globules, which can be considered as fat-filled protein particles, into the new structural matrix may be considered to increase the effective protein concentration (c.f. van Vliet and Dentener-Kikkert, 1982; Marchesseau *et al.*, 1997). The positive correlations between the degree of emulsification and firmness or elasticity, and the inverse relationship between the degree of emulsification and flowability of PCPs support this suggestion (Rayan *et al.*, 1980; Caríc *et al.*, 1985; Savello *et al.*, 1989).

20.4 Effects of different ingredients on properties of processed cheese products

These properties are controlled by formulation, types of ingredients, processing conditions, the gross composition of the final processed cheese, and their manipulation by the state-of-the-art knowledge and know-how of manufacturers. They have been extensively investigated and several comprehensive reviews are available on this topic (Meyer, 1973; Zehren and Nusbaum, 1992; Carić and Kaláb, 1993; Guinee *et al.*, 2004; Kapoor and Metzger, 2008). The following discussion focuses on the effects of different ingredients.

20.4.1 Effect of characteristics of natural cheese used in processing

Little published information is available on the effect of casein hydration or the concentration of intact paracasein in the unheated cheese on the rheological characteristics of the resultant processed cheese.

Cheese type and level of intact protein

Cheese is the major ingredient in all PCPs, with the use of many different varieties, differing with respect to composition (for example moisture, calcium level, pH) and degree of maturity. During maturation of rennet-curd cheeses such as Cheddar or Gouda, the paracasein is increasingly hydrolysed into peptides and free amino acids by various enzymatic activities, including residual coagulant and the proteinase and/or peptidases systems of milk, starter culture lactic acid bacteria, non starter lactic acid bacteria, secondary cultures and/or exogenous enzyme preparations (Upadhyay *et al.*, 2004). The level of intact protein, as measured by the level of nitrogen insoluble at pH 4.6 (casein isoelectric pH), decreases concomitantly with paracasein hydrolysis. The level of intact protein (the fraction of non-degraded protein as a percentage total protein) in the natural cheese has a major effect on the properties of the unheated and heated processed cheese. As the level of intact casein in the cheese used for processing decreases, there is a significant decrease in the firmness of the unheated processed cheese and increase in meltability of the heated processed cheese, as evidenced by increases in flowability and loss tangent (on heating between 20 and 82°C) and a decrease in the elastic shear modulus, G', at 20°C (Olson *et al.*, 1958; Brickley *et al.*, 2007). In corollary, the flowability of processed cheeses in which Cheddar cheese was partially replaced by an ultrafiltred (UF) retentate/cream pre-cheese blend significantly improved when casein hydrolysis in the blend was increased by treatment with fungal proteinases (from *Aspergillus oryzae*, *Candida cylindracea*) and rennet. Similarly, it has been observed that model processed cheese prepared from acidified curd became excessively soft when a high concentration of peptides of molecular mass <10 kDa was present following treatment of the curd by a fungal proteinase (from *A. oryzae*) prior to processing (Mahoney *et al.*, 1982).

The effects of the degree of cheese maturity, and hence intact protein, has long been recognised and reflected in industrial practices. Thus, the use of cheese age as a major selection criterion for blend formulation at commercial level is common practice. Block processed cheeses with good sliceability and elasticity are generally formulated with a high proportion of young cheese (85–95% intact protein) in the blend; whereas predominantly medium ripe cheese (60–85% intact casein) is used for cheese spreads. It is also generally recognised in practice that interior, bacterial-ripened hard and semi-hard cheese varieties, such as Cheddar, Gouda and Emmental, give firmer, longer-bodied processed products than surface ripened cheeses (e.g. mould-ripened: Camembert and Blue cheese. Bacterial-ripened: Ersom, Limburger). The latter cheeses generally have a higher pH and undergo more extensive proteolyis than the former during ripening (Reps, 1993; Brennan and Cogan, 2004; Spinnler and Gripon, 2004), and therefore are likely to have a relatively low degree of intact casein. However, the level of intact casein in any variety will be influenced by a myriad of factors (e.g. age, ripening conditions, milk pre-treatments, composition, types of coagulant/cultures, addition of exogenous enzymes). Hence, the measurement of intact casein content (e.g. by level of protein insoluble in water at pH 4.6, gel electrophoresisis/densiometry, and/or reversed phase HPLC) is being used increasingly as a quality control tool in selection of, and determining the suitability of, natural cheese for specific processed cheese recipes. However, the response of process cheese functionality to intact casein level is influenced by many other factors such as gross composition, calcium level, and pH (Marchesseau *et al.*, 1997; Piska and Štětina, 2004), a fact worthy of consideration when formulating commercially.

Calcium content

In cheese, calcium and phosphorous exist mainly in the form of an insoluble colloidal calcium phosphate complex (associated with the *paracasein* matrix) in equilibrium with soluble phosphate and calcium; typically, the ratio of the insoluble to soluble salts varies, e.g. from ~1.3 for Cheddar (pH ~5.2) to ~3.34 for low moisture Mozzarella (pH ~5.6); and increases with pH (Guinee *et al.*, 2000a). The ratio of total Ca to P is relatively constant (e.g. ~1.3–1.4) for most rennet cheeses at pH values ≈ ≥5.1. Hence, the effects of total calcium content of cheese on its properties or processability should be expressed more precisely as the effects of total calcium plus phosphorous, rather than of calcium *per se*. The levels of total calcium in natural cheeses exhibit marked inter- (e.g. from ~15.2 $mg \cdot g^{-1}$ protein in Bavarian Blue cheese to 35.6 in Emmental) and intra- (e.g. 26.5 to 30 $mg \cdot g^{-1}$ protein in Cheddar cheese) variety variation. It has a major impact on the degree of protein hydration and the physical properties of natural cheese

(Keller *et al.*, 1974; Lawrence *et al.*, 1984; Kindstedt *et al.*, 2004; Kapoor and Metzger, 2008).

Yet, the calcium content has seldom been considered in the selection of natural cheese for processing or for its effects on the physical properties of the final processed cheese products. Kapoor *et al.* (2007) and Biswas *et al.* (2008) compared the effects of increasing the calcium content of Cheddar cheese (21 and 27 $mg \cdot g^{-1}$ protein) on the physical properties of processed cheese food (15.0 and 18.5 mg $Ca \cdot g^{-1}$ protein) formulated using cheese, non-fat dry milk powder and butter oil, and containing high or low levels of lactose (~1.52 and 0.3 $g \cdot 100\ g^{-1}$ Cheddar) and salt-in-moisture (S/M; ~6.4 and 5.0 $g \cdot 100\ g^{-1}$ Cheddar). Reducing calcium level resulted in softer and more meltable processed cheeses, but only when the S/M levels of the cheddar cheeses used were high; however, the results were somewhat complicated by random variation in levels of intact casein (79 to 89% of total), pH (5.7 to 6.2), and protein level in the two-month old Cheddar cheeses used. In a recent study, Guinee and O'Kennedy (2009) found that reducing the levels of calcium (from 29.8 through 24.4 to 19.6 $mg \cdot g^{-1}$ casein) and phosphorous (from 20 to 16 $mg \cdot g^{-1}$ casein) in Cheddar cheese (intact casein level, 89% of total; protein, ~18 $g \cdot 100\ g^{-1}$), had a marked impact on the functionality of the resultant pasteurised processed Cheddar cheese, leading to significant reductions in fracture stress, fracture strain and firmness of the unheated product, and increases in the extent of flow (Fig. 20.2) and fluidity (loss tangent) of the melted cheese. Regression analysis indicated that the functionality of the processed Cheddar cheese was strongly correlated with the high levels of water insoluble calcium and phosphate (Guinee and O'Kennedy, 2009). A tentative suggestion for this relationship is that the high levels of water-insoluble calcium and phosphate may form an amorphous structure or gel (Relyveld, 1977; Becker *et al.*, 2003) that interferes with cheese functionality to an extent dependent on formulation, composition, type of emulsifying salt used, and processing conditions.

Cheese pH

pH has a major effect on the structure and functional properties of processed cheese products (Gupta *et al.*, 1984; Cavalier-Salou and Cheftel, 1991; Marchesseau *et al.*, 1997; Lee and Klostermeyer, 2001; Piska and Štětina, 2004). Processed cheeses with low pH (4.8–5.2), as affected by the sole use of monosodium citrate, monosodium phosphate or sodium hexametaphosphate as emulsifying salt, are short, dry, crumbly and prone to oiling-off during formation, storage and subsequent heating, whereas products with high pH (>6.0) tend to be very soft and flow excessively on heating (Gupta *et al.*, 1984).

Increasing pH in the region 5.7 to 6.2 is accompanied by decreases in hardness and storage modulus (G′), and by increases in the flowability and fluidity (loss tangent) of the melted processed cheese (Fig. 20.3; Marchesseau

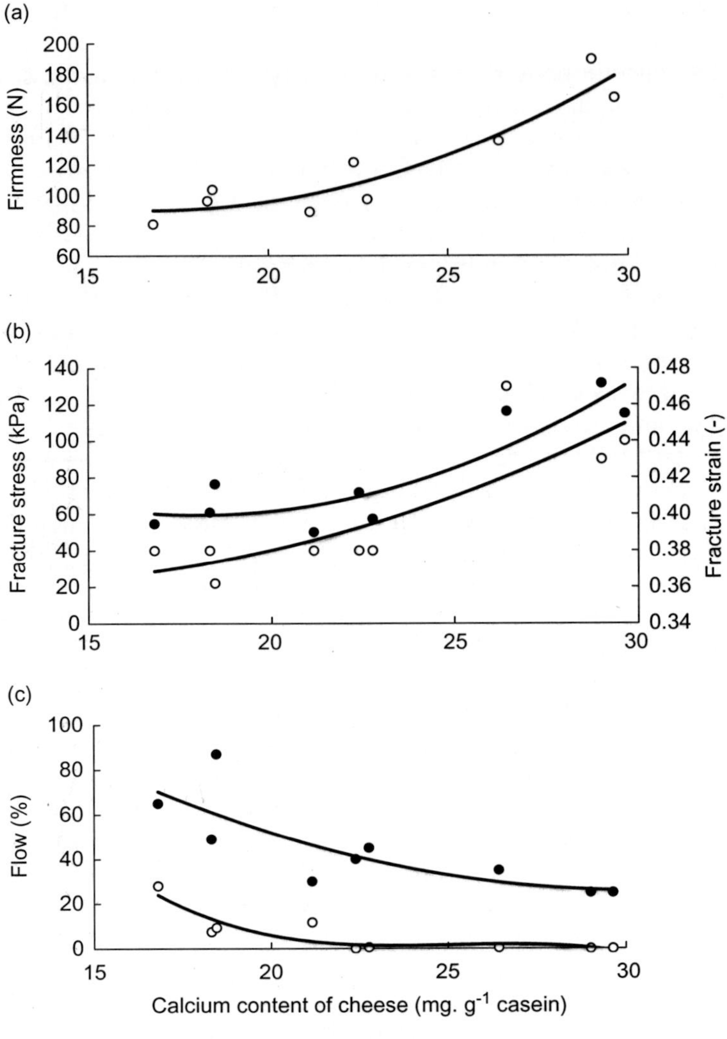

Fig. 20.2 Effect of calcium level (expressed as mg·g^{-1} casein) in natural cheese on: a) firmness (○); b) fracture stress (●) and fracture strain (○); and c) flowability as measured using modified versions of Price-Olson (●) or Schreiber (○) methods. Presented data show the results of nine trials using different Cheddar cheeses varying in calcium level.

et al., 1997; Lee and Klostermeyer, 2001). These changes coincide with increases in casein hydration (Fig. 20.4), a lower degree of paracasein aggregation, a finer gel network and a more uniform structure. However, it is difficult to study the effect of pH *per se* because of indirect effects of the methods used to control pH (different emulsifying salts; pH adjustment before or after product formation using a given type of emulsifying salt) on

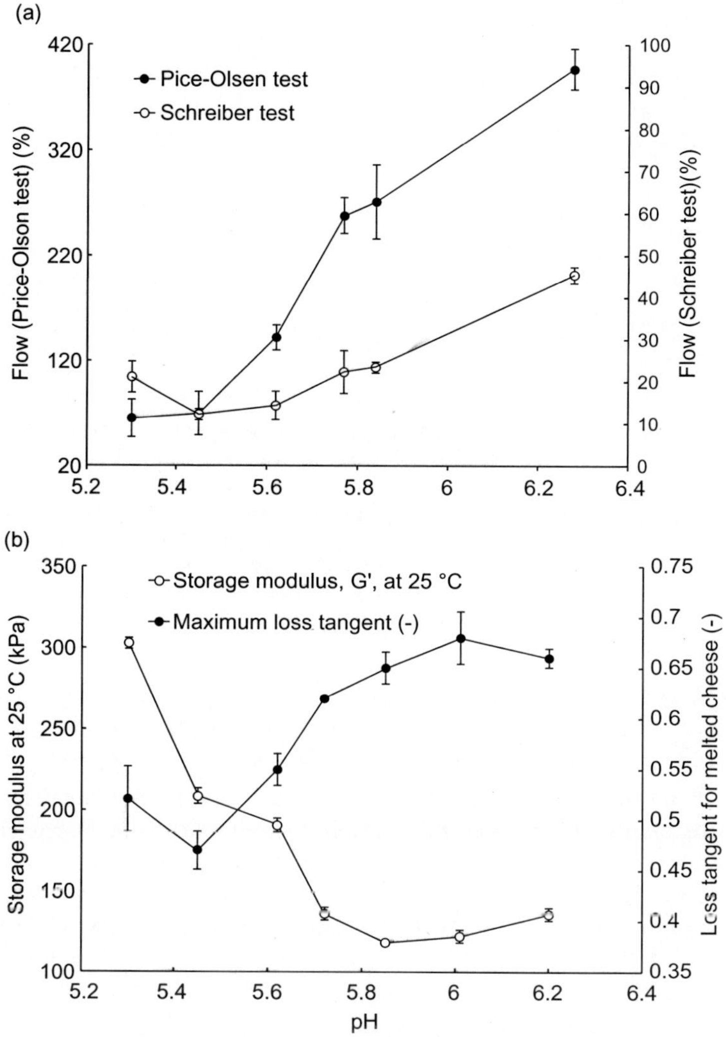

Fig. 20.3 Effect of pH on the cooking characteristics of processed cheese: a) heat-induced flowability, as measured using modified versions of Price-Olson or Schreiber methods, and b) changes in viscoelastic parameters (storage modulus at 25°C and maximum value of loss tangent). Data are the means of three replicate trials, and error bars show standard deviations.

calcium sequestration during product formation and cooling. Adjusting pH before processing affects the calcium sequestering ability of different emulsifying salts to a degree depending on their dissociation constants, while pH adjustment on the completion of processing allows time for the emulsifying salt to sequester calcium during mixing (of acidulant) and cooling. A more precise measurement of the effect of pH may be obtained by the approach

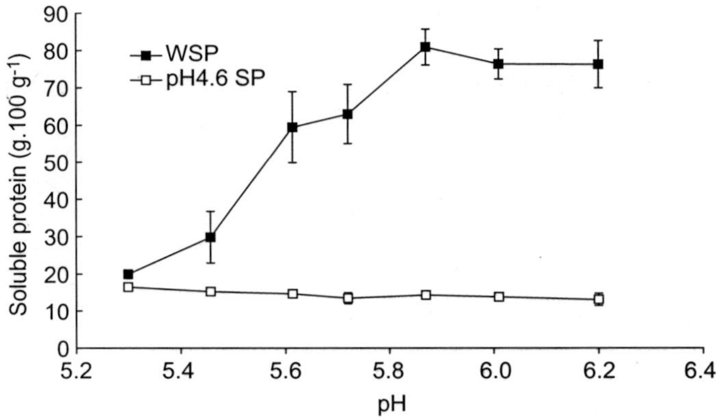

Fig. 20.4 Effect of pH on levels of pH 4.6 soluble-protein (pH 4.6SP) and water-soluble protein (WSP), expressed as percentages of total protein. For all pH values, blocks of the same batch of Cheddar cheese, with similar levels of pH 4.6SP and WSP (~14 $g \cdot 100\ g^{-1}$ of total protein) were used. The pH of the hot molten processed cheese blend (pH 5.9) was adjusted at the end of manufacture by adding the appropriate quantities of concentrated sodium hydroxide or hydrochloric acid, and mixing for 30 s. Data are the means of three replicate trials, and error bars show standard deviations.

of Gigante *et al.* (2006), whereby slices of the cooled product were exposed to vapours of ammonia or acetic acid (Almena-Aliste *et al.*, 2006).

The author is not aware of any study that has systematically examined the effect of cheese pH on processed cheeses, presumably because of difficulties of changing pH without changing other parameters (e.g. calcium level, intact casein level) in the natural cheese, which otherwise cannot be standardised during the manufacture of the process cheese. Increasing the pH of Cheddar cheese from 5.05 to 5.35 increased the firmness and reduced the meltability (flow) of the resultant processed cheese food (Kapoor *et al.*, 2007); however, the levels of calcium and intact casein in the Cheddar cheese increased simultaneously with pH.

20.4.2 Milk proteins

Attempts to reduce formulation costs of PCP and improve end-product consistency have led to extensive investigation on the development of, and study of, the effects of ingredients that are more cost-effective than cheese (Mann, 1993, 1997). In this regard, it has been attempted to replace blend cheese by other dairy ingredients, including milk proteins, milk ultrafiltrate (Sood and Kosikowski, 1979; Anis and Ernstrom, 1984) and cheese base (Ernstrom *et al.*, 1980; Park *et al.*, 1992; Simbuerger *et al.*, 1997).

Milk proteins (e.g. casein, whey protein) and their derivatives (e.g. caseinates, co-precipitates) are used widely in processed cheese foods and spreads, where significant quantities may be added (e.g. $10\ g \cdot 100\ g^{-1}$), depending on the fat-in-dry matter and dry matter levels of the final product (FAO/WHO, 2008). Depending on their solubility, structure and chemistry, added proteins have a marked influence on the physico-chemical, rheological, stability and usage appeal characteristics of pasteurised processed cheese products (PCPs) and analogue cheese products (ACPs) (Savello *et al.*, 1989; Abou-El-Nour *et al.*, 1996). The functional properties may be defined as those rheological, physico-chemical, micro-structural and organoleptic (i.e. flavour and texture) characteristics that affect the behaviour of the cheese in food systems during preparation, processing, storage, cooking and/or consumption. Hence, protein has a major influence on processed cheese quality and on its industrial applications.

Caseins and caseinates

In milk, casein which comprises ~80 $g \cdot 100\ g^{-1}$ of the total protein and is the main structural protein in acid-induced milk gels, exists in the form of spherical-shaped colloid particles (~40–300 nm diameter), casein micelles (Fox and Brodkorb, 2008; McMahon and Oommen, 2008). The micelles on a dry weight basis consist of 7 $g \cdot 100\ g^{-1}$ ash (mainly Ca and P), 92 $g \cdot 100\ g^{-1}$ casein, and 1 $g \cdot 100\ g^{-1}$ minor compounds . The casein is heterogeneous, comprising four main types:- α_{s1}, α_{s2}, β and κ, present at a ratio of ~4:1:4:1.2. Model studies in dilute dispersions indicate that the individual caseins vary in the content and distribution of phosphate, and consequently have different calcium binding properties. Generally, α_{s1}-, α_{s2}-, β- caseins bind calcium strongly, and precipitate at relatively low calcium concentrations (e.g. ~0.005–0.1 M $CaCl_2$ solutions), inclusive of the calcium level in milk (30 mM); in contrast, κ-casein is not sensitive to these calcium concentrations and can, in fact, stabilise up to ten times its mass of the calcium-sensitive caseins. Owing to the differences in degree of phosphorylation, the arrangement of casein within the micelle is such that the core is occupied by the calcium-sensitive caseins and κ-casein is principally located at the surface, with the hydrophilic c-terminal region (caseino macropeptide) oriented outwards toward the serum phase in the form of protruding negatively-charged hairs which confer stability to the micelle by electrostatic repulsion, Brownian movement and steric effect.

The production of casein powders, rennet casein and acid casein, from milk has been recently reviewed (Mulvihill and Ennis, 2003). They are formed by coagulation of the casein, as induced by selective hydrolysis of κ-casein using rennet (rennet casein) or pH adjustment to the isolectric point (acid casein). Following coagulation and whey separation, the insoluble curds are washed, pasteurised and dried to moisture levels of ~10 $g \cdot 100\ g^{-1}$. In both cases, the casein is more aggregated and dehydrated compared to casein in rennet-curd or acid-curd cheeses owing to the relatively high temperatures during curd pasteurisation and drying, and the lower moisture levels (~0.12 $g \cdot g^{-1}$ protein compared ~1.25 to 6.25 $g \cdot g^{-1}$ protein in cheese, depending on the variety). While both caseins are insoluble in water, they also differ in composition. Similar to rennet-curd cheese, the protein in rennet casein is calcium phosphate paracaseinate (at pH ~6.8–7.1), where the caseino-macropeptide region of the casein micelle-stabilising protein (κ-casein) has been cleaved by the addition of rennet (e.g. chymosin); the remainder of the caseins (referred to as paracasein) coagulate via calcium binding and associated charge suppression, and retain the full compliment of micellar calcium and phosphate. In contrast, acid casein contains the full compliment of caseins (including κ-casein), but all the micellar calcium and phosphate are solubilised during acidification (to pH ~4.6). Owing to the differences in casein and mineral composition, the potential of acid casein to hydrate and bind water is much higher than that of rennet casein when the pH is reversed to the pH of native milk, i.e. ~6.7) (from 4.6) because of its low calcium level (Sood *et al.*, 1979), the presence of the highly hydrated caseino-macropeptide, and high pH.

To enhance its functionality, acid casein can be converted to sodium or calcium caseinate during (casein) manufacture; this essentially involves treatment of the curd with alkali solution and readjustment of the pH of the casein back to values (~6.8–7.0) similar to or slightly higher than that of the milk. Manufacture involves: physical disintegration of the dewatered acid casein curd (~40% moisture) mass by agitation; addition of water at 40°C to give a dry matter level of 25 $g \cdot 100\ g^{-1}$; formation of a slurry by passing through a colloid mill; addition of alkali which, depending on the type of caseinate, may differ, e.g. with NaOH for sodium caseinate, $Ca(OH)_2$ for calcium caseinate, or NH_4OH for ammonium caseinate; pH adjustment to pH ~6.8; solubilisation of the casein on heating to ~75–90°C while vigorously agitating; and drying.

While rennet and acid caseins are insoluble, sodium caseinate is highly dispersible and forms highly viscous solutions at concentrations >~15 $g \cdot 100\ g^{-1}$ protein; similarly, calcium caseinate is readily dispersable but not as soluble as sodium caseinate.

Rennet casein, despite its insolubility, is the preferred protein type for use in block processed cheese or analogue/substitute pizza cheese. In both products, it behaves like fresh natural rennet-curd cheese, conferring a high degree of elasticity and firmness to the unheated processed cheese, and moderate meltability. Partial replacement of young Cheddar cheese (intact casein content, ~93 $g \cdot 100\ g^{-1}$ protein) with commercial rennet casein (intact casein, ~99 $g \cdot 100\ g^{-1}$ protein) results in firmer processed cheese that melts to a lower extent on heating. This effect may be attributed to the higher calcium phosphate content of rennet casein, ~33 $mg \cdot g^{-1}$ casein compared to Cheddar cheese (~27 $mg \cdot g^{-1}$ casein). Rennet casein also contributes to the desired degrees of stringiness and stretchability in the hot processed mass during the manufacture of analogue pizza cheese, an attribute that is highly desired by the manufacturer as it gives the appearance of natural low moisture Mozzarella cheese to the latter. However, despite this, commercial analogue pizza cheeses made from rennet casein generally possess inferior stringiness compared to natural Mozzarella cheese when subsequently melted (Guinee *et al.*, 2000a). This probably reflects differences in the degree of protein hydration and microstructure. Recent work in the author's laboratory has shown the level of protein hydration (as measured by the degree of water extractable protein when using a 2:1 water:cheese slurry) in analogue cheese made with a very low level of disodium orthophosphate (~0.75 $g \cdot 100\ g^{-1}$) was much higher than that in freshly-made Mozzarella cheese (~40 versus ~5 $g \cdot 100\ g^{-1}$); the level of water-soluble protein in PCPs and ACPs is generally much higher (~60–80 $g \cdot 100\ g^{-1}$) owing to the higher level of emulsifying salts used (e.g. 1–2 $g \cdot 100\ g^{-1}$). Moreover, microstructural studies have shown that the aligned calcium phosphate paracasein fibres (with interspersed fat pools/columns showing similar orientation) which give natural molten Mozzarella its stringiness are absent in analogue pizza cheese, which consists of discrete rounded fat droplets of varying size in a continuous protein phase (Guinee *et al.*, 2000a; Kindstedt *et al.*, 2004).

Caseinates (especially sodium) find most application in processed cheese spreads (PCSs), where their high water-binding capacity and good emulsifying properties promote a desired *creaming effect*. Gouda *et al.* (1985) reported that full replacement of non-fat cheese solids by calcium caseinate caused deterioration in spreadability of Cheddar PCS, probably due an excessive creaming effect. However, partial replacement in a formulation (with skim milk powder, calcium caseinate, ripe Cheddar, butter oil and emulsifying salt at respective levels of 6–8, 5–7, 15, 14, and 3%, w/w) improved the meltability of the PCS, suggesting a desirable level of creaming. Caseinates may be used in spreadable imitation cheese products (Marshall, 1990; Hokes *et al.*, 1989).

Acid casein is generally not used in processed cheese manufacture because of its insolubility, low pH, and its depressing effect on the pH of the processed cheese blend, which in turn reduces the degree of dissociation of the emulsifying salts and their ability to sequester calcium from other proteins (e.g. cheese protein, or rennet casein). Consequently, the use of acid casein at quantities of 1 to 3% (w/w) of the processed cheese formulation can markedly extend the product make time, unless the pH of the blend is increased to its normal value (e.g. ~5.8 to 6.0) via addition of alkali and/or an emulsifying salt blend with the desired pH-buffering effect. A comparison of the functionality of acid casein and rennet casein on the melt properties of model processed cheeses (made from the casein powder, vegetable fat, water and emulsifying salts) was made by Savello *et al.* (1989). The response of meltability to casein type depended on the type of emulsifying salt and the pH to which the acid casein was adjusted (upwards) during processing. Acid casein (pH adjusted to ~7.0 to 7.7) gave model processed cheeses with higher meltability than those made with rennet casein, using disodium hydrogen orthophosphate or tetrasodium pyrophosphate, and an opposite effect when using sodium aluminium phosphate or trisodium citrate.

Casein hydrolysates may be used successfully as a means of partially replacing sodium orthophosphates in processed cheese manufacture (Kwak *et al.*, 2002), despite the fact that the pH of the processed cheeses containing the hydrolysate is significantly lower than the control (~5.7–5.9 compared with 6.3). This was attributed to the increased emulsifying capacity of the hydrolysate over the parent protein, sodium caseinate. However, when added at a level of 3%, w/w, to replace the emulsifying salt completely, the hydrolysate gave a PCP that on cooking had a high flowability but excessive oiling-off.

Whey proteins

In practice, whey proteins are added to processed cheese formulations mostly in the form of whey powder (e.g. sweet whey powder with ~12–15 $g \cdot 100\ g^{-1}$ protein), which is used as a cost-effective filler to impart a mild sweetish taste and a smooth consistency, especially desirable in high-moisture processed cheese spreads and dips. Whey protein concentrates (WPCs) or whey protein isolates (WPIs), though used less frequently, are included in formulations for specific functions such as (i) to give low-flow or flow-resistance in 'controlled–melt' PCPs targeted to specific cooking applications (e.g. cheese insets/pieces in meat-based products, fried cheese), and (ii) to confer 'body' (stiffness and viscosity) to high moisture spreadable PCPs.

Numerous studies have investigated the effects of added whey proteins on the texture and cooking characteristics of PCPs (Schulz, 1976; Savello *et al.*, 1989; Hill and Smith, 1992; Gupta and Reuter, 1993; Kaminarides and Stachtiaris, 2000; Gigante *et al.*, 2001; Mleko and Foegeding, 2000; 2001; French *et al.*, 2002; Mleko and Lucey, 2003; Mounsey *et al.*, 2007). These studies have generally shown that the use of whey proteins, even at relatively low levels (1–2 $g \cdot 100\ g^{-1}$), as a substitute for cheese protein or casein at similar levels, reduces the flowability of the heated PCP to an extent that increases as the level of substitution increases (Savello *et al.*, 1989; Gupta and Reuter, 1993; Gigante *et al.*, 2001; Mleko and Foegeding, 2001; Mounsey *et al.*, 2007). Indeed, an early patent (Schulz, 1976) describes a process for the manufacture of PCP that is resistant to flow on cooking, based on the addition of a heat-coagulable protein (3–7%, w/w, lactalbumin, egg albumen), at a temperature <70°C, to the PCP formulation near the completion of processing. The loss of flowability observed with the addition of whey proteins was dependent on concentration of whey proteins but independent of whether the whey protein was denatured (heated at pH 6.6 at 85°C for 1 hour) or undenatured (Savello *et al.*, 1989) or of the pH (7 through 10) at which the whey protein was denatured (at 80°C for 30 min) (Mleko and Foegeding, 2001). The latter surmised that on heating, processed cheese with added whey proteins behaved like a two-component system, consisting of a melting casein network and a non-melting whey protein network. Such a network may be comprised of whey protein particles/aggregates that interact to form larger particles or a network in the processed cheese at the high temperatures (80 to 100°C) reached during manufacture and subsequent cooking of the product in the particular food application (e.g. in sandwich, pizza). The high calcium level and relatively low pH (~6.0) of the processed cheese environment would be conducive to a high degree of interaction of heat-denatured whey proteins by thiol–thiol, thiol–disulphide, hydrophobic and electrostatic interactions. A recent study showed that heating of β-lactoglobulin at pH 4.6 in dilute solutions ($1\ g \cdot 100\ g^{-1}$) with medium ionic strength at ~80°C is conducive to the formation of precipitable aggregates (O'Kennedy and Mounsey, 2009). The fact that the inclusion of casein-co-precipitates formed by preheating skim milk at pH 9.5 yielded meltable processed cheese (when included at ~$2\ g \cdot 100\ g^{-1}$) (Mounsey *et al.*, 2007), suggests that the presence of casein when preheating whey proteins somehow reduces their subsequent reactivity in the processed cheese environment.

The use WPCs as a replacement of cheese solids has also been found to accelerate storage-related flavour deterioration, which increased with the level of WPC added in the range 0 to 20%, w/w (Thapa and Gupta, 1992).

Casein–Whey Protein Co-precipitates (CWPts)

Conventionally, co-participates, also referred to as total milk proteinates, are protein products containing casein and whey proteins, and they are formed by heat treatment of the milk and subsequent precipitation of the protein complex by acidification to pH 4.6 and calcium addition (Mulvihill and Ennis, 2003). Generally, their inclusion in PCPs has been found to reduce the heat-induced flowability to an extent that increases with level (Abou-El-Nour *et al.*, 1996; French *et al.*, 2002; Mounsey *et al.*, 2007).

Depending on the level of $CaCl_2$ added, three types variants of CWPts may be obtained, namely, high, medium and low calcium CWPts containing 2.5–3.0, 1.0–2.0 and 0.5–0.8%, w/w, calcium, respectively. The use of various CWPts at levels up to $5\ g \cdot 100\ g^{-1}$ formulation, increased the firmness and sliceability, and reduced the meltability/flowability of the resultant PCPs (Thomas, 1970). Owing to their excellent emulsifying capacity, Thomas and Hyde (1972) concluded that the level of emulsifying salts could be reduced from 3.0 to $2.0\text{–}2.5\ g \cdot 100\ g^{-1}$, if calcium CWPt was added at a level of $2\text{–}3\ g \cdot 100\ g^{-1}$ to the processed cheese formulation. However, high levels ($>3\ g \cdot 100\ g^{-1}$) significantly reduced the flowability of the PC, especially when a high proportion of young cheese was used. The depressing effect of CWPt on meltability of PCPs may be associated with a number of factors including *inter alia*:

- a higher emulsion stability (Thomas, 1970) and/or degree of emulsification (Gupta and Reuter, 1993);
- the formation of large whey–protein casein aggregates in the CWPt (O'Kennedy and Mounsey, 2009) which, on inclusion in processed

cheese, impede slippage of the protein layers and thereby restrict heat-induced flow, similar to the melt-inhibitory effect observed by McMahon *et al.* (1996) in low-fat Mozzarella cheese made from milk with added Novogel (a proprietary blend of microcrystalline cellulose and guar gum); and/or
- the formation of a non-melting whey protein network (Mleko and Lucey, 2003).

Co-precipitates are not widely used in the food industry owing to their generally poor functionality (low solubility, emulsification) as an ingredient, and adverse effects on the meltability/flowability of processed cheese products. However, new approaches aimed at controlling heat-induced denaturation and aggregation of whey protein, and the size/gelation capacity of the resultant reaction products may improve the functionality of CWPts (Mleko and Foegeding, 1999; Donato and Guyomarc'h, 2009). Mounsey *et al.* (2007) investigated the effect of varying the pH of skim milk to values greater (7.5, 9.5) or lower (3.5) than the control (~6.6), prior to heating and re-acidification to pH 4.6, on the functionality of the resultant CWPts in model processed cheese products. The CWPts were added at $8\ g \cdot 100\ g^{-1}$ as a substitute for acid casein powder and the level of whey protein ($g \cdot 100\ g^{-1}$) in the processed cheese varied from 1.37 (for pH 7.5 and 9.5) to 0.96 (pH 3.0) due to the effect of pH at heat treatment on whey protein denaturation and recovery. Increasing the pH at heating yielded CWPts that markedly improved the meltability, which, for the processed cheese made from the pH 9.5, CWPt was similar to the control processed cheese containing $8\ g \cdot 100\ g^{-1}$ acid casein. Conversely, reducing the pH had the opposite effect.

Micellar Caseins (MC) and Milk Protein Concentrates (MPCs)

Micellar caseins are produced from skim milk, which is subjected to microfiltration and diafiltration to remove whey proteins, lactose and soluble salts (Kelly *et al.*, 2000). MPCs are similarly manufactured except that the skim milk is ultrafiltered instead of microfiltered so as to recover the whey protein along with the micellar casein (Guinee *et al.*, 2006). For both ingredients the diafiltered retentates are spray dried to yield powders with protein levels of $\sim \leq 82\ g \cdot 100\ g^{-1}$, being comprised of casein micelles and colloidal ash in MCs and micellar casein, whey proteins and ash in MPCs.

The use of MC or MPC at a level of $6.3\ g \cdot 100\ g^{-1}$, as a substitute for young Cheddar cheese protein (intact casein $\sim 90\ g \cdot 100\ g^{-1}$ protein) in the formulation reduced flowability of the melted processed cheese, with the effect being more pronounced for the MPC (Fig. 20.5). The lower flow with the MPC is most likely due to heat-induced aggregation and/or gelation of whey proteins, present at a level of $\sim 1.3\ g \cdot 100\ g^{-1}$ in the processed cheese, even though other factors (such as the higher intact casein content of the added MPC) may also contribute.

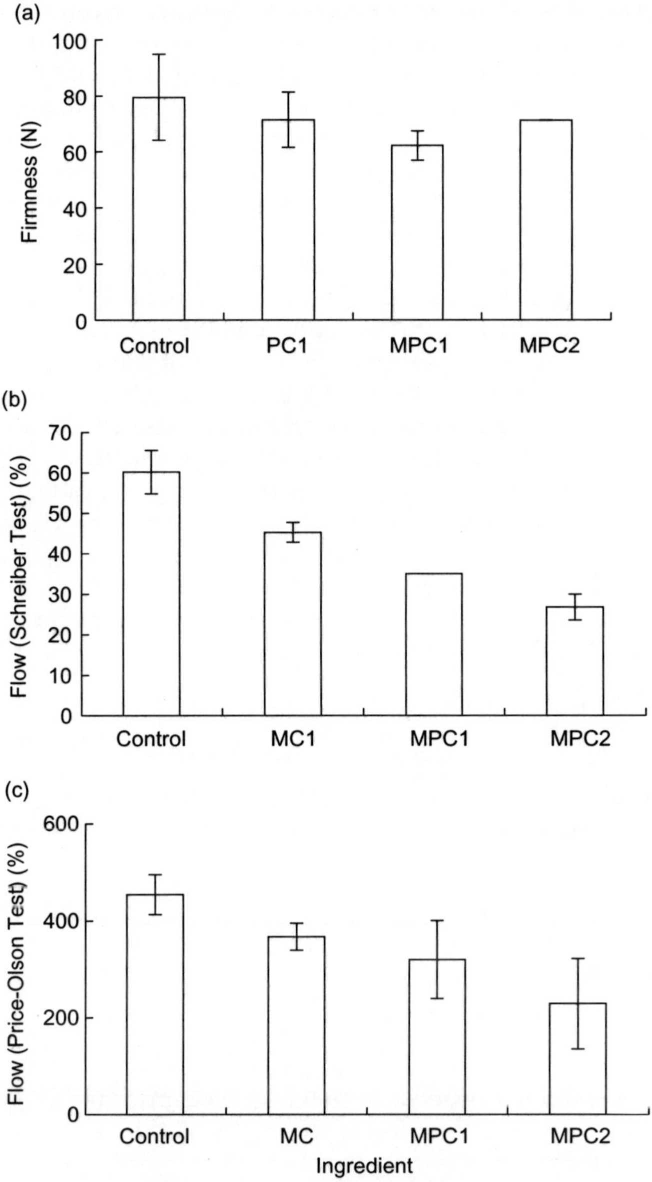

Fig. 20.5 Effect of partial replacement of Cheddar cheese with different ingredients (MC and MPC) on the physical properties processed cheese. The ingredients were added to replace 25% of the protein content of the Cheddar cheese. To ensure similar levels ($g \cdot 100\ g^{-1}$) of moisture (~47), fat (~25) and protein (20.5) in all processed cheeses, anhydrous butter oil and water were added along with the ingredients. Treatment processed were made using Cheddar cheese only (control), and using Cheddar cheese plus micellar casein powder (MC), milk protein concentrate powder 1 (MPC1), or milk protein concentrate powder 2 (MPC2). All powders had $80\ g \cdot 100\ g^{-1}$ protein, and the MPC powders differed with respect to heat treatment applied during manufacture (low heat, MPC1; high heat, MPC 2). Data are the means of three replicate trials, and error bars show standard deviations.

20.4.3 Cheese Base (CB) and Ultrafiltered Milk Retentate (UFMR)

Production of CB generally involves ultrafiltration and difiltration of skim milk (typically to 20–25 $g \cdot 100\ g^{-1}$ dry matter), inoculation of the retentate with lactic culture, incubation to a set pH (5.2–5.8), pasteurisation and scraped-surface evaporation to ~60 $g \cdot 100\ g^{-1}$ dry matter (Ernstrom *et al.*, 1980; Sutherland, 1991; Ganguli, 1991). However, rennet may be added to the retentate to form a curd from which a small quantity of whey is removed (compared to that in natural cheese manufacture) and which is dry-salted salted and pressed, and stored as natural cheese. The retentate may also be treated with lipase to enhance the flavour of the final PC; it was claimed that up to 80%, w/w, of Ras cheese solids could be replaced by the lipase-treated retentate, with the resultant PCs having flavour and consistency considered to be superior to those of the control (Aly *et al.*, 1995). Hyde *et al.* (2002) describes the preparation of cheese base by the acidification and cooling of a blend comprising one or more powdered milk protein ingredients, milk fat, NaCl, food-grade acid and/or preservative.

Increasing the level of substitution of natural cheese by CB, made in the conventional manner, or UFMR, normally results in a 'longer-bodied,' firmer PCP which is less flowable on heating (Collinge and Ernstrom, 1988; Tamime *et al.*, 1990; Younis *et al.*, 1991). The lower flowability may be attributed to a number of factors including:

- a higher degree of intact casein in the cheese base,
- the presence of whey proteins in the cheese base (~8.7%, w/w), which are denatured and complex with para-κ-casein to form a pseudo-gel at the high processing temperature (85–90°C for 3 min) (cf. Doi *et al.*, 1983a,b, 1985).

The tendency of whey proteins to aggregate and gel is probably accentuated by the high levels of protein and soluble calcium in the cheese (Doi *et al.*, 1983a,b; O'Kennedy and Mounsey, 2009; Donato and Guyomarc'h, 2009). However, the effects on flowability vary depending on the method of preparation of the CB and UFMR, and the subsequent heat treatment during processing:

(i) Reducing the pH of milk, from 6.6 to ≥5.2, prior to UF resulted in CBs with lower calcium levels and PCPs with improved meltability (Anis and Ernstrom, 1984);
(ii) Rennet treatment of the UF retentate results in poorer meltability (Anis and Ernstrom, 1984), an effect which may be attributed to the higher degree of interaction between β-lactoglobulin and para-κ-casein (than with native casein) during subsequent processing (Doi *et al.*, 1983 a,b);
(iii) Treatment of retentate with exogenous proteinases (e.g. Savorase-A, and enzymes from *Aspergillus oryzae* and *Candida cylindracea*), which increase the level of proteolysis in the CB, yields PCPs that are

softer and more meltable than those made with untreated CB (Sood and Kosikowski, 1979; Tamime *et al.*, 1990, 1991);

(iv) Increasing the processing temperature in the range 66 to 82°C results in PCPs with reduced meltability, an effect attributed to heat-denaturation and gelation of whey proteins at the higher temperatures, especially when rennet-treated CB is used (Collinge and Ernstrom, 1988).

20.4.4 Other dairy ingredients

Skim milk powder

Addition of skim milk powder to PCP blends at a level of 3–5%, w/w, results in softer, more spreadable products (Kairyukshtene and Zakhrova, 1982). However, higher levels (7–10%, w/w) lead to textural defects such as crumbliness and lack of cohesiveness (Thomas and Hyde, 1972; Kairyukshtene and Zakhrova, 1982) and the powder may remain undissolved. However, a high level may be added if the skim-milk powder is first reconstituted and then precipitated by proteolytic enzymes or citric acid, and the curd added to the blend (Thomas, 1970).

Lactose

Added lactose, in the range of 0 to 5%, w/w, resulted in lower spreadability, lower water activity and increased propensity to non-enzymatic browning in PCPs during processing (especially at a high temperature) and storage (Piergiovanni *et al.*, 1989; Kombila-Moundounga and Lacroix, 1991). Hong (1990) found that replacement of experimental cheeses by lactose at levels of 5 to 20%, w/w, reduced the firmness of PCP. Excess lactose may also increase the propensity to crystallisation in PCPs during storage, with the formation of mixed crystals containing various species, e.g. Ca, P, Mg, Na, tyrosine and/or citrate. Owing to the relatively high level of bound water in PCPs (a maximum of 1.6 g/g solids-non-fat; Csøk, 1982), the effective lactose concentration in the free moisture phase may easily exceed its solubility limit (~15 g/100 g H_2O at 21°C). This may result in the formation of lactose crystals, which could serve as nuclei for crystallisation of mineral species that are supersaturated (Guinee *et al.*, 2004).

20.5 Conclusions

Processed cheese products (PCPs) are composites formulated from natural cheese, water, emulsifying salts and optional ingredients, the type and level of which depend on the category of the PCP, e.g. named variety processed cheese, processed cheese or processed cheese preparation. The formulation is blended and heated to ~75–85°C while shearing continuously to give a hot uniform molten PCP which is hot-packed and stored chilled. While natural cheese is the major ingredient, accounting for $\geq$51 $g \cdot 100\ g^{-1}$ in all categories of PCPs, dairy ingredients can be added at substantial levels (e.g. 10 $g \cdot 100\ g^{-1}$) to some categories such as processed cheese preparations, as defined by the Codex Alimentarius Commission (FAO/WHO, 2008).

Calcium phosphate paracasein is the main protein in rennet curd cheeses (e.g. Cheddar, Gouda), which are the main varieties used in PCPs. This protein is inherently insoluble because of intermolecular calcium and/or calcium phosphate mediated linkages. Hence, heating cheese to a high temperature (e.g. 90°C) while shearing, is conductive to protein aggregation, destabilisation of the fat globule membrane, fat coalescence, and leakage of free fat and moisture. However, the addition of emulsifying salts, such as sodium orthophosphate and/or citrate, prevents such an occurrence and promotes the formation of stable uniform PCPs. The salts do this by increasing the pH of the blend (e.g. ~5.7 to 6.0, compared with <5.5 in natural cheese) and competing with the casein for calcium. They thereby mediate the transfer of most of the calcium and phosphate from the protein and its conversion to a predominantly 'sodium' paracaseinate, as confirmed by the large proportions of insoluble calcium and phosphate (~70% of total) and water-soluble protein (~70–90% of total) in PCPs. The paracaseinate is the primary stabiliser in PCPs, binding free water and emulsifying free fat released during processing of the cheese. The degree of paracaseinate hydration and the size distribution of emulsified oil droplets are major determinants of the rheology and heat-induced functionality of the resultant PCPs.

Casein-based ingredients (and butter fat), when added as a partial substitute for cheese protein, similarly contribute to the formation and stabilisation of PCPs. Nevertheless, differences in initial solubility (e.g. rennet casein versus MC), pH (e.g. acid casein versus sodium caseinate) and mineral composition lead to diversity in their functionality (e.g. water binding, emulsification) during processing and in the characteristics of the resultant PCPs. For example, for similar processing conditions and product composition, rennet casein is generally more suited to the manufacture of elastic block-type PCPs with moderate heat-induced melt/flow than acid casein or caseinates, which are ideal for the manufacture of high moisture, spreadable PCPs.

Whey protein is generally added as a cheap substitute for cheese protein. It is not as amphiphilic as casein and therefore not as effective as an emulsifier of free fat during processing. Whey proteins are prone to heat-induced denaturation and aggregation/gelation *via* disulphide and other (e.g. hydrophobic) interactions. Consequently, they markedly restrict the ability of PCPs to melt and flow when heated by the consumer. The increased melt/flow resistance is undesirable in most cooking applications, but can be desirable in others (such as fried cheese).

Dairy ingredients containing whey protein and casein (e.g. MPCs, CWPts, CB) impart the properties of both protein components, but the relative effects on the properties of PCPs can be influenced by the type and level of ingredient, product pH, processing conditions (e.g. temperature, homogenisation) and cooling rate.

20.6 References

ABOU-EL-NOUR, A.M., SCHEURER, G.J., OMAR, M.M. and BUCHHEIM, W., 'Physicochemical and rheological properties of block-type processed cheese analogue made from rennet casein and total milk protein', *Milchwissenschaft*, 1996, 51, 684–687.

ALMENA-ALISTE, M., GIGANTE, M.L. and KINDSTEDT, P.S., 'Impact of pH on the texture of cultured cream cheese: Firmness and water phase characteristics', *Milchwissenschaft*, 2006, 61, 400–404.

ALY, M.E., ABDEL-BAKY, A.A., FARAHAT, S.M. and HANA, U.B.B., 'Quality of processed cheese spread made using ultrafiltered retentates treated with some ripening agents', *Int. Dairy J.*, 1995, 5, 191–209.

ANIS, S.M.K. and ERNSTROM, C.A., 'Calcium and the meltability of process cheese food made from ultrafiltered retentate', *J. Dairy Sci.*, 1984, 67 (Suppl. 1), 79.

AUTY, M.A.E., TWOMEY, M., GUINEE, T.P. and MULVIHILL, D.M., 'Development and application of confocal scanning laser microscopy methods for studying the distribution of fat and protein in selected dairy products', *J. Dairy Res.*, 2001, 68, 417–427.

BECKER, P., BUHRMEISTER, M. and NEUMANN, H.G., *Method for coating a substrate with calcium phospha*, Patent No WO03052164 (A2), 2003.

BISWAS, A.C., MUTHUKUMARAPPAN, K. and METZGER, L.E., 'Dynamic rheological properties of process cheese: Effect of Ca and P content, residual lactose, salt-to-moisture ratio and cheese temperature', *Int. J. Food Properties*, 2008, 11, 282–295.

BRENNAN, N.M. and COGAN, T.M. 'Bacterial surface-ripened cheeses' in Fox, P.F., Mcsweeney, P.L.H., Cogan, T.M. and Guinee, T.P., *Cheese Chemistry, Physics and Microbiology, Vol. 2 Major Cheese Groups, 3rd edn.*, Amsterdam, Elsevier Academic Press, 2004, 197–225.

BRICKLEY, C.A., GOVINDASAMY-LUCEY, S., JAEGGI, J.J., JOHNSON, M.E., MCSWEENEY, P.L.H. and LUCEY, J.A., 'Influence of emulsifying salts on the textural properties of nonfat process cheese made from direct acid cheese bases', *J. Dairy Sci.*, 2008, 91, 39–48.

BRICKLEY, C.A., AUTY, M.A.E., PIRAINO, P. and MCSWEENEY, P.L.H., 'The effect of natural Cheddar cheese ripening on the functional and textural properties of the processed cheese manufactured therefrom', *J. Food Science*, 2007, 72, C483–C490.

CARIĆ, M. and KALÁB, M. 'Processed cheese products', in Fox, P.F., *Cheese, Chemistry, Physics and Microbiology, Vol. 2. Major Cheese Groups*, London, Chapman and Hall, 1993, 467–505.

CARIĆ, M., GANTAR, M. and KALÁB, M., 'Effects of emulsifying agents on the microstructure and other characteristics of process cheese – A review', *Food Microstruc*, 1985, 4, 297–312.

CAVALIER-SALOU, C. and CHEFTEL, J.C., 'Emulsifying salts influence on characteristics of cheese analogs from calcium caseinate', *J. Food Sci.*, 1991, 56, 1542–1547, 1551.

CODE OF FEDERAL REGULATIONS (2008). Part 133: Cheese and related products, in *Food and Drugs 21. Code of Federal Regulations, Parts 100–169*, Washington, D.C., US Government Printing Office, (www.gpoaccess.gov/CFR/INDEX.HTML)

COLLINGE, S.K. and ERNSTROM, C.A., 'Relationship between soluble N at pH 4.6 and meltability of pasteurized process cheese food made from UF curd', *J. Dairy Sci.*, 1988, 71 (Suppl. 1), 71.

CSØK, J. The effect of holding time on free- and bound-water contents of processed cheese, in *Proc. XXI Intern. Dairy Congr., Brief Communications Volume 1, Book 1*, Moscow, Mir Publishers, 1982, 475–476.

DOI, H., HIRMAMATUS, M., IBUKI, F. and KANAMORI, M., 'Gelation of the heat-induced complex between κ-casein and α-lactalbumin', *J Nutri. Sci Vitaminol.*, 1985, 31, 77–87.

DOI, H., IDENO, S., HUANG KUO, F., IBUKI, F. and KANAMORI, M., 'Gelation of the complex between κ-casein and β-lactoglobulin', *J Nutrit. Sci Vitaminol*, 1983a, 29, 679–689.

DOI, H., IDENO, S., IBUKI, F., KANAMORI, M., 'Participation of the hydrophobic bond in complex formation between κ-casein and β-lactoglobulin', *Agri Biol Chem*, 1983b, 47, 407–409.

DONATO, L. and GUYOMARC'H, F., 'Formation and properties of the whey protein/κ-casein complexes in heated skim milk – a review', *Dairy Sci Technol*, 2009, 89, 1–27.

ERNSTROM, C.A., SUTHERLAND, B.J. and JAMESON, G.W., 'Cheese base for processing. A high yield product from whole milk by ultrafiltration', *J Dairy Sci*, 1980, 63, 228–234.

FAO/WHO (2008). *Codex General Standard for Process(ed) Cheese Preparations, Process(ed) Cheese Food and Process(ed) Cheese Spread*: Codex Stan 287–1978, Amended 2008. www.codexalimentarius.net/web/standard/list

FOX, K.K., HARPER, M.K., HOLSINGER, V.H. and PALLANSCH, M.J., 'Gelation of milk solids by orthophosphate', *J Dairy Sci*, 1965, 179–185.

FOX, P.F. and BRODKORB, A., 'The casein micelle: Historical aspects, current concepts and significance', *Int Dairy J*, 2008, 18, 677–684.

FRENCH, S.J., LEE, K.M., DECASTRO, M. and HARPER, W.J., 'Effects of different protein concentrates and emulsifying salt conditions on the characteristics of a processed cheese product', *Milchwissenschaft*, 2002, 57, 79–83.

GANGULI, N.C., 'Cheesebase – a base for new foods in its own right', *Indian Dairyman*, 1991, 43, 449–451.

GAUCHER, I., PIOT, M., BEAUCHER, E. and GAUCHERON, F., 'Physico-chemical characterization of phosphate-added skim milk', *Int Dairy J*, 2007, 17, 1375–1383.

GE, O., ALMENA-ALISTE, M. and KINDSTEDT, P.S., 'Reversibility of pH-induced changes in the calcium distribution and melting characteristics of mozzarella cheese', *Aus. J Dairy Technol*, 2002, 57, 3–9.

GIGANTE, M.L., ALMENA-ALISTE, M., KINDSTEDT, P.S., 'Effect of cheese pH and temperature on serum phase characteristics of cream cheese during storage', *J Food Sci*, 2006, 71, C7–C11.

GIGANTE, M.L., ANTUNES, A.J., PETENATE, A.J. and ROIG, S.M., 'Firmness and melting properties of Requeijão cheese', *Milchwissenschaft*, 2001, 56, 560–562.

GOUDA, A., EL-SHABRAWY, S.A., EL-ZAYAT, A.I. and EL-BAGOURY, E., 'Use of calcium caseinate in processed cheese spreads making', *Egypt J Dairy Sci*, 1985, 13, 115–119.

GREEN, M.L., 'Effect on the composition and properties of casein micelles of interaction with ionic material', *J Dairy Res*, 1982, 49, 87–98.

GUINEE, T.P., CARIĆ, M., KALÁB, M. and FOX, P.F., 'Pasteurized processed cheese and substitute/imitation cheese products' in Fox, P.F., Mcsweeney, P.L.H., Cogan, T.M. and Guinee, T.P., *Cheese Chemistry, Physics and Microbiology, Vol. 2 Major Cheese Groups, 3rd edn.*, Amsterdam, Elsevier Academic Press, 2004, 349–394.

GUINEE, T.P., AUTY, M.A.E., MULLINS, C., CORCORAN, M.O. and MULHOLLAND, E.O., 'Preliminary observations on effects of fat content and degree of fat emulsification on the structure–functional relationship of Cheddar-type cheese', *J Text Studies*, 2000b, 31, 645–663.

GUINEE, T.P., FEENEY, E.P., AUTY, M.A.E. and FOX, P.F., 'Effect of pH and calcium concentration on some textural and functional properties of Mozzarella cheese', *J Dairy Sci*, 2002, 85, 1655–1659.

GUINEE, T.P., HARRINGTON, D., CORCORAN, M.O., MULHOLLAND, E.O. and MULLINS, C., 'The composition and functional properties of commercial Mozzarella, Cheddar and analogue pizza cheese', *Int J Dairy Technol*, 2000a, 53, 51–56.

GUINEE, T.P., O'KENNEDY, B.T. and KELLY, P.K., 'Effect of Milk Protein Standardization Using Different Methods on the Composition and Yields of Cheddar Cheese', *J Dairy Sci*, 2006, 89, 468–482.

GUINEE, T.P. and O'KENNEDY, B.T., 'The effect of calcium content of Cheddar-style cheese on the biochemical and rheological properties of processed cheese', *Dairy Sci. and Technol.* 2009 (In press). Available at www.dairy-journal.org

GUPTA, S.K., KARAHADIAN, C. and LINDSAY, R.C., 'Effect of emulsifier salts on textural and flavour properties of processed cheeses', *J Dairy Sci*, 1984, 67, 764–778.

GUPTA, V.K. and REUTER, H., 'Firmness and melting quality of processed cheese foods with added whey protein concentrates', *Lait,* 1993, 73, 381–388.

HASSAN, A., JOHNSON, M.E. and LUCEY, J.A., 'Changes in the proportions of soluble and insoluble calcium during the ripening of Cheddar cheese', *J Dairy Science*, 2004, 87, 854–862.

HEERTJE, I., BOSKAMP, M.J., VAN KLEEF, F. and GORTEMAKER, F.H., 'The microstructure of processed cheese', *Neth Milk Dairy J*, 1981, 35, 177–179.

HILL, A.R. and SMITH, A.K., 'Texture and ultrastructure of process cheese spreads made from heat precipitated proteins', *Milchwissenschaft,* 1992, 47, 71–74.

HOKES, J.C., HANSEN, P.M.T and MANGINO, M.E., 'Functional properties of commercial calcium caseinates for use in imitation cheese', *Food Hydrocoll*, 1989, 3, 19–31.

HONG, Y.H., 'Physicochemical and textural characteristics of process cheese manufactured with different kinds and quantities of ingredients', *Korean J Animal Sci*, 1990, 32, 406–412.

HYDE, M.A., CARRUTHERS, L.M., MCGEOWN, J.J. and TREKER, G.W., *Method for preparing cheese products and process cheese bases*, United States Patent 6773740 B2, 2002.

KAIRYUKSHTENE, I. and ZAKHROVA, N., Investigation on the consistency of low-fat cheese, in *Proc. XXI Int. Dairy Congr. Vol. 1, Book 1*, Moscow, Mir Publishers, 1982, 495.

KALÁB, M., Practical aspects of electron microscopy in cheese research, in E.L. Malin and M.H. Tunick, *Chemistry of Structure–Function Relationships in Cheese*, New York, Plenum Press, 1995, 247–276.

KAMINARIDES, S. and STACHTIARIS, S., 'Production of processed cheese using Kasseri cheese and processed cheese analogues incorporating whey protein concentrate and soybean oil', *Int J Dairy Technol*, 2000, 53, 69–74.

KAPOOR, R. and METZGER, L.E., 'Process Cheese: Scientific and Technological Aspects–A Review', *Compr Rev Food Sci Food Safety*, 2008, 7, 194–214.

KAPOOR, R., METZGER, L.E., BISWAS, A.C. and MUTHUKUMMARAPPAN, K. 'Effect of natural cheese characteristics on process cheese properties', *J. Dairy Sci.*, 2007, 90, 1625–1634.

KELLER, B., OLSON, N.F. and RICHARDSON, T., 'Mineral retention and rheological properties of Mozzarella cheese made by direct acidification', *J. Dairy Sci.*, 1974, 57, 174–180.

KELLY, P.M., KELLY, J., MEHRA, R., OLDFIELD, D.J., RAGGETT, E. and O'KENNEDY, B.T., 'Implementation of integrated membrane processes for pilot scale development of fractionated milk components', *Lait,* 2000, 80, 139–153.

KIMURA, T., TANEYA, S. and FURUICHI, E., Electron microscopic observation of casein particles in processed cheese, in *Proc 20th Int Dairy Congr.*, Paris, 1979, E239–240.

KINDSTEDT, P.S., CARIĆ, M. and MILANOVÍC, S., Pasta-filata cheeses, in Fox, P.F., Mcsweeney, P.L.H., Cogan, T.M. and Guinee, T.P., *Cheese Chemistry, Physics and Microbiology, Vol. 2 Major Cheese Groups, 3rd edn.*, Elsevier Academic Press, 2004, 251–277.

KOMBILA-MOUNDOUNGA, E. and LACROIX, C., 'Effect des combinations de sel, lactose et glycérol sur l'activité de l'eau (Aw) des fromages fondus à tartiner', *Can Inst Food Sci.Technol J*, 1991, 24, 239–251.

KWAK, H.S., CHOI, S.S., AHN, J. and LEE, S.W., 'Casein hydrolysate fractions act as emulsifiers in process cheese', *J. Food Sci.*, 2002, 67, 821–825.

LAWRENCE, R.C., HEAP, H.A. and GILLES, J., 'A controlled approach to cheese technology', *J Dairy Sci*, 1984, 67, 1632–1645.

LEE, B.O., KILBERTUS, G. and ALAIS, C., 'Ultrastructural study of processed cheese. Effect of different parameters', *Milchwissenschaft*, 1981, 36, 343–348.

LEE, B.O., PAQUET, D. and ALAIS, C., 'Etude biochimique de la fonte des fromages IV. Effect du type de sels de fonte et de la nature de la matière protéique sur la peptisation. Utilisation d'un système modèle.' *Le Lait*, 1986, 66, 257–267.

LEE, S.K. and KLOSTERMEYER, H, 'The effect of pH on the rheological properties of reduced-fat model processed cheese spreads', *Lebensm Wissens und Technol*, 2001, 34, 288–292.

LU, Y., SHIRASHOII, N. and LUCEY, J.A., 'Effects of pH on the textural properties and meltability of pasteurized process cheese made with different types of emulsifying salts', *J Food Sci*, 2008, 73, E363–E369.

MAHONEY, R.R., LAZARIDIS, H.N. and ROSENAU, J.R., 'Protein size and meltability in enzyme-treated, direct-acidified cheese products', *J. Food Sci.*, 1982, 47, 670–671.

MANN, E.J., 'Processed cheese', *Dairy Ind Int*, 1993, 58 (i) 14–15.

MANN, E.J., 'Processed cheese', *Dairy Ind Int*, 1997, 62 (2), 16–17.

MARCHESSEAU, S. and CUQ, J.L., 'Water-holding capacity and characterization of protein interactions in processed cheese', *J Dairy Res*, 1995, 62, 479–489.

MARCHESSEAU, S., GASTALDI, E., LAGAUDE, A. and CUQ, J.-L., 'Influence of pH on protein interactions and microstructure of process cheese', *J. Dairy Sci.*, 1997, 80, 1483–1489.

MARSHALL, R.J., 'Composition, structure, rheological properties and sensory texture of processed cheese analogues', *J Sci Food Agric*, 1990, 50, 237–252.

MCMAHON, D.J. and OOMMEN, B.S., 'Supramolecular structure of the casein micelle', *J. Dairy Sci.*, 2008, 91, 1709–1721.

MCMAHON, D.J., ALLEYNE, M.C., FIFE, R.L. and OBERG, C.J., 'Use of fat replacers in low fat Mozzarella cheese', *J Dairy Sci*, 1996, 79, 1911–1921.

MCMAHON, D.J., FIFE, R.L. and OBERG, C.J., 'Water partitioning in Mozzarella cheese and its relationship to cheese meltability', *J Dairy Sci*, 1999, 82, 1361–1369.

MCMAHON, D.J., OBERG C.J. and MCMANUS, W., 'Functionality of Mozzarella cheese', *Aus J Dairy Technol*, 1993, 48, 99–104.

MEYER, A., Processed Cheese Manufacture, London, Food Trade Press Ltd., 1973.

MLEKO, S. and FOEGEDING, E.A., 'Formation of whey protein polymers: Effects of a two-step heating process on rheological properties', *J Texture Studies*, 1999, 30, 137–149.

MLEKO, S. and FOEGEDING, E.A., 'Physical properties of rennet casein gels and processed cheese analogs containing whey proteins', *Milchwissenschaft*, 2000, 55, 513–516.

MLEKO, S. and FOEGEDING, E.A., 'Incorporation of polymerized whey proteins into processed cheese analogs', *Milchwissenschaft*, 2001, 56, 612–615.

MLEKO, S. and LUCEY, J.A., 'Production and properties of processed cheese with reduced lactose whey', *Milchwissenschaft*, 2003, 58, 498–502.

MOUNSEY, J.S., O'KENNEDY, B.T. and KELLY, P.M., 'Effect of the aggregation-state of whey protein-based ingredients on processed cheese functionality', *Milchwissenschaft*, 2007, 62, 44–47.

MULVIHILL, D.M. and ENNIS, M.P., 'Functional milk Proteins: Production and utilization, in Fox, P.F., and McSweeney, P.L.H., *Advanced Dairy Chemistry, Volume 1, Protein, third edition, Part B.*, 2003, 1175–1228.

NAKAJIMA, I., KAWANISHI, G. and FURUICHI, E., 'Reaction of melting salts upon casein micelles and their effects on calcium, phosphorus and bound water', *Agric Biol Chem*, 1975, 39, 979–987.

O'KENNEDY, B.T. and MOUNSEY, J.S., 'The dominating effect of ionic strength on the heat-induced denaturation and aggregation of β-lactoglobulin in simulated milk ultrafiltrate', *Int Dairy J*, 2009, 19, 123–128.

OLSON, N.F., VAKALERIS, D.G. and PRICE, W.V., 'Acidity and age of natural cheese as factors affecting the body of process cheese spread', *J Dairy Sci*, 1958, 41, 1005–1016.

ORTEGA-FLEITAS, F.O., REINERY, P.L., ROCAMORA, Y., REAL-DEL-SOL, E., CABRERA, M.C. and CASALS, C., 'Effect of homogenization pressure on texture properties of soy cream cheese', *Alimentaria*, 2000, 313, 125–127.

PANOUILLÉ, M.T., NICOLAI, T. and DURAND, D., 'Heat-induced aggregation and gelation of casein submicelles', *Int Dairy J*, 2004, 14, 297–303.

PARK, J.N., LEE, K.T. and YU, J.H., 'The effect of emulsifying salts on the texture and flavour of block processed cheese made with ultrafiltered cheese base', *Korean J Dairy Sci*, 1992, 14, 234–249.

PIERGIOVANNI, L., DE NONI, I., FAVA, P. and SCHIRALDI, A., 'Nonenzymatic browning in processed cheeses. Kinetics of the Maillard reaction during processing and storage', *Ital J Food Sci*, 1989, 1, 11–20.

PISKA, I. and ŠTĚTINA, J., 'Influence of cheese ripening and rate of cooling of the processed cheese mixture on rheological properties of processed cheese', *J Food Engineering*, 2004, 61, 551–555.

RAYAN, A.A., KALÁB, M. and ERNSTROM, C.A., 'Microstructure and rheology of process cheese', *Scanning Electron Microsc*, 1980, III, 635–643.

RELYVELD, E., *Calcium phosphate gel for absorbing vaccines*, United States Patent 4016252, 1977.

REPS, A., 'Bacterial surface-ripened cheeses', in Fox, P.F., *Cheese: Chemistry, Physics and Microbiology, Volume 2, Major Cheese Groups*, London, Chapman & Hall, 1993, 137–172.

RUEGG, M. and MOOR, U., 'Effect of calcium on the hydration of casein. I. Water vapour sorption and fine structure of calcium caseinates compared with sodium caseinates in the pH range 4.6–8.0', *J Dairy Res*, 1984, 51, 103–111, 29.

SAVELLO, ERNSTROM, C.A. and KALÁB, M., 'Microstructure and meltability of model process cheese made with rennet and acid casein', *J Dairy Sci*, 1989, 72, 1–11.

SCHULZ, M.E., *Preparation of melt-resistant processed cheese*, United States Patent 3962483, 1976.

SIMBUERGER, S., WOLFSCHOON, A.F., KEMPTER, K., ROSE, M. and MARDER, U., *Acidified ultrafiltration concentrates as raw material for the production of processed cheese* Eur. Patent Application EP 0 755 630 A1, 1997.

SNOEREN, T.H.M., KLOK, H.J., VAN HOOYDONK, A.C.M. and DAMMAN, A.J., 'The voluminosity of casein micelles', *Milchwissenschaft*, 1984, 39, 461–463.

SOOD, S.M., GAIND, D.K. and DEWAN, R.K., 'Correlation between micelle solvation and calcium content', *NZ J Dairy Sci Technol*, 1979, 14, 32–34.

SOOD, V.K. and KOSIKOWSKI, F.V., 'Process Cheddar cheese from plain and enzyme treated retentates', *J Dairy Sci*, 1979, 62, 1713–1718.

SØRENSEN, H.H. and PEDERSEN, E., 'The world market for Cheese, Sixth Edition', International Dairy Federation Bulletin 402, 2005.

SPINNLER, H.-E. and GRIPON, J.-C. 'Surface mould-ripened cheeses' in Fox, P.F., Mcsweeney, P.L.H., Cogan, T.M. and Guinee, T.P., *Cheese Chemistry, Physics and Microbiology, Vol. 2 Major Cheese Groups, 3rd edn.*, Amsterdam, Elsevier Academic Press, 2004, 155–174.

SUTHERLAND, B.J., 'New cheese products as food ingredients', *CSIRO–Food Res Quart*, 1991, 51, 114–119.

TAMIME, A.Y., KALÁB, M., DAVIES, G. and YOUNIS, M.F., 'Microstructure and firmness of processed cheese manufactured from cheddar cheese and skim milk powder cheese base', *Food Struct*, 1990, 9, 23–37.

TAMIME, A.Y., YOUNIS, M.F. and DAVIES, G., 'Production of processed cheese using Cheddar cheese and cheese base. 2. Production of a cheese base from skim milk powder', *Milchwissenschaft*, 1991, 46, 495–499.

TANEYA, S., KIMURA, T., IZUTSU, T. and BUCHHEIM, W., 'The submicroscopic structure of processed cheese with different melting properties', *Milchwissenschaft*, 1980, 35, 479–481.

THAPA, T.B. and GUPTA, V.K., 'Changes in sensoric and rheological characteristics during storage of processed cheese foods prepared with added whey protein concentrates', *Indian J Dairy Sci*, 1992, 45, 140–145.

THOMAS, M.A., 'Use of calcium co-precipitates in processed cheese', *Aust J Dairy Technol*, 1970, 25, 23–26.

THOMAS, M.A. and HYDE, K.A., *The Manufacture of Processed Cheese – A Working Manual*. Bulletin D40, Division of Dairying, New South Wales Department of Agriculture, Sydney, 1972.

THOMAS, M.A., NEWELL, G. and TURNER, A.D., 'Effect of emulsifying salts on objective and subjective properties of processed cheese', *J Food Sci*, 1980, 45, 458–459, 466.

TUNICK, M.H., MACKEY, K.L., SHIEH, J.J., SMIT, P.W., COOKE, P. and MALIN, E.L., 'Rheology and microstructure of low-fat Mozzarella cheese', *Int Dairy J*, 1993, 3, 649–662.

UPADHYAY, V.K., MCSWEENEY, P.L.H., MAGBOUL, A.A.A. and FOX, P.F., 'Proteolysis in cheese during ripening', in Fox, P.F., McSweeney, P.L.H., Cogan, T.M. and Guinee, T.P., *Cheese Chemistry, Physics and Microbiology, Vol. 1 General Aspects, 3rd edn.*, 2004, Amsterdam, Elsevier Academic Press, 391–433.

VAN VLIET, T. and DENTENER-KIKKERT, A., 'Influence of the composition of the milk fat globule membrane on the rheological properties of acid milk gels', *Neth Milk Dairy J*, 1982, 35, 261–265.

YOUNIS, M.F., TAMIME, A.Y., DAVIES, G., HUNTER, E.A. and ABD-EL-HADY, S.M., 'Production of processed cheese using Cheddar cheese and cheese base. 5. Rheological properties', *Milchwissenschaft*, 1991, 46, 701–705.

ZEHREN, V.L. and NUSBAUM, D.D., *Process Cheese*. Madison, Wisconsin, Cheese Reporter Publishing Company, Inc., 1992.

21

Emulsions and nanoemulsions using dairy ingredients

P. J. Wilde, Institute of Food Research, UK

Abstract: Milk is often used as a source of ingredients to stabilize emulsions as such ingredients are readily available and easily purified. The most widely used are proteins, of which there are two classes in milk – caseins and whey proteins. This chapter will briefly review the use and functionality of emulsions and the benefits of using dairy proteins as emulsifying agents. The chapter will then discuss how the different components are utilized as emulsifiers, the effect of processing on functionality, interactions with other components such as polysaccharides, and novel methods to improve functionality. The chapter will then discuss future trends, together with sources of further information.

Key words: milk, emulsifiers, emulsions, proteins.

21.1 Introduction

21.1.1 Emulsions and nanoemulsions

Making an emulsion is a fundamentally natural thing for a dairy ingredient to do. Dairy products are derived from milk, and milk is an emulsion. The whole purpose of milk is to deliver nutrients to offspring and it contains all the macro and micronutrients required for their survival. One important nutrient is fat, and that is delivered in milk in the form of an emulsion. The emulsion has to be formed of small droplets that are stable to coalescence to facilitate effective secretion, but are susceptible to subsequent breakdown during digestion and uptake.

An emulsion is defined as a mixture of at least two immiscible fluids or phases. One phase is formed into droplets (known as the dispersed phase), and is suspended in another fluid (the continuous phase). The fact that the continuous and dispersed phases are immiscible ensures that emulsions are a unique class of structured fluids. Their rheological properties are often completely different from those of their component parts. For example, a traditional mayonnaise is a viscoelastic solid, but it is made up of two phases that are viscous liquids. The rheological properties of the mayonnaise are due to the colloidal interactions between the droplets. This is why emulsions are a popular method for imparting texture to food products. The problem is to make them robust and stable enough to retain the desired functionality in the final product, and this will be discussed in more detail throughout this chapter.

The two phases are normally liquid when formed, to allow the droplets to be broken up and dispersed. However, changes in temperature can often be used to solidify or crystallize one or both of the phases to change the texture and/or functionality of the emulsion. For example, in emulsions made from vegetable oils, for example, such as mayonnaise, salad dressings and sauces, both phases are liquid. For milk and yoghurt, when prepared at temperatures >35°C, the fat is liquid, but upon cooling and storage, the dispersed fat droplets will be solid. Butter and margarine are examples of liquid water droplets in a solid fat phase; these are special cases in terms of emulsified dairy products and will not be discussed in this chapter.

21.1.2 Microemulsions and macroemulsions

Emulsions can be divided into two main classes based on their thermodynamic stability. Microemulsions are thermodynamically stable, and macroemulsions are metastable. Macro-emulsions (commonly referred to simply as 'emulsions'), as stated previously, are dispersions of droplets in an immiscible fluid. Breaking up the dispersed phase into droplets requires energy to overcome the surface tension forces between the two phases, and hence they are thermodynamically unstable and will eventually collapse back into their component phases. The droplet diameter is generally in the range 0.1–10 μm. Microemulsions overcome the thermodynamic barrier by reducing the interfacial tension to near-zero. This is achieved by high concentrations of surfactants (and often co-surfactants, mixtures of oils and oil soluble compounds). The result is that when the two phases come into contact, they spontaneously emulsify to form a microemulsion with very small droplet diameters (<30 nm), and are often transparent. The high concentration of surfactant required means that they are usually restricted to non-food applications. The ambiguously named 'nanoemulsions', (Mason *et al.*, 2006a) are a class of macroemulsions. They have the same thermodynamic stability as macroemulsions, but they are formed under much higher shear forces, using techniques such as the microfluidizer (Mason *et al.*, 2006a and 2006b) or ultrasound (Kentish *et al.*, 2008). This results in much smaller droplets (typically 30–100 nm diameter), giving enhanced functionality such as rheology and flow, and enhanced delivery (Kotyla *et al.*, 2007; Wang *et al.*, 2008) due to the massive increase in surface area compared with a conventional emulsion.

Over one third of the food we consume contains emulsified fats at some level. The fat is present to impart texture, pleasurable mouthfeel, flavour and also to deliver lipid-soluble nutrients such as vitamin A and β-carotenes (or drugs in pharmaceuticals). Humans have evolved to desire a certain degree of fat in our diets to give us energy that can be stored during times of low food availability. Fat content in foods varies depending on the product type and functional properties. These could be at high levels in full cream (40–50%), traditional mayonnaise (50–80%), salad dressings (20–50%), whipping cream (30–40%); intermediate levels in cream cheese (20–25%), single cream (20%), ice cream (10–15%); and at lower levels in sauces (2–10%), yoghurts (3–10%), soup (3–5%), milk: whole (3–4%), semi skimmed (2%), skimmed (<0.5%), and low fat versions of many of the above products. The sensory perception of emulsified fats in food is a complex combination of physical, chemical and even psychological stimuli, thus making the reduction of fat such a difficult task.

21.1.3 Functionality

Oil in water (OW) emulsions are used to suspend lipids or lipid-soluble compounds (nutrients, drugs, chemically active agents) in an aqueous medium. For example, to deliver an oil soluble drug, the delivery medium has to be an aqueous one in order for the medium to be dispersed effectively in the gut, or bloodstream, but the drug is soluble only in oil. Therefore, dissolving the drug into small oil droplets allows homogeneous dispersion of the medium and allows the drug to partition slowly out of the oil phase during digestion or lipolysis. In foods, emulsified fats not only directly affect texture (cream, mayonnaise), but can deliver lipid soluble flavours and aromas to impart the unique sensory quality of a particular food product. Emulsions can also be used to carry oil-soluble pigments in paints and photographic films, and reactive and bio-active components in applications such as cosmetics, detergents and explosives. Emulsifiers derived from dairy ingredients are largely used in food applications due to their wide availability, relatively low cost, and natural, wholesome appeal; there are legislative limitations on types and quantities of emulsifiers for use in foods.

21.2 Emulsion formation and stability

To understand the functionality of dairy-based emulsifiers, it is important to understand the basic generic mechanisms of emulsion formation and stability. As stated earlier, emulsions are usually (except for microemulsions) thermodynamically unstable, and thus require the presence of emulsifiers to aid their formation and stability.

To create an emulsion normally requires energy for applying shear forces to break up the dispersed phase into small droplets in the continuous phase. The simplest way of doing this is by using a conventional domestic blender containing the continuous phase, and adding drops of the dispersed phase at a slow rate. However, simply dispersing, for example, pure oil into water, even under high energy conditions, will form only a transient emulsion. The droplets will be large, and they will soon merge (coalesce) with others to form increasingly larger droplets, and eventually separate into a second phase. Therefore, to efficiently create and stabilize the oil droplets against coalescence, a protective, adsorbed layer of molecules must be formed around them (by an emulsifier). This layer helps prevent coalescence, and hence allows the formation of a foam or emulsion that may be stable for long periods, depending on the surface properties of the adsorbed layer. We will now discuss the separate processes of emulsion formation and stability.

21.2.1 Emulsion formation

To form an emulsion requires energy to break up one phase into small droplets. This results in a huge increase in interfacial area of the system. The energy barrier to this surface expansion is the interfacial free energy or interfacial tension. The interfacial tension is defined as the force per unit length acting on an interface. Therefore, a lower interfacial tension will require less energy to produce a specific increase in surface area. For pure water, strong hydrogen bonding between water molecules results in a high interfacial tension, therefore requiring high energy to disperse droplets or expand the interfacial area. Figure 21.1 shows how the deformability of an oil droplet will determine the size distribution of an emulsion during homogenization. Essentially, interfacial tension is the force acting on a

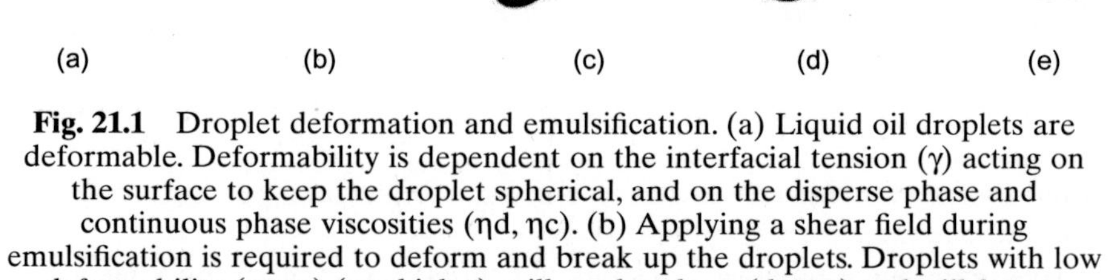

Fig. 21.1 Droplet deformation and emulsification. (a) Liquid oil droplets are deformable. Deformability is dependent on the interfacial tension (γ) acting on the surface to keep the droplet spherical, and on the disperse phase and continuous phase viscosities (ηd, ηc). (b) Applying a shear field during emulsification is required to deform and break up the droplets. Droplets with low deformability (c-top) (e.g. high γ), will not break up (d- top) and will form a coarse emulsion (e- top). A highly deformable droplet (c- bottom) will be deformed in the shear field and break up into smaller droplets (d- bottom), and hence produce a fine emulsion with many small droplets (e- bottom).

surface that is trying to minimize surface area and make the droplet spherical. If this force is reduced, the droplets become more deformable, and hence easier to break up. Reducing the interfacial tension is achieved through the use of emulsifiers, which adsorb at the interface between the two phases and lower the interfacial tension by disrupting the hydrogen bonds between the water molecules at the interface. Emulsifiers can be simple, lipid-like molecules or complex macromolecules such as proteins; however, they are all amphiphilic (i.e. contain both hydrophilic and hydrophobic parts) so that they naturally align themselves at the interface. The interfacial tension is determined by the structure of the emulsifier and the interfacial concentration; therefore, the most effective emulsifiers are those whose structure allows the formation of a dense, close packed interfacial layer.

The dynamics are also very important because the homogenization process is rapid; so it is the dynamic interfacial tension over timescales of $<<1$ second that is most important, not the equilibrium value. Rapid adsorption of the emulsifier to lower the interfacial tension will thus result in more effective emulsification of the dispersed phase during the homogenization process. Lower interfacial tensions also result in smaller droplets for a given energy input, as the droplets are more easily broken up. Simple emulsifiers are smaller than proteins; therefore, in general, they will diffuse to the interface more quickly and reduce the interfacial tension more rapidly than proteins. Also, simple emulsifiers are able to pack more effectively at interfaces, so the resultant interfacial tension will also be lower than that for proteins. Hence, simple emulsifiers are generally more effective at forming emulsions than amphiphilic macromolecules such as proteins, requiring lower shear rates and resulting in smaller droplet diameters. However, it is not a simple relationship, as there are many limiting factors such as the method of emulsification.

21.2.2 Emulsification methods

Emulsification is the process by which the dispersed phase is broken up into small droplets. Normally a coarse premix is created by rapid mixing of the ingredients. This is sufficient to break up the dispersed phase into large droplets, and allow adsorption of the emulsifiers prior to final emulsification. There are two main methods/principles commonly used to homogenize emulsions in the food industry, mechanical (e.g. rotor–stator) and high pressure. (Walstra, 1983; Robins and Wilde, 2003). Mechanical methods induce high shear fields to break up droplets. High pressure homogenizers are now very common, and simply force the premix through a narrow orifice or valve at high pressures (typically 10–100 MPa). Forcing the emulsion through a valve at high pressure creates turbulence and very high shear forces, thus breaking up the droplets. Shear forces are not the most efficient way of breaking up droplets, as the shear field tends to lose energy rotating the droplet. Forcing the droplets through a simple orifice induces elongational flow forces which break up droplets more efficiently. The droplet size produced is, within certain boundaries, generally inversely proportional to the energy input during homogenization. More accurately, it is the energy dissipated into droplet break-up. This is affected by the interfacial tension as described previously, and also the viscosities of the dispersed and continuous phases. As continuous phase viscosity increases, droplet size will decrease, yet as dispersed phase viscosity increases, the droplets become more difficult to break and droplet size consequently increases (Walstra, 1993).

Another homogenization method receiving increased attention, and which is the most energy efficient process, is crossflow membrane emulsification (XME). Here, the dispersed phase, or a premix, is forced through a porous substrate, into the continuous phase (Charcosset and Fessi, 2005). The droplet size produced is dependent on the interfacial tension and the pore sizes. These microporous methods are still being developed and have not yet been widely utilized in the food industry, due to the low throughput.

21.2.3 Emulsion stability

Once formed, a stable emulsion is normally defined as a homogeneous distribution of droplets in a continuous phase. As most emulsions are thermodynamically unstable, they will ultimately want to phase separate into their primary components of oil and water. There are three main instability mechanisms in emulsions as shown in Fig. 21.2, namely flocculation, coalescence and creaming (or sedimentation).

Flocculation is where droplets aggregate together and form clumps or networks. It is a result of the imbalance between attractive and repulsive forces between droplets, resulting in a net attractive force. This can be driven by various mechanisms (which are described briefly in Fig. 21.3), the most common of which is through electrostatic forces. Emulsions are often stabilized by electrostatic repulsion from a net surface potential on the surface of droplets repelling them from close contact, and hence maintaining a homogeneous droplet distribution. Electrostatic repulsion can be negated by changing the salt concentration, composition or pH to counteract or screen the surface charges on the droplets. The droplets will become more attractive through the Van der Waals attractive forces (Fig. 21.3b), get stuck in the potential energy minimum and begin to aggregate. This flocculation is often reversible by changing the ionic strength, pH, or using highly charged surfactants. Bridging flocculation occurs when a large, surface active polymeric molecule is able to form a bridge between two droplets (Fig. 21.3c). If the polymer has multiple hydrophobic regions and is large enough, the hydrophobic regions could adsorb two different droplets simultaneously, thus forming a strong interaction between them.

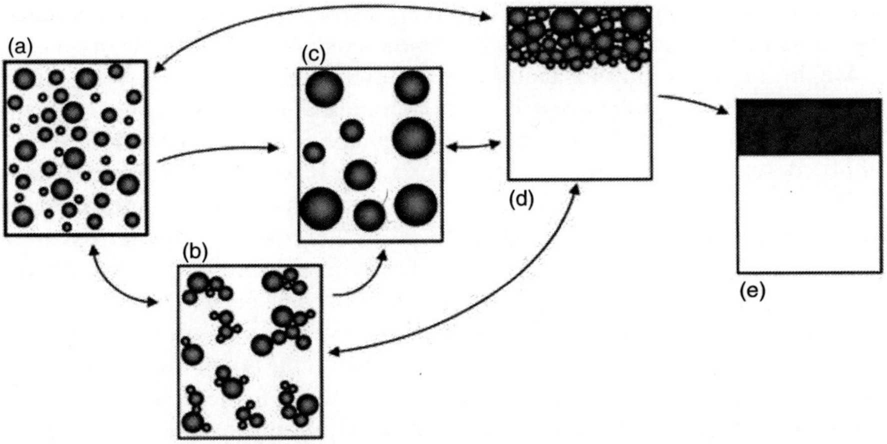

Fig. 21.2 Emulsion stability pathways. Most emulsions are metastable and begin with (a) a homogenous dispersion of oil droplets in water. The three main instability processes are (b) flocculation, (c) coalescence and (d) creaming (sedimentation). (e) Finally, the two phases will completely separate or break into their component parts. The arrows denote possible routes and indicate where reversibility is potentially achievable through gentle re-dispersion.

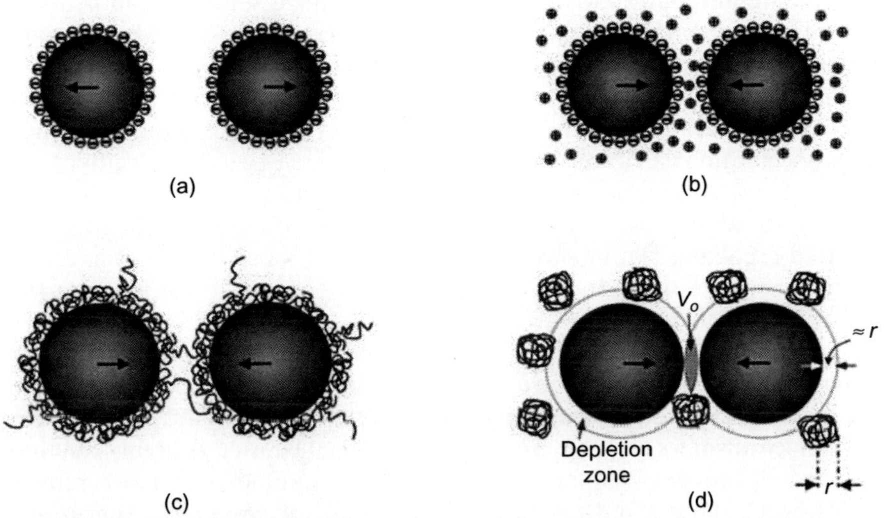

Fig. 21.3 Flocculation mechanisms. Under normal, stable conditions (a) emulsion droplets are repelled, often by a surface charge (electrostatic repulsion) to maintain stability. Added counter ions (b) can screen electrostatic repulsion, inducing flocculation. Adsorbing polymers can induce bridging flocculation (c) by adsorbing to more than one droplet. Depletion flocculation (d) is caused by non-adsorbing polymers or particles. Here a zone exists around the droplets where the polymer concentration is reduced (depletion zone) or is even zero (V_o), inducing an osmotic gradient that increases the attraction between droplets.

Bridging flocculation is often irreversible, but may be prevented by increasing the concentration of the polymer to increase adsorption.

Depletion flocculation is caused by the presence of non-adsorbing polymers or particles. The process is quite complex and is dealt with in detail by other authors (e.g. Jenkins and Snowden, 1996; McClements, 1999), but will be described here only briefly (Fig. 21.3d). If we consider an emulsion which contains non-adsorbing polymers or particles in the continuous phase which are smaller than the emulsion droplets, there will be a depletion zone close to the droplet surface which contains a reduced concentration of polymer (Fig. 21.3d). When two droplets approach each other, they will reach a separation distance that is smaller than the size of the depletion zone, hence this region will contain no polymer, and is sometimes referred to as the depleted or overlap volume. In this situation, there is a difference in the polymer concentration between the continuous phase and the depletion volume. This effectively creates an osmotic potential difference, driving solvent molecules away from the depleted volume into the bulk continuous phase, thus driving the droplets even further together. The depletion zone, and hence the depletion interactions, is dependent on the size and concentration of the depleting polymer/particle (McClements, 1999 and 2000), the size and concentration of the emulsion droplets, and the osmotic potential of the polymer (Jenkins and Snowden, 1996). Optimum polymer concentrations are required, as low concentrations will have too weak an osmotic potential, and high polymer concentrations can actually stabilise the droplets and prevent flocculation.

The effect of flocculation is to create larger flocs of droplets, which will cream and separate more quickly. Sometimes, at higher droplet concentrations, flocculation can induce a space-spanning network to form, rather than discrete flocs, thus inducing stability into the structure. However, this structure can be weak, and often suffers catastrophic collapse events, followed by rapid creaming (Poon *et al.*, 1999). The other consequence of flocculation is that by bringing the droplets into close contact, it will increase the probability of coalescence.

Coalescence is the merger of two or more droplets together to form a larger droplet. This can occur only when droplets come into close contact, so can be accelerated by creaming and flocculation. There are several ways in which emulsifiers and proteins can prevent coalescence. A highly charged surface will prevent droplets from coming into contact and hence reduce coalescence. Steric stabilization through the use of large molecular moieties, often proteins and polymers attached to the surface of the emulsion droplets, will physically or sterically repel droplets. Proteins often form a highly viscoelastic surface layer on droplets which can resist mechanical disruption. Droplets which come into contact must disrupt this film before coalescence takes place. Low molecular weight emulsifiers often use the Gibbs-Marangoni mechanism of stabilization. This involves the rapid diffusion of surfactants at the surface to negate surface tension gradients that arise due to deformation or collision events between droplets. If the surface diffusion is fast enough, the action will drag solvent back into the gap between the droplets, counteract the deformation, and hence prevent coalescence.

Creaming is driven by the density difference between the dispersed and continuous phases, resulting in buoyancy forcing the less dense phase to the top of the emulsion. The terminal velocity V_s of a droplet radius r in a fluid of viscosity η_c medium is estimated by Stokes law:

$$V_s = \frac{2g\Delta\rho r^2}{9\eta_c} \qquad [21.1]$$

where g is the acceleration due to gravity and $\Delta\rho$ is the density difference between the droplet and continuous phases. Although this is valid for solid particles, the velocity can be affected by the viscosity of the dispersed phase and other factors, which are reviewed by McClements (1999). Clearly, therefore, by referring to Equation 21.1, creaming can be reduced by decreasing the droplet size, minimizing the density difference between the dispersed and continuous phases, or increasing the viscosity of the continuous phase. All these can be used to slow down the rate of creaming and hence extend the shelf-life of the emulsion.

A further instability mechanism is *Ostwald ripening*, whereby the increased Laplace pressure in small droplets forces the mass transfer of oil from small to larger droplets leading to a coarsening of the emulsion and increasing the rate of creaming. However, in dairy-based emulsions, Ostwald ripening is not normally a problem, as it requires the oil phase to have some finite solubility in water in order to transfer between droplets, and generally, significant Ostwald ripening has been observed only in hydrocarbon oil systems (Hemar and Horne, 1999).

This section has briefly reviewed the generic principles involved in the formation and stability of emulsions, and the role of emulsifiers (simple or protein). In the next section we will discuss the role of specific dairy ingredients to form and stabilize emulsions.

21.3 Dairy ingredients as emulsifiers

To be effective emulsifiers, as discussed previously, the molecule has to be amphiphilic, i.e. contain both hydrophobic and hydrophilic parts. Milk contains two classes of ingredients that fulfil this criterion – proteins and polar lipids. Milk also contains two sources of surface active material: the aqueous phase and the milk fat globule membrane (MFGM). The aqueous phase of milk contains the water-soluble components; of those that are capable of stabilizing emulsions, proteins are the main class, comprising approximately 3.5 wt%. The MFGM is formed during secretion of the fat droplets in the mammary gland, and is designed to stabilize the fat droplets; being extremely stable (Corredig and Dalgleish, 1998). The MFGM comprises about 40–50% protein; 30–40% neutral lipids and 15–25% phospholipids, depending on source, breed, diet and season. As emulsifying agents, the milk proteins are the most widely utilized, so most of the next section will describe their functionality; however, we will also review the literature concerning the uses of MFGM-derived material.

21.3.1 Milk proteins

As stated previously, the protein content of milk approximately 3.5%, depending on type, source and processing. The two main fractions are caseins and whey proteins, about which there is more information in other chapters and in the literature (Wong *et al.*, 1996; Farrell *et al.*, 2004).

Caseins

Caseins comprise about 80% of the total milk protein content. They are phosphor-proteins and their main function is to form complex aggregates in milk known as casein micelles, which are between 50 and 600 nm in diameter (Fox and Brodkorb, 2008) and encapsulate insoluble minerals (calcium and phosphate) plus small amounts of other minerals, to prevent crystallization in the milk, so aiding secretion and digestion. There are four main types, α_{s1}- α_{s2}-, β-, and κ-casein, which have molecular weights between 19 and 24 kD (Wong *et al.*, 1996) and they all assemble to form the complex micelle structures. These are colloidal aggregates whose core is primarily α and β-caseins, which are highly phosphorylated, and hence chelate the calcium and phosphate, and they are thought to be coated mainly by an outer so called 'hairy layer' of κ-casein, which confers colloidal stability to the aggregates. They are insoluble around the iso-electric point (pH 4–5) and are therefore quite simple to concentrate and extract from milk through isoelectric precipitation at pH 4.6.

The main surface active casein protein is β-casein. It forms approximately 30–35% of the casein proteins (Wong *et al.*, 1996) and is the most surface active; therefore it tends to dominate the interfacial properties of casein-based products, although there is some evidence of interactions with other casein proteins at the interface. Bovine β-casein is a rheomorphic protein in that it is mainly composed of disordered secondary structure (Creamer *et al.*, 1981), has no internal disulphide bonds, and is therefore a very flexible molecule. This aids the adsorption of the protein because, following adsorption, proteins try to rearrange to adopt low energy conformations at the surface, and the highly flexible nature of the protein accelerates this process, allowing rapid reductions in interfacial tension, hence promoting good emulsification properties (Mitchell, 1986). For a protein it is quite surfactant-like in some respects. The hydrophobic residues are largely concentrated at the C-terminal half of the molecule; the N-terminus is more hydrophilic and contains a concentration of five charged phospho-serine groups (Wong *et al.*, 1996). Most globular proteins tend to bury hydrophobic amino acids in the core of their structure, to keep the entire molecule more soluble. However, as β-casein has very little secondary structure, it tends to adopt a similar approach to a surfactant and is thought to form micelle-like aggregates in solution (Walstra, 1999), with a number of proteins aggregating together with their hydrophobic parts in the centre and the hydrophilic residues on the surface. The structure of the interfacial film of β-casein has been studied extensively by a range of techniques, such as x-ray, neutron and light scattering and in-silico mathematical modelling. It is generally agreed that β-casein forms a dense layer less than 2 nm thick, close to the surface, and a very diffuse layer extending up to 10 nm into the continuous phase (Dickinson, 2001). This less dense outer layer is composed largely of the N-terminal part of the protein containing the phosphorserine groups that are highly charged, and is repelled from the rest of the protein, forming a tail protruding into the continuous phase. This layer is very effective at stabilizing emulsion droplets through a combination of charge and steric hindrance (Husband *et al.*, 1997). However, because casein is so flexible, it does not form strong viscoelastic structures at interfaces (Graham and Phillips, 1980); it tends to be weaker and easier to disrupt by other surface active components such as lipids and surfactants (Mackie *et al.*, 1999). Also, because the stability relies on the extension of the phosphorylated region away from the surface, anything that collapses that region back into the dense surface layer, such as low pH, high ionic strength, presence of calcium, or dephosphorylation, can result in a dramatic reduction in the thickness of the layer, and consequently a major loss of stability, resulting in flocculation (Schokker and Dalgleish, 2000) and/or coalescence (Husband *et al.*, 1997).

The other main casein protein that may play a role at the interface is α_{s1}-casein. It is the most dominant casein, forming up to 40% of the casein protein fraction (Wong *et al.*, 1996). It is amphiphilic, like β-casein, but because it is less flexible, its adsorption rate at hydrophobic interfaces is generally slower than β-casein. Therefore α_{s1}-casein is not thought to play a major role in emulsion stabilization. α_{s1}-Casein has a higher net charge and a greater number of phosphoserines than β-casein. The molecular structure of α_{s1} means that it tends to lay flatter on the surface than β-casein (Dickinson, 1999), and so does not sterically stabilize emulsion droplets to the same extent. Its higher charge means that it is more sensitive to ionic strength, and can therefore easily lead to flocculation in the presence of salts (Dickinson *et al.*, 1987).

The functionality of caseins in terms of emulsion formation and stability is largely dominated by β-casein; however, as purifying β-casein is an expensive process, cruder fractions composed of mixtures of caseins, known as caseinates, are normally used. Caseinates are produced by solubilizing the casein fraction obtained by isoelectric precipitation by raising the pH to around 6.6–7. This is done by adding various salts, such as sodium hydroxide, sodium phosphate, or potassium or calcium hydroxide, thus producing sodium, calcium or potassium caseinates. Sodium caseinate contains small protein aggregates together with a significant amount of monomeric protein (Lucey *et al.*, 2000). Because caseins are sensitive to calcium concentration, calcium caseinate contained much larger aggregates, some more than 10 μm in diameter (Moughal *et al.*, 2000). This can have a significant effect on emulsification properties (Srinivasan *et al.*, 1999 and 2003) as the larger aggregates of calcium caseinate diffuse more slowly and lower interfacial tension less rapidly than sodium caseinate. Hence, sodium caseinate can form finer, more stable emulsions at lower protein concentrations, compared with calcium caseinate (Srinivasan *et al.*, 1999).

Whey proteins

Whey proteins are effectively all the other soluble, non-casein proteins. Whey is the milk serum that remains following isoelectric precipitation of 'curds' for cheesemaking or casein production. Upon removal of the precipitated protein or curd fraction, the remaining liquid is known as whey and contains around 0.7% protein, 5% lactose, plus some other minor components (Smithers, 2008). In the 17th and 18th centuries, the medicinal benefits of whey were recorded; however, whey was still, until recently, considered a waste product of the cheese manufacturing process. With the advent of modern, cheap ultrafiltration methods and stricter environmental laws on waste disposal, it became more effective to extract the protein from whey (Smithers, 2008). Today, whey proteins have a huge market as a highly functional, cheap source of protein, and claim a wide variety of health benefits. Whey proteins form about 20% of the total milk proteins, and, depending on the source and the process, comprise approximately 50–60% β-lactoglobulin, 15–20% α-lactalbumin, 15% glycomacropeptide (in renneted whey only), and about 15–20% minor protein/peptide components (e.g. immunoglobulins, lactoferrin, lactoperoxidase, serum albumin, lysozyme, and growth factors). Regarding emulsification properties, the two most abundant proteins, β-lactoglobulin and α-lactalbumin, dominate the functionality of whey.

The major protein in whey is β-lactoglobulin (BLG), and this is probably the most surface active. It has a propensity to bind hydrophobic ligands and has been linked with the binding of components such as retinol, but its physiological role is not known. Nevertheless, its propensity to bind hydrophobic ligands makes BLG highly surface active. BLG has two internal disulphide bonds which confer some degree of structural stability (Wong *et al.*, 1996); hence, it is not as flexible as β-casein, so does not reduce the interfacial tension as quickly, and tends to be slightly less effective at emulsion formation. However, once formed, BLG has a much greater structural integrity at the emulsion interface. BLG is a globular protein, and has a tendency to form intermolecular β-sheet upon adsorption; thus it tends to form highly viscoelastic films at interfaces (Kragel *et al.*, 1995), conferring high degrees of stability to emulsions. This strong interfacial film is thought to be strengthened by intermolecular disulphide bonds due to the existence of a free thiol (Dickinson and Matsumura, 1991). The interfacial viscoelasticity of BLG can make it quite difficult to disrupt and displace by surfactants (Mackie *et al.*, 1999). The protein is quite sensitive to pH. Although it does not precipitate to the same degree as the caseins around its isoelectric point (pH 5.1), it does form more ordered octamers (Pessen *et al.*, 1985). Thus, between pH 4 and 5, the emulsifying properties of BLG are poor and emulsions will tend to flocculate extensively due to the lack of charge repulsion between emulsion droplets. Above pH 6 it is largely a dimer, but dissociates into monomers as the pH increases, and above pH 8 it undergoes an irreversible denaturation. It is acid stable, and below pH 3.5 begins to dissociate into monomers, it is still capable of forming stable emulsions at this pH; however, it has increased competition from α-lactalbumin as will be discussed in the next section.

α-Lactalbumin (ALA) is the next most abundant protein in whey, forming 15–20% of whey protein. It is smaller than BLG at 14.2 kD, and is much more structurally stable, containing mainly α-helix, stabilized by four internal disulphide bonds (Wong *et al.*, 1996). The physiological role of ALA is well defined, and it is involved in enzyme pathways for lactose synthesis. ALA binds calcium, and the structural stability is dependent on the presence of calcium. It is less hydrophobic than BLG, so is not as surface active in its native state. Therefore, considering the relative abundance of BLG compared with ALA in whey and their relative surface activities, it is not surprising that the emulsification of whey protein at neutral pH is dominated by BLG. However, at lower pH, ALA does become more surface active (Gao *et al.*, 2008). Initially, this is due to reduced charge repulsion at the interface between the proteins as the pH approaches the iso-electric point (pH 4.2). Below this point, ALA begins to unfold as the protein approaches its metastable 'molten globule' state (Kronman *et al.*, 1965). This unfolding, due to disulphide bond breakage/reshuffling, leads to the protein being more flexible and surface active. Hence, it has been shown that for whey protein-based emulsions at low pH, ALA becomes more dominant at the interface than BLG (Hunt and Dalgleish, 1994). Thus, in combination, ALA and BLG make a highly flexible and adaptive food ingredient for stabilizing emulsions.

The other whey proteins do not appear to play a significant role in emulsion formation. Although bovine serum albumin (BSA) is surface active and is able to form stable emulsions when used in isolation (Dickinson and Semenova, 1992), it is not generally present at high enough concentrations to have a significant impact. Similarly to caseins, whey proteins are rarely used commercially as individual proteins, although some fractions can be enriched in either ALA or BLG. They are more widely used in the form of total whey protein fractions, whey protein concentrates (WPC) or whey protein isolates (WPI).

Whey proteins are concentrated using a range of filtration and separation techniques, and dried to form WPC or WPI powders (Smithers, 2008). WPC is produced using mainly ultrafiltration techniques and, depending on the process, contains anything between 30 to 75% protein, 2–4% fat, and the remainder made up mainly of lactose and salts. WPI tends to have a much higher protein content (≥90 wt%), as the separation process often involves ion-exchange and is therefore more effective in separating the protein from the non-protein whey components. The proteins are thought to be less denatured, although pasteurization and drying processes can induce significant structural changes. WPI certainly contains lower levels of fat, and therefore the emulsions are less susceptible to lipid destabilization; hence WPI tends to be more effective at forming and stabilizing emulsions than WPC, but is usually more expensive as an ingredient.

21.3.2 Phospholipids

Milk contains amphiphilic polar lipids, mainly in the milk fat globule membrane (MFGM), at low levels (around 300 ppm in whole milk and 12 ppm in skimmed milk) (Rombaut *et al.*, 2007). There is a lot of interest in the health benefits of phospholipids from milk. Composition of the MFGM is up to about 70% protein and 25% phospholipids, with small amounts of cerebrosides and cholesterol (DeWettinck *et al.*, 2008). The proteins are mainly membrane proteins, are not found elsewhere in milk in any quantity, and form only 1–2% of the total milk proteins. The extraction of the MFGM is more convoluted than for protein purification, but there is an interest in the polar lipid fraction from MGFM due to their proposed health benefits. Due to processing of milk products into, for example, butter, the MFGM is partitioned into the serum phase; hence buttermilk can contain up to 10 times more phospholipids than fresh milk, and butter serum up to five times more than that (Dewettinck *et al.*, 2008). MFGM contains mixture of phospholipids (Lopez *et al.*, 2008) and sphingomyelin (SM), both of which seem to form very stable emulsions (Malmsten *et al.*, 1994). This is probably due to the presence of high melting point, mainly saturated long chain fatty acids (Dewettinck *et al.*, 2008).

Untreated MFGM material can have excellent foam and emulsifiying properties (Kanno *et al.*, 1991). Other workers have shown that processing (e.g. heating) induced changes to the composition of MFGM material significantly decreases functionality (Corredig and Dalgleish, 1998), probably due to denaturation of the membrane proteins and possible association of whey proteins with the MFGM material. The MFGM can be modified during cheesemaking to improve the stabilization of oil and water droplets in cheese to increase yields. Here, the use of phospholipase is thought to liberate soluble polar lipids lyso-PC from the MFGM to improve stabilization of oil and water droplets and hence increase the yield (Hoier *et al.*, 2006; Lilbaek *et al.*, 2007). There is little evidence that milk phospholipids are used commercially to stabilize emulsions. Current research is focused on cost effective isolation of phospholipids from milk to produce liposomes for use as delivery vehicles in the cosmetic and pharmaceutical industries. The health benefits of phospholids are attracting increased interest (Dewettinck *et al.*, 2008) and thus may drive the incorporation of MFGM phospholipids into food products.

21.4 Effects of food processing

The principal methodologies for producing caseinates, whey protein fractions and MFGM material have now been described, together with how different processes may influence functionality. In this section we will deal with common effects encountered when milk-based emulsifiers are exposed to common food processing conditions during the production of emulsified foods. The most common processes are those of heat and shear, but high pressure processing is also being encountered. As proteins are the main emulsifying agent in milk, this discussion will concentrate on them. The main physicochemical properties that define the ability of a protein to form and stabilize emulsions are size, solubility, hydrophobicity, charge and flexibility. Any processing scheme which affects these parameters is likely to have an impact upon emulsion properties. However, there is a complex interplay between many of these attributes. For example, to remain soluble, proteins generally hide their hydrophobic residues in their core to reduce contact with the aqueous solvent. To expose these residues will make the protein more surface active, but will also make them more susceptible to aggregation via hydrophobic interactions. This will reduce their solubility, and as a consequence, reduce their ability to diffuse to and stabilize interfaces. There is a general scheme that describes the input of processing energy (whether heat, shear or high pressure) and the impact on protein structure and solubility, and hence on functional properties, as shown in Fig. 21.4. Commercially-produced protein ingredients have usually been exposed to heat and shear treatments, and so they will often form aggregates when rehydrated in preparation for use. Protein aggregates are slower to adsorb and less surface active, and hence are less effective at stabilizing emulsions. Moderate energy processing (e.g. medium shear mixing) could break these aggregates and increase the amount of protein available for adsorption (Husband *et al.*, 1994). Further processing (e.g. shear or thermal) will induce unfolding, leading to increased surface hydrophobicity and hence surface activity (Mitchell, 1986). The level of unfolding for optimal surface activity will vary between proteins. As the energy is increased, the protein will unfold to an extent that the exposed hydrophobicity will render it insoluble and it will tend to aggregate, resulting in less protein available for emulsification and a loss of functionality (Pittia *et al.*, 1996b). Further treatment will result in degradation of the protein into smaller peptides.

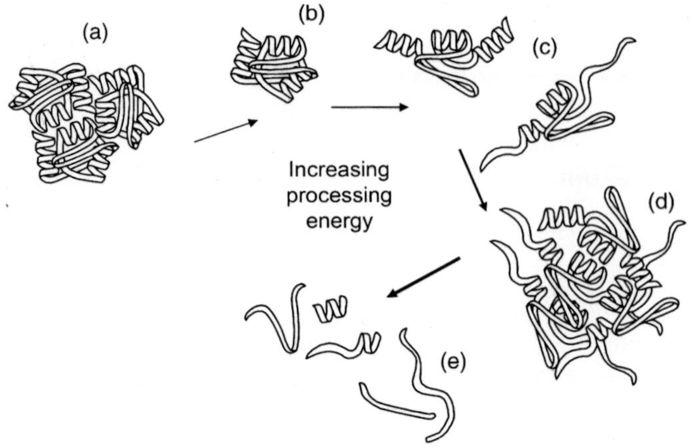

Fig. 21.4 Generalised scheme of the effect of processing energy on protein structure and functionality. (a) protein aggregates/quaternary structures can be disrupted to (b) release more surface active protein monomers. (c) Further processing causes unfolding of tertiary structure to give optimal hydrophobicity and surface activity. (d) Overprocessing will result in hydrophobic aggregation and (e) eventually, degradation of primary structure, thus reducing functionality.

Some of these peptides may be highly surface active and effective emulsifiers, but due to their low molecular weight, are unlikely to prevent flocculation or coalescence and so functionality tends to be poor (Phillips and Beuchat, 1981).

Thermal treatment for cooking and sterilization is the most common processing condition encountered during food processing. Increasing temperature induces changes in hydration, particularly for the more hydrophobic residues, which reduces hydrogen bonding, changes the volume and steric parameters, and hence disrupts the local structure. The weaker or more accessible bonds will be affected at the lowest temperatures, resulting in loss of quaternary structure (i.e. dissociation of oligomers) followed by disruption of some tertiary structure. Hydrogen and covalent bonds require far more energy to break (Messens *et al.*, 1997) and will therefore require much higher temperatures to disrupt. Because of the infinite number of combinations of amino acids and interactions available, no two protein molecules will have the same thermal denaturation profile. Some proteins denature very rapidly at relatively low temperatures, others with many internal disulphide bonds (such as soybean 11 s globulin which has 21) (Kinsella, 1982) are stable even at high temperatures (Sorgentini *et al.*, 1995). Some proteins even have an intermediate metastable denatured state, often referred to as the molten globule state. For example, it is well known that above 70°C β-lactoglobulin exists in a state which is intermediate between the native and the totally unfolded conformations (Qi *et al.*,

1997). Mild heat treatments, up to 100°C for short periods, have resulted in increased surface hydrophobicity in a range of protein systems (Townsend and Nakai, 1983). An increase in surface hydrophobicity has also been associated with improved emulsification properties (Kato *et al.*, 1981). Therefore, by increasing the surface hydrophobicity without drastically reducing the solubility of a protein will increase the affinity of a protein for the interface, as found by Corredig and Dalgleish (1996) who showed that heating β-lactoglobulin between 65 and 75°C increased its affinity for adsorption to the milk fat globules. Similarly with ALA, its greater structural stability means that temperature treatments at 80°C for two hours markedly improved surface activity (Wijesinha-Bettoni *et al.*, 2007). However, similar heat treatments of skim milk powder significantly decreased foaming and emulsifying capacity (Elsamragy *et al.*, 1993). Parris *et al.* found that heating skim milk powder stimulated interactions between whey proteins and casein micelles resulting in more hydrophilic aggregates with poor solubility, but with improved foaming and emulsifying activity (Parris *et al.*, 1991). This demonstrates that in complex protein isolates, process induced interactions between different classes of proteins within the system can severely complicate any relationships between structural and functional properties of individual proteins. Combining heat treatments with pH changes can have cumulative effects. Lieske and Konrad (1994) showed that the solubility of Na-caseinate between pH 4 and 5 can be markedly improved by heating to 90°C at pH 10. This resulted in large improvements in the formation and stability of both foams and emulsions, particularly between pH 4 and 5 where caseinate is normally insoluble.

High shear fluid treatments such as homogenization and the microfluidizer technique do not always have a big impact on protein structure, but may have effects on larger structures and protein aggregates. The microfluidizer technique has been used to prepare liposomes from MFGM lipids (Thompson and Singh, 2006; Thompson *et al.*, 2007); this avoids the use of solvents and the impact they may have on the phase behaviour of the phospholipids.

High pressure treatment is being increasingly used in the food industry as a sterilization process (Ledward *et al.*, 1995). Unlike temperature, when pressure is applied to a system, the change in pressure is instant and homogenous throughout the material; therefore there are no untreated areas where complete sterilization may not occur. Inactivation of microorganisms by high pressure can take place at room temperature, so there is often little or no effect on the flavour, colour or texture of the product (Balny and Masson, 1993). Yet applying high pressure to proteins and polysaccharides can result in molecular changes which can cause gelation and aggregation (Dumay *et al.*, 1996); therefore it is potentially a powerful tool for modifying the functionality of proteins. Electrostatic and hydrophobic interactions which stabilize quaternary structures can be affected at pressures around 1.5 kbar (150 MPa) (Messens *et al.*, 1997), and these dissociations can be reversible. At higher pressures (>4 kbar), the hydrophobic interactions between non-polar amino acids within the protein molecule are broken and the molecule rearranges its tertiary structure (Balny and Masson, 1993). As pressure increases, the protein is compressed, which re-orders the amino acid into a more efficiently packed conformation within the volume constraints created by the increase in pressure. This will affect any intramolecular interactions as the amino acids are forced in to a non-native conformation, away from the structure held together by the delicate balance of forces within the primary structure and between the solvent. It is at pressures above 7 kbar that these effects become important, and protein secondary structure is disturbed. High pressure treatment of skimmed milk was shown to reduce casein micelle size (Anema, 2008), which could explain the observed changes in surface activity (Pittia *et al.*, 1996a). In contrast, the aggregation of whey proteins (Gracia-Julia *et al.*, 2008) and specifically of BLG (Pittia *et al.*, 1996b) was found to increase upon high pressure treatment. This resulted in an increase in surface hydrophobicity. However, rather than increasing surface activity, the result was increased aggregation, even under mild treatments, resulting in poorer emulsification properties.

Regarding processing, there has been much research on the effect of processing on the structure of dairy proteins, as described above. However, less attention has been given to the structures formed by dairy ingredients, which are then subsequently exposed to processing conditions. This is clearly highly relevant, as many food emulsions are processed and cooked after the emulsion has been formed. Interfacial structures are sensitive to heat treatment following adsorption. Xu *et al.* (2008) showed how heating β-lactoglobulin adsorbed films resulted in increased surface elasticity, probably due to increased levels of interfacial aggregation, and greater susceptibility to interfacial perturbation. This increased strength is likely to lead to more stable emulsions (Mitchell, 1986). However, whey protein stabilized emulsions are very sensitive to heat induced flocculation (Livney *et al.*, 2003; Dickinson and Parkinson, 2004), due to increased interactions caused by structural changes in the protein. In fact, freeze–thaw cycles can also induce whey protein stabilized emulsions to flocculate (Cortes-Munos *et al.*, 2009). However, casein stabilized emulsions seem to be less susceptible to heat induced flocculation so the incorporation of small amounts of casein proteins can help prevent flocculation of whey protein stabilized emulsions by thermal processing (Parkinson and Dickinson, 2007).

Another processing variable that is often overlooked in the production of dairy products is water quality. Emulsions stabilized by dairy proteins are normally oil in water emulsions, and hence the major component in these systems is water. As discussed above, dairy proteins, particularly the caseins, are very sensitive to ionic strength and particularly calcium levels. Hence water hardness can have an enormous impact on the stability of emulsions stabilized by caseins. Elevated concentrations of calcium can lead to extensive flocculation of caseinate stabilized emulsions (Schokker and Dalgleish, 2000; Dickinson *et al.*, 2003). Therefore, water quality needs to be monitored and controlled by either deionization or addition of chelating agents.

21.5 Improving functionality

The key physicochemical properties of proteins that enable them to form emulsions are size, solubility, hydrophobicity, charge and flexibility. Their ability to stabilize emulsions depends on the charge and the steric and viscoelastic properties of the interfacial film that they form. In order to improve functionality, aspects of a protein's performance that are restricted by a lack of one or more of these attributes must be addressed.

Lower molecular weight proteins, if soluble, will diffuse more quickly to the interface and often result in increased functionality (Grunden *et al.*, 1974). Reducing molecular weight by hydrolyzing the protein can achieve a similar effect (Horiuchi *et al.*, 1978; Phillips and Beuchat, 1981). However, hydrolyzed milk proteins tend to have mixed results. In general, low levels of hydrolysis can lead to some improvements in emulsion formation (Haque and Mozaffar, 1992; Tuncturk and Zorba, 2006) but further hydrolysis leads to reduced emulsion properties (Tuncturk and Zorba, 2006; Sinha *et al.*, 2007), and emulsion stability is usually reduced (Ven *et al.*, 2001) probably due to lower molecular weight peptides not having the steric or viscoelastic properties necessary to stabilize the interface. This goes back to the general vision of processing effects that an optimum level of modification is required to achieve optimal levels of adsorption and stability.

Similarly, hydrophobic proteins, if soluble, will lower the interfacial tension more effectively, and improve functionality (Kato and Nakai, 1980; Townsend and Nakai, 1983; Horiuchi *et al.*, 1978; Mitchell, 1986). Increasing hydrophobicity and hence functionality has been achieved by thermal treatment (Corredig and Dalgleish, 1996), and chemical modification such as succinylation (Pearce and Kinsella, 1978) and alkylation (Toledano and Magdassi, 1998).

A big problem with milk proteins, and in particular caseins, is the solubility between pH 4 and 6. Any treatment that could improve solubility and hence functionality of caseins over this pH range would be of clear benefit. Santos and Tomasula (2000) showed that acetylation and succinylation can increase the solubility of caseinate at pH 5, as these are common ways of increasing protein solubility. Reducing the sensitivity to calcium by dephosphorylation can also be used to increase solubility (Hekken and Strange, 1993). However, as explained earlier, this can reduce the ability of the casein to stabilize emulsions (Husband *et al.*, 1997) and hence the effect is generally negative on emulsification properties (Hekken and Strange, 1993). Relatively simple treatments such as heating to 90°C at high pH

(Lieske and Konrad, 1994) dramatically improved casein solubility at low pH, resulting in large improvements in emulsion properties.

21.6 Future trends

The future for the use of dairy ingredients is an interesting one. Traditionally, milk was widely regarded as a cheap source of functional ingredients, particularly with the advent of cost effective extraction and preparation technologies. The health benefits of, for example, whey proteins have contributed to increased costs of these as emulsifiers, and food manufacturers are constantly looking at cheaper sources of functional proteins. Additionally, as the apparent incidences of food allergies and intolerances increase, more consumers are turning away from traditional ingredients based on wheat and dairy products.

However, the positive health-promoting aspects of dairy ingredients, whey, caseins and MFGM could actually increase their use in foods. Current opinions regarding the future of food is largely concerned with the ability of the food to deliver improved health and quality of life. Proteins, peptides and phospholipids can all have impacts on human health through availability of essential amino acids, bioactivity and signalling of homeostatic processes such as satiety, and improving membrane and nervous function. The evidence supporting the health benefits of milk-based ingredients such as whey proteins (Frestedt *et al.*, 2008; Krissansen, 2007), casein derived peptides (Moller *et al.*, 2008) and milk phospholipids (Noh and Koo, 2004; DeWettinck *et al.*, 2008), mean that their incorporation into foods can potentially deliver improved health.

More relevant to the subject of this chapter, is the use of milk-based ingredients in emulsions. Emulsified foods are of course a source of dietary fat. Certainly within westernized populations, increased fat intake and reduced exercise are thought to be mainly responsible for the increased incidence of obesity. Therefore, novel strategies are required to reduce fat intake whilst maintaining consumer desirability for a particular food. The influence of whey proteins on energy intake and fat metabolism are beginning to attract interest (Pichon *et al.*, 2008). Whey proteins could also affect the rheology of emulsions compared to conventional emulsifiers, probably due to their interfacial rheology and charge interactions (Mackie *et al.*, 2007), and hence protein stabilized emulsions have been shown to enhance the perception of fat in emulsions (Moore *et al.*, 1998). The digestion of emulsified fat is also likely to attract a great deal of interest in the future, as modulating fat digestion can be used to promote satiety (Maljaars *et al.*, 2008; Mei *et al.*, 2006) through the ileal break mechanism (Mela, 2006). Lipase hydrolysis of triglyceride fat is an interfacial mechanism, and lipase activity is sensitive to the interfacial composition, so changing the interfacial composition to reduce lipase activity could induce satiety. Current research is underway to try to modify the interfacial properties of milk proteins, either enzymatically or using heat treatments, to be able to slow down lipid digestion and hence promote satiety.

Hence milk-based components can be used as multifunctional ingredients to provide both physical stability but also to deliver nutritional and health benefits.

21.7 Sources of further information and advice

Advanced Dairy Chemistry. Volume 1. Dairy Proteins. Eds. P.F. Fox and P.L.H. McSweeney. Kluwer Academic/Plenum Publishers, New York, 2003. Everything you ever needed to know about dairy proteins, including some aspects concerning emulsification.

Encyclopedia of Emulsion Technology, Volume 1, Basic theory. Ed. Paul Becher, Marcel Dekker, New York, 1983. Basic principles of emulsion science and technology.

Polymers at Interfaces: G.J. Fleer, M.A. Cohen Stuart, J.M.H.M. Scheutjens, T. Cosgrove, B. Vincent. Chapman & Hall, London, 1993. Basic principles underlying polymer adsorption at interfaces.

Food Emulsions: Principles, Practice, and Techniques, D. J. McClements, CRC Press, New York 1999.

Understanding and Controlling the Microstructure of Complex Foods. Ed. D.J. McClements, Woodhead Publishing, Cambridge, UK, 2007.

21.8 References

ANEMA S G (2008) 'Heat and/or high-pressure treatment of skim milk: Changes to the casein micelle size, whey proteins and the acid gelation properties of the milk', *International Journal of Dairy Technology*, 61(3), 245–252.

BALNY C and MASSON P (1993) 'Effects of high-pressure on proteins', *Food Rev Int*, 9(4), 611–628.

CHARCOSSET C and FESSI H (2005) 'Membrane emulsification and microchannel emulsification processes', *Reviews in Chemical Engineering*, 21(1), 1–32.

CORREDIG M and DALGLEISH D G (1996) 'Effect of different heat treatments on the strong binding interactions between whey proteins and milk fat globules in whole milk', *J Dairy Res*, 63(3), 441–449.

CORREDIG M and DALGLEISH D G (1998) 'Characterization of the interface of an oil-in-water emulsion stabilized by milk fat globule membrane material', *Journal of Dairy Research*, 65(3), 465–477.

CORTES-MUNOZ M, CHEVALIER-LUCIA D and DUMAY E (2009) 'Characteristics of submicron emulsions prepared by ultra-high pressure homogenisation: Effect of chilled or frozen storage', *Food Hydrocolloids*, 23(3), 640–654.

CREAMER L K, RICHARDSON T and PARRY D A D (1981) 'Secondary Structure of Bovine Alpha-S1-Casein and Beta-Casein in Solution', *Arch Biochem Biophys*, 211(2), 689–696.

DEWETTINCK K, ROMBAUT R, THIENPONT N, LE T T, MESSENS K and VAN CAMP J (2008) 'Nutritional and technological aspects of milk fat globule membrane material', *International Dairy Journal*, 18(5), 436–457.

DICKINSON E (1999) 'Caseins in emulsions: Interfacial properties and interactions', *International Dairy Journal*, 9(3–6), 305–312.

DICKINSON E (2001) 'Milk protein interfacial layers and the relationship to emulsion stability and rheology', *Colloids and Surfaces B-Biointerfaces*, 20(3), 197–210.

DICKINSON E and MATSUMURA Y (1991). 'Time dependent polymerisation of β-lactoglobulin through disulphide bonds at the oil–water interface in emulsions' *Int J Biol Macromol*, 13(1), 26–30.

DICKINSON E and SEMENOVA M G (1992) 'Emulsifying behavior of protein in the presence of polysaccharide under conditions of thermodynamic incompatibility', *J Chem Soc Faraday Trans*, 88(6), 849–854.

DICKINSON E and PARKINSON E L (2004) 'Heat-induced aggregation of milk protein-stabilized emulsions: Sensitivity to processing and composition', *International Dairy Journal*, 14(7), 635–645.

DICKINSON E, WHYMAN R H and DALGLEISH D G (1987). 'Colloidal properties of model oil-in-water food emulsions stabilized separately by α_{s1}-casein, β-casein and κ-casein. In E. Dickinson, *Food Emulsions and Foams* (pp. 40–51), Royal Society of Chemistry, London.

DICKINSON E, RADFORD S J and GOLDING M (2003) 'Stability and rheology of emulsions containing sodium caseinate: Combined effects of ionic calcium and non-ionic surfactant', *Food Hydrocolloids*, 17(2), 211–220.

DUMAY E, LAMBERT C, FUNTENBERGER S and CHEFTEL J C (1996) 'Effects of high pressure on the physico-chemical characteristics of dairy creams and model oil/water emulsions', *Food Sci Technol-Lebensmittel-Wissenschaft Technol*, 29(7), 606–625.

ELSAMRAGY Y A, HANSEN C L and MCMAHON D J (1993) 'Production of ultrafiltered skim milk retentate powder. 2. Functional-properties', *J Dairy Sci*, 76(10), 2886–2890.

FARRELL H M, JIMENEZ-FLORES R, BLECK G T, BROWN E M, BUTLER J E, CREAMER L K, HICKS C L, HOLLAR C M, NG-KWAI-HANG K F and SWAISGOOD H E (2004) 'Nomenclature of the proteins of cows' milk – Sixth revision', *Journal of Dairy Science*, 87(6), 1641–1674.

FOX P F and BRODKORB A (2008) 'The casein micelle: Historical aspects, current concepts and significance', *International Dairy Journal*, 18(7), 677–684.

FRESTEDT J L, ZENK J L, KUSKOWSKI M A, WARD L S and BASTIAN E D (2008) 'A whey-protein supplement increases fat loss and spares lean muscle in obese subjects: A randomized human clinical study', *Nutrition & Metabolism*, 5.

GAO C, WIJESINHA-BETTONI R, WILDE P J, MILLS E N C, SMITH L J and MACKIE A R (2008) 'Surface properties are highly sensitive to small pH induced changes in the 3-D structure of alpha-lactalbumin', *Biochem,* 47(6), 1659–1666.

GRACIA-JULIA A, RENE M, CORTES-MUNOZ M, PICART L, LOPEZ-PEDEMONTE T, CHEVALIER D and DUMAY E (2008) 'Effect of dynamic high pressure on whey protein aggregation: A comparison with the effect of continuous short-time thermal treatments', *Food Hydrocolloids*, 22(6), 1014–1032.

GRAHAM D E and PHILLIPS M C (1980) 'Proteins at liquid interfaces. V. Shear properties', *J Colloid Interface Sci*, 76(1), 240–250.

GRUNDEN L P, VADEHRA D V and BAKER R C (1974) 'Effects of proteolytic enzymes on the functionality of chicken egg albumin', *J Food Sci*, 39 (39), 841–843.

HAQUE Z U and MOZAFFAR Z (1992) 'Casein Hydrolysate. 2. Functional-Properties of Peptides', *Food Hydrocolloids*, 5(6), 559–571.

HEKKEN D and STRANGE E (1993) 'Functional properties of dephosphorylated bovine whole casein', *Journal of Dairy Science*, 76(11).

HEMAR Y and HORNE D S (1999) 'A diffusing wave spectroscopy study of the kinetics of Ostwald ripening in protein-stabilised oil/water emulsions', *Colloids and Surfaces B-Biointerfaces*, 12(3–6), 239–246.

HOIER E, LILBAEK H, BROE M L and SORENSEN N K (2006) 'Enhancing cheese yield by phospholipase treatment of cheese milk', *Australian Journal of Dairy Technology*, 61(2), 179–182.

HORIUCHI T, FUKISHIMA D, SUGIMOTO H and HATTORI T (1978) 'Studies on enzyme-modified proteins as foaming agents: Effect of structure on foam stability', *Food Chem*, 3 (3), 35–42.

HUNT J A and DALGLEISH D G (1994) 'Effect of pH on the Stability and Surface-Composition of Emulsions Made with Whey-Protein Isolate', *Journal of Agricultural and Food Chemistry*, 42(10), 2131–2135.

HUSBAND F A, WILDE P J, CLARK D C, RAWEL H M and MUSCHIOLIK G (1994) 'Foaming properties of modified faba bean protein isolates', *Food Hydrocolloids* 8(5), 455–468.

HUSBAND F A, WILDE P J, MACKIE A R and GARROOD M J (1997) 'A comparison of the functional and interfacial properties of beta-casein and dephosphorylated beta-casein', *J Colloid Interface Sci*, 195, 77–85.

JENKINS P and SNOWDEN M (1996) 'Depletion flocculation in colloidal dispersions', *Advances in Colloid and Interface Science*, 68, 57–96.

KANNO C, SHIMOMURA Y and TAKANO E (1991) 'Physicochemical Properties of Milk-fat Emulsions Stabilized with Bovine Milk Fat Globule Membrane', *Journal of Food Science*, 56(5), 1219–1223.

KATO A and NAKAI S (1980) 'Hydrophobicity determined by a fluorescent probe method and its correlation with surface properties of proteins'. *Biochim Biophys Acta*, 624(1), 13–20.

KATO A, TSUTSUI N, MATSUDOMI N, KOBAYASHI K and NAKAI S (1981) 'Effects of partial denaturation on surface properties of ovalbumin and lysozyme', *Agric Biol Chem*, 45(12), 2755–2760.

KENTISH S, WOOSTER J, ASHOKKUMAR A, BALACHANDRAN S, MAWSON R and SIMONS L (2008) 'The use of ultrasonics for nanoemulsion preparation', *Innovative Food Science & Emerging Technologies*, 9(2), 170–175.

KINSELLA J E (1982) 'Relationship between structural and functional properties of food proteins', in *Food Proteins* (eds. P.F. Fox and J.J. Condon) Applied Science Publishers, Barking, 1982, pp 51–104.

KOTYLA T, KUO F, MOOLCHANDANI V, WILSON T and NICOLOSI R (2007) 'Increased bio-availability of a transdermal application of a delta tocopherol nanoemulsion preparation', *Faseb Journal*, 21(5), A266.

KRAGEL J, WUSTNECK R, CLARK D C, WILDE P J and MILLER R (1995) 'Dynamic surface-tension and surface shear rheology studies of mixed beta-lactoglobulin Tween-20 systems', *Colloids Surfaces A-Physicochem Eng Aspects,* 98(1), 127–135.

KRISSANSEN G W (2007) 'Emerging health properties of whey proteins and their clinical implications', *Journal of the American College of Nutrition*, 26(6), 713S–723S.

KRONMAN M J, CERANKOWSKI L and HOLMES L G (1965) 'Inter and intramolecular interactions of alpha-lactalbumin. 3. Spectral changes at acid pH', *Biochem,* 4(3), 518–525.

LEDWARD D A, JOHNSTON D E, EARNSHAW R G and HASTING A P M (eds.) *High Pressure Processing of Foods*, Nottingham University press, Nottingham, 1995.

LIESKE B and KONRAD G (1994) 'Thermal modification of sodium-caseinate.1. Influence of temperature and pH on selected physicochemical and functional-properties', *Milchwissenschaft-Milk Sci Int*, 49(1), 16–20.

LILBAEK H M, FATUM T M, IPSEN R and SORENSEN N K (2007) 'Modification of milk and whey surface properties by enzymatic hydrolysis of milk phospholipids', *Journal of Agricultural and Food Chemistry*, 55(8), 2970–2978.

LIVNEY Y D, CORREDIG M and DALGLEISH D G (2003) 'Influence of thermal processing on the properties of dairy colloids', *Current Opinion in Colloid & Interface Science*, 8(4–5), 359–364.

LOPEZ C, BRIARD-BION V, MENARD O, ROUSSEAU F, PRADEL P and BESLE J M (2008) 'Phospholipid, sphingolipid, and fatty acid compositions of the milk fat globule membrane are modified by diet', *Journal of Agricultural and Food Chemistry*, 56(13), 5226–5236.

LUCEY J A, SRINIVASAN M, SINGH H and MUNRO P A (2000) 'Characterization of commercial and experimental sodium caseinates by multiangle laser light scattering and size-exclusion chromatography', *Journal of Agricultural and Food Chemistry*, 48(5), 1610–1616.

MACKIE A R, GUNNING A P, WILDE P J and MORRIS V J (1999) 'Orogenic displacement of protein from the air/water interface by competitive adsorption', *J Colloid Interface Sci*, 210(1), 157–166.

MACKIE A R, RIDOUT M J, MOATES G, HUSBAND F A and WILDE P J (2007) 'Effect of the interfacial layer composition on the properties of emulsion creams', *J Agric Food Chem*, 55(14), 5611–5619.

MALJAARS J, FOLTZ M, PETERS H P F, WISEMAN S and MASCLEE A A M (2008) 'The intragastric behaviour of dietary fat strongly influences fat absorption: Implications on cholecystokinin release and appetite', *Gastroenterology*, 134(4), A315–A316.

MALMSTEN M, BERGENSTAHL B, NYBERG L and ODHAM G (1994) 'Sphingomyelin from Milk – Characterization of Liquid-crystalline Liposome and Emulsion Properties', *Journal of the American Oil Chemists Society*, 71(9), 1021–1026.

MASON T G, WILKING J N, MELESON K, CHANG C B and GRAVES S M (2006a) 'Nanoemulsions: Formation, structure, and physical properties', *Journal of Physics-Condensed Matter*, 18(41), R635–R666.

MASON T G, GRAVES S M, WILKING J N and LIN M Y (2006b) 'Extreme emulsification: Formation and structure of nanoemulsions', *Condensed Matter Physics*, 9(1), 193–199.

MCCLEMENTS DJ (1999) *Food Emulsions: Principles, Practice, and Techniques*, CRC Press, New York 1999.

MCCLEMENTS D J (2000) 'Comments on viscosity enhancement and depletion flocculation by polysaccharides', *Food Hydrocolloids*, 14(2), 173–177.

MEI J, LINDQVIST A, KRABISCH L, REHFELD J F and ERLANSON-ALBERTSSON C (2006) 'Appetite suppression through delayed fat digestion', *Physiology & Behavior*, 89(4), 563–568.

MELA D J (2006) 'Novel Food Technologies: Enhancing Appetite Control in Liquid Meal Replacers', *Obesity Res*, 14, S179–S181.

MESSENS W, VAN CAMP J and HUYGHEBAERT A (1997) 'The use of high pressure to modify the functionality of food proteins', *Trends Food Sci Technol*, 8(4), 107–112.

MITCHELL J R (1986) Foaming and emulsifying properties of proteins. In *Developments in Food Proteins – 4* (Ed. Hudson, B J F) Elsevier, London. pp. 291–338.

MOLLER N P, SCHOLZ-AHRENS K E, ROOS N and SCHREZENMEIR J (2008) 'Bioactive peptides and proteins from foods: Indication for health effects', *European Journal of Nutrition*, 47(4), 171–182.

MOORE P B, LANGLEY K, WILDE P J, FILLERY-TRAVIS A and MELA D J (1998) 'Effect of emulsifier type on sensory properties of oil-in-water emulsions', *J Sci Food Agric*, 76, 469–476.

MOUGHAL K I, MUNRO P A and SINGH H (2000) 'Suspension stability and size distribution of particles in reconstituted, commercial calcium caseinates', *International Dairy Journal*, 10(10), 683–690.

NOH S K and KOO S I (2004) 'Milk sphingomyelin is more effective than egg sphingomyelin in inhibiting intestinal absorption of cholesterol and fat in rats', *Journal of Nutrition*, 134(10), 2611–2616.

PARKINSON E L and DICKINSON E (2007) 'Synergistic stabilization of heat-treated emulsions containing mixtures of milk proteins', *International Dairy Journal*, 17(2), 95–103.

PARRIS N, WOYCHIK J H, COOKE P H (1991) 'Effect of preheat temperature on the hydrophobic properties of milk proteins', *ACS Symp Series*, 454(1) 25–41.

PEARCE K N and KINSELLA J E (1978) 'Emulsifying properties of proteins: Evaluation of a turbidimetric technique', *J Agric Food Chem*, 26 (3), 716–723.

PESSEN H, PURCELL J M and FARRELL H M (1985) 'Proton relaxation rates of water in dilute-solutions of beta-lactoglobulin – determination of cross relaxation and correlation with structural-changes by the use of two genetic-variants of a self-associating globular protein', *Biochim Biophys Acta,* 828(1), 1–12.

PHILLIPS R D and BEUCHAT L R (1981). 'Enzyme modification of proteins', *ACS Symp Ser,* 147, 275–298.

PICHON L, POTIER M, TOME D, MIKOGAMI T, LAPLAIZE B, MARTIN-ROUAS C and FROMENTIN G (2008) 'High-protein diets containing different milk protein fractions differently influence energy intake and adiposity in the rat', *British Journal of Nutrition*, 99(4), 739–748.

PITTIA P, WILDE P J and CLARK D C (1996a) 'the foaming properties of native and pressure treated beta-casein', *Food Hydrocolloids* 10(1), 335–342.

PITTIA P, WILDE P J, HUSBAND F and CLARK D C (1996b) 'functional and structural properties of beta-lactoglobulin as affected by high pressure treatment', *J Food Sci,* 61(6), 1123–1128.

POON W C K, STARRS L, MEEKER S P, MOUSSAID A, EVANS R M L, PUSEY P N and ROBINS M M (1999) 'Delayed sedimentation of transient gels in colloid-polymer mixtures: Dark-field observation, rheology and dynamic light scattering studies', *Faraday Discussions*, (112), 143–154.

QI X L, HOLT C, MCNULTY D, CLARKE D T, BROWNLOW S and JONES G R (1997) 'Effect of temperature on the secondary structure of beta-lactoglobulin at pH 6.7, as determined by CD and IR spectroscopy: A test of the molten globule hypothesis', *Biochem J*, 324(1), 341–346.

ROBINS M M and WILDE P J (2003) 'Colloids and emulsions'. In *Encyclopaedia of Food Science and Nutrition*, Eds. Caballero B, Trugo L and Finglas PM, Elsevier, London, pp 1517–1524.

ROMBAUT R, DEWETTINCK K and VAN CAMP J (2007) 'Phospho- and sphingolipid content of selected dairy products as determined by HPLC coupled to an evaporative light scattering detector (HPLC-ELSD)', *Journal of Food Composition and Analysis*, 20(3–4), 308–312.

SANTOS C V and TOMASULA P M (2000) 'Acylation and solubility of casein precipitated by carbon dioxide', *Journal of Food Science*, 65(2), 227–230.

SCHOKKER E P and DALGLEISH D G (2000) 'Orthokinetic flocculation of caseinate-stabilized emulsions: Influence of calcium concentration, shear rate, and protein content', *J Agric Food Chem*, 48(1), 198–203.

SINHA R, RADHA C, PRAKASH J and KAUL P (2007) 'Whey protein hydrolysate: Functional properties, nutritional quality and utilization in beverage formulation', *Food Chemistry*, 101(4), 1484–1491.

SMITHERS G W (2008) 'Whey and whey proteins – From "gutter-to-gold" ', *International Dairy Journal*, 18(7), 695–704.

SORGENTINI D A, WAGNER J R and ANON M C (1995) 'Effects of thermal-treatment of soy protein isolate on the characteristics and structure–function relationship of soluble and insoluble fractions', *J Agric Food Chem*, 43(9), 2471–2479.

SRINIVASAN M, SINGH H and MUNRO P A (1999) 'Adsorption behaviour of sodium and calcium caseinates in oil-in-water emulsions', *International Dairy Journal*, 9(3–6), 337–341.

SRINIVASAN M, SINGH H and MUNRO P A (2003) 'Influence of retorting (121 degrees C for 15 min), before or after emulsification, on the properties of calcium caseinate oil-in-water emulsions', *Food Chemistry*, 80(1), 61–69.

THOMPSON A K and SINGH H (2006) 'Preparation of liposomes from milk fat globule membrane phospholipids using a Microfluidizer', *Journal of Dairy Science*, 89(2), 410–419.

THOMPSON A K, MOZAFARI M R and SINGH H (2007) 'The properties of liposomes produced from milk fat globule membrane material using different techniques', *Lait*, 87(4–5), 349–360.

TOLEDANO O and MAGDASSI S (1998) 'Emulsification and foaming properties of hydrophobically modified gelatin', *J Colloid Interface Sci*, 200, 235–240.

TOWNSEND A A and NAKAI S (1983) 'Relationships between hydrophobicity and foaming characteristics of food proteins', *J Food Sci*, 48(2), 588–594.

TUNCTURK Y and ZORBA O (2006) 'The effects of enzymatic hydrolysis of casein on apparent yield stress and some emulsion properties', *Food Hydrocolloids*, 20(4), 475–482.

VEN C, GRUPPEN H, BONT D and VORAGEN A (2001) 'Emulsion properties of casein and whey protein hydrolysates and the relation with other hydrolysate characteristics', *Journal of Agricultural and Food Chemistry*, 49(10), 5005–5012.

WALSTRA P (1983) 'Formation of emulsions'. In *Encyclopedia of Emulsion Technology, Volume 1, Basic Theory*, Paul Becher, Ed., Marcel Dekker, New York, 1983.

WALSTRA P (1993) 'Principles of emulsion formation' *Chem Eng Sci*, 48(2), 333–349.

WALSTRA P (1999) 'Casein sub-micelles: Do they exist?', *Int Dairy J*, 9(3–6), 189–192.

WANG X Y, JIANG Y, WANG Y W, HUANG M T, HO C T and HUANG Q R (2008) 'Enhancing anti-inflammation activity of curcumin through O/W nanoemulsions', *Food Chemistry*, 108, 419–424.

WIJESINHA-BETTONI R, GAO C, JENKINS J A, MACKIE A R, WILDE P J, MILLS E N C and SMITH L J (2007) 'Heat treatment of bovine alpha-lactalbumin results in partially folded, disulfide bond shuffled states with enhanced surface activity', *Biochem*, 46(34), 9774–9784.

WONG D W S, CAMIRAND W M and PAVLATH A E (1996) 'Structures and functionalities of milk proteins', *Crit Rev Food Sci Nutr*, 36(8), 807–844.

XU R, DICKINSON E and MURRAY B S (2008) 'Morphological changes in adsorbed protein films at the oil-water interface subjected to compression, expansion, and heat processing', *Langmuir*, 24(5), 1979–1988.

22

Using dairy ingredients for encapsulation

C. M. Oliver and M. A. Augustin, CSIRO Division of Food and Nutritional Sciences, Australia

Abstract: This chapter discusses the properties of dairy components that allow them to function as effective encapsulants in a range of encapsulated systems. The key principles that underlie the delivery of encapsulated food ingredients and bioactives in food are introduced. The formulation and processing of encapsulated ingredients containing dairy components and their applications in foods are covered. The focus is on the use of milk proteins as encapsulants. Examples of the use of milkfat, lactose and milkfat globule membrane in encapsulation are given. The emerging trends in encapsulation of food ingredients and the future potential of dairy ingredients as encapsulant materials are highlighted.

Key words: dairy ingredients, milk proteins, encapsulation, emulsion, food.

22.1 Introduction

Microencapsulation protects an active component (core) within a secondary material (matrix, shell, wall or encapsulant) from the surrounding environment and enables its controlled release at a target site (Kamyshny and Magdassi, 2006). Microencapsulation is well-established within the pharmaceutical, cosmetic, chemical and agricultural industries, but is relatively new to the food industry (Gibbs *et al.*, 1999; Risch and Reineccius, 1995). The early uses of microencapsulation in the food industry included masking undesirable flavours, controlling the release of desirable flavours, and the conversion of liquid streams into powdered formats for convenience and improved shelf-life. Encapsulation technology is increasingly being used in the food industry to protect and control delivery of food ingredients and bioactives (Augustin and Sanguansri, 2008; Champagne and Fustier, 2007; McClements *et al.*, 2007). Examples of food components that may benefit from encapsulation are given in Table 22.1.

Table 22.1 Active cores that benefit from being protected and delivered into foods with encapsulation

Type of core	Examples	Potential benefit of encapsulation
Oils	Milkfat, omega-3 oils	Improved storage stability, target release
Flavours	Mint, orange oil	Preservation of flavours, controlled release in the mouth
Food additives	Leavening agents	Controlled release during baking
Minerals	Iron salts	Avoiding undesirable interactions (e.g. catalyzing fat oxidation), target delivery on ingestion
Phytonutrients	Flavanoids, polyphenols, tocopherols, phytosterols, carotenoids, carotene, lycopene, lutein	Protection of sensitive ingredients from the environment and interactions with the food matrix, target delivery on ingestion
Probiotics	Bifidobacteria, lactobacilli	Improved survival during storage, survival on exposure to stomach acids

The encapsulating materials used in food applications include food proteins, carbohydrates, fats, waxes and food-grade surfactants. The properties that make these materials suitable as encapsulants are given in Table 22.2. Dairy ingredients are readily available as food-grade materials and their properties make them highly desirable as encapsulating materials. Dairy proteins, lactose, milkfat and its fractions, and milkfat globule membrane (MFGM) components may all be used as encapsulating materials. Some dairy components (e.g. milk protein peptides, conjugated linoleic acid) have been encapsulated for a target function. This chapter discusses the properties of dairy components that allow them to function as encapsulating materials in a range of delivery systems designed for food applications. Examples of micro- and nano-encapsulated systems for the delivery of food ingredients and bioactives that have been formulated with dairy ingredients are given. The emerging trends in encapsulation of food ingredients and the future of dairy ingredients in this arena are highlighted.

22.2 Designing encapsulated ingredients

Designing functional micro- and nano-encapsulated ingredients requires an understanding of the solubility and stability of the core to be encapsulated, the physico-chemical properties of the encapsulating materials and the processes that may be utilized for encapsulation. Both the core and the encapsulating materials used in the formulation of encapsulated ingredients

Table 22.2 Examples of encapsulating materials used in encapsulation and their functional properties

Encapsulant materials	Examples	Characteristics of materials useful for encapsulation
Proteins	Caseins, whey proteins, soy proteins, egg proteins, pea protein, gelatin, hydrolyzed proteins	Ability to build viscosity, gelling, emulsifying
Sugars and glucose syrups	Mono-, di- and oligo-saccharides, glucose syrups	Low viscosity at high solids, ability to form glassy solids on dehydration
Polysaccharides	Starch, maltodextrins, gums, carboxymethylcellulose, pectins, alginates, chitosan	Gelling, emulsion stabilization, film forming, ability to form glassy solids on dehydration
Fats and waxes	Animal fats (e.g. milkfat), vegetable fats (e.g. canola oil), waxes (e.g. beeswax)	Solubilization of lipophilic cores, matrix for embedding cores, water barrier properties, film forming
Surfactants	Mono- and di-glycerides, phospholipids (e.g. lecithin), glycolipids, Tweens, Spans	Emulsifying

for the food industry have to be food-grade (Tables 22.1 and 22.2). The technology to be used for encapsulation may be selected from a range of processing methods (Table 22.3), depending on the formulation and the final structure of the capsule. A number of reviews describe the history of and developments in the encapsulation of food ingredients and encapsulation technologies (Augustin and Sanguansri, 2008; Augustin *et al.*, 2001; Gibbs *et al.*, 1999; Gouin, 2004; Madene *et al.*, 2006; Risch and Reineccius, 1995).

Depending on the target application, a variety of capsule structures can be designed to encapsulate ingredients. These can be classified into the classical core–shell structures or as matrix structures in which the active core is entrapped (Fig. 22.1). Variations of these include multi-layered capsules and capsules with multiple cores (Kamyshny and Magdassi, 2006).

Examples of capsules with core–shell structures are conventional oil-in-water (O/W) emulsions, multiple emulsions (W/O/W), multilayer-stabilized emulsions, solid lipid particles (O/W), nano-emulsions made by high pressure homogenization and liposomes. Hydrogels, lipospheres and biopolymeric micro- and nano-particles with active cores are typical of encapsulated systems with matrix structures.

By adjusting the internal structure and properties of the shell or the matrix material, the release patterns of the core can be manipulated. An

Table 22.3 Selected encapsulation methods used in the food industry[1]

Method	Principle	Comments
Spray drying	Core is dispersed into aqueous encapsulant solution and atomized into a drying chamber	Most commonly used method, cost-effective
Spray chilling	Core is dispersed into coating solution and sprayed into a cold environment to solidify the carrier material	Used for protection of water-soluble cores, also suitable for cores that are sensitive to temperature
Extrusion	Emulsion or dispersion containing the core is passed through a die at high temperature and pressure into a bath for solidification of the particle	Used primarily for encapsulation of flavours and volatile cores in glassy matrices
Fluidized bed coating	Particles are suspended in air and a coating is applied	Used for enhancing and achieving finer control over release properties of the core
Inclusion complexation	An inclusion complex is formed between cyclodextrin and the core	Encapsulation of flavours and lipophilic nutrients
Coacervation	Coacervates are formed when two oppositely charged biopolymers associate and phase separate	Can entrap high loadings of cores, has been used in encapsulation of flavours and many nutrients

[1] Augustin *et al.*, 2001; Gouin, 2004.

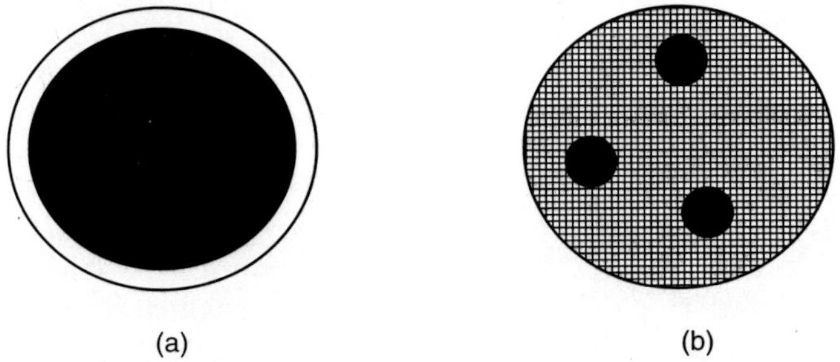

Fig. 22.1 Types of microcapsules (a) core and shell (b) matrix with embedded core.

understanding of the physico-chemical properties of the encapsulating matrix, how it interacts with the core, and its behaviour during processing is therefore required to design an encapsulation system. The core has to be protected until its controlled release in response to a trigger encountered at the appropriate time. Triggers for the release of the core, such as pH, temperature, shear or degradation of the matrix material by enzymes, have been targeted to suit each application. Control of the core release in response to an appropriate trigger is influenced by the encapsulant material, formulation and the processing technique used to produce the encapsulated ingredient.

A recent paper (Ubbink and Krüger, 2006) discusses a retro-design approach to encapsulation. In this approach the functionality in a target application is first established. A systematic evaluation of specific encapsulation technology and materials that can be used to obtain the desired protection and release are based on an understanding of material science, physical chemistry and biophysics.

22.3 Encapsulating properties of dairy ingredients

Some important properties that are desired of encapsulants are their ability to gel, to stabilize an interface, to solubilize actives and to form glassy matrices. These properties are inherent in a range of dairy components. Dairy proteins are commonly chosen as encapsulant materials. Milkfat and milkfat globule membrane (MFGM) material also have applications as encapsulants. A summary of the functional properties of dairy ingredients as encapsulating materials is given in Table 22.4. Different functionalities

Table 22.4 Applications of dairy ingredients in various encapsulation systems

System	Dairy encapsulant material	Function of dairy ingredient
Emulsions	Milk proteins – caseins, whey proteins, milk protein isolates, hydrolyzed milk proteins	Film forming, stabilization of emulsion
	Milkfat, milkfat fractions	Carrier of active cores
Dried emulsions	Milk proteins – casein, whey proteins, hydrolyzed milk proteins	Emulsion stabilization prior to drying, emulsion matrix of dried particle
	Lactose	Formation of glassy matrix on drying
	Milkfat	Carrier of active cores, secondary coating material, barrier to moisture
Hydrogel	Milk proteins – casein, whey proteins	Formation of gel phase, matrix for embedding cores
Lipospheres	Milkfat	Entrapment of active cores
Liposomes	Milk phospholipids, milkfat globule membrane	Stabilization of bilayer vesicles
Coacervates	Milk proteins – casein, whey proteins	Interaction with oppositely charged biopolymers to form a separate phase

are required of encapsulants depending on the target food system and mode of delivery. For example, for core–shell microcapsules, such as emulsion-based systems, dairy proteins are used to lower surface tension and stabilize oil droplets in which an active core is solubilized. In matrix-based systems, such as hydrogels, the ability of dairy proteins to form gels enables active cores to be entrapped. The release of the core from these capsules occurs when there is disintegration of the shell or protein matrix on exposure to the appropriate trigger.

Lactose can exist in the glassy state at low water activities and at ambient temperature (Roos, 2008). This property makes lactose useful as an encapsulating matrix. In the glassy state, there is limited mobility of molecules. When active cores are entrapped in glassy states they are stabilized against adverse reactions. The core can be readily released in response to increased temperature or on exposure to moisture. Lactose may be used on its own to form a glassy matrix. In addition, as lactose readily forms a glass when spray-dried, it is used in combination with proteins to prepare spray-dried emulsion systems (Fäldt and Bergenståhl, 1996).

Bulk milkfat has a wide melting point range and may also be fractionated to produce various melting point fractions (Kaylegian and Lindsay, 1995). Like any other fat, milkfat or its fractions can carry active cores in emulsion-based systems, and in micro- and nano-particles. They can solubilize lipophilic cores (e.g. carotene) or act as carriers for hydrophilic cores (e.g. peptides). MFGM material and milk phospholipids from the MFGM have amphiphilic properties and like other surface-active material have a propensity to self-assemble into supramolecular structures. These properties make them useful in the formulation of micro- and nano-emulsions and liposomes.

22.4 Encapsulated systems utilizing dairy ingredients as encapsulants

Dairy components have been used for formulating encapsulated systems containing a variety of cores.

22.4.1 Milk proteins

The major proteins in milk are caseins and whey proteins. Both these proteins have been utilized in the formulation of a range of encapsulated systems. It is well known that caseins and whey proteins differ in many aspects. For example, there are differences in (a) pI – caseins have a pI of 4.6 while the whey proteins have a pI of ~5.2, (b) conformation – the caseins are random coil proteins while the whey proteins have a globular structure, (c) sensitivity to heat – caseins are relatively heat stable whereas the whey proteins denature at high temperature, (d) emulsifying properties – the globular whey proteins form denser interfaces than the more flexible caseins, and (e) gelling properties – whey protein denaturation generally precedes gelation whereas casein gels may be formed in the presence of calcium ions. It is important to appreciate that protein functionality dictates the selection of the protein in encapsulated systems. The functional properties of milk proteins and how they can be modified has been reviewed extensively (Augustin and Udabage, 2007; Foegeding *et al.*, 2002; Fox, 1997; Kinsella, 1984) and will not be covered here.

22.4.2 Caseins

Particles

The casein micelle is itself a natural nano-particle. It has an average radius of ~100 nm and is made up of nano-clusters of amorphous calcium phosphate surrounded by casein (Holt *et al.*, 2003). It is a natural encapsulation system for the delivery of calcium and phosphate. The casein micelle also has the capacity to carry hydrophobic compounds. Technology developed by Livney and Dalgleish (2007) demonstrated a 5.5-fold increase in vitamin D_2 incorporated into the casein micelle compared to that in the serum phase (Semo *et al.*, 2007). Encapsulation of vitamin D_2 stabilized the bioactive and protected it from degradation (e.g. UV light). Thus the casein micelle-encapsulated bioactive may be incorporated into food to improve its nutritive value. The casein micelle also lends itself to the delivery of other sparingly water-soluble compounds by complexing with the bioactive (Sahu *et al.*, 2008). These studies suggest that there may be opportunities to utilize the inherent properties of the casein micelle nano-structure to deliver other sensitive hydrophobic cores to improve the health benefits of food systems (Semo *et al.*, 2007). Another interesting possibility may be offered by the observation that the binding of hydrophobic components to casein can be modulated by adding an anionic surfactant (Liu and Guo, 2007).

β-Casein is the most hydrophobic casein protein but it also has a relatively high lysine content that predisposes it to conjugation with reducing carbohydrates via the naturally-occurring Maillard reaction. Under controlled conditions, amphiphilic block co-polymers (e.g. casein–polysaccharide conjugates) have the ability to self-assemble into micelles with a hydrophobic core and a hydrophilic shell. The block co-polymers can be used to construct nano-particlulate encapsulant systems. Pan *et al.* (2007a) reported the simultaneous formation and encapsulation of casein–dextran-based nano-particles. The authors took advantage of the hydrophobic interaction between β-carotene and β-casein to form the capsule core. Maillard-derived co-polymers of dextran and β-casein formed the shell, providing stability and dispersibility to the particles over a wide pH range (pH 2–12). Encapsulation protected carotene against oxidation in the presence of ferric chloride, and it could be released on digestion of the protein particle with pepsin or trypsin.

Hydrogels

Casein hydrogels are used to entrap and protect active cores. The colloidal stability of casein hydrogels to acid and heat can be modified by cross-linking within casein micelles using transglutaminase (Huppertz and de Kruif, 2008). This has the potential to widen the application of casein-based hydrogel encapsulation systems. This is because modification of the properties of the encapsulant material provides an additional lever by which the release of the core may be modulated.

Emulsions

Sodium caseinate has been used in the formulation of O/W emulsions to produce a range of spray-dried soy oil powders (up to 75% fat). As was to be expected, higher fat powders had lower microencapsulation efficiency as reflected in the higher levels of solvent extractable fat (Hogan *et al.*, 2001a). Microencapsulation efficiency is improved when proteins are used in combination with carbohydrates to stabilize spray-dried emulsions. Increasing the dextrose equivalent of the carbohydrate further improves the encapsulation efficiencies (Danviriyakui *et al.*, 2002; Hogan *et al.*, 2001b). The molecular assembly of casein affects its encapsulation efficiency. Vega *et al.* (2007) demonstrated that sodium caseinate is superior to micellar casein when spray-dried with sugars for encapsulation of milkfat. This was attributed to the higher mobility and flexibility of casein molecules in sodium caseinate compared to the more aggregated casein micelles. However, when used in combination with lactose as encapsulant matrix, Keogh *et al.* (2001) showed that aggregated caseins provided more protection to fish oil powders during storage due to their low vacuole volume.

For the encapsulation of oils that are very prone to oxidation (e.g. fish oils), Maillard reaction products formed by heating caseins and reducing carbohydrates have been found to be superior encapsulants to corresponding physical blends (Augustin *et al.*, 2006). The molecular weight profile and oxygen permeability of the glycated encapsulant matrix play a key role in the oxidative stability of the encapsulated fish oil (Drusch *et al.*, 2009).

Casein-carbohydrate conjugates can also provide a stable and protective environment for probiotics that are encapsulated by spray drying an O/W emulsion (Crittenden *et al.*, 2006). The probiotic cells *(Bifidobacterium infantis)* resisted digestion in simulated gastric fluid and were released in simulated intestinal fluid.

Casein–dextran Maillard conjugates provide enhanced stability to double emulsions. Fechner *et al.* (2007) showed that diffusion of vitamin B_{12} from the inner aqueous phase of a W/O/W emulsion to the outer phase was significantly decreased by replacing caseinate with a casein–dextran conjugate, as the secondary emulsifier. Caseinate–dextran conjugates were also used as a secondary emulsifier in a W/O/W emulsion containing immunoglobulin Y (IgY) (Shimizu and Nakane, 1995). However, in this application the activity of IgY was compromised, with inactivation of IgY being dependent on its extent of adsorption at the oil–water interface.

Milk proteins, such as casein, can be cross-linked to provide additional strength to the shell of emulsion-based systems. Lee *et al.* (2006) showed that sodium caseinate emulsion gels cross-linked by transglutaminase provided enhanced thermal stability and improved the retention of aroma compounds. The unique textural properties of cross-linked emulsion gels enhance control of the core's release (e.g. aroma compounds) and have the potential to create dairy products with different sensory characteristics.

Coacervates

Coacervates are formed when two oppositely charged biopolymers associate and phase separate (de Kruif *et al.*, 2004). Coacervates of casein and anionic polysaccharides are formed at pH below the pI of casein. Complex coacervates of casein and pectin were employed to microencapsulate probiotics (*Bifidobacterium lactis* and *Lactobacillus acidophilus*) and dried by spouted bed drying. While there was little loss of viability of the probiotics during drying (0.3 log loss), the microencapsulated bacteria were not protected at low pH (Oliveira *et al.*, 2007).

Electrostatic β-casein–lysozyme complexes have been shown to self-assemble in aqueous solution to produce complex micelles. With subsequent heat treatment, the lysozyme gels and thereby entraps β-casein within the gel network (Pan *et al.*, 2007b). This approach provides further opportunities to formulate and design delivery systems. The technology may be useful to encapsulate β-casein that is complexed with hydrophobic molecules, for example.

22.4.3 Whey proteins

Particles

β-Lactoglobulin (β-LG) is the major whey protein of bovine milk and is a natural molecular nano-carrier for hydrophobic molecules. Retinoic acid, cholesterol, vitamin D, various aroma compounds (e.g. aldehydes, ketones), and fatty acids (e.g. conjugated linoleic acid, palmitate, oleate), for example, bind within its hydrophobic calyx (Kontopidis *et al.*, 2004). It has been shown that several hydrophobic bioactives (e.g. cholesterol, vitamin D) improve the protein's resistance to thermal denaturation (Puyol *et al.*, 1994). This may provide a means of enhancing the stability of β-LG during food manufacturing processes and offer greater control over its ability to deliver specific bioactives in food systems. In its native conformation, β-LG is resistant to degradation by gastric proteases (Morr and Ha, 1993) and hence may be expected to remain largely intact following transport through the stomach.

Schmitt *et al.* (2007) have described a process for preparing aggregated whey protein micelles which provide improved delivery and controlled release of the encapsulated core, compared to traditional encapsulation systems. The micelles are prepared by mixing the whey proteins and the core. The whey proteins are then denatured to form whey protein micelles and an aggregate of the whey protein micelles is produced. The hydrophilic parts of the proteins are oriented towards the outer part of the agglomerate and the hydrophobic parts of the proteins are oriented towards the inner 'core' of the micelle. The micellar aggregates have been described as effective stabilizers for both water-soluble and lipophilic compounds, and improve bioavailability of the encapsulated core.

Hydrogels

Like most globular proteins, whey proteins possess excellent gelation properties. The gelation conditions (e.g. pH, temperature, protein concentration, presence of ions) determine the physico-chemical properties of the gel, including its strength, pH sensitivity, permeability and mechanical properties. Thus, manipulation of the gelation conditions provides a means of modulating the release profiles of core agents entrapped within the gel matrices. Release properties of the gels can be further modulated by cross-linking (Lee and Rosenberg, 1999) or coating with an appropriate biopolymer (Gunasekaran *et al.*, 2006, 2007).

The gelation process produces hydrogels either as continuous networks or as discrete gel particles (microspheres). Continuous gel networks are effective delivery systems for encapsulated water-soluble core substances. For example, Gunasekaran *et al.* (2006) entrapped the water-soluble molecule caffeine (as a model compound) within a heat-induced whey protein concentrate (WPC) hydrogel network. The gel matrices displayed pH-sensitive swelling properties which determined the release profile of caffeine. Core release profiles could be readily modified by adding external layers of alginate (Gunasekaran *et al.*, 2006).

Another important property of whey proteins is their ability to gel under cold conditions. This is particularly useful for encapsulation of heat-labile molecules. In this approach, the whey proteins are pre-denatured by heat treatment and then cooled in the presence of a divalent cation. Cold-gelation has conventionally been carried out in the presence of Ca ions (Barbut and Foegeding, 1993). Recently, however, Remondetto *et al.* (2002) showed that ferrous (Fe) ions could also produce cold-induced gelation of β-LG. Gels with different microstructure and Fe release properties were produced depending on the iron concentration used. At low iron concentration (10 mM) 'filamentous' gels were formed, whereas at high iron concentration (30 mM) the gels were 'particulate'. Based on *in vitro* digestion and cell-line (CaCO-2) model studies, 'filamentous' gels displayed maximal Fe bioavailability (Remondetto *et al.*, 2004). Inadequate iron intake is a global nutritional issue. Hence, the ability to deliver iron within whey protein hydrogels is a promising approach to improve Fe intake.

Various methods have been developed to prepare whey protein hydrogels as microspheres. Rosenberg and Lee (2004) developed an 'all aqueous' encapsulation system by which relatively large (1–2 mm diameter) alginate-coated, water-insoluble, whey protein-based microspheres were produced which protected encapsulated actives from oxidative degradation.

Subirade *et al.* (2002) have described an encapsulation process based on emulsification of denatured whey protein with oil in the presence of a bioactive. The emulsion was then contacted with salt solution to induce formation of protein microspheres. Fat-soluble vitamins (e.g. retinol) (Beaulieu *et al.*, 2002) and water-soluble vitamins (e.g. riboflavin) (Chen and Subirade, 2006), have been encapsulated using a combination of emulsification and Ca ion-induced cold-gelation. In all these studies, the microspheres demonstrated controlled or sustained release of the active core. There was delayed release of the active cores under gastric conditions and complete release of the active core in intestinal fluid.

Encapsulation systems containing whey proteins have been used to deliver probiotic bacteria. Reid *et al.* (2005) employed direct extrusion of a denatured whey protein isolate (WPI)–concentrated probiotic (*Lactobacillus rhamnosus*) cell suspension into a Ca ion solution. The microcapsules (3 mm diameter) showed good encapsulation efficiency (96% of the probiotic cells encapsulated) but only 23% cell viability after the gelation process. The microcapsules were subsequently (Reid *et al.*, 2007) incorporated into vegetable juice, where ~33% of the cells were viable after storage (2 wk) compared to ~7% for the non-encapsulated cells. Although the gelation process requires further optimization to improve cell viability, this study demonstrates the feasibility of this approach for delivering actives in food systems. Guérin *et al.* (2003) encapsulated bifidobacteria in whey protein–polysaccharide (alginate–pectin) gel beads coated with and without a membrane, and formed by the transacylation reaction. Their results suggest that entrapment of the bacteria in membrane-coated beads enhances their survival during gastrointestinal transit.

Emulsions

Whey proteins are effective wall matrices for the encapsulation of milkfat (Keogh and O'Kennedy, 1999; Lee and Rosenberg, 2000; Moreau and Rosenberg, 1996; Young *et al.*, 1993a, b) and other sensitive lipids such as soy oil (Hogan *et al.*, 2001a, b), conjugated linoleic acid (Jimenez *et al.*, 2006), flaxseed oil (Partanen *et al.*, 2008) and orange oil (Kim and Morr, 1996). Microencapsulation efficiencies of >90% anhydrous milkfat (AMF) have been reported for spray-dried emulsions. Incorporation of carbohydrates, such as lactose, increased encapsulation efficiency (95%), especially for high oil loads (75%) (Young *et al.*, 1993a). Furthermore, the whey protein matrices protected the sensitive lipids against oxidation during storage (Jimenez *et al.*, 2006; Moreau and Rosenberg, 1996; Partanen *et al.*, 2008).

Spray-dried, water-soluble whey proteins or whey protein–carbohydrate microcapsules also provide excellent retention of volatiles used to enhance food flavour and aroma (Baranauskienė *et al.*, 2006). The encapsulation is achieved by exploiting the strong binding affinity of whey proteins, including bovine serum albumin (BSA), α-lactalbumin (α-LA) and β-LG, to flavour compounds (Kühn *et al.*, 2006).

In the case of bioactives, it is critical that the wall system not only encapsulates and protects the active core during food processing and storage but that it also provides protection to the core during digestion until its release at the target site. Interestingly, a recent study (Sandra *et al.*, 2008) indicated that heat-induced protein cross-linking of β-LG-stabilized O/W emulsions did not delay the rate and extent of lipid digestion during simulated digestion *in vitro.* Pancreatic lipase was still able to access the emulsified oil despite cross-linking of the interfacial protein. In other studies, Lee and Rosenberg (2001) showed that composite wall systems based on whey protein and AMF delayed the release of water-soluble core materials (theophylline was used as a model compound) into simulated gastrointestinal fluids (Lee and Rosenberg, 2001). Lee and Rosenberg (2000) used an alternative approach to show that heat gelation of whey protein-based emulsions also offers potential for controlled release of core materials, due to the limited water-solubility of the microcapsules (Lee and Rosenberg, 2000). Indeed, by implementing this approach, Cho *et al.* (2005) showed that IgG was more stable against degradation by acid and heat, and displayed improved resistance against digestion with pepsin. In another study, calcium alginate coating of whey protein-stabilized O/W emulsions containing paprika oleoresin was employed to improve core stability and alter release properties (Rosenberg and Lee, 2004).

Whey proteins are suitable wall materials for the encapsulation of beneficial bacteria. Picot and Lacroix (2004) encapsulated probiotics by emulsification of milkfat with heat-denatured whey protein prior to spray drying. The microencapsulated probiotics were subsequently incorporated into yoghurt. Viable cell counts (~28%) were significantly higher (+2.6 log cycles) than that of the non-encapsulated probiotics during refrigerated storage (28 d, 4°C), as well as after *in vitro* digestion (+2.7 log cycles) (Picot and Lacroix, 2004).

A strategy that has been shown to be particularly effective for improving the interfacial properties of lipid droplets uses electrostatic complexation. This can be achieved either by formation of an electrostatic complex between two oppositely charged biopolymers prior to emulsification or by a technique described as layer-by-layer electrostatic deposition. In both cases, the core is surrounded by multicomponent interfacial coatings. For example, double W/O/W emulsions formulated with WPI–xanthan complexes were used to entrap vitamin B_1 (thiamine). The WPI–xanthan complex formed a thick, protective barrier against premature release of the thiamine by increasing the emulsion stability under various pH and temperature conditions. The stability was enhanced at low pH due to the strong electrostatic interactions between the two biopolymers at pH < pI (Benichou *et al.*, 2007). Using the layer-by-layer technique, de Lorenzis *et al.* (2008) showed the feasibility of delivering lactoferrin using chitosan–lecithin emulsions. Hong and McClements (2007a) prepared double emulsions comprising chitosan-coated β-LG-stabilized lipid droplets. The stability of the emulsions was increased at pH ~ pI of the protein (pH 4.5–5.5). The authors suggest the formulation may be useful as a delivery system in acidic foods and/or for improving core stability during gastric digestion. Another advantage of polysaccharide coatings is their ability to improve freeze–thaw and freeze-drying stability of protein-coated lipid droplets (Mun *et al.*, 2008).

In additional to traditional O/W/O and W/O/W multiple emulsions, novel microstructures can also be created, potentially increasing the range and applications of bioactive delivery systems. For example, Kim *et al.* (2006) combined a heat-denatured WPI-pectin O/W emulsion with an aqueous, phase-separated, heat-denatured WPI-pectin solution to create O/W/W emulsions, which were subsequently gelled in the presence of Ca ions. More recently, Gu *et al.* (2007) developed an innovative technology to prepare 'colloidosomes' that used electrostatic layer-by-layer deposition. This process involved mixing an O/W emulsion comprising large, anionic droplets of pectin-coated β-LG corn oil droplets with a second O/W emulsion comprising small, cationic β-LG-stabilized oil droplets. At pH 4 pectin adsorbs to β-LG particles. By increasing the concentration of the pectin, the charge of the droplets is changed from cationic to anionic. Gu *et al.* (2007) suggested that, with further optimization, this technology may have important implications for improving emulsion stability, for developing unique controlled or trigger-released delivery systems and for compartmentalizing active components.

Coacervates

Coacervates formed between whey proteins and polyelectrolytes have been utilized to entrap various core materials. Several studies focused on the formation of coacervates between whey proteins and pectin. In acidic conditions (pH 3.5), complexation between heat-denatured WPI and low methoxyl pectin was used as a matrix for thiamine entrapment. Although the entrapment efficiency was low (4.5–7%), it was sufficient to provide the recommended daily intake of the vitamin for an adult male based on just 70–80 mg of the dry complex (Bédié *et al.*, 2008). A similar approach was taken by Santipanichwong *et al.* (2008) to produce core–shell biopolymer nanoparticles from heat-denatured β-LG-pectin complexes. The particles showed good stability over pH 4–6 and in the presence of salt. The complex formed between native β-LG and pectin was an effective encapsulant for docosahexaneoic acid (DHA), as demonstrated by the fact that the concentration of DHA was ~166-fold that in the surrounding medium (Zimet and Livney, 2009). Furthermore, encapsulation of DHA protected it against oxidation during storage. It was suggested that these coacervates may be used to deliver omega-3 oils into low-pH beverages.

Complexes formed between proteins and polyelectrolytes involve weak interactions and are stable only within a narrow pH range. The coacervate shell can be further modified to control core release. Chen and Subirade (2005) refined the stability of native β-LG–chitosan coacervates by cold-induced gelation in the presence of tripolyphosphate. An alternative approach, described by Hong and McClements (2007b), involved the heating of β-LG-chitosan coacervates to form hydrogel particles. De Kruif and Weinbreck (2005) illustrated the application of complex coacervates of β-LG and gum arabic to encapsulate lemon oil used to flavour cheese. In this study, cross-linking of the coacervates with gluteraldehyde was used to modulate the release profile of the core (de Kruif and Weinbreck, 2005).

The electrostatic interaction between proteins and polyelectrolytes is not limited to polysaccharides. Pouzot *et al.* (2008) recently patented a process to produce charged, liposome-like structures containing a denatured supramolecular protein core, which can entrap various core materials. Additionally, the lipid-containing shell surrounding the protein core is claimed to improve the solubility of proteins at pH ~ pI, to provide a protective barrier against adverse environmental conditions (e.g. oxygen, humidity, agglomeration during drying) and to reduce protein astringency.

22.4.4 Milk protein hydrozylates

Controlled proteolysis of α-LA results in the formation of self-assembled α-LA nano-tubes (~110 × 20 nm) via association of the protein hydrozylates in the presence of Ca ions (Graveland-Bikker and de Kruif, 2006; Ipsen and Otte, 2007). The inner cavity of the tubes (8 nm diameter) is a putative site for encapsulation of molecules such as vitamins or enzymes (Graveland-Bikker and de Kruif, 2006; Ipsen and Otte, 2007). Furthermore, the nano-tubes are heat-stable (72°C, 40 s) and able to withstand the freeze-drying process. The stability of the nano-tubes may be further enhanced by cross-linking (e.g. using transglutaminase). The nano-tubes can also be fitted with lipid caps that open and close to allow controlled release of the encapsulated molecule (Graveland-Bikker and de Kruif, 2006). Further research is required to establish their role as encapsulants and delivery vehicles.

Milk protein hydrozylates can also be used as encapsulants in emulsion-based systems. Numerous studies have shown that proteolysis of the milk proteins modifies their functionality, including their surface-active and gelation properties. The effect of proteolysis on the physical functionality of proteins depends on the type of milk protein and the hydrolysis conditions (Chobert, 2003; Foegeding *et al.*, 2002; Kilara and Panyam, 2003). For example, Chobert *et al.* (1987, 1988) showed that limited hydrolysis of casein (2–10%) could increase its emulsifying activity. Euston *et al.* (2001) showed hydrolysis (10–27%) improved the emulsifying capacity of WPC. Treatment of whey proteins with a protease has also been shown to induce gel formation more rapidly during heat treatment compared to the intact proteins (Chen *et al.*, 1994). Gel properties can also be manipulated by modulating hydrolysis conditions (Otte *et al.*, 1996). These properties allow milk protein hydrozylates to function as effective encapsulants and may lead to the development of unique delivery systems.

22.4.5 Milkfat

The major component of milkfat is triglycerides (98%). The composition of milkfat is highly complex with over 400 fatty acids and 200 triglycerides identified (Gresti *et al.*, 1993). Milkfat melts over the temperature range ~−40°C to 40°C. The melting temperature of milkfat is a function of the crystalline state (Walstra and Jenness, 1984) and the composition of the triglycerides. The fractionation process provides milkfat with various triglyceride compositions and is performed to differentiate the physical and functional properties of milkfat, such as crystallization behaviour and melting point (Kaylegian and Lindsay, 1995). In the area of encapsulation, milkfat fractions with a defined melting point can be exploited; for example, to facilitate finer control over the release of core materials that have been embedded in a fat matrix.

Milkfat is an obvious carrier for lipophilic cores. Braun and Olson (1986) encapsulated proteins and peptides in milkfat and investigated their functionality. Butter oil capsules demonstrated good freeze–thaw dispersibility. However, when the capsules were held at high incubation temperatures (>32°C), such as those that may be encountered during cheesemaking, the microcapsules showed poor temperature stability.

Active cores may also be embedded in a solid fat matrix. A high-melting point fraction of milkfat was used to encapsulate (39% encapsulation efficiency) protease enzymes (Kailasapathy and Lam, 2005), which were subsequently incorporated (73.5%) into cheese curd. Encapsulation of the proteases accelerated cheese ripening compared to the control cheese. Jackson and Lee (1991) prepared lipid-coated microcapsules of Fe to fortify cheese. However, neither a high-melting point fraction of milkfat nor hydrogenated milkfat was a suitable coating for this application.

Milkfat can also be used in emulsion-based systems. Encapsulation of α-tocopherol in a stearin-rich milkfat fraction was prepared as a caseinate O/W emulsion to protect the antioxidant against degradation during storage (Relkin *et al.*, 2008). Al-Nabulsi *et al.* (2006) microencapsulated lactoferrin in a polyglycerol condensed ricinoleate (PGPR)-stabilized butterfat-corn oil W/O emulsion. The fat encapsulant improved the antimicrobial activity of lactoferrin in cured meat by protecting it against adverse environmental conditions.

Magee and Olson (1981) developed a method to produce microencapsulated enzyme-substrate flavouring systems as a W/O/W emulsion using milkfat. Encapsulation protected the enzyme-substrate concoction from pH and ionic environments (Magee *et al.*, 1981). The concentration of the enzyme and its substrate, by encapsulation, increased the concentrations of the flavour components and reduced the loss of enzymes during cheese manufacture (Magee *et al.*, 1981).

22.4.6 Milkfat globule membrane

The MFGM is a complex mixture of glycolipids, phospholipids and proteins (Singh, 2006). MFGM phospholipids constitute 26–31% of its total lipid content, and together with proteins constitute more than 90% of the membrane dry weight (Singh, 2006). It has a high nutritional value (Dewettinck *et al.*, 2008; Spitsberg, 2005) and acts as a natural barrier against lipolysis of milkfat in milk. MFGM material is a natural emulsifier and behaves differently from milk proteins (caseins, β-LG) when used for stabilizing emulsions. Corredig and Dalgleish (1998) revealed that MFGM-stabilized O/W emulsions are less susceptible to interfacial displacement by low molecular weight surfactants than milk protein-stabilized emulsions, and are not affected by the presence of other milk proteins (caseins, whey). These properties have been attributed to a strong interfacial interaction between MFGM and the oil. The phospholipids may also contribute by lowering the interfacial tension. However, the source of the raw MFGM, the isolation procedure, and the conditions during emulsion formation are critical, in order to maximize MFGM functionality (Singh, 2006).

Liposomes prepared from MFGM phospholipids possess several advantages over their soy phospholipid counterparts, including a thicker membrane, lower membrane permeability and higher phase transition temperature (Thompson *et al.*, 2006a). These properties afford the MFGM-derived phospholipid liposomes with improved environmental stability (e.g. heat, pH, presence of divalent cations) compared to liposomes prepared from soy-derived phospholipids (Thompson *et al.*, 2006b). Furthermore, Thompson and Singh (2006) showed that MFGM-derived phospholipid liposomes can be produced at high throughput in a continuous and reproducible manner using microfluidization. The liposomes were relatively stable but tended to be highly polydisperse, with various lamellarities and structures. Liposomes may function as a putative delivery system for various hydrophobic and hydrophilic molecules (e.g. Se). Potential applications of liposomes in the food industry include protecting sensitive molecules, masking undesirable flavours, and improving the efficacy of food bioactives. Furthermore, *in vivo* studies show MFGM enhances intestinal drug absorption (Liu *et al.*, 1995), indicating opportunities for the use of MFGM for bioactive delivery.

22.5 Future trends

Dairy components play a valuable role as encapsulant materials, due to their many unique and versatile physico-chemical properties. Micro- and nano-encapsulation systems formulated with dairy components can be designed in various delivery formats (e.g. colloidosomes, liposomes, emulsion-based systems, hydrogels, micro- and nano-particles) and produced using a range of technologies and processing conditions. As encapsulant materials, dairy components can protect a wide range of both hydrophilic and hydrophobic core materials and enable their controlled release in response to an appropriate trigger.

It is important to remember that encapsulation deals with multiple components. Consideration of the core(s), and effect of the encapsulant(s), formulation, process design and process conditions are all crucial to achieve the optimum result for a defined purpose. In this respect it is anticipated that a greater focus will be given to enhancing our understanding of the effect of core materials on encapsulation processes and encapsulant properties. Knowledge that provides fundamental information on the nature of the interaction between the dairy ingredients and the core materials in order to anticipate core release properties will be required to formulate and design delivery systems with a greater range of core materials, and core release profiles. Future challenges will include a focus on delivering a 'cocktail' of cores, either within a given delivery system or by combining dairy ingredient delivery systems that have different functions.

It is anticipated that the trend to exploit the properties of milk ingredients to develop encapsulation formulations and delivery systems will continue, particularly those designed for food systems. Exploitation of the unique properties of milk protein hydrozylates, of combining dairy ingredients (e.g. caseins and whey) in new ways and/or by modifying their properties to further improve their encapsulation properties, via traditional manufacturing and/or emerging technologies or by using 'smart' chemistry, can be expected.

A challenging but poorly investigated area remains the application of dairy ingredients as encapsulants in real food systems. The time has come to focus on the impact of encapsulation in foods. This is of paramount importance for encapsulated bioactives. The bioavailability and bioactivity of encapsulated bioactives and their targeted delivery in food systems to promote health and assist in treatment of diseases requires extensive research.

22.6 Sources of further information and advice

There are excellent reviews on microencapsulation and its application in the food industry, which provide additional information on the subject (Augustin and Hemar, 2009; Gibbs *et al.*, 1999; Gouin, 2004; Risch and Reineccius, 1995). An overview of encapsulation techniques (Gouin, 2004) including spray drying (Gharsallaoui *et al.*, 2007), micro-emulsions (Flanagan and Singh, 2006) and coacervates (Turgeon *et al.*, 2003) are recommended.

Further information on bioactives and their encapsulation and delivery in foods and nutraceuticals is available (Augustin and Sanguansri, 2008; Champagne and Fustier, 2007; Garti, 2008; Shi, 2006; Wildman, 2007).

Fox (1997) is recommended for detailed information on the physico-chemical properties of dairy ingredients, including the milk proteins, milkfat and lactose. Kinsella (1984) provides a comprehensive overview of the functional properties of milk proteins in particular.

Milk protein-based delivery systems (including hydrogels, microspheres and emulsions) are reviewed by Chen *et al.* (2006) and Vega and Roos (2006).

22.7 References

AL-NABULSI AA, HAN JH, LIU Z, RODRIGUES-VIEIRA ET, HOLLEY RA (2006), 'Temperature-sensitive microcapsules containing lactoferrin and their action against *Carnobacterium viridans* on bologna', *J Food Sci*, 71 (6), M208–M214. DOI: 10.1111/j.1750-3841.2006.00103.x

AUGUSTIN MA, HEMAR Y (2009), 'Nano- and micro-structured assemblies for encapsulation of food ingredients', *Royal Soc Chem*, 38, 902–912.

AUGUSTIN MA, SANGUANSRI L (2008), 'Encapsulation of bioactives', in Aguilera JM and Lillford PJ, *Food Materials Science – Principles and Practice*, New York, Springer, 577–601.

AUGUSTIN MA, SANGUANSRI L, BODE O (2006), 'Maillard reaction products as encapsulants for fish oil powders', *J Food Sci*, 71 (2), E25–E32. DOI: 10.1111/j.1365-2621.2006.tb08893.x

AUGUSTIN MA, SANGUANSRI L, MARGETTS C, YOUNG B (2001), 'Microencapsulation of food ingredients', *Food Australia,* 53 (6), 220–223.

AUGUSTIN MA, UDABAGE P (2007), 'Influence of processing on functionality and milk and dairy proteins', *Advances Food Sci Nutr*, 53, 1–38. DOI: 10.1016/S1043-4526(07)53001-9

BARANAUSKIENĖ R, VENSKUTONIS PR, DEWETTINCK D, VERHÉ R (2006), 'Properties of oregano (*Origanum vulgare* L.), citronella (*Cymbopogon nardus* G.) and majoram (*Majorana hortensis* L.) flavors encapsulated into milk protein-based matrices', *Food Res Int*, 39 (4), 413–425. DOI: 10.1016/j.foodres.2005.09.005

BARBUT S, FOEGEDING EA (1993), 'Ca^{2+}-induced gelation of pre-heated whey protein isolate', *J Food Sci*, 58 (4), 867–871.

BEAULIEU L, SAVOIE L, PAQUIN P, SUBIRADE M (2002), 'Elaboration and characterization of whey protein beads by an emulsification/cold gelation process: Application for the protection of retinol', *Biomacromolecules*, 3 (2), 239–248. DOI: 10.1021/bm010082z

BÉDIÉ GK, TURGEON SL, MAKHLOUF J (2008), 'Formation of native whey protein isolate – low methoxyl pectin complexes as a matrix for hydro-soluble food ingredient entrapment in acidic foods', *Food Hydrocolloids*, 22 (5), 836–844. DOI: 10.1016/j.foodhyd.2007.03.010

BENICHOU A, ASERIN A, GARTI N (2007), 'W/O/W double emulsions stabilized with WPI–polysaccharide complexes', *Colloids Surf A Physicochem Eng Asp*, 294 (1–3), 20–32. DOI: 10.1016/j.colsurfa.2006.07.056

BRAUN SD, OLSON NF (1986), 'Encapsulation of proteins and peptides in milkfat: Encapsulation efficiency and temperature and freezing stabilities', *J Microencapsul*, 3 (2), 115–126.

CHAMPAGNE CP, FUSTIER P (2007), 'Microencapsulation for the improved delivery of bioactive compounds into foods', *Curr Opin Biotechnol*, 18 (2), 184–190. DOI: 10.1016/j.copbio.2007.03.001

CHEN L, SUBIRADE M (2005), 'Chitosan/β-lactoglobulin core–shell nanoparticles as nutraceutical carriers', *Biomaterials*, 26 (30), 6041–6053. DOI: 10.1016/j.biomaterials.2005.03.011

CHEN L, SUBIRADE M (2006), 'Alginate – whey protein granular microspheres as oral delivery vehicles for bioactive compounds', *Biomaterials*, 27 (26), 4646–4654. DOI: 10.1016/j.biomaterials.2006.04.037

CHEN SX, SWAISGOOD HE, FOEGEDING EA (1994), 'Gelation of β-lactoglobulin treated with limited proteolysis by immobilized trypsin', *J Agric Food Chem*, 42 (2), 234–239. DOI: 10.1021/jf00038a002

CHEN L, REMONDETTO GE, SUBIRADE M (2006), 'Food protein-based materials as nutraceutical delivery systems', *Trends Food Sci Technol*, 17 (5), 272–283. DOI: 10.1016/j.tifs.2005.12.011

CHO Y-H, LEE J-J, PARK I-B, HUH C-S, BAEK Y-J, PARK J (2005), 'Protective effect of microencapsulation consisting of multiple emulsification and heat gelation processes on immunoglobulin in yolk', *J Food Sci*, 70 (2), E148–E151.

CHOBERT J-M (2003), 'Milk protein modification to improve functional and biological properties', *Adv Food Nutr Res*, 47, 1–71. DOI: 10.1016/S1043-4526(03)47001-0

CHOBERT J-M, SITOHY MZ, WHITAKER JR (1987), 'Specific limited hydrolysis and phosphorylation of food proteins for improvement of functional and nutritional properties', *J Am Oil Chem Soc*, 64 (12), 1704–1711. DOI: 10.1007/BF02542507

CHOBERT J-M, SITOHY MZ, WHITAKER JR (1988), 'Solubility and emulsifying properties of caseins modified enzymatically by *Staphylococcus aureus* V8 protease', *J Agric Food Chem*, 36 (1), 220–224. DOI: 10.1021/jf00079a055

CORREDIG M, DALGLEISH DG (1998), 'Characterization of the interface of an oil-in-water emulsion stabilized by milk fat globule membrane material', *J Dairy Res*, 65 (3), 465–477. DOI: 10.1017/S0022029998002982

CRITTENDEN R, WEERAKKODY R, SANGUANSRI L, AUGUSTIN MA (2006), 'Synbiotic microcapsules that enhance microbial viability during nonrefrigerated storage and gastrointestinal tract', *Appl Environ Microbiol*, 72 (3), 2280–2282. DOI: 10.1128/AEM.72.3.2280-2282.2006

DANVIRIYAKUI S, MCCLEMENTS DJ, DECKER E, NAWAR WW, CHINACHOTI P (2002), 'Physical stability of spray-dried milk fat emulsion as affected by emulsifiers and processing conditions', *J Food Sci*, 67 (6), 2183–2189. DOI: 10.1111/j.1365-2621.2002.tb09524.x

DE KRUIF CG, WEINBRECK F (2005), 'Complex coacervates of whey proteins and anionic polysaccharides, and their use for encapsulation', *Food New Zealand*, 5 (3), 23–30.

DE KRUIF CG, WEINBRECK F, DE VRIES R (2004), 'Complex coacervation of proteins and anionic polysaccharides', *Curr Opin Colloid Interface Sci*, 9 (5), 340–349. DOI: 10.1016/j.cocis.2004.09.006

DE LORENZIS E, SEMERARO C, DE BLASI MD, MITA G, POLTRONIERI P (2008), 'Emulsions based on the interactions between lactoferrin and chitosans', *Food Biophysics*, 3 (2), 169–173. DOI: 10.1007/s11483-008-9063-x

DEWETTINCK K, ROMBAUT R, THIENPONT N, LE TT, MESSENS K, CAMP JV (2008), 'Nutritional and technological aspects of milk fat globule membrane material', *Int Dairy J*, 18 (5), 436–457. DOI: 10.1016/j.idairyj.2007.10.014

DRUSCH S, BERG S, SCAMPICCHIO M, SERFERT Y, SOMOZA V, MANNINO S, SCHWARZ K (2009), 'Role of glycated caseinate in stabilisation of microencapsulated lipophilic functional ingredients', *Food Hydrocolloids*, 23, 942–948. DOI:10.1016/j.foodhyd.2008.07.004

EUSTON SR, FINNIGAN SR, HIRST RL (2001), 'Heat-induced destabilization of oil-in-water emulsions formed from hydrolyzed whey protein', *J Agric Food Chem*, 49 (11), 5576–5583. DOI: 10.1021/jf0102620

FÄLDT P, BERGENSTÅHL B (1996), 'Spray-dried whey protein/lactose/soybean oil emulsions. 1. Surface composition and particle structure', *Food Hydrocolloids*, 10 (4), 421–429.

FECHNER A, KNOTH A, SCHERZE I, MUSCHIOLIK G (2007), 'Stability and release properties of double-emulsions stabilised by caseinate – dextran conjugates', *Food Hydrocolloids*, 21 (5–6), 943–952. DOI: 10.1016/j.foodhyd.2006.10.021

FLANAGAN J, SINGH H (2006), 'Microemulsions: A potential delivery system for bioactives in foods', *Crit Rev Food Sci Nutr*, 46 (3), 221–237. DOI: 10.1080/10408690590956710

FOEGEDING EA, DAVIS JP, DOUCET D, MCGUFFEY MK (2002), 'Advances in modifying and understanding whey protein functionality', *Trends Food Sci Technol*, 13 (5), 151–159.

FOX PF (1997), *Advanced Dairy Chemistry. Volumes 1–3*, New York, Elsevier.

GARTI N (2008), *Delivery and Controlled Release of Bioactives in Foods and Nutraceuticals*, New York, CRC Press.

GHARSALLAOUI A, ROUDAUT G, CHAMBIN O, VOILLEY A, SAUREL R (2007), 'Applications of spray-drying in microencapsulation of food ingredients: An overview', *Food Res Int*, 40 (9), 1107–1121. DOI: 10.1016/j.foodres.2007.07.004

GIBBS BF, KERMASHA S, ALLI I, MULLIGAN CN (1999), 'Encapsulation in the food industry: A review', *Int J Food Sci Nutr*, 50 (3), 213–224. DOI: 10.1080/096374899101256

GOUIN S (2004), 'Microencapsulation: Industrial appraisal of existing technologies and trends', *Trends Food Sci Technol*, 15 (7–8), 330–347. DOI:10.1016/j.tifs.2003.10.005

GRAVELAND-BIKKER JF, DE KRUIF CG (2006), 'Unique milk protein based nanotubes: Food and nanotechnology meet', *Trends Food Sci Technol*, 17 (5), 196–203. DOI: 10.1016/j.tifs.2005.12.009

GRESTI J, BUGAUT M, MANIONGUI C, BEZARD J (1993), 'Composition of molecular species of triacylglycerols in bovine milk fat', *J Dairy Sci*, 76 (7), 1850–1869.

GU Y-S, DECKER EA, MCCLEMENTS DJ (2007), 'Formation of colloidosomes by adsorption of small charged oil droplets onto the surface of large oppositely charged oil droplets', *Food Hydrocolloids*, 21 (4), 516–526. DOI: 10.1016/j.foodhyd.2006.05.011

GUÉRIN D, VUILLEMARD J-C, SUBIRADE M (2003), 'Protection of bifidobacteria encapsulated in polysaccharide–protein gel beads against gastric juice and bile', *J Food Prot*, 66 (11), 2076–2084.

GUNASEKARAN S, XIAO L, ELEYA MMO (2006), 'Whey protein concentrate hydrogels as bioactive carriers', *J Appl Polym Sci*, 99 (5), 2470–2476. DOI: 10.1002/app.22838

GUNASEKARAN S, KO S, XIAO L (2007), 'Use of whey proteins for encapsulation and controlled delivery applications', *J Food Eng*, 83 (1), 31–40. DOI:10.1016/j.jfoodeng.2006.11.001

HOGAN SA, MCNAMEE BF, O'RIORDAN ED, O'SULLIVAN M (2001a), 'Microencapsulating properties of sodium caseinate', *J Agric Food Chem*, 49 (4), 1934–1938. DOI: 10.1021/jf000276q S0021-8561(00)00276-4

HOGAN SA, MCNAMEE BF, O'RIORDAN ED, O'SULLIVAN M (2001b), 'Emulsification and microencapsulation properties of sodium caseinate/carbohydrate blends', *Int Dairy J*, 11 (3), 137–144. DOI: S0958-6946(01)00091-7

HOLT C, DE KRUIF CG, TUINIER R, TIMMINS PA (2003), 'Substructure of bovine casein micelles by small-angle X-ray and neutron scattering', *Colloids Surf A Physicochem Eng Aspects*, 213 (2–3), 275–284.

HONG Y-H, MCCLEMENTS DJ (2007a), 'Modulation of pH sensitivity of surface charge and aggregation stability of protein-coated lipid droplets by chitosan addition', *Food Biophysics*, 2 (1), 46–55. DOI: 10.1007/s11483-007-9028-5

HONG Y-H, MCCLEMENTS DJ (2007b), 'Formation of hydrogel particles by thermal treatment of β-lactoglobulin – chitosan complexes', *J Agric Food Chem*, 55 (14), 5653–5660. DOI: 10.1021/jf070564n

HUPPERTZ T, DE KRUIF CG (2008), 'Structure and stability of nanogel particles prepared by internal cross-linking of casein micelles', *Int Dairy J*, 18 (5), 556–565. DOI:10.1016/j.dairyj.2007.10.009

IPSEN R, OTTE J (2007), 'Self-assembly of partially hydrolysed α-lactalbumin', *Biotechnol Adv*, 25 (6), 602–605. DOI: 10.1016/j.biotechadv.2007.07.006

JACKSON LS, LEE K (1991), 'Microencapsulated iron for food fortification', *J Food Sci*, 56 (4), 1047–1050.

JIMENEZ M, GARCIA HS, BERISTAIN CI (2006), 'Spray-dried encapsulation of conjugated linoleic acid (CLA) with polymeric matrices', *J Sci Food Agric*, 86 (14), 2431–2437. DOI: 10.1002/jsfa.2636

KAILASAPATHY K, LAM SH (2005), 'Application of encapsulated enzymes to accelerate cheese ripening', *Int Dairy J*, 15 (6–9), 929–939. DOI: 10.1016/j.idairyj.2004.11.006

KAMYSHNY A, MAGDASSI S (2006), 'Microencapsulation', in Somasundaran P, *Encyclopedia of Surface and Colloid Science*, Boca Raton, Taylor & Francis, CRC Press, 3957–3969. DOI: 10.1081/E-ESCS-120023308

KAYLEGIAN KE, LINDSAY RC (1995), *Handbook of Milkfat Fractionation Technology and Applications*, Champaign, Illinois, AOCS Press. DOI: 10.1002/food.19960400316

KEOGH MK, O'KENNEDY BT (1999), 'Milk fat microencapsulation using whey proteins', *Int Dairy J*, 9 (9), 657–663.

KEOGH MK, O'KENNEDY BT, KELLY J, AUTY MA, KELLY PM, FUREBY A, HAAHR A-M (2001), 'Stability to oxidation of spray-dried fish oil powder microencapsulated using milk ingredients', *J Food Sci*, 66 (2), 217–224. DOI: 10.1111/j.1365-2621.2001.tb11320.x

KILARA A, PANYAM D (2003), 'Peptides from milk proteins and their properties', *Crit Rev Food Sci Nutr*, 43 (6), 607–633. DOI: 10.1080/10408690390251138

KIM YD, MORR CV (1996), 'Microencapsulation properties of gum Arabic and several food proteins: Spray-dried orange oil emulsion particles', *J Agric Food Chem*, 44 (5), 1314–1320. DOI: 10.1021/jf9503927

KIM H-J, DECKER EA, MCCLEMENTS DJ (2006), 'Preparation of multiple emulsions based on thermodynamic incompatibility of heat-denatured whey protein and pectin solutions', *Food Hydrocolloids*, 20 (5), 586–595. DOI:10.1016/j.foodhyd.2005.06.007

KINSELLA JE (1984), 'Milk proteins: Physicochemical and functional properties', *Crit Rev Food Sci Nutr*, 21, 197–262.

KONTOPIDIS G, HOLT C, SAWYER L (2004), 'Invited review: Beta-lactoglobulin: Binding properties, structure and function', *J Dairy Sci*, 87 (4), 785–796.

KÜHN J, CONSIDINE T, SINGH H (2006), 'Interactions of milk proteins and volatile flavor compounds: Implications in the development of protein foods', *J Food Sci*, 71 (5), R72–R82. DOI:10.1111/j.1750-3481.2006.00051.x

LEE SJ, ROSENBERG M (1999), 'Preparation and properties of glutaraldehyde cross-linked whey protein-based microcapsules containing theophylline', *J Control Release*, 61 (1–2), 123–136.

LEE SJ, ROSENBERG M (2000), 'Whey protein-based microcapsules prepared by double emulsification and heat gelation', *Lebensm -Wiss u- Technol*, 33 (2), 80–88. DOI: 10.1006/fstl.1999.0619

LEE SJ, ROSENBERG M (2001), 'Microencapsulation of theophylline in composite wall system consisting of whey proteins and lipids', *J Microencapsul*, 18 (3), 309–321. DOI: 10.1080/02652040010019523

LEE HA, CHOI SJ, MOON TW (2006), 'Characteristics of sodium caseinate- and soy protein isolate-stabilized emulsion-gels formed by microbial transglutaminase', *J Food Sci*, 71 (6), C352–C357. DOI:10.1111/j.1750-3841.2006.00110.x

LIU Y, GUO R (2007), 'Interaction between casein and sodium dodecyl sulfate', *J Colloid Interface Sci*, 315 (2), 685–692. DOI: 10.1016/j.jcis.2007.07.018

LIU H-X, ADACHI I, HORIKOSHI I, UENO M (1995), 'Mechanism of promotion of lymphatic drug absorption by milk fat globule membrane', *Int J Pharmaceutics*, 118 (1), 55–64. DOI: 10.1016/0378-5173(94)00336-4

LIVNEY YD, DALGLEISH DG (2007), *Casein micelles for nanoencapsulation of hydrophobic compounds*, Patent WO2007122613A1.

MADENE A, JACQUOT M, SCHER J, DESOBRY S (2006), 'Flavour encapsulation and controlled release – a review' *Int J Food Sci Technol*, 41 (s1), 1–21. DOI: 10.1111/j.1365-2621.2005.00980.x

MAGEE JR EL, OLSON NF (1981), 'Microencapsulation of cheese ripening systems: Formation of microcapsules', *J Dairy Sci*, 64 (4), 600–610.

MAGEE JR EL, OLSON NF, LINDSAY RC (1981), 'Microencapsulation of cheese ripening system: Production of diacetyl and acetoin in cheese by encapsulated bacterial cell-free extract', *J Dairy Sci*, 64 (4), 616–621.

MCCLEMENTS DJ, DECKER EA, WEISS J (2007), 'Emulsion-based delivery systems for lipophilic bioactive components', *J Food Sci*, 72 (8), R109–R124. DOI: 10.1111/j.1750-3841.2007.00507.x

MOREAU DL, ROSENBERG M (1996), 'Oxidative stability of anhydrous milkfat encapsulated in whey proteins', *J Food Sci*, 61 (1), 39–43.

MORR CV, HA EYW (1993), 'Whey protein concentrates and isolates: Processing and functional properties', *Food Sci Nutr*, 33 (6), 431–476.

MUN S, CHO Y, DECKER EA, MCCLEMENTS DJ (2008), 'Utilization of polysaccharide coatings to improve freeze-thaw and freeze-dry stability of protein-coated lipid droplets', *J Food Eng*, 86 (4), 508–518. DOI: 10.1016/j.foodeng.2007.11.002

OLIVEIRA AC, MORETTI TS, BOSCHINI C, BALIERO JCC, FREITAS LAP, FREITAS O, FAVARO-TRINDADE CS (2007), 'Microencapsulation of *B. lactis* (BI 01) and *L. acidophilus* (LAC 4) by complex coacervation followed by spouted-bed drying', *Drying Technol*, 25 (10–12), 1687–1693. DOI:10.1080/07373930701.590939

OTTE J, JU ZY, FAEGERMAND M, LOMHOLT SB, QVIST KB (1996), 'Protease-induced aggregation and gelation of whey proteins', *J Food Sci*, 61 (5), 911–915, 923.

PAN X, YAO P, JIANG M (2007a), 'Simultaneous nanoparticle formation and encapsulation driven by hydrophobic interaction of casein-*graft*-dextran and β-carotene', *J Colloid Interface Sci*, 315 (2), 456–463. DOI:10.1016/j.jcis.2007.07.015

PAN X, YU S, YAO P, SHAO Z (2007b), 'Self-assembly of β-casein and lysozyme', *J Colloid Interface Sci*, 316 (2), 405–412. DOI:10.1016/j.jcis.2007.09.007

PARTANEN R, RAULA J, SEPPĀNEN R, BUCHERT J, KAUPPINEN E, FORSSELL P (2008), 'Effect of relative humidity on oxidation of flaxseed oil in spray dried whey protein emulsions', *J Agric Food Chem*, 56 (14), 5717–5722. DOI: 10.1021/jf8005849

PICOT A, LACROIX C (2004), 'Encapsulation of bifidobacteria in whey protein-based microcapsules and survival in simulated gastrointestinal conditions and in yoghurt', *Int Dairy J*, 14 (6), 505–515. DOI: 10.1016/j.idairyj.2003.10.008

POUZOT M, SCHMITT CJE, MEZZENGA R (2008), *Food protein and charged emulsifier interaction*, Patent WO2008025784.

PUYOL P, PEREZ MD, PEIRO JM, CALVO M (1994), 'Effect of binding of retinol and palmitic acid to bovine β-lactoglobulin on its resistance to thermal denaturation', *J Dairy Sci*, 77 (6), 1494–1502.

REID AA, VUILLEMARD JC, BRITTEN M, ARCAND Y, FARNWORTH E, CHAMPAGNE CP (2005), 'Microentrapment of probiotic bacteria in a Ca^{2+}-induced whey protein gel and effects on their viability in a dynamic gastro-intestinal model', *J Microencapsul*, 22 (6), 603–619. DOI: 10.1080/02652040500162840

REID AA, CHAMPAGNE CP, GARDNER N, FUSTIER P, VUILLEMARD JC (2007), 'Survival in food systems of *Lactobacillus rhamnosus* R011 microentrapped in whey protein gel particles', *J Food Sci*, 72 (1), M31–M37. DOI: 10.1111/j.1750-3841.2006.00222.x

RELKIN P, YUNG J-M, KALNIN D, OLLIVON M (2008), 'Structural behaviour of lipid droplets in protein-stabilized nano-emulsions and stability of α-tocopherol', *Food Biophysics*, 3 (2), 163–168. DOI: 10.1007/s11483-008-9064-9

REMONDETTO GE, PAQUIN P, SUBIRADE M (2002), 'Cold gelation of β-lactoglobulin in the presence of iron', *J Food Sci*, 67 (2), 586–594. DOI: 10.1111/j.1365-2621.2002.tb10643.x

REMONDETTO GE, BEYSSAC E, SUBIRADE M (2004), 'Iron availability from whey protein hydrogels: An *in vitro* study', *J Agric Food Chem*, 52 (26), 8137–8143. DOI: 10.1021/jf040286h

RISCH SJ, REINECCIUS GA (1995), *Encapsulation and Controlled Release of Food Ingredients*, Washington DC, ACS Symposium Series 590, American Chemical Society.

ROOS YH (2008), 'The glassy state', in Aguilera JM and Lillford PJ, *Food Materials Science – Principles and Practice*, New York, Springer, 7–81.

ROSENBERG M, LEE S-J (2004), 'Water-insoluble, whey protein-based microspheres prepared by an all-aqueous process', *J Food Sci*, 69 (1), FEP50–FEP58. DOI: 10.1111/j.1365-2621.2004.tb17867.x

SAHU A, KASOJU N, BORA U (2008), 'Fluorescence study of the curcumin–casein micelle complexation and its application as a drug nanocarrier to cancer cells', *Biomacromolecules*, 9 (10), 2905–2912. DOI: 10.1021/bm800683f

SANDRA S, DECKER EA, MCCLEMENTS DJ (2008), 'Effect of interfacial protein cross-linking on the *in vitro* digestibility of emulsified corn oil by pancreatic lipase', *J Agric Food Chem*, 56 (16), 7488–7494. DOI: 10.1021/jf800741w

SANTIPANICHWONG R, SUPHANTHARIKA M, WEISS J, MCCLEMENTS DJ (2008), 'Core–shell biopolymer nanoparticles produced by electrostatic deposition of beet pectin onto heat-denatured β-lactoglobulin aggregates', *J Food Sci*, 73 (6), N23–N29. DOI: 10.1111/j.1750-3841.2008.00804.x

SCHMITT CJE, BOVETTO LJR, BOVETTO L (2007), *Whey protein vehicle for active agent delivery*, Patent WO2007110422A2.

SEMO E, KESSELMAN E, DANINO D, LIVNEY YD (2007), 'Casein micelle as a natural nano-capsular vehicle for nutraceuticals', *Food Hydrocolloids*, 21 (5–6), 936–942. DOI:10.1016/j.foodhyd.2006.09.006

SHI J (2006), *Functional Food Ingredients and Nutraceuticals: Processing Technologies*, Boca Raton, Taylor & Francis, CRC Press.

SHIMIZU M, NAKANE Y (1995), 'Encapsulation of biologically active proteins in a multiple emulsion', *Biosci Biotechnol Biochem*, 59 (3), 492–496.

SINGH H (2006), 'The milk fat globule membrane – a biophysical system for food applications', *Curr Opin Colloid Interface Sci*, 11 (2–3), 154–163. DOI: 10.1016/j.cocis.2005.11.002

SPITSBERG VL (2005), 'Bovine milk fat globule membrane as a potential nutraceutical', *J Dairy Sci*, 88 (7), 2289–2294.

SUBIRADE M, BEAULIEU L, PAQUIN P (2002), *Process for making protein delivery matrix and uses thereof*, Patent WO200280881A2.

THOMPSON AK, SINGH H (2006), 'Preparation of liposomes from milk fat globule membrane phospholipids using a microfluidizer', *J Dairy Sci*, 89 (2), 410–419.

THOMPSON AK, HINDMARSH JP, HAISMAN D, RADES T, SINGH H (2006a), 'Comparison of the structure and properties of liposomes prepared from milk fat globule membrane and soy phospholipids', *J Agric Food Chem*, 54 (10), 3704–3711. DOI: 10.1021/jf052859b

THOMPSON AK, HAISMAN D, SINGH H (2006b), 'Physical stability of liposomes prepared from milk fat globule membrane and soy phospholipids', *J Agric Food Chem*, 54 (17), 6390–6397. DOI: 10.1021.jf0605695

TURGEON SL, BEAULIEY M, SCHMITT C, SANCHEZ C (2003), 'Protein – polysaccharide interactions: Phase-ordering kinetics, thermodynamic and structural aspects',

Curr Opin Colloid Interface Sci, 8 (4–5), 401–414. DOI: 10.1016/j.cocis.2007.07.007

UBBINK J, KRÜGER J (2006), 'Physical approaches for the delivery of active ingredients in foods', *Trends Food Sci Technol*, 17 (5), 244–254. DOI:10.1016/j.tifs.2006.01.007

VEGA C, ROOS YH (2006), 'Invited review: Spray-dried dairy and dairy-like emulsions – compositional considerations', *J Dairy Sci*, 89 (2), 383–401.

VEGA C, GOFF HD, ROOS YH (2007), 'Casein molecular assembly affects the properties of milk fat emulsions encapsulated in lactose or trehalose matrices', *Int Dairy J*, 17 (6), 683–695. DOI:10.1016/j.dairyj.2006.08.004

WALSTRA P, JENNESS R (1984), 'Outline of milk composition and structure', in Walstra P, Jenness R, Badings HT, *Dairy Chemistry and Physics*, New York, Wiley, 1–11.

WILDMAN REC (2007), *Handbook of Nutraceuticals and Functional Foods*, Boca Raton, CRC Press.

YOUNG SL, SARDA X, ROSENBERG M (1993a), 'Microencapsulating properties of whey proteins. 1. Microencapsulation of anhydrous milk fat', *J Dairy Sci*, 76 (10), 2868–2877.

YOUNG SL, SARDA X, ROSENBERG M (1993b), 'Microencapsulating properties of whey proteins. 2. Combinations of whey proteins with carbohydrates, *J Dairy Sci*, 76 (10), 2878–2885.

ZIMET P, LIVNEY YD (2009), 'Beta-lactoglobulin and its nano-complexes with pectin as vehicles for ω-3 polyunsaturated fatty acids', *Food Hydrocolloids*, 23, 1120–1126. DOI: 10.1016/j.foodhyd.2008.10.008

23

Using dairy ingredients to produce edible films and biodegradable packaging materials

P. M. Tomasula, United States Department of Agriculture, USA

Abstract: Major food retailers and consumers are concerned about the waste that packaging generates and the scarce natural resources and energy used in its manufacture. Edible films and coatings made from food-grade proteins and carbohydrates are an untapped source of renewable material that, while compostable and biodegradable, can also be consumed with the food product. This chapter focuses on films and coatings made from dairy proteins with an emphasis on those based on casein and whey, the major proteins found in milk, and the research efforts that have been undertaken to improve their mechanical and barrier properties.

Key words: edible films and coatings, packaging, renewable materials, dairy proteins, casein, whey, barrier properties.

23.1 Introduction

23.1.1 Food packaging films

In an effort to reduce food packaging waste, manufacturers are turning their attention to renewable materials to replace petroleum-based packaging films. Edible films and coatings made from food-grade proteins and carbohydrates are an untapped source of renewable material that, while compostable and biodegradable, can also be consumed with the food product. This chapter focuses on edible films and coatings made from dairy proteins with an emphasis on those based on casein and whey, the major proteins found in milk, and the research efforts that have been undertaken to improve their mechanical and barrier properties so that they may be used in a variety of applications.

23.1.2 Edible films and coatings

Edible protein-based *films* are usually formed as free-standing thin sheets, while edible *coatings* are thin films formed directly on the food product. Edible films have the potential for use as food wrapping, or as part of the food itself, to extend its shelf-life and enhance its properties. Edible films and coatings also have the potential to replace one or more polymeric film layers in multilayer packaging systems, being placed next to the food to protect it from contact with the polymeric packaging or to facilitate recycling.

In edible films and coatings, the protein functions as the foundation and matrix of the film. Most films or coatings made from proteins are strong, but are brittle and not crease resistant, have limited ability to elongate or stretch, and swell or dissolve under humid conditions. Most are excellent oxygen barriers.

Besides being biodegradable and compostable, edible films or coatings based on proteins are desirable because they also offer a lucrative outlet for surplus agricultural materials. Edible films or coatings may be used to inhibit migration between components in foods, to evenly distribute food ingredients and additives throughout a food product, to modify the appearance of foods, and to improve the mechanical integrity and handling characteristics of foods. They also may be used as mass transfer barriers between the food and the environment, controlling transfer of moisture, oxygen, carbon dioxide, and aromas (Kester and Fennema, 1986). Edible films and coatings are not a replacement for synthetic packaging materials for prolonged storage of food but are to be used to improve food quality, extend shelf-life and possibly improve the efficiency of packaging materials (Kester and Fennema, 1986). The films may also carry antimicrobials and be used at the surface of foods to prevent bacterial growth. They may also be formed into pouches or bags for individual or small portions of food, or used for microencapsulation of food ingredients to control their addition and release into foods (Robertson, 1993).

23.2 The milk proteins used for film formation

23.2.1 Casein and caseinate films

Casein and whey are the major milk proteins, with casein comprising approximately 80% of the milk proteins. Casein exists as a micelle in milk and is comprised of the amphipathic αs_1-, αs_2-, β- and κ-caseins in the ratio of 40:10:35:12. The caseins are bound by calcium-phosphate linkages. αs_1-, αs_2- and β-caseins are within the hydrophobic core of the micelle and are calcium sensitive, while κ-casein is distributed over the micelle, stabilizing it through steric stabilization.

αs_1- and β-casein do not contain cysteine residues and, unlike the whey proteins, the caseins do not form disulfide bonds. Because of a large number of proline residues, casein assumes a random-coil structure. Its loose, open structure, which exposes the hydrophobic groups, is sensitive to environmental changes, such as pH changes or changes in temperature, which are exploited to precipitate casein from milk.

Casein may be precipitated from milk by (i) enzyme treatment with chymosin or rennet, as in cheese making, to make para-κ-casein, more popularly known as rennet casein; (ii) through the action of lactic acid bacteria, to make lactic acid casein; or (iii) by addition of acids to make acid casein. In rennet casein production, rennet cleaves the Phe (105)-Met (106) linkage of κ-casein, destabilizing the micelle. Para-κ-casein remains with the micelle and is hydrophobic. The rest of the peptide, known as casein macropeptide, is hydrophilic and is no longer part of the micelle. The casein micelles aggregate due to hydrophobic interactions and are usually heated to about 40°C to firm the precipitated curd. The calcium (Ca) content of the resulting curd ranges from 2.6 to 3.0% and the pH of the curd is 7.3 to 7.7 (Southward and Walker, 1980). Rennet casein is insoluble in water.

Acid casein may be produced either through the addition of lactic acid bacteria, which lowers the pH of milk to 4.6 after an incubation period of 14 to 16 hours (Southward and Walker, 1980), or by the addition of a mineral acid such as HCl or H_2SO_4 until the pH of milk is 4.6. As pH of milk is lowered from pH 6.7 to 5.4 during acid casein production, the internal Ca-phosphate linkages stabilizing the micelle are solubilized into the whey. Destabilization of the micelle begins. As pH is decreased further, the surface of the micelle, which is negative, is neutralized and aggregation begins because of hydrophobic interactions. The resulting acid curd is heated to 40°C to firm the curd. The Ca content of acid casein is approximately 0.1% (Southward and Walker, 1980). Acid caseins are insoluble in water.

Because acid and rennet caseins are insoluble in water, they are converted to water soluble forms to make films. Acid caseins are reacted with bases to form the water soluble caseinates. The most common caseinates are calcium caseinates (CaCAS) and sodium caseinates (NaCAS), which are the products of the reaction of acid casein with either $Ca(OH)_2$ or NaOH, respectively. The micellar structure of casein is not restored upon addition of the base. The Ca content of CaCAS is approximately 1.3% and the Na content of NaCAS ranges from approximately 1.2 to 1.4%. The random coil nature of the caseinates and their ability to form hydrogen, electrostatic and hydrophobic bonds makes them excellent film and coating formers (Avena-Bustillos and Krochta, 1993). The caseinates show good adhesion to different substrates due to their high amount of polar groups and their hydrophilicity makes them excellent barriers to substances such as oxygen, carbon dioxide, and aromas (Audic *et al.*, 2003). However, their hydrophilic nature makes them a weak barrier to moisture. Much of the research on edible casein films to date is directed at improving their water vapor barrier properties. The caseinates can also participate in enzymatic and chemical reactions to modify their properties for food or nonfood use (Li Chan, 2004; Audic *et al.*, 2003; Santos *et al.*, 1999; Santos and Tomasula, 2000) but these types of modifications have rarely been explored to modify the properties of edible films. Also, the film forming properties of the individual caseins such as αs1-casein and β-casein, which are less commercially available, have been studied to a very limited extent.

In a novel process (Jordan *et al.*, 1987; Tomasula *et al.*, 1995), casein has also been precipitated from milk using high-pressure carbon dioxide (CO_2). Addition of CO_2 to milk causes carbonic acid to form, which in turn, lowers the pH of the milk. With CO_2 pressures greater than 5.2 MPa at 38°C, the pH is reduced to about 5.4 and CO_2–casein (CO_2–CAS) precipitates. The pH of the curd ranges from 5.4 to 5.8, varying with CO_2 pressure. CO_2–CAS has some of its micellar structure intact, which imparts a haze to films made from this casein. A continuous process (Tomasula *et al.*, 1997) was also developed to produce CO_2–CAS. It has a Ca content of approximately 1%, is 7% soluble in water and its other properties are between those of acid and rennet casein. The performance of the CO_2–CAS films though is similar to that of CaCas films.

23.2.2 Whey protein films

Sweet whey, once considered a waste of the cheese making process, is now an important product of cheese manufacture. Using technologies such as ultrafiltration and ion exchange, whey protein concentrates (WPC) with protein contents ranging from 35 to 85% and whey protein isolates (WPI) with protein contents greater than 90% are commercially available (Bonnaillie and Tomasula, 2008). Whey protein-based edible films are usually made from commercial WPI, though some have been made using WPC. Commercial WPI, depending on manufacturing process, is comprised of approximately 30% alpha-lactalbumin (α-LA) and 60% beta-lactoglobulin (β-LG), 7% of other whey proteins, and small amounts of lactose, ash and fat. WPI is chosen over other whey products in order to exploit the gelling and solubility properties of β-LG (Kilara and Vaghela, 2004).

The whey proteins are globular proteins, soluble over a wide range of pH, and are denatured by heat. Unlike casein, the hydrophobic, polar, and charged amino acids are uniformly distributed over the whey protein. The proteins fold so that most of the hydrophobic groups are buried inside the structure. At temperatures between 67 and 82°C and pH of 7.5 (the conditions important for making water insoluble whey protein films), the globular structure of β-LG opens, exposing the free thiol group at CYS 121 (Dangaran and Krochta, 2008) and the hydrophobic groups. Polymerization then occurs through intermolecular disulfide bonding by thiol – disulfide interchange and thiol oxidation reactions.

23.3 Edible films and coatings made from casein or whey proteins

23.3.1 Edible film and coatings preparation

Edible films and coatings are formed by simple coacervation, complex coacervation, or by thermal gelation or precipitation (Kester and Fennema, 1986). The performance of the films in an application depends on the conformation of proteins in the film, which is influenced by the composition of the film, the solvent environment used to dissolve the protein, the influence of the ambient environment, and the techniques used to dry and process the films. Post-treatment of the films to modify surface texture is also used to modify the properties of the films.

For edible film and coating applications, the proteins are dissolved in either water or food-grade ethanol or mixtures of the two solvents and stirred or homogenized. The solution is placed under mild vacuum to remove dissolved air. There are several methods available for preparing edible films, and solvent casting is the one mostly used in laboratories. In solvent casting, a weighed amount of the film solution, usually containing about 2 to 10% (w/w) of protein, is pipetted into a polystyrene or PTFE petri dish, or spread onto any other flat surface or substrate that can contain the solution. The film preparation is then allowed to dry overnight under ambient conditions. The dried film is then peeled from the dish or other surface and is stored at 50% relative humidity (RH) in a dessicator at room temperature. Thickness is measured using a micrometer at up to 10 points across the film to determine an average film thickness. Film thickness for a particular protein may be varied by varying the amount of solids in the film-forming solution. For edible coatings, the protein solution is sprayed directly onto the food, onto other films, or used as a dipping solution.

Kozempel *et al.* (2003) determined that the optimal concentration for preparing edible films made from CaCAS or CO_2–CAS was 10% (w/w). Surface tension, a measure of the inward forces that must be overcome to expand the surface area of a liquid (Whitten *et al.*, 1992), was essentially constant for solution concentrations in the range from 4 to 10% (w/w), then increased rapidly with concentrations greater than 10% (w/w). Above 10% solids, the high viscosity of the protein solution made it difficult to remove air bubbles before casting films and entrapped bubbles, and other defects were noted in the dried films.

The choice of substrate is also important for film casting. The film solution must uniformly spread over, or wet, the substrate and be readily released when the film is dry (Churaev, 2003; Kozempel and Tomasula, 2004). The substrate is an environmental factor that influences the properties of the finished film; film defects and properties may be attributed to the temperature and type of substrate (Kester and Fennema, 1986).

Few studies have examined the effects of drying techniques or parameters on the barrier and mechanical properties of films. Drying techniques may also affect the appearance of the films (Perez-Gago and Krochta, 2000). The kinetic mechanism for ambient air-drying of films or drying of films in a hot-air dryer proceeds in three-stages (Kozempel *et al.*, 2003). The first drying stage is a constant rate period in which surface evaporation is the rate-controlling step. The second drying stage is a falling rate period in which the solvent can no longer move to the surface of the film to saturate the surface and diffusion begins to predominate as the rate-controlling step. The third stage is a second falling rate period that is unlikely to be observed graphically for thin edible films. At this stage, the evaporating surface of the film is located deeper within the film and the film itself restricts diffusion. For hot air drying of caseinate films, the optimal drying temperature for clear colorless films was 34°C. Higher temperatures led to slightly yellow and progressively darker films.

Cast films tend to be shiny on the side that faced the substrate and matte on the side facing the air. The barrier properties of the films appear to be independent of the side facing the highest concentration of permeant, with the exception of emulsion films as discussed below.

23.3.2 Plasticizers

Protein films tend to shrink and become brittle as they dry, making removal of a single intact film from the substrate difficult. Addition of a plasticizer, a low molecular weight compound, is required to impart flexibility to the films. Plasticizers reduce the internal hydrogen bonding and electrostatic interactions that operate within the protein chains and reduce the glass transition temperature (Krochta and De Mulder-Johnson, 1997). They also increase the molecular spacing in films or, in the case of low molecular weight compounds, occupy the free volume. Proteins tend to have a large free volume relative to that of synthetic polymers because they are less linear with many amino acid side groups. Internal water also acts as a plasticizer and its effects may be difficult to control because protein films are sensitive to the RH of the environment (Coupland *et al.*, 2000).

Various food-grade plasticizers have been used to impart flexibility and elongation to edible films. Polyols such as glycerol (GLY), which is hydrophilic, are added to the protein solution before solvent casting or other types of processing. GLY may also establish hydrogen bonding with amino acid residues of casein (Tomasula *et al.*, 1998) and electrostatic attraction between calcium and the hydroxyl groups of GLY is possible. Other plasticizers that have been used in edible films are propylene glycol (PG), polypropylene glycol (PPG), sorbitol (SOR), sucrose and various blends of the polyols.

The choice of a plasticizer depends on its size and shape as well as its compatibility with the protein. While plasticizers increase the flexibility of the films, they modify their barrier and mechanical properties by decreasing the strength of the films while increasing their permeability. Dangaran and

Krochta (2007) demonstrated that, over time, the plasticizer itself may crystallize, as is the case with sucrose, and affect the properties of the film. Addition of a crystallization inhibitor such as raffinose or lactose is necessary to prevent this change in the plasticizer.

23.3.3 Barrier and mechanical properties of films

Water vapor permeability (WVP) and oxygen permeability (OP) are the most commonly reported barrier properties for edible films. The mechanism by which water vapor or a gas such as oxygen permeates a protein film is assumed to occur as follows: the permeant is absorbed into the film matrix at the high concentration side, dissolves and diffuses into the film according to the concentration gradient, and is desorbed and evaporates from the other side of the film. Dissolution and diffusion of the permeant depends on the structure of the protein film matrix and the properties of the permeant. The structural aspects of the film such as its hydrogen bonding, van der Waals interactions, degree of cross-linking, and crystallinity, as well as the amount and type of plasticizer have been implicated (Kumins, 1965; Siew *et al.*, 1999; Coupland *et al.*, 2000). The properties of the permeant such as its temperature, size, shape and polarity and its solubility in the film are also of importance (Kester and Fennema, 1986). Because of the many variables affecting barrier properties of edible films, reporting of their properties requires that the thickness of the films, RH gradient across the films, and temperature be reported as well.

For protein films, permeability is the product of the diffusion coefficient and the solubility, and is dependant on the cross-sectional area of the film, A, and the thickness of the film, L. The permeability, P, is defined by the following equation (Kester and Fennema, 1986) based on Fick's First Law of Diffusion and Henry's Law of Solubility:

$$P = DS = QL/(At\Delta p) \tag{23.1}$$

where D is the diffusion coefficient, S is the solubility coefficient, Q is the quantity of permeant passing through the film, and Δp is the partial pressure difference across the film. This equation assumes that D and S are independent of the concentration of the permeant. For hydrophilic protein films, the water sorption isotherms are nonlinear and anomalous effects in film properties have been observed compared to polymer-based films. For example, WVP has been shown to vary with the thickness of NaCAS films but is a constant value for polymer films. McHugh *et al.* (1993) demonstrated that WVP increases with increasing film thickness because of the associated increase in RH of these films as the absorbed water vapor interacts with the hydrophilic, polar films and imparts a plasticizing effect. The amount of hydrophilic plasticizer incorporated into the film is also responsible for increases in the water vapor permeability.

A model was developed by Buonocore *et al.* (2003) to gain insight into the phenomena controlling water solubilization and diffusion into casein–starch-based edible coatings in which solubility and diffusion processes were described separately. The model successfully fitted the experimental data.

Water Vapor Permeability

Standard E96-95 (ASTM, 1995) is usually applied for the determination of WVP of films although instrument methods are also available. Most researchers apply the correction factor of McHugh *et al.* (1993) for hydrophilic films, which accounts for the water vapor partial pressure gradient in the stagnant air layer of the test cups used for testing. The WVP of several milk-based films are reported in Table 23.1. In general, WVP of milk protein films are several times greater than those of synthetic films.

A comparison of the WVP of films based on the type of protein or plasticizer used is difficult because of the variation in experimental conditions employed to test the films and the variations in film thicknesses. Lower WVP are apparent for the thinner films as is the case for the CaCAS:GLY and CO_2–CAS:GLY films (Tomasula *et al.*, 1998) and WVP increases with increasing film thickness. However, the increase noted in the case of WPI films plasticized with GLY is probably insignificant (Longares *et al.*, 2004).

Instead of adding plasticizer to films to reduce the intermolecular forces between protein chains, Sothornvit and Krochta (2000b) hypothesized that using WPI with some degree of hydrolysis would increase the number of polymer chain end groups and the free volume and, as a consequence, reduce the amount of plasticizer needed. It was shown that the whey protein molecular weight had little effect on WVP, and that decreasing the molecular weight increased the solubility of the films and improved their flexibility relative to WPI films.

Oxygen permeability

Most milk protein films are excellent oxygen barriers, especially at low RH, compared to synthetic films such as low density polyethylene (LDPE) and high density polyethylene (HDPE) and approach or equal the properties of films such as ethylene vinyl alcohol (EVOH) and poly(vinylidene chloride) (PVDC) due to their high degree of hydrogen bonding (McHugh and Krochta, 1994b). Even though milk protein films have relatively high WVP which would prevent their use in many applications, their low oxygen permeabilities (OP) make them excellent candidates to consider for a particular oxygen barrier application or to use as part of a composite system.

Instruments are available for the determination of OP over wide ranges of temperature, typically from 10 to 40°C and RH at 0 and from 30 to 75%. The instruments operate according to Standard Method D-3985-95 (ASTM, 1995a). A sample of the film is clamped into a diffusion cell, is purged using pure oxygen, and then pure oxygen is introduced to the cell. Molecules of oxygen that diffuse through the film are then sensed by an oxygen sensor. RH probes are also located on both sides of the film for control of RH.

Table 23.1 Water vapor permeability (WVP) properties of solvent cast protein films and synthetic films

Dairy protein	Composition	Weight ratio (protein/ plasticizer)	Thickness (mm)	Test conditions T°C; RH %*	WVP g.mm/ m^2.h.KPa	Reference
NFDM	NFDM:potassium sorbate	9:1	0.110	25; 0/74	3.32	Maynes & Krochta (1994)
	NFDM:hydrolyzed lactose	–	0.120	25; 0/72	3.76	Maynes & Krochta (1994)
	NFDM (ultrafiltered):GLY	4:1	0.071	25; 0/65	2.93	Maynes & Krochta (1994)
β-casein	β-casein:GLY	2:1	0.11–0.12	22.5; 53/11	0.654	Mauer *et al.* (2000)
Lactic acid casein	LAC:GLY	0.6:1	0.203	37.8; 90	2.29	Chick & Ustunol (1998)
	LAC:GLY	1:1	0.203	37.8; 90	2.47	Chick & Ustunol (1998)
	LAC:GLY	1.4:1	0.203	37.8; 90	2.28	Chick & Ustunol (1998)
	LAC:SOR	0.6:1	0.203	37.8; 90	1.87	Chick & Ustunol (1998)
	LAC:SOR	1:1	0.203	37.8; 90	1.88	Chick & Ustunol (1998)
	LAC:SOR	1.4:1	0.203	37.8; 90	1.42	Chick & Ustunol (1998)
	LAC:SOR	1:1	0.104	37.8; 50	0.081	Chick & Hernandez (2002)
	LAC:SOR	1:1	0.104	37.8; 70	0.667	Chick & Hernandez (2002)
	LAC:SOR:carnauba wax	3.33:2.33:1	0.104	37.8; 50	0.039	Chick & Hernandez (2002)
	LAC:SOR:carnauba wax	3.33:2.33:1	0.104	37.8; 70	0.495	Chick & Hernandez (2002)
	LAC:SOR:candelilla wax	3.33:2.33:1	0.104	37.8; 50	0.025	Chick & Hernandez (2002)
	LAC:SOR:candelilla wax	3.33:2.33:1	0.104	37.8; 70	0.185	Chick & Hernandez (2002)
Rennet casein	REN:GLY	0.6:1	0.203	37.8/90	2.41	Chick & Ustunol (1998)
	REN:GLY	1:1	0.203	37.8/90	2.43	Chick & Ustunol (1998)
	REN:GLY	1.4:1	0.203	37.8/90	1.88	Chick & Ustunol (1998)
	REN:GLY	4:1	0.091	25; 0/77	2.33	McHugh & Krochta (1994d)
	REN:SOR	0.6:1	0.203	37.8/90	2.07	Chick & Ustunol (1998)
	REN:SOR	1:1	0.203	37.8/90	2.07	Chick & Ustunol (1998)
	REN:SOR	1.4:1	0.203	37.8/90	1.65	Chick & Ustunol (1998)
MgCAS	MgCAS:GLY	4:1	0.077	25; 0/77	1.83	McHugh & Krochta (1994d)

Table 23.1 *Cont'd*

Dairy protein	Composition	Weight ratio (protein/ plasticizer)	Thickness (mm)	Test conditions T°C; RH %*	WVP g.mm/ m^2.h.KPa	Reference
Calcium caseinate	CaCAS	–	0.082	25; 0/85	1.17	Avena-Bustillos & Krochta (1993)
	CaCAS:BW	1.7:1	0.075	25; 0/97	0.15	Avena-Bustillos & Krochta (1993)
	CaCAS:GLY	2:1	0.105	23; 0/72	7.9	Banerjee & Chen (1995)
	CaCAS:GLY	2.33:1	0.171	30; 0/87	3.18	Tomasula *et al.* (1998)
	CaCAS:GLY	2.33:1	0.222	30; 0/86	4.45	Tomasula *et al.* (1998)
	CaCAS:GLY	2.33:1	0.100	30; 0/85	2.40	Tomasula *et al.* (2003)
	CaCAS:GLY	9:1	0.331	25; 0/82	3.54	Dangaran *et al.* (2006)
	CaCAS:GLY	4:1	0.301	25; 0/82	5.04	Dangaran *et al.* (2006)
	CaCAS:GLY	2.33:1	0.280	25; 0/82	6.69	Dangaran *et al.* (2006)
	CaCAS:GLY:PPG	2.33:0.75:0.25	0.100	30; 0/85	2.00	Tomasula *et al.* (2003)
CO_2 casein	CO_2–CAS:GLY	2.33:1	0.112	30; 0/86	2.22	Tomasula *et al.* (1998)
	CO_2–CAS:GLY	2.33:1	0.163	30; 0/88	2.58	Tomasula *et al.* (1998)
	CO_2–CAS:GLY	2.33:1	0.184	30; 0/88	3.21	Tomasula *et al.* (1998)
	CO_2–CAS:GLY	2.33:1	0.277	30; 0/90	3.80	Tomasula *et al.* (1998)
	CO_2–CAS:GLY	2.33:1	0.100	30; 0/89	2.30	Tomasula *et al.* (2003)
	CO_2–CAS:GLY (Particle size = 111.8 μm)	9:1	0.236	25; 0/82	2.32	Dangaran *et al.* (2006)
	CO_2–CAS:GLY (Particle size = 111.8 μm)	4:1	0.229	25; 0/82	3.47	Dangaran *et al.* (2006)
	CO_2–CAS:GLY (Particle size = 111.8 μm)	2.33:1	0.100	25; 0/82	3.57	Dangaran *et al.* (2006)
	CO_2–CAS:GLY:PPG	2.33:0.75:0.25	0.100	30; 0/89	1.90	Tomasula *et al.* (2003)

AcCAS	Acylated casein	2.33:1	0.175	30; 0/90	2.40	Tomasula *et al.* (2003)
Sodium caseinate	NaCAS	–	0.083	25; 0/81	1.53	Aveno-Bustillos & Krochta (1993)
	NaCAS	–	0.3364	23; 0/90	1.9	Schou *et al.* (2005)
	NaCAS (Buffer-treated pH 4.6 with Ca Ascorbate)	–	0.072	25; 0/86	0.87	Aveno-Bustillos & Krochta (1993)
	NaCAS:GLY	2:1	0.109	23; 0/55	12.9	Banerjee *et al.* (1996)
	NaCAS:GLY	0.89:1	0.104	20; 45/0	0.473	Siew *et al.* (1999)
	NaCAS:GLY	1.67:1	0.072	20; 45/0	0.225	Siew *et al.* (1999)
	NaCAS:GLY	2.125:1	0.3577	23; 0/90	3.0	Schou *et al.* (2005)
	NaCAS:GLY	4:1	0.10	20; 0/90	0.303	Khwaldia *et al.* (2004a)
	NaCAS:GLY	4:1	0.065	25; 75	0.25	Chambi & Grosso (2006)
	NaCAS:PEG	0.81:1	0.099	20; 45/0	0.936	Siew *et al.* (1999)
	NaCAS:PEG	1.32:1	0.071	20; 45/0	0.612	Siew *et al.* (1999)
	NaCAS:Lauric Acid	4:1	0.074	25; 0/92	0.40	McHugh & Krochta (1994d)
	NaCAS:AM	1:4	0.04	25; 0/84	0.66	Aveno-Bustillos & Krochta (1993)
	NaCAS:AM	1:1	0.088	25; 0/88	0.92	Avena-Bustillos & Krochta (1993)
	NaCAS:BW	1.66:1	0.103	25; 0/95	0.40	Avena-Bustillos & Krochta (1993)
	NaCAS:GLY:AMF	10:2.5:1	0.100	20; 0/90	0.300	Khwaldia *et al.* (2004a)
	NaCAS:GLY:AMF	3.33:0.83:1	0.100	20; 0/90	0.506	Khwaldia *et al.* (2004a)
	NaCAS:Gelatin:GLY	3:1:1	0.065	25; 0/75	0.29	Chambi & Grosso (2006)
	w/transglutaminase	3:1:1	0.065	25; 0/75	0.21	Chambi & Grosso (2006)

Table 23.1 *Cont'd*

Dairy protein	Composition	Weight ratio (protein/ plasticizer)	Thickness (mm)	Test conditions T°C; RH %*	WVP g.mm/ m^2.h.KPa	Reference
Whey protein isolate	Native WPI:GLY	2.33:1	0.139	25; 0/71	5.06	Perez-Gago *et al.* (1999)
	WPI:GLY	2.33:1	0.139	25; 0/71	4.96	Perez-Gago *et al.* (1999)
	WPI:GLY	1.6:1	0.106	25; 0/11	0.275	McHugh *et al.* (1994)
	WPI:GLY	1.6:1	0.121	25; 0/65	5.00	McHugh *et al.* (1994)
	WPI:GLY	4:1	0.129	25; 0/77	2.93	McHugh *et al.* (1994)
	WPI:GLY	15:1	0.11–0.18	26.3; 0/88	1.8918	Shellhammer & Krochta (1997a)
	WPI:SOR	1.6:1	0.13	25; 0/79	2.58	McHugh *et al.* (1994)
	WPI:BW:SOR (Mean Diameter = 1.97 μm)	3.5:1.8:1	0.19	25; 0/93	1.17	McHugh & Krochta (1994a)
	WPI:BW:SOR (Mean Diameter = 0.82 μm)	3.5:1.8:1	0.14	25; 0/98	0.221	McHugh & Krochta (1994a)
	Hydrolyzed WPI (5.5%):GLY	2.33:1	–	25; 0/90	4.2	Sothornvit & Krochta (2000b)
	Hydrolyzed WPI (10%):GLY	2.33:1	–	25; 0/90	4	Sothornvit & Krochta (2000b)
Synthetic films	HDPE	–	–	38; 90/0	.00083	McHugh & Krochta (1994d)
	LDPE	–	–	27.6; 0/100	.0013	Shellhammer & Krochta (1997)
	PVDC	–	–	27.6; 0/100	.0008	Shellhammer & Krochta (1997)
	PVC	–	–	27.6;0/100	0.0257	Shellhammer & Krochta (1997)
	Beeswax	–	–	25.9; 0/100	.0037	Shellhammer & Krochta (1997)
	Anhydrous Milkfat	–	–	27.5;0/100	0.5993	Shellhammer & Krochta (1997)
	Carnauba Wax	–	–	27.5;0/100	0.0041	Shellhammer & Krochta (1997)
	Candelilla Wax	–	–	24.9;0/100	0.0005	Shellhammer & Krochta (1997a)

* %RH = relative humidities outside and inside of test cup if used, (outside/inside), otherwise, a single RH% is reported.
NFDM = nonfat dry milk; GLY = glycerol; SOR = sorbitol; BW = beeswax; PPG = polypropylene glycol; AM = acetylated monoglyceride; PEG = polyethylene glycol; MgCas = Magnesium caseinate; AMF = anhydrous milk fat; LDPE = low density polyethylene; HDPE = high density polyethylene; PVDC = poly(vinylidene chloride); PVC = poly(vinyl chloride).

The OP of various milk protein films are listed in Table 23.2 along with those of synthetic films for comparison. The OP is a function of plasticizer content, type of protein, RH and temperature. Examination of the table shows that for any given milk protein, films plasticized with SOR, which is a solid at room temperature, are more effective oxygen barriers than films plasticized with GLY. The improved OP of the films plasticized with SOR may be due to the creation of impenetrable crystalline domains within the film that lower permeability, although it was expected that the crystalline domains would negatively affect the tensile properties (Dangaran and Krochta, 2008). The OP of films increased with increasing amounts of hydrophilic plasticizer.

The protein type also affects the oxygen barrier properties of the films. Lactic acid casein and rennet films had lower OP but were tested under different conditions from many of the other films. For films tested at 23°C and 50% RH, CaCAS and WPI plasticized with 30% GLY have OP of 86 and 76.1, respectively, but the OP of CO_2–casein is 144 cc-µm/KPa.d.m^2, almost twice as large. OP is related to the aggregation of the protein, its conformation due to environmental variables, and its impact on the free volume.

Mechanical properties

The mechanical properties of milk-based films are usually reported in terms of the tensile strength (TS), a measure of the strength of the film, and percent elongation (%E), a measure of the distance that the film will stretch from its initial length before breaking, or of its flexibility. The tensile properties are measured according to Standard D882-01 (ASTM, 2001) for plastic films. Prior to measurement of TS, the milk protein based films are conditioned at 23°C and 50% RH for at least 40 hours. Overall, the milk protein films are strong due to the extensive intermolecular forces operating among the protein chains, but plasticizers reduce these forces and increase protein chain mobility, which improves the flexibility of the films. The tensile properties of milk-based films are reported in Table 23.3 and tend to vary according to types of protein and plasticizer used and film thickness.

In general, the TS values for most films approach those of the synthetic films but %E are much lower. Higher TS values of films tend to correlate with greater protein content. Higher %E values are desirable because the films are more durable when handled during processing or during use by the consumer. Addition of plasticizer weakens the milk-based films but at the lowest values of plasticizer, 10–15% GLY, the milk protein based films are approximately as strong as the LDPE and HDPE films. Increasing GLY plasticizer weakens the films further but improves the %E. Whey and caseinate films plasticized with SOR are approximately as strong as those plasticized with GLY but exhibit lower %E because the crystalline structure of SOR inhibits flexibility.

Table 23.2 Oxygen permeabilities of solvent cast protein films compared with synthetic films. Test conditions were 23°C and 50% RH unless otherwise noted

Dairy protein	Composition	Weight ratio (protein/ plasticizer)	Thickness (mm)	Oxygen permeability (cc. μm/kPa.d.m^2)	Reference
Lactic acid casein	LAC:GLY[a]	0.6:1	0.203	0.88	Chick and Ustunol (1998)
	LAC:GLY[a]	1:1	0.203	2.18	Chick and Ustunol (1998)
	LAC:GLY[a]	1.4:1	0.203	0.73	Chick and Ustunol (1998)
	LAC:SOR[a]	0.6:1	0.203	0.65	Chick and Ustunol (1998)
	LAC:SOR[a]	1:1	0.203	0.73	Chick and Ustunol (1998)
	LAC:SOR[a]	1.4:1	0.203	0.81	Chick and Ustunol (1998)
	LAC:SOR	1:1	0.104	0.414	Chick and Hernandez (2002)
	LAC:SOR:carnauba wax	3.33:2.33:1	0.104	0.544	Chick and Hernandez (2002)
	LAC:SOR:candelilla wax	3.33:2.33:1	0.104	0.613	Chick and Hernandez (2002)
Rennet casein	REN:GLY[a]	0.6:1	0.203	7.06	Chick and Ustunol (1998)
	REN:GLY[a]	1:1	0.203	5.55	Chick and Ustunol (1998)
	REN:GLY[a]	1.4:1	0.203	1.84	Chick and Ustunol (1998)
	REN:SOR[a]	0.6:1	0.203	0.71	Chick and Ustunol (1998)
	REN:SOR[a]	1:1	0.203	1.02	Chick and Ustunol (1998)
	REN:SOR[a]	1.4:1	0.203	0.96	Chick and Ustunol (1998)
Calcium caseinate	CaCAS:GLY	2.33:1	0.1	86	Tomasula *et al.* (2003)
	CaCAS:GLY:PPG	2.3:0.75:0.25	0.1	68	Tomasula *et al.* (2003)
Sodium caseinate	NaCas:GLY	4:1	0.100	44.6	Khwaldia *et al.* (2004a)
	NaCas:GLY:AMF	10:2.5:1	0.100	38.4	Khwaldia *et al.* (2004a)
	NaCas:GLY:AMF	3.33:0.83:1	0.100	33.6	Khwaldia *et al.* (2004a)

CO_2–casein	CO_2–CAS:GLY	2.3:1	0.1	144	Tomasula *et al.* (2003)
	CO_2–CAS:GLY:PPG	2.3:0.75:0.25	0.1	74	Tomasula *et al.* (2003)
AcCas	Acylated casein	2.33:1	0.175	48	Tomasula *et al.* (2003)
Whey protein isolate	Hydrolyzed WPI:GLY 5.5% DH	3.1–1.8:1	0.0130	42.2–111.9	Sothornvit and Krochta (2000a)
	Hydrolyzed WPI:GLY 10% DH	3.1–1.8:1	0.0130	35.5–89.1	Sothornvit and Krochta (2000a)
	Unhydrolyzed WPI:GLY	3.1–0.8:1	0.013	41.3–333.1	Sothornvit and Krochta (2000a)
	WPI:GLY	2.3:1	0.11	76.1	McHugh and Krochta (1994b)
	WPI:GLY	5.7:1	0.11	18.5	McHugh and Krochta (1994b)
	WPI:SOR	2.3:1	0.11	4.3	McHugh and Krochta (1994b)
	WPI:SOR	1:1	0.11	8.3	McHugh and Krochta (1994b)
	WPI:SOR	3.5:1	0.11	2.6	McHugh and Krochta (1994b)
	WPI:SOR[b]	3.5:1	0.11	0.7	McHugh and Krochta (1994b)
	WPI:SOR[c]	3.5:1	0.11	43.3	McHugh and Krochta (1994b)
Synthetic films	LDPE	–	0.0254	1870	Salame (1986)
	HDPE	–	–	427	Salame (1986)
	PVDC-based films	–	–	0.38–5.1	Salame (1986)
	EVOH (70% VOH)[d]	–	–	12	Salame (1986)

[a] Test conditions, 23°C, 0% RH; [b] test conditions 23°C 40% RH; [c] test conditions 23°C, 70% RH; [d] test conditions 23°C, 95% RH.
GLY = glycerol; SOR = sorbitol; PPG = polypropylene glycol; AMF = anhydrous milk fat; LDPE = low density polyethylene; HDPE = high density polyethylene; PVDC = poly(vinylidene chloride).

Tensile properties are difficult to determine for milk protein films without added plasticizer because of their brittleness; however, Schou *et al.* (2005) report TS for a NaCAS film without added plasticizer which surpasses that of LDPE and HDPE films. On the other hand, %E of the NaCAS film is only 2.1% compared to values of 500 and 300%, respectively, for the LDPE and HDPE films.

The lactic acid and rennet casein films (Chick and Ustunol, 1998), which were made by adjusting pH to 10 using NaOH to solubilize the caseins in water prior to film making, were plasticized with greater amounts of GLY and SOR than for the other films reported in Table 23.3. The TS of these films plasticized with about 40% GLY are about as strong as the whey and casein films plasticized with 30% GLY yet have %E approaching that of the synthetic films, possibly due to the greater plasticizer content.

Little information is available on the performance of individual caseins in films. Mauer *et al.* (2000) showed that films made from the most hydrophobic component of casein, β-casein, demonstrated marginal improvement in WVP and TS, but % E was about 250% compared to approximately 80% obtained for both CO_2–CAS and CaCAS films (Tomasula *et al.*, 1998). Motoki *et al.* (1987) made films from the hydrophilic α_{s1}-casein cross-linked with transglutaminase. The resulting films were water insoluble but the tensile properties were comparable to those of CO_2–CAS or CaCAS films.

Rapid drying of solvent cast films may limit the mobility of the protein chains, preventing the development of intermolecular interactions in films as the solvent is removed (Banker, 1966). This was demonstrated by Alcantara *et al.* (1998) for the drying of whey protein isolate films. Water vapor permeabilities were lower for films dried at 95°C and 30% RH than for films dried at 21°C and 50% RH. The films dried at 95°C were stronger but less extensible than films dried at 21°C. Kaya and Kaya (2000) examined the effects of microwave drying on the properties of whey protein isolate films containing 50% WPI and 50% GLY. Drying time was reduced to 5 minutes compared to 18 hours of drying under room conditions. TS and %E improved by approximately 15 and 40%, respectively, with microwave drying but WVP did not improve.

Aroma barrier properties and appearance of milk protein films

Information on the aroma barrier properties of casein films is scarce but there are a few studies that report the aroma barrier properties of whey protein films, such as those reported in Dangaran and Krochta (2008).

Caseinates and whey proteins form films that are transparent and glossy. These characteristics are important for using the films as coatings or in laminates with synthetic films. CO_2–CAS films (Tomasula *et al.*, 1998) are nearly opaque and hazy because of the partial micellar character of the films. Standards D523-89 (ASTM, 1989) and D4039-93 (ASTM, 1999) are used to measure gloss and haze of films. Dangaran *et al.* (2006) improved the gloss of CO_2–CAS films from 55.3 gloss units to 73 gloss units by

Table 23.3 Tensile properties of protein films compared with synthetic films. Test conditions were 23°C and 50% RH unless otherwise noted

Dairy protein	Composition	Weight ratio (protein/ plasticizer)	Thickness (mm)	Tensile strength (MPa)	Percent elongation (%)	References
NFDM	NFDM (Lactose free):GLY	4:1	0.069	5.1	12.2	Maynes and Krochta (1994)
	NFDM (ultrafiltered):GLY	4:1	0.071	10.0	5.2	Maynes and Krochta (1994)
α_{s1}-casein	α_{s1}-casein:GLY	49:1		4.1	38	Motoki *et al.* (1987)
	α_{s1}-casein:GLY w/ transglutaminase	49:1		10.6	77	Motoki *et al.* (1987)
β-casein	Beta-casein:GLY	2:1	0.11–0.12	6.0	274	Mauer *et al.* (2000)
Lactic acid casein	LAC:GLY	0.6:1	0.216	0.42	121.4	Chick and Ustunol (1998)
	LAC:GLY	1:1	0.216	1.24	253.6	Chick and Ustunol (1998)
	LAC:GLY	1.4:1	0.216	2.51	194.1	Chick and Ustunol (1998)
	LAC:SOR	0.6:1	0.216	2.43	170.7	Chick and Ustunol (1998)
	LAC:SOR	1.4:1	0.216	11.65	50.6	Chick and Ustunol (1998)
	LAC:SOR	1:1	0.216	7.48	156.0	Chick and Ustunol (1998)
	LAC:SOR	1:1	0.104	6.2	156	Chick and Hernandez (2002)
	LAC:SOR[a]	1:1	0.104	1.1	167	Chick and Hernandez (2002)
	LAC:SOR:carnauba wax	3.3:2.3:1	0.104	7.9	31	Chick and Hernandez (2002)
	LAC:SOR:carnauba wax[a]	3.3:2.3:1	0.104	1.9	88	Chick and Hernandez (2002)
	LAC:SOR:candelilla wax	3.3:2.3:1	0.104	8.3	37	Chick and Hernandez (2002)
	LAC:SOR:candelilla wax	3.3:2.3:1	0.104	1.8	74	Chick and Hernandez (2002)
Rennet casein	REN:GLY	0.6:1	0.216	0.83	123.2	Chick and Ustunol (1998)
	REN:GLY	1:1	0.216	2.42	185.4	Chick and Ustunol (1998)
	REN:GLY	1.4:1	0.216	4.5	223.5	Chick and Ustunol (1998)
	REN:SOR	0.6:1	0.216	3.83	4.9	Chick and Ustunol (1998)
	REN:SOR	1:1	0.216	9.53	7.6	Chick and Ustunol (1998)
	REN:SOR	1.4:1	0.216	15.12	17.9	Chick and Ustunol (1998)
Calcium caseinate	CaCAS:GLY	2:1	0.105	4.25	1.45	Banerjee and Chen (1995)
	CaCAS:GLY	2.33:1	0.11	1.6	66.6	Tomasula *et al.* (1998)
	CaCAS:GLY	2.33:1	0.15	1.9	76	Tomasula *et al.* (1998)
	CaCAS:GLY	2.33:1	0.10	7.0	66	Tomasula *et al.* (2003)
	CaCAS:GLY:PPG	2.33:0.75:0.25	0.10	12	88	Tomasula *et al.* (2003)

Table 23.3 *Cont'd*

Dairy protein	Composition	Weight ratio (protein/ plasticizer)	Thickness (mm)	Tensile strength (MPa)	Percent elongation (%)	References
AcCAS	AcCAS:GLY	2.33:1	0.10	4	43	Tomasula *et al.* (2003)
Sodium caseinate	NaCAS	–	0.3364	37.5	2.1	Schou *et al.* (2005)
	NaCAS:GLY	2:1	0.109	13.9	30.8	Banerjee and Chen (1995)
	NaCAS:GLY	2.125:1	0.3577	9.0	13.6	Schou *et al.* (2005)
	NaCAS:GLY	4:1	0.085	17.4–26.7	10.5	Siew *et al.* (1999)
	NaCAS:GLY	2:1	0.085	10.9–11.7	73.7–84.2	Siew *et al.* (1999)
	NaCAS:GLY	4:1	0.10	12.8	24	Khwaldia *et al.* (2004a)
	NaCAS:PEG	4.54:1	0.085	10.9–16.35	5.3	Siew *et al.* (1999)
	NaCAS:PEG	1.9:1	0.085	10.9–13.9	25.4	Siew *et al.* (1999)
	NaCAS:GLY:AMF	10:2.5:1	0.100	11.5	10	Khwaldia *et al.* (2004a)
	NaCAS:GLY:AMF	3.33:0.83:1	0.100	4.8	6	Khwaldia *et al.* (2004a)
CO_2–casein	CO_2–CAS:GLY	2.33:1	0.11	1.2	50.2	Tomasula *et al.* (1998)
	CO_2–CAS:GLY	2.33:1	0.15	3	74.2	Tomasula *et al.* (1998)
	CO_2–CAS:GLY	2.33:1	0.10	5	56	Tomasula *et al.* (2003)
	CO_2–CAS:GLY:PPG	2.33:0.75:0.25	0.10	7	88	Tomasula *et al.* (2003)
Whey protein isolate	Native WPI:GLY	2.3:1	0.139	3.1	7	Perez-Gago *et al.* (1999)
	WPI:GLY	2.3:1	0.139	6.9	41	Perez-Gago *et al.* (1999)
	WPI:GLY	5.7:1	0.110	29	4	McHugh and Krochta (1994b)
	WPI:GLY	2.3:1	0.110	13.9	30.8	McHugh and Krochta (1994b)
	WPI:SOR	2.3:1	0.110	14	1.6	McHugh and Krochta (1994b)
	WPI:SOR	1:1	0.110	14.7	8.7	McHugh and Krochta (1994b)
	WPI:GLY, 5.5% DH	2.33:1	0.1344	1	40	Sothornvit and Krochta (2000a)
	WPI:GLY, 10% DH	2.33:1	0.1344	2	4	Sothornvit and Krochta (2000a)
Synthetic films	LDPE	–	–	13	500	Salome (1986)
	HDPE	–	–	26	300	Salome (1986)
	Polystyrene	–	–	35–55	1	Houston (1986)
	Plasticized PVC (wrap film)	–	–	15–30	150–350	Audic *et al.* (2003)

[a] Test conditions, 23°C; 75% RH.

NFDM = nonfat dry milk; GLY = glycerol; SOR = sorbitol; PPG = polypropylene glycol; PEG = polyethylene glycol; AMF = anhydrous milk fat; LDPE = low density polyethylene; HDPE = high density polyethylene; PVC = poly(vinyl chloride).

reducing the particle size of granular dry CO_2 casein from 126 μm to 111 μm. The films were still hazy. Reducing the particle size to less than 86 μm resulted in films that were glossy and also transparent. However, reducing particle size decreased the TS of the films and increased WVP.

The gloss and appearance properties of whey protein films are reported in detail in Dangaran and Krochta (2008). Compared to the CO_2 casein films, whey protein films exhibit gloss units near that of shellac (92.9).

23.4 Improvements to edible films

Although the TS and OP of milk-based edible films with added plasticizer approach that of the synthetic films, especially at low %RH, the low %E, the high WVP and in some cases the high solubility of the films in water, precludes their use in most applications. While the amount of added plasticizer tends to improve the %E of the films, it also weakens the TS of the films. Several studies have focused on improving the WVP properties of the films by making various modifications that include the addition of lipids and waxes, or cross-linking agents, or treatment with ultrasound and irradiation.

23.4.1 Addition of lipids or waxes or polysaccharides

Hydrophobic lipids or waxes have been added to milk-based edible films either by creating a stable emulsion with the lipid or forming a bilayer or composite film with the lipid. The hydrophobic lipids or waxes do not allow water to diffuse through them.

Introducing acetylated monoglyceride (AM) or beeswax (BW) to NaCAS in emulsion films resulted in significant improvements in WVP (Avena-Bustillos and Krochta, 1993) with significant reductions compared with the caseinate film. Addition of BW to CaCAS films also resulted in significant reductions in WVP. Increasing concentrations of BW resulted in thicker films but did not improve WVP because the wax was not dispersed effectively. The lipid type and concentration was also found to influence the WVP of the caseinate films. BW, due to its crystalline structure was more effective than stearic acid and AM in reducing WVP of NaCAS films.

Addition of carnauba or candelilla waxes resulted in little improvement of the WVP of lactic acid casein-based emulsion films plasticized with SOR (Chick and Hernandez, 2002). The WVP decreased with increasing wax content but the reduction in WVP was low despite the added wax. The OP of the films with the added wax was comparable to that of films without wax, while TS improved slightly and %E was reduced. Scanning electron microscopy showed that the wax was partially distributed within the protein–water–SOR matrix.

Films made from an emulsion of anhydrous milk fat (AMF), comprising 10–30% of the NaCAS by weight and plasticized with GLY (25%), were weaker than NaCAS films without AMF, had a decreased % E, and showed no improvement in WVP (Khwaldia *et al.*, 2004a). The loss of mechanical properties of the films was attributed to a loss of continuity of the protein matrix because of the presence of lipid globules. At AMF levels greater than 30%, it was hypothesized that the distribution of lipid globules with high particle size affected the structural cohesion of the polymeric matrix and the emulsion stability, leading to an increase in WVP and a decrease in the film tortuosity, which increased water diffusion. Water does not diffuse through the lipid. OP was found to decrease with increasing amounts of AMF up to 20% and then increased again at AMF levels greater than 20%.

The addition of carrageenans can improve the mechanical properties of CO_2–CAS films plasticized with GLY (20%). CO_2–CAS films were blended in a 1:1 ratio with either lambda (λ-), iota (ι-), or kappa (κ-) carrageenan (Unpublished data, Dangaran and Tomasula, 2007). The carrageenans did not lower WVP, but TS increased from 13.2 MPa to 22.4–23.3 MPa with the addition of κ-carrageenan. The TS of blends containing the carrageenans was approximately the average of the TS of the casein film and that of the carrageenan alone. %E was significantly increased from 5.6% to 42.9% with λ-carrageenan. The improvement in %E appears to be due to the presence of the carrageenan alone which had a %E of 40%. ι- and κ-carrageenan approximately doubled %E. DSC analysis showed that blends of the carrageenans with the protein were miscible as indicated by a single peak on thermograms. The differences in charge densities of the carrageenans compared to the proteins appear to affect their interactions.

Unlike the results for the casein films, the addition of lipids and waxes was found to improve the WVP of whey protein films, but decreased their tensile properties. Films made from an emulsion of WPI and BW, and plasticized with SOR exhibited lower WVP than the WPI films (McHugh and Krochta, 1994a). Decreasing the lipid particle size which increased the number of lipid particles and film tortuousity resulted in a further decrease in WVP and improvement in TS (Perez-Gago and Krochta, 2000) Heat denaturation of the WPI resulted in intermolecular disulfide bonds through thiol-oxidation and thiol-disulfide interchange reactions that further improved the WVP. Reducing the pH of the emulsion below that of the pI of the protein caused WVP to increase due to a sharp change in viscosity with an increase in protein aggregation, which possibly lowered the lipid mobility and reduced the interconnectivity among the lipid droplets.

Lipid concentration and lipid type also influenced WVP of WPI films (McHugh and Krochta, 1994c). Increasing the lipid content of the films resulted in a downward trend in WVP for WPI films with added BW or palmitic acid, but there was a lesser trend with stearyl alcohol. Emulsion films comprised of WPI, lipid, and SOR exhibited a low WVP if the lipid was a fatty acid (palmitic acid, myristic acid) or BW but was twice as high if the lipid was stearyl alcohol, hexadecanol or tetradecanol.

Shellhammer and Krochta (1997) determined the viscoelastic properties of four lipids and waxes, carnauba wax, candelia wax, milkfat fraction and BW, and showed that the viscoelastic milkfat and BW improved WVP in whey films more than carnauba or candelia waxes. The milkfat and BW deform to a greater extent during drying to form an intact lipid network due to their more plastic nature.

In order to enhance or reduce lipid deformability and the ability to yield an internal interconnecting lipid network in the films, Perez-Gago and Krochta (2000) dried WPI films in an emulsion with either BW, anhydrous milkfat fraction, or candelilla wax. Significantly lower WVP was observed for films dried at 80°C and 40% RH compared to films dried at 25 or 40°C and 40% RH. The mechanical properties of the films were not modified by the drying temperature.

23.4.2 Cross-linking of milk protein films

For synthetic polymers, cross-linking is used to decrease polymer chain mobility to increase resistance to vapor and gas transport (Kumins, 1965), by joining two or more functional groups of the polymer by a covalent bond. pH adjustment, ionic cross-linking, through the introduction of calcium ion, and cross-linking using heating, γ-irradiation or food grade enzymes have been explored to improve the mechanical and barrier properties of edible films.

pH treatment

The WVP of NaCAS films, soaked for 1 minute in sodium acetate, sodium ascorbate or calcium ascorbate buffers, and adjusted to pH 4.6 or soaked for 1 minute in calcium chloride solution and then adjusted to pH 9.6, was reduced by approximately 40% (Avena-Bustillos and Krochta, 1993). The WVP was lowered possibly due to increased protein–protein interaction, as indicated by the decreased film thickness of the treated films.

γ-irradiation

Treatment of solutions of NaCAS and CaCAS, with or without added GLY, by γ-irradiation formed edible films with improved mechanical properties, such as puncture strength and puncture deformation (Brault *et al.*, 1997). γ-irradiation results in the formation of bityrosine, a covalently bound biphenol, produced by the reaction of two tyrosyl radicals or a tyrosyl radical plus a tyrosine molecule. GLY increased the formation of cross-links within protein chains and was found to improve the mechanical strength and film flexibility. The greatest effects of irradiation dose on the mechanical properties of the films occurred at GLY:caseinate ratios of 1:2 and 2:3, leading to branching of polypeptide chains to form a 3-D network. The addition of $CaCl_2$ to CaCAS solution was found to facilitate a reduction in the salt bridges and electrostatic bonds facilitating the formation of bityrosine during irradiation. Gels were formed at irradiation doses greater than 16 kGy (Ressouany *et al.*, 1998).

Vachon *et al.* (2000) combined thermal treatment and γ-irradiation and showed that the molecular weight of CaCAS increased, the mechanical strength of the films increased and solubility of the films in water decreased. WPI did not show similar behavior because cross-links were not generated due to the low tyrosine content of the protein. However, WPI could be substituted for CaCAS in films, resulting in an increase of the puncture strength of the treated films. γ-irradiation may induce more intermolecular interactions and the formation of inter- and/or intra-molecular covalent cross-links in the film-forming solutions (Ouattara *et al.*, 2002). A reduction in WVP from 2.07 (unirradiated) to 1.38 g mm/m^2 d/mm Hg (irradiated) was noted for CaCAS-WPI films in the ratio of 25:75.

Transglutaminase

Transglutaminase (TGase) is a food-grade enzyme that uses the acyltransferase mechanism, in the presence of Ca ion, to catalyze the transfer of gamma-carboxyamide (acyl donor) of a peptide-bound glutamine residue to the gamma-amine (acyl acceptor) of lysine residues along protein chains (Mahmoud and Savello, 1992). Intramolecular and intermolecular ε-(γ-glutaminyl)lysine cross-links are generated when the ε-primary amine of lysine is bound to a glutamine-containing protein. It was shown to effectively cross-link whey-based formulations containing a modified whey powder (35% protein) (Aboumahmoud and Savello, 1990) and decrease the solubility of 1:1 mixtures of α-LA and β-LG with added GLY in buffers at various pH and temperature (Mahmoud and Savello, 1992). The films were protease-digestible (Mahmoud and Savello, 1993).

The addition of TGase to WPI films, plasticized with GLY (Yildirim and Hettiarachchy, 1998), did not improve WVP compared to the control, possibly due to the creation of additional, or an increase in size of, existing pores. TS increased from 5.64 MPa for the untreated whey protein film to 12.53 MPa.

The globular nature of the whey proteins, especially β-LG, prevents cross-linking by TGase, except if denatured. Cross-linking is further limited by the formation of non-covalent aggregates between calcium and β-LG, induced by electrostatic interaction (Faergemand *et al.*, 1998). Sharma *et al.* (2001) determined the susceptibility to cross-linking by TGase of the individual milk proteins in skim milk. In both heated and unheated milk, there was a reduction in the monomeric forms of κ- and β-caseins, indicating that these proteins were most susceptible to TGase-induced cross-linking. Only preheated β-LG was susceptible to the same amount of cross-linking as α-LA, preheated or not. Nieuwenhuizen *et al.* (2004) demonstrated the modification of the lysine and glutamine residues in native β-LG to improve the accessibility of the residues to reaction with TGase under nonreducing and nondenaturing conditions, with a possible application for the creation of novel foods.

In another study, Sharma *et al.* (2002) showed that commercial α-LA was susceptible to more cross-linking by TGase than that from the study by Aboumahmoud and Savello (1990). Reduction of disulfide bonds in the commercial protein was found to be unnecessary because structural modifications of the protein occurred during manufacture exposing available sites for TGase reaction.

Chambi and Grosso (2006) cross-linked NaCAS, gelatin, and mixtures of the two, using TGase to produce edible films that showed improvement in the mechanical and barrier properties of the films, depending on the ratio of casein to gelatin in the film. The films contained 25 g GLY/100 g protein. TS of the NaCAS (~12 MPa), gelatin film(~36 MPa) and of the caseinate–gelatin films, which increased with increasing gelatin content, were not significantly different from that of the untreated films. On the other hand, % E of the untreated films increased from 9% for NaCAS film to a maximum value (~27%) at caseinate:gelatin ratios of 75:25 and 50:50, and then decreased to ~9% for the gelatin film. With TGase treatment, %E of the NaCAS film approximately doubled while that of the gelatin film was not significantly different. The greatest increase in %E was noted for blends containing 75:25 caseinate:gelatin, showing % E of 56.59% compared to a value of 27.2% for the untreated film. This film also showed an anomalous decrease in WVP with TGase treatment, while the others showed increases in WVP. It was concluded that cross-linking may have increased the proportion of hydrophobic segments on the surface of the film, which are mostly provided from the NaCAS, thus leading to a lower WVP.

O'Connell and de Kruif (2003) showed that TGase treatment at either 0 or 35°C affects the association behavior of β-casein. TG appeared to freeze the micellization process, indicating that the associative state of β-casein should be considered prior to treatment with the enzyme. The changes appeared to be due to charge modification. This finding may be of interest in the design of future experiments involving edible films.

Building on the observations that ultra-high temperature (UHT) treatment of milk seemed to produce a more open, looser casein micelle structure, Bönisch *et al.* (2004) heated NaCAS solutions to 140°C and immediately cooled them using ice water. The UHT-treated NaCAS solutions and controls were then treated with TGase. TGase treatment of NaCAS resulted in a degree of protein polymerization of about 40% after 90 min compared to 60% for TGase treatment of NaCAS that was heated at 140°C for 20 s. This finding may have application in the development of stronger films with improved WVP.

Tyrosinase

While the caseins are more susceptible to cross-linking using TGase than the whey proteins, mushroom tyrosinase (E.C.1.14.18.1), a polyphenol oxidase, has been shown to be an effective cross-linking agent for the whey proteins, α-LA and β-LG (Thalman and Lotzbeyer, 2002), but its ability to improve the mechanical and barrier properties of dairy-based films has not yet been reported. Tyrosinase first induces the formation of *o*-quinones in the presence of low molecular weight (MW) phenolic compounds or from the tyrosine residues. The low MW compounds, such as caffeic or chlorogenic acid, act as a substrate for the cross-linking process and are oxidized to the *o*-quinone structure. The *o*-quinones may then interact with each other or react with amino and sulphydryl groups or pyrrolidine side chains of the proteins (Thalman and Lotzbeyer, 2002; Seo *et al.*, 2003; Lantto *et al.*, 2005). α-LA and β-LG polymers with a MW greater than 300 kDa were produced from the proteins with initial MW of 14 kDa and 18 kDa, respectively, after treatment with tyrosinase (Thalman and Lotzbeyer, 2002). For α-LA, addition of caffeic acid was not necessary but increasing temperatures up to 50°C alone resulted in polymers with molecular weight greater than 300. The polymerization of β-LG required use of an external phenolic source (Thalman and Lotzbeyer, 2002). Onwulata and Tomasula (2008) used tyrosinase to investigate the gelling properties of microparticulated WPI and CaCAS slurries in the presence of alginic acid. Stable gels were obtained with increasing amounts of tyrosinase. While the gels were designed as replacements for carbohydrates in low-fat foods, the slurries have potential as films.

Genipin

Genipin, obtained from the iridoid glucoside, geniposide, present in the fruit of *Gardenia jasminoides* Ellis, is another natural cross-linking agent which has potential use in improving the mechanical and barrier properties of casein and whey protein films. However, it characteristically gives a deep blue color to primary amino acids, such as the lysine residues of proteins. Bigi *et al.* (2002) showed that the extensibility, the swelling properties, and the enthalpy of denaturation, of gelatin films cross-linked with genipin were similar to that using glutaraldehyde but showed a greater thermal stability. The mechanical properties and stability of chitosan and chitosan blend films in water were also improved when cross-linked with genipin, suggesting potential applications in biomaterials (Jin *et al.*, 2004). Under basic conditions, genipin undergoes a ring-opening polymerization prior to cross-linking with chitosan. The crosslinked bridge consists of polymerized genipin macromers or oligomers (7–88 monomer units) (Mi *et al.*, 2005). Under acidic or neutral conditions, genipin reacts with the primary amino groups on chitosan to form heterocyclic amines. The heterocyclic amines are further associated to form cross-linked networks with short chains of dimer, trimer, and tetramer bridges.

Preliminary work in our laboratory to investigate the use of genipin for improving the properties of CO_2–CAS, CaCAS, and whey protein films, shows that reaction of the proteins with genipin is favored under neutral and basic conditions, the solubility of the films in water is decreased and TS increases.

Other treatments
Banerjee *et al.* (1996) demonstrated that ultrasound treatment of NaCAS and WPC solutions with a 1:2 ratio of GLY:protein improved the TS of the films compared to controls. However, WVP and %E were not affected. Ultrasound may have created greater molecular order in the films, enabling greater inter-molecular attraction, and therefore enhanced TS. WPC films showed smaller particle and lipid droplet sizes, which may also have enabled greater molecular attraction.

23.5 Milk proteins in composite films

Paper is the most commonly used packaging material and it is coated with wax or laminated to synthetic films or aluminum foil to improve its resistance to moisture, oil or grease, and to give oxygen barrier properties. Researchers have tested the properties of edible milk protein-based films in composite structures with the intention of placing the edible film side of the composite in contact with the food and to facilitate composting or recycling.

WPI solutions containing GLY spread on pulp paper improved the printability of the paper (Han and Krochta, 1999). WVP of the paper was also improved because WPI impregnated the porous paper structure, but this also increased the maximum amount of water absorbed by the paper. However, the WPI did not significantly change the mechanical and optical properties of the paper (Han and Krochta, 2001) and paper coated with 18 g/m^2 of WPI demonstrated increased oil resistance. Both denatured and native WPI showed grease barrier properties similar to that of commercial polyvinyl alcohol and fluorocarbon coatings and were highly impermeable to grease penetration after four hours when coated on paperboard (Chan and Krochta, 2001). Lin and Krochta (2003) found similar results for WPC-80 coatings on paperboard. Using sucrose as a plasticizer, the grease resistance properties of WPC-80 were retained after ambient storage and during extended testing times of 16 hours.

The oxygen barrier properties of WPI with various plasticizers have been exploited to create a laminated structure with polypropylene (PP). An important consideration in designing a laminate structure is the surface energy of the synthetic polymer and surface tensions of the liquids and their contact angle when coating synthetic substrates with the protein formulations (Hong *et al.*, 2004). Non-polar PP was first treated with corona discharge to improve its adherence to the WPI coating. OP of the laminated PP films was reduced 4-fold at 40°C with the addition of a WPI:GLY coating at 50% RH. OP of the laminated films was sensitive to RH and increased significantly with RH in the range from 30–85% range, approaching that of the uncoated PP film at 80% RH (Hong and Krochta, 2003). In another study (Hong and Krochta, 2006), OP was significantly reduced using either WPI or WPC coating at 40°C and 50% RH on both PE and PP laminates. OP of the PP laminates were significantly lower than that of the PE laminates due to differences in the intrinsic characteristics of the base films. Excellent OP properties were noted for both laminates at low to intermediate RH. WPI, plasticized with GLY, PEG, SOR or sucrose, and coated onto PP films resulted in films with TS of up to 70 MPa and % E as high as 135%.

The mechanical and WVP properties of edible films containing a whey powder and NaCAS mixture, with whey powder as the principal component, were improved when laminated to corn-zein based films (Cho *et al.*, 2002); however, %E was significantly reduced. Whey powder, containing 0.11 g protein/g powder is significantly less expensive than WPC or WPI. Addition of stearic acid to the corn-zein layer resulted in a reduced WVP compared to the whey-caseinate film.

A study conducted to determine the mechanical and WVP properties of blends or bilayers of NaCAS–pullulan films plasticized with SOR (Kristo and Biliaderis, 2006; Kristo *et al.*, 2007) showed that varying the ratio of the polymers changed the mechanical properties of the film but not WVP. Increasing the pullulan content decreased TS and increased %E, showing that the pullulan imparts flexibility and NaCAS stiffness to the films. Application of BW improved the WVP of the films.

23.6 Modifying the properties of edible films through processing

Most of the studies to determine the properties of edible milk protein films and coatings have been conducted on films prepared through solvent casting on the laboratory-scale. Larger scale production is needed to produce samples for applications studies.

There are several examples in the patent literature for the production of casein films using extrusion methods. However, little information is provided on the mechanical and barrier properties of these films. Kozempel and Tomasula (2004) developed a continuous solvent casting process to make casein films from CaCAS or CO_2–CAS. Hot air drying was used in the process. The mechanical properties of the films were similar to those made on the laboratory-scale. Nonfat dry milk was substituted for up to 20% of the CaCAS or CO_2–CAS content with no loss in physical properties.

Casein films have also been made through the wet-spinning process. In this method, acid casein is extruded with water and sodium hydroxide to form sodium caseinate, discharged into a coagulating bath, and then collected on a roller (Frinault *et al.*, 1997). Because of the chemicals used in the bath to harden the casein films, the films were rendered inedible. However, the films had low solubility in water, WVP was approximately 1.40 g-mm/m^2 hkPa, TS was 4.5 MPa and %E was 68.6%, greater than that reported for solvent cast films, which may be due to shearing effects of extrusion on the molecular structure of casein.

Müller-Buschbaum *et al.* (2006) prepared casein films using a spin-coating technique, a method that is used to prepare thin and ultrathin films from synthetic polymers, on base-treated glass surfaces. The glass surfaces were treated to assure wetability of the casein solution. Films dried in less than 30 s and the moderate pressures introduced by spin-coating force the micelles to rearrange into a more compact structure. This method is still in the experimental stages but has potential for the preparation of advanced coatings of casein in laminate structures or on foods.

Sheets made by compression molding of WPI:GLY mixtures were compared to solvent cast films having the same concentration of WPI and GLY (Sothornvit *et al.*, 2003, 2007). Operating temperature and pressure of the process did not affect the mechanical properties of the films but increasing amounts of GLY resulted in films with decreased TS and no change in %E. Compression molded films had higher WVP than solvent cast films because the compression molded films demonstrated less protein cross-linking, were thicker and were more soluble in water.

Dangaran and Krochta (2008) report details of extrusion studies which demonstrate that extrusion operating conditions provide sufficient heat-denaturing and cross-linking to produce transparent, flexible whey protein sheets with improved tensile properties relative to cast films. Hernandez-Izquierdo *et al.* (2008) showed that extruded WPI plasticized with GLY had tensile properties comparable to solvent cast and compression molded films but exhibited greater %E. A WPI film containing 50% GLY had a %E of 132% compared with 68% for a solvent cast WPI film containing 55% GLY. It was concluded that extrusion created a greater alignment of the polymer molecules in the machine direction as the sheets exited the extruder. Onwulata *et al.* (2003, 2006) and Tunick and Onwulata (2006) have demonstrated that shearing of the protein from the extrusion process leads to creation of aligned whey protein polymers.

23.7 Potential Applications

The film-forming properties of casein have long been exploited in non-food applications (Audic *et al.*, 2003). While research on investigation of the mechanical and barrier properties of casein and whey proteins is active, research lags on applications of casein and whey films in food packaging and preservation, to improve the appearance and quality of foods, and as a barrier layer in packaging systems.

Some of the potential applications of casein and whey films and coatings as moisture and gas barriers for fresh fruits and vegetables, meats, cereals, nuts and frozen foods are reported by Khwaldia *et al.* (2004) and Vargas *et al.* (2008). A more recent review on applications of whey protein based films and coatings includes nuts and peanuts, eggs, confectionary products, meat product, fruits and vegetables (Dangaran and Krochta, 2008). Yildirim and Hettiarachchy (1998) suggested that TGase-cross-linked WPI films could be used as a wrapping to prevent quality changes in products such as meat pies and high-moisture cakes requiring films permeable to water vapor. However, the addition of milk-based films and coatings to fresh fruits and vegetables would require additional labeling on the package to alert consumers with milk allergies, intolerances, religious-based dietary restrictions, or who maintain a vegetarian lifestyle (Gennadios, 2004). Limiting applications to dairy-based applications, such as films and coatings for cheese, or fresh, baked and frozen products expected to be consumed with or to contain milk-based proteins is a more logical approach and applications research should focus on these types of products (Albert and Mittal, 2002; Guillard *et al.*, 2003; Schou *et al.*, 2005). Casein films have also been explored as tablet coatings (Abu Diak *et al.*, 2007).

Recent research efforts have focused on the development of active whey protein films for packaging applications. Active packaging interacts directly with a food or headspace of the product (Ozdemir and Floros, 2004). Whey-based films alone have the advantage of being able to be in direct contact with food and in active packaging systems could contain additional flavors, natural oxygen scavengers and antimicrobials (Lee *et al.*, 2008; Dangaran and Krochta, 2008). Cagri *et al.* (2004) have reviewed the various types of protein-based and lipid-based edible films and a wide range of antimicrobial agents that could be used or are currently used to enhance the safety and shelf-life of ready-to-eat foods.

Huppertz and de Kruif (2008) demonstrated the creation of stable nanogel particles by cross-linking all caseins within the casein micelle using TGase from which all micellar calcium phosphate can be removed. The particles may have applications in the encapsulation of minerals, vitamins and pharmaceuticals, as well as to add flavors, additives and other components to foods.

23.8 Future trends

Current milk-protein based research efforts on edible films arose from the need to utilize surplus dairy proteins and was directed to improving the mechanical and barrier properties of films made with these proteins. This research needs to continue, especially for applications research, but more attention needs to be directed to structuring the films through consideration of the conformation of the protein in the film-forming solution as affected by environmental variables (Lent *et al.*, 1998; Anker *et al.*, 2000; Farrell *et al.*, 2003a, 2003b; O'Connell and de Kruif, 2003; Lefèvre *et al.*, 2005; Lencki, 2005; Qi, 2007). Just as importantly, future research needs to extend to the production of the films on a large-scale basis with pilot-plant scale equipment used by the food industry to produce large quantities of the films for food applications and shelf-life studies. Little is known about the endurance of the films during refrigeration, freezing, baking and microwaving. Cast films may have immediate application as reinforcements of foil packaging and for food protection; e.g. to reinforce the foil lids of dairy food containers while protecting the contents, or to protect entrees in microwaveable meals. Little is known about the heat-sealing properties of the films.

The impact of processing on the mechanical and barrier properties of the films also needs to be established. Some work has been done to date using a continuous process, compression molding, injection molding and extrusion. These techniques have been shown to improve the mechanical properties of the films. However, the proteins are heat sensitive which can cause color and sensory changes; this needs to be addressed. Wet spinning is another area that deserves more attention, possibly investigating the use of tyrosinase, TGase, genipin and other natural cross-linkers as replacements for the chemicals typically used in the process.

Coatings of foods with milk-based coatings on a commercial basis may be easier to achieve than film production at the present time. Techniques used by the food industry to coat foods, such as panning, fluidized bed coating, spray coating, and dipping, require little modification for use with edible milk-based coating solutions. However, research is still needed to establish operation parameters and processing protocols for each food product.

The use of milk-based films as part of multilayered packaging systems has not yet begun in earnest. However, this may prove to be the best outlet for edible films and coatings. Research has demonstrated that application of milk-based protein films to synthetic packaging improves the oxygen barrier properties of the film significantly. Milk-based films have the potential to reduce the amount of synthetic films or aluminum foil used, and facilitate biodegradability (Li and Chen, 2000; de Vlieger, 2003) However, research is needed to determine compatible film systems, the stability of the protein film layer in contact with the food, and resistance to microbial degradation as a function of shelf-life, for example. While costs are part of the drivers for selection of a packaging system, the costs of synthetic films are rising. Kozempel *et al.* (2003) have estimated that the hot-air drying costs for a 0.23 mm milk-protein film are $0.26/m^2, assuming a 10% initial solids concentration in the film-forming solution. In addition, industry has incorporated sustainability of their operations as part of their business planning, and is committed to reducing packaging.

Research is also scarce on the creation of new blended films made from synthetic film material, or the biodegradable polylactic acid (PLA), with milk-based proteins and other materials. The blends would reduce the use of the synthetic feedstock. Prototype cups (not films) made in our laboratory contained up to 25% WPI as a replacement for some of the PLA and PE material. Information on the properties of synthetic films that should be targeted in blends is found in texts such as Massey (2003).

Research on the use of nanoparticles to strengthen edible milk protein films for multi-layer packaging options is currently underway in our laboratory and is an unexplored area. Inclusion of nanoparticles such as montmorillonite clay or calcium phosphate may improve the handling of the films for manufacturing. Inclusion of whey (Bonnaillie and Tomasula, 2008) or casein fractions, pending commercial availability, may have similar impact by imparting their individual functionality to film properties.

23.9 Sources of further information and advice

The use of films or coatings made from proteins and other biobased materials has been the subject of many review articles and book chapters (Kester and Fennema, 1986; Guilbert, 1986; Chen, 1995; Krochta and De Mulder-Johnson, 1997; Debeaufort *et al.*, 1998; Korhonen, 2002; Tharanathan, 2003; Gennadios, 2004; Dangaran and Krochta, 2008).

General information on edible films, coatings, and applications in commercial use may be found at several vendor websites. Examples include: Watson, Inc., a manufacturer of edible films http://www.watson-inc.com/film_edible.php.; Ascona Ingredients Ltd, http://www.ediblefilms.com; a manufacturer of edible films, and Mantrose-Haeuser, http://www.mbzgroup.com, another manufacturer of edible films and coatings. Several use USDA/ARS technology and that developed by other research institutions in the field. Another website providing information on edible films is that sponsored by the California Dairy Research Board, http://www.ediblefilms.org. Information on commercial packaging use is found at industry websites such as http://www.packworld.com and http://www.innoviafilms.com. General information on new commercial developments in the dairy foods area may be found at http://www.dairyfoods.com. The above list is not meant to be exhaustive.

New research on milk protein edible films and other edible films is published in The *Journal of Food and Agricultural Food Chemistry*, the *Journal of Food Science*, and the *Journal of Dairy Science*. Use of search engines and software for publishing and managing bibliographies will yield most, if not all, research conducted to date on milk protein edible films not included in this chapter, as well as patents and information on films for non-food use. Research is also presented at annual meetings of the American Chemical Society (ACS), the American Institute of Chemical Engineers (AICHE), the Institute of Food Technologists (IFT), the American Dairy Science Association (ADSA), as well as at International meetings such as that held by the International Dairy Federation (IDF).

23.10 References

ABOUMAHMOUD, R and SAVELLO, P (1990), 'Crosslinking of whey protein by transglutaminase', *J Dairy Sci* 73:256–263.

ABU DIAK, O, BANI-JABER, A, AMRO, B, JONES, D and ANDREWS, G P (2007), 'The manufacture and characterization of casein films as novel tablet coatings', *Trans IChemE, Part C, Food and Bioproducts Processing*, 85, C3, 284–290.

ALBERT, S and MITTAL, G S (2002), 'Comparative evaluation of edible coatings to reduce fat uptake in a deep-fried cereal product', *Food Research International*, 35, 445–458.

ALCANTARA, C R, RUMSEY, T R and KROCHTA, J M (1998), 'Drying rate effect on the properties of whey protein films', *J Food Process Engineering*, 21, 387–405.

ANKER, M, STADING, M and HERMANSSON, A-M (2000), 'Relationship between the microstructure and the mechanical and barrier properties of whey protein films', *J Agric Food Chem*, 48, 3806–3816.

ASTM (1989) Standard D523-89, *Standard test methods for specular gloss D523-89*, ASTM International, West Conshohocken, PA.

ASTM (1995) Standard E96-95, *Standard test methods for water vapor transmission of materials*, ASTM International, West Conshohocken, PA.

ASTM (1995a) Standard D3985-95, *Standard test method for oxygen gas transmission rate through plastic film and sheeting using a coulometric sensor*, ASTM International, West Conshohocken, PA.

ASTM (1999) Standard D4039-93, *Standard test method for reflection haze of high-gloss surfaces*, ASTM International, West Conshohocken, PA.

ASTM (2001) Standard D882-01, *Standard test method for tensile properties of thin plastic sheeting*, ASTM International, West Conshohocken, PA.

AUDIC, J-L, CHAUFER, B and DAUFIN, G (2003), 'Non-food applications of milk components and dairy co-products: A review', *Lait*, 83, 417–438.

AVENA-BUSTILLOS, R J and KROCHTA, J M (1993), 'Water vapor permeability of caseinate-based edible films as affected by pH, calcium crosslinking and lipid content'. *J Food Sci*, 58(4), 904–907.

BANERJEE, R and CHEN, H (1995), 'Functional properties of edible films using whey protein concentrate', *J Dairy Sci*, 78, 1673–1683.

BANERJEE, R, CHEN, H and WU, J (1996), Milk protein-based edible film mechanical strength changes due to ultrasound process, *J Food Sci*, 61(4), 824–828.

BANKER, G S (1966), 'Film coating theory and practice', *J Pharm Sci* 55(1), 81–89.

BIGI, A, COJAZZI, G, PANZAVOLTA, S, ROVERI, N, and RUBINI, K (2002), 'Stabilization of gelatin films by crosslinking with genipin', *Biomaterials*, 23, 4827–4832.

BÖNISCH, M P, LAUBER, S and KULOZIK, U (2004), 'Effect of ultra-high temperature treatment on the enzymatic cross-linking of micellar casein and sodium caseinate by transglutaminase', *J Food Sci*, 69, E398–E404.

BONNAILLIE, L M and TOMASULA, P M (2008), Whey processing and fractionation, in Onwulata C and Huth P, *Whey Processing, Functionality and Health Benefits*, Institute of Food Technologists Series, Wiley-Blackwell, 15–38.

BRAULT, D, D'APRANO, G and LACROIX, M (1997), 'Formation of free-standing sterilized edible-films from irradiated caseinates', *J Agric Food Chem*, 45, 2964–2969.

BUONOCORE, G G, DEL NOBILE, M A, DI MARTINO, C, GAMBACORTA, G, LA NOTTE, E and NICOLAIS, L (2003), 'Modeling the water transport properties of casein-based edible coating', *J Food Engineering*, 60, 99–106.

CAGRI, A, USTUNOL, Z and RYSER, E T (2004), 'Antimicrobial edible films and coatings', *J Food Protection*, 67, 833–848.

CHAMBI, H and GROSSO, C (2006), 'Edible films produced with gelatin and casein cross-linked with transglutaminase', *Food Research International*, 39, 458–466.

CHAN, M A and KROCHTA, J M (2001), 'Grease and oxygen barrier properties of whey protein isolate-coated paperboard', *TAPPI*, 84, 57.
CHEN, H (1995), 'Functional properties and applications of edible films made of milk proteins', *J Dairy Sci*, 78, 2563–2583.
CHICK, J and USTUNOL, Z (1998), 'Mechanical and barrier properties of lactic acid and rennet precipitated casein-based edible films', *J Food Sci*, 63, 1024–1027.
CHICK, J and HERNANDEZ, R J (2002), 'Physical, thermal, and barrier characterization of casein–wax-based edible films', *J Food Sci*, 67, 1073–1075.
CHO, S Y, PARK, J-W and RHEE, C (2002), 'Properties of laminated films from whey powder and sodium caseinate mixtures and zein layers', *Lebensm.-Wiss. U-Technol*, 35, 135–139.
CHURAEV, N V (2003), 'Surface forces in wetting films', *Advances in Colloid and Interface Science*, 103, 197–218.
COUPLAND, J N, SHAW, N B, MONAHAN, F J, O'RIORDAN, E D and O'SULLIVAN, M (2000), 'Modeling the effect of glycerol on the moisture sorption behavior of whey protein edible films', *J Food Engineering*, 43, 25–30.
DANGARAN, K L, COOKE, P and TOMASULA, P M (2006), 'The effect of protein particle size reduction on the physical properties of CO_2-precipitated casein films', *J Food Sci*, 71, E196–201.
DANGARAN, K L and KROCHTA, J M (2007), 'Preventing the loss of tensile, barrier and appearance properties caused by plasticiser crystallisation in whey protein films', *International Journal of Food Science and Technology*, 42, 1094–1100.
DANGARAN, K and KROCHTA, J M (2008), Whey protein films and coatings, in Onwulata C and Huth P, *Whey Processing, Functionality and Health Benefits*, Institute of Food Technologists Series, Wiley-Blackwell, 133–168.
DEBEAUFORT, F, QUEZADA-GALLO, J-A and VOILLEY, A (1998), Edible films and coatings: Tomorrow's packagings: A review. *Critical Reviews in Food Sci*, 38, 299–313.
DE VLIEGER, J J (2003) 'Green plastics for food packaging'. In Ahvenainen, R, *Novel Food Packaging Technologies*, Cambridge, Woodhead, 519–534.
FAERGEMAND, M, OTTE, J and QVIST, K B (1998), 'Crosslinking of whey proteins by enzymatic oxidation', *J. Agricultural and Food Chemistry*, 46, 1326–1333.
FARRELL, H M JR, BROWN, E M, MALIN, E L and HOAGLAND, P D (2003a), Higher order structures in casein: A paradox. In Fox, P F and McSweeny, P H, *Advanced Dairy Chemistry: Proteins, 3rd ed*, New York, Plenum-Kluwar Inc, 96–116.
FARRELL, H M, COOKE, P H, WICKHAM, E D, PIOTROWSKI, E G and HOAGLAND, P D (2003b), 'Environmental influences on bovine κ–casein: Reduction and conversion to fibrillar (amyloid) structures', *J Protein Chem*, 22, 259–273.
FRINAULT, A, GALLANT D J, BOUCHET, B and DUMONT, J P (1997), 'Preparation of casein films by a modified wet spinning process', *J. Food Sci*, 62, 744–747.
GENNADIOS, A (2004), 'Edible films and coatings from proteins'. In Yada R Y, *Proteins in Food Processing*, Cambridge, Woodhead, 442–467.
GUILBERT, S (1986), 'Technology and application of edible protective films'. In Mathiouthi M, *Food Packaging and Preservation*, London, Elsevier Applied Science Publishers, 371–394.
GUILLARD, V, BROYART, B, BONAZZI, C, GUILBERT, S and GONTARD, N (2003), 'Preventing moisture transfer in a composite food using edible films: Experimental and mathematical study', *J Food Science*, 68, 2267–2277.
HAN, J H and KROCHTA, J M (1999), 'Wetting properties and water vapor permeability of whey-coated paper', *Transactions of the ASAE*, 42, 1375–1382.
HAN, J H and KROCHTA, J M. (2001) 'Physical properties and oil absorption of whey protein coated paper', *J Food Science*, 66, 294–299.
HERNANDEZ-IZQUIERDO, V M, REID, D S., MCHUGH, T H., BERRIOS, DE J and KROCHTA, J M (2008), 'Thermal transitions and extrusion of glycerol-plasticized whey protein mixtures', *J Food Science* 73, E169–E175.

HONG, S-I and KROCHTA, J M (2003), 'Oxygen barrier properties of whey protein isolate coatings on polypropylene films', *J Food Science*, 68, 224–228.

HONG, S-I, HAN, J H and KROCHTA, J M (2004), 'Optical and surface properties of whey protein isolate coatings on plastic films as influenced by substrate, protein concentration, and plasticizer type', *J Applied Polymer Sci*, 92, 335–343.

HONG, S-I and KROCHTA, J M (2006), 'Oxygen barrier performance of whey protein-coated plastic films as affected by temperature, relative humidity, base film and protein type', *J Food Engineering*, 77, 739–745.

HOUSTON, J S (1986), 'Polystyrene'. In: *The Wiley Encyclopedia of Packaging Technology*; Bakker, M, John Wiley & Sons., New York.

HUPPERTZ, T and DE KRUIF, C G (2008), 'Structure and stability of nanogel particles prepared by internal cross-linking of casein micelles', *International Dairy Journal*, 18, 556–565.

JIN, J, SONG, M and HOURSTON, D J (2004), 'Novel chitosan-based films cross-linked by genipin with improved physical properties', *Biomacromolecules*, 5, 162–168.

JORDAN, P J, LAY, K, NGAN, N and RODLEY, G F (1987), 'Casein precipitation using high pressure carbon dioxide', *N Z J Dairy Sci Technol*, 22, 247–256.

KAYA, S and KAYA, A (2000), 'Microwave drying effects on properties of whey protein isolate edible films', *J Food Engineering*, 43, 91–96.

KESTER, J J and FENNEMA, O R (1986), 'Edible films and coatings: A review', *Food Technol*, 40, 47–59.

KHWALDIA, K, PEREZ, C, BANON, S, DESOBRY, S and HARDY, J (2004) 'Milk proteins for edible films and coatings', *Critical Reviews in Food Science and Nutrition*, 44, 239–251.

KHWALDIA, K, BANON, S, DESOBRY, S and HARDY, J (2004a), 'Mechanical and barrier properties of sodium caseinate–anhydrous milk fat edible films', *Int. J. Food Sci. Tech*, 39, 403–411.

KILARA, A and VAGHELA, M N, (2004), Whey proteins. In Yada R Y, *Proteins in Food Processing*, Cambridge, Woodhead, 72–99.

KORHONEN, H (2002), 'Technology options for new nutritional concepts', *Int. J. Dairy Tech*, 55, 79–88.

KOZEMPEL, M, TOMASULA, P M and MCALOON, A (2003), 'Drying kinetics of calcium caseinate', *J Agric Food Chem*, 51, 773–776.

KOZEMPEL, M and TOMASULA, P M (2004), 'Development of a continuous process to make casein films', *J Agric Food Chem*, 52, 1190–1195.

KRISTO, E and BILIADERIS, C G (2006), 'Water sorption and thermo-mechanical properties of water/sorbitol-plasticized composite biopolymer films: Caseinate–pullulan bilayers and blends', *Food Hydrocolloids*, 20, 1057–1071.

KRISTO, E, BILIADERIS, C G and ZAMPRAKA, A (2007), 'Water vapour barrier and tensile properties of composite-pullulan films: Biopolymer composition effects and impact of beeswax lamination', *Food Chemistry*, 101, 753–764.

KROCHTA, J M and DE MULDER-JOHNSON, C (1997), 'Edible and biodegradable polymer films', *Food. Technol*, 51, 61–74.

KUMINS, C A (1965), 'Transport through polymer films', *J Polymer Sci*, Part C, 10, 1–9.

LANTTO, R, HEINE, E, FREDDI, G, LAPPALAINEN, A, MIETTINEN-OINONEN, A, NIKU-PAAVOLA, M-L and BUCHERT, J (2005), 'Enzymatic modification of wool with tyrosinase and peroxidase', *J Textile Institute*, 96, (2), 109–116.

LEE, J-W, SON, S-M and HONG, S-I (2008), 'Characterization of protein-coated polypropylene films as a novel composite structure for active food packaging application', *J Food Engineering*, 86, 484–493.

LEFÈVRE, T, SUBIRADE, M and PÈZOLET, M (2005), 'Molecular description of the formation and structure of plasticized globular protein films', *Biomacromolecules* 6, 3209–3219.

LENCKI, R W (2007), 'Evidence for fibril-like structure in bovine casein micelles', *J Dairy Sci* 90, 75–89.

LENT, L E, VANASUPA, L S and TONG, P S (1998), 'Whey protein edible film structures determined by atomic force microscope', *J Food Science*, 63, 1–4.

LI, J and CHEN, H (2000), 'Biodegradation of whey protein-based edible films', *J Polymers and the Environment'*, 8, 135–143.

LI-CHAN, E C Y (2004), Properties of proteins in food systems: An introduction. In Yada R Y, *Proteins in Food Processing*, Cambridge, Woodhead, 2–28.

LIN, S-Y and KROCHTA, J M (2003), 'Plasticizer effect on grease barrier and color properties of whey-protein coatings on paperboard', *J Food Science*, 68, 229–233.

LONGARES, A, MONAHAN, F J, O'RIORDAN, E D and O'SULLIVAN, M (2004), 'Physical properties and sensory evaluation of WPI films of varying thickness', *Lebensm.-Wiss. U-Technol*, 37, 545–550.

MAHMOUD, R and SAVELLO, P A (1992), 'Mechanical properties of and water vapor transferability through whey protein films', *J Dairy Sci*, 75, 942–946.

MAHMOUD, R and SAVELLO, P A (1993), 'Solubility and hydrolyzability of films produced by transglutaminase catalytic crosslinking of whey protein', *J Dairy Sci*, 76, 29–35.

MASSEY, L.K. 2003. *Permeability properties of plastics and elastomers. A guide to packaging and barrier materials*. Plastics Design Library. Norwich, NY.

MAUER, L J, SMITH, D E and LABUZA, T P (2000), 'Water vapor permeability, mechanical, and structural properties of edible β–casein films', *International Dairy J*, 10, 353–358.

MAYNES, J R and KROCHTA, J M (1994), Properties of edible films from total milk protein. *J. Food Sci*, 59, 909–911.

MCHUGH, T H, AVENA-BUSTILLOS, R and KROCHTA, J M (1993), 'Hydrophilic edible films: Modified procedure for water vapor permeability and explanation of thickness effects', *J Food Science*, 58, 899–903.

MCHUGH, T H and KROCHTA, J M (1994a), 'Dispersed phase particle size effects on water vapor permeability of whey protein–beeswax edible emulsion films', *J Food Processing and Preservation*, 18, 173–188.

MCHUGH, T H and KROCHTA, J M (1994b), Sorbitol- vs glycerol-plasticized whey protein edible films: Integrated oxygen permeability and tensile property evaluation', *J Agric Food Chem*, 42, 841–845.

MCHUGH, T H and KROCHTA, J M (1994c), 'Water vapor permeability properties of edible whey protein–lipid emulsion films', *JAOCS*, 71(3), 307–312.

MCHUGH, T H and KROCHTA, J M (1994d), 'Milk protein-based edible films and coatings', *Food Technology*, January, 97–103.

MCHUGH, T H, AUJARD, J F and KROCHTA, J M (1994), 'Plasticized whey protein edible films: Water vapor permeability properties', *J Food Sci*, 59, 416–419, 423.

MI, F-L, SHYU, S-S and PENG, C-K (2005), 'Characterization of ring-opening polymerization of genipin and pH-dependent cross-linking reactions between chitosan and genipin', *J Polymer Sci*, Part A, 43, 1985–2000.

MOTOKI, M, ASO, H, SEGURO, K and NIO, N (1987), 'Alpha s1 casein film prepared using transglutaminase', *Agric Biol Chem*, 52, 993–996.

MÜLLER-BUSCHBAUM, P, GEBHARDT, R, MAURER, E, BAUER, E, GEHRKE, R and DOSTER, W (2006), 'Thin casein films as prepared by spin-coating: Influence of film thickness and of pH', *Biomacromolecules*, 7, 1773–1780.

NIEUWENHUIZEN, W F, DEKKER, H L, GRONEVELD, T, DE KOSTER, C G and DE JONG, G A H (2004), 'Transglutaminase-mediated modification of glutamine and lysine residues in native bovine lactoglobulin', *Biotechnology and Bioengineering*, 85, 248–258.

O'CONNELL, J E and DE KRUIF, C G (2003), 'β-casein micelles: Cross-linking with transglutaminase', *Colloids and Surfaces A: Physicochem. Eng. Aspects*, 216, 75–81.

ONWULATA, C I, KONSTANCE, R P, COOKE, P H and FARRELL, H M (2003), 'Functionality of extrusion-texturized whey proteins', *J Dairy Sci*, 86, 3775–3782.

ONWULATA, C I, ISOBE, S, TOMASULA, P M and COOKE, P H (2006), 'Properties of whey protein isolates extruded under acidic and alkaline conditions', *J Dairy Sci*, 89, 71–81.

ONWULATA, C I and TOMASULA, P M (2008), 'Gelling properties of tyrosinase-treated dairy proteins', Accepted for publication to *Food Bioprocess Technol.* DOI: 10.1007/s11947-008-0124-4.

OUATTARA, B, CANH, LT, VACHON, C, MATEESCU, M A and LACROIX, M (2002), 'Use of γ-irradiation cross-linking to improve the water vapor permeability and the chemical stability of milk protein films', *Radiation Physics and Chemistry*, 63, 821–825.

OZDEMIR, M and FLOROS, J D (2004), 'Active food packaging technologies', *Critical Reviews in Food Science and Nutrition*, 44, 185–193.

PEREZ-GAGO, M B, NADAUD, P and KROCHTA, J M (1999), 'Water vapor permeability, solubility, and tensile properties of heat-denatured versus native whey protein films', *J Food Science*, 64, 1034–1037.

PEREZ-GAGO, M B and KROCHTA, J M (2000), 'Drying temperature effect on water vapor permeability and mechanical properties of whey protein–lipid emulsion films', *J Agric Food Chem*, 48, 2687–2692.

QI, P X (2007), 'Casein micelle structure: The past and the present', *Le Lait*, 87, 363–383.

RESSOUANY, M, VACHON, C and LACROIX, M (1998), 'Irradiation dose and calcium effect on the mechanical properties of cross-linked caseinate films', *J Agric Food Chem*, 46, 1618–1623.

ROBERTSON, G L (1993), Processing and converting of thermoplastic polymers, in Robertson, G L, *Food Packaging, Principles and Practice*, New York, Marcel Dekker, 1–8; 111–143.

SALAME, M (1986), 'Barrier Polymers', in *The Wiley Encyclopedia of Packaging Technology*; Bakker, M, John Wiley & Sons, New York, 48–54.

SANTOS, C V, TOMASULA, P M and KURANTZ, M J (1999), 'Deamidation kinetics of casein precipitated by carbon dioxide compared with commercial caseinates', *J Food Science*, 64, 400–414.

SANTOS, C V and TOMASULA, P M (2000), 'Acylation and solubility of casein precipitated by carbon dioxide', *J Food Science*, 65, 227–230.

SCHOU, M, LONGARES, A, MONTESINOS-HERRERO, C, MONAHAN, F J, O'RIORDAN, D and O'SULLIVAN, M (2005), 'Properties of edible sodium caseinate films and their application as food wrapping', *LWT*, 38, 605–610.

SEO, S-Y, SHARMA, V K and SHARMA, N (2003), 'Mushroom tyrosinase: Recent prospects', *J. Agric. Food Chem*, 51, 2837–2853.

SHARMA, R, LORENZEN, P C and QUIST, K B (2001), 'Influence of transglutaminase treatment of skim milk on the formation of ε-(γ-glutamyl)lysine and the susceptibility of individual proteins towards crosslinking', *International Dairy J.*, 11, 785–793.

SHARMA, R, ZAKORA, M and QVIST, K B (2002), 'Susceptibility of an industrial α-lactalbumin concentrate to cross-linking by microbial transglutaminase', *International Dairy J*, 12, 1005–1012.

SHELLHAMMER, T H and KROCHTA, J M (1997), 'Water vapor barrier and rheological properties of simulated and industrial milk fat fractions,' *Transactions of the ASAE*, 40, 1119–1127.

SHELLHAMMER, T H and KROCHTA, J M (1997a), 'Whey protein emulsion film performance as affected by lipid type and amount', *J Food Science*, 62, 390–394.

SIEW, D C W, HEILMANN, C, EASTEAL, A J and COONEY, R P (1999), 'Solution and film properties of sodium caseinate/glycerol and sodium caseinate/polyethylene glycol edible coating systems', *J Agri Food Chem*, 47, 3432–3440.

SOTHORNVIT, R and KROCHTA, J M (2000a), 'Oxygen permeability and mechanical properties of films from hydrolyzed whey protein', *J Ag Food Chem*, 48, 3913–3916.

SOTHORNVIT, R and KROCHTA, J M (2000b), 'Plasticizer effect on oxygen permeability of beta-lactoglobulin films', *J Ag Food Chem*, 48, 6298–6302.

SOTHORNVIT, R and KROCHTA, J M (2000b), 'Water vapor permeability and solubility of films from hydrolyzed whey protein', *J Food Science*, 65, 700–703.

SOTHORNVIT, R, OLSEN, C W, MCHUGH, R H and KROCHTA, J M (2003), 'Formation conditions, water-vapor permeability and solubility of compression-molded whey protein films', *J Food Science*, 68, 1985–1989.

SOTHORNVIT, R, OLSEN, C W, MCHUGH, T H and KROCHTA, J M (2007), 'Tensile properties of compression-molded whey protein sheets: Determination of molding condition and glycerol-content effects and comparison with solution-cast films', *J Food Engineering*, 78, 855–860.

SOUTHWARD, C R and WALKER, N J (1980), 'The manufacture and industrial use of casein', *N Z J Dairy Sci and Technol*, 15, 201–217.

THALMAN, C R and LOTZBEYER, T (2002), 'Enzymatic cross-linking of proteins with tyrosinase', *Eur Food Res Technol*, 214, 276–281.

THARANATHAN, R N (2003), Biodegradable films and composite coatings: Past, present, and future, *Trends Food Sci Technol*, 14, 71–78.

TOMASULA, P M, CRAIG, J C JR, BOSWELL, R T, COOK, R D, KURANTZ, M J and MAXWELL, M (1995), 'Preparation of casein using carbon dioxide', *J Dairy Sci*, 78 (3), 506–514.

TOMASULA, P M, CRAIG, J C and BOSWELL, R T (1997), 'A continuous process for casein production using high-pressure carbon dioxide', *J Food Eng*, 33, 405–419.

TOMASULA, P M, PARRIS, N, YEE, W and COFFIN, D (1998), 'Properties of films made from CO_2-precipitated casein', *J Agric Food Chem*, 46, 4470–4474.

TOMASULA, P M, YEE, W C and PARRIS, N (2003), 'Oxygen permeability of films made from CO_2-precipitated casein and modified casein', *J Agric Food Chem*, 51, 634–639.

TUNICK, M and ONWULATA, C I (2006), 'Rheological properties of extruded milk powders', *International J Food Properties*, 9, 835–844.

VACHON, C, YU, H-L, YEFSAH, R, ALAIN, R, ST-GELAIS, D and LACROIX, M (2000), 'Mechanical and structural properties of milk protein edible films cross-linked by heating and γ-irradiation', *J Agric Food Chem*, 48, 3202–3209.

VARGAS, M, PASTOR, C, CHIRALT, A, MCCLEMENTS, D J and GONZALEZ-MARTINEZ, C (2008), 'Recent advances in edible coatings for fresh and minimally processed fruits', *Critical Reviews in Food Science and Nutrition*, 48, 496–511.

WHITTEN, K W, GAILE, K D and DAVIS, R E (1992), *General Chemistry*, 4th ed.; Saunders College Publishing, Harcourt Brace Jovanovich College Publishers: Philadelphia, PA.

YILDIRIM, M and HETTIARACHCHY, N S (1998), 'Properties of films produced by cross-linking whey proteins and 11S globulin using transglutaminase', *J Food Sci*, 63, 248–252.

24

Transformation of lactose for value-added ingredients

A. H. J. Paterson*, Massey University, New Zealand; S. J. Kellam, Hilmar Cheese Company, USA

Abstract: Lactose, as a raw material in whey, is currently available in quantities that far exceed the demand for the product. This means that lactose is inexpensive relative to the costs of other dairy products. This chapter looks at the sources of lactose, the reasons behind why there was a recent spike in lactose price and the range of different derivatives that people have considered as ways of using lactose. Tagatose seems to be the product with the most potential to develop into a significantly large market that will use significant amounts of lactose.

Key words: lactose, oligosaccharides, lactulose, lactitol, lactobionic acid, lactitol, hydrolyzed lactose, tagatose, ethanol.

24.1 Introduction

24.1.1 Sources of lactose

Lactose is a disaccharide sugar made up of one glucose and one galactose group (Fig. 24.1.) and it is the main energy source in mammalian milk. As such, it is found in the milk of most mammals. The level of lactose in human milk is around 7%, but the level in bovine milk is only about 4.5%. When cows' milk is fed to human infants, there is a deficiency in the level of energy in the formula. This applies whether the milk is fresh milk or is reconstituted from whole milk powder. Henry Nestlé was the first person who realized this and he formulated an infant milk powder from cows' milk boosted by the addition of more lactose (Anon., 2007a). This became the first major use of lactose and led to the development of lactose plants to recover the lactose from cheese whey.

The overall production of milk has been increasing over the last ten years at a rate of between 1.4 and 3% per annum. It had reached a level of 551 million tonnes of cows' milk in 2007 and 104 million tonnes of other animal

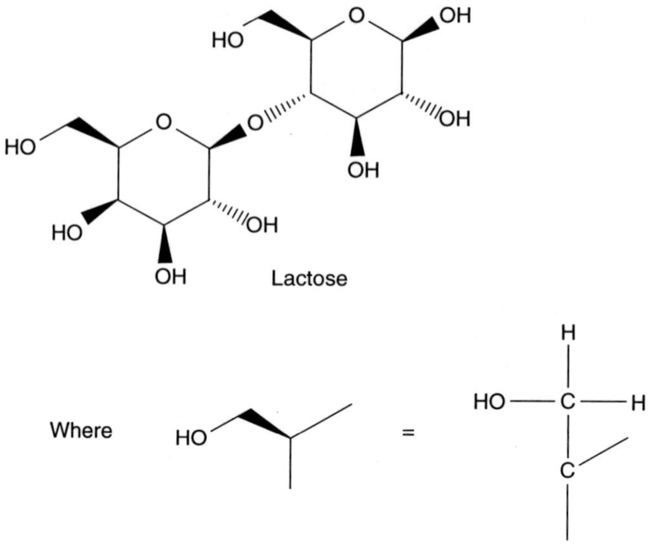

Fig. 24.1 Structure of lactose.

milk (Anon., 2007b). From, this the overall cheese production was 17 million metric tonnes (Anon., 2007b) and hence the production of whey is estimated to be about 157 million tonnes per annum (Fox *et al.*, 2000). The total lactose available for production is estimated to be 6.3 million tonnes per annum. Affertsholt (2008) quotes the world production of lactose at 890 thousand metric tones for 2007 and that this is projected to rise to between 940 to 1000 thousand metric tonnes by 2010. This shows that only 15% of the potentially available lactose is currently being processed into lactose. Currently, most of the excess is dried into whey powder and used in swine production or otherwise disposed of.

The lactose market is broadly split into two segments: edible grade at 780000 metric tonnes and 110000 metric tonnes of pharmaceutical grade (Affertsholt, 2008). Edible grade lactose is defined in CODEX as: 'Lactose: A natural constituent of milk normally obtained from whey with an anhydrous lactose content of not less than 99.0%, w/w, on a dry basis. It may be anhydrous or contain one molecule of water of crystallization or be a mixture of both forms.' http://www.codexalimentarius.net/download/standards/338/CXS_212e.pdf (accessed 11 February 2009).

Descriptions of the edible lactose production method and the methods that can be used to produce pharmaceutical grade lactose are given in Paterson (2009).

The main use of lactose has traditionally been as an addition to infant formula and this is still the main use of lactose in the USA today. Table 24.1 shows the split in market share for lactose between the market in the

Table 24.1 Market split of lactose in the USA and in EU (Affertsholt, 2007)

Uses of lactose	Market share USA (%)	Market share EU (%)
Infant formula	66	18
Neutraceuticals	5	28
Pharmaceuticals		
Confectionery	16	16
Bakery		5
Processed foods including meat		30
All other uses	13	3

EU and USA. This table shows that lactose uses in the EU are more diversified than in the USA. The excess of available lactose from whey over the current market size shows that there is still a surplus of lactose world wide. This is reflected in the price of lactose. Figure 24.2 shows the price of lactose in the USA since 2004 (Anon., 2008b). The increase in lactose price in 2007 has been attributed to two effects. The first was the overall high prices being paid for all dairy products world wide and the second was the effect of protein standardization of milk powders which saw a lot of lactose being taken out of the traditional markets and instead being used in milk powders. The price of lactose started to approach the price of skim milk powder as this became the default use. However, it did not take long for more lactose plants to come on stream and bring the price back down to its traditional level. It is an interesting debate on the laws of supply and demand as to what the true value of lactose is.

When added to skim milk powder or whole milk power to standardize the powders' protein content (lactose is added so that the protein content matches the specified minimum for the product), its inherent value is up to the price of SMP or WMP. However, if the dairy factory has to sell any excess lactose production, the market value is only a fraction of this value. This leads to many debates as to what price lactose should be priced at within a company when deciding whether to invest in capital to recover lactose.

The large world wide over-supply of lactose and the unacceptability of discharging lactose-rich streams with their high biological oxygen demand into waste waters means that alternative uses for lactose are actively being sought. People have been turning to the derivatives of lactose to find useful ones that will give a high return. We will now look at the derivatives of lactose, how they are formed and why they might prove to be useful as alternative products.

Figure 24.3, adapted from Gänzle *et al.* (2008), shows the relationship of the various derivatives that can be made from lactose. Table 24.2 shows the derivative market size for 2007 and that predicted for 2010 by Affertsholt

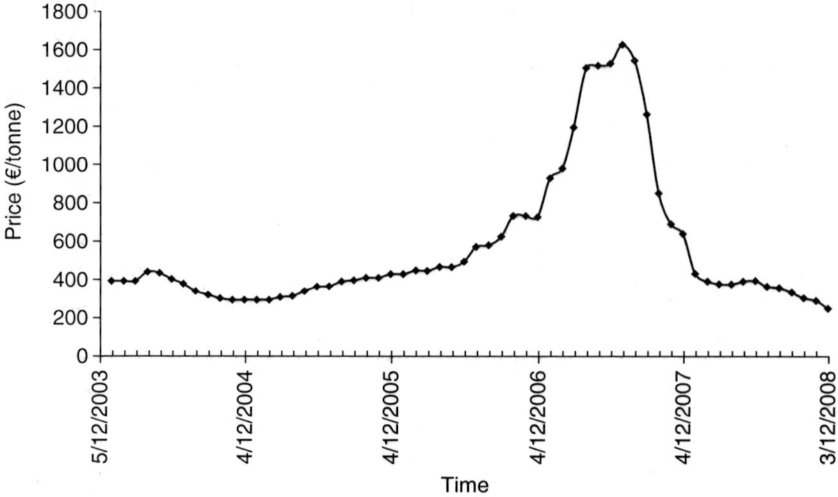

Fig. 24.2 International price of edible grade lactose from 2003 to 2008 in euros/tonne.

(2008) with the exception of tagatose which was not included in the report. Independent estimates for tagatose from the authors have been included as we see this product, along with the galacto oligosaccharides as being the main products with potential to develop into substantial markets.

The first improvements to milk powder to turn it into an infant formula were made by Henry Nestlé by the addition of lactose to make infant formula closer to that of human milk. There have been studies done which demonstrate that babies do better on human breast milk than on infant formula in terms of brain development and general health in that their ability to fight off gastrointestinal upsets appears to be better (Benno and Mitsuoka, 1986; Bouhnik *et al.*, 1994). This has been traced back to the levels of oligosaccharides in human breast milk, especially during the first few weeks when the colostrum is being expressed. Research has shown that there are over 130 different oligosaccharides, including sialic acid, N-acetylglucosamine, L-fucose, D-glucose and D-galactose in human milk (Kunz *et al.*, 2000; Kunz and Rudloff, 2006; MacFarlane *et al.*, 2006). These oligosaccharides have been shown to be responsible for the production of a healthy lower intestine flora of bacteria rich in bifidobacteria and lactobacillus strains which are now known to restrict the growth of enteropathogens.

In an effort to try to make the infant formulas more like human milk, there has been a lot of research conducted looking at the effectiveness of different oligosaccharides at mimicking these effects (Hawkins, 1993; Rowland and Tanaka, 1993; Van Loo *et al.*, 1999; Kunz *et al.*, 2000; Gopal

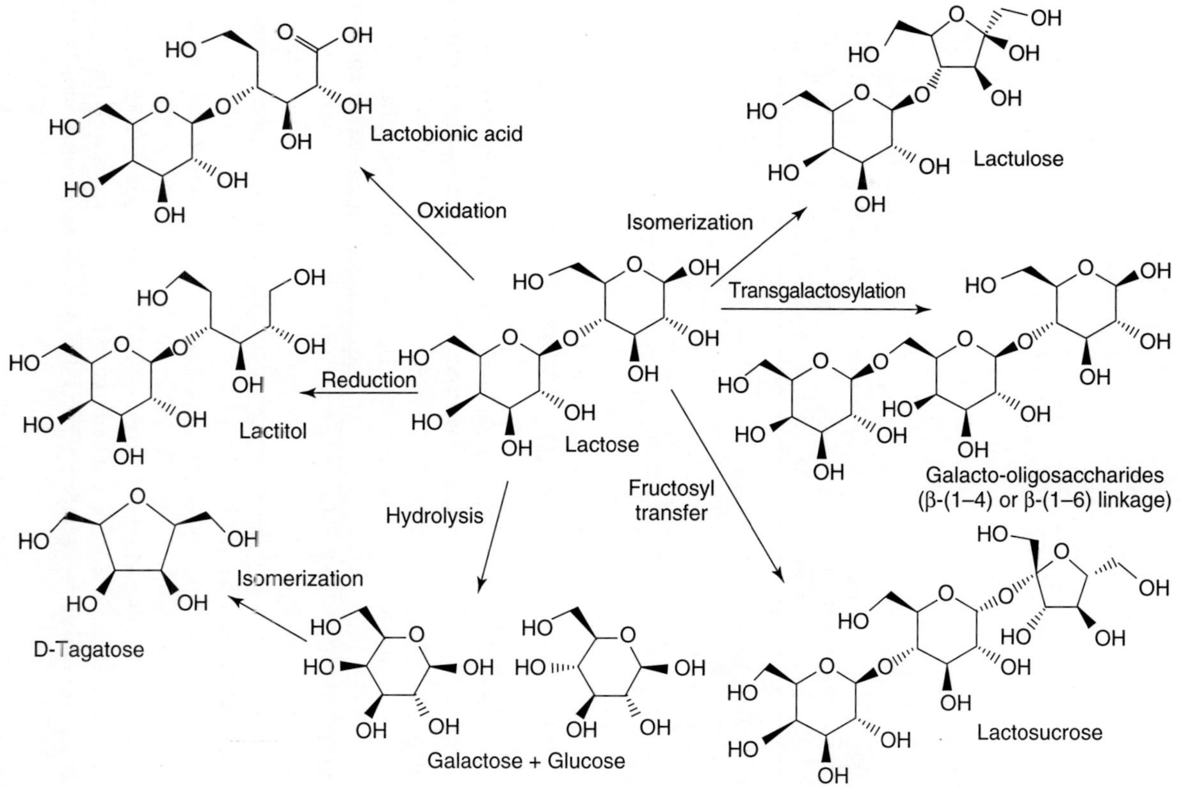

Fig. 24.3 Lactose and its relationship to its derivatives, adapted from Gänzle *et al.* (2008).

Table 24.2 Lactose derivatives market for 2007 and predictions for 2010 (Affertsholt, 2008)

	2007		2010	
	Volume MT	Value MUSD	Volume MT	Value MUSD
Galactose	10	0.45–0.6	10–11	0.45–0.7
Lactulose	30–40 000	160–176	36–42 000	180–190
Lactitol	10–11 000	40–50	11–12 000	44–55
Lactobionic acid	N/A	N/A	N/A	N/A
GOS	25 000	135–170	33–43 000	180–290
Lactic acid	170 000	238–425	196–226 000	270–560
Lactosucrose	3 500	28	4–4 600	32–37
Tagatose	900	1.3	20–30 000*	37.5

* Author's estimate by 2015 rather than 2010.
MT = metric tonne.
MUSD = million US dollars.

et al., 2001; Rivero-Urgell and Santamaria-Orleans, 2001; Playne *et al.*, 2003; Rastall, 2003; Tuohy *et al.*, 2005; Veereman-Wauters, 2005; Bode, 2006; Kunz and Rudloff, 2006; MacFarlane *et al.*, 2006, 2008; Moro *et al.*, 2006; Shoaf *et al.*, 2006; Boehm, 2008; Gänzle *et al.*, 2008; Schaafsma, 2008). These oligosaccharides come from many sources, such as lactose, corn syrups, soya beans and sugar beet, among others. The forms vary from gluco-oligosaccharides, galacto-oligosaccharides (GOS), fructo-oligosaccharides (FOS) and xylo-oligosaccharides. These oligosaccharides have all been shown to have prebiotic benefits to greater or lesser extents.

There have been several reviews written recently in this area that are well worth a read for anyone seeking a background in the advantages of using oligosaccharides as prebiotics in general. (Hawkins, 1993; Rowland and Tanaka, 1993; Sako *et al.*, 1999; Kunz *et al.*, 2000; Rivero-Urgell and Santamaria-Orleans, 2001; Rastall, 2003; Tuohy *et al.*, 2005; Bode, 2006; Kunz and Rudloff, 2006; MacFarlane *et al.*, 2006, 2008; Boehm, 2008; Gänzle *et al.*, 2008; Schaafsma, 2008). Tomomatsu (1994) has reviewed the early work, which has included trials on humans as well as *in vitro* work. The benefits attributed to the addition of GOS and FOS types include; the preferential growth of bifidobacteria leading to a reduction of detrimental bacteria in the gut which in turn leads to less toxic metabolites and cancerous substances, protection of the liver (due to less toxic material for the liver to deal with). The increased numbers of bifidobacteria has also been associated with the prevention of constipation, a reduction of serum cholesterol, the reduction of blood pressure and the production of nutrients.

The review of Rivero-Urgell and Santamaria-Orleans (2001) covers in detail the research into the advantges that adding oligosaccharides to infant formulas.

24.2 Galacto-oligosaccharides

In general, galacto-oligosaccharides have been shown to have mainly beneficial effects on the human gut flora and most of the benefits can be traced back to the fact that they are preferred as a food source by the bifidobacteria and they are not consumed by the harmful bacteria. GOS are usually made from lactose by enzymatic hydrolysis (see Fig. 24.3) using β-galactosidases which come from many different sources including bacteria and fungi. Each β-galactosidase behaves differently and produces a different mixture of GOS, glucose and galactose, and the mixture also depends on how long and under what conditions the hydrolysis is carried out. Most commercially available β-galactosidases have been selected to give high yields of glucose and galactose, but recent research has shown that high levels of GOS can be produced using high concentrations of lactose at the start of the reaction and higher temperatures (60°C), and stopping the reaction when 70 to 80% complete. Note: if left to continue, the β-galactosidases generally start to break down the GOS structure into galatose units.

As an example of the mixtures of GOS that are generated by different β-galactosidases, Boon *et al.* (2000) states 'Clear differences between the β-galactosidases were found concerning amount, size, and type of oligosaccharides produced. The β-galactosidases from *Bacillus circulans* produced the most abundant amount, the most different, and largest-sized oligosaccharides. The β-galactosidases from *Kluyveromyces* spp. produced mainly trisaccharides.' There is a large literature body on this subject and little is gained by discussing it in detail here. As an example, Onishi and Tanaka (1996) screened 19 different β-galactosidases from different organisms and looked at the yields that could be obtained from lactose as well as the mixtures of GOS formed. This whole area is an active area of research as dairy companies try to find the best mix of GOS to add to infant formulas and the right β-galactosidase or mixture of β-galactosidases to make the desired mixture of GOS.

24.3 Lactulose

Lactulose is the first lactose derivative that was commercially produced. As seen in Fig. 24.3, lactulose is made by isomerization of lactose. This is usually carried out in alkaline solutions with boric acid added to prevent side reactions (Kozempel and Kurantz, 1993; Kozempel *et al.*, 1995, 1999). The conversion yield has always been low at less than 30% and there are problems with the degradation of the lactulose to galactose and isosaccharinic acids (Harju, 1992). An alternative enzyme route has been reported by several authors (Lee *et al.*, 2004; Mayer *et al.*, 2004; Kim *et al.*, 2006). Lactulose purification normally requires several different steps though a simpler chromatographic separation method for the purification of lactulose has also been patented (Heikkilä *et al.*, 2007). The product is normally sold in a syrup form due to problems encountered when crystallizing and drying it (Harju, 1992).

It was used as a treatment of chronic portal-systemic encephalopathy by Conn *et al.* (1977) who found it to be quite effective. This was confirmed in a full review of lactulose health effects given by Havernaar and van Dokkam (2001). It appears that lactulose behaves in the gut in a similar manner to GOS, in that it is not digested in the upper intestine and has been shown to increase the levels of bifidiobacteria in the lower intestine. Studies have been carried out using lactulose as an additive to infant formula with positive results. The other main use of lactulose is in the treatment of constipation. Lactulose has been shown to have a safe laxative effect when used at about 20 gm per day dosage. It is especially good for pregnant women as it is not adsorbed through the intestinal lining. Consequently, it does not reach the foetus and it has not been found in breast milk. Higher dosage levels can lead to diarrhoea.

24.4 Lactitol

The first probable description of lactitol dihydrate was reported by Senderens (1920) with the anhydrous (Karrer and Büchi, 1937; Wolfrom *et al.*, 1938), monohydrate (van Velthuijsen, 1979) and trihydrate (Kawashima *et al.*, 1991) forms appearing subsequently in the literature. It was not until the 1980s that lactitol was developed into a commercial product, with applications as a sugar alcohol for use in confectionery, medical treatments and pharmaceutical formulations.

Lactitol is prepared by the catalytic reduction of lactose using hydrogen under high pressure. A lactose solution is hydrogenated at 100°C for 6 hrs at 8825 kPa using a Raney nickel catalyst. The catalyst is then filtered off and the resulting lactitol solution is purified by ion-exchange and activated carbon steps. Evaporation and crystallization of the liquor yields the monohydrate product (van Velthuijsen, 1979). Manipulation of the crystallization conditions allows production of lactitol as the monohydrate, dihydrate (Blankers, 1995) or anhydrous form (Myers *et al.*, 2005). Extensive lactitol technical data is provided by van Velthuijsen (1979).

Lactitol is not a reducing sugar and is relatively stable to heat, acids, and bases, and to Maillard-type reactions. The relative sweetness is about 35% as compared with sucrose, although the exact value varies with concentration (van Velthuijsen, 1979). Lactitol has been approved by the FAO/WHO Expert Committee of Food Additives, with its accepted daily intake being recorded as 'not specified' (IDF, 1983).

Lactitol has a probiotic effect in the human gut in that it reaches the colon intact and is fermented by specific bacterial species (Gibson and Roberfroid, 1995). Over 95% of ingested lactitol passes through to the colon where it is fermented primarily by *Lactobacillus* and *Bifidobacter* species, thus increasing their numbers. Such changes in the microflora composition positively affect health through specific alterations of the colonic environment that may include a reduction in pH and an inhibition in the production of ammonia (Kitler *et al.*, 1992). Lactitol does not cause hyperglycaemia and hyperinsulinaemia. The maximum metabolic energy derived from lactitol reaching the colon is 8 kJ/g (2 kcal/g) which is half that of carbohydrates.

A significant use of lactitol is in the prevention of dental caries. *In vitro* studies have shown much reduced demineralization of dental enamel mineral compared with other sugars (Grenby *et al.*, 1989). The same workers studied the cariogenicity of lactitol in rats and concluded that substitution of lactitol for sucrose significantly reduced dental decay. These studies demonstrated the suitability of lactitol for use in chewing gums and allied confectionery. There are many other applications of lactitol in chocolate, baked goods and other products and these have been recently reviewed (Young, 2006). It is notable that of the new products destined for market in recent years, almost three-quarters were aimed at the low-calorie sector (Affertsholt, 2007).

Lactitol may also be used as a pharmaceutical excipent, although it is typically more expensive than pharmaceutical grades of lactose. Consequently, its use is normally restricted to applications that exploit the low hydroscopicity of this sugar alcohol: namely, tablet preparations containing moisture-sensitive active compounds. In parallel with lactose, crystalline forms of lactitol produce tablets with poor hardness and friability characteristics. To overcome this, methods have been devised to create directly compressible lactitol forms that produce acceptable tablets (Oliger and Pearson, 1998).

Current production stands at about 10 000 tons per annum (Affertsholt, 2007).

24.5 Lactobionic acid

Lactobionic acid (4-*O*-ß D-galactopyranosyl-D-gluconic acid) is rarely found in nature. It was first described from a chemical synthesis (Fischer and Meyer, 1889) as an oxidation product of the lactose free aldehyde group (Harju *et al.*, 1993). A number of other preparation methods have since been described, including the use of lactose oxidase systems from various bacterial species, an example of which is the glucose/fructose oxidoreductase system from *Zymomonas mobilis* (Satory *et al.*, 1997), but there are many other lactose oxidase systems from other bacterial species.

Lactobionic acid is a strong metal chelator. Consequently, a primary use of lactobionic acid is as a component of solutions used to stabilize organs prior to transplantation. The acid's strong metal chelating properties means it acts to reduce the oxidative damage during storage and preservation (Southard and Belzer, 1995). It is also used for preparing the lactobionate salts of macrolide antibiotics, most notably erythromycin, thereby increasing their solubilities.

Though a number of product applications have been demonstrated for lactobionic acid, none have had significant commercial uptake. A small number of companies manufacture about 1000MT/yr (Affertsholt, 2007). This limited tonnage of lactobionic acid in commercial products makes little impact on global lactose stocks.

24.6 Hydrolyzed lactose

Over the last 100 years there has been a massive increase in cheese production and with it, vast volumes of whey. The development of ultrafiltration membrane systems has allowed the protein from this stream to be captured and manufactured into whey protein concentrate grades. The permeate from this process is suitable for lactose production. The problem facing dairy companies is that lactose markets are frequently out of balance with the cheese markets, so finding derivatives from lactose has become necessary.

The hydrolysis of lactose to glucose and galactose to make sweetening syrups can be easily envisaged as a process and a use. In principle, cleavage of the glycosidic linkage is all that is required and, although it is simple to achieve, lactose hydrolysate has proven to be commercially elusive as a product. Lactose may be hydrolyzed either chemically using mineral acids/strong cation resins in the hydrogen form, or enzymatically with β-galactosidase, yielding the monosaccharides in equimolar proportions. However, due to the processing conditions employed, oligosaccharides may be formed and Maillard-type side reactions can occur. Many reports have been published and patents filed covering this subject matter and so the reader is referred to comprehensive reviews for details on the history of this product and hydrolysis processes (IDF, 1993; Gänzle *et al.*, 2008).

The problem with hydrolyzed lactose sweeteners is that, given a product price comparison with corn-derived high fructose syrups, it is unlikely that the former will ever gain any significant market presence. It is however conceivable that with sweetener markets trending towards more natural sources, bulk lactose hydrolysates may eventually become viable. The most significant use of lactose hydrolysis technologies is in the production of low-lactose milks, in which the lactose is partially removed and the remainder broken down by free or immobilized β-galactosidase to give a product acceptable to the lactose intolerant/malabsorbing community.

Although there are very few stand-alone plants operating anywhere around the world, lactose hydrolysis processes are found integrated into other higher-value lactose derivative processes.

24.7 Tagatose

The last few decades have seen a considerable increase in rates of both adult and child obesity in the Western World due to the ready availability of foods and sedentary lifestyles. This is of major concern because of the strains it places on the health of the patient, medical services and the social cost. A reduction of calorific intake is desirable for some of these individuals and low-calorie sweeteners have a clear part to play. There are many of these sweeteners currently on the market though new entrants are of particular interest if they impart additional health, functionality and/or sensory advantages.

D-Tagatose is a ketohexose sugar, a stereoisomer of D-fructose. It was first detected in a gum from the cacao tree, *Sterculia setigera* (Hirst *et al.*, 1949) and subsequently isolated from many other sources. The range of these sources is varied but primarily encompasses fruit at the higher end of the scale and dairy products at the other (Skytte, 2006), with the notable exception of sterilized milk (Troyano *et al.*, 1996). Clearly, the consumption of tagatose has always been a natural part of human and animal diets.

Tagatose was recognized as a promising candidate during a search for alternative sweeteners by the American company Biospherix. Patents covering its production soon followed (Beadle *et al.*, 1991, 1992). The devised production scheme is revealed in these patents and a submission made to the UK Advisory Committee on Novel Foods and Processes (Bär, 2004). In essence, lactose is hydrolyzed using β-galactosidase and the resulting saccharides separated by simulated moving bed chromatography (SMBC), a purification technique that can achieve high yield and purity on a continuous running basis. The galactose fraction is isomerized by the addition of calcium hydroxide, which raises the solution pH, complexes with the tagatose formed and, as a result, moves the equilibrium towards product formation. This complex precipitates and can be harvested. The tagatose is then released by sparging the system with carbon dioxide gas, calcium carbonate being produced as a by-product. Purification of the tagatose is achieved using a further SMBC elution, ion exchange and activated carbon treatment. Evaporation to an elevated solids level, crystallization and centrifugal extraction of the crystals complete the purification process.

In 1996, Arla Food Ingredients acquired the tagatose production rights from Biospherix. SweetGredients, an Arla/Nordzucker joint company, built a large pilot plant at Nordstemmen, Germany that by 2004 was producing enough tagatose to provide customers with trial quantities. In 2006 production ceased due to the unfavorable economics of this process.

Simultaneously, other parties had been working on an enzymatic method for tagatose production. Arabinose isomerase (EC 5.3.1.4) normally catalyzes the conversion of L-arabinose to L-ribulose but it can also mediate the conversion of D-galactose into D-tagatose. Particular interest has been taken in arabinose isomerase enzymes from thermophilic bacterial species

(Lim *et al.*, 2007), and two notable patents have appeared using enzymes from *Thermoanaerobacter mathranii* (Bertelsen *et al.*, 2006) and *Thermotoga neapolitana* (Pyun *et al.*, 2005) respectively.

Since 2007, the Belgian company Nutrilab has been small-scale manufacturing tagatose using an undisclosed enzymatic process and expects to have a plant producing some 10000 tonnes annually by September 2009 (Halliday, 2008). Biospherix have formed an association with Inalco (Italy) covering the immediate provision and long-term supply of tagatose (Anon., 2008a). As previously mentioned, tagatose has been advocated as an alternative sweetener. The US Food and Drug Administration has acknowledged 'Generally Regarded As Safe' (GRAS) status for tagatose as a food additive, with specified use levels (Rulis, 2001). On consumption, about 20% of the tagatose is absorbed through the ileum and is metabolized in the liver by the same metabolic pathway as for fructose. The majority, however, reaches the colon, where tagatose has been shown to demonstrate probiotic activity and be fermented ultimately into short-chain fatty acids (Venema *et al.*, 2005). The metabolic energy value has been determined as 6 kJ/g (1.5 kcal/g), a value that was accepted by the US FDA with 'no objections'. The US FDA has also amended a directive on dental caries to include tagatose (FDA 2003).

Tagatose has 92% of the sweetness of sucrose when compared as a 10% solution. It has a slightly faster taste onset but the flavour is sucrose-like, with no after-taste. In common with the polyols, tagatose is a probiotic and non-carcinogenic but does not have a laxative effect when consumed in moderate amounts (Levin *et al.*, 1995). Tagatose is also known to enhance flavours in confectionery systems (Rosenplenter and Mende, 2004) and, as a reducing sugar, will undergo Maillard type browning reactions when combined in food systems.

Application patents have appeared relating to the use of tagatose (Howling and Callagan, 2002; Vigh and Andersen, 2002, 2007). The wide range of applications indicated include biscuits and cakes, ready-to-eat snacks, chocolate and wider confectionery products, but given tagatose supply constraints, few products are currently on the market.

A more recent development is the use of tagatose as a treatment for Type 2 diabetes due to its beneficial effects on postprandial hyperglycaemia and hyperinsulinaemia (Lu *et al.*, 2008). Studies also showed promise in inducing weight loss and raising high-density lipoprotein cholesterol. If a current phase 3 clinical trial that is scheduled for completion at the end of 2009 demonstrates the efficacy of D-tagatose, Spherix plans to source further material from Inalco (Italy) over the longer term. Spherix is seeking approval from the US Food and Drug Administration (FDA) in order to market D-tagatose as a drug in the USA. If approval of the New Drug Application is granted, an application for approval of D-tagatose to treat diabetes in the European Union (EU) may follow (Spherix, 2008).

Of all the lactose derivatives, tagatose has the potential to use large tonnages of lactose. The chemical properties of this compound coupled with excellent biological acceptance and wide applicability in foodstuffs and specific medical treatments explain the strong interest by academic and industrial groups.

24.8 Lactose as a fermentation substrate

Lactose is used to commercially produce ethanol, with the potable alcohol grade having been produced from lactose in New Zealand for over twenty years. When the price of lactose rose in 2007, this became a marginal operation and consideration was given to stopping the process, but with the price of lactose dropping again during 2008/9, potable alcohol has returned to being a profitable process. The low price of lactose means that it can be considered as a viable substrate for the production of bioethanol and several other fermented products. In these cases, the lactose must compete against other cheap carbohydrate sources such as corn syrup and sucrose. It will be an interesting market over the next ten years as the demand for bioethanol climbs. This will affect the price of all carbohydrate substrates either directly as a competitor substrate or through the cost of the raw materials as the price of food rises in response to less food being grown. It is unlikely that lactose can ultimately compete as a fermentation substrate for many products as it is cheaper to produce glucose substrates from other sources, such as corn syrups. For this reason, for lactose to be used as a preferred carbon source, the product being produced needs to be made using bacteria that preferentially utilize lactose. The only commercial product that falls into this category at present is nisin.

Nisin is a naturally occurring antibiotic that consists of a group of polypeptides that have antibiotic properties. Its structure has been given by (Gross and Morell, 1971) and is shown in Fig. 24.4, while (Liu and

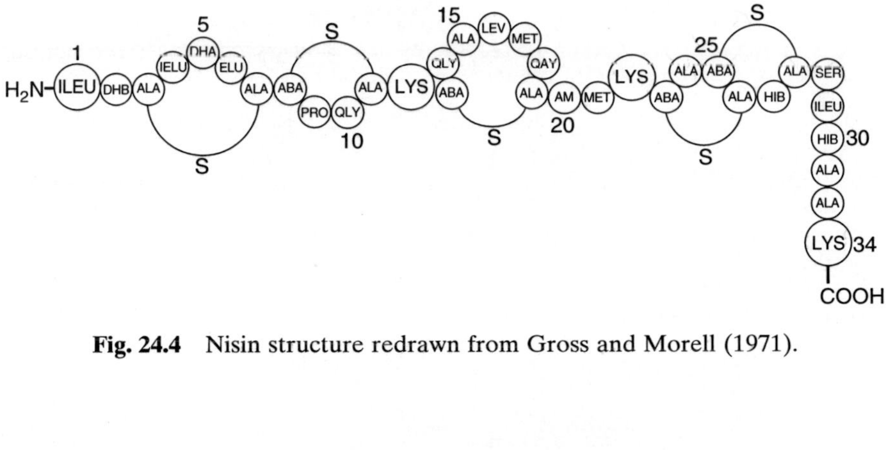

Fig. 24.4 Nisin structure redrawn from Gross and Morell (1971).

Hansen, 1990) have covered the chemical and physical properties of nisin. Nisin is made by growing *Lactococcus lactis* on milk and then concentrating and drying the nisin from the solution. The resultant dried product is milled and standardized with salt (NaCl) so that the nisin is about 2.5%, with 77.5% salt and non-fat dry milk (12% protein and 6% carbohydrate) and less than 3% moisture (Chen and Hoover, 2003). Nisin is the only antibiotic that is allowed to be added to foods in the USA and it has also been approved for use in at least 50 other countries (Delves-Broughton *et al.*, 1996; Fennema, 1996). The use of nisin has been researched for many applications. It has been found to be active against gram positive bacteria but it is not effective against gram negative bacteria or some strains of clostridia. It does stop spore formation in the gram positive bacteria and it has found use in high-moisture processed cheeses to stop the outgrowth of *Clostridium botulinum* spores. (Scott and Taylor, 1981; Stevens *et al.*, 1991, 1992; Cutter and Siragusa, 1995; Delves-Broughton *et al.*, 1996; Fennema, 1996; Chavasirikunton *et al.*, 2006; Rydlo *et al.*, 2006) Other uses include: a wide range of heat-processed foods such as fresh milk, processed cheese and other dairy products, pasteurized liquid eggs, processed meat, seafood, canned food, fruit drinks, plant protein drinks, baked products, instant food, beer, wine, a preservative in gelatin processing, cosmetics, drugs and health products. It has been found to be non-toxic to humans and, because it is currently the only naturally occurring antimicrobial agent that has been approved to be added to food, it is expected that the production of nisin is likely to rise with time, although there are alternative peptide products looking to break into the market (Rydlo *et al.*, 2006). It is currently estimated that the world market is 60 to 70 tonnes per year at about $350 US per kg.

The only drawback for nisin is that it is not effective when applied to meat. It is thought to be inactivated by the enzymatic reaction with glutathione (Rayman *et al.*, 1983; Rose *et al.*, 1999, 2002).

24.9 References

AFFERTSHOLT, T. (2007). Market developments and industry challenges for lactose and lactose derivatives, *IDF Symposium 'Lactose and its Derivatives'*. Moscow, Russia.

AFFERTSHOLT, T. (2008, 5 December). *World Report on Milk Production and Population*. 3A Business Consulting

ANON. (2008a). *D-tagatose to be supplied by Inalco s.p.a* from http://www.biospace.com/news_story.aspx?NewsEntityId=94678

ANON. (2007a). *Nestlé home page*. Retrieved 5 November, 2007, from http://www.nestle.com/AllAbout/AllAboutNestle.htm

ANON. (2007b). The world dairy situation 2007. *Bulletin of the International Dairy Federation*, 423, 1–91.

ANON. (2008b). *USA lactose*. Retrieved 9 December, 2008, from http://www.clal.it/en/index.php?section=lattosio_usa

BÄR, A. (2004). *D-tagatose, dossier prepared and submitted by Bioresco on behalf of Arla Food Ingredients Amba, Viby, Denmark for evaluation pursuant to EU Novel Foods Regulation (EC) 258/97 by the UK Advisory Committee on Novel Foods and Processes, dated August 2, 2004.*

BEADLE, J. R., SAUNDERS, J. P. and WAJDA JR., T. J. (1991). *Process for manufacturing tagatose*, Biospherics Inc. US Patent Number 5,002,612

BEADLE, J. R., SAUNDERS, J. P. and WAJDA JR., T. J. (1992). *Process for manufacturing tagatose*, Biospherics Inc. US Patent Number 5,078,796

BENNO, Y. and MITSUOKA, T. (1986). Development of intestinal microflora in humans and animals. *Microflora*, 5, 13–25.

BERTELSEN, H., ERIKNAUER, K., BOTTCHER, K., CHRISTENSEN, H. J., STOUGAARD, P., HANSEN, O. C. and JORGENSEN, F. (2006). *Process for manufacturing tagatose*. Arla Foods Amba. US Patent Number 6,991,923

BLANKERS, I. (1995). Properties and applications of lactitol. *Food Technol.*, 49, 66–68.

BODE, L. (2006). Recent advances on structure, metabolism, and function of human milk oligosaccharides. *Journal of Nutrition*, 136, 2127–2130.

BOEHM, G. (2008). Structural and functional aspects of prebiotics used in infant nutrition. *Journal of Nutrition*, 138, 1818–1828.

BOON, M. A., JANSSEN, A. E. M. and VAN T RIET, K. (2000). Effect of temperature and enzyme origin on the enzymatic synthesis of oligosaccharides. *Enzyme and Microbial Technology*, 26, 271–281.

BOUHNIK, Y., FLOURIE, B., OUARNE, N., BISETTI, F., BORNET , J. C. and RAMBAUD, J. C. (1994). Effects de l'administration prolongée de fructo-oligosaccharides (Actilight1) sur la flore bactériene colique de l'homme (journées francophones de pathologie digestive). *Gastroenterol Clin Biol*, 18:A50.

CHAVASIRIKUNTON, V., VATANYOOPAISARN, S. and PHALAKORNKULE, C. (2006). Bacteriocin-like activity from weissella confusa and pediococcus acidilactici isolated from traditional Thai fermented sausages. *Journal of Culture Collections*, 5, 64–72.

CHEN, H. and HOOVER, D. G. (2003). Bacteriocins and their food applications. *Comprehensive Reviews in Food Science and Food Safety*, 2, 82–100.

CONN, H. O., LEEVY, C. M., VLAHCEVIC, Z. R., RODGERS, J. B., MADDREY, W. C., SEEFF, L. and LEVY, L. L. (1977). Comparison of lactulose and neomycin in the treatment of chronic portal-systemic encephalopathy. A double blind controlled trial *Gastroenterology*, 72, 573–583.

CUTTER, C. N. and SIRAGUSA, G. R. (1995). Population reductions of gram-negative pathogens following treatments with nisin and chelators under various conditions. *Journal of Food Protection*, 58, 977–983.

DELVES-BROUGHTON, J., BLACKBURN, P., EVANS, R. J. and HUGENHOLTZ, J. (1996). Applications of the bacteriocin, nisin. *Antonie van Leeuwenhoek*, 69, 193–202.

FDA (2003). Food labeling: Health claims; d tagatose and dental caries, *Federal Registar, Vol:68, No.128.*

FENNEMA, O. R. (1996). *Food Chemistry*. New York: Marcel Dekker Inc.

FISCHER, E. and MEYER, J. (1898). Oxidation des Milchzuckers. *Ber. dt. chem. Ges.*, 23, 361–364.

FOX, P. F., GUINEE, T. P., COGAN, T. M. and MCSWEENEY, P. L. H. (2000). *Fundamentals of Cheese Science*. Gaithersburg, USA: Aspen Publishers, Inc.

GÄNZLE, M. G., HAASE, G. and JELEN, P. (2008). Lactose: Crystallization, hydrolysis and value-added derivatives. *International Dairy Journal*, 18, 685–694.

GIBSON, G. R. and ROBERFROID, M. B. (1995). Dietry modulation of the human colonic microbiota. Introducing the concept of prebiotics. *J. Nutrition*, 125, 1401–1412.

GOPAL, P. K., SULLIVAN, P. A. and SMART, J. B. (2001). Utilisation of galacto-oligosaccharides as selective substrates for growth by lactic acid bacteria including bixdobacterium lactis dr10 and lactobacillus rhamnosus dr20. *International Dairy Journal*, 11, 19–25.

GRENBY, T. H., PHILLIPS, A. and MISTRY, M. (1989). Studies of the dental properties of lactitol compared with five other bulk sweeteners *in vitro*. *Caries Research*, 23, 315.

GROSS, E. and MORELL, J. L. (1971). Structure of nisin. *Journal of the American Chemical Society*, 93, 4634–4635.

HALLIDAY, J. (May 14 2008). *Nutrilab works towards a bright future for tagatose*, from http://www.nutrilab.be/content.php?hmID=28&smID=123

HARJU, M. (1992). Production and properties of lactulose, lactitol and lactobionic acid. *IDF workshop on lactose hydrolysis, 3–4 Sep 1992*. Munich (Germany).

HARJU, M., VESANEN, K., PERTTILA, M. and JALKANEN, M. (1993). *Production and properties of lactose, lactitol and lactobionic acid*. Brussels: International Dairy Federation.

HAVERNAAR, R. and VAN DOKKAM, W. (2001). Lactulose: A review on effects, clinical results, and safety aspects in relation to its influence on the colonic environment. In Cho, S. and Dreher, M. L. (Eds.), *Handbook of Dietary Fiber* (pp. 179–194): Marcel Dekker

HAWKINS, S. M. (1993). Bifidobacteria in dairy products. *Cultured Dairy Products*, 28, 16–20.

HEIKKILÄ, H., LEWANDOWSKI, J. and KUISMA, J. (2007). *Chromatographic separation method*. Danisco Sweeteners Oy. US Patent Number 7229558 from http://www.patentstorm.us/patents/7229558.html

HIRST, E. L., HOUGH, L. and JONES, J. K. N. (1949). Composition of the gum of sterculia setigera: Occurance of d-tagatose in nature. *Nature*, 163, 177.

HOWLING, D. and CALLAGAN, J. (2002). *Rte cereals and other foods presweetened with d-tagatose*. US Patent Number 6,475,540

IDF (1993). *Lactose Hydrolysis*. Brussels, Belgium: International Dairy Federation.
INCHEM. (1983). Lactitol.

KARRER, P. and BÜCHI, J. (1937). Reaktionsprodukte von disacchariden: Maltit, lactit, cellobit. *Helvetica Chimica Acta*, 20, 86–90.

KAWASHIMA, S., IDE, H., KATO, K., MAGARA, M. and ISHII, Y. (1991). *Lactitol trihydrate crystal, crystalline mixture solid containing it and process for preparing them*. Japan. Patent Number 4,999,058

KIM, Y.-S., PARK, C.-S. and OH, D.-K. (2006). Lactulose production from lactose and fructose by a thermostable [beta]-galactosidase from sulfolobus solfataricus. *Enzyme and Microbial Technology*, 39, 903–908.

KITLER, M. E., LUGINBUHL, M., LANG, O., WUHL, F., WYSS, A. and LEBEK, G. (1992). Lactitol and lactulose – an *in vivo* and *in vitro* comparison of their effects on the human intestinal flora. *Drug Investigation*, 4, 73–82.

KOZEMPEL, M. and KURANTZ, M. (1993). A continuous reactor system for production of lactulose. *Journal of Chemical Technology & Biotechnology*, 59, 265–269.

KOZEMPEL, M. F., HICKS, K. B., KURANTZ, M. J. and CRAIG, J. C. (1999). *Continuous process for the production of lactulose from lactose using boric acid as a complexation agent*. US Patent Number 5895814, from http://www.freepatentsonline.com/5895814.html

KOZEMPEL, M. F., KURANTZ, M. J., CRAIG, J. C. and HICKS, K. B. (1995). Development of a continuous lactulose process: Separation and purification. *Biotechnology Progress*, 11, 592–595.

KUNZ, C. and RUDLOFF, S. (2006). Health promoting aspects of milk oligosaccharides. *International Dairy Journal*, 16, 1341–1346.

KUNZ, C., RUDLOFF, S., BAIER, W., KLEIN, N. and ROBEL, S. S. (2000). Oligosaccharides in human milk: Structural, functional, and metabolic aspects. *Annual Reviews Nutrition*, 20, 699–722.

LEE, Y. L., KIM, C. S. and OH, D. K. (2004). Lactulose production by b-galactosidase in permeabilized cells of kluyveromyces lactis. *Applied Microbiology and Biotechnology*, 64, 787–793.

LEVIN, G. V., ZEHNER, L. R., SAUNDERS, J. P. and BEADLE, J. R. (1995). Sugar substitutes: Their energy values, bulk characteristics, and potential health benefits. *Am J Clin Nutr*, 62, 1161S–1168S.

LIM, B.-C., KIM, H.-J. and OH, D.-K. (2007). High production of d-tagatose by the addition of boric acid. *Biotechnology Progress*, 23, 824.

LIU, W. and HANSEN, J. N. (1990). Some chemical and physical properties of nisin, a small-protein antibiotic produced by lactococcus lactis. *Applied and Environmental Microbiology*, 56, 2551–2558.

LU, Y., LEVIN, G. V. and DONNER, T. W. (2008). Tagatose, a new antidiabetic and obesity control drug. *Diabetes, Obesity & Metabolism*, 10, 109.

MACFARLANE, G. T., STEED, H. and MACFARLANE, S. (2008). Bacterial metabolism and health-related effects of galacto-oligosaccharides and other prebiotics. *Journal of Applied Microbiology*, 104, 305–344.

MACFARLANE, S., MACFARLANE, G. T. and CUMMINGS, J. H. (2006). Review article: Prebiotics in the gastrointestinal tract. *Alimentary Pharmacology & Therapeutics*, 24, 701–714.

MAYER, J., CONTRAD, J., KLAIBER, I., LUTZ-WAHL, S., BEIFUSS, U. and FISCHER, L. (2004). Enzymatic production and complete nuclear magnetic resonance assignment of the sugar lactulose. *Journal of Food and Agricultural Chemistry*, 52, 6983–6990.

MORO, G., ARSLANOGLU, S., STAHL, B., JELINEK, J. and WAHN, U. (2006). A mixture of prebiotic oligosaccharides reduces the incidence of atopic dermatitis during the first six months of age *Archives of Disease in Childhood*, 91, 814–819.

MYERS, C., HEIKKILA, H., NIKANDER, H., NURMI, J., NYGRÉN, J., PERKKALAINEN, P., PEPPER, T., PITKANEN, I. and VALKONEN, J. (2005). *Anhydrous lactitol crystals, a product containing the same and a process for the preparation thereof as well as use thereof.* Xyrofin Oy. US Patent Number 6872414

OLIGER, P. M. and PEARSON, J. (1998). *Directly compressible lactitol and method.* Xyrofin Oy. US Patent Number 5846568

ONISHI, N. and TANAKA, T. (1996). Purification and properties of a galacto- and gluco-oligosaccharide-producing [beta]-glycosidase from rhodotorula minuta ifo879. *Journal of Fermentation and Bioengineerlng*, 82, 439–443.

PATERSON, A. H. J. (2009). Production and uses of lactose. In Fox, P. F. and McSweeney, P. L. H. (Eds.), *Advanced Dairy Chemistry* (Vol. 3): Springer Science+Business Media, LLC.

PLAYNE, M. J., BENNETT, L. E. and SMITHERS, G. W. (2003). Functional dairy foods and ingredients. *Australian Journal of Dairy Technology*, 58, 242–264.

PYUN, Y. R., KIM, B. C., LEE, H. S., LEE, D. W. and LEE, Y. H. (2005). *Thermostable l-arabinose isomerase and process for preparing d-tagatose.* CJ Corp. US Patent Number 6,933,138

RASTALL, R. (2003). Chapter 12: Enhancing the functionality of prebiotics and probiotics. In *Functional Dairy Products*, T. Mattila-Sandholm and M. Saarela (eds) (pp. 1–16): CRC Press LLC.

RAYMAN, K., MALIK, N. and HURST, A. (1983). Failure of nisin to inhibit outgrowth of clostridium botulinum in a model cured meat system. *Appl. Environ. Microbiol.*, 46, 1450–1452.

RIVERO-URGELL, M. and SANTAMARIA-ORLEANS, A. (2001). Oligosaccharides: Application in infant food. *Early Human Development*, 65, S43–S52.

ROSE, N. L., PALCIC, M. M., SPORNS, P. and MCMULLEN, L. M. (2002). Nisin: A novel substrate for glutathione s-transferase isolated from fresh beef. *Journal of Food Science*, 67, 2288–2293.

ROSE, N. L., SPORNS, P., STILES, M. E. and MCMULLEN, L. M. (1999). Inactivation of nisin by glutathione in fresh meat. *Journal of Food Science*, 64, 759–762.

ROSENPLENTER, K. and MENDE, K. (2004). *Use of d-tagatose for improving aroma and taste*. Patent Number WO/2004/073419

ROWLAND, I. R. and TANAKA, R. (1993). The effects of transgalactosylated oligosaccharides on gut flora metabolism in rats associated with a human faecal microflora. *Journal of Applied Microbiology*, 74, 667–674.

RULIS, A. M. (2001). United States Food and Drug Administration response letter. GRAS Notice No. Grn 000078.

RYDLO, T., MILTZ, J. and MOR, A. (2006). Eukaryotic antimicrobial peptides: Promises and premises in food safety. *Journal of Food Science*, 71, 125–135.

SAKO, T., MATSUMOTO, K. and TANAKA, R. (1999). Recent progress on research and applications of non-digestible galacto-oligosaccharides. *International Dairy Journal*, 9, 69–80.

SATORY, M., FURLINGER, M., HALTRICH, D., KULBE, K. D., PITTNER, F. and NIDETZKY, B. (1997). Continuous enzymatic production of lactobionic acid using glucose–fructose oxidoreductase in an ultrafiltration membrane reactor. *Biotechnology Letters*, 19, 1205–1208.

SCHAAFSMA, G. (2008). Lactose and lactose derivatives as bioactive ingredients in human nutrition. *International Dairy Journal*, 18, 458–465.

SCOTT, V. N. and TAYLOR, S. L. (1981). Effect of nisin on the outgrowth of *clostridium botulinum* spores. *Journal of Food Science*, 46, 117–126.

SENDERENS, J. B. (1920). *Comptes Rendus Hebdomadaires des Seances de l'Academie des Sciences*, 170, 47–50.

SHOAF, K., MULVEY, G. L., ARMSTRONG, G. D. and HUTKINS, R. W. (2006). Prebiotic galactooligosaccharides reduce adherence of enteropathogenic escherichia coli to tissue culture cells. *Infection and Immunity*, 74, 6920–6928.

SKYTTE, U. P. (2006). Tagatose. In Mitchell, H. (Ed.), *Sweeteners and Sugar Alternatives in Food Technology* (pp. 262–294). Oxford, U.K.: Blackwell Publishing Ltd.

SOUTHARD, J. H. and BELZER, F. O. (1995). Organ preservation. *Annual Review of Medicine*, 46, 235–247.

SPHERIX. (2008). Spherix d-tatgatose to be supplied by Inalco s.P.A. http://www.spherix.com/Press Release/pr043008

STEVENS, K. A., SHELDON, B. W., KLAPES, N. A. and KLAENHAMMER, T. R. (1991). Nisin treatment for inactivation of salmonella species and other gram-negative bacteria. *Applied and Environmental Microbiology*, 57, 3613–3615.

STEVENS, K. A., SHELDON, B. W., KLAPES, N. A. and KLAENHAMMER, T. R. (1992). Effect of treatment conditions on nisin inactivation of gram-negative bacteria. *Journal of Food Protection*, 55, 763–766.

TOMOMATSU, H. (1994). Health effects of oligosaccharides. *Food Technology*, 48, 61–65.

TROYANO, E., VILLAMIEL, M., OLANO, A., SANZ, J. and MARTINEZCASTRO, I. (1996). Monosaccharides and myo-inositol in commercial milks. *Journal of Agricultural and Food Chemistry*, 44, 815–817.

TUOHY, K. M., ROUZAUD, G. C. M., BRÜCK, W. M. and GIBSON, G. R. (2005). Modulation of the human gut microflora towards improved health using prebiotics – assessment of efficacy. *Current Pharmaceutical Design*, 11, 75–90.

VAN LOO, J., CUMMINGS, J., DELZENNE, N., ENGLYST, H., FRANCK, A., HOPKINS, M., KOK, N., MACFARLANE, G., NEWTON, D., QUIGLEY, M., ROBERFROID, M., VLIET, T. V. and HEUVEL, E. V. D. (1999). Functional food properties of non-digestible oligosaccharides: A consensus report from the endo project (dgxii airii-ct94-1095). *British Journal of Nutrition*, 81, 121–132.

VAN VELTHUIJSEN, J. A. (1979). Food additives derived from lactose: Lactitol and lactitol palmitate. *J. Agric. Food Chem.*, 27, 680–686.

VEEREMAN-WAUTERS, G. (2005). Application of prebiotics in infant foods. *British Journal of Nutrition*, 93, S57–S60.

VENEMA, K., VERMUNT, S. H. F. and BRINK, E. J. (2005). D-tagatose increases butyrate production by the colonic microbiota in healthy men and women. *Microbial Ecology in Health and Disease*, 17, 47–57.

VIGH, M. L. and ANDERSEN, H. (2002). *Synergistic combination of sweeteners including d-tagatose*. US Patent Number 6,432,464

VIGH, M. L. and ANDERSEN, H. (2007). *Use of d-tagatose as a prebiotic food component*. US Patent Number 7,202,219

WOLFROM, M. L., BURKE, W. J., BROWN, K. R. and ROSE JR., R. S. (1938). *J. Am. Chem. Soc.*, 60, 571–573.

YOUNG, H. (2006). Lactitol. In Mitchell, H. (Ed.), *Sweeteners and Sugar Alternatives in Food Technology* (pp. 205–218). Oxford: Blackwell Publishing Ltd.

25

Protein interactions and functionality of milk protein products

H. Singh, Massey University, New Zealand

Abstract: Milk proteins are nutritionally important and provide a wide range of dynamic functional properties which are widely exploited by the food industry. Several methods for the industrial-scale production of milk proteins have been developed over the last 40 years. As a result, a vast range of milk protein products, specifically designed for particular applications, is manufactured by the dairy industry. These products include the traditional milk protein products, such as skim milk powder and whey powders, and higher protein products, such as caseins and caseinates, whey protein concentrates and isolates and milk protein concentrates and isolates. The processes used in the manufacture of these products can modify the native structures of proteins which can lead to further protein–protein interactions, consequently impacting on the protein functionality. This chapter provides an overview of the manufacture, composition and functionality of milk protein products and milk powders. It also considers possible interactions of proteins during the manufacture of milk protein products and their consequences for the functional properties and applications of the products.

Key words: milk proteins, milk powders, functionality, interactions, processing, food systems.

25.1 Introduction

The major component of milk is water; the remainder consists of fat, lactose and protein (casein and whey proteins). Milk also contains smaller quantities of minerals, specific blood proteins, enzymes and small intermediates of mammary synthesis. The composition of milk varies with the breed of the cow, the progress of lactation, the kind and quantity of feed, and mastitis (Walstra and Jenness, 1984).

Normal bovine milk contains about 3.5% protein, which can be fractionated into caseins and whey proteins. Caseins can be fractionated into four distinct proteins, α_{s1}-, α_{s2}-, β- and κ-casein, all of which are phosphoproteins with the phosphate groups being esterified to the serine residues. The phosphate groups bind large amounts of calcium, and are important to the structure of casein micelles. Casein micelles are coarse colloidal particles, with diameters ranging from 80 to 300 nm (average ~150 nm). On a dry weight basis, the micelles consist of ~94% protein and ~6% small ions, principally calcium, phosphate, magnesium and citrate, referred to collectively as colloidal calcium phosphate (CCP).

The precise structure of the casein micelle is debatable; a number of models have been proposed over the past 45 years. These models include coat–core models, which postulate that the interior of the micelle is composed of proteins that are different from those on the exterior, and sub-unit structure models, to which the term sub-micelle is attached (see Rollema, 1992; Fox and McSweeney, 1998, 2003; de Kruif and Holt, 2003; Horne, 2009). Other models consider the micelle as a porous network of proteins (of no fixed conformation); the calcium phosphate nanoclusters are responsible for cross-linking the protein and holding the network together (Holt, 1992). A model proposed by Horne (1998) assumes that the assembly of the casein micelle is governed by a balance of electrostatic and hydrophobic interactions between casein molecules; α_{s1}-, α_{s2}- and β-caseins consist of distinct hydrophobic and hydrophilic regions. Two or more hydrophobic regions from different molecules form a bonded cluster. Growth of these polymers is inhibited by the protein charge residues, the repulsion of which pushes up the interaction free energy. Neutralization of the phosphoserine clusters by incorporation into the CCP diminishes that free energy and produces a second type of cross-linking bridge. κ-Casein acts as a terminator for both types of growth, as it contains neither a phosphoserine cluster nor another hydrophobic anchor point.

A common factor in all models is that most of the κ-casein appears to be present on the surface of the casein micelles. The hydrophilic, C-terminal, part of κ-casein is assumed to protrude 5–10 nm from the micelle surface into the surrounding solvent, giving the micelle a 'hairy' appearance. The highly charged flexible 'hairs' physically prevent the approach and interactions of hydrophobic regions of the casein molecules. Removal of the hairs by cleavage with rennet or their collapse in ethanol destroys the stabilization effect of κ-casein, allowing micelles to interact and aggregate (see de Kruif and Holt, 2003).

Field emission scanning electron microscopy has shown a complex surface structure of the casein micelles, with cylindrical or tubular protrusions between 10 and 20 nm in diameter extending from the surface (Dalgleish *et al.*, 2004).

The principal fractions of whey proteins are β-lactoglobulin (β-LG), bovine serum albumin (BSA), α-lactalbumin (α-LA) and immunoglobulins (Igs). Major reviews covering the structures and properties of the whey proteins have been published (see Kinsella and Whitehead, 1989; Hambling *et al.*, 1992; Edwards *et al.*, 2009). β-LG is the most abundant whey protein, represents about 50% of the total whey protein in bovine milk, has a molecular weight of 18000 Da and contains two internal disulphide bonds and a single free thiol group.

Milk salts consist mainly of chlorides, phosphates, citrates and bicarbonates of sodium, potassium, calcium and magnesium. Some of the milk salts (i.e. the chlorides of sodium and potassium) are soluble and are present almost entirely as ions dissolved in milk whey. Others, calcium and phosphate in particular, are much less soluble and, at the normal pH of milk, exist partly in dissolved form and partly in insoluble (i.e. colloidal) form, in close association with the casein micelles (Walstra and Jenness, 1984).

Lactose is found in cows' milk at levels of about 5% and makes a major contribution to the colligative properties of milk (osmotic pressure, freezing point depression, boiling point elevation). For example, it accounts for about 50% of the osmotic pressure of milk. Compared with many other sugars, lactose is relatively less soluble in water; its solubility at 25°C is only 17.8 g per 100 g solution. This relatively low solubility can cause some manufacturing problems because lactose crystals are gritty in texture. Crystallization of lactose is also responsible for caking and lumping of dried milk during storage, particularly if moisture is absorbed from the air. Lactose, like other reducing sugars, can react with free amino groups of proteins to give products that are brown in colour (Walstra and Jenness, 1984).

The macro-constituents of milk (proteins, lipids and lactose) can be fractionated readily and are used widely as food ingredients. The proteins of milk are the most valuable components, and several methods for the industrial-scale production of milk proteins have been developed. Milk proteins are used in nutritional, nutraceutical and functional applications. This chapter provides an overview of the manufacture, composition and functionality of milk protein products and milk powders. The chapter also considers possible interactions of proteins during the manufacture of milk protein products and their consequences for the functional properties and applications of the products.

25.2 Milk protein products and powders

A milk protein product can be defined as a dried material that can be added as an ingredient in dairy products or other foods to achieve certain nutritional and/or functional properties. A vast range of milk protein products, specifically designed for particular applications, is manufactured by the dairy industry. Many different grades and types of caseins and caseinates, whey protein concentrates and isolates, milk protein concentrates and isolates, and hydrolyzed proteins can be manufactured, by altering processing steps during manufacture, as shown in Fig. 25.1. These protein products vary with respect to the concentration of protein, minerals, lipids and lactose.

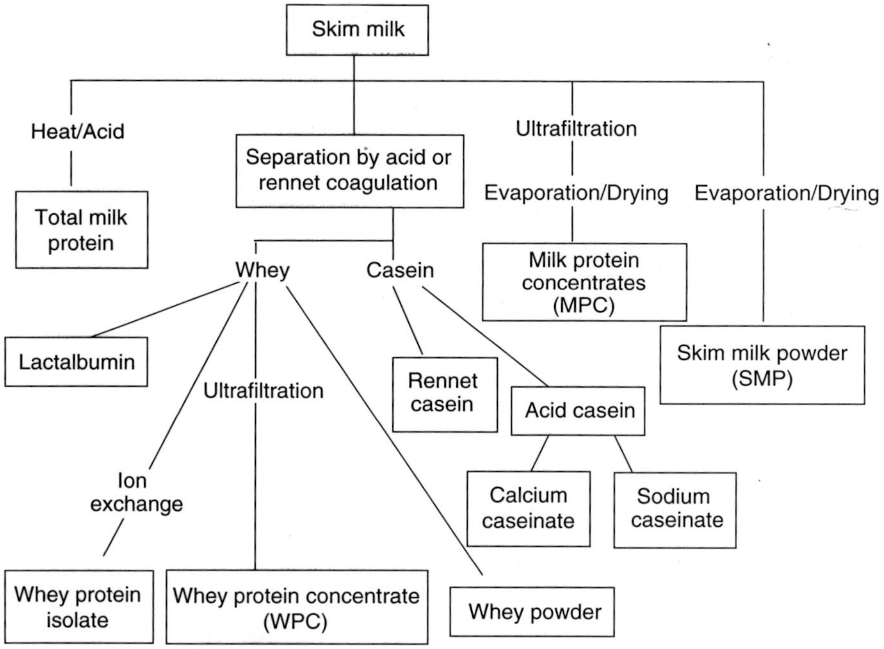

Fig. 25.1 Functional protein powders from milk.

The traditional milk protein products, skim milk powder (SMP) and whey powders, have a long history of use as ingredients in a vast range of food systems from beverages to meat, confectionery, ice cream and bakery products. SMPs are commonly classified as low, medium or high heat powder on the basis of their whey protein nitrogen index (WPNI). High protein milk powders with protein content in the range 50–85% are commonly referred to as milk protein concentrates (MPCs) and may be classified as MPC56, MPC70 and MPC85; the number denotes the approximate protein concentration. The compositions of various types of milk powders are given in Table 25.1.

The manufacturing processes for SMP consist of several distinct stages (Fig. 25.2): (i) separation and pasteurization of the raw milk into desired components for final product specifications, (ii) preheat treatment of the milk for a set time/temperature combination, (iii) concentration of the milk by multiple-effect evaporator, (iv) heat treatment of the concentrate to reduce the viscosity prior to its entry into the spray drier, and (v) spray drying of the concentrate.

Milk powders are used in a number of applications (Baldwin and Pearce, 2005; Oldfield and Singh, 2005; Singh, 2007). In recombining applications, SMP and anhydrous milk fat are mixed together by dissolving them in water (reconstitution) and homogenizing. The products that can be made include

Table 25.1 Typical compositions of milk powders

Product	SMP	MPC56	MPC70	MPC85
Milk fat	0.8	1.3	1.4	1.0
Protein	34.0	57.1	70.0	85.0
Lactose	53.5	30.1	17.0	3.0
Minerals (ash)	7.9	7.7	7.2	7.0
Moisture	3.8	3.8	4.4	4.0

SMP = skim milk powder; MPC = milk protein concentrate powder.

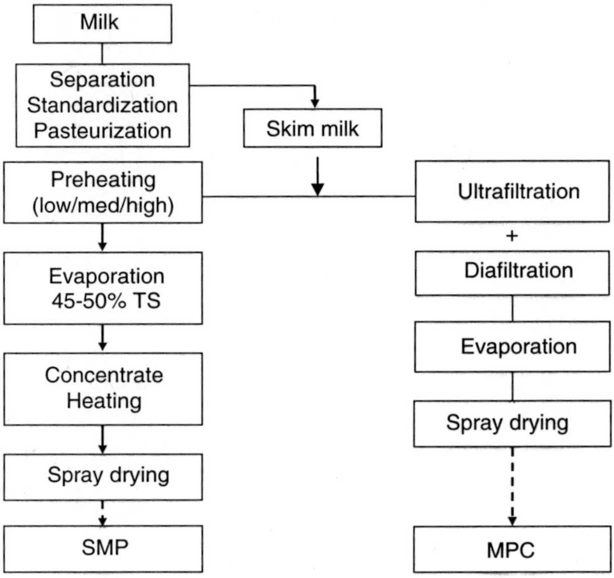

Fig. 25.2 Manufacturing processes for a typical skim milk powder (SMP) or milk protein concentrate powder (MPC).

recombined evaporated milk (REM), recombined milks, recombined sweetened condensed milk, yoghurt, cheese and fermented milk beverages. Milk powders can also be used as functional ingredients in a wide variety of foods, e.g. chocolate, bakery, beverages and confectionery. Milk powders may be used by consumers directly at home by mixing milk powder and water. Alternatively, milk powder can be used as a whitener by adding it directly to hot coffee or tea. For household use, the powders are instantized during spray drying by lecithination and agglomeration, so that they dissolve rapidly in water at room temperature.

In the manufacture of MPC, skim milk is concentrated by ultrafiltration (UF) to increase the protein content and to remove lactose and salts

(Fig. 25.2). In the manufacture of higher protein powders (normally above 70% protein on a dry powder basis), diafiltration (DF) is usually applied. In general, no preheating step is involved. After UF and DF of the skim milk, the retentate is evaporated to remove more water and then spray dried.

MPCs are commonly added to milk or cheese formulations to enhance the protein content and/or the yield of the final product. MPC can be used to enhance the textural characteristics of yoghurts. The use of MPCs in nutritional drinks is growing. In this application, MPC provides both casein and whey proteins in the same ratio as in milk, but without the high lactose content (Baldwin and Pearce, 2005).

The manufacture of whey protein products involves combinations of several processes, such as UF, DF, reverse osmosis, ion exchange, evaporation and drying (Hobman, 1992). Commercial whey protein products include various whey powders, whey protein concentrates (WPCs), whey protein isolates (WPIs) and fractionated proteins, such as α-LA and β-LG (Mulvihill, 1992). The most important commercial whey protein products are WPC (up to about 85% protein) and WPI (approximately 95% protein). In addition to α-LA, β-LG, BSA and Igs, these products contain various other components, including lipids, lactose and minerals. Typical values for the composition of industrial WPCs are shown in Table 25.2.

These products have applications in meat products, beverages, baked products and infant foods (Kinsella, 1984; de Wit, 1989). The important functional properties of whey protein products include water binding, emulsification, foaming and whipping, and gelation. The functional properties of whey protein products depend not only on their composition but also on various processing treatments during manufacture (de Wit *et al.*, 1996).

Casein, the principal protein in milk, can be extracted via precipitation in three ways: (i) by the addition of mineral acid, (ii) by the action of starter bacteria which utilize lactose and release lactic acid, and (iii) by the action of chymosin or an appropriate microbial protease (see Mulvihill, 1992; Mulvihill and Ennis, 2003). The manufacture of mineral acid casein involves the acidification of pasteurized skim milk (72°C for 15 s) using sulphuric

Table 25.2 Typical compositions of whey protein concentrate powders (WPCs)

Component	Range (%)
Moisture	3.30–4.50
Protein	73.0–86.0
Lactose	0.80–7.60
Total lipids	3.30–7.40
Phospholipids	0.80–1.55
Ash	2.50–5.30

acid or hydrochloric acid to achieve a pH of approximately 4.6 at 20°C. In the manufacture of lactic acid casein, pasteurized skim milk is mixed with selected strains of starter bacteria followed by incubation at 22–26°C. The bacteria grow in the milk producing lactic acid, which lowers the pH, eventually causing isoelectric precipitation. In rennet casein manufacture, calf rennet or microbial rennet is added to skim milk, followed by incubation at about 30°C. The action of chymosin is to cleave one of the bonds in κ-casein, resulting in the aggregation of casein particles.

For all three types of casein, the curd obtained is heated, e.g. by direct steam injection, to 50–55°C in order to further promote aggregation of casein particles and cause shrinkage. The curd is then washed with water, followed by a dewatering process to reduce the moisture content to 50–60%. Drying of casein curd is most commonly carried out on horizontal vibrating fluidized bed driers.

Sodium caseinate is usually prepared by solubilizing acid casein with NaOH by increasing the pH to 6.7–7.0 to produce a solution of about 23% solids. This solution is then spray dried. A similar process is used in calcium caseinate manufacture, except that calcium hydroxide is used to neutralize the acid casein curd.

A protein product including both casein and whey proteins, known as total milk proteinate, may be manufactured by a variation of the process described above.

The casein protein products described have a wide range of food ingredient and industrial applications, many utilizing the emulsifying, water and fat binding; and nutritional properties of the proteins (Kinsella, 1984).

25.3 Protein interactions during the manufacture of skim milk powder and milk protein concentrate powder

The process of the conversion of milk into powder alters the nature and the behaviour of the milk components. The modification of protein structures, and consequently the interactions of different proteins with each other and with non-protein components, are critical in determining the functional behaviour of powders. Process-induced changes that occur in proteins during powder manufacture have been reviewed recently by Singh (2007) and are summarized in Table 25.3.

25.3.1 Preheat treatment

Preheat treatment commonly refers to the heat treatment given to milk prior to evaporation and spray drying. The main purpose of preheating is to alter the functional properties of the powder, and to raise the milk temperature high enough to ensure that the milk boils in the first effect of the evaporator. Heating temperatures range from 72°C for 15 s to 120°C for

Table 25.3 Process-induced changes in protein during the manufacture of SMP and MPC

Process	Effect on milk proteins
Preheating	Denaturation of whey proteins; association of whey proteins with casein micelles; changes in micelle size, transfer of the soluble salts to the colloidal phase; pH decrease, lactosylation of proteins
Evaporation	Increase in colloidal salts; increase in casein micelle size; decrease in pH
Concentrate heating	Reduction in concentrate viscosity; increase in colloidal salts; further aggregation of whey proteins and their association with casein micelles
Ultrafiltration/ Diafiltration	Swelling of casein micelles, loss of colloidal calcium and phosphate into the serum, dissociation of casein micelles, aggregation of dissociated caseins during diafiltration
Spray drying	Rapid removal of water; relatively minor changes in proteins

2 min. These heat treatments cause a number of competitive and often interdependent reactions in milk proteins, and this subject has been discussed in a number of reviews and books (Fox, 1982; Singh and Creamer, 1992; Singh and Newstead, 1992; Singh, 1995, 2004; O'Connell and Fox, 2003; Anema, 2009).

The most important reactions during the preheating of milk proteins are the denaturation of the whey proteins, interactions of the whey proteins with the casein micelles, and the aggregation/dissociation of the casein micelles. The whey proteins in milk vary widely in their susceptibility to denaturation; β-LG and α-LA show significant denaturation only at temperatures >70–75°C, whereas BSA and Igs begin to denature at ~65°C. The order of sensitivity of various whey proteins to heat-induced denaturation has been reported to be: Igs > BSA > β-LG A > β-LG B > α-LA (Dannenberg and Kessler, 1988; Oldfield *et al.*, 1998a).

The kinetics of denaturation of the whey proteins are quite complex, with the reaction characteristics showing marked changes above 80–100°C. The denaturation reaction shows a non-linear Arrhenius plot; there is a marked change in temperature dependence at about 80–90°C for both α-LA and β-LG. The apparent activation energies are in the range 260–280 kJ/mol for β-LG and 270–280 kJ/mol for α-LA at temperatures below 90°C (Dannenberg and Kessler, 1988; Oldfield *et al.*, 1998a). At higher temperatures, the activation energy is lower, ranging from 54 to 60 kJ/mol for β-LG and from 55 to 70 kJ/mol for α-LA, indicating chemical (aggregation) reactions.

In the milk powder industry, the extent of denaturation of the whey proteins is expressed as the whey powder nitrogen index (WPNI). SMP is dissolved in water and then saturated with NaCl solution, followed by filtration, which retains denatured whey proteins and casein. The filtrate is analyzed for its protein content and the value is expressed as the WPNI, which is the quantity of undenatured whey proteins per gram of powder. Thus, as the extent of denaturation of the whey proteins increases, the WPNI decreases. On the basis of the WPNI, SMPs can be classified into low, medium and high heat types.

Following denaturation, the whey proteins can interact with the casein micelles, more specifically with κ-casein in the micelles. Two disulphide bridges and a free thiol group in the native structure of β-LG play an important role in these interactions. The interaction between β-LG and κ-casein has been shown to involve the formation of disulphide bonds. Heat treatment causes native β-LG to interchange into non-native monomer and non-native disulphide-bonded dimer, which then interact with κ-casein.

β-LG can also form complexes with α_{s2}-casein and α-LA (Patel *et al.*, 2006). Through their interactions with micellar κ-casein at the micelle surface, some of the denatured whey proteins become associated with the casein micelles. Some remain in the serum, where they may form aggregates with other whey proteins or with serum κ-casein. When the temperature of milk is increased gradually above 70°C, as in an indirect heating system, most of the β-LG and most of the α-LA associate with the micelles, whereas, when milk is heated rapidly, as in a direct heating system (e.g. direct steam injection), only about half of the β-LG and half of the α-LA associate with the micelles (Corredig and Dalgleish, 1996; Oldfield *et al.*, 1998b).

Oldfield *et al.* (1998b) suggested that there are at least three possible species of denatured β-LG that could associate with the micelles: (i) unfolded monomeric β-LG, (ii) self-aggregated β-LG and (iii) β-LG/α-LA aggregates. The relative rates of association of these species with the casein micelles depend on temperature and heating rate, which in turn affect the relative rates of unfolding and the formation of the various aggregated species. At high temperatures and fast heating rates, all whey proteins begin to unfold in a short period of time, thus presenting more opportunity for unfolded monomeric β-LG to self-aggregate, which consequently is likely to associate with the casein micelles less efficiently. These β-LG aggregates could protrude from the micelle surface, providing steric effects for further β-LG association. In addition, these aggregates may have their reactive sulphydryl group buried within the interior of the aggregate, and therefore unavailable for sulphydryl–disulphide interchange reactions with micellar κ-casein. The formation of unfolded β-LG may be promoted by long heating times at low temperatures, or by heating at a slow rate to the required temperature. These unfolded monomeric β-LG molecules would be expected to penetrate the κ-casein hairy layer with greater ease and have a readily accessible sulphydryl group.

In addition to time and temperature of heating, several other factors influence the extent of association of denatured whey protein with the casein micelles; these include the pH of the milk prior to heating, levels of soluble calcium and phosphate, and the milk solids concentration (Singh, 1995). Heating at pH values less than 6.7 results in a greater quantity of denatured whey proteins associating with the casein micelles, whereas at higher pH values, whey protein κ-casein complexes dissociate from the micelle surface, apparently because of dissociation of κ-casein (Singh and Fox, 1985).

The interactions of whey proteins with the casein micelles causes changes in the casein micelle size during heating at temperatures up to 100°C, but the change in micelle size is dependent on the pH of the milk at heating, which also influences the extent of association of denatured whey proteins with the micelles. At pH 6.5, where about 70% of the denatured whey proteins are associated with the micelles, the average micelle size increases by about 35 nm. At pH 6.7, where the level of whey protein association is about 30%, the micelle size increases by only about 5 nm (Anema and Li, 2003).

Association of whey proteins modifies the surface of the casein micelles and is likely to influence the functionality of the reconstituted milk powders, particularly rennet clotting behaviour, solubility and heat stability.

25.3.2 Evaporation

Falling film evaporators are predominant in the dairy industry and are designed to remove as much water from the milk as possible without detrimentally affecting the quality of the powder. A falling film evaporator has low residence times (typically 3–6 min) and operates under vacuum so that boiling temperatures range from 40 to 70°C. About 80% of the water can be removed from milk, concentrating it from 9–14% total solids up to 44–50% total solids.

Evaporation causes a number of changes in the milk system (Singh and Newstead, 1992). The pH of the milk decreases during concentration from an average initial value of 6.7 to approximately 6.3 at 45% total solids. This is partly due to changes in salt equilibria as more calcium phosphate is transferred from the soluble state to the colloidal state, with a concomitant release of hydrogen ions. There is no significant denaturation of β-LG or Ig G during the evaporation of skim milk to 45% total solids, whereas some denaturation of α-LA may occur (Oldfield *et al.*, 2005). This is considered to be partly due to the increased stability of whey proteins at high total solids concentrations, because of the stabilizing effect of lactose on the protein structure. However, there is an increase in the amounts of β-LG and α-LA associated with the micelles after evaporation, which can be attributed mainly to the decrease in pH during evaporation (Oldfield *et al.*, 2005).

Casein micelles may increase in size because of the increase in CCP or increased association of the whey proteins with the casein micelles. Unfortunately, no reliable data on the size distribution of casein micelles in highly concentrated systems are available, mainly because of the non-availability of measurement techniques. In addition, concentrated milk is very difficult to handle because of its high viscosity and its tendency towards age gelation.

Under normal conditions with the evaporation temperatures below 70°C, the alterations to the proteins are likely to be minor compared with those incurred during preheating (Singh and Creamer, 1991a).

25.3.3 Concentration of skim milk by ultra- and diafiltration

Ultrafiltration (UF) is a pressure-driven filtration process in which porous membranes are used to separate the components of a solid–liquid mixture on the basis of size and shape. UF membranes are available in many forms but tubular membranes are most popular in modern dairy UF plants. These membranes are available with wide ranges of porosity and molecular mass: UF of skim milk is commonly carried out using membranes with a molecular weight cut-off of 10000 Da. By this process, a concentrate of high molecular weight (retentate) and a permeate fraction of low molecular weight are obtained. UF processing of skim milk allows water, lactose, non-protein nitrogen compounds and soluble salts to pass through the membrane while retaining casein micelles and whey proteins. Some calcium, magnesium, phosphate and citrate are associated with the casein micelles in milk, and hence are retained in the concentrate. The temperature of UF has an impact on the composition of the concentrate, with higher temperatures resulting in higher permeation rates and lower retention of lower molecular weight components. Diafiltration (DF) is employed to produce milk powders with high protein contents, typically over 80%. Purified water is commonly used. The number of DF stages and the volume of water used are dependent on the design and operation of the UF plant and the final product specification.

Relatively little work on the behaviour of the casein micelles during UF/DF has been reported. Srilaorkul *et al.* (1991), using electron microscopy, showed that the size of the highest proportion of casein micelles (80–100 nm) in milk was reduced (to 60–80 nm) at a volume concentration factor of about 5. These changes in micelle size were considered to be due mainly to the changes in the levels of milk salts, particularly calcium and phosphate. In contrast, McKenna (2000) found that there was little change in particle size in the early stages of UF, but that the average micelle size increased significantly at the end of the UF process, and particularly during DF. McKenna (2000) observed a progressive swelling of the casein micelles and the formation of non-micellar material.

Figure 25.3 shows a freeze fracture transmission electron micrograph of skim milk; roughly spherical particles, i.e. casein micelles, of different sizes

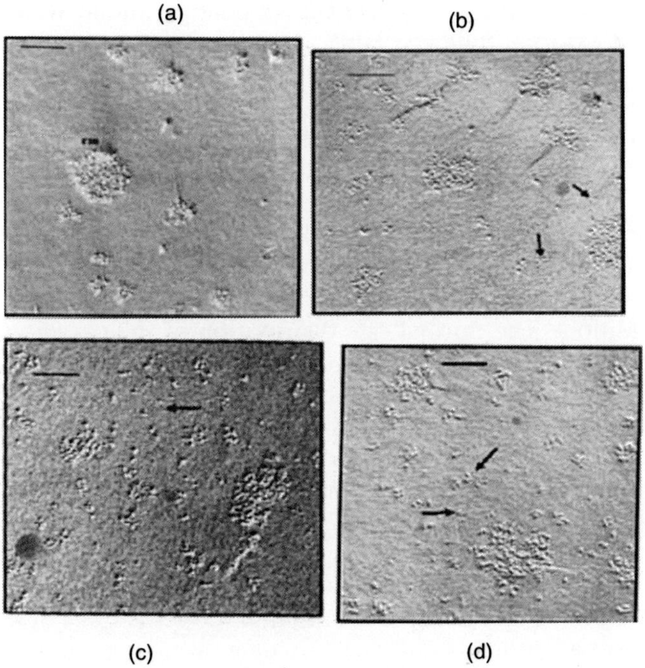

Fig. 25.3 Freeze fracture transmission electron microscopy of (a) skim milk, (b) ultrafiltrated skim milk (mid-way), (c) ultrafiltrated skim milk and (d) diafiltrated skim milk (From McKenna, 2000).

are clearly observed (Fig. 25.3a). Mid-way through and after UF, there appears to be no considerable change in casein micelle size, but more non-micellar material can be seen in the micrographs (arrows) (Fig. 25.3b,c). After DF, a significant change in the structure of the casein micelles appears to take place, with an increase in non-micellar material, and the casein micelle appears to be swollen and less dense (Fig. 25.3d). The non-micellar material appears to link these micelles together. The dissolution of CCP results in loosening of the casein micelle structure and is thus responsible for the swelling of the micelles.

25.3.4 Spray drying

Spray drying removes the remaining water in the concentrate to produce a powder with 3–5% moisture. Milk concentrate is atomized into a fine spray that gives a large surface area over which drying can take place. In the drying chamber, the droplets are intimately mixed with the incoming hot air (150–250°C). Initially, milk droplets dry rapidly at their wet bulb temperature (45–60°C); however, as the droplets progress down the drying chamber, they become drier and the particle temperature eventually approaches the air temperature at the drier outlet (70–90°C).

There are a number of different drier configurations, with or without internal and external fluidized beds. Two-stage and multi-stage driers are more energy efficient and less damaging to the properties of milk powder than single-stage driers.

The drying conditions influence powder properties: moisture content, solubility, bulk density and powder flowability (Pisecky, 1997). The outlet temperature has a marked effect on powder solubility. Increasing the outlet temperature results in poor powder solubility. However, little is known about the changes to proteins that occur during spray drying. The native properties of the milk components are essentially unmodified by moderate drying conditions. The normal size distribution of the casein micelles and their heat stabilities and renneting characteristics are substantially recovered on the reconstitution of spray-dried milk.

Under typical drying conditions, whey proteins are not denatured to any great extent. Oldfield *et al.* (2005) showed that the denaturation of whey proteins during spray drying was minimal with no apparent loss of Ig G and only a small loss of BSA (3–7%). Varying the inlet/outlet drier air temperature (200/100°C–160/89°C) appears to have no significant effect on whey protein denaturation. However, high outlet temperature conditions (>80°C) can cause lactosylation of milk proteins (part of the Maillard reaction between proteins and lactose), decreasing their nutritional value (Guyomarc'h *et al.*, 2000).

Spray drying would be expected to cause some changes in the salt equilibria, possibly an increase in CCP and a decrease in pH. It has been shown that the concentrations of soluble calcium and soluble phosphate in reconstituted skim milk are about 20% lower than those in the original milk (Le Graet and Brule, 1982).

Roller driers can also be used to produce milk powders, but they are limited to producing whole milk powders for confectionary, especially chocolate, manufacture (Augustin, 2001). Roller-dried whole milk powders have a high level of free fat and low solubility, which make them ideal for use in chocolate.

25.4 Protein interactions during the manufacture of whey protein and casein products

25.4.1 Whey protein products

Sweet or acid whey, derived from cheese or casein production, is used for the manufacture of WPCs and WPIs. WPIs can also be manufactured directly from skim milk using microfiltration technology.

There is a range of whey pretreatment processes, such as clarification, preheating and pasteurization, that may be carried out to improve flux and to reduce fouling during UF, and to manipulate the chemical composition of the final product. The purpose of centrifugal clarification is to remove particulate matter such as fat, casein fines and microorganisms. Preheating of whey to at least the temperature of UF is carried out to improve the efficiency of the UF process. These processes are unlikely to have an effect on the native structures of the whey proteins.

The clarified whey, particularly in the case of cheese whey, is commonly pasteurized at 72°C for 15 s, which is likely to cause denaturation of minor, heat-sensitive whey proteins, such as Igs and BSA. In comparison, acid whey is generally not pasteurized because such a heat treatment at the natural pH of acid whey (pH ~4.6) can result in protein aggregation (Hobman, 1992).

UF and DF processes of whey are commonly carried out at ~50°C. This temperature is generally considered to be optimum; acceptable fluxes are achieved and thermal denaturation of the whey proteins is avoided. The evaporation of whey retentate is carried out using high vacuum, low boiling temperature (e.g. 50°C) falling film evaporators, which enable effective concentration of the retentate up to 40% total solids (Nielsen, 1988), while avoiding thermal denaturation of the whey proteins. Drying of the retentate is usually performed using a spray drier fitted with nozzle atomization. Typically, the inlet and outlet air temperatures used are 160–180°C and less than 80°C, respectively (Hobman, 1992). As in the case of milk powder drying, the temperature of the drying droplets does not exceed 70°C, ensuring no damage to the proteins.

de la Fuente *et al.* (2002) examined the structural changes in proteins during the major steps of WPC manufacture, using size exclusion chromatography combined with multi-angle laser light scattering (SEC–MALLS). Samples of acid whey and cheese whey were obtained from commercial plants, throughout the WPC process as follows: whey after clarification; retentate mid-way through UF/DF; retentate after DF; concentrate after evaporation; and WPC after spray drying.

There were some differences in the protein compositions between the two types of whey, with the acid casein whey containing lower proportions of 'soluble' aggregates than the cheese whey. UF/DF processes caused a removal of non-protein, low molecular weight components, but had no significant effect on protein composition and aggregation. Evaporation and spray drying of the retentates caused no significant changes in protein denaturation and aggregation. The presence of small amounts of protein aggregates in WPC products results from heat treatments applied to milk or whey (mainly pasteurization) prior to UF/DF and drying. Therefore, it appears that variability in protein denaturation between WPCs is likely to arise from heat treatments used in casein and cheese manufacture rather than from the WPC manufacturing process.

25.4.2 Casein products

The manufacture of acid casein involves the isoelectric precipitation of casein from skim milk, either by direct addition of mineral acid or by acidification by bacterial cultures that ferment lactose to lactic acid. During the acidification of milk, many of the physico-chemical properties of casein micelles undergo considerable changes. As the pH of skim milk is reduced, CCP is dissolved and the caseins are dissociated into the milk serum phase (Roefs *et al.*, 1985; Dalgleish and Law, 1988). The extent of dissociation of the caseins is dependent on the temperature of acidification; at 30°C, a decrease in pH causes virtually no dissociation; at 4°C, about 40% of the caseins are dissociated in the serum at pH ~5.5 (Dalgleish and Law, 1988). Apparently little change in the average hydrodynamic diameter of the casein micelles occurs during acidification of the milk to pH ~5.0 (Roefs *et al.*, 1985). The lack of change in the size of the micelles on reducing the pH of milk to 5.5 may be due to concomitant swelling of the particles as CCP is solubilized.

Aggregation of casein occurs as the isoelectric point (pH 4.6) is approached. The pH and the temperature of the curd and whey at separation determine the properties of the curd (Jablonka and Munro, 1985). These effects appear to be related to the level of residual calcium in the curd. Below about pH 4.4, almost all of the calcium is lost from the micelle at any temperature but above pH 4.4, the loss of calcium is temperature sensitive with more calcium being retained at higher temperatures. The curd strength follows the same trend, with high calcium being comparable with high cohesiveness of the curd. There is an ideal cohesiveness and an ideal calcium concentration, which give springy non-sticky curd that holds together. If the curd is produced at too acid a pH, it lacks cohesiveness but this can be improved to some degree by using a higher cooking temperature. If not enough acid is added, the curd will be too cohesive (tough and unwashable and may be sticky) but this can be improved to some degree by using a lower cooking temperature.

In the manufacture of caseinate, fresh acid casein curd or dry acid casein is solubilized in a dilute solution of alkali and then spray dried, as described earlier. Manufacturing processes for sodium caseinate can vary from producer to producer. Differences in the chemical composition between batches of sodium caseinate from the same and different manufacturers have been found. Lynch *et al.* (1997) reported some differences in the profiles obtained by SEC on different batches of sodium caseinate. Lucey *et al.* (2000) examined a range of commercial and laboratory-made sodium caseinate samples using a MALLS system. Sodium caseinate solutions, analyzed using a MALLS system alone, gave weight-average molar mass (M_w) values in the range 1200–4700 kDa and the z-average root-mean-square radius (R_g) values ranged from ~50 to 120 nm. When these solutions were ultracentrifuged at 90000 g for 1 h, and the subnatants analyzed by SEC–MALLS, the M_w values were found to be in the range ~30–575 kDa and the R_g values ranged from ~22 to 49 nm. Laboratory-made (never dried) sodium caseinate samples had the smallest M_w values, suggesting some aggregation of casein molecules during the drying process.

In the manufacture of rennet casein, the coagulation of milk occurs in phases: a primary enzymatic phase, a secondary non-enzymatic phase and a less clearly defined tertiary phase (Dalgleish, 1992). During the primary phase, rennet causes specific hydrolysis of κ-casein in the region of the phenylalanine$_{105}$–methionine$_{106}$ bond, Phe$_{105}$ supplying the carboxyl group and Met$_{106}$ supplying the amino group, resulting in the formation of two peptides of contrasting physical and chemical properties. The glycomacropeptide (GMP) moiety, which comprises amino acid residues 106–169, is hydrophilic and diffuses away from the casein micelle after splitting from κ-casein and into the serum. The second peptide, para-κ-casein, which consists of amino acid residues 1–105, is strongly hydrophobic and remains attached to the micelles.

κ-casein hydrolysis during the primary phase alters the properties of the casein micelles, rendering them susceptible to aggregation and this marks the onset of the secondary phase. Loss of GMP during the primary phase of renneting results in loss of about half the negative charge as well as surface steric repulsion of κ-casein (Dalgleish, 1984). Consequently, the micelle surface becomes more hydrophobic because of the accumulation of para-κ-casein, and micelle aggregation becomes possible. There is a critical value of κ-casein hydrolysis (86–88%) below which the micelles cannot aggregate. The action of rennet can be seen as providing 'hot spots' (areas on the casein micelle surfaces from which the protective GMP moiety has been 'shaved off') via which the micelles can aggregate, these reactive areas being produced by the removal of κ-casein from a sufficiently large area (Dalgleish, 1992). As the last of the stabilizing surface is removed (i.e. during the destruction of the final 10% of the κ-casein), the concentration of micelles capable of aggregation and the rate at which they aggregate increase rapidly.

The process of gel assembly during the secondary phase of rennet coagulation appears to involve the formation of small aggregates in which the micelles tend to be joined in chains that link together randomly to form a network (Green *et al.*, 1978). Initially, linkage of aggregated micelles is through bridges, which slowly contract with time, forcing the micelles into contact and eventually causing them to partly fuse. This process probably progressively strengthens the links between micelles, giving an increase in curd firmness after coagulation.

The tertiary stage involves the rearrangement of the network and processes such as syneresis and the non-specific proteolysis of the caseins in the rennet gel (Dalgleish, 1992).

25.5 Functional properties and applications of milk protein products

25.5.1 Skim milk powders

SMPs are used as ingredients in a range of other dairy foods, in particular in recombined milk products. SMP is generally mixed with water to the required concentration and other ingredients are added, followed by further processing and packaging. Various types of recombined milk products, including recombined pasteurized milk, recombined evaporated milk (REM), recombined sweetened condensed milk, yoghurt, cheese and ice cream, are manufactured using SMP. In fact, recombined analogues of all dairy products made from fresh milk can be produced. Among others, the functional properties of SMP in these products include solubility, heat stability, rennet coagulation, acid gelation and thickening. These properties are determined by the physical and chemical properties of the primary constituents, namely proteins and lactose, both individually and in combination, which in turn are affected by the extent of protein interactions during powder manufacture.

As the heat stability of SMP is one of the most important functional properties in most applications, the following section discusses some key aspects of this property.

Heat stability

The heat stability of milk refers to the ability of milk to withstand high processing temperatures without visible coagulation or gelation or a large increase in viscosity. The heat stability characteristics of reconstituted SMP are important when milk powders are used in the manufacture of REM or as ingredients in hot beverages, such as custards and soups. However, most of the commercial heat stability problems are concerned with recombined concentrated products (e.g. REM). The total solids content of REM ranges from 26 to 31% and these milks are typically sterilized at 120°C for 12–23 min. Milk powders reconstituted to normal milk solids (9%) and subjected to UHT-type heat treatments or even sterilization heat treatments (120°C) are generally stable. The heat stability aspects of skim milk have been studied by several workers and have been reviewed extensively (Fox, 1982; Singh and Creamer, 1992; O'Connell and Fox, 2003). A number of factors influence the heat stability of milk (both normal and concentrated). The factors affecting the heat stability of fresh milk and reconstituted and recombined milks are very similar and hence the vast amount of information available on fresh milk can be applied directly to recombined milks.

The heat stability of milk is affected by the mineral balance, the lactose, urea and protein contents and the milk processing parameters, such as preheating and homogenization. Fat *per se* has no effect on the heat stability because skim milk is as stable to heat as the whole milk from which it is prepared. However, when fat-containing milk is subsequently homogenized, a clear decrease in heat stability is observed. Some of the important factors, particularly those relevant in the manufacture of recombined milks, are described next.

The heat coagulation time (HCT) of fresh skim milk at 140°C is ~14–19 min and is markedly affected by pH. A plot of HCT versus pH

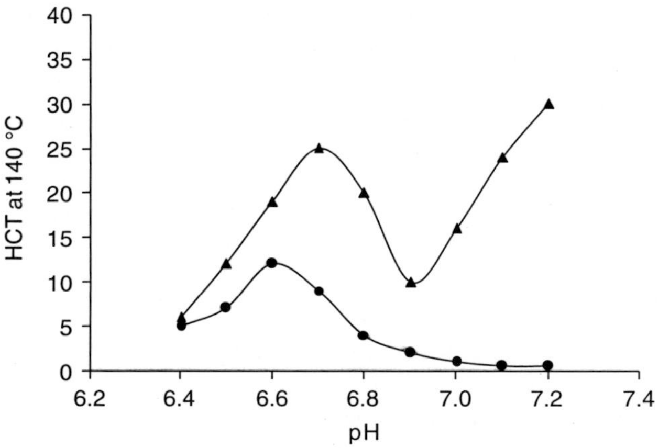

Fig. 25.4 Heat coagulation time (HCT) versus pH profile for normal skim milk (▲) or concentrated milk (20% total solids) (•) heated at 140°C.

typically shows a maximum at around pH 6.6–6.7 and a minimum in the pH region 6.8–6.9 (Fig. 25.4). The heat stability of fresh or reconstituted concentrated milk is much lower than that of unconcentrated milk but the effect of pH is similar, with the maximum heat stability usually occurring in the pH range 6.4–6.6. However, concentrated milk is unstable at all pH values above 6.8 (Fig. 25.4).

It is widely recognized that preheating improves the heat stability of milk, especially when the product is subsequently concentrated. Preheat treatments of 80–90°C with holding times of 10–20 min are commonly used. However, it has been shown that preheating temperatures in excess of 100°C with shorter holding times provide better stability in the concentrated product. Studies in New Zealand on the effects of preheat treatment on the heat stability of REM have shown that heat treatments of 120°C for 120 s, prior to evaporation and drying of the SMP, result in a heat-stable REM (Newstead *et al.*, 1975). Longer periods of preheating at 120°C can further improve the heat stability during the early and late parts of the season, but decrease the heat stability during the mid-season (Singh and Tokley, 1990).

Soluble salts, particularly calcium and phosphate, play an important role in the heat stability of milk. Generally, the addition of calcium salts to milk decreases the heat stability, whereas a reduction in calcium has a stabilizing effect. Concentrated milks with lower levels of soluble salts (e.g. concentrates prepared by UF) are markedly more stable that those prepared by standard evaporation. The addition of sodium or potassium salts of orthophosphoric and citric acids is permitted in many countries and increases the heat stability of concentrated milk. The addition of commercial phosphates (mixtures of NaH_2PO_4, Na_2HPO_4 and Na_3PO_4) improves the heat stability of concentrated milk and is a common practice in the recombined milk industry (Augustin, 2000).

Urea concentration is a major factor affecting the heat stability of skim milk. Addition of urea improves the heat stability of unconcentrated milks but has no effect on the heat stability of concentrated milks. Addition of formaldehyde to normal milk and concentrated milk increases the heat stability.

The heat stability of recombined milk may be improved by incorporating buttermilk powder (BMP) and lecithin during homogenization (Singh and Tokley, 1990; McCrae and Muir, 1992). BMP improved the heat stability of REM during most of the dairying season (Singh and Tokley, 1990). Lecithin can be added at different stages of the manufacturing process: (i) before preheating, (ii) after preheating but before homogenization, and (iii) immediately after homogenization. Polyphenol-rich extracts of tea, coffee, cocoa, wine, pine bark and oak leaves and several pure phenolic compounds increase the heat stability of both unconcentrated milk and concentrated milk (O'Connell *et al.*, 1998). However, the effects of these compounds on the sensory properties of the milks are not known.

Mechanisms of heat coagulation

The coagulation of milk on extended heating at high temperatures (120–140°C) is a consequence of the loss of casein micelle stability, as a result of numerous physical and chemical changes in its components. Casein micelles are stabilized primarily by the surface layer of κ-casein, which provides steric stabilization and contributes the overall negative charge. Inside the casein micelle, the individual casein molecules are associated by hydrophobic and electrostatic bonds in which CCP also plays an important role (Holt, 1992). Heat treatment markedly changes the environment around the casein micelles (i.e. serum phase) (e.g. change in pH and soluble minerals, in particular calcium ions, breakdown of lactose and urea) as well as the casein micelles themselves (association of whey proteins, changes in CCP, dephosphorylation, casein dissociation). It is not known exactly which particular changes are directly responsible for coagulation, predispose the milk to coagulation or are a consequence of the coagulation process. For more detailed discussion on this topic, refer to Singh and Creamer (1992), O'Connell and Fox (2003) and Singh (2004).

The pH dependence of heat stability has received considerable attention, and it is now believed that the occurrence of a maximum (pH around 6.7) in the HCT–pH profile of normal milk is essentially caused by whey protein interactions, through the formation of a complex between κ-casein and β-LG on the surface of the casein micelles, thus altering the steric and electrostatic interactions (Singh and Fox, 1987; van Boekel *et al.*, 1989; Singh and Creamer, 1992; Singh, 2004). Although the exact mechanism of the coagulation of whey-protein-coated micelles is not clear, it is likely that further heating causes lowering of the pH, dephosphorylation, covalent bond formation and other reactions. As a result, the altered micelles are able to make more frequent contact and probably form covalent cross-links, thus promoting micelle aggregation. At pH below 6.7, the protective effect of the complex is insufficient to overcome the effects of increasing acidity and high calcium ion concentrations.

At pH above 6.9, the stability decreases due to dissociation of micellar κ-casein, thus reducing the stabilizing effect of κ-casein (Singh and Fox, 1987; van Boekel *et al.*, 1989). The κ-casein-depleted micelles are sensitive to calcium ion concentrations. Therefore, the minimum in the HCT–pH profile is a result of coagulation of κ-casein-depleted micelles; coagulation is essentially salt induced, caused by calcium ions. At higher pH, although the dissociation of micellar κ-casein increases, the HCT increases because of the increase in protein charge and the low calcium ion activity.

When concentrated milk is heated at its normal pH, there is initial association of whey proteins with the micellar κ-casein. This is followed by dissociation of κ-casein-rich protein from the micelles, which creates sites on the micelles that are favourable for micelle–micelle interactions. Thus, in concentrated milk, coagulation is due to the sensitivity of κ-casein-depleted micelles to calcium ions, a situation that is somewhat similar to that existing for the coagulation of normal milk within the region of minimum stability. At high pH, the micelles are highly dissociated and coagulation is more like a gelation, with the dissociated material acting as a bridging agent (Nieuwenhuijse *et al.*, 1991; Singh and Creamer, 1991b; Singh *et al.*, 1995).

Production of heat-stable milk powders – practical considerations

The problems relating to the production of heat-stable milk powders (i.e. milk powders that are stable to heat sterilization after reconstitution to form REM) largely relate to seasonal variations and batch-to-batch variations in heat stability. Practical solutions to the heat stability problems include the following:

Manipulation of preheat treatments – Production of heat-stable SMP is primarily controlled by providing a sufficient preheat treatment, before evaporation, to denature the whey proteins. High heat treatments are typically of the order of 115–125°C for 1–3 min. However, high heat treatments are not a guarantee of producing heat-stable SMP. In countries with seasonal variations in milk composition (e.g. New Zealand, Australia, Ireland), milk from the middle of the season is used, as milk at the beginning and the end of the season (late lactation milk) tends to have lower heat stability (Kelly *et al.*, 1982). It is possible to make heat-stable SMP from late lactation milk by careful screening of powders and control of processing conditions (Singh and Newstead, 1992).

Addition of different levels of phosphate salts – Sodium and potassium salts of orthophosphate and citrate are used to improve the heat stability of

REM. Care should be taken as over-use of salts can cause the heat stability to become worse. Thus, heat stability tests are used to select the right combination and level of salts. Food regulations limit the amount of stabilizing salt that can be added – besides the limit of not producing a salty product. In order for stabilizing salts to work, the overall heat stability maximum needs to be increased and, more importantly, the pH of the REM should be positioned on or close to this maximum. For example, di-sodium hydrogen orthophosphate (DSP) and mono-sodium di-hydrogen orthophosphate (MSP) are commonly used to improve heat stability. Addition of either salt shifts the heat stability curve to a more acidic pH and increases the heat stability maximum. The sodium salts also change the pH: DSP increases the natural pH, whereas MSP reduces the natural pH. Thus, salt selection depends on which side of the maximum the pH of REM is. Often, small pH shifts are required to position the pH of REM at the maximum, so that combinations of MSP and DSP are used. It is also possible to incorporate appropriate amounts of stabilizing phosphate salts into the milk prior to drying.

Addition of buttermilk and phospholipids at appropriate levels – BMP can be used to replace up to 10% of the SMP in REM. Phospholipids in the buttermilk improve the heat stability, contribute to the creamy flavour, and help to emulsify the fat (Singh and Tokley, 1990). Alternatively, buttermilk can be incorporated into milk powders before or after evaporation, to improve their heat stability.

25.5.2 Milk protein powders

The milk protein products, such as WPC, WPI, MPC, are added to a formulation for a particular property, either nutritional or functional. The functional properties of proteins in a vast range of food products include water absorption, solubility, viscosity, gelation, emulsification, foaming and film formation. Some important functional properties of milk protein products are shown in Table 25.4.

The functional properties of milk protein products may be considered to be the consequence of the modifications of molecular structures and their interactions. Therefore, the modification of native protein structures and their consequent interactions with other food components, such as lipids, non-dairy proteins and carbohydrates, can exert a negative or positive effect on the functional properties of milk protein products. The functional properties of milk proteins vary with pH, temperature, ionic strength, and concentration of calcium and other polyvalent ions, sugars and hydrocolloids, as well as with processing treatments. Some of the key functional properties, such as solubility, emulsification and gelation, are discussed in this section. For more extensive information on the functional properties of milk proteins, the reader is referred to de Wit (1989), Kinsella and Whitehead (1989), Mulvihill (1992) and Mulvihill and Ennis (2003).

Table 25.4 Key functional properties of different milk protein products

	Caseinates		Whey protein products		Milk protein concentrates
	Sodium	Calcium	WPC	WPI	MPC
Solubility	xxx	x	xxx	xxx	xx
Emulsification	xxx	x	xx	xx	x
Foaming	xxx	x	xx	xx	x
Water binding	xxx	x	x	x	x
Viscosity	xxx	x	x	x	x
Gelation	–	–	xxx	x	–
Heat stability	xxx	x	x	x	x
Acid stability	x	x	xxx	xxx	x
Freeze/thaw stability	xxx	x	x	x	x

x: poor, xx: good, xxx: excellent.

Solubility
Solubility is generally defined as the amount of protein in a sample that goes into solution or into colloidal dispersion under specified conditions and is not sedimented by low centrifugal forces. The most common methods for determining solubility involve the preparation of a protein solution of known concentration (0.5–5.0%), centrifugation at forces ranging from 500 to 65 000 g for 5–40 min and then protein analysis of the supernatant.

The solubility of milk protein products depends on the nature of the proteins in the product, the processing treatments used, the physical form of the powder, pH, and the temperature and ionic concentration of the solution. In general, high temperature treatments during the manufacturing process result in low solubility of the final product. Salts may enhance the solubility of protein up to a point and then decrease it. The pH affects the charge and electrostatic interactions between protein molecules. Above and below the isoelectric point, proteins have a net negative or positive charge and enhanced solubility. At the isoelectric point, attractive forces predominate, causing protein molecules to associate, resulting in a loss of solubility.

SMPs generally have excellent solubility in water under a wide range of conditions, except at close to the isoelectric point of casein. MPCs with up to 70% protein have good solubility at pH 7.0, but MPC powders with very high protein content (e.g. MPC85) are generally known to have poor solubility upon reconstitution in water at 20°C. The solubility of these powders improves at higher reconstitution temperatures, but decreases with storage time at elevated temperatures (McKenna, 2000). The major factor affecting the solubility behaviour of MPC85 appears to be related to the rate of water transfer into the powder particle rather than to the thermal processes during manufacture (Schuck *et al.*, 1994). Various methods for manufacturing MPC powders with improved solubility in cold water have been proposed, involving the addition of a monovalent salt to the ultrafiltered retentate prior to drying (Carr, 2002) or partial replacement (~30%) of the calcium content of ultrafiltered retentate by sodium ions (Bhaskar *et al.*, 2001).

The insoluble material found in reconstituted MPC powders has been characterized by McKenna (2000) and Havea (2006). Electron microscopy showed that the insoluble material in MPC85 consists of large particles (up to 100 μm) formed by the fusion of casein micelles, involving some kind of protein–protein interactions (McKenna, 2000). These fused casein micelles appear to form a skin-like structure on the outside of the powder particle, inhibiting the movement of water into the particle. Upon reconstitution in water, large parts of these particles remain intact. Sodium dodecyl sulphate–polyacrylamide gel electrophoresis (SDS-PAGE) analysis showed that the insoluble material consisted predominantly of α_s- and β-caseins, and that this material was held together by weak non-covalent interactions that were easily disrupted under the PAGE conditions (Havea, 2006). Some disulphide-linked protein aggregates consisting of κ-casein and β-LG were present in the MPC powders, but these aggregates were not considered to play an important role in the formation of insoluble material. Further work is required to understand the nature of the protein–protein interactions involved in the formation of insoluble material, and to develop methods that can be used to minimize protein interactions during the manufacture of MPC.

Rennet and acid caseins are insoluble in water but caseinates (sodium, potassium and ammonium) have excellent solubility. However, caseinates are insoluble in the isoelectric (pH 4.0–5.0) region, but they are soluble at pH below 3.5. Calcium caseinate forms a coarse colloidal dispersion under these conditions (see Mulvihill and Ennis, 2003).

Whey proteins show excellent solubility over the entire pH range, but heat-denatured whey proteins are insoluble in the pH range 4–5. Processing treatments used in the manufacture of WPCs and WPIs may sometimes cause small amounts of denaturation, which tends to reduce their solubility (de Wit and Klarenbeek, 1984; de Wit *et al.*, 1996). The solubility of whey proteins is generally not altered by UF or spray drying, although pasteurization may cause up to 20% denaturation with a consequent loss of solubility.

Emulsification

As milk proteins, both caseinates and whey proteins, are surface active, they are adsorbed rapidly to the oil–water interface, forming stable oil-in-water emulsions. Sodium caseinate and WPC/WPI show excellent emulsifying abilities. In emulsions formed with sodium caseinate or WPC, the surface protein concentration increases with an increase in protein concentration until it reaches a plateau value of about 2.0–3.0 mg/m^2 (Singh, 2005; Singh and Ye, 2009). The emulsifying ability of 'aggregated' milk protein products, such as MPC and calcium caseinate, is much lower than that of WPC or sodium caseinate, i.e. much higher concentrations of protein are required to make stable emulsions and larger droplets are formed under similar homogenization conditions (Euston and Hirst, 1999). The surface protein concentration of emulsions formed with MPC is in the range 5–20 mg/m^2 depending on the protein concentration used in making the emulsions. When the ratio of protein to oil is low in these emulsions, protein aggregates are shared by adjacent droplets, resulting in bridging flocculation and consequently a marked increase in droplet size.

In sodium-caseinate-based emulsions, there is competition between individual proteins for adsorption on to the oil–water interface during emulsification. β-Casein appears to adsorb in preference to α_{s1}-casein at low protein concentrations (<2.0%), whereas a larger amount of α_{s1}-casein than β-casein is present at the interface at high protein concentrations. At all concentrations, κ-casein from sodium caseinate appears to be less readily adsorbed (Srinivasan *et al.*, 1996, 1999). The concentration dependence of the competitive adsorption of α_{s1}-casein and β-casein in sodium caseinate emulsions appears to be related to the formation of complexes in solution. At low concentrations, β-casein exists as a monomer, but, with increasing protein concentration, β-casein aggregates to form micelles, which results in the loss of its competitive ability.

When casein is in the highly aggregated form of casein particles, as in calcium caseinate or MPC, there is very little competitive adsorption and protein exchange (Euston and Hirst, 1999; Srinivasan *et al.*, 1999). In these systems, the average surface composition is probably determined by the adsorption of protein aggregates of fixed composition. The whey proteins β-LG and α-LA also show no preferential adsorption regardless of the protein-to-oil ratio in the emulsion (Hunt and Dalgleish, 1994; Dalgleish, 1996).

Emulsions stabilized by sufficient quantities of milk proteins are generally very stable to coalescence over prolonged storage, but they can be susceptible to different types of flocculation, which in turn leads to enhanced creaming or serum separation. When there is insufficient protein to fully cover the oil–water interface during homogenization, bridging flocculation and/or coalescence of droplets may occur during or immediately after emulsion formation. Bridging flocculation is commonly observed in emulsions formed with aggregated milk protein products, such as calcium caseinate or MPC, in which the droplets are bridged by casein aggregates or micelles. Optimum stability can generally be attained at protein concentrations high enough to allow full saturation coverage at the oil–water interface. However, at very high protein-to-oil ratios, the presence of excess, unadsorbed protein may lead to depletion flocculation in some emulsions. Both depletion flocculation and bridging flocculation cause an emulsion to cream more rapidly (Dalgleish, 1997; Dickinson, 1998; Singh, 2005).

Sodium-caseinate-based emulsions show depletion flocculation (Dickinson and Golding, 1997; Srinivasan *et al.*, 2001; Ye and Singh, 2001) when the protein content is increased to above 3.0 wt% in a 30% oil emulsion. This flocculation results in a marked decrease in creaming stability, but is reversible by shearing or dilution with water. Emulsions formed with calcium caseinate, MPC or whey proteins do not show depletion flocculation and are stable against flocculation, coalescence and creaming for several weeks.

Gelation

Milk proteins have the ability to form heat-induced irreversible gels that hold water and fat, and provide structural support in food systems. Gelation induced by the addition of acid or rennet to milk has been discussed in previous sections. WPC and WPI products can provide gelation under a wide range of conditions and are used in a variety of food systems.

Gelation of whey proteins occurs upon heat treatments and involves a series of steps, starting with the unfolding of protein molecules, followed by aggregation. A gel is formed when the extent of aggregation exceeds some critical level; a three-dimensional, self-supporting network that traps the solvent in the system is formed (Clark and Ross-Murphy, 1987; Doi, 1993). It has been shown that the interactions between unfolded protein molecules lead to the formation of 'soluble' aggregates (McSwiney *et al.*, 1994a, 1994b; Havea *et al.*, 1998; Schokker *et al.*, 1999), which are the basic building blocks leading to the formation of gels. It is generally assumed that the driving forces for aggregation are non-specific interactions between the hydrophobic regions of unfolded polypeptide chains, but sulphydryl–disulphide interchange reactions and ionic interactions are likely to participate as well (Clark and Ross-Murphy, 1987; McSwiney *et al.*, 1994a, 1994b).

When the extent of aggregation is below some critical minimum, soluble aggregates or a precipitate will form. Therefore, gel formation and the properties of the gel depend on the type and the number of protein–protein interactions, which in turn are affected by variables such as type of protein, protein concentration, temperature, pH, ionic strength and the presence of other ingredients such as lactose (de Wit and Klarenbeek, 1984; Kinsella, 1984). The structure of protein gels can vary widely depending on these conditions, and has an impact on gel properties, such as rheological properties, sensory qualities and water-holding capacity.

The most outstanding microstructural feature of the gels formed from whey proteins is the presence of a homogeneous network of connected protein particles, usually referred to as 'fine-stranded' gels, or aggregates that form a three-dimensional matrix with the interstices filled by a liquid or aqueous solution, usually referred to as 'particulated' gels (Langton and Hermansson, 1992; Stading *et al.*, 1993). The gels that are formed on heating whey protein solutions at pH 4–6 are opaque, whereas translucent gels are formed above and below this pH range. Opaque gels have been described as soft and creamy and they tend to lose water during compression. The transparent gels formed at pH below 4 are weak (low values for fracture stress) and brittle (low values for fracture strain) whereas those formed at pH above 6.0 are strong and rubbery, with high fracture stress and strain values. The variations in gel properties with pH are attributed to variations in electrostatic interactions and disulphide bonding (Stading and Hermansson, 1990).

Generally, heating a whey protein solution above the minimum denaturation temperature of the constituent proteins is required for gel formation. The strength of whey protein gels is affected by the concentration and purity of the protein. A protein concentration of 7.5% or greater is needed to form a strong gel from WPC at pH 7.0 upon heating for 10 min at 100°C. Pure solutions of β-LG and BSA can form gels at 5 and 4% protein, respectively, after heating for 15 min at 90°C. When other factors are maintained, gel hardness increases with increasing heating temperature and time. The heating rate also affects the gelation process; slow heating allows the proteins enough time for unfolding and aggregation, resulting in much stronger gels.

Salts have a major effect on the properties of whey protein gels, especially at pH values far from the isoelectric point where the proteins carry a large net charge. Addition of NaCl or $CaCl_2$ to a dialysed solution of WPC or WPI results in an increase in gel strength until maximum values are reached, and the gel strength then decreases at higher salt concentrations (Kuhn and Foegeding, 1991; Tang *et al.*, 1993). Calcium ions influence protein–protein interactions by shielding electrostatic repulsion and also by forming calcium bridges between protein molecules. In general, divalent calcium ions have a much greater effect on gel properties than monovalent sodium or potassium ions (Havea *et al.*, 2002).

Calcium caseinate solutions at protein concentrations greater than 15% form reversible gels upon heating to 50–60°C. The gelation temperature increases with protein concentration from 15 to 20% and with pH in the range 5.2–6.0. Upon cooling, these gels are changed into viscous liquids.

25.6 References

ANEMA S G (2009), 'The whey proteins in milk: Thermal denaturation, physical interactions and effects on the functional properties of milk', in Thompson A, Boland M and Singh H, *Milk Proteins: From Expression to Food*, New York, Academic Press, 239–273.

ANEMA S G and LI Y (2003), 'Effect of pH on the association of denatured whey proteins with casein micelles in heated reconstituted skim milk', *J Dairy Res*, 51, 73–83.

AUGUSTIN M A (2000), 'Mineral salts and their effect on milk functionality', *Aust J Dairy Technol*, 55, 61–64.

AUGUSTIN M A (2001), 'Dairy ingredients in chocolate – chemistry and ingredient interactions', *Food Aust*, 53(a), 389–391.

BALDWIN A and PEARCE D (2005), 'Milk powder', in Onwulata C, *Encapsulated and Powdered Foods*, Boca Raton, Florida, Taylor & Francis, 387–434.

BHASKAR G V, SINGH H and BLAZEY N D (2001), 'Milk protein concentrate products and process', *International Patent Specification WO01/41578*, Palmerston North, New Zealand, New Zealand Dairy Research Institute.

CARR A J (2002), 'Milk protein concentrate products and uses thereof', *International Patent Specification WO02/196208A2*, Wellington, New Zealand, New Zealand Dairy Board.

CLARK A H and ROSS-MURPHY S B (1987), 'Structural and mechanical properties of bio-polymer gels', *Adv Polym Sci*, 83, 60–192.

CORREDIG M and DALGLEISH D G (1996), 'Effect of temperature and pH on the interactions of whey proteins with casein micelles in skim milk', *Food Res Int*, 29, 49–55.

DALGLEISH D G (1984), 'Measurement of electrophoretic mobilities and zeta potentials of particles from milk using laser Doppler electrophoresis', *J Dairy Res*, 51, 425–438.

DALGLEISH D G (1992), 'The enzymatic coagulation of milk', in Fox P F, *Advanced Dairy Chemistry, Volume 1, Proteins*, 2nd edn, Barking, Essex, Elsevier Applied Science, 579–619.

DALGLEISH D G (1996), 'Conformations and structures of milk proteins adsorbed to oil–water interfaces', *Food Res Int*, 29, 541–547.

DALGLEISH D G (1997), 'Adsorption of protein and the stability of emulsions', *Trends Food Sci Technol*, 8, 1–6.

DALGLEISH D G and LAW A J R (1988), 'pH-induced dissociation of casein micelles. 1. Analysis of liberated caseins', *J Dairy Res*, 55, 529–538.

DALGLEISH D G, SPAGNUOLO P A and GOFF H D (2004), 'A possible structure of the casein micelle based on high-resolution field-emission scanning electron microscopy', *Int Dairy J*, 14, 1025–1031.

DANNENBERG F and KESSLER H G (1988), 'Reaction kinetics of the denaturation of whey proteins in milk', *J Food Sci*, 53, 258–263.

DE KRUIF C G and HOLT C (2003), 'Casein micelle structure, functions and interactions' in Fox P F and McSweeney P L H, *Advanced Dairy Chemistry, Volume 1, Proteins*, 3rd edn, New York, Kluwer Academic/Plenum Publishers, 233–276.

DE LA FUENTE M A, HEMAR Y, TAMEHANA M, MUNRO P A and SINGH H (2002), 'Process-induced changes in whey proteins during the manufacture of whey protein concentrates', *Int Dairy J*, 12, 361–369.

DE WIT J N (1989), 'Functional properties of whey proteins', in Fox P F, *Developments in Food Proteins 4*, London, Elsevier Science Publishers, 285–321.

DE WIT J N and KLARENBEEK G (1984), 'Effects of various heat treatments on structure and solubility of whey proteins', *J Dairy Sci*, 67, 2701–2710.

DE WIT J N, KLARENBEEK G and ADAMSE M (1996), 'Evaluation of functional properties of whey protein concentrates and whey protein isolates. 2. Effects of processing history and composition', *Neth Milk Dairy J*, 40, 41–56.

DICKINSON E (1998), 'Proteins at interfaces and in emulsions: Stability, rheology and interactions', *J Chem Soc Faraday Trans*, 94, 1657–1669.

DICKINSON E and GOLDING M (1997), 'Depletion flocculation of emulsions containing unadsorbed sodium caseinate', *Food Hydrocolloids*, 11, 13–18.

DOI E (1993), 'Gels and gelling of globular proteins', *Trends Food Sci Technol*, 4, 1–5.

EDWARDS P B, CREAMER L K and JAMESON G B (2009), 'Structure and stability of whey proteins', in Thompson A, Boland M and Singh H, *Milk Proteins: From Expression to Food*, New York, Academic Press, 163–190.

EUSTON S R and HIRST R L (1999), 'Comparison of the concentration-dependent emulsifying properties of protein products containing aggregated and non-aggregated milk protein', *Int Dairy J*, 9, 693–701.

FOX P F (1982), 'Heat induced coagulation of milk', in Fox P F, *Developments in Dairy Chemistry – I. Proteins*, London, Applied Science Publishers, 189–228.

FOX P F and MCSWEENEY P L H (1998), *Dairy Chemistry and Biochemistry*, London, Chapman & Hall.

FOX P F and MCSWEENEY P L H (2003), *Advanced Dairy Chemistry, Volume 1, Proteins*, 3rd edn, New York, Kluwer Academic/Plenum Publishers.

GREEN M L, HOBBS, D G, MORANT S V and HILL V A (1978), 'Intermicellar relationships in rennet-treated separated milk. II. Process of gel assembly', *J Dairy Res*, 45, 413–422.

GUYOMARC'H F, WARIN F, MUIR D D and LEAVER J (2000), 'Lactosylation of milk proteins during the manufacture and storage of skim milk powders', *Int Dairy J*, 10, 863–872.

HAMBLING S G, MCALPINE A S and SAWYER L (1992), 'β-Lactoglobulin', in Fox P F, *Advanced Dairy Chemistry, Volume 1, Proteins*, 2nd edn, Barking, Essex, Elsevier Applied Science, 141–190.

HAVEA P (2006), 'Protein interactions in milk protein concentrate powders', *Int Dairy J*, 16, 415–422.

HAVEA P, SINGH H, CREAMER L K and CAMPANELLA O H (1998), 'Electrophoretic characterization of the protein products formed during heat treatment of whey protein concentrate solutions', *J Dairy Res*, 65, 79–91.

HAVEA P, SINGH H and CREAMER L K (2002), 'Heat-induced aggregation of whey proteins: Comparison of acid and cheese WPC and relevance of mineral environment', *J Agric Food Chem*, 50, 4674–4681.

HOBMAN P G (1992), 'Ultrafiltration and manufacture of whey protein concentrates', in Zadow J G, *Whey and Lactose Processing*, London, Elsevier Applied Science, 195–230.

HOLT C (1992), 'Structure and stability of bovine casein micelles', *Adv Protein Chem*, 43, 64–151.

HORNE D S (1998), 'Casein interactions: Casting light on the black boxes, the structure in dairy products', *Int Dairy J*, 8, 171–177.

HORNE D S (2009), 'Casein micelle structure and stability', in Thompson A, Boland M and Singh H, *Milk Proteins: From Expression to Food*, New York, Academic Press, 133–157.

HUNT J A and DALGLEISH D G (1994), 'Adsorption behaviour of whey protein isolate and caseinate in soya oil-water emulsions', *Food Hydrocolloids*, 8, 175–187.

JABLONKA M S and MUNRO P A (1985), 'Particle size distribution and calcium content of batch-precipitated acid casein curd: Effect of precipitation temperature and pH', *J Dairy Res*, 52, 419–428.

KELLY P M, O'KEEFE A M, KEOGH M K and PHELAN J A (1982), 'Studies of milk composition and its relationship to some processing criteria. III. Seasonal variation in heat stability of milk', *Ir J Food Sci Technol*, 6, 29–38.

KINSELLA J E (1984), 'Milk proteins: Physico-chemical and functional properties', *CRC Crit Rev Food Sci Nutr*, 21, 197–262.

KINSELLA J E and WHITEHEAD D M (1989), 'Proteins in whey: Chemical, physical, and functional properties', *Adv Food Nutr Res*, 33, 343–438.

KUHN P R and FOEGEDING E A (1991), 'Mineral salt effects on whey protein gelation', *J Agric Food Chem*, 39, 1013–1016.

LANGTON M and HERMANSSON A-M (1992), 'Fine-stranded and particulate gels of β-lactoglobulin and whey protein at varying pH', *Food Hydrocolloids*, 5, 523–539.

LE GRAET Y and BRULE G (1982), 'Effect of concentration and drying on mineral equilibria of skim milk and retentates', *Lait*, 62, 113–125.

LUCEY J A, SRINIVASAN M, SINGH H and MUNRO P A (2000), 'Characterization of commercial and experimental sodium caseinates by multi-angle laser light scattering and size exclusion chromatography', *J Agric Food Chem*, 48, 1610–1616.

LYNCH A G, MULVIHILL D M, LAW A J R, LEAVER J and HORNE D S (1997), 'Chromatographic elution profiles, electrophoretic properties and free amino and sulphydryl group contents of commercial sodium caseinates', *Int Dairy J*, 7, 213–220.

MCCRAE C H and MUIR D D (1992), 'Heat stability of recombined milk: Influence of lecithins on the heat coagulation time–pH profile', *J Dairy Res*, 59, 177–185.

MCKENNA A B (2000), '*Effect of processing and storage on reconstitution properties of whole milk and ultrafiltered skim milk powders*', PhD thesis, Palmerston North, New Zealand, Massey University.

MCSWINEY M, SINGH H and CAMPANELLA O H (1994a), 'Thermal aggregation and gelation of bovine β-lactoglobulin', *Food Hydrocolloids*, 8, 441–453.

MCSWINEY M, SINGH H, CAMPANELLA O H and CREAMER L K (1994b), 'Thermal gelation and denaturation of bovine ß-lactoglobulins A and B', *J Dairy Res*, 61, 221–232.

MULVIHILL D M (1992), 'Production, functional properties and utilization of milk protein products', in Fox P F, *Advanced Dairy Chemistry, Volume 1, Proteins*, 2nd edn, Barking, Essex, Elsevier Applied Science, 369–404.

MULVIHILL D M and ENNIS M P (2003), 'Functional milk proteins: Production and utilization', in Fox P F and McSweeney P L H, *Advanced Dairy Chemistry, Volume 1, Proteins*, 3rd edn, New York, Kluwer Academic/Plenum Publishers, 1175–1228.

NEWSTEAD D F, SANDERSON W B and BAUCKE A G (1975), 'The effect of heat treatment and pH on the heat stability of recombined evaporated milk', *NZ J Dairy Sci Technol*, 10, 113–118.

NIELSEN P S (1988), 'Membrane filtration for whey proteins concentrate', *Marketing Bulletin*, Aarhus, Denmark, APV Pasilac A/S.

NIEUWENHUIJSE J A, SJOLLEMA A, VAN BOEKEL M A J S, VAN VLIET T and WALSTRA P (1991), 'The heat stability of concentrated milk', *Neth Milk Dairy J*, 45, 193–224.

O'CONNELL J E and FOX P F (2003), 'Heat induced coagulation of milk', in Fox P F and McSweeney P L H, *Advanced Dairy Chemistry, Volume 1, Proteins*, 3rd edn, New York, Kluwer Academic/Plenum Publishers, 879–930.

O'CONNELL J E, FOX P D, TAN-KINTIA R and FOX P F (1998), 'Effects of tea, coffee and cocoa extracts on the colloidal stability of milk and concentrated milk', *Int Dairy J*, 8, 689–693.

OLDFIELD D and SINGH H (2005), 'Functional properties of milk powders', in Onwulata C, *Encapsulated and Powdered Foods*, Boca Raton, Florida, Taylor & Francis, 365–386.

OLDFIELD D J, SINGH H, TAYLOR M W and PEARCE K N (1998a), 'Kinetics of denaturation and aggregation of whey proteins in skim milk heated in an ultra-high temperature (UHT) pilot plant', *Int Dairy J*, 8, 311–318.

OLDFIELD D J, SINGH H and TAYLOR M W (1998b), 'Association of β-lactoglobulin and α-lactalbumin with the casein micelles in skim milk heated in an ultra-high temperature plant', *Int Dairy J*, 8, 765–770.

OLDFIELD D J, SINGH H and TAYLOR M W (2005), 'Effect of preheating and other process parameters on whey protein reactions during skim milk powder manufacture', *Int Dairy J*, 15, 501–511.

PATEL H A, SINGH H, ANEMA S G and CREAMER L K (2006), 'Effects of heat and high hydrostatic pressure treatments on disulfide bonding interchanges among the proteins in skim milk', *J Agric Food Chem*, 54, 3409–3420.

PISECKY J (1997), *Handbook of Milk Powder Manufacture*, Copenhagen, Denmark, Niro A/S.

ROEFS S P F M, WALSTRA P, DALGLEISH D G and HORNE D S (1985), 'Preliminary note on the change in casein micelles caused by acidification', *Neth Milk Dairy J*, 39, 119–122.

ROLLEMA H S (1992), 'Casein association and micelle formation', in Fox P F, *Advanced Dairy Chemistry, Volume 1, Proteins*, 2nd edn, Barking, Essex, Elsevier Applied Science, 111–140.

SCHOKKER E P, SINGH H, PINDER D N, NORRIS G E and CREAMER L K (1999), 'Characterization of intermediates formed during heat-induced aggregation of β-lactoglobulin A, B at neutral pH', *Int Dairy J*, 9, 791–800.

SCHUCK P, PIOT M, MEJEAN S, LE GRAET Y, FAUQUANT J, BRULE G and MABOUIS J (1994), 'Spray drying of native phosphocaseinate obtained by membrane microfiltration', *Lait*, 74, 375–388.

SINGH H (1995), 'Heat-induced changes in caseins including interactions with whey proteins', in Fox P F, *Heat-induced Changes in Milk*, Special Issue 9501, Brussels, International Dairy Federation, 86–99.

SINGH H (2004), 'Heat stability of milk', *Int J Dairy Technol*, 57, 111–119.

SINGH H (2005), 'Milk protein functionality in food colloids', in Dickinson E, *Food Colloids: Interactions, Microstructure and Processing*, Cambridge, The Royal Society of Chemistry, 179–193.

SINGH H (2007), 'Interactions of milk proteins during the manufacture of milk powders', *Lait,* 87, 413–423

SINGH H and CREAMER L K (1991a), 'Denaturation, aggregation and the stability of milk protein during the manufacture of skim milk powder', *J Dairy Res*, 58, 269–283.

SINGH H and CREAMER L K (1991b), 'Influence of concentration of milk solids on the dissociation of micellar κ-casein on heating reconstituted skim milk at 120°C', *J Dairy Res*, 58, 99–105.

SINGH H and CREAMER L K (1992), 'Heat stability of milk', in Fox P F, *Advanced Dairy Chemistry, Volume 1, Proteins*, 2nd edn, Barking, Essex, Elsevier Applied Science, 621–656.

SINGH H and FOX P F (1985), 'Heat stability of milk: pH-dependent dissociation of micellar κ-casein on heating milk at ultrahigh temperatures', *J Dairy Res*, 52, 529–538.

SINGH H and FOX P F (1987), 'Heat stability of milk: Role of β-lactoglobulin in the pH-dependent dissociation of micellar κ-casein', *J Dairy Res*, 54, 509–521.

SINGH H and NEWSTEAD D F (1992), 'Aspects of proteins in milk powder manufacture', in Fox P F, *Advanced Dairy Chemistry, Volume 1, Proteins*, 2nd edn, Barking, Essex, Elsevier Applied Science, 735–765.

SINGH H and TOKLEY R P (1990), 'Effects of preheat treatments and buttermilk addition on the seasonal variations in the heat stability of recombined evaporated milk and reconstituted concentrated milk', *Aust J Dairy Technol*, 45, 10–16.

SINGH H, CREAMER L K and NEWSTEAD D F (1995), 'Heat stability of concentrated milk', in Fox P F, *Heat-induced Changes in Milk*, Special Issue 9501, Brussels, International Dairy Federation, 256–274.

SINGH H and YE A (2009), 'Interactions and functionality of milk proteins in food emulsions', in Thompson A, Boland M and Singh H, *Milk Proteins: From Expression to Food*, New York, Academic Press, 321–340.

SRILAORKUL S, OZIMEK L, OORAIKUL B, HADZIYEV D and WOLFE F (1991), 'Effect of ultrafiltration of skim milk on casein micelle size distribution in retentate', *J Dairy Sci*, 74, 50–57.

SRINIVASAN M, SINGH H and MUNRO P A (1996), 'Sodium caseinate-stabilized emulsions: Factors affecting coverage and composition of surface proteins', *J Agric Food Chem*, 44, 3807–3811.
SRINIVASAN M, SINGH H and MUNRO P A (1999), 'Adsorption behaviour of sodium and calcium caseinates in oil-in-water emulsions', *Int Dairy J*, 9, 337–341.
SRINIVASAN M, SINGH H and MUNRO P A (2001), 'Creaming stability of oil-in-water emulsions formed with sodium and calcium caseinates', *J Food Sci*, 66, 441–446.
STADING M and HERMANSSON A-M (1990), 'Viscoelastic behaviour of β-lactoglobulin gel structures', *Food Hydrocolloids*, 4, 121–135.
STADING M, LANGTON M and HERMANSSON A-M (1993), 'Microstructure and rheological behaviour of particulated ß-lactoglobulin gels', *Food Hydrocolloids*, 7, 195–212.
TANG Q, MCCARTHY O J and MUNRO P A (1993), 'Oscillatory rheological study of the gelation mechanism of whey protein concentrate solutions: Effects of physicochemical variables on gel formation', *J Dairy Res*, 60, 543–555.
VAN BOEKEL M A J S, NIEUWENHUIJSE J A and WALSTRA P (1989), 'The heat coagulation of milk: Mechanisms', *Neth Milk Dairy J*, 43, 97–127.
YE A and SINGH H (2001), 'Interfacial composition and stability of sodium caseinate emulsions as influenced by calcium ions', *Food Hydrocolloids*, 15, 195–207.
WALSTRA P and JENNESS R (1984), *Dairy Chemistry and Physics*, New York, Wiley.

Index